ALGEBRA

PURE AND APPLIED MATHEMATICS

A Wiley-Interscience Series of Texts, Monographs, and Tracts

Founded by RICHARD COURANT
Editor Emeritus: PETER HILTON and HARRY HOCHSTADT
Editors: MYRON B. ALLEN III, DAVID A. COX, PETER LAX,
JOHN TOLAND

A complete list of the titles in this series appears at the end of this volume.

ALGEBRA

PIERRE ANTOINE GRILLET

A Wiley-Interscience Publication

JOHN WILEY & SONS, INC.

New York · Chichester · Weinheim · Brisbane · Singapore · Toronto

Copyright © 1999 by John Wiley & Sons, Inc. All rights reserved.

Published simultaneously in Canada.

Library of Congress Cataloging-in-Publication Data:

Grillet, P. A. (Pierre Antoine), 1941–
    Algebra/Pierre Antoine Grillet.
        p.   cm.   –   (Pure and applied mathematics)
    "A Wiley-Interscience publication."
    Includes bibliographical references and index.
    ISBN 0-471-25243-3 (cloth: alk. paper)
    1. Algebra.   I. Title.   II. Series: Pure and applied mathematics
    (John Wiley & Sons: Unnumbered)
    QA154.2.G736   1999
    512–dc21   98-41793

Printed in the United States of America

10 9 8 7 6 5 4 3 2 1

To Gail, with all my love

CONTENTS

RINGS AND MODULES

APPENDICES

PREFACE

This book is intended for graduate students (or for bright undergraduates) as a basic text for a one-year course in algebra. Except for linear algebra and elementary set theory, the material is self-contained. However, a previous semester of abstract algebra is highly recommended. I have assumed that the reader can do simple proofs with sets and elements.

Writing such a textbook requires some guiding principles.

Besides the basic material, with enough exercises to develop research skills, I have covered a diversity of topics, to convey some of the scope and thrust of contemporary algebra.

Moreover, students should sense the unity and usefulness of algebra and see that results in one area often have fruitful applications to another. I have included as many such connections as possible, and avoided loose ends (results that don't have such connections).

Diagrams and universal properties are introduced in the first chapters, to assist the transition from "element" thinking to "arrow" thinking. The last chapters give more abstract views of algebra as a whole, for further conceptual understanding.

Finally, students should to be prepared for additional courses, particularly in homological algebra, which many will need.

All this amounts to rather more material than can be covered in one year. I expect students to read the rest, and have provided complete proofs throughout. To give instructors more flexibility in selecting a syllabus, I have indicated the interdependence of chapters, and partitioned sections into numbered parts, so that sections do not need to be covered in full.

A basic course might cover the following: Sections 1.1 through 1.6; Sections 2.1, 2.2, 2.5, and 2.6; Sections 3.1 through 3.5; Chapter 4 (quickly); Sections 5.1, 5.4, and 5.5; Sections 6.1, 6.2, and 6.3; Sections 7.1, 7.2, 7.4, and 7.6; Sections 10.1 through 10.8; Sections 14.1, 14.2, 14.3, 14.5, and 14.6; Sections 15.1, 15.2; Sections 17.1, 17.2, 17.3, 17.4, 17.6, and 17.7; and, at some point, Section B.2 and some of Section B.3. Parts of some sections may be omitted.

I wish to thank Professors Laszlo Fuchs, Thomas R. Shemanske of Dartmouth College, and Lynne Walling of the University of Colorado for their comments, and Jessica Downey, Steve Quigley, and the staff at Wiley for their help during the preparation of the manuscript.

PIERRE GRILLET

New Orleans, May 1998

OVERALL ORGANIZATION

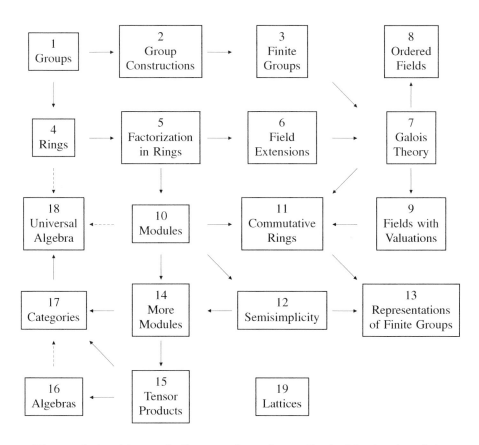

These relationships, and all quotes from Appendix A (Numbers) and Appendix B (Sets and Order), are listed in more detail at the beginning of each chapter.

ALGEBRA

1

GROUPS

"Classical" algebra dealt with "concrete" objects: real or complex numbers, polynomials with complex coefficients, or specific groups of transformations. "Modern" algebra replaced these concrete objects by elements of a set, whose nature is irrelevant and whose relationships to each other are specified by axioms.

This new "abstract" algebra studied sets endowed with one or more operations (and perhaps with an order relation) whose properties are deduced from axioms. As a result algebra attained unprecendented abstraction, clarity, and generality, enabling Artin, Krull, Noether, Schreier, and others to reach unprecedented depth as well. A similar transformation began late in the nineteenth century and occurred in all of mathematics.

Group theory was one of the first areas to be affected. By 1897 enough was known about abstract groups for Burnside to publish the first book on the subject, *Theory of Groups of Finite Order*. Group theory has prospered since then and now has many applications throughout mathematics. Even more applications are expected from one of the major accomplishments of this century, the Classification Theorem for simple groups, whose proof was completed in the early 1980s.

This chapter contains basic properties of groups and homomorphisms, subgroups, normal subgroups, quotient groups, and cyclic groups. The first section has general properties of binary operations; I have included Light's associativity test (which has long been used in semigroup theory) for its usefulness in constructing groups from presentations.

In preparation for later chapters, I have paid more attention than usual to homomorphisms. Readers who have seen this material before will thus be exposed to new ideas.

Required results: elementary set theory (Section B.1) and integer division from Section A.3; other results from Section A.3 may be used but are not required.

1.1. OPERATIONS

1.1.a. Operations. The general definition of operations is the following. An *n*-**ary operation** on a set X is a mapping of X^n into X, where $X^n = X \times X \times \cdots \times X$ is the cartesian product of n copies of X. The number n is the **arity** of the operation. A **ternary** operation has arity $n = 3$. A **binary** operation has arity $n = 2$. A **unary** operation is an operation of arity $n = 1$ and is a mapping of $X^1 = X$ into X. By convention, X^0 is your favorite one-element set (popular choices are $\{1\}$ and $\{\emptyset\}$); an operation of arity $n = 0$ merely selects one element of X and is called a **constant** operation.

Most operations in this book are binary, but we will also encounter unary and constant operations.

When X is finite, a binary operation μ on X can be specified by a **table** which lists all values of m; for example,

μ	a	b	c	d
a	a	b	c	b
b	b	c	a	c
c	c	a	b	a
d	b	c	a	c

An operation table

It is understood that $\mu(x,y)$ is found in the row of x and in the column of y. In the example above, $X = \{a,b,c,d\}$ and $\mu(a,d) = b$, $\mu(b,c) = a$, and so on.

1.1.b. Products. When $\mu : X \times X \longrightarrow X$ is a binary operation on X, it is cumbersome to always denote by $\mu(x,y)$ the typical value of μ. In practice, $\mu(x,y)$ is most often denoted by xy or $x \cdot y$ (the **multiplicative notation**) or by $x + y$ (the **additive notation**). The additive notation is generally reserved for abelian (= commutative) groups. In this chapter we generally use the multiplicative notation. Then a table of values of μ is a **multiplication table**, and $\mu(x,y) = xy$ is referred to as the **product** of x and y.

Successive applications of μ yield more complex products such as $(xy)z$ or $(ab)(c(de))$. Parentheses are necessary here since μ cannot accommodate more than two variables at a time. However, most operations used in algebra have the property that such long products may be written without parentheses. This can be achieved as follows.

First we need a precise definition of long products. Given a binary operation on X (denoted multiplicatively) and a finite nonempty sequence $x_1,\ldots,x_m$ of elements of X, **products** of $x_1,\ldots,x_m$ (in that order) are defined recursively (by induction on m) as follows. If $m = 1$, there is one product of x_1, namely x_1. If $m > 1$, then p is a product of $x_1,\ldots,x_m$ (in that order) if and only if there exists $1 \leqslant k < m$ and $q,r \in X$ such that $p = qr$, q is a product of $x_1,\ldots,x_k$, and r is a product of $x_{k+1},\ldots,x_m$; parentheses may be placed around q and r for clarity. The elements $x_1,\ldots,x_m$ are the **terms** of the product p.

If, for example, $m = 2$, then necessarily $k = 1$, and there is just one product of x_1, x_2, namely $x_1 x_2$. There are two products of x_1, x_2, x_3 (in that order): one product with $k = 1$, namely $x_1(x_2 x_3)$, and one product with $k = 2$, namely $(x_1 x_2) x_3$.

In the additive notation, products become **sums**.

A binary operation on a set X (denoted multiplicatively) is **associative** in case

$$x(yz) = (xy)z \qquad \text{for all} \quad x, y, z \in X.$$

Proposition 1.1.1. *An associative binary operation (denoted multiplicatively) has the property that all products of $x_1, \ldots, x_m$ (in that order) are equal.*

Proof. Associativity yields this property for $m \leqslant 3$. For $m > 3$ we show by induction on the number of terms that every product p of $x_1, \ldots, x_m$ equals the product $x_1(x_2(\ldots(x_{m-1} x_m)\ldots))$. By definition, $p = qr$, where q is a product of $x_1, \ldots, x_k$, r is a product of $x_{k+1}, \ldots, x_m$, and $1 \leqslant k < m$.

If $k = 1$, then $r = x_2(x_3(\ldots(x_{m-1} x_m)\ldots))$ by the induction hypothesis, so that $p = x_1(x_2(\ldots(x_{m-1} x_m)\ldots))$.

If $k > 1$, then $q = x_1(x_2(\ldots(x_{k-1} x_k)\ldots))$ by the induction hypothesis. Associativity, applied to $x = x_1$, $y = x_2(x_3(\ldots(x_{k-1} x_k)\ldots))$, and $z = r$, yields $p = qr = (x_1 y)r = x_1(yr)$. Now $yr = x_2(x_3(\ldots(x_{m-1} x_m)\ldots))$, by the induction hypothesis, and $p = x_1(x_2(\ldots(x_{m-1} x_m)\ldots))$. $\qquad \square$

By Proposition 1.1.1, associativity ensures that there is only one product of $x_1, \ldots, x_m$ (in that order). This product is usually written $x_1 x_2 \ldots x_m$ (without parentheses, since there is no need to specify which product is meant).

1.1.c. Commutativity. Even with associativity the order of the terms in a product $x_1 x_2 \ldots x_m$ must in general be respected; for instance, this is the case in products of matrices. However, many operations used in algebra have the property that such long products do not change when their terms are permuted.

A binary operation on a set X (denoted multiplicatively) is **commutative** in case

$$yx = xy \qquad \text{for all} \quad x, y \in X.$$

Proposition 1.1.2. *A commutative and associative binary operation (denoted multiplicatively) has the property that all products of $x_1, \ldots, x_m$ (in any order) are equal. More generally, an associative binary operation (denoted multiplicatively) has the following property: when $x_1, \ldots, x_m$ all commute with each other ($x_i x_j = x_j x_i$ for all i, j), then all products of $x_1, \ldots, x_m$ (in any order) are equal.*

The proof of Proposition 1.1.2 consists in showing, by induction on m, that $x_{\sigma 1} x_{\sigma 2} \ldots x_{\sigma m} = x_1 x_2 \ldots x_m$ for every permutation σ of $\{1, 2, \ldots, m\}$. We leave this to the reader as an exercise.

1.1.d. Powers. Powers are a particular case of products. Given an associative binary operation on a set X (denoted multiplicatively) and a positive integer m, the m-**th power** of $x \in X$ is the product $x^m = x_1 x_2 \ldots x_m$ in which $x_1 = x_2 = \cdots = x_m = x$. Associativity is required in this definition to avoid ambiguity.

In the additive notation, $x^n = xx \ldots x$ becomes $nx = x + x + \cdots + x$ and is an **integer multiple** of x.

Proposition 1.1.3. *Given an associative binary operation on a set X (denoted multiplicatively), powers have the following properties: for all $x, y \in S$ and $m, n > 0$:*

(1) $x^1 = x$;

(2) $x^m x^n = x^{m+n} = x^m x^n$;

(3) $(x^m)^n = x^{mn}$;

(4) *if $xy = yx$, then $(xy)^m = x^m y^m$.*

The proof is left to the reader.

1.1.e. Products of Subsets. Every binary operation μ on a set X induces a binary operation on the set 2^X of all subsets of X. If μ is denoted multiplicatively, the **product** of two subsets A and B of X is

$$AB = \{ ab \,;\, a \in A,\ b \in B \}.$$

AB is denoted by $A + B$ if μ is written additively.

It is customary to denote $\{a\}B$ by aB, and $A\{b\}$ by Ab. This creates no ambiguity since $\{a\}\{b\} = \{ab\}$. The reader will show:

Proposition 1.1.4. *If the operation on X (denoted multiplicatively) is associative, then the product of subsets of X is associative. If the operation on X is commutative, then the product of subsets of X is commutative.*

Other properties will be found in the exercises.

1.1.f. Congruences. Let $\mathscr{E}$ be an equivalence relation on the set X. As in Section B.1 we denote by E_x the equivalence class (the $\mathscr{E}$-class) of x. We show how an operation on the set X can induce an operation on the quotient set $X/\mathscr{E}$ (the set of all $\mathscr{E}$-classes).

Proposition 1.1.5. *Given a binary operation on X (denoted multiplicatively), and an equivalence relation $\mathscr{E}$ on X, the following are equivalent:*

(1) *There is an operation on $X/\mathscr{E}$ such that $E_x E_y = E_{xy}$ for all $x, y \in X$;*

(2) *$a \,\mathscr{E}\, b$, $c \,\mathscr{E}\, d$ implies that $ac \,\mathscr{E}\, bd$ for all $a, b, c, d \in X$.*

In (1) *the operation on* $X/\mathscr{E}$ *is unique; the product* E_xE_y *in* $X/\mathscr{E}$ *is the equivalence class which contains the product of* E_x *and* E_y *as subsets of* X.

Proof. (1) implies (2). If there is an operation on $X/\mathscr{E}$ that satisfies (1), then $a \mathrel{\mathscr{E}} b$, $c \mathrel{\mathscr{E}} d$ implies that $E_a = E_b$, $E_c = E_d$, $E_{ac} = E_aE_c = E_bE_d = E_{bd}$, and $ac \mathrel{\mathscr{E}} bd$.

(2) implies (1). Assume that $a \mathrel{\mathscr{E}} b$, $c \mathrel{\mathscr{E}} d$ implies $ac \mathrel{\mathscr{E}} bd$. Let $x, y \in E$. Then $a \in E_x$, $b \in E_y$ implies $a \mathrel{\mathscr{E}} x$, $b \mathrel{\mathscr{E}} y$, $ab \mathrel{\mathscr{E}} xy$, and $ab \in E_{xy}$. Thus the product E_xE_y of E_x and E_y as subsets of X is contained in the single $\mathscr{E}$-class E_{xy}. Therefore there is a binary operation on $X/\mathscr{E}$ under which the product of two $\mathscr{E}$-classes is the $\mathscr{E}$-class which contains their product as subsets of X, and this operation satisfies (1). The proof also shows that this operation is the only operation that satisfies (1).   □

An equivalence relation is called a **congruence** (relative to some binary operation) in case it satisfies the conditions in Proposition 1.1.5.

1.1.g. Light's Test. We conclude this section with a technique to verify the associativity of an operation μ on a set X from its multiplication table. Testing for associativity can be a lengthy task: for a six-element set, one must consider 216 triples. **Light's associativity test** provides an efficient way to do just that. Associativity means that, for each $t \in X$, the two binary operations on X, $(x, y) \longmapsto (xt)y$ and $(x, y) \longmapsto x(ty)$, are identical. In Light's test each element t is tested for associativity by comparing these two operations. Tables of the two operations are readily constructed from the multiplication table (T) of μ: for the first operation the row of x coincides with the row of xt in (T); for the second operation the column of y coincides with the column of ty in (T).

Here is a practical procedure which tests the element t. Using the given multiplication table (T), construct the table (t) of the operation $(x, y) \longmapsto (xt)y$, which we call the **Light's table** of t, as follows: i) write the column of t on the left; ii) for each entry $u = xt$ in the column of t, the row of u in table (t) is the row of u in table (T); iii) write the row of t at the top. The element t **passes Light's test** $((xt)y = (xt)y$ for all $x, y)$ if and only if, for every entry $v = ty$ in the row of t, the column of v in table (t) equals the column of v in table (T).

Example 1.1.6. Given the multiplication table (T) on the left,

T	a	b	c	d		d	b	c	a	c
a	a	b	c	b		b	b	c	a	c
b	b	c	a	c		c	c	a	b	a
c	c	a	b	a		a	a	b	c	b
d	b	c	a	c		c	c	a	b	a

the Light's table (d) of d is on the right. Inspection shows that in this case, for each entry $v = dy$ in the row of d, the column of v in table (d) equals the column of v in table (T); hence d passes Light's test.

To establish associativity, one should test every element of X. However, the following result states that it suffices to test the generators of X. **Generators** are elements of X such that every element of S is a product of [a sequence of] generators. Note that there may be more than one set of generators. Also generators of groups or modules have somewhat different definitions, but we will see (Corollary 1.4.10) that the generators of a *finite* group can be defined as above.

Proposition 1.1.7. *In Light's associativity test, only the generators need be tested.*

Proof. If one of the generators fails the test, then we don't have associativity. Now assume that every generator passes. We prove by induction on m that every product of m generators must also pass. This is true if $m = 1$ and implies that every element of X passes, and associativity holds.

Let p be a product of $m > 1$ generators. Then $p = qr$, where q is a product of $k < m$ generators and r is a product of $m - k < m$ generators. By the induction hypothesis, q and r both pass Light's test. Hence

$$x(py) = x((qr)y) = x(q(ry)) = (xq)(ry) = ((xq)r)y = (x(qr))y = (xp)y$$

for all x, y, and p passes Light's test. $\square$

For instance, the operation in Example 1.1.6 is associative. We already saw that the element d passes Light's test. Now $dd = c$, $cd = a$, and $cc = b$; hence $\{d\}$ is a set of generators, and associativity follows from Proposition 1.1.7.

Other examples will be found in the exercises.

Exercises

1. List all products of a,b,c,d (in that order). (Do not assume associativity!)
2. List all products of a,b,c,d,e (in that order). (Do not assume associativity.)
3. Prove Proposition 1.1.2.
4. Prove Proposition 1.1.3.
5. Prove Proposition 1.1.4.
6. Let $A,B \subseteq X$. Show that $AB = \emptyset$ if and only if $A = \emptyset$ or $B = \emptyset$.
7. Show that the product of subsets distributes unions:

$$\left(\bigcup_{i \in I} A_i \right) B = \bigcup_{i \in I} (A_i B) \quad \text{and} \quad A \left(\bigcup_{j \in J} B_j \right) = \bigcup_{j \in J} (AB_j)$$

for all $A, B, A_i, B_j \subseteq X$.

8. Is the following operation associative?

	a	b	c	d
a	a	b	a	b
b	a	b	a	b
c	c	d	c	d
d	c	d	c	d

9. Is the following operation associative?

	a	b	c	d
a	a	b	a	b
b	b	a	d	c
c	a	b	c	d
d	d	c	d	c

1.2. DEFINITION AND EXAMPLES OF GROUPS

1.2.a. Definition. A **group** $G = (G, \bullet)$ is a set G together with a binary operation $\bullet$ on G (denoted multiplicatively in this chapter) with the following properties:

(1) The operation is associative.

(2) There exists an **identity element** (an element $e \in G$ such that $ex = xe = x$ for all $x \in G$), relative to which:

(3) each element $x \in G$ has an **inverse** (an element $y \in G$ such that $xy = yx = e$).

A **semigroup** $S = (S, \bullet)$ is a set S together with an associative binary operation $\bullet$ on S. A semigroup with an identity element is also called a **monoid**. Thus a group is a monoid in which every element has an inverse.

It is customary to denote a group $(G, \bullet)$ and its **underlying set** G by the same letter. Older abstract algebra texts begin the definition of a group with the axiom:

every ordered pair (x, y) of elements of G has exactly one product xy in G,

which precisely means that there is a binary operation on G. This is still a good property to consider first when proving that some object is a group.

The **order** of a group G is the [cardinal] number $|G|$ of elements of G.

1.2.b. Examples. A **trivial** group is a group with just one element e (then $ee = e$).

Finite groups with few elements can be specified by their multiplication table.

Example 1.2.1. The **Klein's four-group** is the group $V_4 = \{1,a,b,c\}$ with the multiplication table

$$
\begin{array}{c}
V_4 \\
\hline
\begin{array}{cccc}
1 & a & b & c \\
a & 1 & c & b \\
b & c & 1 & a \\
c & b & a & 1
\end{array}
\end{array}
$$

Note that the identity element is listed first, so that we can omit the column headings, which would duplicate the first row, and the row headings, which would duplicate the first column. Associativity follows from Light's test and is left as an exercise. There is an obvious identity element, and in V_4 each element is its own inverse.

We see that V_4 has order 4 and is an **abelian** group (its operation is commutative). Abelian groups play a major role in algebra as part of richer structures such as rings, fields, and modules. They will be studied in some detail at a later time.

Our next example is denoted additively. It is the abelian group $\mathbb{Z}$ whose elements are all integers, and whose operation is addition. We know from Proposition A.2.4 that the addition of integers is associative ($x + (y + z) = (x + y) + z$) and commutative ($x + y = y + x$), the integer 0 is an identity element ($0 + x = x + 0 = x$ for all x), and each integer x has an inverse (generally called its **opposite**) $-x$, such that $x + (-x) = (-x) + x = 0$. Thus $\mathbb{Z}$ is an abelian group.

The group $\mathbb{Z}_n$ of integers modulo n in Section A.3 is a similar example, for each integer $n > 0$. Number systems provide other examples of groups (see the exercises).

The simplest example of a group which need not be abelian or finite is the group of all permutations of a set.

Example 1.2.2. Given a set X, the **symmetric group** on X is the group S_X whose elements are all bijections of X into itself (also called **permutations** of X), and whose operation is composition of mappings. The symmetric group on $\{1,2,\ldots,n\}$ is denoted by S_n; it has order $n!$.

We write permutations as left operators (so that the image of $x \in X$ under $\sigma \in S_X$ is σx). Composition is given by $(\sigma\tau)x = \sigma(\tau x)$ (with τ applied first). Properties of bijections in Section B.1 imply that S_X is a group: if σ and τ

are permutations of X, then so is $\sigma\tau$; composition is associative; the identity mapping on X is an identity element of S_X; each permutation of X has an inverse bijection, which is also its inverse in S_X.

Many interesting examples of groups consist of permutations. We give two such examples.

Example 1.2.3. Given a vector space V over a field F, the **general linear group** is the group $GL(V)$ whose elements are all invertible (= bijective) linear transformations of V into itself, and whose operation is composition of mappings. A related example is the group $GL(n,F)$ whose elements are all invertible $n \times n$ matrices with coefficients in F, and whose operation is the multiplication of matrices.

As above, properties of linear transformations and matrices, which skeptical readers will prove for themselves, imply that $GL(V)$ and $GL(n,F)$ are groups.

The next example arises from any euclidean space E. An **isometry** of E is a permutation σ of E which preserves distances (if A and B are two points of E, then $d(\sigma A, \sigma B) = d(A,B)$).

Given a bounded subset S of an euclidean space E, the **group of all rotations and symmetries of S** is the group whose elements are all isometries of E which leave S invariant ($\sigma S = S$), and whose operation is composition of mappings. Readers who know nothing of euclidean spaces can still verify that this is a group. It can be shown that an isometry which leaves a bounded subset invariant must be a rotation or a reflection, hence the name of this group.

Example 1.2.4. Given a positive integer $n \geqslant 2$, the **dihedral group** D_n of index n is the group of all rotations and symmetries of a regular polygon with n vertices in an euclidean plane.

We note that a regular polygon with n vertices has a center and has n axes of symmetry, which are the straight lines joining the center to a vertex or to the midpoint of an edge. Hence D_n contains n reflections; D_n also contains n rotations about the center, by angles $2k\pi/n$ ($1 \leqslant k \leqslant n$). Our intuition tells us that D_n contains only these $2n$ isometries, and so D_n has order $|D_n| = 2n$.

To set up a multiplication table for D_n, let ρ_k denote the rotation by $2k\pi/n$ about the center, let L_0 be the straight line joining the center to any vertex, let L_k be the straight line obtained by rotating L_0 by $k\pi/n$ about the center, and let σ_k be the reflection about L_k. It is convenient to let k be any integer so that $\rho_{k+n} = \rho_k$ and $\sigma_{k+n} = \sigma_k$. The distinct elements of D_n are $\rho_0, \ldots, \rho_{n-1}, \sigma_0, \ldots, \sigma_{n-1}$. We see that $\rho_i \circ \rho_j = \rho_{i+j}$. Next geometry shows that $\sigma_i \circ \sigma_j$ is a rotation by twice the angle from L_j to L_i, that is, $\sigma_i \circ \sigma_j = \rho_{i-j}$. Composing with σ_i and σ_j yields $\sigma_j = \sigma_i \circ \rho_{i-j}$ so that $\sigma_i \circ \rho_k = \sigma_{i-k}$, and $\sigma_i = \rho_{i-j} \circ \sigma_j$ so that $\rho_k \circ \sigma_j = \sigma_{k+j}$. Thus

$$\rho_i \rho_j = \rho_{i+j}, \qquad \rho_i \sigma_j = \sigma_{i+j}, \qquad \sigma_i \sigma_j = \rho_{i-j}, \qquad \sigma_i \rho_j = \sigma_{i-j}.$$

1.2.c. Properties. We now return to the definition of a group and point out some general consequences of it.

Proposition 1.2.5. *In a group, the identity element is unique, and the inverse of an element is unique.*

This is fun to prove; we'll let the reader do it. In a group (denoted multiplicatively) the identity element is generally denoted by 1 and the inverse of x by x^{-1}. Thus a trivial group is generally denoted by $\{1\}$ (or by 1).

Proposition 1.2.5 shows that a group has in fact three basic operations: its binary operation, the constant operation which selects the identity element, and the unary operation $x \longmapsto x^{-1}$.

Proposition 1.2.6. *In a group, the **cancellation laws** hold: $ca = cb$ implies $a = b$; $ac = bc$ implies $a = b$. Moreover the equations $ax = b$, $ya = b$, have unique solutions $x = a^{-1}b$, $y = ba^{-1}$.*

Proof. If $ca = cb$, then $a = 1a = c^{-1}ca = c^{-1}cb = 1b = b$. Similarly $ac = bc$ implies $a = b$. Moreover $ax = b$ implies $x = a^{-1}ax = a^{-1}b$; conversely, $a(a^{-1}b) = b$. Similarly $ya = b$ if and only if $y = ba^{-1}$. $\qquad\square$

More products and powers can be defined in a group than with associative operations in general. In a group (more generally, in a monoid) the **empty product** (the product of the empty sequence, $x_1, \ldots, x_m$ with $m = 0$) is 1. In particular, $x^0 = 1$. This definition ensures that if we multiply, say, xy by the empty product, which adds no new terms, the result is still xy.

Proposition 1.2.7. *In a group*

$$(x^{-1})^{-1} = x \qquad and \qquad (x_1 x_2 \ldots x_m)^{-1} = x_m^{-1} \ldots x_2^{-1} x_1^{-1}.$$

Proof. The equalities $x^{-1}x = xx^{-1} = 1$ show that x is the inverse of x^{-1}.

The second equality in the statement is true if $m = 0$ or $m = 1$. If $m > 1$, it is proved by induction on m: the equality $x_m x_m^{-1} = 1$, and the induction hypothesis $(x_1 x_2 \ldots x_{m-1})^{-1} = x_{m-1}^{-1} \ldots x_2^{-1} x_1^{-1}$ yield

$$x_1 \ldots x_{m-1} x_m \, x_m^{-1} x_{m-1}^{-1} \ldots x_1^{-1} = x_1 \ldots x_{m-1} x_{m-1}^{-1} \ldots x_1^{-1} = 1.$$

Since $(x_1 \ldots x_{m-1} x_m)(x_1 \ldots x_{m-1} x_m)^{-1} = 1$, the cancellation law (Proposition 1.2.6) then yields $(x_1 x_2 \ldots x_m)^{-1} = x_m^{-1} \ldots x_2^{-1} x_1^{-1}$. $\qquad\square$

If X is a subset of a group, then $X^{-1} = \{x^{-1}; x \in X\}$ denotes the set of all inverses of elements in X. Proposition 1.2.7 implies $(X_1 X_2 \ldots X_m)^{-1} = X_m^{-1} \ldots X_2^{-1} X_1^{-1}$ for all subsets $X_1, X_2, \ldots, X_m$.

1.2.d. Powers. It follows from Proposition 1.2.7 that $(x^n)^{-1} = (x^{-1})^n$ for all $n > 0$. Hence negative powers can be defined in a group as follows: if n is a positive integer, then

$$x^{-n} = (x^n)^{-1} = (x^{-1})^n.$$

With $n = 1$ this defines x to power -1 as the inverse of $x^1 = x$; no confusion arises when both operations are denoted by x^{-1}.

In a group, x^n is now defined for every integer n. This was achieved with no loss of properties:

Proposition 1.2.8. *Powers in a group G have the following properties: for all $x, y \in G$ and $m, n \in \mathbb{Z}$:*

(1) $x^0 = 1^n = 1$;

(2) $(x^n)^{-1} = (x^{-1})^n$;

(3) $x^1 = x$;

(4) $x^m x^n = x^{m+n} = x^m x^n$;

(5) $x^m (x^n)^{-1} = (x^n)^{-1} x^m = x^{m-n}$;

(6) $(x^m)^n = x^{mn}$;

(7) *if $xy = yx$, then $(xy)^m = x^m y^m$.*

This is no fun to prove; we'll leave it to the reader.

It follows from Proposition 1.2.8 that equalities between powers of one element follow simple patterns.

Proposition 1.2.9. *Let a be an element of a group. Either $a^n \neq 1$ for all $n \neq 0$, and then $a^p = a^q$ if and only if $p = q$; or there exists a smallest integer $n > 0$ such that $a^n = 1$, and then $a^p = a^q$ if and only if n divides $p - q$.*

Proof. By Proposition 1.2.8, $a^p = a^q$ if and only if $a^{p-q} = a^p(a^q)^{-1} = 1$. If $a^n \neq 1$ for all $n \neq 0$, then $a^p = a^q$ if and only if $p - q = 0$.

Now assume that $a^n = 1$ for some $n \neq 0$. Then $a^{-n} = 1$, so that $a^n = 1$ for some $n > 0$. Let n denote the smallest such positive integer. If $a^m = 1$, then dividing m by n yields $m = nq + r$, where $0 \leqslant r < n$, $a^r = 1$ (since $a^m = a^{nq} = 1$), $r = 0$ (since $a^r = 1$ with $0 < r < n$ would contradict the choice of n), and n divides m. Thus $a^m = 1$ if and only if n divides m. (This also follows from Proposition A.3.3, since $I = \{ m \in \mathbb{Z}; a^m = 1 \}$ is an ideal of $\mathbb{Z}$.) Then $a^p = a^q$ if and only if $a^{p-q} = 1$, if and only if n divides $p - q$. $\square$

The **order** of each element a of a group is defined as follows: if $a^n \neq 1$ for all $n \neq 0$, then a **has infinite order** and the order of a is the infinite cardinal $\aleph_0 = |\mathbb{Z}|$; otherwise, a **has finite order** and the order of a is the smallest integer $n > 0$ such that $a^n = 1$.

It should be remembered that an equality $a^m = 1$ with $m > 0$ does not imply that a has order m; by Proposition 1.2.9, it implies only that a has finite order and that the order of a divides m.

When a has finite order $n > 0$, the powers of a arrange themselves in a cyclic pattern. First, $a^0 = 1$, $a^1 = a$, $a^2, \ldots, a^{n-1}$ are all distinct, since n cannot divide $i - j$ when $i \neq j$ and $0 \leqslant i, j < n$. Thereafter $a^n = a^0$, $a^{n+1} = a$, $a^{n+2} = a^2$, and so on. Similarly $a^{-1} = a^{n-1}$, $a^{-2} = a^{n-2}$, and so on. Thus the order of a is also the number of distinct powers of a. (This also holds if a has infinite order.)

In a finite group it is not possible for all the powers of a to be distinct. Hence:

Corollary 1.2.10. *In a finite group G, every element a has finite order. In particular, a^{-1} is a positive power of a.*

Exercises

1. Let G be a semigroup in which there is a **left identity element** (an element $e \in G$ such that $ex = x$ for all $x \in G$) relative to which every element x has a **left inverse** (an element $y \in G$ such that $yx = e$). Prove that G is a group.

2. Let G be a nonempty semigroup in which the equations $ax = b$ and $ya = b$ can be solved for every $a, b \in G$. Prove that G is a group.

3. Let G be a finite nonempty semigroup in which the cancellation laws hold. Prove that G is a group. Show by an example that this result need not be true if G is infinite.

4. Verify that V_4 is a group.

5. Draw a multiplication table for S_3.

6. Prove that in a monoid the identity element is unique and that the inverse of an element (if it exists) is unique.

7. Prove Proposition 1.2.8.

8. List the order of every element of D_3.

9. List the elements of S_4 and find the order of each element.

10. When $X = \mathbb{N}$, $\mathbb{Z}$, or $\mathbb{Z}_n$, when is $(X, +)$ a group? when is $(X, \cdot)$ a group? when is $(X \backslash \{0\}, \cdot)$ a group?

11. Can you think of subsets of $\mathbb{R}$ that are groups under the multiplication on $\mathbb{R}$? What about $\mathbb{C}$?

12. Let P be the set of all subsets of a nonempty set X. Show that P is a monoid, but not a group, under the binary operation $\cup$.

1.3. HOMOMORPHISMS

Homomorphisms are the means by which groups relate to each other. These relationships between groups are important, since what happens inside a group is often determined by how it relates to other groups.

1.3.a. Definition. A **homomorphism** of a group G into a group G' is a mapping φ of G into G' such that

$$\varphi(xy) = \varphi(x)\varphi(y) \qquad \text{for all} \quad x, y \in G;$$

G is the **domain** of φ and G' is the **codomain** of φ.

A homomorphism of groups must also preserve identity elements, inverses, and all products and powers:

Proposition 1.3.1. *Every homomorphism φ of a group G into a group G' has the following properties: for all $x \in G$ and $n \in \mathbb{Z}$:*

(1) $\varphi(1) = 1;$
(2) $\varphi(x^{-1}) = \varphi(x)^{-1};$
(3) $\varphi(x_1 x_2 \ldots x_m) = \varphi(x_1)\varphi(x_2) \ldots \varphi(x_m)$ *for all* $x_1, \ldots, x_m \in G;$
(4) $\varphi(x^n) = \varphi(x)^n.$

Proof

(1) By definition, $\varphi(1)\varphi(1) = \varphi(1 \cdot 1) = \varphi(1) = \varphi(1)1$; hence $\varphi(1) = 1$.
(2) Similarly $\varphi(x)\varphi(x^{-1}) = \varphi(1) = 1 = \varphi(x)\varphi(x)^{-1}$; hence $\varphi(x^{-1}) = \varphi(x)^{-1}$.
(3) This follows from the definition if $m > 0$ and from (1) if $m = 0$.
(4) This follows from (3) if $n > 0$, from (1) if $n = 0$, and from (2) if $n < 0$. $\qquad\square$

1.3.b. Composition. The **composition** of two homomorphisms is a homomorphism: if $\varphi : G \longrightarrow G'$ and $\psi : G' \longrightarrow G''$ are homomorphisms, then so is $\psi \circ \varphi : G \longrightarrow G''$. Composition of homomorphisms is, like the composition of mappings, associative: $\chi \circ (\psi \circ \varphi)$ is defined if and only if $(\chi \circ \psi) \circ \varphi$ is defined, and then $\chi \circ (\psi \circ \varphi) = (\chi \circ \psi) \circ \varphi$.

The **identity** mapping $1 = 1_G : G \longrightarrow G$ on a group G is a homomorphism (defined by: $1_G(x) = x$ for all $x \in G$). This identity homomorphism has the property that $\varphi \circ 1 = \varphi$ for all $\varphi : G \longrightarrow G'$ and $1 \circ \psi = \psi$ for all $\psi : G'' \longrightarrow G$. The identity homomorphism $1 : G \longrightarrow G$ is distinguished from the identity element $1 \in G$ only from context; we will avoid confusing expressions such as $1(x) = x$.

An injective homomorphism φ is often called **one-to-one** or a **monomorphism**; it has the property that $\varphi \circ \psi = \varphi \circ \chi$ implies $\psi = \chi$. A surjective homomorphism φ is often called **onto** or an **epimorphism**; it has the property that $\psi \circ \varphi = \chi \circ \varphi$ implies $\psi = \chi$. The proofs are exercises.

1.3.c. Isomorphisms. An **isomorphism** is a bijective homomorphism.

Let $\theta : G \longrightarrow G'$ be an isomorphism. The reader will verify that the inverse bijection $\theta^{-1} : G' \longrightarrow G$ is also an isomorphism (the **inverse** of θ). Then $\theta^{-1} \circ$

$\theta = 1_G$ and $\theta \circ \theta^{-1} = 1_{G'}$. If, conversely, $\theta : G \longrightarrow G'$ and $\zeta : G' \longrightarrow G$ are homomorphisms such that $\zeta \circ \theta = 1$ and $\theta \circ \zeta = 1$, then θ is an isomorphism and $\zeta = \theta^{-1}$.

Two groups G and G' are **isomorphic** when there exists an isomorphism of G onto G'; this relationship (also called isomorphism) is denoted by $G \cong G'$.

For example, the groups

V_4

1	a	b	c
a	1	c	b
b	c	1	a
c	b	a	1

G

1	x	y	z
x	1	z	y
y	z	1	x
z	y	x	1

are isomorphic: there is an isomorphism $\theta : V_4 \longrightarrow G$ which sends $1 \in V_4$ to $1 \in G$, a to x, b to y, and c to z.

In general, $G \cong G$ (since 1_G is an isomorphism), $G \cong G'$ implies $G' \cong G$ (since the inverse of an isomorphism is an isomorphism), and $G \cong G' \cong G''$ implies $G \cong G''$ (since the composition of two isomorphisms is an isomorphism). Thus isomorphism is like an equivalence relation. It would be an actual equivalence relation if only groups would cooperate and form a set. Nevertheless, one speaks of the **isomorphism class** of a group G, which consists of all groups isomorphic to G.

Philosophical considerations give isomorphism a particular importance. Abstract algebra studies very carefully how the elements of a group relate to each other but cares little what its elements are like. Accordingly, isomorphic groups are treated as is they were the same "abstract" group. For example, all the dihedral groups D_3 of index 3 (as defined in Section 1.2) are isomorphic to each other; we regard them all as instances of one abstract group, *the* dihedral group D_3. Likewise we speak of *the* trivial group.

1.3.d. Endomorphisms and Automorphisms. An **endomorphism** of a group G is a homomorphism of G into G. An **automorphism** of a group G is an isomorphism of G onto G. The **group of automorphisms** of a group G is the group $\text{Aut}(G)$ whose elements are all automorphisms of G and whose operation is composition.

General properties of isomorphisms ensure that $\text{Aut}(G)$ is a group. The following construction will show that $\text{Aut}(G)$ is nontrivial when G is nonabelian:

Proposition 1.3.2. *Let G be a group. For each $a \in G$, $\alpha_a : x \longmapsto axa^{-1}$ is an automorphism of G. Furthermore $\alpha : a \longmapsto \alpha_a$ is a homomorphism of G into $\text{Aut}(G)$.*

Proof. We have $\alpha_a(x)\alpha_a(y) = axa^{-1}aya^{-1} = axya^{-1} = \alpha_a(xy)$ (since $a^{-1}a = 1$). Also $\alpha_a(\alpha_b(x)) = a(bxb^{-1})a^{-1} = \alpha_{ab}(x)$ (since $(ab)^{-1} = b^{-1}a^{-1}$), and

$\alpha_a \circ \alpha_b = \alpha_{ab}$. Hence $\alpha_a \circ \alpha_{a^{-1}} = \alpha_1 = 1_G = \alpha_{a^{-1}} \circ \alpha_a$, so that α_a is an automorphism, and $\alpha : G \longrightarrow \mathrm{Aut}(G)$ is a homomorphism. $\qquad\qquad\square$

The automorphisms of a group G constructed in Proposition 1.3.6 are the **inner automorphisms** of G.

If $\alpha_a = 1_G$, then $axa^{-1} = x$ and $ax = xa$ for all $x \in G$. If $\alpha_a = 1_G$ for all $a \in G$, then G is abelian. If G is not abelian, then $\alpha_a \neq 1_G$ for some $a \in G$, and therefore $\mathrm{Aut}(G)$ is nontrivial.

Exercises

1. Let $\theta : G \longrightarrow G'$ be an isomorphism. Verify that the inverse bijection θ^{-1} is an isomorphism.
2. Show that an injective homomorphism φ of groups has the property that $\varphi \circ \psi = \varphi \circ \chi$ implies $\psi = \chi$.
3. Show that a surjective homomorphism φ of groups has the property that $\psi \circ \varphi = \chi \circ \varphi$ implies $\psi = \chi$.
4. Show that a homomorphism φ of groups is injective if and only if $\varphi(x) = 1$ implies $x = 1$.
5. Find all homomorphisms of D_2 into D_3.
6. Find all homomorphisms of D_3 into D_2.
7. Let V be a vector space of dimension n over a field K. Show that $GL(V,K) \cong GL(n,K)$. (See Example 1.2.3 for definitions.)
8. Show that $D_2 \cong V_4$.
9. Show that $D_3 \cong S_3$.
10. Find all automorphisms of V_4. Draw their multiplication table. Then prove that $\mathrm{Aut}(V_4) \cong S_3$.
11. Find all automorphisms of D_3. Then find $\mathrm{Aut}(D_3)$.
12. Prove that $G \cong G'$ implies $\mathrm{Aut}(G) \cong \mathrm{Aut}(G')$.
13. Show that $\mathrm{Aut}(G) \cong \mathrm{Aut}(G')$ does not imply $G \cong G'$.

1.4. SUBGROUPS

1.4.a. Definition. Recall that for each subset $Y \subseteq X$ of a set X there is an **inclusion mapping** $\iota : Y \longrightarrow X$ defined by: $\iota(y) = y$ for every $y \in Y$; ι is injective and $\iota(Y) = Y$.

A **subgroup** of a group G is a group H such that

(1) as a set, $H \subseteq G$, and
(2) the inclusion mapping $\iota : H \longrightarrow G$ ($\iota(x) = x \in G$ for all $x \in H$) is a homomorphism (the **inclusion homomorphism**).

We denote this relationship by $H \leqslant G$. (The more traditional notation $H < G$ seems to imply that $H \neq G$.) We write $H \lneqq G$ when $H \leqslant G$ and $H \neq G$.

In the definition of a subgroup, the condition that the inclusion mapping $\iota : H \longrightarrow G$ be a homomorphism means that xy is the same in H and G whenever $x, y \in H$. Hence a subgroup of G is completely determined by its underlying set. The latter is also called a subgroup. Subgroups in this sense are characterized by the following straightforward result:

Proposition 1.4.1. *A subset H of a group G is [the underlying set of] a subgroup of G if and only if:*

(1) $1 \in H$;

(2) *H is closed under inverses ($x \in H$ implies $x^{-1} \in H$);*

(3) *H is closed under multiplication ($x, y \in H$ implies $xy \in H$).*

Proof. A subgroup has properties (1), (2), and (3) by definition of homomorphisms and Proposition 1.3.1. Conversely, let H be a subset of G with properties (1), (2), and (3). By (3), the operation $G \times G \longrightarrow G$ can be restricted to an operation $H \times H \longrightarrow H$ on H. Associativity is inherited from G, and properties (1) and (2) ensure that H is a group. The inclusion mapping $\iota : H \longrightarrow G$ is now a homomorphism. ☐

Proposition 1.4.1 is often used as a definition of subgroups. It is interesting that condition (3) by itself does not characterize subgroups and that a subgroup must be closed under all three basic group operations (identity, inverse, and multiplication). A subgroup H is also closed under products (if $x_1, \ldots, x_m \in H$, then $x_1 x_2 \ldots x_m \in H$) and closed under powers (if $x \in H$, then $x^n \in H$ for all $n \in \mathbb{Z}$). Proposition 1.4.1 also implies that $H^{-1} = \{ x^{-1} ; x \in H \} = H$ and that $HH = H$ (by (3) and (1), $HH \subseteq H = 1H \subseteq HH$).

It is useful to have more compact characterizations of subgroups. We leave the proofs as exercises.

Lemma 1.4.2. *A subset H of a group G is a subgroup of G if and only if $H \neq \emptyset$, and $x, y \in H$ implies $xy^{-1} \in H$.*

Lemma 1.4.3. *A subset of a* finite *group G is a subgroup of G if and only if it is nonempty and closed under multiplication.*

Lemma 1.4.3 follows from Corollary 1.2.10 and does not extend to infinite groups. Skeptical readers will readily find counterexamples.

1.4.b. Examples. Every group G has subgroups: the **trivial** subgroup $\{1\}$ and the group G itself are always subgroups of G. Subgroups $H \neq \{1\}$ of G are called **nontrivial**; subgroups $H \lneq G$ are called **proper**.

It follows from Proposition 1.2.12 that the powers of an element a of G constitute a subgroup

$$\langle a \rangle = \{ a^n ; n \in \mathbb{Z} \}.$$

This subgroup is the **cyclic** subgroup of G **generated by** a. We saw that the order of a cyclic subgroup $\langle a \rangle$ (the number of distinct powers of a) equals the order of its generator a: if a has finite order n, then $\langle a \rangle = \{1, a, a^2, \ldots, a^{n-1}\}$, by Proposition 1.2.9.

A group G is **cyclic** in case $G = \langle a \rangle$ for some $a \in G$. The following result is proved in Appendix A (Proposition A.3.3); we give a direct proof:

Proposition 1.4.4. *Every subgroup of $\mathbb{Z}$ is cyclic.*

Proof. In the additive notation used for $\mathbb{Z}$, products xy become sums $x + y$, and the inverse x^{-1} becomes an opposite $-x$; hence $H \subseteq \mathbb{Z}$ is a subgroup of $\mathbb{Z}$ if and only if $0 \in H$, and $x, y \in H$ implies $x - y \in H$. Similarly a positive power $x^n = xx \ldots x$ becomes $x + x + \cdots + x = nx$; $x^0 = 1$ becomes $0x = 0$; a negative power $x^{-n} = (x^n)^{-1}$ (with $n > 0$) becomes $-(nx) = (-n)x$.

Let H be a subgroup of $\mathbb{Z}$. If $H = \{0\}$, then $H = \langle 0 \rangle$ is cyclic. Now assume that $H \neq \{0\}$. Since $x \in H$ implies $-x \in H$, H contains a positive integer. Let n be the smallest positive integer contained in H. Then $\langle n \rangle \subseteq H$. Conversely, let $x \in H$; dividing x by n yields $x = nq + r$, where $0 \leqslant r < n$, $r = x - nq \in H$, $r = 0$ (since $r \in H$ with $0 < r < n$ would contradict the choice of n), and $x \in \langle n \rangle$. Thus $H = \langle n \rangle$ is cyclic. $\square$

In particular, $\mathbb{Z}$ is a cyclic group, with $\mathbb{Z} = \langle 1 \rangle = \langle -1 \rangle$.

The name "cyclic" (circular) may have originated with the complex n-th roots of unity $\omega_k = \cos(2k\pi/n) + i\sin(2k\pi/n) = \omega_1^k$ ($k = 0, 1, \ldots, n - 1$), which lie on the unit circle $|z| = 1$ and form a cyclic group G under multiplication (generated by ω_1). The generators of G are called **primitive** n-th roots of unity; ω_k is primitive (generates G) if and only if g.c.d. $(k, n) = 1$ (see the exercises). Proposition 1.2.9 shows that every cyclic group of order n is isomorphic to the cyclic group of n-th roots of unity.

Subgroups of cyclic groups are themselves cyclic (see Proposition 1.4.16 below).

Proposition 1.4.5. *When* $\varphi : G \longrightarrow G'$ *is a homomorphism of groups, then*

$$\operatorname{Im} \varphi = \varphi(G) = \{\varphi(x); x \in G\}$$

is a subgroup of G'.

Proof. By Proposition 1.3.2, $1 \in \operatorname{Im} \varphi$, and $u = \varphi(x) \in \operatorname{Im} \varphi$ implies $u^{-1} = \varphi(x^{-1}) \in \operatorname{Im} \varphi$. By definition of a homomorphism, $u = \varphi(x)$, $v = \varphi(y) \in \operatorname{Im} \varphi$ implies $uv = \varphi(x)\varphi(y) \in \operatorname{Im} \varphi$. $\square$

Other examples of subgroups will be found in the exercises.

1.4.c. Properties

Proposition 1.4.6. *Every intersection of subgroups of a group G is a subgroup of G.*

The proof is an exercise.

A union of subgroups is not in general a subgroup (see the exercises). We note the following exception. A **chain** of subgroups is a family $(H_i)_{i \in I}$ of subgroups such that $H_i \subseteq H_j$ or $H_j \subseteq H_i$ holds for each $i, j \in I$.

Proposition 1.4.7. *The union of a nonempty chain of subgroups of a group G is a subgroup of G.*

Proof. Let $(H_i)_{i \in I}$ be a chain of subgroups, with $I \neq \emptyset$, and $H = \bigcup_{i \in I} H_i$. We have $1 \in H$, since $I \neq \emptyset$, and $1 \in H_i$ for some i. Let $x, y \in H$. Then $x \in H_i$ and $y \in H_j$ for some $i, j \in I$. If $H_i \subseteq H_j$, then $x, y \in H_j$ and $xy^{-1} \in H_j$. If $H_j \subseteq H_i$, then $x, y \in H_i$ and $xy^{-1} \in H_i$. In either case, $xy^{-1} \in H$. $\square$

Proposition 1.4.8. *Let $\varphi : G \longrightarrow G'$ be a homomorphism. If $H \leqslant G$, then $\varphi(H) \leqslant G'$. If $H' \leqslant G'$, then $\varphi^{-1}(H') \leqslant G$.*

The proof is an exercise. Recall that $\varphi(H) = \{ \varphi(x)\,;\ x \in H \}$ and that $\varphi^{-1}(H') = \{ x \in G\,;\ \varphi(x) \in H' \}$; the notation $\varphi^{-1}(H)$ does not imply that φ is an isomorphism or that there is a mapping $\varphi^{-1} : G' \longrightarrow G$.

1.4.d. Generators. Let G be a group and X be a subset of G. By Proposition 1.4.6, there exists a smallest subgroup of G containing X (= a subgroup of G which contains X and is contained in every other subgroup which contains X): namely, the intersection of all the subgroups of G which contain X. If, for instance, $X = \{a\}$, then a subgroup which contains a must contain all powers of a; hence $\langle a \rangle$ is the smallest subgroup containing a.

In general, the smallest subgroup of a group G which contains a subset X of G is the subgroup of G **generated by** X, which we denote by $\langle X \rangle$. The group G itself is **generated by** X in case $G = \langle X \rangle$; then X **generates** G and the elements of X are the **generators** of G. Note that $\langle X \rangle$ depends on G, not just on X. Moreover, being a generator is not an intrinsic property and depends on the subset X. For instance, it is clear that $G = \langle G \rangle$; hence every element of G belongs to a generating set.

The next result gives a general description of $\langle X \rangle$.

Proposition 1.4.9. *Let G be a group and X be a subset of G. The subgroup $\langle X \rangle$ of G generated by X is the set of all products of elements of X and inverses of elements of X.*

REMARK. We were careful to allow products to be empty or have just one term.

Proof. Let H be the set of all products of elements of X and inverses of elements of X. Clearly a subgroup which contains X contains every element of H. It now suffices to prove that H is a subgroup containing X. Since products may be empty or have just one term, $1 \in H$ and $X \subseteq H$. Also H is closed under multiplication, and under inverses by Proposition 1.2.7. $\square$

Corollary 1.4.10. *Let G be a finite group and X be a subset of G. The subgroup $\langle X \rangle$ of G generated by X is the set of all products of elements of X.*

Proof. This follows from Corollary 1.2.10. $\square$

Proposition 1.4.11. *If $\varphi, \psi : G \longrightarrow G'$ are homomorphisms that agree on a generating subset X of G (if $\varphi(x) = \psi(x)$ for all $x \in X$), then $\varphi = \psi$.*

Proof. Let $H = \{ a \in G \, ; \, \varphi(a) = \psi(a) \}$. Since φ and ψ are homomorphisms, H is a subgroup of G. (H is the **equalizer** of φ and ψ.) If φ and ψ agree on X, then $X \subseteq H$; if X generates G, this implies $H = G$ and $\varphi = \psi$. $\square$

Many groups can be characterized by properties of suitably selected generators. For instance, a group is cyclic with n elements if and only if it is generated by one element of order n.

Proposition 1.4.12. *A group G is isomorphic to D_n if and only if G is generated by two elements a and b such that a has order n, $b \neq a$ has order 2, and $bab^{-1} = a^{-1}$.*

Proof. As in Example 1.2.4, let $\rho = \rho_1$ be the rotation by $2\pi/n$ and $\sigma = \sigma_0$. Then $\rho_i = \rho^i$ and $\sigma_i = \rho_i \sigma_0 = \rho^i \sigma$. We see that ρ has order n, σ has order 2, $\sigma \neq \rho$, and $\sigma \rho \sigma^{-1} = \sigma_0 \rho_1 \sigma_0 = \sigma_0 \sigma_1 = \rho_{-1} = \rho^{-1}$.

Now let G be a group generated by an element a of order n and an element $b \neq a$ of order 2 such that $bab^{-1} = bab = a^{-1}$. By Corollary 1.4.10, an element of G is a product of positive powers of a and b. Since $\alpha_b : x \longmapsto bxb = bxb^{-1}$ is an automorphism, we have $ba^k b = a^{-k}$ and $ba^k = a^{-k} b$. Therefore any product of powers of a and b can be rearranged so that all the a's are at the beginning and all the b's at the end. Hence $G = \{ 1, a, \ldots, a^{n-1}, b, ab, \ldots, a^{n-1}b \}$. As above, we have $a^i a^j = a^{i+j}$ ($= a^{i+j-n}$ if $i + j \geqslant n$); $a^i (a^j b) = a^{i+j} b$ ($= a^{i+j-n}b$ if $i + j \geqslant n$); $(a^i b)(a^j b) = a^{i-j}$ ($= a^{i-j+n}$ if $i - j < 0$); and $(a^i b)a^j = a^{i-j}b$ ($= a^{i-j+n}b$ if $i - j < 0$).

We show that $1, a, \ldots, a^{n-1}, b, ab, \ldots, a^{n-1}b$ are all distinct. First $1, a, \ldots, a^{n-1}$ are all distinct, since a has order n. By the cancellation law, $b, ab, \ldots, a^{n-1}b$ are all distinct. Finally assume that $a^i = a^j b$ for some i, j. Then $b = a^k$ for some $k = i - j$. If $n > 2$, this forces b to commute with a, contradicting $bab^{-1} = a^{-1} \neq a$. If $n = 2$, $b = a^k$ implies $b = 1$ or $b = a$, contradicting the hypothesis that $b \neq a$ has order 2. Therefore $\{ 1, a, \ldots, a^{n-1} \}$ and $\{ b, ab, \ldots, a^{n-1}b \}$ are disjoint, and G has exactly $2n$ elements.

We can now define $\theta : G \longrightarrow D_n$ by $\theta(a^k) = \rho^k$ and $\theta(a^k b) = \rho^k \sigma$. It is clear that θ is an isomorphism. $\square$

Proposition 1.4.12 may be taken as a purely algebraic definition of D_n. Another characterization will be given in Chapter 2.

1.4.e. Cosets. It turns out that equivalence relations can be constructed from any subgroup. The following result is straightforward:

Lemma 1.4.13. *When H is a subgroup of G:*

(1) *the binary relation $\mathscr{R}$ on G defined by*

$$x \mathscr{R} y \qquad \text{if and only if} \quad xy^{-1} \in H$$

is an equivalence relation; the $\mathscr{R}$-class of $x \in G$ is Hx.

(2) *the binary relation $\mathscr{L}$ on G defined by*

$$x \mathscr{L} y \qquad \text{if and only if} \quad y^{-1}x \in H$$

is an equivalence relation; the $\mathscr{L}$-class of $x \in G$ is xH. $\square$

The **right cosets** of a subgroup H of a group G are the sets Hx ($x \in G$); the **left cosets** of H are the sets xH ($x \in G$). The **right coset of** $x \in G$ modulo H is Hx; the **left coset of** $x \in G$ modulo H is xH.

For example, $H = H1 = 1H$ is a right coset of itself, and also a left coset of itself. Lemma 1.4.13 implies that $H = Hx = xH$ for all $x \in H$.

By Lemma 1.4.13, the right cosets of a subgroup H of G constitute a partition of G, and so do the left cosets. We will see in Section 1.5 that these two partitions are in general different. However:

Proposition 1.4.14. *The number of right cosets of a subgroup equals the number of its left cosets.*

Proof. If Hx is a right coset of H, then $(Hx)^{-1} = x^{-1}H^{-1} = x^{-1}H$ is a left coset of H. If conversely xH is a left coset of H, then $(xH)^{-1} = Hx^{-1}$ is a right coset of H. This constructs a one-to-one correspondence (a pair of mutually inverse bijections) between right cosets and left cosets. $\square$

The **index** of a subgroup is the [cardinal] number $[G : H]$ of its right cosets, and also the number of its left cosets.

1.4.f. Lagrange's Theorem. Our next result shows that the order of a finite group has a strong influence on its structure.

Theorem 1.4.15 (Lagrange's Theorem). *When H is a subgroup of a group G, $|G| = |H| [G : H]$. If G is finite, then the order of every subgroup of G divides the order of G, and the index of every subgroup of G divides the order of G.*

Proof. Right multiplication by x is a bijection of H onto Hx (right multiplication by x^{-1} provides the inverse bijection). Hence $|Hx| = |H|$. Now G is partitioned into $[G : H]$ right cosets, all of which have $|H|$ elements; therefore G has $[G : H] |H|$ elements. □

Corollary 1.4.16. *In a finite group G, the order of every element of G divides the order of G.*

Proof. The order of $a \in G$ is also the order of $\langle a \rangle \leqslant G$. □

The converse of Lagrange's Theorem states that every divisor of $|G|$ is the order of some subgroup of G. This is true if G is cyclic (see below) or dihedral (see exercises). However, we will construct in Chapter 3 a permutation group in which the converse of Lagrange's Theorem does not hold. Chapter 3 also contains a number of partial converses.

Proposition 1.4.17. *Let $G = \langle a \rangle$ be a cyclic group of order n. For each divisor d of n, G has exactly one subgroup of order d, namely $\langle a^{n/d} \rangle = \{ x \in G ; x^d = 1 \}$.*

Proof. Since a has order n, $a^{n/d}$ has order d, and G has a subgroup $\langle a^{n/d} \rangle$ of order d.

Let $D = \{ x \in G ; x^d = 1 \}$. Every subgroup H of G of order d is contained in D, since by Corollary 1.4.16 the order of any $x \in H$ must divide d. On the other hand, $a^k \in D$ is equivalent to $a^{kd} = 1$, $n \mid kd$, and $n/d \mid k$; hence D consists of $1, a^{n/d}, a^{2n/d}, \ldots, a^{(d-1)n/d}$ and has d elements. Therefore D is the only subgroup of G of order d. In particular, $D = \langle a^{n/d} \rangle$. □

Proposition 1.4.17 implies the following result, whose proof is an exercise:

Corollary 1.4.18. *A group $G \neq 1$ has no subgroup $H \neq 1, G$ if and only if G is a cyclic group of prime order.*

As a last application of cosets we show:

Proposition 1.4.19. *When A and B are subgroups of G, then $|AB| = |A| |B| / |A \cap B|$.*

Proof. AB is a union of right cosets of A, namely the right cosets of the elements of B. Now $x, y \in B$ have the same right coset modulo A if and only if $xy^{-1} \in A$, if and only if $xy^{-1} \in A \cap B$, if and only if, in the group B, x and y are in the same right coset modulo $A \cap B$. Therefore AB is the union of $[B : A \cap B]$ distinct right cosets of A, and has $(|B|/|A \cap B|)|A|$ elements. □

Exercises

1. Prove Lemma 1.4.2.

2. Prove Lemma 1.4.3.

3. Give an example of a nonempty subset of an infinite group which is closed under multiplication but not a subgroup.

4. Find all subgroups of V_4.

5. Find all subgroups of D_3.

6. Show that the complex n-th roots of unity form a cyclic group under complex multiplication.

7. Show that $\omega_k = \cos(2k\pi/n) + i\sin(2k\pi/n)$ is a **primitive** complex n-th root of unity (generates the group in the previous exercise) if and only if g.c.d. $(k,n) = 1$.

8. Show that $\mathbb{Z}_n$ is cyclic (generated by $\overline{1}$).

9. Describe the multiplication on a cyclic group of order n.

10. Prove that any two cyclic groups of order n are isomorphic.

11. Find all subgroups of a cyclic group of order 6.

12. Find all subgroups of a cyclic group of order 8.

13. Let $G = \langle a \rangle$ be a cyclic group of order n. Show that a^k generates G if and only if $(k,n) = 1$.

14. Prove Proposition 1.4.6.

15. Let $H, K \leqslant G$. Prove that $H \cup K \leqslant G$ if and only if $H \subseteq K$ or $K \subseteq H$. Give an example where $H \cup K$ is not a subgroup.

16. Prove Proposition 1.4.8.

17. Prove that every divisor of $|D_n|$ is the order of a subgroup of D_n.

18. Prove that a group $G \neq 1$ has no subgroup $H \neq 1, G$ if and only if G is a cyclic group of prime order.

1.5. QUOTIENT GROUPS

1.5.a. Definition. When $\mathscr{E}$ is an equivalence relation on a set X, the **quotient set** $X/\mathscr{E}$ is the set of all equivalence classes modulo $\mathscr{E}$. The quotient set comes with a **projection** mapping $\pi : X \longrightarrow X/\mathscr{E}$ which assigns to each $x \in X$ its equivalence class and induces on X the given equivalence relation $\mathscr{E}$ ($\pi(x) = \pi(y)$ if and only if $x \mathscr{E} y$).

We define a **quotient group** of a group G as a group G' such that:

(1) as a set, G' is a quotient set of G;

(2) the projection mapping $G \longrightarrow G'$ is a homomorphism, the **projection** homomorphism.

1.5.b. Normal Subgroups. It turns out that all quotient groups of a group G can be constructed from subgroups of G.

To understand this, we look at the equivalence relation $\mathscr{E}$ induced by a projection homomorphism, or more generally, by any homomorphism $\varphi : G \longrightarrow G'$ of groups. Observe that $x \mathscr{E} y$, that is, $\varphi(x) = \varphi(y)$, is equivalent to $\varphi(x)\varphi(y)^{-1} = 1$ and to $\varphi(xy^{-1}) = 1$. Also $\varphi(x) = \varphi(y)$ is equivalent to

$\varphi(y)^{-1}\varphi(x) = 1$ and to $\varphi(y^{-1}x) = 1$. Thus $\mathscr{E}$ is determined very simply by the set

$$K = \{x \in G; \varphi(x) = 1\};$$

namely $x \mathscr{E} y$ if and only if $xy^{-1} \in K$, if and only if $y^{-1}x \in K$, as in Lemma 1.4.16. This suggests the following definitions.

The **kernel** of a homomorphism $\varphi : G \longrightarrow G'$ of groups is the set

$$\mathrm{Ker}\,\varphi = \{x \in G; \varphi(x) = 1\}.$$

Lemma 1.5.1. *When* $\varphi : G \longrightarrow G'$ *is a homomorphism of groups, the kernel* $K = \mathrm{Ker}\,\varphi$ *of* φ *is a subgroup of* G, *and* $\varphi(x) = \varphi(y)$ *if and only if* $xy^{-1} \in K$, *if and only if* $y^{-1}x \in K$. *Furthermore* $xKx^{-1} \subseteq K$ *for all* $x \in G$.

Proof. It follows from Proposition 1.3.1 that $1 \in K$; also $x, y \in K$ implies $\varphi(xy^{-1}) = \varphi(x)\varphi(y)^{-1} = 1$ and $xy^{-1} \in K$. Hence K is a subgroup. Also $k \in K$ implies $\varphi(xkx^{-1}) = \varphi(x)\varphi(k)\varphi(x)^{-1} = \varphi(x)\varphi(x)^{-1} = 1$ and $xkx^{-1} \in K$, for all $x \in G$. □

Lemma 1.5.1 shows that the kernel K of a homomorphism $\varphi : G \longrightarrow G'$ is not an arbitrary subgroup of G but must satisfy $xKx^{-1} \subseteq K$ for all $x \in G$.

A **normal** subgroup of a group G is a subgroup $N \leqslant G$ such that $xNx^{-1} \subseteq N$ for all $x \in G$. We denote this relationship by $N \trianglelefteq G$. (The more traditional notation $N \triangleleft G$ seems to imply that $N \neq G$.) For example, $G \trianglelefteq G$ and $1 \trianglelefteq G$; the exercises contain less trivial examples, and examples of subgroups that are not normal. Normal subgroups are also called **invariant**, which is short for "invariant under inner automorphisms."

We saw (Lemma 1.4.13) that the right cosets of a subgroup H of G constitute a partition of G, and so do the left cosets. The following result shows that these two partitions coincide precisely when H is normal; then the set $Hx = xH$ is the **coset** of x (a right or left side need not be specified).

Proposition 1.5.2. *For a subgroup* N *of a group* G, *the following conditions are equivalent:*

(1) N *is a normal subgroup of* G *($xNx^{-1} \subseteq N$ for all $x \in G$)*;
(2) $xNx^{-1} = N$ *for all* $x \in G$;
(3) $xN = Nx$ *for all* $x \in G$;
(4) *for all* $x, y \in G$, $xy^{-1} \in N$ *if and only if* $y^{-1}x \in N$.

Proof. (1) implies (2). If $xNx^{-1} \subseteq N$ for all $x \in G$, then $x^{-1}Nx \subseteq N$; applying the inner automorphism α_x yields $N = xx^{-1}Nxx^{-1} \subseteq xNx^{-1}$, so that $xNx^{-1} = N$.

(2) implies (3). Multiply on the right by x.

(3) is equivalent to (4). Note that $xy^{-1} \in N$ if and only if $x \in Ny$ and $y^{-1}x \in N$ if and only if $x \in yN$.

(3) implies (1). Multiply on the right by x^{-1}. $\square$

Proposition 1.5.2 implies that every subgroup of an abelian group is normal.

Proposition 1.5.3. *Every intersection of normal subgroups of G is a normal subgroup of G.*

The proof is an exercise.

1.5.c. Construction. Now let G' be a quotient group of G so that, as a set, G' is the quotient set $G/\mathscr{E}$ of G by an equivalence relation $\mathscr{E}$ on G, and the projection $\pi : G \longrightarrow G'$ is a homomorphism. By Lemma 1.5.1, $N = \operatorname{Ker} \pi$ is a normal subgroup of G, and $x \mathscr{E} y$ if and only if $Nx = Ny$, since $\mathscr{E}$ is induced by π. Thus $\mathscr{E}$ is the partition on G into cosets of N, and these are the elements of G'. Moreover the product of Nx and Ny in G' is $\pi(x)\,\pi(y) = \pi(xy) = Nxy$.

Given this generous hint, we now construct a quotient group of G from any normal subgroup N of G. By the above, every quotient group can be constructed in this fashion.

Proposition 1.5.4. *When N is a normal subgroup of G, the [right] cosets of N are the elements of a group G/N whose operation is well-defined by $(Nx)(Ny) = Nxy$. The projection $\pi : x \longmapsto Nx$ is a homomorphism of G onto G/N, and $\operatorname{Ker} \pi = N$.*

Proof. When we multiply Nx and Ny as subsets of G, we obtain $NxNy = NNxy = Nxy$, since $xN = Nx$. (In particular, $Na = Nb$ and $Nc = Nd$ implies $Nac = Nbd$, as in Proposition 1.1.5.) Thus multiplication of subsets induces a binary operation on the set G/N, which is associative by Proposition 1.1.4. The coset $N = N1$ is the identity element of G/N, since $NxN = NNx = Nx$ for all Nx. The coset of Nx^{-1} is the inverse of Nx in G/N, since by the above $NxNx^{-1} = Nxx^{-1} = N$ and $Nx^{-1}Nx = Nx^{-1}x = N$. Thus G/N is a group. The equality $(Nx)(Ny) = Nxy$ shows that the projection π is a homomorphism. Finally $\operatorname{Ker} \pi = N$, since $Nx = N$ if and only if $x \in N$. $\square$

It follows from Proposition 1.5.4 and Lemma 1.5.1 that a subgroup is the kernel of some homomorphism if and only if it is normal.

The **quotient group** or **factor group of** a group G **by** a normal subgroup N of G is the group G/N in Proposition 1.5.4. The order of G/N is the index $|G/N| = [G : N]$ of N. The projection $\pi : G \longrightarrow G/N$ is a **canonical** homomorphism, which means that it depends only upon G and N; it is surjective and has kernel N. Proposition 1.5.4 is often used as the definition of quotient groups.

1.5.d. Examples. We saw that $\{1\}$ and G are always normal subgroups of G. The corresponding quotients are $G/\{1\} \cong G$ (since $\{1\}x = \{x\}$) and $G/G \cong \{1\}$.

For a less trivial example, let $G = \mathbb{Z}$. We saw (Proposition 1.4.4) that every nontrivial subgroup of $\mathbb{Z}$ has the form $n\mathbb{Z} = \{nz\,;\, z \in \mathbb{Z}\}$ for some $n > 0$; this is a normal subgroup, since $\mathbb{Z}$ is abelian. The quotient group $\mathbb{Z}_n = \mathbb{Z}/n\mathbb{Z}$ is the **group of integers modulo** n. We verify that this is the same group of integers modulo n as in Section A.3.

By definition of quotient groups, $\mathbb{Z}_n$ is the group whose elements are all $\bar{x} = x + n\mathbb{Z}$ with $x \in \mathbb{Z}$, and whose operation (denoted additively) is $\bar{x} + \bar{y} = \overline{x + y}$. Division by n shows that every element of $\mathbb{Z}_n$ can be written in the form $\bar{x}$ for some $0 \leqslant x < n$. Moreover $\bar{x} = \bar{y}$ implies $x = y$ when $0 \leqslant x, y < n$: indeed $-n < x - y < n$, so that n divides $x - y$ only if $x - y = 0$. Thus $\mathbb{Z}_n$ has n elements, namely $\bar{0}, \bar{1}, \ldots, \overline{n-1}$. When $0 \leqslant x, y < n$,

$$\bar{x} + \bar{y} = \begin{cases} \overline{x+y} & \text{if } x + y < n, \\ \overline{x+y-n} & \text{if } x + y \geqslant n. \end{cases}$$

In particular, $\mathbb{Z}_n$ is a cyclic group of order n, with $\mathbb{Z}_n = \langle \bar{1} \rangle$.

Other examples will be found in the exercises.

1.5.e. Subgroups of Quotient Groups. All subgroups of G/N can be obtained from subgroups of G. This follows from a refinement of Proposition 1.4.8, whose proof is an exercise:

Proposition 1.5.5. *Let* $\varphi : G \longrightarrow G'$ *be a homomorphism of groups. There is a one-to-one, order-preserving correspondence between the subgroups of* $\operatorname{Im} \varphi$ *and the subgroups of* G *which contain* $\operatorname{Ker} \varphi$. *The subgroup that corresponds to* $\operatorname{Ker} \varphi \leqslant H \leqslant G$ *is* $H' = \varphi(H)$; *the subgroup that corresponds to* $H' \leqslant \operatorname{Im} \varphi$ *is* $H = \varphi^{-1}(H')$. *Furthermore* $H \trianglelefteq G$ *if and only if* $H' \trianglelefteq \operatorname{Im} \varphi$.

Let $N \trianglelefteq G$ and $\pi : G \longrightarrow G/N$ be the projection. Proposition 1.5.5 provides a one-to-one correspondence between the subgroups of $\operatorname{Im} \pi = G/N$ and the subgroups of G which contain $\operatorname{Ker} \pi = N$. The subgroup that corresponds to $N \leqslant H \leqslant G$ is $H' = \pi(H)$. Thus H' is the set of all the cosets of elements of H, multiplied (as in G/N) by $Nx\,Ny = Nxy$; in other words, $H' = H/N$ (note that $N \trianglelefteq H$ by Proposition 1.5.2). This proves:

Proposition 1.5.6. *Let N be a normal subgroup of a group G. Every subgroup of G/N is of the form H/N for some unique subgroup H of G which contains N. Furthermore $H/N \trianglelefteq G/N$ if and only if $H \trianglelefteq G$.*

Proposition 1.5.5 implies that the subgroups of $\mathbb{Z}_n$ correspond to subgroups of $\mathbb{Z}$ which contain $n\mathbb{Z}$, and to divisors of n. (Compare with Proposition 1.4.17.)

Exercises

1. Show that a subgroup N of a group G satisfies $NxNy \subseteq Nxy$ for all $x, y \in G$ if and only if it is normal (in other words, the decomposition of G into right cosets of N is a congruence on G if and only if N is normal).

2. Find all subgroups of D_3. Which are the normal subgroups? What are the quotient groups (i.e., the quotient groups are isomorphic to which known examples of groups)?

3. Find all subgroups of D_4. Which are the normal subgroups? What are the quotient groups?

4. Let D_n be generated by elements a and b as in Proposition 1.4.15. Show that $\langle a \rangle \trianglelefteq D_n$.

5. Show that $G = D_4$ contains subgroups A and B such that $A \trianglelefteq B$ and $B \trianglelefteq G$ but not $A \trianglelefteq G$.

6. Prove that every subgroup of index 2 is normal.

7. Prove that every intersection of normal subgroups of a group G is a normal subgroup of G.

8. Prove that the union of a nonempty chain of normal subgroups of G is a normal subgroup of G.

9. Let $\varphi : G \longrightarrow G'$ be a homomorphism of groups with kernel K. Show that $\varphi^{-1}(\varphi(H)) = HK$ for every $H \leqslant G$ and that $\varphi(\varphi^{-1}(H')) = H' \cap \operatorname{Im} \varphi$ for every $H' \leqslant G'$.

10. Prove Proposition 1.5.5; you may use the previous exercise.

1.6. THE ISOMORPHISMS THEOREMS

This section contains additional properties of subgroups and homomorphisms: results that allow a homomorphism to factor through another, the Homomorphism Theorem, and the First and Second Isomorphism Theorems. (These three results are also called the First, Second, and Third Isomorphism Theorems.)

Relationships between groups are often summarized by diagrams of groups and homomorphisms. Much about groups is proved by elementary means; yet diagrams are also used and proofs by diagram will become more and more important as we progress from groups to rings to modules. Factoring homomorphisms through each other provides a basic and eventually indispensable technique for constructing diagrams.

1.6.a. Factoring through the Domain. We begin by factoring mappings through each other.

Every mapping $\varphi : X \longrightarrow Y$ has a **range** or **image** $\operatorname{Im} \varphi = \varphi(X) = \{\varphi(x); x \in X\}$ which is a subset of its codomain Y. Images determine when one mapping $\varphi : X \longrightarrow Z$ factors through another mapping $\psi : Y \longrightarrow Z$ with the same codomain.

Proposition 1.6.1. *Let X, Y, Z be sets and $\varphi : X \longrightarrow Z$, $\psi : Y \longrightarrow Z$ be mappings.*

(1) *φ factors through ψ (i.e., $\varphi = \psi \circ \chi$ for some mapping $\chi : X \longrightarrow Y$) if and only if $\operatorname{Im} \varphi \subseteq \operatorname{Im} \psi$.*

(2) *If ψ is injective and $\operatorname{Im} \varphi \subseteq \operatorname{Im} \psi$, then φ factors uniquely through ψ (i.e., there is only one mapping $\chi : X \longrightarrow Y$ such that $\varphi = \psi \circ \chi$).*

(3) *If φ and ψ are injective and $\operatorname{Im} \varphi = \operatorname{Im} \psi$, then the mapping $\chi : X \longrightarrow Y$ such that $\varphi = \psi \circ \chi$ is bijective.*

Proof. We prove (1) and leave (2) and (3) to our tireless reader. If $\varphi = \psi \circ \chi$, then $\varphi(x) = \psi(\chi(x)) \in \operatorname{Im} \psi$ for every $x \in X$ and $\operatorname{Im} \varphi \subseteq \operatorname{Im} \psi$. Conversely, assume that $\operatorname{Im} \varphi \subseteq \operatorname{Im} \psi$. For every $x \in X$ we have $\varphi(x) \in \operatorname{Im} \psi$ and $\varphi(x) = \psi(y)$ for some $y \in Y$. Choose one such $y \in Y$ for each x, and let $\chi(x)$ be this chosen element. This defines a mapping $\chi : X \longrightarrow Y$. By construction, $\psi(\chi(x)) = \varphi(x)$ for all $x \in X$. $\square$

Factoring one homomorphism through another is more difficult than factoring one mapping through another. To factor one mapping $\varphi : X \longrightarrow Z$ through another mapping $\psi : Y \longrightarrow Z$, we only need a mapping $\chi : X \longrightarrow Y$ such that $\varphi = \psi \circ \chi$. To factor one homomorphism $\varphi : A \longrightarrow G$ through another homomorphism $\psi : B \longrightarrow G$, we need a *homomorphism* $\chi : A \longrightarrow B$ such that $\varphi = \psi \circ \chi$. In the proof of Proposition 1.6.1, we can no longer choose for $\chi(x)$ any old y such that $\varphi(x) = \psi(y)$; we must ensure that $\chi(x_1 x_2) = \chi(x_1) \chi(x_2)$ for all x_1, x_2. This may be impossible (see the exercises).

There is one general case where Proposition 1.6.1 extends to homomorphisms: when ψ is injective.

Theorem 1.6.2. *Let G, A, B be groups and $\varphi : A \longrightarrow G$, $\psi : B \longrightarrow G$ be homomorphisms.*

(1) *If ψ is injective, then φ factors through ψ (i.e., $\varphi = \psi \circ \chi$ for some homo-morphism $\chi : A \longrightarrow B$) if and only if $\operatorname{Im} \varphi \subseteq \operatorname{Im} \psi$, and then φ factors uniquely through ψ (i.e., χ is unique).*

(2) *If φ and ψ are injective and $\operatorname{Im} \varphi = \operatorname{Im} \psi$, then the homomorphism $\chi : A \longrightarrow B$ such that $\varphi = \psi \circ \chi$ is an isomorphism.*

Proof. If $\operatorname{Im} \varphi \subseteq \operatorname{Im} \psi$, then Proposition 1.6.1 provides a unique mapping $\chi : A \longrightarrow B$ such that $\varphi = \psi \circ \chi$. If $x, y \in A$, then

$$\psi(\chi(xy)) = \varphi(xy) = \varphi(x)\varphi(y) = \psi(\chi(x))\psi(\chi(y)) = \psi(\chi(x)\chi(y))$$

and $\chi(xy) = \chi(x)\chi(y)$, since ψ is injective. Hence χ is a homomorphism. $\square$

1.6.b. Factoring through the Codomain. Every mapping $\varphi : X \longrightarrow Y$ in-duces an equivalence relation $\mathscr{E}$ on its domain X, namely $x \mathrel{\mathscr{E}} y$ [or, $(x, y) \in \mathscr{E}$] if and only if $x, y \in X$ and $\varphi(x) = \varphi(y)$. These equivalence relations de-termine when one mapping $\varphi : X \longrightarrow Y$ factors through another mapping $\psi : X \longrightarrow Z$ with the same domain.

Proposition 1.6.3. *Let X, Y, Z be sets, with $Y \neq \varnothing$, and $\varphi : X \longrightarrow Y$, $\psi : X \longrightarrow Z$ be mappings. Let $\mathscr{E}_\varphi$ and $\mathscr{E}_\psi$ be the equivalence relations on X induced by φ and ψ.*

(1) *φ factors through ψ (that is, $\varphi = \chi \circ \psi$ for some mapping $\chi : Z \longrightarrow Y$) if and only if $\mathscr{E}_\psi \subseteq \mathscr{E}_\varphi$.*

(2) *If ψ is surjective and $\mathscr{E}_\psi \subseteq \mathscr{E}_\varphi$, then φ factors uniquely through ψ (i.e., there is only one mapping $\chi : Z \longrightarrow Y$ such that $\varphi = \chi \circ \psi$).*

(3) *If φ and ψ are surjective and $\mathscr{E}_\varphi = \mathscr{E}_\psi$, then the mapping $\chi : Z \longrightarrow Y$ such that $\varphi = \chi \circ \psi$ is bijective.*

Proof. If $\varphi = \chi \circ \psi$, then $\psi(x) = \psi(y)$ implies $\varphi(x) = \varphi(y)$, $(x, y) \in \mathscr{E}_\psi$ im-plies $(x, y) \in \mathscr{E}_\varphi$, and $\mathscr{E}_\psi \subseteq \mathscr{E}_\varphi$.

Conversely assume $\mathscr{E}_\psi \subseteq \mathscr{E}_\varphi$. Construct a mapping $\xi : \operatorname{Im} \psi \longrightarrow Y$ as fol-lows. Formally

$$\xi = \{ (\psi(x), \varphi(x)) \in \operatorname{Im} \psi \times Y ; x \in X \}.$$

For every $z \in \operatorname{Im} \psi$, we have $z = \psi(x)$ for some $x \in X$ and $(z, y) \in \xi$ for some $y = \varphi(x) \in Y$. If $(\psi(x_1), \varphi(x_1)) \in \xi$, $(\psi(x_2), \varphi(x_2)) \in \xi$, and $\psi(x_1) = \psi(x_2)$, then $\varphi(x_1) = \varphi(x_2)$ (since $\mathscr{E}_\psi \subseteq \mathscr{E}_\varphi$). Thus ξ is a mapping. (Informally we say that $\xi : \operatorname{Im} \psi \longrightarrow Y$ is *well defined* by $\xi(\psi(x)) = \varphi(x)$.)

We have $\xi(\psi(x)) = \varphi(x)$ for all $x \in X$. For any $y \in Y$ we can now define $\chi : Z \longrightarrow Y$ as follows: if $z \in \operatorname{Im} \psi$, then $\chi(z) = \xi(z)$; otherwise, $\chi(z) = y$. Then $\varphi = \chi \circ \psi$. This proves (1).

We leave (2) to the reader but this time prove (3). If φ and ψ are surjective and $\mathscr{E}_\varphi = \mathscr{E}_\psi$, then by (1) φ and ψ factor through each other, and there exist mappings $\chi : Z \longrightarrow Y$ and $\zeta : Y \longrightarrow Z$ such that $\varphi = \chi \circ \psi$ and $\psi = \zeta \circ \varphi$. Then $\zeta \circ \chi \circ \psi = \zeta \circ \varphi = \psi = 1_Z \circ \psi$; since ψ is surjective, it factors through itself in only one way, by (2); therefore $\zeta \circ \chi = 1_Z$. Similarly $\chi \circ \zeta \circ \varphi = \chi \circ \psi = \varphi = 1_Y \circ \varphi$; since φ is surjective, it factors through itself in only one way, by (2); therefore $\chi \circ \zeta = 1_Y$. Thus χ and ζ are mutually inverse bijections. $\square$

With homomorphisms, we can use Lemma 1.5.1 to replace equivalence relations by kernels. But factoring a group homomorphism $\varphi : G \longrightarrow A$ through another group homomorphism $\psi : G \longrightarrow B$ is still more difficult that factoring mappings: we are no longer content with mere mappings and want a homomorphism $\chi : B \longrightarrow A$ such that $\chi \circ \psi = \varphi$. For this the condition $\mathscr{E}_\psi \subseteq \mathscr{E}_\varphi$ no longer suffices (see the exercises).

There is one general case where Proposition 1.6.3 extends to homomorphisms: when ψ is surjective.

Theorem 1.6.4. *Let G, A, B be groups and $\varphi : G \longrightarrow A$, $\psi : G \longrightarrow B$ be homomorphisms.*

(1) *If ψ is surjective, then φ factors through ψ (i.e., $\varphi = \chi \circ \psi$ for some homomorphism $\chi : B \longrightarrow A$) if and only if $\operatorname{Ker} \psi \subseteq \operatorname{Ker} \varphi$, and then φ factors uniquely through ψ (i.e., χ is unique).*

(2) *If φ and ψ are surjective and $\operatorname{Ker} \varphi = \operatorname{Ker} \psi$, then the homomorphism $\chi : B \longrightarrow A$ such that $\varphi = \chi \circ \psi$ is an isomorphism.*

Proof. If $\varphi = \chi \circ \psi$ factors through ψ, then $\psi(x) = 1$ implies $\varphi(x) = \chi(\psi(x)) = \chi(1) = 1$, and $\operatorname{Ker} \psi \subseteq \operatorname{Ker} \varphi$.

If conversely $\operatorname{Ker} \psi \subseteq \operatorname{Ker} \varphi$, then $\psi(x) = \psi(y)$ implies $xy^{-1} \in \operatorname{Ker} \psi$, $xy^{-1} \in \operatorname{Ker} \varphi$, and $\varphi(x) = \varphi(y)$. Hence Proposition 1.6.3 provides a unique mapping $\chi : B \longrightarrow A$ such that $\varphi = \chi \circ \psi$. If $x, y \in B$, then $x = \psi(u)$, $y = \psi(v)$ for some $u, v \in G$, since ψ is surjective,

$$\chi(\psi(u)\,\psi(v)) = \chi(\psi(uv)) = \varphi(uv) = \varphi(u)\,\varphi(v) = \chi(\psi(u))\,\chi(\psi(v)),$$

and $\chi(xy) = \chi(x)\,\chi(y)$. Thus χ is a homomorphism. $\square$

A **homomorphic image** of a group G is a group G' such that there exists a surjective homomorphism φ of G onto G'. By part (2) of Theorem 1.6.4, a group is a homomorphic image of G if and only if it is isomorphic to a quotient group of G: if indeed there is a surjective homomorphism $\varphi : G \longrightarrow G'$, then the homomorphism φ and the projection $\pi : G \longrightarrow G/\mathrm{Ker}\,\varphi$ have the same kernel, so there is an isomorphism $\theta : G/\mathrm{Ker}\,\varphi \longrightarrow G'$ such that $\varphi = \theta \circ \pi$.

1.6.c. The Homomorphism Theorem. Set theory provides that any mapping can be decomposed into a projection to a quotient set, followed by a bijection, followed by an inclusion mapping. Indeed let $\varphi : X \longrightarrow Y$ be a mapping. Let $\mathscr{E}$ be the equivalence relation induced by φ. By Proposition 1.6.1, φ factors through the inclusion mapping $\iota : \mathrm{Im}\,\varphi \longrightarrow Y$: there is a mapping $\psi : X \longrightarrow \mathrm{Im}\,\varphi$ such that $\varphi = \iota \circ \psi$. Then Proposition 1.6.3 provides a bijection $\theta : X/\mathscr{E} \longrightarrow \mathrm{Im}\,\varphi$ such that $\psi = \theta \circ \pi$, so that $\varphi = \iota \circ \theta \circ \pi$.

Any homomorphism of groups has the same general structure: it can be decomposed into a projection to a quotient group, followed by an isomorphism, followed by an inclusion homomorphism. This constructs all homomorphisms!

Theorem 1.6.5 (Homomorphism Theorem). *When* $\varphi : G \longrightarrow G'$ *is a homomorphism of groups,* $\mathrm{Im}\,\varphi$ *is a subgroup of* G', $\mathrm{Ker}\,\varphi$ *is a normal subgroup of* G, *and there exists an isomorphism* $\theta : G/\mathrm{Ker}\,\varphi \longrightarrow \mathrm{Im}\,\varphi$ *such that the following diagram commutes (i.e.,* $\varphi = \iota \circ \theta \circ \pi$):

$$
\begin{array}{ccc}
G & \xrightarrow{\ \varphi\ } & G' \\
{\scriptstyle \pi}\downarrow & & \uparrow{\scriptstyle \iota} \\
G/\mathrm{Ker}\,\varphi & \xrightarrow[\theta]{} & \mathrm{Im}\,\varphi
\end{array}
$$

where $\pi : G \longrightarrow G/\mathrm{Ker}\,\varphi$ *is the projection and* $\iota : \mathrm{Im}\,\varphi \longrightarrow G'$ *is the inclusion homomorphism. In particular,* $\mathrm{Im}\,\varphi \cong G/\mathrm{Ker}\,\varphi$.

Proof. $\mathrm{Im}\,\varphi$ is a subgroup of G' and $\mathrm{Ker}\,\varphi$ is a normal subgroup of G, by Proposition 1.4.5 and Lemma 1.5.1. By Theorem 1.6.2, φ factors through the inclusion homomorphism $\iota : \mathrm{Im}\,\varphi \longrightarrow G'$:

$$
\begin{array}{ccc}
G & \xrightarrow{\ \varphi\ } & G' \\
{\scriptstyle \pi}\downarrow & {\scriptstyle \psi}\searrow & \uparrow{\scriptstyle \iota} \\
G/\mathrm{Ker}\,\varphi & \dashrightarrow[\theta]{} & \mathrm{Im}\,\varphi
\end{array}
$$

there is a homomorphism $\psi : X \longrightarrow \mathrm{Im}\,\varphi$ such that $\varphi = \iota \circ \psi$. We see that ψ is surjective and that $\mathrm{Ker}\,\psi = \mathrm{Ker}\,\varphi$ (since ι is injective). Since the projection $\pi : G \longrightarrow G/\mathrm{Ker}\,\varphi$ also has these properties, Theorem 1.6.4 provides an isomorphism $\theta : X/\mathscr{E} \longrightarrow \mathrm{Im}\,\varphi$ such that $\psi = \theta \circ \pi$. Then $\varphi = \iota \circ \theta \circ \pi$. $\qquad\square$

1.6.d. Cyclic Groups. We illustrate the Homomorphism Theorem with another look at cyclic groups. Cyclic groups are homomorphic images of $\mathbb{Z}$, and they are isomorphic to quotient groups of $\mathbb{Z}$.

Proposition 1.6.6. *Let G be a group and $a \in G$. If a has infinite order, then $\langle a \rangle \cong \mathbb{Z}$. If a has finite order n, then $\langle a \rangle \cong \mathbb{Z}_n$.*

Proof. By Proposition 1.2.8, the mapping $\pi : k \longmapsto a^k$ is a homomorphism of $\mathbb{Z}$ into G. If a has infinite order, then $\operatorname{Ker} \pi = \{0\}$ and it follows from Theorem 1.6.5 that $\operatorname{Im} \pi = \langle a \rangle \cong \mathbb{Z}$. If a has finite order n, then n is the smallest positive element of $\operatorname{Ker} \pi$, so that $\operatorname{Ker} \pi = n\mathbb{Z}$ and Theorem 1.6.5 yields $\operatorname{Im} \pi = \langle a \rangle \cong \mathbb{Z}_n$. $\square$

The above provides another proof of Proposition 1.2.9.

1.6.e. First Isomorphism Theorem. Let G be a group and A be a normal subgroup of G. By Proposition 1.5.6, every normal subgroup of G/A has the form B/A for some $A \lhd B \lhd G$. The First Isomorphism Theorem states that this implies $(G/A)/(G/B) \cong G/B$:

Theorem 1.6.7 (First Isomorphism Theorem). *Let G be a group and $A \lhd G$, $B \lhd G$. If $A \subseteq B$, then $A \lhd B$, $B/A \lhd G/A$, and*

$$(G/A)/(B/A) \cong G/B;$$

more precisely, there exists an isomorphism $\theta : (G/A)/(B/A) \longrightarrow G/B$ such that the following diagram commutes (i.e., $\theta \circ \sigma \circ \pi = \rho$):

where π, ρ, σ are the projections.

Proof. We have $A \lhd B$, since $A \lhd G$. Since $A \subseteq B$, Theorem 1.6.4, applied to the projections $\pi : G \longrightarrow G/A$ and $\rho : G \longrightarrow G/B$, provides a homomorphism $\omega : G/A \longrightarrow G/B$ such that $\omega \circ \pi = \rho$.

In other words, $\omega(Ax) = Bx$ for all $x \in G$. Then

$$\operatorname{Ker} \omega = \{ Ax \in G/A;\ Bx = B \} = \{ Ax \in G/A;\ x \in B \} = B/A.$$

Hence it follows from the Homomorphism Theorem that $B/A \lhd G/A$ and that there is an isomorphism $\theta : (G/A)/(B/A) \longrightarrow G/B$ such that $\theta \circ \sigma = \omega$. Then $\theta \circ \sigma \circ \pi = \rho$. $\qquad\square$

1.6.f. Second Isomorphism Theorem

Theorem 1.6.8 (Second Isomorphism Theorem). *Let G be a group and $A \leqslant G$, $N \lhd G$. Then $AN \leqslant G$, $N \lhd AN$, $A \cap N \lhd A$, and*

$$AN/N \cong A/A \cap N;$$

more precisely, there exists an isomorphism $\theta : A/A \cap N \longrightarrow AN/N$ such that the following diagram commutes (i.e., $\rho \circ \iota = \theta \circ \pi$):

$$
\begin{array}{ccc}
A & \xrightarrow{\ \iota\ } & AN \\
{\scriptstyle \pi}\downarrow & & \downarrow{\scriptstyle \rho} \\
A/A \cap N & \xrightarrow{\ \theta\ } & AN/N
\end{array}
$$

where π and ρ are the projections and ι is the inclusion homomorphism.

Proof. We have $1 = 11 \in AN$. Since $Nx = xN$ for all $x \in G$, we also have $NA = AN$. Hence AN is closed under multiplication, since $ANAN = AANN = AN$, and is closed under inverses, since $(AN)^{-1} = N^{-1}A^{-1} = NA = AN$ (recall that $H^{-1} = \{ h^{-1};\ h \in H \}$). Thus $AN \leqslant G$. Also $N = 1N \subseteq AN$ and hence $N \lhd AN$. This yields a homomorphism $\varphi = \rho \circ \iota$; by definition, $\varphi(a) = \rho(a) = Na\ (= aN)$ for all $a \in A$.

$$
\begin{array}{ccc}
A & \xrightarrow{\ \iota\ } & AN \\
{\scriptstyle \pi}\downarrow & {\scriptstyle \varphi}\searrow & \downarrow{\scriptstyle \rho} \\
A/A \cap N & \dashrightarrow[\theta] & AN/N
\end{array}
$$

Now φ is surjective. Also $\varphi(a) = 1 \in AN/N$ if and only if $Na = N$, if and only if $a \in N$; hence $\operatorname{Ker} \varphi = A \cap N$. By the Homomorphism Theorem, $A \cap N \lhd A$, and there is an isomorphism $\theta : A/A \cap N \longrightarrow AN/N$ such that $\varphi = \theta \circ \pi$. Then $\theta \circ \pi = \rho \circ \iota$. $\qquad\square$

When $A \leqslant G$ and $N \lhd G$, Theorem 1.6.8 implies $|AN|/|N| = |A|/|N \cap A|$, as in Proposition 1.4.19.

Exercises

1. Show that the identity homomorphism $1 : \mathbb{Z}_2 \longrightarrow \mathbb{Z}_2$ does not factor through the projection $\pi : \mathbb{Z} \longrightarrow \mathbb{Z}_2$, even though $\mathrm{Im}\, 1 \subseteq \mathrm{Im}\, \pi$. (Of course, π is not injective.)

2. Show that the identity homomorphism $1 : 2\mathbb{Z} \longrightarrow 2\mathbb{Z}$ does not factor through the inclusion homomorphism $\iota : 2\mathbb{Z} \longrightarrow \mathbb{Z}$, even though $\mathrm{Ker}\, \iota \subseteq \mathrm{Ker}\, 1$. (Of course, ι is not surjective.)

3. Show that the group $\mathbb{R}/\mathbb{Z}$ (under addition) is isomorphic to the **circle group** C, which is the group of all complex numbers of modulus 1, under multiplication.

4. Show that the group $\mathbb{Q}/\mathbb{Z}$ (under addition) is isomorphic to the multiplicative group U of all complex roots of unity (complex numbers u such that $u^k = 1$ for some $k \in \mathbb{N}$).

5. Let $n \in \mathbb{N}$ and $d \in \mathbb{N}$ be a divisor of n. Let C_n be a cyclic group of order n and C_d be its subgroup of order d. Show that C_n/C_d is cyclic of order n/d.

6. Let V be a finite-dimensional vector space over a field K. The **special linear group** is the set $SL(V)$ of all linear transformations $T : V \longrightarrow V$ of determinant 1. Show that $SL(V) \lhd GL(V)$, with $GL(V)/SL(V) \cong K^*$, the multiplicative group of all nonzero elements of K.

7. Let G be a finite group, H be a subgroup of G, and N be a normal subgroup of G such that $|N|$ and $[G : N]$ are relatively prime. Show that $H \subseteq N$ if and only if $|H|$ divides $|N|$. (Hint: consider HN.)

8. Let G be a group. Show that for each $a \in G$ the mapping $\lambda_a : G \longrightarrow G$ defined by: $\lambda_a(x) = ax$ is a permutation of G. Use this construction to prove **Cayley's Theorem**: every group is isomorphic to a subgroup of some permutation group S_X.

1.7. GROUPS WITH OPERATORS

Groups with operators are groups with an additional external operation. This structure is of interest to us largely because some theorems about groups (specifically, the Krull-Schmidt Theorem and the Jordan-Hölder Theorem) extend readily to groups with operators and therefore hold in less general structures, such as modules.

In this section we extend to groups with operators the basic properties of subgroups and homomorphisms from previous sections.

1.7.a. Definition. A left **action** of a set S on a set X is a mapping $S \times X \longrightarrow X$, typically denoted by $(s,x) \longmapsto s \bullet x$ or by $(s,x) \longmapsto sx$; then S **acts on** X. This provides the set X with a unary operation $x \longmapsto s \bullet x$ for each $s \in S$ (the **action** of s). For example, a vector space over a field K is an abelian group V with a well-behaved action of K on V.

When S is a set (the **set of operators**), an S-**group** (also called a **group with operators**) is a group G together with an action $S \times G \longrightarrow G$ of S onto G, which we denote by $(s,g) \longmapsto {}^s g$ (using the **left exponential notation**), such that

$$^s(xy) = {}^s x \, {}^s y$$

for all $x \in S$ and $x, y \in G$. This condition states that S acts on G **by endomor-**
phisms, which means that the action $x \longmapsto {}^s x$ of each $s \in S$ is an endomorphism of G.

The main examples are groups and modules. Letting $S = \emptyset$ makes every
group a group with operators (which is in fact a group without operators). A
vector space over a field K is a special kind of abelian K-group. Modules (a
more general structure than vector spaces) will be defined in Chapter 10.

1.7.b. Homomorphisms. Let S be a set. An S-**homomorphism** (also called
a **homomorphism**) of an S-group G into an S-group H is a group homomorphism $\varphi : G \longrightarrow H$ such that

$$\varphi({}^s g) = {}^s \varphi(g)$$

for all $s \in S$ and $g \in G$. Note that G and H must have the same set of operators S.

The composition of two S-homomorphisms is an S-homomorphism. The
identity mapping on an S-group is an S-homomorphism.

An S-**endomorphism** (also called an **endomorphism**) of an S-group G is
an S-homomorphism of G into G. An S-**isomorphism** (also called an **isomor-**
phism) is a bijective S-homomorphism; then the inverse bijection is also an
S-isomorphism.

1.7.c. Subgroups. An S-**subgroup** of an S-group G is an S-group H such
that, as a set, $H \subseteq G$, and the inclusion mapping $H \longrightarrow G$ is an S-homomor-
phism. We denote this relationship by $H \leqslant_S G$ (or by $H \leqslant G$ if it is clear
that S-subgroups are meant rather than subgroups). The following result is
straightforward, and is often taken as the definition of S-subgroups:

Proposition 1.7.1. *H is an S-subgroup of G if and only if H is a subgroup
of G and ${}^s h \in H$ for all $s \in S$ and $h \in H$.*

A **normal** S-**subgroup** of an S-group G is an S-subgroup of G which is a
normal subgroup of G. We denote this relationship by $H \trianglelefteq_S G$ (or by $H \trianglelefteq G$
if it is clear that S-subgroups are meant rather than subgroups).

S-subgroups arise from S-homomorphisms in two ways:

Proposition 1.7.2. *When $\varphi : G \longrightarrow H$ is an S-homomorphism, the image
$\operatorname{Im} \varphi$ of φ is an S-subgroup of H, and the kernel $\operatorname{Ker} \varphi$ of φ is a normal
S-subgroup of G.*

1.7.d. Quotient Groups

Proposition 1.7.3. *When N is a normal S-subgroup of an S-group G, the
[right] cosets of N are the elements of an S-group G/N, whose operations*

are well defined by $(Nx)(Ny) = Nxy$ and $^s(Nx) = N(^sx)$. The projection $\pi :$ $x \longmapsto Nx$ is an S-homomorphism of G onto G/N, and $\operatorname{Ker} \pi = N$.

Proof. The similar result for groups is Proposition 1.5.4.

If $s \in S$, $n \in N$, and $x \in G$, then $^sn \in N$ and $^s(nx) = {}^sn\,{}^sx \in N(^sx)$. Thus $\{^sy; y \in Nx\}$ is contained in the single coset $N(^sx)$ of N. Hence there is a mapping $Nx \longmapsto {}^s(Nx)$ of G/N into G/N that assigns to each coset Nx the coset which contains $\{^sy; y \in Nx\}$. We see that $^s(Nx) = N(^sx)$; in particular, the projection $\pi : G \longrightarrow G/N$ is an S-homomorphism. $\qquad\square$

The extension of Theorems 1.6.2 and 1.6.4 to groups with operators is left as an exercise. We extend Theorem 1.6.5:

Theorem 1.7.4 (Homomorphism Theorem). *When $\varphi : G \longrightarrow H$ is an S-homomorphism:*

(1) *$\operatorname{Im} \varphi$ is an S-subgroup of H;*

(2) *$K = \operatorname{Ker} \varphi$ is a normal S-subgroup of G;*

(3) *$\varphi(x) = \varphi(y) \iff Kx = Ky \iff xK = yK$;*

(4) *$\operatorname{Im} \varphi \cong G/\operatorname{Ker} \varphi$; in fact there exists an S-isomorphism $\theta : G/\operatorname{Ker} \varphi \longrightarrow$ $\operatorname{Im} \varphi$ such that the following diagram commutes:*

$$
\begin{array}{ccc}
G & \xrightarrow{\ \varphi\ } & H \\
{\scriptstyle\pi}\downarrow & & \uparrow{\scriptstyle\iota} \\
G/\operatorname{Ker} \varphi & \xrightarrow[\ \theta\ ]{} & \operatorname{Im} \varphi
\end{array}
$$

(where $\pi : G \longrightarrow G/\operatorname{Ker} \varphi$ is the projection and $\iota : \operatorname{Im} \varphi \longrightarrow H$ is the inclusion homomorphism).

Proof. (1) and (2) follow from Proposition 1.7.2; by Theorem 1.6.5, we need only prove that the isomorphism θ in the diagram is an S-isomorphism. Since the diagram commutes, we have $\theta(Kx) = \varphi(x)$ for all $x \in G$. Hence

$$\theta(^s(Kx)) = \theta(K(^sx)) = \varphi(^sx) = {}^s\varphi(x) = {}^s(\theta(Kx)). \qquad\square$$

1.7.e. Isomorphism Theorems. The extension of Theorems 1.6.7 and 1.6.8 to groups with operators is also left as an exercise.

Theorem 1.7.5 (First Isomorphism Theorem). *Let G be an S-group and $A \trianglelefteq_S G$, $B \trianglelefteq_S G$. If $A \subseteq B$, then $A \trianglelefteq_S B$, $B/A \trianglelefteq_S G/A$, and*

$$(G/A)/(B/A) \cong G/B;$$

more precisely, there exists an S-isomorphism $\theta : (G/A)/(B/A) \longrightarrow G/B$ such that the following diagram commutes (i.e., $\theta \circ \sigma \circ \pi = \rho$):

$$G \xrightarrow{\;\pi\;} G/A \xrightarrow{\;\sigma\;} (G/A)/(B/A)$$

$$\rho \searrow \qquad \swarrow \theta$$

$$G/B$$

where π, ρ, σ are the projections.

Theorem 1.7.6 (Second Isomorphism Theorem). *Let G be an S-group and $A \leqslant_S G$, $N \trianglelefteq_S G$. Then $AN \leqslant_S G$, $N \trianglelefteq_S AN$, $A \cap N \trianglelefteq_S A$, and*

$$AN/N \cong A/A \cap N ;$$

more precisely, there exists an S-isomorphism $\theta : A/A \cap N \longrightarrow AN/N$ such that the following diagram commutes (i.e., $\rho \circ \iota = \theta \circ \pi$):

$$
\begin{array}{ccc}
A & \xrightarrow{\;\iota\;} & AN \\
\pi \downarrow & & \downarrow \rho \\
A/A \cap N & \xrightarrow[\;\theta\;]{} & AN/N
\end{array}
$$

where π and ρ are the projections and ι is the inclusion homomorphism.

Exercises

1. If vector spaces over a field K are regarded as abelian K-groups, what are the K-subgroups? the K-homomorphisms?

2. Every group G acts on itself by inner automorphisms ($g \bullet x = gxg^{-1}$). What are the G-subgroups of G? the G-endomorphisms of G?

3. Let G, A, B be S-groups and $\varphi : A \longrightarrow G$, $\psi : B \longrightarrow G$ be S-homomorphisms. Assume that ψ is injective. Prove that φ factors through ψ (i.e., $\varphi = \psi \circ \chi$ for some S-homomorphism $\chi : A \longrightarrow B$) if and only if Im $\varphi \subseteq$ Im ψ, and then φ factors uniquely through ψ (i.e., χ is unique).

4. Let G, A, B be S-groups and $\varphi : G \longrightarrow A$, $\psi : G \longrightarrow B$ be S-homomorphisms. Assume that ψ is surjective. Prove that φ factors through ψ (i.e., $\varphi = \chi \circ \psi$ for some S-homomorphism $\chi : B \longrightarrow A$) if and only if Ker $\psi \subseteq$ Ker φ, and then φ factors uniquely through ψ (i.e., χ is unique).

5. Let G be an S-group and $A \trianglelefteq_S G$, $B \trianglelefteq_S G$, $A \subseteq B$. Prove that $A \trianglelefteq_S B$, $B/A \trianglelefteq_S G/A$, and $(G/A)/(B/A) \cong G/B$; more precisely, there exists an S-isomorphism $\theta : (G/A)/(B/A) \longrightarrow G/B$ such that $\theta \circ \sigma \circ \pi = \rho$, where $\pi : G \longrightarrow G/A$, $\rho : G \longrightarrow G/B$, and $\sigma : G/A \longrightarrow (G/A)/(G/B)$ are the projections.

6. Let G be an S-group and $A \leqslant_S G$, $N \trianglelefteq_S G$. Prove that $AN \leqslant_S G$, $N \trianglelefteq_S AN$, $A \cap N$ $\trianglelefteq_S A$, and $AN/N \cong A/A \cap N$; more precisely, there exists an S-isomorphism $\theta :$ $A/A \cap N \longrightarrow AN/N$ such that $\rho \circ \iota = \theta \circ \pi$, where $\pi : A \longrightarrow A/A \cap N$ and $\rho : AN$ $\longrightarrow AN/N$ are the projections and $\iota : A \longrightarrow AN$ is the inclusion homomorphism.

2

GROUP CONSTRUCTIONS

The constructions in this chapter are direct sums and products, semidirect products, group extensions, free groups, presentations, and free products. These constructions produce new groups from old ones, or from thin air, and yield a number of useful examples. Some constructions (group extensions, free products, and perhaps semidirect products) can be skipped at first reading.

Required results: Chapter 1 is used throughout. However, some of the more specialized results from Chapter 1 are needed only as follows. We use Light's associativity test from Section 1.1 in examples of presentations. The characterization of D_n in Section 1.4 helps explain its presentation. Proposition 1.4.17 is used twice in Section 2.2. Chain conditions from Section B.2 are used in Section 2.9.

2.1. DIRECT SUMS AND PRODUCTS

Direct products are the simplest way of constructing new groups from old ones. Applications of this construction are given in the next section.

2.1.a. Direct Products. The **direct product** of a family of groups $(G_i)_{i \in I}$ is the cartesian product $\prod_{i \in I} G_i$, whose elements are all families $x = (x_i)_{i \in I}$ such that $x_i \in G_i$ for all $i \in I$; x_i is the i **component** of x. The operation on $\prod_{i \in I} G_i$ is the **componentwise** operation defined (in the multiplicative notation) by

$$(x_i)_{i \in I} \ (y_i)_{i \in I} = (x_i y_i)_{i \in I} \ .$$

It is immediate that $\prod_{i \in I} G_i$ is a group, of order $|\prod_{i \in I} G_i| = \prod_{i \in I} |G_i|$. If $I = \varnothing$, then $\prod_{i \in I} G_i$ has only one element (the empty family) and $\prod_{i \in I} G_i = 1$. If $I = \{1\}$, then $\prod_{i \in I} G_i = G_1$. If I is finite, then $\prod_{i \in I} G_i$ is a **finite** direct product (even though the groups G_i may be infinite). If $I = \{1, 2, \ldots, n\}$, then $\prod_{i \in I} G_i$ is also denoted by $G_1 \times G_2 \times \cdots \times G_n$, or by $G_1 \oplus G_2 \oplus \cdots \oplus G_n$, and then $(x_i)_{i \in I}$ is denoted by $(x_1, x_2, \ldots, x_n)$.

In this section we pay special attention to finite direct products, which are the only ones used in Section 2.2. Section 11.4 gives a more complete treatment of direct products and sums.

The direct product $\prod_{i \in I} G_i$ comes with **projections** $\pi_i : \prod_{i \in I} G_i \longrightarrow G_i$ (one for each $i \in I$); π_i assigns to $(x_i)_{i \in I} \in \prod_{i \in I} G_i$ its i component x_i. The projections are surjective, and the operation on $\prod_{i \in I} G_i$ is defined precisely so that every projection is a homomorphism.

The direct product $\prod_{i \in I} G_i$ also comes with **injections** $\iota_j : G_j \longrightarrow \prod_{i \in I} G_i$ (one for each $j \in I$); when $x_j \in G_j$, $x = \iota_j(x_j)$ is defined by

$$x_i = \begin{cases} x_j & \text{if } i = j, \\ 1 \in G_i & \text{if } i \neq j. \end{cases}$$

We see that every ι_i is an injective homomorphism. Moreover $\iota_j(x_j)$ and $\iota_k(x_k)$ commute whenever $j \neq k$: indeed $\iota_j(x_j)\, \iota_k(x_k)$ and $\iota_k(x_k)\, \iota_j(x_j)$ have the same i component for every i, namely x_j if $i = j$, x_k if $i = k$, and 1 if $i \neq j, k$. (On the other hand, $\iota_j(x_j)$ and $\iota_j(y_j)$ do not commute unless x_j and y_j commute in G_j.)

2.1.b. Homomorphisms. In order to build diagrams, we construct homomorphisms whose domain or codomain is a direct product.

If $\varphi : G \longrightarrow \prod_{i \in I} G_i$ is a homomorphism, then $\varphi_i = \pi_i \circ \varphi : G \longrightarrow G_i$ is a homomorphism for each $i \in I$.

$$
\begin{array}{ccc}
G & \xrightarrow{\ \varphi\ } & \prod_{i \in I} G_i \\
& \searrow{\scriptstyle \varphi_i} & \downarrow{\scriptstyle \pi_i} \\
& & G_i
\end{array}
$$

It turns out that the homomorphisms $\varphi_i : G \longrightarrow G_i$ can be chosen arbitrarily:

Proposition 2.1.1. *Let* G *and* $(G_i)_{i \in I}$ *be groups and* $\varphi_i : G \longrightarrow G_i$ *be a homomorphism for each* $i \in I$. *There exists a homomorphism* $\varphi : G \longrightarrow \prod_{i \in I} G_i$ *unique such that* $\pi_i \circ \varphi = \varphi_i$ *for all* $i \in I$; *namely* $\varphi(x) = (\varphi_i(x))_{i \in I}$ *for all* $x \in G$.

Proof. The condition $\pi_i \circ \varphi = \varphi_i$ for all $i \in I$ requires that $\varphi(x) = (\varphi_i(x))_{i \in I}$ for all $x \in G$; hence there is at most one homomorphism φ with this property. On the other hand, the mapping $\varphi : G \longrightarrow \prod_{i \in I} G_i$ defined by $\varphi(x) = (\varphi_i(x))_{i \in I}$ for all $x \in G$, is a homomorphism:

$$\varphi(x)\varphi(y) = (\varphi_i(x))_{i \in I}\, (\varphi_i(y))_{i \in I} = (\varphi_i(x)\, \varphi_i(y))_{i \in I} = (\varphi_i(xy))_{i \in I} = \varphi(xy)$$

since every φ_i is a homomorphism. $\qquad\square$

Proposition 2.1.1 is an example of **universal property**. Generally, a universal property states that certain entities can be obtained from unique homomorphisms; a general definition is given in Section 18.7. In Proposition 2.1.1 every family of homomorphisms $\varphi_i : G \longrightarrow G_i$ is obtained from a unique homomorphism $\varphi : G \longrightarrow \prod_{i \in I} G_i$. Universal properties are generally used to construct homomorphisms.

Another universal property yields homomorphisms from direct products when I is finite. (The general case is considered at the end of the section.) If $\varphi : \prod_{i \in I} G_i \longrightarrow G$ is a homomorphism, then $\varphi_j = \varphi \circ \iota_j : G_j \longrightarrow G$ is a homomorphism for each $j \in I$.

These homomorphisms have one additional property: every $\varphi_j(x_j)$ commutes with every $\varphi_k(x_k)$ when $j \neq k$, since $\iota_j(x_j)$ and $\iota_k(x_k)$ commute in $\prod_{i \in I} G_i$. Conversely, homomorphisms with this property induce a homomorphism of $\prod_{i \in I} G_i$ into G, provided that I is finite.

Proposition 2.1.2. *Let G and $G_1, G_2, \ldots, G_n$ be groups and $\varphi_1 : G_1 \longrightarrow G$, $\varphi_2 : G_2 \longrightarrow G, \ldots, \varphi_n : G_n \longrightarrow G$ be homomorphisms such that $\varphi_j(x_j)$ commutes with $\varphi_k(x_k)$ whenever $x_j \in G_j$, $x_k \in G_k$, and $j \neq k$. There exists a homomorphism $\varphi : G_1 \times G_2 \times \cdots \times G_n \longrightarrow G$ unique such that $\varphi \circ \iota_i = \varphi_i$ for all i; namely $\varphi(x_1, x_2, \ldots, x_n) = \varphi_1(x_1)\,\varphi_2(x_2) \ldots \varphi_n(x_n)$ for all $(x_1, x_2, \ldots, x_n) \in G_1 \times G_2 \times \cdots \times G_n$.*

Proof. Assume that $\varphi : G_1 \times \cdots \times G_n \longrightarrow G$ is a homomorphism such that $\varphi \circ \iota_i = \varphi_i$ for all i. For each $(x_1, \ldots, x_n) \in G_1 \times \cdots \times G_n$, we have

$$(x_1, x_2, \ldots, x_n) = \iota_1(x_1)\, \iota_2(x_2) \ldots \iota_n(x_n)$$

in $G_1 \times \cdots \times G_n$. (This does not work for infinite direct products.) Hence

$$\varphi(x_1, x_2, \ldots, x_n) = \varphi(\iota_1(x_1))\, \varphi(\iota_2(x_2)) \ldots \varphi(\iota_n(x_n))$$
$$= \varphi_1(x_1)\, \varphi_2(x_2) \ldots \varphi_n(x_n).$$

Thus φ is unique and can be constructed as in the statement.

To prove that there exists a homomorphism $\varphi : G_1 \times G_2 \times \cdots \times G_n \longrightarrow G$ such that $\varphi \circ \iota_i = \varphi_i$ for all i, define a mapping $\varphi : G_1 \times \cdots \times G_n \longrightarrow G$ by

$$\varphi(x_1, x_2, \ldots, x_n) = \varphi_1(x_1)\, \varphi_2(x_2) \ldots \varphi_n(x_n).$$

We see that $\varphi \circ \iota_i = \varphi_i$ for all i. To show that φ is a homomorphism, we compute

$$\varphi((x_1,x_2,\ldots,x_n)(y_1,y_2,\ldots,y_n)) = \varphi(x_1y_1,x_2y_2,\ldots,x_ny_n)$$

$$= \varphi_1(x_1y_1)\,\varphi_2(x_2y_2)\ldots\varphi_n(x_ny_n)$$

$$= \varphi_1(x_1)\,\varphi_1(y_1)\,\varphi_2(x_2)\,\varphi_2(y_2)\ldots\varphi_n(x_n)\,\varphi_n(y_n)$$

$$= \varphi_1(x_1)\,\varphi_2(x_2)\,\varphi_1(y_1)\,\varphi_2(y_2)\ldots\varphi_n(x_n)\,\varphi_n(y_n)$$

$$= \cdots$$

$$= \varphi_1(x_1)\,\varphi_2(x_2)\ldots\varphi_n(x_n)\,\varphi_1(y_1)\,\varphi_2(y_2)\ldots\varphi_n(y_n)$$

$$= \varphi(x_1,x_2,\ldots,x_n)\,\varphi(y_1,y_2,\ldots,y_n)$$

using the hypothesis that $\varphi_j(x_j)$ commutes with $\varphi_k(y_k)$ when $j \neq k$ to move $\varphi_2(x_2),\ldots,\varphi_n(x_n)$ in front of $\varphi_1(y_1),\ldots,\varphi_n(y_n)$. $\qquad\square$

The reader will verify that Propositions 2.1.1 and 2.1.2 characterize direct products up to isomorphism (see the exercises).

2.1.c. Direct Sums. Using Proposition 2.1.2, finite direct products can also be characterized in terms of subgroups rather than homomorphisms.

Lemma 2.1.3. *Let G, G_1, ..., G_n be groups. Then $G \cong G_1 \times \cdots \times G_n$ if and only if G contains subgroups $A_1 \cong G_1,\ldots,A_n \cong G_n$ such that:*

(a) *$a_i \in A_i$ and $a_j \in A_j$ commute whenever $i \neq j$;*

(b) *every $x \in G$ can be written uniquely as a product $x = a_1a_2\ldots a_n$ in which $a_i \in A_i$ for all i.*

Then the homomorphism $A_1 \times \cdots \times A_n \longrightarrow G$, $(a_1,a_2,\ldots,a_n) \longmapsto a_1a_2\ldots a_n$ induced by the inclusion homomorphisms is an isomorphism.

Proof. Let $\iota_i : G_i \longrightarrow G_1 \times \cdots \times G_n$ $(i \in J)$ be the injections. The subgroups $\mathrm{Im}\,\iota_i$ have all the properties in the statement: $\mathrm{Im}\,\iota_i \cong G_i$, since ι_i is injective; we saw that $\iota_i(x_i)$ and $\iota_j(x_j)$ commute whenever $i \neq j$; and every $(x_1,x_2,\ldots,x_n) \in G_1 \times G_2 \times \cdots \times G_n$ can be written uniquely in the form

$$(x_1,x_2,\ldots,x_n) = \iota_1(t_1)\,\iota_2(t_2)\ldots\iota_n(t_n)$$

with $t_1 \in G_1, t_2 \in G_2,\ldots, t_n \in G_n$, since

$$\iota_1(t_1)\,\iota_2(t_2)\ldots\iota_n(t_n) = (t_1,t_2,\ldots,t_n).$$

If now $\theta : G_1 \times \cdots \times G_n \longrightarrow G$ is an isomorphism, then the subgroups $A_i = \theta(\operatorname{Im} \iota_i) \cong \operatorname{Im} \iota_i \cong G_i$ of G also have properties (a) and (b).

Conversely, assume that G contains subgroups $A_i \cong G_i$ with properties (a) and (b). Applying Proposition 2.1.2 to the inclusion homomorphisms $A_i \longrightarrow G$ yields a homomorphism $\theta : A_1 \times \cdots \times A_n \longrightarrow G$, which sends $(a_1, \ldots, a_n) \in A_1 \times \cdots \times A_n$ to the product $a_1 \ldots a_n$ in G. Property (b) shows that θ is bijective. Hence $G \cong A_1 \times \cdots \times A_n$, and $A_1 \times \cdots \times A_n \cong G_1 \times \cdots \times G_n$, since $A_i \cong G_i$ for all i (see the exercises). $\qquad\square$

A group G is the **internal direct sum** $G = A_1 \oplus \cdots \oplus A_n$ of subgroups $A_1, \ldots, A_n$ in case properties (a) and (b) in Lemma 2.1.3 are satisfied (*external* direct sums are defined below). There is a close relationship between internal direct sums and finite direct products, which justifies the notation $G_1 \oplus \cdots \oplus G_n$ often used for direct products: every direct product $G_1 \times \cdots \times G_n$ is the internal direct sum of $\operatorname{Im} \iota_1 \cong G_1, \ldots, \operatorname{Im} \iota_n \cong G_n$; conversely, every internal direct sum $G = A_1 \oplus \cdots \oplus A_n$ is isomorphic to the direct product $A_1 \times \cdots \times A_n$. The group G can then be constructed from smaller groups. In the next section we use this method to construct abelian groups.

The following result gives a more convenient characterization of direct sums, which is often used as definition:

Proposition 2.1.4. *A group G is the internal direct sum of subgroups $A_1, A_2, \ldots, A_n$ if and only if:*

   (i) $A_i \trianglelefteq G$ *for all* i;

   (ii) $(A_1 A_2 \ldots A_i) \cap A_{i+1} = 1$ *for all* $i < n$; *and*

   (iii) $G = A_1 A_2 \ldots A_n$.

Condition (ii) implies that $A_i \cap A_j = 1$ whenever $i \neq j$ (since $A_i \subseteq A_1 A_2 \ldots A_{j-1}$ when, say, $i < j$), but this weaker condition is not sufficient (see the exercises). Also note that Proposition 2.1.4 holds (trivially) if $n = 0$ (then $G = 1$) or if $n = 1$ (then $G = A_1$).

Proof. Assume $G = A_1 \oplus \cdots \oplus A_n$, so that (a) and (b) hold. Then $\theta : (a_1, \ldots, a_n) \longmapsto a_1 \ldots a_n$ is an isomorphism of $A_1 \times \cdots \times A_n$ onto G. Let $\iota_i : A_i \longrightarrow A_1 \times \cdots \times A_n$ be the i-th injection. For every $a \in A_i$, θ sends $\iota_i(a)$ onto a; hence $A_i = \theta(\operatorname{Im} \iota_i)$. Now $\operatorname{Im} \iota_i$ is a normal subgroup of $A_1 \times \cdots \times A_n$: if $x = (x_1, \ldots, x_n) \in A_1 \times \cdots \times A_n$ and $a = (a_1, \ldots, a_n) \in \operatorname{Im} \iota_i$, then $a_j = 1$ for all $j \neq i$, $x_j a_j x_j^{-1} = 1$ for all $j \neq i$, and $xax^{-1} \in \operatorname{Im} \iota_i$. Therefore A_i is a normal subgroup of G.

If $x \in (A_1 \ldots A_i) \cap A_{i+1}$, then $x = a_1 \ldots a_i = a_{i+1}$ for some $a_1 \in A_1$, ..., $a_{i+1} \in A_{i+1}$ and applying (b) to the equality

$$a_1 \ldots a_i \, 1 \, 1 \ldots 1 = 1 \ldots 1 \, a_{i+1} \, 1 \ldots 1$$

yields $a_1 = \cdots = a_i = a_{i+1} = 1$ and $x = 1$. Property (b) also yields $G = A_1 \ldots A_n$. Thus (i), (ii), and (iii) hold.

Conversely, assume that G contains subgroups $A_1, \ldots, A_n$ with properties (i), (ii), (iii). If $a_i \in A_i$, $a_j \in A_j$, and, say, $i < j$, then $a_i a_j a_i^{-1} \in A_j$ by (i), $a_j a_i a_j^{-1} \in A_i$, $a_i a_j a_i^{-1} a_j^{-1} \in A_i \cap A_j \subseteq (A_1 \ldots A_{j-1}) \cap A_j$, $a_i a_j a_i^{-1} a_j^{-1} = 1$ by (ii), and $a_i a_j = a_j a_i$; thus (a) holds.

By (iii), every element x of G can be written in the form $x = a_1 \ldots a_n$ with $a_i \in A_i$ for all i. Assume that $a_1 \ldots a_n = b_1 \ldots b_n$, with $a_i, b_i \in A_i$ for all i, and $a_i \neq b_i$ for some i. Let j be the greatest i such that $a_i \neq b_i$. Then $a_j \neq b_j$ and $a_i = b_i$ for all $i > j$. Hence $a_1 \ldots a_j = b_1 \ldots b_j$. Let

$$t = a_j b_j^{-1} = (a_1 \ldots a_{j-1})^{-1} (b_1 \ldots b_{j-1}).$$

Then $t = a_j b_j^{-1} \in A_j$, and

$$t = (a_1 \ldots a_{j-1})^{-1} (b_1 \ldots b_{j-1}) = a_1^{-1} b_1 \ldots a_{j-1}^{-1} b_{j-1} \in A_1 \ldots A_{j-1}$$

by (a). Hence (ii) yields $t = 1$ and $a_j = b_j$. This contradiction shows that (b) holds. $\qquad \square$

Corollary 2.1.5. *A group G is the direct sum of two subgroups A and B if and only if* (i) $A \trianglelefteq G$ *and* $B \trianglelefteq G$, (ii) $A \cap B = 1$, *and* (iii) $G = AB$.

2.1.d. Infinite Direct Sums. We now extend Proposition 2.1.2 to infinite families $(G_i)_{i \in I}$ of groups, leaving most of the details to the reader. Infinite direct sums are most useful with abelian groups; this case is treated in more detail in Section 11.4.

Inspection suggests that the proof of Proposition 2.1.2 can be saved if we only consider families $(x_i)_{i \in I}$ in which all but finitely many x_i are equal to 1. This leads to the following definitions.

Let I be a set and $P(i)$ be any property of $i \in I$ (a statement about i which is either true or false, depending on $i \in I$). We say that $P(i)$ is true **for almost all** $i \in I$ in case $P(i)$ is true for all but finitely many $i \in I$; that is, in case the set $\{ i \in I \,; P(i)$ is false$\}$ is finite.

Lemma 2.1.6. *When $(G_i)_{i \in I}$ is a family of groups,*

$$\bigoplus_{i \in I} G_i = \{ (x_i)_{i \in I} \,; x_i = 1 \text{ for almost all } i \}$$

is a subgroup of $\prod_{i \in I} G_i$.

Proof. We see that $1 = (1)_{i \in I} \in \bigoplus_{i \in I} G_i$. If $x, y \in \bigoplus_{i \in I} G_i$, then $x_i y_i^{-1} \neq 1$ implies $x_i \neq 1$ or $y_i \neq 1$, and the set

$$\{ i \in I \, ; \, x_i y_i^{-1} \neq 1 \} \subseteq \{ i \in I \, ; \, x_i \neq 1 \} \cup \{ i \in I \, ; \, y_i \neq 1 \}$$

is finite; thus $xy^{-1} \in \bigoplus_{i \in I} G_i$. $\qquad\qquad\qquad\qquad\qquad\qquad\qquad\qquad\qquad\square$

The subgroup $\bigoplus_{i \in I} G_i$ of $\prod_{i \in I} G_i$ is the **external direct sum** of the groups $(G_i)_{i \in I}$. If I is finite, then $\bigoplus_{i \in I} G_i = \prod_{i \in I} G_i$. If I is infinite, then it is to $\bigoplus_{i \in I} G_i$, rather than to $\prod_{i \in I} G_i$, that the universal property in Proposition 2.1.2 can be extended (see the exercises in Section 2.8).

We see that $\iota_j(x_j) \in \bigoplus_{i \in I} G_i$ for all $j \in J$, $x_j \in G_j$. This provides an **injection** $\iota_j : G_j \longrightarrow \bigoplus_{i \in I} G_i$ for each $j \in I$. As before, $\iota_j(x_j)$ and $\iota_k(x_k)$ commute whenever $j \neq k$. The reader will show:

Proposition 2.1.7. *Let G and $(G_i)_{i \in I}$ be groups and $\varphi_i : G_i \longrightarrow G$ be a homomorphism for each $i \in I$. If $\varphi_j(x_j)$ commutes with $\varphi_k(x_k)$ whenever $x_j \in G_j$, $x_k \in G_k$, and $j \neq k$, then there exists a unique homomorphism $\varphi : \bigoplus_{i \in I} G_i \longrightarrow G$ such that $\varphi \circ \iota_i = \varphi_i$ for all i.*

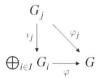

Proposition 2.1.8. *Let G and $(G_i)_{i \in I}$ be groups. Then $G \cong \bigoplus_{i \in I} G_i$ if and only if G contains subgroups $A_i \cong G_i$ (one for each $i \in I$) such that:*

(a) *$a_i \in A_i$ and $a_j \in A_j$ commute whenever $i \neq j$;*
(b) *every $x \in G$ is a product of unique elements $a_i \in A_i$ (with $a_i = 1$ for almost all i).*

A group G is the **internal direct sum** $G = \bigoplus_{i \in I} A_i$ of subgroups $(A_i)_{i \in I}$ in case properties (a) and (b) in Lemma 2.1.3 are satisfied. There is a close relationship between internal and external direct sums: every external direct sum $\bigoplus_{i \in I} G_i$ is the internal direct sum of its subgroups $\operatorname{Im} \iota_i \cong G_i$ $(i \in I)$; conversely, every internal direct sum $G = \bigoplus_{i \in I} A_i$ is isomorphic to the external direct sum $\bigoplus_{i \in I} A_i$. The same notation $\bigoplus$ is used for both; it should be clear from context whether a direct sum is internal or external.

Exercises extend Proposition 2.1.4 to all internal direct sums, in the case of abelian groups.

Exercises

1. Prove that $V_4 \cong \mathbb{Z}_2 \oplus \mathbb{Z}_2$.

2. Show that $K \lhd A$ and $L \lhd B$ implies $K \oplus L \lhd A \oplus B$, and $(A \oplus B)/(K \oplus L) \cong (A/K) \oplus (B/L)$.

3. Show a normal subgroup of $A \oplus B$ need not equal $K \oplus L$ with $K \lhd A$, $L \lhd B$.

4. Prove that direct products are commutative, in the sense that there is a canonical isomorphism $A \oplus B \cong B \oplus A$ for every groups A and B.

5. Prove that direct products are associative, in the sense that there are canonical isomorphisms $(A \times B) \times C \cong A \times (B \times C) \cong A \times B \times C$, for every groups A, B, and C.

6. Prove the following assiciativity property of direct products: if $I = \bigcup_{j \in J} I_j$ is a partition of I, then $\prod_{i \in I} G_i \cong \prod_{j \in J} (\prod_{i \in I_j} G_i)$.

7. Show that $A \cong A'$, $B \cong B'$ implies $A \times B \cong A' \times B'$.

8. Show that $A \times B \cong A' \times B'$ does not imply $A \cong A'$ even when $B \cong B'$. (Use an infinite direct sum.)

9. Show that Proposition 2.1.1 characterizes direct products up to isomorphism. In detail, let P and $(G_i)_{i \in I}$ be groups and $\rho_i : P \longrightarrow G_i$ be a homomorphism for each $i \in I$. Assume the same universal property: for every homomorphisms $\varphi_i : G \longrightarrow G_i$, there exists a unique homomorphism $\varphi : G \longrightarrow P$ such that $\rho_i \circ \varphi = \varphi_i$ for all i. Prove that $P \cong \prod_{i \in I} G_i$.

10. Show that Proposition 2.1.2 characterizes finite direct products up to isomorphism. In detail, let $G_1, \ldots, G_n$ and P be groups and $\kappa_1 : G_1 \longrightarrow P, \ldots, \kappa_n : G_n \longrightarrow P$ be homomorphisms such that $\kappa_j(x_j)$ and $\kappa_k(x_k)$ commute whenever $j \neq k$. Assume the same universal property: if $\varphi_1 : G_1 \longrightarrow G, \ldots, \varphi_n : G_n \longrightarrow G$ are homomorphisms such that $\varphi_j(x_j)$ and $\varphi_k(x_k)$ commute whenever $j \neq k$, then there exists a unique homomorphism $\varphi : P \longrightarrow G$ such that $\varphi \circ \kappa_i = \varphi_i$ for all i. Prove that $P \cong G_1 \times \cdots \times G_n$.

11. Give an example of a group G with subgroups A, B, C such that (a) $A, B, C \lhd G$; (b) $A \cap B = A \cap C = B \cap C = 1$; (c) $G = ABC$; but (d) $G \neq A \oplus B \oplus C$.

12. Prove Proposition 2.1.7.

13. Prove Proposition 2.1.8.

14. Let G be an abelian group (denoted multiplicatively) and $(A_i)_{i \in I}$ be subgroups of G. Show that every family $(a_i)_{i \in I} \in \bigoplus_{i \in I} A_i$ has a product $\prod_{i \in I} a_i$ in G. ($\bigoplus_{i \in I} A_i$ is the external direct sum.) Show that the set

$$\prod_{i \in I} A_i = \{ \textstyle\prod_{i \in I} a_i \, ; \, (a_i)_{i \in I} \in \bigoplus_{i \in I} A_i \}$$

is a subgroup of G. ($\prod_{i \in I} A_i$ is an internal product of subgroups and should not be confused with the direct product.)

15. Using internal products of subgroups from the previous exercise, prove that an abelian group G is the internal direct sum of subgroups $(A_i)_{i \in I}$ if and only if:

(ii) $(\prod_{i \neq j} A_i) \cap A_j = 1$ for all $j \in I$; and

(iii) $G = \prod_{i \in I} A_i$.

If I is totally ordered, show that (ii) can be replaced by:

(ii') $(\prod_{i<j} A_i) \cap A_j = 1$ for all $j \in I$.

16. How would you define the (internal) product of an infinite family of nontrivial subgroups of any group? (Restrictions on the subgroups are appropriate.)

2.2. DIRECT SUM DECOMPOSITIONS

Now that we have direct sums and products, we can try to construct groups as direct sums or products of simpler groups. This method works best with abelian groups.

2.2.a. Indecomposable Groups. A group G is **indecomposable** in case $G \neq 1$, and $G = A \oplus B$ implies $A = 1$ or $B = 1$. For example, $\mathbb{Z}$ is indecomposable: if $A = a\mathbb{Z}$ and $B = b\mathbb{Z}$ are nontrivial subgroups of $\mathbb{Z}$, then $A \cap B = m\mathbb{Z}$ is nontrivial [with $m = \mathrm{lcm}(a,b)$] and Corollary 2.1.5 shows that $\mathbb{Z} \neq A \oplus B$. When p is a prime number, $\mathbb{Z}_{p^n}$ is indecomposable (see the exercises).

Proposition 2.2.1. *Every finite group is a direct sum of indecomposable groups.*

Recall that we allowed direct sums to be empty or to have just one term.

Proof. Let G be a finite group; we proceed by induction on $|G|$. If $|G| = 1$, then G is an empty direct sum. Now let $|G| > 1$, and assume that the result holds for all groups of order less than $|G|$. If G is indecomposable, then G is a one-term direct sum of indecomposable groups. Otherwise, there are groups $A \neq 1$ and $B \neq 1$ such that $G = A \oplus B$. Since $|G| = |A||B|$, we have $|A| < |G|$ and $|B| < |G|$. By the induction hypothesis, A and B are direct sums of indecomposable groups; hence so is $G = A \oplus B$. $\square$

2.2.b. Abelian Groups. Direct sum decompositions are most useful for abelian groups (which in this section we denote multiplicatively). In this case we can find all finite indecomposable groups. Thus direct sums yield a complete construction of all finite abelian groups.

When p is a prime number, a p-**group** is a group in which the order of every element is a power of p. The following result implies that a finite abelian group, which is indecomposable, is a p-group for some prime p.

Proposition 2.2.2. *Every finite abelian group G is a direct sum of p-groups. In detail, if G has order $n = p_1^{k_1} p_2^{k_2} \ldots p_r^{k_r}$, where $p_1, p_2, \ldots, p_r$ are distinct primes,* *then $G = G(p_1) \oplus G(p_2) \oplus \cdots \oplus G(p_r)$, where*

$$G(p) = \{ x \in G ; \text{ the order of } x \text{ is a power of } p \}.$$

for every prime p.

Proof. Let p be prime. We have $1 \in G(p)$. If $x^{p^k} = y^{p^\ell} = 1$, and $m = \max(k, \ell)$, then $(xy^{-1})^{p^m} = x^{p^m} y^{-p^m} = 1$. Therefore $G(p)$ is a subgroup of G, which is normal, since G is abelian. We show that $G = G(p_1) \oplus G(p_2) \oplus \cdots \oplus G(p_r)$, using Proposition 2.1.4.

Let $i < r$ and $x \in (G(p_1) \oplus \cdots \oplus G(p_i)) \cap G(p_{i+1})$. Since $x \in G(p_1) \oplus \cdots \oplus G(p_i)$, x is a product of elements of orders $p_1^{j_1}, \ldots, p_i^{j_i}$ for some $j_1, \ldots, j_i \geq 0$, and $x^\ell = 1$, where $\ell = p_1^{j_1}, \ldots, p_i^{j_i}$. Since $x \in G(p_{i+1})$ we also have $x^m = 1$, where $m = p_{i+1}^j$ for some $j \geq 0$. Since $p_1, \ldots, p_r$ are distinct, ℓ and m are relatively prime, and there exist integers u, v such that $u\ell + vm = 1$ (Proposition A.3.8). Hence $x = (x^\ell)^u (x^m)^v = 1$. Thus $(G(p_1) \oplus \cdots \oplus G(p_i)) \cap G(p_{i+1}) = 1$.

Let $x \in G$. Since $|G| = n$, the order m of x divides n, and $m = p_1^{j_1} \ldots p_r^{j_r}$, where $0 \leq j_i \leq k_i$ for all i. Let $m_i = m/p_i^{j_i}$. Then $x^{m_i} \in G(p_i)$. Also $\gcd(m_1, \ldots, m_r) = 1$; hence there exist integers $u_1, \ldots, u_r$ such that $u_1 m_1 + \cdots + u_r m_r = 1$, and $x = (x^{m_1})^{u_1} \ldots (x^{m_r})^{u_r} \in G(p_1) \ldots G(p_r)$. Thus $G = G(p_1) \ldots G(p_r)$. $\square$

Corollary 2.2.3. *If $n = p_1^{k_1} p_2^{k_2} \ldots p_r^{k_r}$, where $p_1, \ldots, p_r$ are distinct primes, then $\mathbb{Z}_n \cong \mathbb{Z}_{p_1^{k_1}} \oplus \mathbb{Z}_{p_2^{k_2}} \oplus \cdots \oplus \mathbb{Z}_{p_r^{k_r}}$.*

Proof. If the order of $x \in \mathbb{Z}_n$ is a power of p_i, then the order of x, which must divide n, is at most $p_i^{k_i}$. By Proposition 1.4.17, $\mathbb{Z}_n(p_i)$ is the cyclic subgroup of $\mathbb{Z}_n$ of order $p_i^{k_i}$, and $\mathbb{Z}_n(p_i) \cong \mathbb{Z}_{p_i^{k_i}}$. $\square$

As another application of Proposition 2.2.2 we show:

Proposition 2.2.4. *Every finite multiplicative subgroup of a field is cyclic.*

Proof. Let K be a field, G be a finite multiplicative subgroup of K, and $n = |G|$. (Then $x^n = 1$ for all $x \in G$, so that G consists of n-th roots of unity.) For each prime divisor p of n, the finite subgroup $G(p)$ of G contains an element a whose order p^k is maximal. Then $x^{p^k} = 1$ for all $x \in G(p)$. This implies $|G(p)| \leq p^k$, since in a field the equation $x^m = 1$ has at most m solutions. Since $\langle a \rangle \subseteq G(p)$ has order p^k, we have $G(p) = \langle a \rangle$, and $G(p)$ is cyclic. If now $p_1, \ldots, p_r$ are the distinct prime divisors of n, Corollary 2.2.3 yields

$$G = G(p_1) \oplus \cdots \oplus G(p_r) \cong \mathbb{Z}_{p_1^{k_1}} \oplus \cdots \oplus \mathbb{Z}_{p_r^{k_r}} \cong \mathbb{Z}_n. \qquad \square$$

The next result implies that a finite abelian p-group which is indecomposable is cyclic. Conversely, a cyclic p-group is indecomposable (since $\mathbb{Z}_{p^n}$ is indecomposable).

Lemma 2.2.5. *Let G be a finite abelian p-group. Let $a \in G$ have maximal order p^k. Then $G = \langle a \rangle \oplus B$ for some subgroup B.*

Proof. First we show that there is a subgroup B of G such that $B \cap \langle a \rangle = 1$ but $C \cap \langle a \rangle \neq 1$ for all $B \lneq C \leqslant G$. Otherwise, for every subgroup B with $B \cap \langle a \rangle = 1$, there is a larger subgroup $B \lneq C \leqslant G$ with $C \cap \langle a \rangle = 1$. Starting with $B_1 = 1$, we can then find infinitely many subgroups $B_1 \lneq B_2 \lneq B_3 \lneq \cdots$ such that $B_i \cap \langle a \rangle = 1$; this contradicts the finiteness of G.

We show that $x^p \in B$ implies $x \in \langle a \rangle B$. We may assume that $x \notin B$. Then $B \lneq B\langle x \rangle \leqslant G$, $B\langle x \rangle \cap \langle a \rangle \neq 1$ by the choice of B, and $bx^\ell = a^m \neq 1$ for some $b \in B$ and $\ell, m > 0$. If $p \mid \ell$, then $a^m = bx^\ell \in B$, since $x^p \in B$, contradicting $B \cap \langle a \rangle = 1$. Hence $\gcd(\ell, p) = 1$, $u\ell + vp = 1$ for some integers u, v, and $x = (x^\ell)^u (x^p)^v \in \langle a \rangle B$, since $x^\ell = a^m b^{-1}$ and x^p are in $\langle a \rangle B$.

Next we show that $x^p \in \langle a \rangle B$ implies $x \in \langle a \rangle B$. Assume that $x^p = a^m b$ for some $m \geqslant 0$ and $b \in B$. We have $x^{p^k} = 1$ by the choice of a. Then $a^{mp^{k-1}} b^{p^{k-1}} = x^{p^k} = 1$, $a^{mp^{k-1}} = b^{-p^{k-1}} \in B \cap \langle a \rangle$, $a^{mp^{k-1}} = 1$, and $p \mid m$, since a has order p^k. Hence $y = xa^{-m/p}$ satisfies $y^p = b \in B$; by the above, $y \in \langle a \rangle B$, and $x = ya^{m/p} \in \langle a \rangle B$.

We can now show that $G = \langle a \rangle \oplus B$. Let $x \in G$. Then $x^{p^k} = 1$ by the choice of a, and there is a least $t \geqslant 0$ such that $x^{p^t} \in \langle a \rangle B$. If $t > 0$, then $y = x^{p^{t-1}}$ satisfies $y^p \in \langle a \rangle B$ and the above yields $y \in \langle a \rangle B$, contradicting the choice of t. Therefore $t = 0$ and $x \in \langle a \rangle B$. Thus $G = \langle a \rangle B$; since $\langle a \rangle \cap B = 1$ by the choice of B, Corollary 2.1.5 yields $G = \langle a \rangle \oplus B$. $\square$

Combining Proposition 2.2.1 (every finite group is a direct sum of indecomposable groups), Proposition 2.2.2 (a finite abelian group which is indecomposable is a p-group for some prime p), and Lemma 2.2.5 (a finite abelian p-group which is indecomposable is cyclic), we obtain:

Theorem 2.2.6. *Every finite abelian group is a direct sum of cyclic p-groups (for various primes p).*

More generally, a group is **finitely generated** in case it is generated by a finite subset. Theorem 2.2.6 is a particular case of a more general result which will be proved in Chapter 10 by a different method:

Theorem 2.2.7 (Fundamental Theorem of Finitely Generated Abelian Groups). *Every finitely generated abelian group G is a direct sum of cyclic groups. More precisely, G is a direct sum of infinite cyclic groups and cyclic* *p-groups (for various primes p); in this direct sum the number of infinite cyclic groups, and the orders of the cyclic p-groups, are unique.*

Corollary 2.2.8. *In Theorem 2.2.6 the orders of the cyclic p-groups are unique: if G is the direct sum of cyclic groups of orders $p_1^{k_1}, \ldots, p_r^{k_r}$, and the direct sum of cyclic groups of orders $q_1^{\ell_1}, \ldots, q_s^{\ell_s}$ (where $p_1, \ldots, p_r$ are distinct primes and $q_1, \ldots, q_s$ are distinct primes), then $r = s$ and $q_1^{\ell_1}, \ldots, q_s^{\ell_s}$ can be reindexed so that $p_i^{k_i} = q_i^{\ell_i}$ for all i.*

A more general uniqueness statement applies to all finite groups; the proof is rather difficult and is tucked away in Section 2.9.

Theorem 2.2.9 (Krull-Schmidt). *Let*

$$G = A_1 \oplus \cdots \oplus A_m = B_1 \oplus \cdots \oplus B_n$$

be direct sum decompositions of a finite group G into indecomposable groups. Then $m = n$ and $B_1, \ldots, B_n$ can be reindexed so that $A_i \cong B_i$ for all i.

Using Theorem 2.2.6 and Corollary 2.2.8, all abelian groups of small order n are readily constructed (up to isomorphism). We give two examples.

There are three ways to write the number 8 as a product of positive powers of primes: $8 = 2 \cdot 2 \cdot 2$, $8 = 2 \cdot 4$, and $8 = 8$. This yields three abelian groups: $A = \mathbb{Z}_2 \oplus \mathbb{Z}_2 \oplus \mathbb{Z}_2$, $B = \mathbb{Z}_2 \oplus \mathbb{Z}_4$, and $C = \mathbb{Z}_8$. By Theorem 2.2.6, every abelian group of order 8 is isomorphic to A, B, or C. No two of A, B, and C are isomorphic: if, say, $A \cong B$, then some poor misguided abelian group is the direct sum of three cyclic groups of order 2, and also the direct sum of a cyclic group of order 2 and a cyclic group of order 4, contradicting Corollary 2.2.8. Thus there are, up to isomorphism, exactly three abelian groups of order 8.

There are four ways to write the number 36 as a product of positive powers of primes: $36 = 2 \cdot 2 \cdot 3 \cdot 3$, $36 = 4 \cdot 3 \cdot 3$, $36 = 2 \cdot 2 \cdot 9$, and $36 = 4 \cdot 9$. Therefore there are, up to isomorphism, exactly four abelian groups of order 36: $\mathbb{Z}_2 \oplus \mathbb{Z}_2 \oplus \mathbb{Z}_3 \oplus \mathbb{Z}_3$, $\mathbb{Z}_4 \oplus \mathbb{Z}_3 \oplus \mathbb{Z}_3$, $\mathbb{Z}_2 \oplus \mathbb{Z}_2 \oplus \mathbb{Z}_9$, and $\mathbb{Z}_4 \oplus \mathbb{Z}_9$. (By Corollary 2.2.3, $\mathbb{Z}_4 \oplus \mathbb{Z}_9 \cong \mathbb{Z}_{36}$.)

2.2.c. Euler's ϕ Function. As a consequence of the preceding results, we prove some basic properties of Euler's ϕ function. Interestingly, the results are purely number theoretical.

By definition, **Euler's ϕ function** $\phi(n)$ is the number of integers $1 \leqslant k \leqslant n$ which are relatively prime to n.

Lemma 2.2.10. *A cyclic group of order $n > 1$ has exactly $\phi(n)$ elements of order n.*

Proof. Let $G = \langle a \rangle$ have order n, so that $a^i = 1$ in G if and only if $n \mid i$. We have $G = \{a^1, a^2, \ldots, a^n = 1\}$. The order of a^k divides n, since $(a^k)^n = 1$. If $\gcd(k, n) = 1$, then $uk + vn = 1$ for some $u, v \in \mathbb{Z}$, so that $(a^k)^i = 1$ implies $a^i = a^{uki + vni} = 1$ and $n \mid i$; hence a^k has order n. If $\gcd(k, n) = d > 1$, then $(a^k)^{n/d} = 1$, since $d \mid k$, and a^k has order less than n. $\qquad\square$

Proposition 2.2.11. *If m and n are relatively prime, then $\phi(mn) = \phi(m)\phi(n)$.*

Proof. Let C_m (C_n, C_{mn}) be a cyclic group of order m (n, mn). By Corollary 2.2.3, $C_{mn} \cong C_m \oplus C_n$. We show that $(x, y) \in C_m \oplus C_n$ has order mn if and only if x has order m and y has order n; then the result follows from Lemma 2.2.10. If x has order m and y has order n, then $(x, y)^{mn} = 1$, and $(x, y)^k = 1$ implies $x^k = 1$, $y^k = 1$, $m \mid k$, $n \mid k$, and $mn = \mathrm{lcm}(m, n) \mid k$; hence (x, y) has order mn. Similarly, if x has order $k \leqslant m$ and y has order $\ell \leqslant n$, then (x, y) has order $k\ell$; if $k\ell = mn$, then $k = m$ and $\ell = n$. $\qquad\square$

If p is prime and $r > 0$, then $\gcd(k, p^r) > 1$ if and only if $p \mid r$; therefore $\phi(p^r) = p^r - p^{r-1} = p^r(1 - 1/p)$. By induction, Proposition 2.2.11 yields:

Corollary 2.2.12. $\phi(n) = n \prod_{p \text{ prime}, \ p \mid n}(1 - 1/p)$.

Proposition 2.2.13. $\sum_{d \mid n} \phi(d) = n$.

Proof. Let G be cyclic of order n and $d \mid n$. By Proposition 1.4.17, $\{x \in G; x^d = 1\}$ is cyclic of order d. Therefore G contains exactly $\phi(d)$ elements of order d. Since every element of G has an order which is some $d \mid n$, it follows that $n = \sum_{d \mid n} \phi(d)$. $\qquad\square$

Exercises

1. Prove that $\mathbb{Z}_{p^n}$ is indecomposable when p is a prime and $n > 0$.
2. Prove that S_3 is indecomposable.
3. Prove that D_5 is indecomposable.
4. Prove that D_4 is indecomposable.
5. Find all abelian groups of order 15.
6. Find all abelian groups of order 16.
7. Find all abelian groups of order 360.
8. Prove directly that $\mathbb{Z}_{mn} \cong \mathbb{Z}_m \oplus \mathbb{Z}_n$ whenever $\gcd(m, n) = 1$.
9. Let G be a finite abelian group in which $x^n = 1$ for all x (e.g., a finite abelian group of order n). Assume that $n = k\ell$, where k and ℓ are relatively prime. Prove directly that $G = A \oplus B$, where $A = \{x \in G; x^k = 1\}$ and $B = \{x \in G; x^\ell = 1\}$.
10. Prove directly that $\mathbb{Z}_2 \oplus \mathbb{Z}_2 \oplus \mathbb{Z}_2$ is not isomorphic to $\mathbb{Z}_2 \oplus \mathbb{Z}_4$.
11. Prove directly that $\mathbb{Z}_2 \oplus \mathbb{Z}_4$ is not isomorphic to $\mathbb{Z}_8$.
12. Prove that $\mathrm{Aut}(\mathbb{Z}_p)$ is cyclic of order $p - 1$ when p is prime.

2.3. SEMIDIRECT PRODUCTS

Semidirect products generalize the direct product of two groups. They are of interest since nonabelian groups decompose more readily into semidirect products than into direct sums. For instance, the Schur-Zassenhaus Theorem, proved in Chapter 3, decomposes a finite group into a semidirect product whenever there is a normal subgroup whose order and index are relatively prime.

2.3.a. Definition. Let A and B be groups (denoted multiplicatively) and $\varphi : B \longrightarrow \text{Aut}(A)$ be a homomorphism of B into the group of automorphisms of A. It is convenient to denote $\varphi(b)(a)$ by ${}^b a$ (this is the **left exponential** notation). For all $a, a' \in A$ and $b, b' \in B$, we have

$$ {}^1 a = a \qquad \text{and} \qquad {}^b({}^{b'} a) = {}^{bb'} a, $$

since φ is a homomorphism, and

$$ {}^b 1 = 1 \qquad \text{and} \qquad {}^b(aa') = {}^b a \; {}^b a', $$

since $a \longmapsto {}^b a$ is an automorphism.

When $\varphi : B \longrightarrow \text{Aut}(A)$ is a homomorphism, the **semidirect product** $A \times_\varphi B$ of A by B (relative to φ) is the cartesian product $A \times B$ with multiplication

$$ (a,b)(a',b') = (a \; {}^b a', bb'). $$

If, for example, $\varphi(b) = 1_A$ for all $b \in B$, then ${}^b a = a$ and $A \times_\varphi B = A \times B$ is the direct product of A and B. The following general result is straightforward:

Lemma 2.3.1. *When A and B are groups, then $A \times_\varphi B$ is a group.*

2.3.b. Internal Characterization. Corollary 2.1.5 extends to semidirect products.

Proposition 2.3.2. *A group G is isomorphic to a semidirect product of G_1 by G_2 if and only if G contains subgroups $A \cong G_1$ and $B \cong G_2$ such that $A \lhd G$, $A \cap B = 1$, and $G = AB$.*

Proof. Let $G = G_1 \times_\varphi G_2$. Let $A = \{(a,1); a \in G_1\}$ and $B = \{(1,b); b \in G_2\}$. It is immediate that $A \lhd G$, $B \leqslant G$, $A \cong G_1$, and $B \cong G_2$. Moreover $A \cap B = 1$ and $AB = G$ (since $(a,b) = (a,1)(1,b)$).

Conversely, assume that $A \lhd G$ and $B \leqslant G$ satisfy $A \cap B = 1$, and $G = AB$. For each $b \in B$ there is an inner automorphism $\alpha_b : x \longmapsto bxb^{-1}$ of G. Then $\alpha_b(A) \subseteq A$, since $A \lhd G$, and α_b has a restriction $\varphi(b)$ to A. We see that $\varphi(b)$ is an automorphism of A (with inverse $\varphi(b^{-1})$) and that φ is a homomorphism of B into $\text{Aut}(A)$ (since $\alpha_b \circ \alpha_{b'} = \alpha_{bb'}$). By definition, ${}^b a = \alpha_b(a) = bab^{-1}$.

Define $\theta : A \times_\varphi B \longrightarrow G$ by: $\theta(a,b) = ab$. Then θ is a homomorphism:

$$\theta((a,b)(a',b')) = \theta(a\ ^ba', bb') = aba'b^{-1}bb' = aba'b' = \theta(a,b)\theta(a',b');$$

θ is surjective, since $G = AB$, and injective, since $ab = 1$ implies $a = b^{-1} \in A \cap B$ and $a = b = 1$. Thus $G \cong A \times_\varphi B$. Since $A \cong G_1$ and $B \cong G_2$, it follows that G is isomorphic to a semidirect product of G_1 and G_2. $\qquad\square$

Proposition 2.3.2 readily implies that D_n is a semidirect product of a cyclic group of order n by a cyclic group of order 2 (see the exercises). This provides examples of semidirect products that are not direct products.

2.3.c. Cyclic Groups. As a more general example, we construct all semidirect products of a cyclic group C_n of order n by a cyclic group C_m of order m. The groups Q and T in Section 2.6 are of this type.

First we determine $\mathrm{Aut}(C_n)$. Let U_n denote the group of all $\bar{k} \in \mathbb{Z}_n$ such that $\gcd(k,n) = 1$; the multiplication on U_n is induced by the multiplication on $\mathbb{Z}_n$ $(\bar{k}\,\bar{\ell} = \overline{k\ell})$. U_n has order $\phi(n)$; the following result implies that it is a group (this can also be proved directly).

Lemma 2.3.3. *Let $C_n = \langle a \rangle$ be cyclic of order n. Every automorphism α of C_n has the form $\alpha(a^i) = a^{ki}$, where $\gcd(k,n) = 1$. Hence $\mathrm{Aut}(C_n) \cong U_n$.*

Proof. Recall that $a^k = a^\ell$ if and only if $k \equiv \ell$ (mod n), if and only if $\bar{k} = \bar{\ell}$. Let $\alpha \in \mathrm{Aut}(C_n)$ and $\alpha(a) = a^k$. Then $\alpha(a^i) = a^{ki}$. Also a^k generates C_n, since a generates C_n, so that $\gcd(k,n) = 1$. Hence there is a mapping $\theta : \mathrm{Aut}(C_n) \longrightarrow U_n$ such that $\theta(\alpha) = \bar{k}$, where $\alpha(a) = a^k$.

Let $\alpha, \beta \in \mathrm{Aut}(C_n)$, with $\alpha(a) = a^k$, $\beta(a) = a^\ell$. Then $\alpha(\beta(a)) = \alpha(a^\ell) = (a^k)^\ell = a^{k\ell}$; hence θ is a homomorphism. If $\theta(\alpha)$ is the identity element of U_n, then $k \equiv 1$ (mod n), $\alpha(a) = a^k = a$, and α is the identity automorphism; hence θ is injective. Finally let $\bar{k} \in U_n$. Define $\alpha : C_n \longrightarrow C_n$ by $\alpha(a^i) = a^{ki}$. Then α is an endomorphism of C_n; α is surjective, since $\langle a^k \rangle = C_n$, and therefore is an automorphism of C_n. We see that $\theta(\alpha) = \bar{k}$; thus θ is surjective. $\qquad\square$

Proposition 2.3.4. *A group G is isomorphic to a semidirect product of a cyclic group of order n by a cyclic group of order m if and only if G is generated by two elements a and b such that a has order n, b has order m, $b^j \notin \langle a \rangle$ if $0 < j < m$, and $bab^{-1} = a^r$, where $r^m \equiv 1$ (mod n).*

Proof. Let $C_n = \langle a \rangle$ and $C_m = \langle b \rangle$ be cyclic groups of orders n,m. Let $\varphi : C_m \longrightarrow \mathrm{Aut}(C_n)$ be a homomorphism. As above, $\alpha = \varphi(b) \in \mathrm{Aut}(C_n)$ has the form $\alpha(a^i) = a^{ri}$, where $\gcd(r,n) = 1$. Then $\alpha^2(a^i) = \alpha(a^{ri}) = a^{r^2i}$, and by induction, $\alpha^j(a^i) = a^{r^j i}$. Since $b^m = 1$, we must have $\alpha^m = 1$, $a^{r^m} = \alpha^m(a) = a$, and $r^m \equiv 1 \pmod{n}$. Now $C_n \times_\varphi C_m$ is the group of all pairs $(a^i, b^k) \in C_n \times C_m$, with multiplication

$$(a^i, b^j)(a^k, b^\ell) = (a^{i+r^j k}, b^{j+\ell}),$$

since $\varphi(b^j)(a^k) = \alpha^j(a^k) = a^{r^j k}$. Let $c = (a, 1)$ and $d = (1, b)$. Then $c^i = (a^i, 1)$, $d^j = (1, b^j)$ so that c has order n, d has order m, and $d^j \notin \langle c \rangle$ if $0 < j < m$. Also $(a^i, b^j) = (a^i, 1)(1, b^j) = c^i d^j$ so that $C_n \times_\varphi C_m$ is generated by c and d. Finally $dcd^{-1} = (1, b)(a, b^{-1}) = (a^r, 1) = c^r$. Any group G which is isomorphic to $C_n \times_\varphi C_m$ must then have generators with similar properties.

Conversely, assume that G is generated by two elements a and b such that a has order n, b has order m, $b^j \notin \langle a \rangle$ if $0 < j < m$, and $bab^{-1} = a^r$, where $r^m \equiv 1 \pmod{n}$. This implies $ba^i b^{-1} = (bab^{-1})^i = a^{ri}$ and $ba^i = a^{ri} b$. Then $A = \langle a \rangle$ is cyclic of order n, $B = \langle b \rangle$ is cyclic of order m, and $A \cap B = 1$. Also every element of G is a product of nonnegative powers of a and b; in such a product the equality $ba^i = a^{ri} b$ can be used repeatedly to move powers of a in front of b's until all the a's precede all the b's. This writes every element of G in the form $a^i b^j$; hence $G = AB$. Since $aa^i a^{-1} = a^i$, $ba^i b^{-1} = a^{ri}$, it follows that $A \triangleleft G$. Hence $G \cong A \times_\varphi B$ for some φ. $\square$

Corollary 2.3.5. *When p is prime, a semidirect product of a cyclic group of order p by a cyclic group of order 2 is either cyclic or dihedral.*

Proof. If $r^2 \equiv 1 \pmod{p}$, then p divides $r^2 - 1 = (r-1)(r+1)$, p divides $r - 1$ or $r + 1$, and $r \equiv \pm 1 \pmod{p}$. If $r \equiv 1$, then in the above α is the identity, $\varphi(b^j)$ is the identity for all j, and $C_p \times_\varphi C_2 = C_p \times C_2$ is cyclic of order $2p$ by Corollary 2.2.3. If $r \equiv -1$, then in the above $\alpha(a) = a^{-1}$, $\alpha(x) = x^{-1}$ for all $x \in \langle a \rangle$, and $C_p \times_\varphi C_2 \cong D_p$. $\square$

When p is prime, it will eventually follow from Corollary 2.3.5 that every group of order $2p$ is either cyclic or dihedral.

Exercises

1. Prove Lemma 2.3.1.

2. Let $G = A \times_\varphi B$. Show that there is an injective homomorphism $\kappa : A \longrightarrow G$ and a surjective homomorphism $\rho : G \longrightarrow B$ such that $\operatorname{Im} \kappa = \operatorname{Ker} \rho$.

3. Show that D_n is a semidirect product of a cyclic group of order n by a cyclic group of order 2. If $n > 2$, this is not a direct product.

4. Let p and q be primes such that $q \nmid p - 1$ and C_p, C_q be cyclic groups of order p, q. Show that every semidirect product of C_p by C_q is cyclic.

5. Find all semidirect products of $\mathbb{Z}_4$ by $\mathbb{Z}_2$.

6. Find all semidirect products of $\mathbb{Z}_3$ by $\mathbb{Z}_4$.

2.4. GROUP EXTENSIONS

Group extensions are more difficult than semidirect products but more general, since they occur whenever normal subgroups occur. This section contains Schreier's Theorem, which constructs all extensions of one group by another, and some applications (the theorems of Schur and Hölder). The analysis of groups by group extensions (by normal series) is considered in Chapter 3.

2.4.a. Definition. A **group extension** $G \xrightarrow{\kappa} E \xrightarrow{\rho} Q$ of a group G by a group Q consists of a group E, an injective homomorphism $\kappa : G \longrightarrow E$, and a surjective homomorphism $\rho : E \longrightarrow Q$, such that $\operatorname{Im} \kappa = \operatorname{Ker} \rho$. Equivalently, E is a group with a normal subgroup $N \,(= \operatorname{Im} \kappa = \operatorname{Ker} \rho)$ such that $N \cong G$ and $E/N \cong Q$. For example, $G \times Q$, with the injection $G \longrightarrow G \times Q$ and projection $G \times Q \longrightarrow Q$, is a group extension of G by Q; so is every semidirect product of G by Q.

Group extensions need only be constructed up to isomorphism. In detail, an **equivalence** of group extensions $G \xrightarrow{\kappa} E \xrightarrow{\rho} Q$ and $G \xrightarrow{\lambda} F \xrightarrow{\sigma} Q$ of G by Q is an isomorphism $\theta : E \longrightarrow F$ such that the diagram

$$
\begin{array}{ccccc}
G & \xrightarrow{\;\kappa\;} & E & \xrightarrow{\;\rho\;} & Q \\
\Big\| & & \Big\downarrow{\scriptstyle\theta} & & \Big\| \\
G & \xrightarrow{\;\lambda\;} & F & \xrightarrow{\;\sigma\;} & Q
\end{array}
$$

commutes ($\theta \circ \kappa = \lambda$ and $\sigma \circ \theta = \rho$); then the extensions $G \xrightarrow{\kappa} E \xrightarrow{\rho} Q$ and $G \xrightarrow{\lambda} F \xrightarrow{\sigma} Q$ are **equivalent**. An equivalence of group extensions may be viewed as an isomorphism which preserves the relationship of E to G and Q.

2.4.b. Factor Sets. Let $G \xrightarrow{\kappa} E \xrightarrow{\rho} Q$ be a group extension of G by Q and $N = \operatorname{Im} \kappa = \operatorname{Ker} \rho \trianglelefteq E$. For each $a \in Q$, select any $p_a \in E$ such that $\rho(p_a) = a$; select $p_1 = 1$. Since each set $\rho^{-1}(a)$ is a [right] coset of N, this selects one element in each coset of N. We call $p = (p_a)_{a \in Q}$ a **cross section** of E (actually, p is a cross section of the partition of E into cosets of N). Every element of E can now be written in the form $n p_a$ for some unique $n \in N$, $a \in Q$, and we have proved:

Lemma 2.4.1. *When* $G \xrightarrow{\kappa} E \xrightarrow{\rho} Q$ *is a group extension of G by Q and p is a cross section of E, every element of E can be written in the form*

$$
\kappa(x)\, p_a \tag{2.1}
$$

for some unique $x \in G$ and $a \in Q$.

We now put every product in E in the form (2.1). We start with products $p_a p_b$ and $p_a \kappa(x)$. We have $\rho(p_a p_b) = \rho(p_a) \rho(p_b) = ab$; by Lemma 2.4.1,

$$p_a p_b = \kappa(s_{a,b}) p_{ab} \tag{2.2}$$

for some unique $s_{a,b} \in G$. The mapping $s : (a,b) \longmapsto s_{a,b}$ of $Q \times Q$ into G is the **factor set** of E relative to p. Similarly $\rho(p_a \kappa(x)) = \rho(p_a) \rho(\kappa(x)) = a$, so

$$p_a \kappa(x) = \kappa(^a x) p_a \tag{2.3}$$

for some unique $^a x \in G$. This defines the **action** of the set Q on the group G relative to p. (This is in general not a group action as defined in Chapter 3.)

It turns out that the factor set s and the set action of Q on G suffice to put all products in E in the form (2.1): indeed equations (2.2) and (2.3) yield

$$\kappa(x) p_a \kappa(y) p_b = \kappa(x) \kappa(^a y) p_a p_b = \kappa(x) \kappa(^a y) \kappa(s_{a,b}) p_{ab}$$
$$= \kappa(x^a y \, s_{a,b}) p_{ab} . \tag{2.4}$$

Lemma 2.4.1 and equation (2.4) construct E (up to isomorphism) from G and Q; this is the essence of Schreier's Theorem.

2.4.c. Schreier's Theorem. The factor set s and set action of Q on G have the following properties.

Lemma 2.4.2. When $G \xrightarrow{\kappa} E \xrightarrow{\rho} Q$ is a group extension of G by Q, the factor set s and set action of Q on G (relative to a cross section of E) have the following properties: for all $a, b, c \in Q$ and $x \in G$, $x \longmapsto {}^a x$ is an automorphism of G, and:

(i) $^1 x = x$ and $s_{a,1} = s_{1,a} = 1$;

(ii) $^a(^b x) s_{a,b} = s_{a,b} \, ^{ab} x$;

(iii) $s_{a,b} s_{ab,c} = {}^a s_{b,c} s_{a,bc}$.

Proof. Let $\varphi(a)$ denote the mapping $x \longmapsto {}^a x$ of G into G. By equation (2.2), $\kappa(^a x) = p_a \kappa(x) p_a^{-1}$. Hence the following diagram commutes:

$$
\begin{array}{ccc}
G & \xrightarrow{\kappa} & N \\
{\scriptstyle \varphi(a)} \downarrow & & \downarrow {\scriptstyle \alpha} \\
G & \xrightarrow{\kappa} & N
\end{array}
$$

where $N = \operatorname{Im} \kappa \trianglelefteq E$ and α is the restriction to N of the inner automorphism $\alpha_{p_a} : n \longmapsto p_a n p_a^{-1}$. It follows that $\varphi(a)$ is an isomorphism, since the other three maps in the diagram are isomorphisms.

Part (1) follows from $p_1 = 1$ and equations (2.2) and (2.3). Parts (2) and (3) follow from associativity in E and are left as exercises. ☐

In what follows we denote by $\varphi(a)$ the mapping $x \longmapsto {}^a x$ of G into G. By Lemma 2.4.2, φ is a mapping of Q into $\mathrm{Aut}(G)$. If, conversely, $\varphi : Q \longrightarrow \mathrm{Aut}(G)$ is a mapping, it is convenient to denote $\varphi(a)(x)$ by ${}^a_\varphi x$ or (if φ is known) by ${}^a x$; then ${}^1 x = x$ and ${}^a(xy) = {}^a x\, {}^a y$, for all a, x, y.

Theorem 2.4.3 (Schreier). *Let G and Q be groups and $s : Q \times Q \longrightarrow G$ and $\varphi : Q \longrightarrow \mathrm{Aut}(G)$ be mappings which satisfy conditions* (i), (ii), *and* (iii) *in Lemma 2.4.2. Let $E(s, \varphi) = G \times Q$ with multiplication given by*

$$(x, a)(y, b) = (x\, {}^a_\varphi y\, s_{a,b}, ab). \tag{2.5}$$

Then $E(s, \varphi)$ is a group extension of G by Q. Furthermore every group extension of G by Q is equivalent to $E(s, \varphi)$ for some s and φ.

Proof. Let s and φ satisfy (i), (ii), and (iii). Construct $E = E(s, \varphi)$ as above. Associativity is proved from (ii) and (iii) by a fat but straightforward calculation which is left as an exercise. It follows from (i) that $(1, 1)$ is the identity element of E. Every element (y, b) of E has a left inverse: if $a = b^{-1}$ and $x = ({}^a y\, s_{a,b})^{-1}$, then (2.5) yields $(x, a)(y, b) = (1, 1)$. Then (x, a) has a left inverse (z, c), and associativity in E yields

$$(z, c) = (z, c)((x, a)(y, b)) = ((z, c)(x, a))(y, b) = (y, b);$$

hence $(y, b)(x, a) = (1, 1)$, (x, a) is a true inverse of (y, b), and we have shown that E is a group. By (i) and (2.5), the mappings $G \overset{\lambda}{\longrightarrow} E \overset{\sigma}{\longrightarrow} Q$ defined by $\lambda(x) = (x, 1)$ and $\sigma(x, a) = a$ are homomorphisms. Thus $E = E(s, \varphi)$, together with λ and σ, is a group extension of G by Q.

Conversely, let $G \overset{\kappa}{\longrightarrow} E \overset{\rho}{\longrightarrow} Q$ be any group extension of G by Q. Let s be the factor set and φ be the set action of Q on G, relative to any cross section p of E. Define a mapping $\theta : E(s, \varphi) \longrightarrow E$ by: $\theta(x, a) = \kappa(x)\, p_a$, for all $x \in G$ and $a \in Q$. We see that θ is bijective, and equation (2.4) shows that θ is a homomorphism. Moreover it is clear that the diagram

$$
\begin{array}{ccccc}
G & \overset{\kappa}{\longrightarrow} & E & \overset{\rho}{\longrightarrow} & Q \\
\| & & \downarrow{\scriptstyle \theta} & & \| \\
G & \underset{\lambda}{\longrightarrow} & E(s, \varphi) & \underset{\sigma}{\longrightarrow} & Q
\end{array}
$$

commutes. Thus E is equivalent to $E(s, \varphi)$. ☐

Proposition 2.4.4. *$E(s, \varphi)$ and $E(t, \psi)$ are equivalent if and only if there exists a mapping $a \longmapsto u_a$ of Q into G such that $u_1 = 1$ and, for all $a, b \in Q$,*

$x \in G$:

(iv) $\overset{a}{\varphi}x = u_a \overset{a}{\psi}x\, u_a^{-1}$;

(v) $s_{a,b} = u_a \overset{a}{\psi}u_b\, t_{a,b}\, u_{ab}^{-1}$.

Proof. Let $\theta : E(s,\varphi) \longrightarrow E(t,\psi)$ be an equivalence. Since the diagram

$$
\begin{array}{ccccc}
G & \longrightarrow & E(s,\varphi) & \longrightarrow & Q \\
\| & & \downarrow{\scriptstyle \theta} & & \| \\
G & \longrightarrow & E(s,\varphi) & \longrightarrow & Q
\end{array}
$$

commutes, we have $\theta(x,1) = (x,1)$ and $\theta(1,a) = (u_a,a)$ for some $u_a \in G$; note that $u_1 = 1$, since $\theta(1,1) = (1,1)$. By (i), $\theta(x,a) = \theta((x,1)(1,a)) = (x,1)(u_a,a) = (xu_a,a)$. Hence

$$\theta((x,a)(y,b)) = (x\,\overset{a}{\varphi}y\,s_{a,b}\,u_{ab}\,,\,ab) \qquad \text{and}$$

$$\theta(x,a)\,\theta(y,b) = (x\,u_a\,\overset{a}{\psi}y\,\overset{a}{\psi}u_b\,t_{a,b}\,,\,ab).$$

Since θ is a homomorphism, this implies

$$x\,\overset{a}{\varphi}y\,s_{a,b}\,u_{ab} = x\,u_a\,\overset{a}{\psi}y\,\overset{a}{\psi}u_b\,t_{a,b} \tag{2.6}$$

for all $a,b \in Q$, $x,y \in G$. With $x = y = 1$, (2.6) yields (v). Then (v) and (2.6) yield

$$x\,\overset{a}{\varphi}y\,s_{a,b}\,u_{ab} = x\,u_a\,\overset{a}{\psi}y\,u_a^{-1}\,u_a\,\overset{a}{\psi}u_b\,t_{a,b} = x\,u_a\,\overset{a}{\psi}y\,u_a^{-1}\,s_{a,b}\,u_{ab}\,,$$

which after cancellations reduces to (iv).

Conversely, this last calculation shows that (iv) and (v) imply (2.6). Therefore the bijection $\theta : E(s,\varphi) \longrightarrow E(t,\psi)$ defined by $\theta(x,a) = (xu_a,a)$ is a homomorphism. It is clear that θ is an equivalence of group extensions. $\quad\square$

2.4.d. Split Extensions. Schreier's Theorem becomes simpler in some cases.

Proposition 2.4.5. *For a group extension* $G \xrightarrow{\ \kappa\ } E \xrightarrow{\ \rho\ } Q$ *the following are equivalent:*

(a) *There exists a homomorphism* $\mu : Q \longrightarrow E$ *such that* $\rho \circ \mu = 1_Q$.

(b) *There is a cross section of E relative to which* $s_{a,b} = 1$ *for all* $a,b \in Q$.

(c) *E is equivalent to a semidirect product of G by Q.*

(d) *Relative to any cross section of E, there exists a mapping $a \longmapsto u_a$ of Q into G such that $u_1 = 1$ and, for all $a,b \in Q$, $s_{a,b} = {}^a u_b\, u_a\, u_{ab}^{-1}$.*

Proof. (a) implies (b). If (a) holds, we can select $p_a = \mu(a)$ for all $a \in Q$. Then $s_{a,b} = 1$ for all $a,b \in Q$, by equation (2.2).

(b) implies (c). If $s_{a,b} = 1$ for all $a,b \in Q$, then $\varphi : Q \longrightarrow \mathrm{Aut}(G)$ is a homomorphism, by (ii); $E(s,\varphi)$ is the semidirect product $G \times_\varphi Q$; and E is equivalent to $E(s,\varphi)$, by Schreier's Theorem.

(c) implies (d). A semidirect product $G \times_\psi Q$ of G by Q is a group extension $E(t,\psi)$ in which $t_{a,b} = 1$ for all a,b. If E is equivalent to $G \times_\psi Q$, then, relative to any cross section of E, $E(s,\varphi)$ and $E(t,\psi)$ are equivalent. By (v) and (iv) in Proposition 2.4.4, this implies $s_{a,b} = u_a\, {}^a_\psi u_b\, u_{ab}^{-1} = {}^a_\varphi u_b\, u_a\, u_{ab}^{-1}$, for all $a,b \in Q$.

(d) implies (a). It follows from (d) that

$$u_a^{-1}\, {}^a(u_b^{-1})\, s_{a,b} = u_a^{-1}\, ({}^a u_b)^{-1}\, {}^a u_b\, u_a\, u_{ab}^{-1} = u_{ab}^{-1}.$$

Define $\mu : Q \longrightarrow E$ by $\mu(a) = \kappa(u_a^{-1})\, p_a$. Then $\rho(\mu(a)) = a$ and (2.4) yields

$$\mu(a)\mu(b) = \kappa(u_a^{-1}\, {}^a(u_b^{-1})\, s_{a,b})\, p_{ab} = \kappa(u_{ab}^{-1})\, p_{ab} = \mu(ab). \qquad \square$$

A group extension **splits** (or, is a split extension) in case it satisfies the equivalent conditions in Proposition 2.4.5. Not all extensions split; the exercises give examples.

2.4.e. Extensions of Abelian Groups. When G is abelian, condition (ii) simply states that $\varphi : Q \longrightarrow \mathrm{Aut}(G)$ is a homomorphism, and Schreier's Theorem becomes:

Corollary 2.4.6. *Let G be an abelian group, Q be a group, $s : Q \times Q \longrightarrow G$ be a mapping, and $\varphi : Q \longrightarrow \mathrm{Aut}(G)$ be a homomorphism which satisfy conditions (i) and (iii) in Lemma 2.4.2. Then $E(s,\varphi)$ is a group extension of G by Q. Furthermore every group extension of G by Q is equivalent to $E(s,\varphi)$ for some s and φ.*

Also condition (iv) simply states that $\varphi = \psi$, and Proposition 2.4.4 becomes:

Corollary 2.4.7. *When G is abelian, $E(s,\varphi)$ and $E(t,\psi)$ are equivalent if and only if $\varphi = \psi$ and there exists a mapping $a \longmapsto u_a$ of Q into G such that $u_1 = 1$ and, for all $a,b \in Q$,*

(v) $s_{a,b} = u_a\, {}^a u_b\, u_{ab}^{-1}\, t_{a,b}.$

2.4.f. Schur's Theorem. As an application of group extensions, we prove:

Theorem 2.4.8 (Schur). *Let G be an abelian group and Q be a group. If G and Q are finite and have relatively prime orders, then every group extension of G by Q splits.*

Proof. Let $|G| = n$, $|Q| = m$, and E be an extension of G by Q with factor set s. For each $a \in Q$, let $t_a = \prod_{c \in Q} s_{a,c} \in G$. Then $t_1 = 1$ by (i). Since Q is a group, $\prod_{c \in Q} s_{a,bc} = \prod_{d \in Q} s_{a,d} = t_a$. Since G is abelian, applying $\prod_{c \in Q}$ to property (iii)

$$s_{a,b}\, s_{ab,c} = {}^a s_{b,c}\, s_{a,bc}$$

yields

$$s_{a,b}^m\, t_{ab} = {}^a t_b\, t_a.$$

We also have $s_{a,b}^n = 1$, since $s_{a,b} \in G$, $t_c^n = 1$ for all c, and $qm + rn = 1$ for some $q, r \in \mathbb{Z}$, since $(m,n) = 1$. Hence

$$s_{a,b} = s_{a,b}^{qm+rn} = {}^a t_b^q\, t_a^q\, t_{ab}^{-q},$$

and it follows from Proposition 2.4.5 (with $u_c = t_c^q$) that the extension splits. $\qquad\square$

When G is abelian and $|G| = n$ and $|Q| = m$ are relatively prime, Schur's Theorem implies that every extension of G by Q contains a subgroup of order m. We show that any two such subgroups are conjugate (each is the image of the other under some inner automorphism of G).

Proposition 2.4.9. *Let E be an extension of a finite abelian group G by a finite group Q whose orders $|G| = n$ and $|Q| = m$ are relatively prime. Any two subgroups of E of order m are conjugate.*

Proof. Let A and B be subgroups of E of order m. Since $(m,n) = 1$, we have $A \cap G = B \cap G = 1$ and $AG = BG = E$. Each coset Ga of G intersects A in exactly one element. Hence for each $a \in A$ there exists $u_a \in N$ unique such that $u_a a \in B$. (This is similar to the proof of Proposition 2.4.4; indeed B is a cross section of E, viewed as an extension of G by A.) Let $a, b \in A$. Since B is a subgroup, we have $u_a(au_b a^{-1})ab = (u_a a)(u_b b) \in B$. Hence

$$u_{ab} = u_a a u_b a^{-1} \qquad (2.7)$$

for all $a, b \in A$. (Compare with condition (v) in Corollary 2.4.7.)

Let $v = \prod_{b \in A} u_b \in G$. Since G is abelian, equation (2.7) yields $v = u_a^m a v a^{-1}$. Since $(m,n) = 1$, we have $mq + nr = 1$ for some integers q, r; then $u_a^{mq} = u_a$ and $w = v^q \in G$ satisfies $w = u_a a w a^{-1}$. Hence $u_a a = w a w^{-1}$ for all $a \in A$. Thus $B = wAw^{-1}$ is a conjugate of A. $\qquad\square$

In Chapter 3 we will extend Schur's Theorem 2.4.8 to all finite groups, and Proposition 2.4.9 to all finite solvable groups.

2.4.g. Hölder's Theorem. Hölder's Theorem describes all extensions of a cyclic group by another; this generalizes Proposition 2.3.4.

Proposition 2.4.10 (Hölder's Theorem). *A group G is an extension of a cyclic group of order m by a cyclic group of order n if and only if G contains generators a and b such that a has order m, $b^n = a^t$, $b^i \notin \langle a \rangle$ if $0 < i < n$, and $bab^{-1} = a^r$, where $r^n \equiv 1$ and $rt \equiv t \pmod m$.*

Such a group exists for every choice of integers m,n,r,t such that $m,n > 0$, $r^n \equiv 1$, and $rt \equiv t \pmod m$.

Proof. Let G contain generators a and b such that a has order m, $b^n = a^t$, $b^i \notin \langle a \rangle$ if $0 < i < t$, and $bab^{-1} = a^r$, where $r^n \equiv 1$ and $rt \equiv t \pmod m$. Then a and b have finite order, and every element of G is a product of a's and b's. Since $bab^{-1} \in A$, we have $bAb^{-1} \subseteq A$, and it follows that $A = \langle a \rangle \lhd G$. Also G/A is generated by Ab, since G is generated by a and b; since $b^n \in A$ and $b^i \notin A$ when $0 < i < n$, G/A is cyclic of order n.

Conversely, let G be a group extension of a cyclic group $A = \langle a \rangle$ of order m, which we regard as a normal subgroup of G, by a cyclic group $C = \langle c \rangle$ of order n. Let $b \in G$ be sent to c by the projection $G \longrightarrow C$. Since $C = \langle c \rangle = \{c^i ; i \in \mathbb{Z}\}$, we have $G/A = \{Ab^i ; i \in \mathbb{Z}\}$; hence every element of G can be written in the form $a^i b^j$, and G is generated by a and b. Since c has order n, we have $b^n = a^t \in \langle a \rangle$ for some t and $b^i \notin \langle a \rangle$ when $0 < i < n$. Also $bab^{-1} = a^r \in \langle a \rangle$ for some r, since $A \lhd G$. This implies $a^{rt} = ba^t b^{-1} = bb^n b^{-1} = a^t$ and $rt \equiv t \pmod m$. Also $b^2 ab^{-2} = ba^r b^{-1} = (a^r)^r = a^{r^2}$, and, by induction, $b^i ab^{-i} = a^{r^i}$; since $b^n ab^{-n} = a$, we have $r^n \equiv 1 \pmod m$.

In the above, $1 = p_1$, $b = p_c$, ..., $b^{n-1} = p_{c^{n-1}}$ is a cross section of G. We saw that $b^i a^k b^{-i} = a^{kr^i}$. When $0 \leqslant i, j < n$, we also have $b^i b^j = b^{i+j}$ if $i + j < n$, $b^i b^j = a^t b^{i+j-n}$ if $i + j \geqslant n$. This suggests how Corollary 2.4.6 can be used to show that a group G exists for every suitable choice of m,n,r,t.

Assume that $m,n > 0$, $r^n \equiv 1$, and $rt \equiv t \pmod m$. Let $A = \langle a \rangle$ be cyclic of order m and $C = \langle c \rangle$ be cyclic of order n. Since $r^n \equiv 1$, r and m are relatively prime and an automorphism α of A is defined by $\alpha(a^k) = a^{kr}$. Moreover $\alpha^i(a^k) = a^{kr^i}$ for all i. In particular, α^n is the identity on A. Hence there is a homomorphism $\varphi : C \longrightarrow \mathrm{Aut}(A)$ such that $\varphi(c^i) = \alpha^i$ for all i. With ${}^i a^k = {}^{c^i}_\varphi a^k$ this yields

$$ {}^i a^k = {}^{c^i}_\varphi a^k = a^{kr^i} . $$

Since $rt \equiv t \pmod m$, we have $\alpha(a^t) = a^t$ and ${}^i a^t = a^t$ for all i. Define a factor set by

$$ s_{i,j} = s_{c^i, c^j} = \begin{cases} 1 & \text{if } i + j < n, \\ a^t & \text{if } i + j \geqslant n, \end{cases} $$

for all $0 \leqslant i, j < n$. Condition (i) in Lemma 2.4.2 holds. The proof of condition (iii) falls into the following cases. Let $0 \leqslant i, j, k < n$.

If $i + j + k < n$, then $s_{i,j} s_{i+j,k} = 1 = {}^i s_{j,k} s_{i,j+k}$.

If $i + j < n$, $j + k < n$, and $i + j + k \geqslant n$, then $s_{i,j} s_{i+j,k} = a^t = {}^i s_{j,k} s_{i,j+k}$.

If $i + j < n$ and $j + k \geqslant n$, then $c^j c^k = c^{j+k-n}$, $i + j + k - n < k < n$, and
$s_{i,j} s_{i+j,k} = a^t = {}^i a^t = {}^i s_{j,k} s_{i,j+k-n}$.

If $i + j \geqslant n$ and $j + k < n$, then similarly $c^i c^j = c^{i+j-n}$, $i + j + k - n < i < n$, and $s_{i,j} s_{i+j-n,k} = a^t = {}^i s_{j,k} s_{i,j+k}$.

If $i + j \geqslant n$, $j + k \geqslant n$, and $i + j + k < 2n$, then $i + j + k - n < n$ and
$s_{i,j} s_{i+j-n,k} = a^t = {}^i a^t = {}^i s_{j,k} s_{i,j+k-n}$.

If finally $i + j \geqslant n$, $j + k \geqslant n$, and $i + j + k \geqslant 2n$, then $i + j + k - n \geqslant n$ and
$s_{i,j} s_{i+j-n,k} = a^t a^t = {}^i a^t a^t = {}^i s_{j,k} s_{i,j+k-n}$.

Thus (iii) holds. By Corollary 2.4.6, there is an extension $E(s, \varphi)$ of A by C. $\square$

Exercises

1. Complete the proof of Lemma 2.4.2 (prove (ii) and (iii)).
2. Verify that $E(s, \varphi)$ is associative.
3. Produce a cross section of $E(s, \varphi)$ relative to which s is the factor set and φ the set action of Q on G.
4. Show that $E(s, \varphi)$ and $E(t, \psi)$ are equivalent if and only if there is a group extension E and two cross sections of E relative to which s, φ and t, ψ are the factor set and set action of E.
5. Find all extensions of $\mathbb{Z}_3$ by $\mathbb{Z}_2$ (up to equivalence).
6. Find all extensions of $\mathbb{Z}_4$ by $\mathbb{Z}_2$ (up to equivalence).
7. Find all extensions of $\mathbb{Z}_p$ by $\mathbb{Z}_q$ (up to equivalence) when p and q are distinct primes.
8. Find all extensions of $\mathbb{Z}_3$ by $\mathbb{Z}_4$ (up to equivalence).

2.5. FREE GROUPS

In this section and the next we construct groups which are generated by a given set X.

We saw in Chapter 1 (Proposition 1.4.9) that a group G is generated by a subset X of G if and only if every element of G is a product of elements of X and inverses of elements of X. An element of G can be written in this form in many ways; for instance, $xx^{-1} = x^{-1}x = 1$ for all $x \in X$. Thus there are always nontrivial equalities between our products. The simplest way to construct a group generated by X is to have as few such equalities as possible. The result is the free group on X.

2.5.a. Words. We begin with the similar construction for monoids, which is simpler. (Recall that a monoid is a set with an associative operation for which there is an identity element.)

A monoid M is generated by a subset X of M in case every element of M is a product of elements of X. An element of M can usually be written in this form in many ways. As with groups, the simplest way to construct a monoid generated by X is to have as few equalities as possible between products of elements of X. With monoids, we can arrange to have no equalities at all between products of elements of X, so that products of different sequences are different. The result is the free monoid on X.

Given a set X, we define "formal products" of elements of X as follows. A **word** in the **alphabet** X is a finite sequence $a = (a_1, a_2, \ldots, a_n)$ of elements of X, the **letters** of a; the **length** of a is the nonnegative integer n. (The usual notation for $(a_1, a_2, \ldots, a_n)$ is $a_1 a_2 \ldots a_n$. For instance, the very words you are now reading are really finite sequences of elements of the latin alphabet $\{a, b, c, \ldots, z\}$.) Words are multiplied by **concatenation**:

$$(a_1, a_2, \ldots, a_n)(b_1, b_2, \ldots, b_m) = (a_1, a_2, \ldots, a_n, b_1, b_2, \ldots, b_m).$$

Concatenation may be of German origin, and it is associative. The **empty** word $1 = (\ )$, which is the word $(a_1, \ldots, a_n)$ with length $n = 0$, is an identity element for concatenation. We denote by W_X the set of all words in the alphabet X (this is also denoted by X^*). We see that W_X is a monoid; W_X is the **free monoid** on the set X.

W_X comes with an **injection** of X into W_X which assigns to each $x \in X$ the one letter word $\iota(x) = (x)$. (It is common practice to identify $x \in X$ and $\iota(x) \in W_X$.) Every word $a = (a_1, \ldots, a_n)$ can be recovered by concatenating the one letter words $(a_1), \ldots, (a_n)$:

$$(a_1, a_2, \ldots, a_n) = \iota(a_1)\, \iota(a_2)\, \ldots\, \iota(a_n);$$

hence W_X is generated (as a monoid) by $\iota(X)$, which is a copy of X.

We show that W_X is the "simplest" monoid generated by [a copy of] X. For this, we prove that every monoid M generated by X is a homomorphic image of W_X. More generally, W_X has the following universal property.

Proposition 2.5.1. *Let X be a set. For each mapping φ of X into a monoid M there is a homomorphism ψ of W_X into M unique such that the diagram*

commutes ($\psi \circ \iota = \varphi$); *namely,* $\psi(a_1,\ldots,a_n) = \varphi(a_1)\ldots\varphi(a_n)$ *for all* $(a_1,\ldots,a_n) \in W_X$. *If* $\varphi(X)$ *generates* M, *then* ψ *is surjective.*

Proof. A homomorphism of monoids is a mapping which preserves identity elements and products.

Assume that $\psi : W_X \longrightarrow M$ is a homomorphism such that $\psi(\iota(x)) = \varphi(x)$ for all $x \in X$. For any $a = (a_1,\ldots,a_n) \in W$, we have

$$\psi(a) = \psi(\iota(a_1))\ldots\psi(\iota(a_n)) = \varphi(a_1)\ldots\varphi(a_n).$$

Hence there is at most one homomorphism $\psi : W_X \longrightarrow M$ such that $\psi \circ \iota = \varphi$.

Conversely, define a mapping $\psi : W_X \longrightarrow M$ by

$$\psi(a) = \varphi(a_1)\ldots\varphi(a_n)$$

for all $a = (a_1,\ldots,a_n) \in W$. If $a = 1$ (if $n = 0$), then $\varphi(a_1)\ldots\varphi(a_n)$ is an empty product and $\psi(a) = 1$. Also $\psi(ab) = \psi(a)\psi(b)$ for all $a,b \in W_X$. Thus ψ is a homomorphism (of monoids). We see that $\psi \circ \iota = \varphi$. Also, Im ψ consists of all products of elements of $\varphi(X)$; if M is generated (as a monoid) by $\varphi(X)$, then ψ is surjective. $\quad\square$

2.5.b. Reduced Words. We now construct the free group F_X on a set X. For clarity's sake we have not used the standard notation for F_X (it is indicated in parentheses).

Let $x \longmapsto x'$ be a bijection of X onto a set X' which is disjoint from X. (The usual notation for x' is x^{-1}.) It is convenient to use the same notation $y \longmapsto y'$ for the inverse bijection $X' \longrightarrow X$; then $(x')' = x$ for all $x \in X \cup X'$. Let $Y = X \cup X'$ and $W = W_Y$ be the set of all words in the alphabet Y; thus an element of W is a finite sequence $a = (a_1,\ldots,a_n)$ (usually denoted by $a_1 a_2 \ldots a_n$), where $n \geqslant 0$ and $a_1,\ldots,a_n \in Y$. We can view the elements of W as formal products of elements of X and inverses of elements of X.

Assume for a minute that X is a subset of a group G. To every element of Y there corresponds an element of G (either an element of X or the inverse of an element of X); this defines a mapping $Y \longrightarrow G$. Every word $a = (a_1,\ldots,a_n) \in W$ has a product in G: in W, a is the product of its letters, and the corresponding elements of G have a product in G. In fact this describes the monoid homomorphism $W \longrightarrow G$ provided by Proposition 2.5.1. If G is generated as a group by the elements of X, then G is generated as a monoid by the elements of X and their inverses, and $W \longrightarrow G$ is surjective. In particular, there must be a surjection $W \longrightarrow F_X$; thus every element of F_X comes from a word in W.

Certain words yield all of F_X. A word $a = (a_1,\ldots,a_n)$ is **reduced** in case it does not contain consecutive letters a_i, a_{i+1} such that $a_{i+1} = a_i'$.

Reduction is the removal of subsequences (a_i, a_i') from a word until a reduced word is reached; the detailed process is as follows. A **one-step re-**

duction replaces $a = (a_1, \ldots, a_i, a_{i+1}, \ldots, a_n)$ by $b = (a_1, \ldots, a_{i-1}, a_{i+2}, \ldots, a_n)$ in case $a_{i+1} = a_i'$; we denote this relationship by $a \geqslant_1 b$. A k **step reduction** is a sequence of k one-step reductions; we write $a \geqslant_k b$ when a k step reduction replaces a by b. For instance $a \geqslant_0 a$. In general, $a \geqslant_k b$ if and only if $a \geqslant_1 c_1 \geqslant_1 c_2 \geqslant_1 \cdots \geqslant_1 c_k = b$ for some $c_1, c_2, \ldots, c_k \in W$. We write $a \geqslant b$ (or $b \leqslant a$) when $a \geqslant_k b$ for some k.

If a is reduced, then there is no $a \geqslant_1 b$; hence $a \geqslant_k b$ implies $k = 0$, and $a \geqslant b$ implies $a = b$. (Thus reduced words are minimal under $\leqslant$.) The following result states that every word a can be reduced (= transformed into a reduced word by a k step reduction); this is readily proved by induction on the length of a.

Lemma 2.5.2. *For each word $a \in W$ there is a reduced word $b \leqslant a$.*

Assume for one more minute that X is a subset of a group G. As above, every word $(a_1, \ldots, a_n) \in W$ has a product in G. When b is obtained from a by a one step reduction, then a and b have the same product in G. If therefore $a \geqslant b$, then a and b have the same product in G. Thus reduction gives us words which must have the same product in any group $G \supseteq X$. In particular, every element of the free group come from a reduced word.

2.5.c. Reduction. We now prove that a given word reduces to only one reduced word; that is, in Lemma 2.5.2, b is unique.

Lemma 2.5.3. *If $a \geqslant_1 b$ and $a \geqslant_1 c \neq b$, then $b \geqslant_1 d$ and $c \geqslant_1 d$ for some d.*

Proof. By definition, $b = (a_1, \ldots, a_{i-1}, a_{i+2}, \ldots, a_n)$, where $a_{i+1} = a_i'$, and $c = (a_1, \ldots, a_{j-1}, a_{j+2}, \ldots, a_n)$, where $a_{j+1} = a_j'$. We have $i \neq j$, since $b \neq c$; we may assume that $i < j$. If $j = i + 1$, then $a_{i+2} = a_{j+1} = a_j' = a_{i+1}' = a_i = a_{j-1}$ and $b = c$. Hence $j \geqslant i + 2$. It follows that a_j and a_{j+1} are consecutive letters of b, so that $b \geqslant_1 d$, where

$$d = (a_1, \ldots, a_{i-1}, a_{i+2}, \ldots, a_{j-1}, a_{j+2}, \ldots, a_n)$$

(if $j = i + 2$, then $d = (a_1, \ldots, a_{i-1}, a_{j+2}, \ldots, a_n)$). Similarly a_i and a_{i+1} are consecutive letters of c, so that $c \geqslant_1 d$, with d as above. $\qquad\square$

Lemma 2.5.4. *If $a \geqslant b, c$ in W, then $b, c \geqslant d$ for some $d \in W$.*

Proof. We prove by induction on $k + \ell$ that $a \geqslant_k b$ and $a \geqslant_\ell c$ imply $c \geqslant_i d$ and $b \geqslant_j d$ for some $d \in W$ and $i \leqslant k$, $j \leqslant \ell$. This holds if $k = 0$, if $\ell = 0$, or (by Lemma 2.5.3) if $k = \ell = 1$ (with $d = b = c$ if $b = c$), and thus holds whenever $k + \ell \leqslant 2$.

Now assume that $k + \ell > 2$. We may assume that $k, \ell > 0$. First consider the case where $k > 1$. Then $a \geqslant_1 b_1 \geqslant_{k-1} b$ for some b_1 and $1 + \ell < k + \ell$. By the induction hypothesis, $a \geqslant_1 b_1$ and $a \geqslant_\ell c$ imply $c \geqslant_{i'} d_1$ and $b_1 \geqslant_{j'} d_1$ for some $d_1 \in W$ and $i' \leqslant 1$, $j' \leqslant \ell$. Then $(k-1) + j' < k + \ell$; by the induction hypothesis, $b_1 \geqslant_{k-1} b$ and $b_1 \geqslant_\ell d_1$ imply $d_1 \geqslant_{i''} d$ and $b \geqslant_{j''} d$ for some $d \in W$ and $i'' \leqslant k - 1$, $j'' \leqslant j'$. These relationships may be summarized in the diagram:

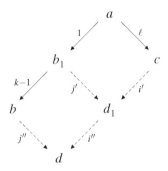

Hence $c \geqslant_{i'+i''} d$ and $b \geqslant_{j''} d$, with $i' + i'' \leqslant k$ and $j'' \leqslant \ell$.

If $k = 1$, then $\ell > 1$; exchanging b and c in the above yields $b \geqslant_j d$ and $c \geqslant_i d$ for some $d \in W$ and $i \leqslant k$, $j \leqslant \ell$. $\qquad\square$

Lemma 2.5.4 implies that b is unique in Lemma 2.5.2: if $a \geqslant b$ and $a \geqslant c$ with b and c reduced, then $b \geqslant d$ and $c \geqslant d$ for some $d \in W$, which implies $b = d = c$ since b and c are reduced. Thus:

Lemma 2.5.5. *For each word $a \in W$, there is a reduced word $b \leqslant a$, and b is unique.*

We denote by red a the reduced word b such that $b \leqslant a$.

2.5.d. Construction. It follows from Lemma 2.5.5 that reduced words can be used directly as elements of the free group on X. Thus we can define the **free group** $F = F_X$ on the set X as the set of all reduced words in W. The operation $\bullet$ on F (usually denoted by juxtaposition) is defined by concatenation and reduction:

$$a \bullet b = \mathrm{red}\,(ab).$$

The **injection** of X into F_X is the injective mapping $\iota : X \longrightarrow F_X$ which assigns to each $x \in X$ the one letter reduced word $\iota(x) = (x)$. (It is common practice to identify $x \in X$ and $\iota(x) \in F$.) It is convenient to extend ι to X' by $\iota(x') = (x')$. (This creates no ambiguity since X' is disjoint from X.)

Proposition 2.5.6. F_X *is a group and is generated by $\iota(X)$.*

Proof. First we show that $\bullet$ is associative. Note that $a \geqslant_1 b$ implies $ac \geqslant_1 bc$. By induction, $a \geqslant b$ implies $ac \geqslant bc$. Similarly $a \geqslant b$ implies $ca \geqslant cb$. It follows

that

$$abc \geqslant (a \cdot b)c \geqslant (a \cdot b) \cdot c \qquad \text{and} \qquad abc \geqslant a(b \cdot c) \geqslant a \cdot (b \cdot c).$$

Since $(a \cdot b) \cdot c$ and $a \cdot (b \cdot c)$ are reduced, Lemma 2.5.5 implies $(a \cdot b) \cdot c = a \cdot (b \cdot c)$.

The empty word 1 is reduced and is the identity element of F. We show that the inverse of $a = (a_1, \ldots, a_n) \in F$ is $a' = (a'_n, a'_{n-1}, \ldots, a'_1)$: indeed a' is reduced, since a is, and

$$aa' = (a_1, \ldots, a_n, a'_n, \ldots, a'_1)$$
$$\geqslant_1 (a_1, \ldots, a_{n-1}, a'_{n-1}, \ldots, a'_1) \geqslant_1 \cdots \geqslant_1 (a_1, a'_1) \geqslant_1 1,$$

so that $a \cdot a' = 1$. Since $(a')' = a$, this implies $a' \cdot a = 1$. Thus F is a group. That $\iota(X)$ generates F follows from the next lemma, which is easily proved by induction on n. □

Lemma 2.5.7. *When $a = (a_1, \ldots, a_n)$ is reduced, then*

$$a = \iota(a_1) \cdot \iota(a_2) \cdot \cdots \cdot \iota(a_n).$$

Also $\iota(x') = \iota(x)^{-1}$ for all $x \in X$.

When x is identified with $\iota(x)$, and x' is written x^{-1}, Lemma 2.5.7 writes every element of F uniquely as a reduced product of elements of X and inverses of elements of X.

2.5.e. Universal Property. We now show that F_X is the "simplest" group generated by [a copy of] X. For this, we prove that every group G generated by X is a homomorphic image of F_X (hence there are at least as many equalities in G between products of elements of X and inverses of elements of X as there are in F_X). More generally, F_X has the following universal property.

Theorem 2.5.8. *Let X be a set. For each mapping φ of X into a group G, there is a homomorphism ψ of F_X into G unique such that the diagram*

commutes ($\psi \circ \iota = \varphi$); namely $\psi(a_1, \ldots, a_n) = \varphi(a_1) \ldots \varphi(a_n)$ for all $(a_1, \ldots, a_n) \in F_X$. If $\varphi(X)$ generates G, then ψ is surjective.

If each $x \in X$ is identified with $\iota(x) \in F_X$ so that $X \subseteq F_X$, then Theorem 2.5.8 states that every mapping of X into a group G extends uniquely to a homomorphism of F_X into G.

Proof. It is convenient to extend φ to X' by $\varphi(x') = \varphi(x)^{-1}$ for all $x \in X$. Then $\varphi(y') = \varphi(y)^{-1}$ for all $y \in Y = X \cup X'$.

Assume that $\psi : F = F_X \longrightarrow G$ is a homomorphism such that $\psi(\iota(x)) = \varphi(x)$ for all $x \in X$. Then $\psi(\iota(x')) = \psi(\iota(x)^{-1}) = \varphi(x)^{-1} = \varphi(x')$ for all $x \in X$. For any $a = (a_1, \ldots, a_n) \in F$, Lemma 2.5.7 implies

$$\psi(a) = \psi(\iota(a_1)) \ldots \psi(\iota(a_n)) = \varphi(a_1) \ldots \varphi(a_n).$$

This shows that there is at most one homomorphism $\psi : F \longrightarrow G$ such that $\psi \circ \iota = \varphi$.

To prove that such a homomorphism exists, define

$$\chi(a) = \varphi(a_1) \ldots \varphi(a_n)$$

for all $a = (a_1, \ldots, a_n) \in W$; thus $\chi : W_Y \longrightarrow G$ is the monoid homomorphism induced by $\varphi : Y \longrightarrow G$ in Proposition 2.5.1. In particular, $\chi(ab) = \chi(a)\chi(b)$ for all $a, b \in W$. If $a \geqslant_1 b$, then $b = (a_1, \ldots, a_{i-1}, a_{i+2}, \ldots, a_n)$, where $a_{i+1} = a_i'$, so that $\varphi(a_{i+1}) = \varphi(a_i)^{-1}$ and

$$\chi(a) = \varphi(a_1) \ldots \varphi(a_{i-1}) \, \varphi(a_i) \, \varphi(a_{i+1}) \, \varphi(a_{i+2}) \ldots \varphi(a_n)$$
$$= \varphi(a_1) \ldots \varphi(a_{i-1}) \, \varphi(a_{i+2}) \ldots \varphi(a_n) = \chi(b).$$

By induction on k, $a \geqslant_k b$ implies $\chi(a) = \chi(b)$. Hence

$$\chi(a \bullet b) = \chi(\mathrm{red}\,(ab)) = \chi(ab) = \chi(a)\chi(b)$$

whenever a and b are reduced. Hence the restriction ψ of χ to F is a homomorphism of F into G. We see that $\psi(\iota(x)) = \varphi(x)$ for all $x \in X$.

If G is generated by $\varphi(X)$, then every element g of G is a product of elements of $\varphi(Y)$ and $g = \chi(a) = \psi(\mathrm{red}\,a)$ for some $a \in W$; hence ψ is surjective. $\square$

The universal property of F_X characterizes F_X (and ι) up to isomorphism (see the exercises) and is often used as a definition of free groups.

2.5.f. Cyclic Groups. We already met one example of free group. The following result is readily proved from the construction of F_X:

Proposition 2.5.9. *When $X = \{x\}$, then F_X is an infinite cyclic group, generated by $\iota(x)$.*

The universal property of $F_{\{x\}} \cong \mathbb{Z}$ implies properties encountered in Chapter 1. Let $X = \{x\}$, $c = \iota(x)$, and G be any group. For each $a \in G$ there is one mapping $\varphi : X \longrightarrow G$, which sends x to a. By Theorem 2.5.8, there is one homomorphism $\psi : F_X \longrightarrow G$ such that $\psi(c) = a$, which is surjective if G is generated by a; namely $\psi(c^n) = a^n$. This implies properties of powers (most parts of Proposition 1.2.8), and implies that every cyclic group is a homomorphic image of an infinite cyclic group (Proposition 1.6.6).

2.5.g. Free Abelian Groups. The free abelian group on a set X, which we denote by FA_X, is similar to the free group on X in that it is the "simplest" abelian group which is generated by X (or by a copy of X).

Let A be an abelian group which is generated by $X \subseteq A$. Every element a of A is (in the multiplicative notation) a product of powers of elements of X and, by commutativity, a product $a = x_1^{k_1} x_2^{k_2} \ldots x_r^{k_r}$ of powers of distinct elements of X. Thus a is determined by the family of integers $(k_x)_{x \in X}$, where k_x is the exponent of x in the product $x_1^{k_1} x_2^{k_2} \ldots x_r^{k_r}$. Since $x_1^{k_1} x_2^{k_2} \ldots x_r^{k_r}$ is a finite product, $k_x = 0$ for almost all $x \in X$. This suggests that FA_X is an external direct sum of copies of $\mathbb{Z}$.

Given a set X, we formally construct the **free abelian group** FA_X on X as the external direct sum $\bigoplus_{x \in X} \mathbb{Z}$ (this is short for $\bigoplus_{x \in X} G_x$, where $G_x = \mathbb{Z}$ for all x). Thus FA_X is the set of all families $a = (a_x)_{x \in X}$, normally written as linear combinations $a = \sum_{x \in X} a_x x$, such that $a_x \in \mathbb{Z}$ for all $x \in X$, and $a_x = 0$ for almost all $x \in X$. The operation on FA_X is componentwise addition:

$$\sum_{x \in X} a_x x + \sum_{x \in X} b_x x = \sum_{x \in X} (a_x + b_x) x .$$

If X is finite, then FA_X consists of all families $(a_x)_{x \in X}$ of integers, under componentwise addition, and FA_X is the direct product $\mathbb{Z}^X$ of $|X|$ copies of $\mathbb{Z}$. If X has just one element, then $FA_X \cong \mathbb{Z} \cong F_X$.

The direct sum injections $\iota_x : \mathbb{Z} \longrightarrow FA_X$ provides the **injection** $\iota : X \longrightarrow FA_X$, which is defined by $\iota(x) = \iota_x(1)$; in other words,

$$\iota(y) = \sum_{x \in X} \delta_{xy} x ,$$

where

$$\delta_{xy} = \begin{cases} 1 & \text{if } x = y, \\ 0 & \text{if } x \neq y; \end{cases}$$

δ is the **Kronecker delta** function.

Our reader will eagerly verify the universal property of FA_X:

Proposition 2.5.10. *FA_X is an abelian group and is generated by $\iota(X)$. Moreover, for every mapping φ of X into an abelian group A, there is a unique homo-*

morphism ψ of FA_X into A such that $\psi \circ \iota = \varphi$. If A is generated by $\varphi(X)$, then ψ is surjective.

2.5.h. Free Commutative Monoids. The free commutative monoid on a set X, which we denote by C_X, is similar to the free monoid on X in that it is the "simplest" commutative monoid which is generated by X (or by a copy of X). It will be used in Section 4.3 to construct polynomial rings with several indeterminates.

Let M be a commutative monoid which is generated by $X \subseteq M$. Every element a of M is a product of elements of X and, by commutativity, a product $a = x_1^{k_1} x_2^{k_2} \ldots x_r^{k_r}$ of positive powers of distinct elements of X. Thus a is determined by the family of nonnegative integers $(k_x)_{x \in X}$, where k_x is the exponent of x in the product $x_1^{k_1} x_2^{k_2} \ldots x_r^{k_r}$, in particular $k_x = 0$ for almost all $x \in X$. This suggests that C_X can be retrieved from FA_X.

Given a set X, we construct the **free commutative monoid** C_X on X as the **nonnegative cone** of FA_X:

$$C_X = \{\, a = (a_x)_{x \in X} \in FA_X \,;\, a_x \geqslant 0 \text{ for all } x \in X \,\}.$$

The **length** or **degree** of $a = (a_x)_{x \in X}$ is the nonnegative integer $\sum_{x \in X} a_x$; this is a finite sum, since $a_x = 0$ for almost all x. The elements of C_X are often written as linear combinations $a = \sum_{x \in X} a_x x$ or as products $a = \prod_{x \in X} x^{a_x}$.

When $a, b \in C_X$, the sum $a + b$ of a and b in FA_X is in C_X; this provides the operation on C_X. The injection $\iota : X \longrightarrow FA_X$ sends X into C_X; this provides the **injection** $\iota : X \longrightarrow C_X$.

Our tireless reader will verify the universal property of C_X:

Proposition 2.5.11. *C_X is a commutative monoid and is generated by $\iota(X)$. Moreover, for every mapping φ of X into a commutative monoid M, there is a unique monoid homomorphism ψ of C_X into M such that $\psi \circ \iota = \varphi$. If M is generated by $\varphi(X)$, then ψ is surjective.*

Exercises

1. Prove Lemma 2.5.2.

2. Prove Lemma 2.5.7.

3. Prove that the universal property in Theorem 2.5.8 characterizes free groups. (If $\kappa : X \longrightarrow F$ is a mapping and for every mapping φ of X into a group G there is a homomorphism $\psi : F \longrightarrow G$ unique such that $\psi \circ \kappa = \varphi$, then $F \cong F_X$.)

4. Use the construction of free groups to prove that $F_X \cong \mathbb{Z}$ when X has just one element.

5. Show that every mapping $\varphi : X \longrightarrow Y$ induces a unique homomorphism $F_\varphi : F_X \longrightarrow F_Y$ of free groups such that $F_\varphi \circ \iota_X = \iota_Y \circ \varphi$. Show that F_φ is the identity on F_X if φ is the identity on X, and that $F_{\psi \circ \varphi} = F_\psi \circ F_\varphi$.

6. Let X be a set. Show that FA_X is an abelian group and is generated by $\iota(X)$. Moreover, for every mapping φ of X into an abelian group A, show that there is a unique homomorphism ψ of FA_X into A such that $\psi \circ \iota = \varphi$, which is surjective if A is generated by $\varphi(X)$.

7. Let X be a set. Show that C_X is a commutative monoid and is generated (as a monoid) by $\iota(X)$. Moreover, for every mapping φ of X into a commutative monoid M, show that there is a unique homomorphism ψ of C_X into M such that $\psi \circ \iota = \varphi$, which is surjective if M is generated by $\varphi(X)$.

2.6. PRESENTATIONS

Presentations are a compact and versatile method of constructing groups.

By Theorem 2.5.8, every group G generated by a set X is a homomorphic image of the free group F_X. Hence G is isomorphic to a quotient group of F_X and is determined (up to isomorphism) by the set X and by a normal subgroup N of F_X (or by generators of N). Such a specification is called a presentation of G, and describes G as a set of products of elements of X and inverses of elements of X, between which hold equalities (called relations) determined by N. Presentations are also called **definitions by generators and relations**.

2.6.a. Relations. In this section we identify $x \in X$ and $\iota(x) \in F_X$ so that $X \subseteq F_X$ and $\iota : X \longrightarrow F_X$ is the inclusion mapping.

Given a set X we formally define a **relation** between elements of X as an ordered pair (u, v) of elements of F_X. A relation (u, v) is normally written like an equality $u = v$, or in the equivalent form $uv^{-1} = 1$; since u and v are reduced words, nontrivial relations are readily distinguished from actual equalities in F_X. When φ is a mapping of X into a group G, the relation $u = v$ **holds in** G (**via** φ) in case $\psi(u) = \psi(v)$ holds in G, where $\psi : F_X \longrightarrow G$ is the homomorphism which extends φ.

This terminology and notation make most sense when X is a subset of G and $\varphi : X \longrightarrow G$ is the inclusion mapping. Then ψ takes a product of elements of X and inverses of elements of X, calculated in F_X, to the same product of elements of X and inverses of elements of X, calculated in G. Thus a relation $u = v$ holds in G if and only if the corresponding products are equal when calculated in G.

In general, the relation $u = v$ holds in G (via φ) if and only if $uv^{-1} \in \operatorname{Ker} \psi \subseteq F_X$ (where ψ is as above). Thus the normal subgroup $\operatorname{Ker} \psi$ of F_X determines which relations hold in G.

2.6.b. Presentations. Let X be a set and R be a set of relations between elements of X (R is a subset of $F_X \times F_X$). Since an intersection of normal subgroups is a normal subgroup (Proposition 1.5.3), there is a smallest normal subgroup $N \lhd F_X$ which contains $\{ uv^{-1} ; (u,v) \in R \}$; we call N the **normal subgroup generated by** R. (This should be read as "(normal subgroup) generated by R" and not as "normal (subgroup generated by R)"; in particular, N should not be confused with the subgroup of F_X generated by $\{ uv^{-1} ; (u,v) \in R \}$.) We denote the quotient group F_X/N by $\langle X;R \rangle$.

For example, if $R = F_X \times F_X$, then $N = F_X$ and $\langle X;R \rangle = F_X/F_X = 1$. If $R = \varnothing$, then $N = 1$ and $\langle X;\varnothing \rangle = F_X/1 \cong F_X$. For a less trivial example, let $X = \{a\}$ so that $F_X \cong \mathbb{Z}$ (Proposition 2.5.9), and let R consist of $a^n = 1$, where $n > 0$. Then $N = \langle a^n \rangle \subseteq F_X$; the isomorphism $F_X \longrightarrow \mathbb{Z}$ takes N to $n\mathbb{Z}$, so that $\langle a; a^n = 1 \rangle \cong \mathbb{Z}/n\mathbb{Z} = \mathbb{Z}_n$ is cyclic of order n. Additional examples will be seen shortly.

$$
\begin{array}{ccc}
 & X & \\
\scriptstyle\subseteq \Big\downarrow & & \searrow\raisebox{2pt}{$\scriptstyle\iota$} \\
F_X & \!\!\!\longrightarrow\!\!\! & F_X/N = \langle X;R \rangle
\end{array}
$$

In general, the inclusion mapping $X \longrightarrow F_X$ induces a mapping $\iota : X \longrightarrow \langle X;R \rangle$ which sends $x \in X$ to $Nx \in \langle X;R \rangle$ (and which the reader will not confuse with the injection $X \longrightarrow F_X$). The homomorphism $F_X \longrightarrow \langle X;R \rangle$ which extends ι is precisely the projection $F_X \longrightarrow F_X/N$. Therefore $\langle X;R \rangle$ has the following properties:

Proposition 2.6.1. $\langle X;R \rangle$ *is generated by* $\iota(X)$, *and every relation* $u = v$ *in* R *holds in* $\langle X;R \rangle$ *(via* ι*)*.

Due to Proposition 2.6.1, $\langle X;R \rangle$ is called the group **generated by** X **subject to** R; the elements of R are the **defining relations** of $\langle X;R \rangle$. This standard terminology is somewhat misleading. First, X is not a subset of $\langle X;R \rangle$; in fact ι need not be injective. Moreover every homomorphic image of $\langle X;R \rangle$ has the properties in Proposition 2.6.1; thus there are many groups generated by $\iota(X)$ in which every relation $u = v$ in R holds. However, $\langle X;R \rangle$ is the "largest" such group, as shown by its universal property:

Proposition 2.6.2. *Let R be a set of relations between elements of a set X. Let G be a group and $\varphi : X \longrightarrow G$ be a mapping such that every relation $u = v$ in R holds in G (via φ). There is a unique homomorphism $\psi : \langle X;R \rangle \longrightarrow G$ such that the diagram*

$$
\begin{array}{ccc}
X & \overset{\iota}{\longrightarrow} & \langle X;R \rangle \\
\scriptstyle\varphi\Big\downarrow & \swarrow\raisebox{-2pt}{$\scriptstyle\psi$} & \\
G & &
\end{array}
$$

commutes ($\psi \circ \iota = \varphi$). If G is generated by $\varphi(X)$, then ψ is surjective.

If $R = \emptyset$, then $\langle X; R \rangle \cong F_X$, and Proposition 2.6.2 reduces to Theorem 2.5.8.

Proof. Let N be the normal subgroup of F_X generated by R (the smallest normal subgroup which contains $\{ uv^{-1} ; (u, v) \in R \}$), so that $\langle X; R \rangle = F_X / N$; let $\pi : F_X \longrightarrow \langle X; R \rangle$ be the projection, so that $\iota : X \xrightarrow{\subseteq} F_X \xrightarrow{\pi} \langle X; R \rangle$. Let $\chi : F_X \longrightarrow G$ be the homomorphism which extends φ. For each relation $u = v$ in R, we have $\chi(u) = \chi(v)$ (since $u = v$ holds in G (via φ)) and $uv^{-1} \in \text{Ker } \chi$. Therefore $\text{Ker } \pi = N \subseteq \text{Ker } \chi$. By Theorem 1.6.4, $\chi = \psi \circ \pi$ for some homomorphism $\psi : \langle X; R \rangle \longrightarrow G$.

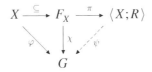

Restriction to X yields $\psi \circ \iota = \varphi$.

If $\psi' : \langle X; R \rangle \longrightarrow G$ is another homomorphism such that $\psi' \circ \iota = \varphi$, then $\psi = \psi'$ by Proposition 1.4.11, since ψ and ψ' agree on a generating subset $\iota(X)$ of $\langle X; R \rangle$.   □

The group $\langle X; R \rangle$ is characterized up to isomorphism by this universal property (see the exercises).

A **presentation** of a group G consists of a set X, a set R of relations between the elements of X, and an isomorphism $G \cong \langle X; R \rangle$. For example, we saw that a cyclic group C_n of order $n > 0$ has the presentation $C_n \cong \langle a; a^n = 1 \rangle$ (with $X = \{a\}$ and $R = \{ a^n = 1 \}$). By Theorem 2.5.8, every group G has a presentation, in which X can be any subset of G which generates G and R contains enough relations to generate the kernel of $\psi : F_X \longrightarrow G$; e.g. $R = \{ (u, 1) ; u \in \text{Ker } \psi \}$.

2.6.c. Examples. Presentations provide a simple and general way to construct specific groups. First we show how to specify known groups by presentations.

Proposition 2.6.3. $D_n \cong \langle x, y ; x^n = y^2 = 1, yxy = x^{-1} \rangle$.

Proof. Let $G = \langle x, y ; x^n = y^2 = 1, yxy = x^{-1} \rangle$. Since x and y have finite order, every element of G is a product of x's and y's. Since $yx = x^{-1}y = x^{n-1}y$, we can move all the y's at the end of the product and write every element of G in the form $x^i y^j$ (with $i, j \geq 0$). Since $x^n = y^2 = 1$, we may further assume $i < n$ and $j < 2$. Hence G has at most $2n$ elements: $1, x, \ldots, x^{n-1}, y, xy, \ldots, x^{n-1}y$.

We saw (Proposition 1.4.12) that D_n contains generators a and b such that $a^n = b^2 = 1$ and $bab = a^{-1}$. This provides a mapping $x \longmapsto a$, $y \longmapsto b$ of $\{ x, y \}$ into D_n via which the defining relations of G hold in D_n. By Proposition 2.6.2, there exists a homomorphism ψ of G onto D_n. Since $|G| \leq 2n = |D_n|$, it follows that $|G| = 2n$ and that ψ is an isomorphism.   □

Note that this presentation of D_n does not contain the conditions "$b \neq a$" and "a has order n" (i.e., $a^i \neq 1$ when $0 < i < n$), which are not relations, and whose purpose in Proposition 1.4.12 was to ensure that $1, a, \ldots, a^{n-1}, b, ab, \ldots, a^{n-1}b$ are distinct. Our next result is similarly related to Proposition 2.4.10; its proof is an exercise:

Proposition 2.6.4 (Hölder's Theorem). *A group G is an extension of a cyclic group of order m by a cyclic group of order n if and only if $G \cong \langle a,b \,;\, a^m = 1,\ b^n = a^t,\ bab^{-1} = a^r \rangle$ for some integers r, t such that $r^n \equiv 1$ and $rt \equiv t \pmod{m}$.*

2.6.d. The Quaternion Group. We now use presentations to construct new groups.

Example 2.6.5. The **quaternion group** is

$$Q = \langle a,b \,;\, a^4 = 1,\ b^2 = a^2,\ bab^{-1} = a^{-1} \rangle;$$

Q is named after the quaternion algebra $\mathbf{Q}$, of which it is a subgroup (see exercises).

Since a and b have finite order, every element of Q is a product of a's and b's. Since $ba = a^{-1}b = a^3b$, we can move all the b's at the end of the product and write every element of Q in the form $a^i b^j$ (with $i, j \geqslant 0$). Since $a^4 = b^4 = 1$, we may further assume $i, j < 4$. Since $b^2 = a^2$, we may in fact assume $j < 2$. Hence Q has at most eight elements: $1, a, a^2, a^3, b, ab, a^2b, a^3b$.

From $bab^{-1} = a^{-1}$, we obtain $ba^2b^{-1} = a^{-2}$, $ba^3b^{-1} = a^{-3}$ so that $ba = a^3b$, $ba^2 = a^2b$, $ba^3 = ab$, and $bab = a^3$, $ba^2b = a^2$, $ba^3b = a$. Hence the elements of Q are multiplied as follows:

1	a	a^2	a^3	b	ab	a^2b	a^3b
a	a^2	a^3	1	ab	a^2b	a^3b	b
a^2	a^3	1	a	a^2b	a^3b	b	ab
a^3	1	a	a^2	a^3b	b	ab	a^2b
b	a^3b	a^2b	ab	a^2	a	1	a^3
ab	b	a^3b	a^2b	a^3	a^2	a	1
a^2b	ab	b	a^3b	1	a^3	a^2	a
a^3b	a^2b	ab	b	a	1	a^3	a^2

It is fairly clear at this point that we have found Q. However, we have not established that the possible elements $1, a, a^2, a^3, b, ab, a^2b, a^3b$ of Q are distinct, or that the above is a group, or even that there exists a group of order 8 with generators a, b satisfying the given relations. To prove all this, we need a model of Q.

If we cannot think of an existing group to serve as model for Q, we can always construct one from the multiplication table we just constructed. Let $G = \{1, a_1, a_2, a_3, b_1, a_1b_1, a_2b_1, a_3b_1\}$ with the multiplication

1	a_1	a_2	a_3	b_1	a_1b_1	a_2b_1	a_3b_1
a_1	a_2	a_3	1	a_1b_1	a_2b_1	a_3b_1	b_1
a_2	a_3	1	a_1	a_2b_1	a_3b_1	b_1	a_1b_1
a_3	1	a_1	a_2	a_3b_1	b_1	a_1b_1	a_2b_1
b_1	a_3b_1	a_2b_1	a_1b_1	a_2	a_1	1	a_3
a_1b_1	b_1	a_3b_1	a_2b_1	a_3	a_2	a_1	1
a_2b_1	a_1b_1	b_1	a_3b_1	1	a_3	a_2	a_1
a_3b_1	a_2b_1	a_1b_1	b_1	a_1	1	a_3	a_2

Associativity is verified with Light's test. We see that G is generated (as in Section 1.1) by a_1 and b_1 (with $a_2 = a_1^2$, $a_3 = a_1^3$); hence it suffices to test a_1 and b_1. The Light's tables for a_1 and b_1 are

a_1	a_1	a_2	a_3	1	a_1b_1	a_2b_1	a_3b_1	b_1
a_1	a_1	a_2	a_3	1	a_1b_1	a_2b_1	a_3b_1	b_1
a_2	a_2	a_3	1	a_1	a_2b_1	a_3b_1	b_1	a_1b_1
a_3	a_3	1	a_1	a_2	a_3b_1	b_1	a_1b_1	a_2b_1
1	1	a_1	a_2	a_3	b_1	a_1b_1	a_2b_1	a_3b_1
a_3b_1	a_3b_1	a_2b_1	a_1b_1	b_1	a_1	1	a_3	a_2
b_1	b_1	a_3b_1	a_2b_1	a_1b_1	a_2	a_1	1	a_3
a_1b_1	a_1b_1	b_1	a_3b_1	a_2b_1	a_3	a_2	a_1	1
a_2b_1	a_2b_1	a_1b_1	b_1	a_3b_1	1	a_3	a_2	a_1

b_1	b_1	a_3b_1	a_2b_1	a_1b_1	a_2	a_1	1	a_3
b_1	b_1	a_3b_1	a_2b_1	a_1b_1	a_2	a_1	1	a_3
a_1b_1	a_1b_1	b_1	a_3b_1	a_2b_1	a_3	a_2	a_1	1
a_2b_1	a_2b_1	a_1b_1	b_1	a_3b_1	1	a_3	a_2	a_1
a_3b_1	a_3b_1	a_2b_1	a_1b_1	b_1	a_1	1	a_3	a_2
a_2	a_2	a_3	1	a_1	a_2b_1	a_3b_1	b_1	a_1b_1
a_3	a_3	1	a_1	a_2	a_3b_1	b_1	a_1b_1	a_2b_1
1	1	a_1	a_2	a_3	b_1	a_1b_1	a_2b_1	a_3b_1
a_1	a_1	a_2	a_3	1	a_1b_1	a_2b_1	a_3b_1	b_1

Inspection shows that a_1 and b_1 pass Light's test. Therefore associativity holds in G. The multiplication table of G shows that G has an identity element and that each element of G has an inverse. Hence G is a group. Furthermore the elements a_1 and b_1 of G satisfy $a_1^4 = 1$, $b_1^2 = a_1^2$, and $b_1 a_1 b_1^{-1} = a_1^{-1}$. By Proposition 2.6.2, there exists a homomorphism ψ of Q onto G that sends a and b onto a_1 and b_1. Since $|Q| \leqslant 8 = |G|$, ψ is an isomorphism, and $Q \cong G$. (Alternately, $1, a, a^2, a^3, b, ab, a^2b, a^3b$ are distinct in Q, since they are distinct in G.) $\qquad\square$

The following group of order 12 is similar to Q:

Example 2.6.6. T is the group

$$T = \langle a, b; a^6 = 1, b^2 = a^3, bab^{-1} = a^{-1} \rangle.$$

The determination of T (its element list and multiplication table) is left as an exercise. Hölder's Theorem (Proposition 2.6.4) shows that Q is an extension of a cyclic group of order 4 by a cyclic group of order 2, and that T is an extension of a cyclic group of order 6 by a cyclic group of order 2.

Exercises

1. Show that $\langle X; R \rangle$ is characterized up to isomorphism by its universal property.
2. Find all groups generated by two elements a, b such that $a^4 = 1$, $b^2 = a^2$, and $bab^{-1} = a^{-1}$.
3. Prove that $Q \ncong D_4$.
4. Show that Q is isomorphic to the multiplicative subgroup $\{1, i, j, k, -1, -i, -j, -k\}$ of the quaternion algebra $\mathbf{Q}$.
5. Show that a group G is isomorphic to Q if and only if G is generated by two elements a, b such that a has order 4, $b^2 = a^2$, b is not a power of a, and $bab^{-1} = a^{-1}$.
6. Find $T = \langle a, b; a^6 = 1, b^2 = a^3, bab^{-1} = a^{-1} \rangle$ (list the elements of T, construct a multiplication table, and prove that your guess is correct).
7. Prove that $T \ncong D_6$.
8. Find $A_4 = \langle a, b; a^3 = 1, b^2 = 1, aba = ba^2b \rangle$ (list the elements of A_4, construct a multiplication table, and prove your guess is correct).
9. Prove that $A_4 \ncong D_6, T$.
10. Find $\langle a, b; a^2 = b^2 = 1 \rangle$.
11. Prove Proposition 2.6.4.
12. Use Hölder's Theorem to list all extensions of a cyclic group of order 4 by a cyclic group of order 2. (Use presentations; make sure that the groups in your list are not isomorphic.)
13. Use Hölder's Theorem to list all extensions of a cyclic group of order 6 by a cyclic group of order 2. (Use presentations; make sure that the groups in your list are not isomorphic.)

2.7. FREE PRODUCTS

Free products of groups are used in algebraic topology. For instance, let T_1 and T_2 be topological spaces with fundamental groups G_1 and G_2, and let T be their one point union (the disjoint union, with one point of T_1 identified with one point of T_2); the fundamental group of T is the free product of G_1 and G_2.

In general, the **free product** of a family $(G_i)_{i \in I}$ of groups is a group G with homomorphisms $\iota_i : G_i \longrightarrow G$ $(i \in I)$ having the following universal property: for any family of homomorphisms $\varphi_i : G_i \longrightarrow G'$ $(i \in I)$, there is a homomorphism $\varphi : G \longrightarrow G'$ unique such that $\varphi \circ \iota_i = \varphi_i$ for all i. This is similar to the universal property of $\bigoplus_{i \in I} G_i$ (Proposition 2.1.7), except that there is no restriction on the homomorphisms $(\varphi_i)_{i \in I}$. The free product of $(G_i)_{i \in I}$ is often denoted by $\coprod_{i \in I} G_i$ (by $G_1 \amalg \ldots \amalg G_n$ if $I = \{1, \ldots, n\}$).

In this section we show that the free product exists and can be constructed very simply from the groups $(G_i)_{i \in I}$. The construction is similar to the construction of free groups in Section 5. It turns out that the free product contains pairwise disjoint subgroups $\operatorname{Im} \iota_i \cong G_i$, and is generated by the union of these subgroups; its universal property shows that the free product is the "largest" such group.

2.7.a. Reduced Words. Up to isomorphism we may assume that the groups $(G_i)_{i \in I}$ are **pairwise disjoint**, which for groups means that $G_i \cap G_j = \{1\}$ whenever $i \neq j$. For instance, G_i may be replaced by $G_i' = \{1\} \cup ((G_i \backslash \{1\}) \times \{i\})$, with multiplication carried to G_i' by the bijection $\theta_i : G_i \longrightarrow G_i'$ $(xy = \theta_i(\theta_i^{-1}(x)\theta_i^{-1}(y)))$; we see that $G_i' \cong G_i$ and that $G_i' \cap G_j' = \{1\}$ when $i \neq j$.

Let $W = W_X$ be the set of all nonempty words in the alphabet $X = \bigcup_{i \in I} G_i$. (In the finished product, $(a_1, a_2, \ldots, a_n) \in W$ is written $a_1 a_2 \ldots a_n$.) Words in W are multiplied by concatenation.

A word $(a_1, \ldots, a_n)$ is **reduced** in case it does not contain consecutive letters which belong to the same group (if $a_t \in G_i$ and $a_{t+1} \in G_j$, then $i \neq j$). Since 1 belongs to all G_i, this implies $a_t \neq 1$ for all t if $n \geqslant 2$.

Reduction simplifies a word until a reduced word is reached. The detailed process is as follows. A **one-step reduction** replaces a word $a = (a_1, \ldots, a_t, a_{t+1}, \ldots, a_n)$, in which consecutive letters a_t and a_{t+1} belong to the same group, by the word

$$b = (a_1, \ldots, a_{t-1}, a_t a_{t+1}, a_{t+2}, \ldots, a_n);$$

we denote this relationship by $a \geqslant_1 b$. A k **step reduction** is a sequence of k one-step reductions; we write $a \geqslant_k b$ when a k step reduction replaces a by b. For instance, $a \geqslant_0 a$. In general, $a \geqslant_k b$ if and only if $a \geqslant_1 c_1 \geqslant_1 c_2 \geqslant_1 \cdots \geqslant_1 c_k = b$ for some $c_1, c_2, \ldots, c_k \in W$. We write $a \geqslant b$ (or $b \leqslant a$) when $a \geqslant_k b$ for some k.

The following result states that every word can be reduced (into a reduced word); this is readily proved by induction on the length:

Lemma 2.7.1. *For each word $a \in W$ there is a reduced word $b \leqslant a$.*

Assume for a minute that every G_i is a subgroup of a group G. Then every word $a = (a_1, \ldots, a_n) \in W$ has a product $a_1 a_2 \ldots a_n$ in G. If G is generated by $\bigcup_{i \in I} G_i$ this provides a surjection $W \longrightarrow G$, so that every element of G comes from a word. When b is obtained from a by a one-step reduction, then a and b have the same product in G. If therefore $a \geqslant b$, then a and b have the same product in G. Hence reduced words yield all the elements of G; the same must be true of the free product.

2.7.b. Reduction. We show that b is unique in Lemma 2.7.1.

Lemma 2.7.2. *If $a \geqslant_1 b$ and $a \geqslant_1 c \neq b$, then $b \geqslant_1 d$ and $c \geqslant_1 d$ for some d.*

Proof. By definition,

$$b = (a_1, \ldots, a_{t-1}, a_t a_{t+1}, a_{t+2}, \ldots, a_n),$$
$$c = (a_1, \ldots, a_{u-1}, a_u a_{u+1}, a_{u+2}, \ldots, a_n),$$

where $a_t, a_{t+1} \in G_i$ and $a_u, a_{u+1} \in G_j$ for some $t, u < n$. We have $t \neq u$, since $b \neq c$; we may assume $t < u$. If $u = t + 1$, then $i = j$ and $b \geqslant_1 d$, $c \geqslant_1 d$, where

$$d = (a_1, \ldots, a_{t-1}, a_t a_{t+1} a_{t+2}, a_{t+3}, \ldots, a_n).$$

If $u \geqslant t + 2$, then a_u and a_{u+1} are consecutive letters of b, a_t and a_{t+1} are consecutive letters of c, and $b \geqslant_1 d$, $c \geqslant_1 d$, where

$$d = (a_1, \ldots, a_{t-1}, a_t a_{t+1}, a_{t+2}, \ldots, a_{u-1}, a_u a_{u+1}, a_{u+2}, \ldots, a_n),$$

with $a_{t+2}, \ldots, a_{u-1}$ omitted if $u = t + 2$. $\qquad\square$

Lemma 2.7.3. *If $a \geqslant b, c$ in W, then $b, c \geqslant d$ for some $d \in W$.*

This is proved exactly like Lemma 2.5.4.

Lemma 2.7.3 implies that b is unique in Lemma 2.7.1: if $a \geqslant b$ and $a \geqslant c$ with b and c reduced, then $b \geqslant d$ and $c \geqslant d$ for some $d \in W$, which implies $b = d = c$, since b and c are reduced. Thus:

Lemma 2.7.4. *For each word $a \in W$, there is a reduced word $b \leqslant a$, and b is unique.*

We denote by $\operatorname{red} a$ the reduced word b such that $a \geqslant b$.

2.7.c. Construction. Lemma 2.7.4 implies that reduced words can be used directly as elements of the free product. Thus we construct the free product

$P = \coprod_{i \in I} G_i$ of the pairwise disjoint groups $(G_i)_{i \in I}$ as the set of all reduced words in W. The multiplication $\cdot$ on P (usually denoted by juxtaposition) is defined by concatenation and reduction:

$$a \cdot b = \mathrm{red}(ab).$$

For each $i \in I$, the **injection** $\iota_i : G_i \longrightarrow P$ assigns to each $x \in G_i$ the one letter reduced word $\iota_i(x) = (x)$. It is immediate that ι_i is an injective homomorphism. (It is common practice to identify $x \in G_i$ and $\iota_i(x) \in P$.) It is convenient to combine all ι_i into a single mapping $\iota : \bigcup_{i \in I} G_i \longrightarrow \coprod_{i \in I} G_i$; by definition, $\iota(x) = \iota_i(x)$ whenever $x \in G_i$. We see that $\mathrm{Im}\, \iota = \bigcup_{i \in I} \mathrm{Im}\, \iota_i$.

Proposition 2.7.5. $\coprod_{i \in I} G_i$ *is a group. Furthermore* $\mathrm{Im}\, \iota_i \cong G_i$, $\mathrm{Im}\, \iota_i \cap \mathrm{Im}\, \iota_j = 1$ *whenever* $i \neq j$, *and* $\coprod_{i \in I} G_i$ *is generated by* $\bigcup_{i \in I} \mathrm{Im}\, \iota_i$.

The proof is similar to the proof of Proposition 2.5.6 and is left as an exercise. As in Section 2.5, the inverse of $a = (a_1, \ldots, a_n)$ is $(a_n^{-1}, a_{n-1}^{-1}, \ldots, a_1^{-1})$. That $\coprod_{i \in I} G_i$ is generated by $\bigcup_{i \in I} \mathrm{Im}\, \iota_i$ follows from the next lemma, which is easily proved by induction on n.

Lemma 2.7.6. *When* $a = (a_1, \ldots, a_n)$ *is reduced, then*

$$a = \iota(a_1) \cdot \iota(a_2) \cdot \cdots \cdot \iota(a_n).$$

When x is identified with $\iota(x)$, Lemma 2.7.6 writes every element of the free product $\coprod_{i \in I} G_i$ as a reduced product of elements of the groups G_i.

2.7.d. Universal Property. It remains to show that $P = \coprod_{i \in I} G_i$ (as constructed above) is indeed a free product of the groups $(G_i)_{i \in I}$.

Proposition 2.7.7. *Let* $(G_i)_{i \in I}$ *be a family of pairwise disjoint groups. If G is any group and* $\varphi_i : G_i \longrightarrow G$ *($i \in I$) are any homomorphisms, then there exists a unique homomorphism* $\psi : \coprod_{i \in I} G_i \longrightarrow G$ *such that* $\psi \circ \iota_i = \varphi_i$ *for all* i.

$$
\begin{array}{ccc}
G_i & & \\
\iota_i \downarrow & \searrow^{\varphi_i} & \\
\coprod_{i \in I} G_i & \dashrightarrow{}_{\psi} & G
\end{array}
$$

Proof. It is convenient to combine all the homomorphisms $\varphi_i : G_i \longrightarrow G$ into a single mapping $\varphi : \bigcup_{i \in I} G_i \longrightarrow G$; by definition, $\varphi(x) = \varphi_i(x)$ whenever $x \in G_i$.

Assume that $\psi : P = \coprod_{i \in I} G_i \longrightarrow G$ is a homomorphism such that $\psi(\iota_i(x)) = \varphi_i(x)$ whenever $x \in G_i$. Then $\psi(\iota(x)) = \varphi(x)$ for all $x \in \bigcup_{i \in I} G_i$. For any $a =$

$(a_1, \ldots, a_n) \in P$, Lemma 2.7.6 implies

$$\psi(a) = \psi(\iota(a_1)) \ldots \psi(\iota(a_n)) = \varphi(a_1) \ldots \varphi(a_n).$$

Hence there is at most one homomorphism $\psi : P \longrightarrow G$ such that $\psi \circ \iota_i = \varphi_i$ for all i.

Conversely, define a mapping $\chi : W \longrightarrow G$ by

$$\chi(a) = \varphi(a_1) \ldots \varphi(a_n).$$

for all $a = (a_1, \ldots, a_n) \in W$. Then $\chi(ab) = \chi(a) \chi(b)$ for all $a, b \in W$. If $a = (a_1, \ldots, a_n) \geqslant_1 b$, then, for some $t < n$, a_t and a_{t+1} belong to the same group G_i, and

$$b = (a_1, \ldots, a_{t-1}, a_t a_{t+1}, a_{t+2}, \ldots, a_n).$$

Since $\varphi(a_t) \varphi(a_{t+1}) = \varphi(a_t a_{t+1})$, we have

$$\chi(a) = \varphi(a_1) \ldots \varphi(a_{t-1}) \varphi(a_t) \varphi(a_{t+1}) \varphi(a_{t+2}) \ldots \varphi(a_n) = \chi(b).$$

Thus $a \geqslant_1 b$ implies $\chi(a) = \chi(b)$. Therefore $a \geqslant_k b$ implies $\chi(a) = \chi(b)$. Hence

$$\chi(a \cdot b) = \chi(\mathrm{red}\,(ab)) = \chi(ab) = \chi(a) \chi(b)$$

whenever a and b are reduced. Thus the restriction ψ of χ to P is a homomorphism of P into G. We see that $\psi(\iota_i(x)) = \psi(x) = \varphi(x) = \varphi_i(x)$ whenever $x \in G_i$. $\qquad\square$

We saw that the homomorphisms ι_i are injective, so that each G_i is isomorphic to a subgroup $\mathrm{Im}\,\iota_i$ of the free product P, and that P is generated by $\bigcup_{i \in I} \mathrm{Im}\,\iota_i$. Proposition 2.7.7 implies that every group with these properties is a homomorphic image of P.

Exercises

1. Prove Lemma 2.7.1.

2. Prove Lemma 2.7.6.

3. Prove Proposition 2.7.5.

4. Show that the universal property of free products characterizes $\coprod_{i \in I} G_i$ and its injections up to isomorphism.

5. Show that $\langle a, b; a^2 = b^2 = 1 \rangle \cong \mathbb{Z}_2 \amalg \mathbb{Z}_2$.

6. Give a presentation of $\mathbb{Z}_3 \amalg \mathbb{Z}_2$.

7. Given presentations of G_i $(i \in I)$, find a presentation of $\coprod_{i \in I} G_i$.

8. Let F be the free group on a set X, $(X_i)_{i \in I}$ be a partition of X, and F_i be the free group on X_i. Prove that $F \cong \coprod_{i \in I} F_i$.

9. Prove the following associativity property of free products: if $I = \bigcup_{j \in J} I_j$ is a partition of I, then $\coprod_{i \in I} G_i \cong \coprod_{j \in J} (\coprod_{i \in I_j} G_i)$.

2.8. FREE PRODUCTS WITH AMALGAMATION

Free products of groups with amalgamation are used in algebraic topology. For instance, let T_1 and T_2 be topological spaces with fundamental groups G_1 and G_2 and homeomorphic subspaces $S_1 \cong S_2$ with fundamental group H. Let T be the disjoint union of T_1 and T_2 in which S_1 is identified with S_2. The fundamental group of T is the free product with amalgamation of G_1 and G_2 amalgamating H.

2.8.a. Group Amalgams. In this section, $(G_i)_{i \in I}$ is a family of groups with a common subgroup H such that $G_i \cap G_j = H$ whenever $i \neq j$; the union $\bigcup_{i \in I} G_i$ is called a **group amalgam** and is said to **amalgamate** H.

In this section we show that every group amalgam can be embedded into a single group G (so that each G_i is a subgroup of G). The **free product with amalgamation** of the groups $(G_i)_{i \in I}$ **amalgamating** H is the "largest" such group G and is defined by its universal property: any family of homomorphisms $\varphi_i : G_i \longrightarrow G'$ ($i \in I$) which agree on H ($\varphi_i(h) = \varphi_j(h)$ for all i, j and all $h \in H$) can be extended uniquely to a homomorphism $\varphi : G \longrightarrow G'$. If $H = 1$ is trivial, then G is the free product of the groups G_i.

The construction of free products with amalgamation is similar to that of free products but not as simple. Let G be the free product with amalgamation of the groups G_i amalgamating H. Every element of G is a product of elements of G_i and can be written as the product of a reduced word. However, this reduced word is not in general unique; if for instance $a \in G_i$, $b \in G_j$, and $h \in H$, then the words (ah, b) and (a, hb) yield the same element of G. Accordingly, the elements of G are constructed not as reduced words but as equivalence classes of reduced words.

2.8.b. Reduced Words. Let $(G_i)_{i \in I}$ be a family of groups such that $G_i \cap G_j = H$ whenever $i \neq j$. Let W be the set of all nonempty words in the alphabet $\bigcup_{i \in I} G_i$, multiplied by concatenation.

A word $a = (a_1, \ldots, a_n)$ is **reduced** in case it does not contain consecutive letters that belong to the same group (if $a_t \in G_i$ and $a_{t+1} \in G_j$, then $i \neq j$). Since H is contained in all G_i, this implies $a_t \notin H$ for all t, if $n \geqslant 2$. **Reduction** is the process of simplifying a word until a reduced word is reached. In detail, a **one step reduction** replaces a word $a = (a_1, \ldots, a_t, a_{t+1}, \ldots, a_n)$, in which a_t and a_{t+1} belong to the same group, by

$$b = (a_1, \ldots, a_{t-1}, a_t a_{t+1}, a_{t+2}, \ldots, a_n).$$

We denote this relationship by $a \geqslant_1 b$. A k **step reduction** is a sequence of k one-step reductions; we write $a \geqslant_k b$ when a k step reduction replaces a by b. For instance, $a \geqslant_0 a$. In general, $a \geqslant_k b$ if and only if $a \geqslant_1 c_1 \geqslant_1 c_2 \geqslant_1 \cdots \geqslant_1 c_k = b$ for some $c_1, c_2, \ldots, c_k \in W$. We write $a \geqslant b$ when $a \geqslant_k b$ for some k.

The following result is readily proved by induction on the length of a:

Lemma 2.8.1. *For each word $a \in W$, there is a reduced word $b \leqslant a$.*

Let $a = (a_1, \ldots, a_m)$ and $b = (b_1, \ldots, b_n) \in W$. We write

$$a \equiv_1 b \quad \text{in case} \quad m = n,$$

and there exists $t < n$ and $h \in H$ such that $a_t h = b_t$, $a_{t+1} = h b_{t+1}$, and $a_u = b_u$ for all $u \neq t, t + 1$. We see that $\equiv_1$ is reflexive and symmetric. We write

$$a \equiv_k b \quad \text{in case} \quad a = c_0 \equiv_1 c_1 \equiv_1 c_2 \equiv_1 \cdots \equiv_1 c_k = b$$

for some $k \geqslant 0$ and $c_0, c_1, \ldots, c_k \in W$. We write $a \equiv b$ and call a and b **equivalent** in case $a \equiv_k b$ for some $k \geqslant 0$. Equivalent words must multiply to the same element of the free product.

Lemma 2.8.2. *The following holds:*

(1) $\equiv$ *is an equivalence relation.*
(2) *If a is reduced and $a \equiv b$, then b is reduced.*
(3) *If $a \equiv_1 b \geqslant_1 c$, then $a \geqslant_1 d \equiv_1 c$ for some d.*
(4) *If $a \equiv_1 b \geqslant_k c$, then $a \geqslant_k d \equiv_1 c$ for some d.*
(5) *If $a \equiv b \geqslant c$, then $a \geqslant d \equiv c$ for some d.*

Proof. (1) is clear.

(2) Let $a = (a_1, \ldots, a_m)$ be reduced and $a \equiv_1 b = (b_1, \ldots, b_n)$ so that $m = n$, $a_t h = b_t$, and $a_{t+1} = h b_{t+1}$ for some $h \in H$ and $t < n$, and $a_u = b_u$ for all $u \neq t, t + 1$. In particular, $n \geqslant 2$ and $a_u \notin H$ for all u. Hence $b_u \notin H$ for all u: this is clear if $u \neq t, t + 1$; also $b_t \in H$ would imply $a_t = b_t h^{-1} \in H$, and similarly for b_{t+1}. We see that a_u and b_u belong to the same group G_i. Hence b is reduced.

If a is reduced and $a = c_0 \equiv_1 c_1 \equiv_1 \cdots \equiv_1 c_k = b$, then by induction $c_0, c_1, \ldots$ and $c_k = b$ are reduced.

(3) Assume $a \equiv_1 b \geqslant_1 c$, so that $a = (a_1, \ldots, a_m)$, $b = (b_1, \ldots, b_n)$ satisfy $m = n$, $a_t h = b_t$, $a_{t+1} = h b_{t+1}$ for some $h \in H$ and $t < n$, and $a_u = b_u$ for all $u \neq t, t + 1$, and

$$c = (b_1, \ldots, b_{u-1}, b_u b_{u+1}, b_{u+2}, \ldots, b_n),$$

where b_u and b_{u+1} belong to the same group G_i.

If $u = t - 1$, then $a_{t-1}, a_t \in G_i$ and

$$a \geqslant_1 (a_1, \ldots, a_{t-2}, a_{t-1}a_t, a_{t+1}, \ldots, a_n)$$
$$\equiv_1 (b_1, \ldots, b_{t-2}, b_{t-1}b_t, b_{t+1}, \ldots, b_n) = c.$$

If $u = t$, then $a_t, a_{t+1} \in G_i$, $a_t a_{t+1} = a_t h b_{t+1} = b_t b_{t+1}$, and

$$a \geqslant_1 (a_1, \ldots, a_{t-1}, a_t a_{t+1}, a_{t+2}, \ldots, a_n)$$
$$= (b_1, \ldots, b_{t-1}, b_t b_{t+1}, b_{t+2}, \ldots, b_n) = c.$$

If $u = t + 1$, then $a_{t+1}, a_{t+2} \in G_i$ and

$$a \geqslant_1 (a_1, \ldots, a_t, a_{t+1}a_{t+2}, a_{t+3}, \ldots, a_n)$$
$$\equiv_1 (b_1, \ldots, b_t, b_{t+1}b_{t+2}, b_{t+3}, \ldots, b_n) = c.$$

If finally $u \neq t - 1, t, t + 1$, then b_u, b_{u+1} are consecutive letters of a and

$$a \geqslant_1 (a_1, \ldots, a_{u-1}, a_u a_{u+1}, a_{u+2}, \ldots, a_n) \equiv_1 c.$$

(4) follows from (3), and (5) from (4), by induction.   □

2.8.c. Reduction. We show that the element b in Lemma 2.8.1 is unique up to equivalence.

Lemma 2.8.3. *If $a \geqslant_1 b$ and $a \geqslant_1 c \neq b$, then $b \equiv_1 c$, or $b \geqslant_1 d$ and $c \geqslant_1 d$ for some d.*

Proof. By definition,

$$b = (a_1, \ldots, a_{t-1}, a_t a_{t+1}, a_{t+2}, \ldots, a_n),$$
$$c = (a_1, \ldots, a_{u-1}, a_u a_{u+1}, a_{u+2}, \ldots, a_n),$$

where $a_t, a_{t+1} \in G_i$ and $a_u, a_{u+1} \in G_j$ for some $t, u < n$. We assume that $b \neq c$; then $t \neq u$, and we may assume $u > t$. If $u > t + 1$, then a_u and a_{u+1} are consecutive letters of b, a_t and a_{t+1} are consecutive letters of c, and $b \geqslant_1 d$, $c \geqslant_1 d$, where

$$d = (a_1, \ldots, a_{t-1}, a_t a_{t+1}, a_{t+2}, \ldots, a_{u-1}, a_u a_{u+1}, a_{u+2}, \ldots, a_n).$$

Now let $u = t + 1$. Then $a_{t+1} \in G_i \cap G_j$. If $i = j$, then $b \geqslant_1 d$, $c \geqslant_1 d$, where

$$d = (a_1, \ldots, a_{t-1}, a_t a_{t+1} a_{t+2}, \ldots, a_{t+3}, \ldots, a_n).$$

If $i \neq j$, then $a_{t+1} = a_u \in H$, and we see that $b \equiv_1 c$.   □

Lemma 2.8.4. *If $a \geqslant b$, $a \geqslant c$, and b, c are reduced, then $b \equiv c$.*

Proof. The proof is by induction on the length n of a. The result is trivial if $n = 1$, or if $a = b$, or if $a = c$. Assume that $a \neq b,c$ (hence $n > 1$). Then $a \geqslant_k b$ for some $k > 0$, so that $a = b^{(0)} \geqslant_1 b' \geqslant_1 b'' \geqslant_1 \cdots \geqslant_1 b^{(k)} = b$; also $a \geqslant_1 c' \geqslant c$ for some $c' \in W$.

Construct by induction a sequence $(d) : c' = d' \geqslant_1 d'' \geqslant_1 \cdots \geqslant_1 d^{(j)}$ such that $b^{(i)} \geqslant_1 d^{(i+1)}$ for all $i < j$ and $b^{(j)} \equiv_1 d^{(j)}$. We already have $a = b^{(0)} \geqslant_1 d'$. If $d^{(i)}$ has been constructed, with $b^{(i-1)} \geqslant_1 d^{(i)}$, then $b^{(i-1)} \neq b$, $b^{(i-1)} \geqslant_1 b^{(i)}$, and by Lemma 2.8.3, we have $b^{(i)} \equiv_1 d^{(i)}$, or $b^{(i)} \geqslant_1 d$, $d^{(i)} \geqslant_1 d$ for some d. If $b^{(i)} \equiv_1 d^{(i)}$, then the sequence (d) stops, and $j = i$; otherwise, the sequence (d) continues with $d^{(i+1)} = d$.

The sequence (d) must terminate at some $j \leqslant k$, since $b = b^{(k)}$ is reduced and we cannot have $b^{(k)} \geqslant_1 d^{(k+1)}$. By part (4) of Lemma 2.8.2, $d^{(j)} \equiv_1 b^{(j)} \geqslant b$ then implies $c' \geqslant d^{(j)} \geqslant d \equiv_1 b$ for some d. The word d is reduced by part (2) of Lemma 2.8.2. Since c' is shorter than a, the induction hypothesis yields $d \equiv c$. Hence $b \equiv c$. $\qquad\square$

2.8.d. Construction. Let $(G_i)_{i \in I}$ be a family of groups such that $G_i \cap G_j = H$ whenever $i \neq j$. We construct the free product with amalgamation P of the groups $(G_i)_{i \in I}$ amalgamating H as the set of all equivalence classes of reduced words in W (under $\equiv$).

For each $a \in W$, let $\mathrm{cls}\,a$ denote the $\equiv$-class of $a \in W$, and let $\mathrm{red}\,a$ denote the $\equiv$-class which by Lemma 2.8.4 contains all reduced words $b \leqslant a$. By part (5) of Lemma 2.8.2, $\mathrm{cls}\,a = \mathrm{cls}\,b$ implies $\mathrm{red}\,a = \mathrm{red}\,b$. The operation $\cdot$ on P (usually denoted by juxtaposition) is defined by concatenation and reduction:

Lemma 2.8.5. *A multiplication on P is well defined by*

$$\mathrm{cls}\,a \cdot \mathrm{cls}\,b = \mathrm{red}\,(ab)$$

for all reduced words a and b.

Proof. It is clear that $a \equiv_1 c$ implies $ab \equiv_1 cb$, and that $b \equiv_1 d$ implies $cb \equiv_1 cd$. By induction, $a \equiv c$ and $b \equiv d$ imply that $ab \equiv cb \equiv cd$, $\mathrm{cls}\,(ab) = \mathrm{cls}\,(cd)$, and $\mathrm{red}\,(ab) = \mathrm{red}\,(cd)$. Hence there is a mapping $\cdot$ of $P \times P$ into P which sends $(\mathrm{cls}\,a, \mathrm{cls}\,b)$ to $\mathrm{red}\,(ab)$. $\qquad\square$

There is a canonical mapping $\iota : \bigcup_{i \in I} G_i \longrightarrow P$ which assigns to each $x \in G_i$ the equivalence class of the one letter reduced word (x):

$$\iota(x) = \mathrm{cls}\,(x).$$

Since $a \equiv_1 b$ requires a to have length at least 2, $\mathrm{cls}\,(x) = \{(x)\}$ for all $x \in \bigcup_{i \in I} G_i$, and ι is injective. Hence $\mathrm{Im}\,\iota \subseteq P$ is a copy of the given amalgam $\bigcup_{i \in I} G_i$. [It is common practice to identify $x \in \bigcup_{i \in I} G_i$ and $\iota(x) \in P$.] For each $i \in I$, the **injection** $\iota_i : G_i \longrightarrow P$ is the restriction of ι to G_i. We see that ι_i an injective homomorphism, and that ι_i and ι_j agree on H ($\iota_i(h) = \iota_j(h)$ for all $h \in H$).

Proposition 2.8.6. *P is a group. Furthermore* Im $\iota_i \cong G_i$, ι_i *and* ι_j *agree on* H, Im $\iota_i \cap$ Im $\iota_j = \iota_i(H) = \iota_j(H)$ *whenever* $i \neq j$, *and* P *is generated by* $\bigcup_{i \in I}$ Im ι_i.

The proof is similar to the proof of Propositions 2.5.6 and 2.7.5, and is left as an exercise. The inverse of $\mathrm{cls}\, a = \mathrm{cls}\,(a_1, \ldots, a_n)$ is $\mathrm{cls}\,(a_n^{-1}, a_{n-1}^{-1}, \ldots, a_1^{-1})$. That P is generated by $\bigcup_{i \in I}$ Im ι_i follows from the next lemma, which is readily proved by induction on n.

Lemma 2.8.7. *When* $a = (a_1, \ldots, a_n) \in F$, *then*

$$\mathrm{cls}\, a = \iota(a_1) \bullet \iota(a_2) \bullet \cdots \bullet \iota(a_n).$$

When x is identified with $\iota(x)$, Lemma 2.8.7 writes the elements of P as products of elements of the groups G_i.

Proposition 2.8.6 shows that a group amalgam can be embedded into a single group, the group P above.

2.8.e. Universal Property. It remains to show that our group P is indeed the free product with amalgamation of the groups $(G_i)_{i \in I}$ amalgamating H.

Proposition 2.8.8. *Let* $(G_i)_{i \in I}$ *be a family of groups such that* $G_i \cap G_j = H$ *whenever* $i \neq j$. *If* G *is any group and* $\varphi_i : G_i \longrightarrow G$ ($i \in I$) *are any homomorphisms which agree on* H ($\varphi_i(h) = \varphi_j(h)$ *for all* $h \in H$), *then there exists a unique homomorphism* $\psi : P \longrightarrow G$ *such that* $\psi \circ \iota_i = \varphi_i$ *for all* i.

Proof. Since the homomorphisms $\varphi_i : G_i \longrightarrow G$ agree on H, it is convenient to combine them into a single mapping $\varphi : \bigcup_{i \in I} G_i \longrightarrow G$; by definition, $\varphi(x) = \varphi_i(x)$ whenever $x \in G_i$.

Assume that $\psi : P \longrightarrow G$ is a homomorphism such that $\psi(\iota_i(x)) = \varphi_i(x)$ whenever $x \in G_i$. Then $\psi(\iota(x)) = \varphi(x)$ for all $x \in \bigcup_{i \in I} G_i$. For any $\mathrm{cls}\, a = \mathrm{cls}\,(a_1, \ldots, a_n) \in P$, Lemma 2.8.7 implies

$$\psi(\mathrm{cls}\, a) = \psi(\iota(a_1)) \ldots \psi(\iota(a_n)) = \varphi(a_1) \ldots \varphi(a_n). \tag{2.8}$$

Assume $a \equiv_1 b$, so that $a = (a_1, \ldots, a_m)$ and $b = (b_1, \ldots, b_n)$ satisfy $m = n$, $a_t h = b_t$, $a_{t+1} = h b_{t+1}$ for some $h \in H$ and $t < n$, and $a_u = b_u$ for all $u \neq t, t+1$.

Then

$$\varphi(a_t)\varphi(a_{t+1}) = \varphi(a_t)\varphi(h)\varphi(b_{t+1}) = \varphi(b_t)\varphi(b_{t+1})$$

and $\varphi(a_1)\ldots\varphi(a_n) = \varphi(b_1)\ldots\varphi(b_n)$. By induction,

$$a \equiv b \quad \text{implies} \quad \varphi(a_1)\ldots\varphi(a_n) = \varphi(b_1)\ldots\varphi(b_n).$$

Therefore there is a mapping of P into G which assigns $\varphi(a_1)\ldots\varphi(a_n)$ to $\operatorname{cls} a$ whenever $a = (a_1,\ldots,a_n)$ is reduced. Then equation (2.8) implies that there is at most one homomorphism $\psi : P \longrightarrow G$ such that $\psi \circ \iota_i = \varphi_i$ for all i: namely the mapping $\operatorname{cls} a \longmapsto \varphi(a_1)\ldots\varphi(a_n)$.

It remains to show that the mapping $\psi : \operatorname{cls} a \longmapsto \varphi(a_1)\ldots\varphi(a_n)$ is a homomorphism. Define a mapping $\chi : W \longrightarrow G$ by

$$\chi(a) = \varphi(a_1)\ldots\varphi(a_n)$$

for all $a = (a_1,\ldots,a_n) \in W$. Then $\chi(ab) = \chi(a)\chi(b)$, and we just showed that $a \equiv b$ implies $\chi(a) = \chi(b)$. Also $\psi(\operatorname{cls} a) = \chi(a)$ when a is reduced. We show that $a \geqslant b$ in W implies $\chi(a) = \chi(b)$. If $a = (a_1,\ldots,a_n) \geqslant_1 b$, then a_t and a_{t+1} belong to the same group G_i for some $t < n$,

$$b = (a_1,\ldots,a_{t-1},a_t a_{t+1},a_{t+2},\ldots,a_n),$$

and $\chi(a) = \chi(b)$, since $\varphi(a_t)\varphi(a_{t+1}) = \varphi(a_t a_{t+1})$. Thus $a \geqslant_1 b$ implies $\chi(a) = \chi(b)$. By induction on k, $a \geqslant_k b$ implies $\chi(a) = \chi(b)$.

Now let a and b be reduced and $\operatorname{cls} a \cdot \operatorname{cls} b = \operatorname{cls} d$, where d is reduced and $ab \geqslant d$. By the above,

$$\psi(\operatorname{cls} a \cdot \operatorname{cls} b) = \psi(\operatorname{cls} d) = \chi(d) = \chi(ab) = \chi(a)\chi(b) = \psi(\operatorname{cls} a)\psi(\operatorname{cls} b).$$

Thus ψ is a homomorphism. The definition of χ shows that $\psi(\iota_i(x)) = \psi(x) = \varphi(x) = \varphi_i(x)$ for all $x \in G_i$. □

We saw that the homomorphisms ι_i are injective, so that each G_i is isomorphic to a subgroup $\operatorname{Im} \iota_i$ of the free product with amalgamation P, and that P is generated by $\bigcup_{i \in I} \operatorname{Im} \iota_i$. Proposition 2.8.8 implies that every group with these properties is a homomorphic image of P.

Exercises

1. Prove Lemma 2.8.1.
2. Prove Lemma 2.8.7.
3. Show that the free product with amalgamation is characterized, up to isomorphism, by its universal property.

4. Recall that $\mathbb{Z}_4$ has only one subgroup of order 2. Give a presentation of the free product of $\mathbb{Z}_4$ and $\mathbb{Z}_4$ amalgamating this subgroup.

5. Recall that $\mathbb{Z}_6$ and D_3 have only one subgroup of order 3. Give a presentation of the free product of $\mathbb{Z}_6$ and D_3 amalgamating this subgroup.

6. Let P be the free product of groups $(G_i)_{i \in I}$ amalgamating H. Prove the following: if $H \trianglelefteq G_i$ for all i, then $H \trianglelefteq P$.

7. Prove the following: when G is a group and $H \leq G$, there exist a group G' and two homomorphisms $\varphi, \psi : G \longrightarrow G'$ such that H is the equalizer $H = \{ x \in G ; \varphi(x) = \psi(x) \}$ of φ and ψ.

8. Let $(G_i)_{i \in I}$ be a family of groups. When G is a group and $\varphi_i : G_i \longrightarrow G$ is a homomorphism for each $i \in I$, and $\varphi_j(x_j)$ commutes with $\varphi_k(x_k)$ whenever $x_j \in G_j$, $x_k \in G_k$, and $j \neq k$, then we know by Proposition 2.1.7 that there exists a unique homomorphism $\varphi : \bigoplus_{i \in I} G_i \longrightarrow G$ such that $\varphi \circ \iota_i = \varphi_i$ for all i. Prove that no subgroup $H \gneqq \bigoplus_{i \in I} G_i$ of $\prod_{i \in I} G_i$ has this property. (Use the previous exercise.)

2.9. THE KRULL-SCHMIDT THEOREM

This section contains the proof of the Krull-Schmidt Theorem, stated in full generality for groups with operators of finite length.

2.9.a. Direct Sum Decompositions. Let S be a set. The **direct product** $\prod_{i \in I} G_i$ of a family $(G_i)_{i \in I}$ of S-groups is their direct product as groups on which S acts componentwise:

$$ {}^s(x_i)_{i \in I} = ({}^s x_i)_{i \in I} . $$

The reader will verify that $\prod_{i \in I} G_i$ is an S-group, and that all properties of direct product in Section 1 extend to S-groups. In particular, an S-group G is isomorphic to a finite direct product $G_1 \times G_2 \times \cdots \times G_n$ if and only if G contains S-subgroups $A_1 \cong G_1, \ldots, A_n \cong G_n$ of which G is the internal direct sum $G = A_1 \oplus A_2 \oplus \cdots \oplus A_n$ ($A_i \trianglelefteq_S G$, $A_{i+1} \cap (A_1 A_2 \ldots A_i) = 1$ for all $i < n$, and $G = A_1 A_2 \ldots A_n$ (Proposition 2.1.4).

An S-group G is S-**indecomposable** in case $G \neq 1$, and $G = A \oplus B$ implies $A = 1$ or $B = 1$.

Every finite S-group is a direct sum of finitely many S-indecomposable S-groups, as in Proposition 2.2.1. This property extends to some infinite S-groups. An S-group G has **finite length** in case G satisfies the ascending chain condition and the descending chain condition on normal S-subgroups (see Section B.2); equivalently (by Proposition B.2.3) in case every chain of normal S-subgroups is finite.

Proposition 2.9.1. *Every S-group of finite length is a direct sum of finitely many S-indecomposable S-groups.*

Proof. Assume that G has finite length but is not a direct sum of finitely many S-indecomposable S-groups. Let $\mathscr{B}$ be the set of all normal S-subgroups B of G such that $G = A \oplus B$ for some S-subgroup A but B is not a direct sum of finitely many S-indecomposable S-groups. Then $G \in \mathscr{B}$. By the d.c.c., $\mathscr{B}$ has a minimal element B. Since B is (in particular) not S-indecomposable or trivial, B contains normal S-subgroups $A', B' \neq 1$ such that $B = A' \oplus B'$. Then $A', B' \underset{\neq}{\leqslant} B$ and $G = A \oplus A' \oplus B'$, in particular, $A', B' \trianglelefteq_S G$. Since B is not a direct sum of finitely many S-indecomposable S-groups, the same is true of at least one of A' and B'; then $A', B' \underset{\neq}{\subsetneqq} B$ contradicts the minimality of B. $\qquad\square$

This proof uses only the d.c.c. on normal S-subgroups. The same conclusion can also be obtained from the a.c.c. alone (see the exercises).

2.9.b. Krull-Schmidt Theorem. The Krull-Schmidt Theorem states that the direct sum decomposition in Proposition 2.9.1 is unique, up to isomorphism and the order of the terms. In fact a somewhat stronger statement holds:

Theorem 2.9.2 (Krull-Schmidt). *Let G be a S-group of finite length. If*

$$G = G_1 \oplus G_2 \oplus \cdots \oplus G_m = H_1 \oplus H_2 \oplus \cdots \oplus H_n,$$

where $G_1, \ldots, G_m$ and $H_1, \ldots, H_n$ are S-indecomposable, then $m = n$ and $H_1, \ldots, H_n$ can be indexed so that $G_i \cong H_i$ for all $i \leqslant n$ and the **exchange property** *holds:*

$$G = G_1 \oplus \cdots \oplus G_k \oplus H_{k+1} \oplus \cdots \oplus H_n$$

for all $k < n$.

The above is the internal direct sum version of the theorem. It implies the external direct sum version: if G has finite length and

$$G \cong G_1 \oplus G_2 \oplus \cdots \oplus G_m \cong H_1 \oplus H_2 \oplus \cdots \oplus H_n,$$

where $G_1, \ldots, G_m$ and $H_1, \ldots, H_n$ are S-indecomposable, then $m = n$ and $H_1, \ldots, H_n$ can be indexed so that $G_i \cong H_i$ for all $i \leqslant n$ and the exchange property holds:

$$G \cong G_1 \oplus \cdots \oplus G_k \oplus H_{k+1} \oplus \cdots \oplus H_n$$

for all $k < n$.

We saw in Section 2.2 that the uniqueness part of the structure theorem for finite abelian groups (Corollary 2.2.8) is a particular case of the Krull-Schmidt Theorem.

Theorem 2.9.2 has another well-known particular case. Let V be an abelian group V on which a field K acts so that V is a vector space. Then a K-subgroup is a vector subspace; every K-subgroup is normal; V has finite length if and only if V has a finite basis; V is K-indecomposable if and only if V has dimension 1. Hence the Krull-Schmidt Theorem implies that any two finite bases $e_1, \ldots, e_m$ and $f_1, \ldots, f_n$ of V have the same number $m = n$ of elements; furthermore $f_1, \ldots, f_n$ can be indexed so that $e_1, \ldots, e_k, f_{k+1}, \ldots, f_n$ is a basis of V for all $k < n$ (this is the exchange property for bases of V). The details of this argument are left as an exercise.

Theorem 2.9.2 follows from properties of endomorphisms; the proof requires some patience.

2.9.c. Endomorphisms. In what follows we may assume that S is disjoint from G. We note that G acts on itself by inner automorphisms: ${}^g x = gxg^{-1}$ for all $g, x \in G$. Then the set $T = S \cup G$ acts on G, with ${}^t x = {}^s x$ if $t = s \in S$ and ${}^t x = {}^g x = gxg^{-1}$ if $t = g \in G$; thus G becomes a T-group. The T-subgroups of G are the normal S-subgroups of G. Hence a direct sum decomposition $G = A \oplus B$ of G as an S-group is also a direct sum decomposition $G = A \oplus B$ of G as a T-group.

A T-endomorphism of G, also called a **normal** S-endomorphism, is an S-endomorphism η such that $\eta(gxg^{-1}) = g(\eta x)g^{-1}$ for all $g, x \in G$; equivalently, an S-endomorphism which commutes with all inner automorphisms of the group G. For example, let $G = A \oplus B$ be a direct sum decomposition of the S-group G; the projections $G \longrightarrow A$, $G \longrightarrow B$ viewed as endomorphisms of G, are T-endomorphisms, since $G = A \oplus B$ is also a direct sum decomposition of the T-group G. In general we write T-endomorphisms as left operators. (Hence $\eta \zeta = \eta \circ \zeta$.)

Lemma 2.9.3. *Let $G = A \oplus B$ be the direct sum of two S-subgroups A and B. Every normal S-subgroup of A is a normal S-subgroup of G. If η is a T-endomorphism of G and $\eta(A) \subseteq A$, then $\eta|_A$ is a T-endomorphism of A.*

Proof. Let $a \in A$ and $b \in B$. Since b commutes with every element of A, the inner automorphism $x \longmapsto abxb^{-1}a^{-1}$ of G induces on A the inner automorphism $x \longmapsto axa^{-1}$ of A. Hence a normal S-subgroup of A is a normal S-subgroup of G. Furthermore, if $\eta \in \mathrm{End}(G)$ commutes with every inner automorphism of G and $\eta|_A$ exists, then $\eta|_A$ commutes with every inner automorphism of A. $\square$

Lemma 2.9.4. *Let G be an S-group of finite length. For a T-endomorphism η of G the following conditions are equivalent: (i) η is injective; (ii) η is surjective; (iii) η is bijective (is a T-automorphism).*

Proof. For each $n > 0$, η^n (the composition $\eta \circ \eta \circ \cdots \circ \eta$) is a T-endomorphism; hence $\mathrm{Im}\,\eta^n$ and $\mathrm{Ker}\,\eta^n$ are T-subgroups (normal S-subgroups) of G.

We have $\operatorname{Im} \eta^{n+1} \subseteq \operatorname{Im} \eta^n$ for all n, and

$$G \supseteq \operatorname{Im} \eta \supseteq \cdots \supseteq \operatorname{Im} \eta^n \supseteq \cdots$$

is a descending sequence of normal S-subgroups of G. By the d.c.c., $\operatorname{Im} \eta^n = \operatorname{Im} \eta^{n+1}$ for some n. For each $x \in G$ we have $\eta^n x = \eta^{n+1} y$ for some $y \in G$; if η is injective, this implies $x = \eta y$, showing that η is surjective.

Also $\operatorname{Ker} \eta^n \subseteq \operatorname{Ker} \eta^{n+1}$ for all n, and

$$1 \subseteq \operatorname{Ker} \eta \subseteq \cdots \subseteq \operatorname{Ker} \eta^n \subseteq \cdots$$

is an ascending sequence of normal S-subgroups of G. By the a.c.c., $\operatorname{Ker} \eta^n = \operatorname{Ker} \eta^{n+1}$ for some n. If η is surjective, then for each $x \in \operatorname{Ker} \eta$ we have $x = \eta^n y$ for some $y \in G$, which implies $\eta^{n+1} y = \eta x = 1$, $y \in \operatorname{Ker} \eta^{n+1} = \operatorname{Ker} \eta^n$, and $x = \eta^n y = 1$, showing that η is injective. $\qquad\square$

Lemma 2.9.5. *Let G be an S-group of finite length. For each T-endomorphism η of G, $G = \operatorname{Im} \eta^n \oplus \operatorname{Ker} \eta^n$ holds for some $n \geqslant 1$.*

Proof. As above, the chains of normal S-subgroups

$$G \supseteq \operatorname{Im} \eta \supseteq \cdots \supseteq \operatorname{Im} \eta^n \supseteq \cdots, \qquad 1 \subseteq \operatorname{Ker} \eta \subseteq \cdots \subseteq \operatorname{Ker} \eta^n \subseteq \cdots$$

must terminate, and there exists $n \geqslant 0$ such that $\operatorname{Im} \eta^{2n} = \operatorname{Im} \eta^n$ and $\operatorname{Ker} \eta^{2n} = \operatorname{Ker} \eta^n$.

If $x \in \operatorname{Im} \eta^n \cap \operatorname{Ker} \eta^n$, then $x = \eta^n y$ for some $y \in G$, $\eta^{2n} y = \eta^n x = 1$, $y \in \operatorname{Ker} \eta^{2n} = \operatorname{Ker} \eta^n$, and $x = \eta^n y = 1$. Thus $\operatorname{Im} \eta^n \cap \operatorname{Ker} \eta^n = 1$.

Let $x \in G$. Then $\eta^n x \in \operatorname{Im} \eta^n = \operatorname{Im} \eta^{2n}$ and $\eta^n x = \eta^{2n} y$ for some $y \in G$. Then $x = \eta^n y (x \eta^n (y^{-1}))$, with $\eta^n y \in \operatorname{Im} \eta^n$, and $x \eta^n (y^{-1}) \in \operatorname{Ker} \eta^n$, since $\eta^n (x \eta^n (y^{-1})) = \eta^n x \eta^{2n} (y^{-1}) = 1$. Thus $(\operatorname{Im} \eta^n)(\operatorname{Ker} \eta^n) = G$. $\qquad\square$

A T-endomorphism η is **nilpotent** in case $\operatorname{Im} \eta^n = 1$ for some n.

Lemma 2.9.6. *Let G be an S-indecomposable S-group of finite length. A T-endomorphism of G is either a T-automorphism or nilpotent.*

Proof. By Lemma 2.9.5, $G = \operatorname{Im} \eta^n \oplus \operatorname{Ker} \eta^n$ holds for some $n \geqslant 1$. If G is S-indecomposable, then either $\operatorname{Im} \eta^n = 1$ and η is nilpotent, or $\operatorname{Ker} \eta^n = 1$ and η is injective and is a T-automorphism by Lemma 2.9.4. $\qquad\square$

Finally the group operation on G induces a pointwise operation $\bullet$ on mappings of G into G: if $f, g : G \longrightarrow G$, then $(f \bullet g)(x) = f(x) g(x)$ for all $x \in G$. The operation $\bullet$ is associative, and induces a partial operation, also denoted by $\bullet$, on the set $\operatorname{End}_T (G)$ of all T-endomorphisms of G: if $\eta, \zeta \in \operatorname{End}_T (G)$, then $\eta \bullet \zeta$ is defined in $\operatorname{End}_T (G)$ if and only if the mapping

$$(\eta \bullet \zeta) x = (\eta x)(\zeta x)$$

is a T-endomorphism. The next result is straightforward:

Lemma 2.9.7. $\eta \cdot \zeta$ *is defined in* $\mathrm{End}_T(G)$ *if and only if* ηx *and* ζy *commute for all* $x, y \in G$, *and then* $\eta \cdot \zeta = \zeta \cdot \eta$. *If* $\eta_1, \ldots, \eta_n \in \mathrm{End}_T(G)$ *and* $\eta_i x$ *commutes with* $\eta_j y$ *for all* $x, y \in G$ *and* $i \neq j$, *then* $\eta_1 \cdot \eta_2 \cdot \cdots \cdot \eta_n$ *is defined in* $\mathrm{End}_T(G)$ *and* $\eta_{\sigma 1} \cdots \cdot \eta_{\sigma n} = \eta_1 \cdot \cdots \cdot \eta_n$ *for every permutation* σ.

If G is abelian, and denoted additively, then $\eta \cdot \zeta$ is denoted by $\eta + \zeta$ and is defined (is a T-endomorphism) for all η and ζ, by Lemma 2.9.7. In general, $\varphi(\eta \cdot \zeta) = \varphi\eta \cdot \varphi\zeta$ and $(\eta \cdot \zeta)\varphi = \eta\varphi \cdot \zeta\varphi$ hold whenever $\eta \cdot \zeta$ is defined.

We saw that direct sum decompositions of G give rise to T-endomorphisms. When $G = G_1 \oplus \cdots \oplus G_n$ is a direct sum, the composition $\eta_i : G \longrightarrow G_i \longrightarrow G$ of the projection $G \longrightarrow G_i$ and injection $G_i \longrightarrow G$ is an endomorphism of G, which we call the i **projection endomorphism**. The following properties are straightforward.

Lemma 2.9.8. *When* $G = G_1 \oplus \cdots \oplus G_n$ *is an internal direct sum, the projection endomorphisms* $\eta_1, \ldots, \eta_n$ *are* T-endomorphisms. *For all* i: $\mathrm{Im}\, \eta_i = G_i$; $\eta_i x = x$ *for all* $x \in G_i$; *and* $\eta_i x = 1$ *for all* $x \in G_j$ *if* $j \neq i$. *Also* $\eta_i x$ *and* $\eta_j y$ *commute for all* $x, y \in G$ *and* $i \neq j$, *and* $\eta_1 \cdot \cdots \cdot \eta_n = 1_G$.

Lemma 2.9.9. *Let* G *be an* S-*indecomposable* S-*group of finite length and* $\eta_1, \ldots, \eta_n \in \mathrm{End}_T(G)$. *If* $\eta_i x$ *commutes with* $\eta_j y$ *for all* $x, y \in G$ *and* $i \neq j$, *and each* η_i *is nilpotent, then* $\eta_1 \cdot \eta_2 \cdot \cdots \cdot \eta_n$ *is nilpotent.*

Proof. It suffices to prove this when $n = 2$; then the general case follows by induction on n.

Assume that $\eta, \zeta \in \mathrm{End}_T(G)$ are nilpotent and that $\alpha = \eta \cdot \zeta$ is defined but not nilpotent. By Lemma 2.9.6, α is a T-automorphism. Let $\varphi = \eta\alpha^{-1}$ and $\psi = \zeta\alpha^{-1}$. Then φ and ψ are nilpotent by Lemma 2.9.6, since they are not automorphisms. Also $\varphi \cdot \psi = (\eta \cdot \zeta)\alpha^{-1} = 1_G$; hence

$$(\varphi\varphi) \cdot (\varphi\psi) = \varphi(\varphi \cdot \psi) = \varphi = (\varphi \cdot \psi)\varphi = (\varphi\varphi) \cdot (\psi\varphi),$$

which implies $\varphi\psi = \psi\varphi$. Therefore $(\varphi \cdot \psi)^n(x)$ can be calculated as in the binomial theorem:

$$(\varphi \cdot \psi)^n(x) = \prod_{i+j=n} (\varphi^i \psi^j x)^{\binom{n}{i}}$$

for all $x \in G$. (The product $\prod_{i+j=n} (\varphi^i \psi^j x)^{\binom{n}{i}}$ can be calculated in any order in G, since φx and ψy commute for all $x, y \in G$.) Now $\mathrm{Im}\, \varphi^k = \mathrm{Im}\, \psi^\ell = 1$ for some $k, \ell > 0$. If $n \geq k + \ell$, then $i + j = n$ implies $i \geq k$ or $j \geq \ell$, and $\varphi^i \psi^j x = 1$, so that $(\varphi \cdot \psi)^n x = 1$. However, $\varphi \cdot \psi = 1_G$, so we have proved that $x = 1$ for all $x \in G$, a contradiction since G is S-indecomposable. $\qquad\square$

2.9.d. Proof. We can now prove Theorem 2.9.2. Let G be a S-group of finite length which is an internal direct sum

$$G = G_1 \oplus G_2 \oplus \cdots \oplus G_m = H_1 \oplus H_2 \oplus \cdots \oplus H_n,$$

where $G_1,\ldots,G_m$ and $H_1,\ldots,H_n$ are S-indecomposable. By Lemma 2.9.3, each G_i has finite length; also, if a T-endomorphism η of G has a restriction to G_i, then that restriction $\eta|_{G_i}$ is a T-endomorphism of G_i. The groups H_i have similar properties.

We prove by induction on k that the following hold for all $k \leqslant n$:

(1) $m \geqslant k$;
(2) $H_1,\ldots,H_n$ can be indexed so that $G_i \cong H_i$ for all $i \leqslant k$; and then
(3) $G = G_1 \oplus \cdots \oplus G_k \oplus H_{k+1} \oplus \cdots \oplus H_n$.

With $k = n$, (1) yields $m \geqslant n$; exchanging G_i's and H_i's then yields $m = n$. Parts (2) and (3) yield the other parts of the Krull-Schmidt Theorem.

If $k = 0$, then (1), (2), and (3) hold. We now assume (1), (2), (3) for $k-1$ and prove them for k.

Let $\eta_1,\ldots,\eta_m$ be the projection endomorphisms for the direct sum $G \cong G_1 \oplus \cdots \oplus G_m$ and $\zeta_1,\ldots,\zeta_n$ be the projection endomorphisms for the direct sum $G \cong G_1 \oplus \cdots \oplus G_{k-1} \oplus H_k \oplus \cdots \oplus H_n$ in the induction hypothesis.

By Lemma 2.9.8, $\eta_1 \bullet \cdots \bullet \eta_m = 1_G$. Hence

$$\zeta_k = \zeta_k(\eta_1 \bullet \cdots \bullet \eta_m) = (\zeta_k \eta_1) \bullet \cdots \bullet (\zeta_k \eta_m).$$

Now $\zeta_k \eta_i x = 1$ for all $x \in G$ and $i < k$; if $m < k$, it follows that $\zeta_k x = 1$ for all x, a contradiction since $\operatorname{Im} \zeta_k = H_k$ is S-indecomposable. Therefore $m \geqslant k$ and (1) holds.

Similarly $\zeta_1 \bullet \cdots \bullet \zeta_m = 1_G$, by Lemma 2.9.8, and $\eta_k \zeta_i x = 1$ for all $i < k$ and $x \in G$, since $\zeta_i x \in G_i$ when $i < k$. Hence

$$\eta_k = \eta_k(\zeta_1 \bullet \cdots \bullet \zeta_n) = (\eta_k \zeta_1) \bullet \cdots \bullet (\eta_k \zeta_n) = (\eta_k \zeta_k) \bullet \cdots \bullet (\eta_k \zeta_n).$$

Now the restriction

$$\eta_k|_{G_k} = (\eta_k \zeta_k)|_{G_k} \bullet \cdots \bullet (\eta_k \zeta_n)|_{G_k}$$

of η_k to G_k, which is the identity on G_k, is not nilpotent. By Lemma 2.9.9, there exists $j \geqslant k$ such that $(\eta_k \zeta_j)|_{G_k}$ is not nilpotent. The groups $H_1,\ldots,H_n$ can be indexed so that $j = k$. We show that $G_k \cong H_k$.

By Lemma 2.9.6, $(\eta_k \zeta_k)|_{G_k}$ is an automorphism of G_k. Therefore $\eta_k \zeta_k$ is not nilpotent, $\operatorname{Im} \eta_k(\zeta_k \eta_k)^n \zeta_k = \operatorname{Im}(\eta_k \zeta_k)^{n+1} \neq 1$ for all $n > 0$, $\zeta_k \eta_k$ is not nilpotent, $(\zeta_k \eta_k)|_{H_k}$ is not nilpotent, and $(\zeta_k \eta_k)|_{H_k}$ is an automorphism of H_k by Lemma 2.9.6. Since $(\eta_k \zeta_k)|_{G_k}$ is an automorphism of G_k, it follows that ζ_k is injective on G_k and $\zeta_k G_k = H_k$; hence $\zeta_k|_{G_k}$ is an S-isomorphism of G_k onto H_k. Similarly $\eta_k|_{H_k}$ is an S-isomorphism of H_k onto G_k. This proves (2).

Let $K = G_1 \ldots G_{k-1} H_{k+1} \ldots H_n$. By the induction hypothesis, $G = G_1 \oplus \cdots \oplus G_{k-1} \oplus H_k \oplus \cdots \oplus H_n$; hence $K = G_1 \oplus \cdots \oplus G_{k-1} \oplus H_{k+1} \oplus \cdots \oplus H_n$. Also $\zeta_k K = 1$, since $\zeta_k G_1 = \cdots = \zeta_k G_{k-1} = \zeta_k H_{k+1} = \cdots = \zeta_k H_n = 1$. Since ζ_k is injective on G_k, it follows that $G_k \cap K = 1$. Hence $G_k K = G_k \oplus K$. We show that $G_k \oplus K = G$; this will prove (3).

Since $\eta_k|_{H_k} : H_k \longrightarrow G_k$ is an S-isomorphism, there is an S-isomorphism

$$G_1 \oplus \cdots \oplus H_k \oplus H_{k+1} \oplus \cdots \oplus H_n \longrightarrow G_1 \oplus \cdots \oplus G_k \oplus H_{k+1} \oplus \cdots \oplus H_n$$

of external direct sums, namely $1_{G_1} \oplus \cdots \oplus 1_{G_{k-1}} \oplus \eta_k|_{G_k} \oplus 1_{H_{k+1}} \oplus \cdots \oplus 1_{H_n}$. This yields an S-isomorphism $G = H_k \oplus K \longrightarrow G_k \oplus K$ of internal direct sums. Composing with the inclusion $G_k \oplus K \longrightarrow G$ yields an injective S-endomorphism θ of G. Since η_k is a T-endomorphism, θ is a T-endomorphism. By Lemma 2.9.4, θ is surjective; hence $G_k \oplus K = G$. $\square$

Exercises

1. Formulate and prove a universal property for the direct product of a family of S-groups.

2. Verify that a direct sum of S-groups is an S-group.

3. Formulate and prove a universal property for the direct sum of S-groups.

4. Verify that an S-group G is isomorphic to the external direct sum $\bigoplus_{i \in I} G_i$ if and only if G contains S-subgroups $A_i \cong G_i$ of which it is the internal direct sum.

5. Prove that every S-group which satisfies the a.c.c. on normal S-subgroups is a direct sum of finitely many indecomposable S-groups.

6. Prove Lemma 2.9.7.

7. Prove Lemma 2.9.8.

8. Use the Krull-Schmidt Theorem to prove that two finite bases $e_1, \ldots, e_m$ and $f_1, \ldots, f_n$ of a vector space V have the same number $m = n$ of elements, and that furthermore $f_1, \ldots, f_n$ can be indexed so that $e_1, \ldots, e_k, f_{k+1}, \ldots, f_n$ is a basis of V for all $k < n$.

3

FINITE GROUPS

This chapter is an introduction to the theory of finite groups, and an investigation of their structure. It combines the constructions and examples in Chapter 2 with two main ideas. Group actions yield the class equation, the Sylow Theorems, and the construction of all small groups. Composition series are used to analyze groups into simple groups and to study solvable groups and nilpotent groups.

The study of finite groups is continued in Chapter 13 (using characters).

Required results: all main results of Chapter 1, and the basic properties of integers in Section A.3. Proposition 1.4.16 is used in Section 3.2. The Schur-Zassenhaus Theorem in Section 3.3 requires Schur's Theorem 2.4.8; Section 3.6 also uses Theorem 2.4.8 and Proposition 2.4.9 to prove the Hall Theorems. The classification of small groups requires the classification of finite abelian groups in Section 2.2 and the examples Q and T in Section 2.6. Section 3.7 requires basic properties of matrices, subspaces, and linear transformations of a finite-dimensional vector space.

3.1. ORBITS

Actions of groups on sets are a basic tool of group theory. This section contains general properties, with nice applications to p-groups.

3.1.a. Group Actions. A left **group action** of a group G on a set X is a mapping $(g,x) \longmapsto g \bullet x$ of $G \times X$ into X such that

$$1 \bullet x = x \qquad \text{and} \qquad g \bullet (h \bullet x) = (gh) \bullet x$$

for all $g,h \in G$ and $x \in X$. The group G **acts on** the set X when there is a group action of G on X. Various notations are used for $g \bullet x$. **Right** group actions are defined similarly by $x \bullet 1 = x$ and $(x \bullet g) \bullet h = x \bullet (gh)$.

When G acts on X, then for each $g \in G$ the mapping $\sigma_g : x \longmapsto g \bullet x$ is a permutation of X: indeed $g \bullet (g^{-1} \bullet x) = 1 \bullet x = x$ and $g^{-1} \bullet (g \bullet x) = 1 \bullet x = x$ for

93

all x so that σ_g and $\sigma_{g^{-1}}$ are mutually inverse permutations. One says that G **acts by permutations** of X. Moreover the action of G on X is completely determined by the mapping $\sigma : g \longmapsto \sigma_g$, which is a homomorphism of G into the group S_X of all permutations of X. In fact there is a one-to-one correspondence between group actions of G on X and homomorphisms $G \longrightarrow S_X$ (see the exercises).

For example, every group G acts on itself **by left multiplication**, meaning that the action is defined by $g \bullet x = gx$ for all $g,x \in G$. More generally, every subgroup H of G acts on G by left multiplication ($h \bullet x = hx$ for all $h \in H$, $x \in G$). Every group G also acts on itself **by inner automorphisms**, meaning that the action is defined by $g \bullet x = gxg^{-1}$ for all $g,x \in G$. The symmetric group S_X (of all permutations of a set X) acts on X by evaluation ($\sigma \bullet x$ is the value of σ at x).

3.1.b. Orbits and Stabilizers. Group actions have certain general properties.

Lemma 3.1.1. *When the group G acts on X, the binary relation*

$$x \equiv y \iff x = g \bullet y \quad \text{for some} \quad g \in G$$

is an equivalence relation on X.

Proof. The relation $\equiv$ is reflexive, since $x = 1 \bullet x$; symmetric, since $x = g \bullet y$ implies that $g^{-1} \bullet x = y$; and transitive, since $x = g \bullet y$, $y = h \bullet z$ imply $x = g \bullet (h \bullet z) = (gh) \bullet z$. $\square$

The $\equiv$-class of an element x of X is the **orbit** of x under the action of G. The orbits form a partition of X. For instance, let H be a subgroup of G, acting on G under left multiplication; the orbit of x is the right coset Hx of x.

The **stabilizer** of an element x of X under the action of G is the set

$$S(x) = S_G(x) = \{ g \in G ; g \bullet x = x \}$$

Proposition 3.1.2. *The stabilizer $S(x)$ of x is a subgroup of G. There is a one-to-one correspondence between the left cosets of $S(x)$ and the elements of the orbit of x; hence the order of the orbit of x equals the index of $S(x)$.*

Proof. It is clear that $S(x) \leqslant G$.
When $y = g \bullet x$ belong to the orbit of x, let

$$C(y) = \{ a \in G ; a \bullet x = y \}.$$

We see that $a \in C(y)$ is equivalent to $g \bullet x = a \bullet x$, $x = g^{-1} \bullet (a \bullet x) = (g^{-1}a) \bullet x$, $g^{-1}a \in S(x)$, and $a \in gS(x)$. Thus $C(y) = gS(x)$ is a left coset of $S(x)$.
Conversely, let $C = gS(x) \subseteq G$ be a left coset of $S(x)$. Then $a \in C$ implies $g^{-1}a \in S(x)$, $(g^{-1}a) \bullet x = x$, and $a \bullet x = g \bullet ((g^{-1}a) \bullet x) = g \bullet x$. Hence $C \bullet x$ is a

single element of X. Let $\theta(C)$ denote this element. By definition, $\theta(gS(x)) = g \cdot x$.

The maps C and θ are mutually inverse bijections: if $y = g \cdot x$, then $C(y) = gS(x)$ and $\theta(C(y)) = g \cdot x = y$; if $C = gS(x)$ is a left coset of $S(x)$ and $y = \theta(C) = g \cdot x$, then $C(\theta(C)) = C(y) = gS(x) = C$.   □

When G is finite, Proposition 3.1.2 (and Lagrange's Theorem) implies that the order of each orbit divides the order of G.

3.1.c. The Class Equation. When applied to a subgroup H of G acting on G by left multiplication, every stabilizer is trivial, and Proposition 3.1.2 states that the order of every right coset of H equals the order of H. We now turn to another example where stabilizers are generally not trivial.

In what follows, G acts on itself by inner automorphisms: $g \cdot x = gxg^{-1}$.

Under this action, $x \equiv y$ if and only if $x = gyg^{-1}$ for some $g \in G$. Two such elements are called **conjugate**, and the orbits are the **conjugacy classes** of G. We denote the conjugacy class of $x \in G$ by C_x. By Lemma 3.1.1, conjugacy is an equivalence relation, and the conjugacy classes form a partition of G.

The **centralizer** of x in G is the stabilizer of $x \in G$ under this action:

$$C(x) = C_G(x) = \{ g \in G\,;\, gxg^{-1} = x \} = \{ g \in G\,;\, gx = xg \}.$$

By Proposition 3.1.2:

Corollary 3.1.3. *For each $x \in G$, the centralizer $C_G(x)$ of x is a subgroup of G, and the number of conjugates of x equals the index of its centralizer* ($|C_x| = [G : C_G(x)]$).

Elements with trivial conjugacy classes are of special interest. By Corollary 3.1.3:

Proposition 3.1.4. *For an element x of a group G the following are equivalent:* (1) $|C_x| = 1$; (2) $C_G(x) = G$; (3) $xy = yx$ *for all* $y \in G$.

The **center** of a group G is the set

$$Z(G) = \{ x \in G\,;\, xy = yx \text{ for all } y \in G \}.$$

Proposition 3.1.5. *The center of a group G is a **characteristic** subgroup of G : $\alpha(Z(G)) = Z(G)$ for every automorphism α of G. Every subgroup of $Z(G)$, in particular $Z(G)$ itself, is a normal subgroup of G.*

We leave these properties as exercises.

Theorem 3.1.6. *In any group G,*

$$|G| = |Z(G)| + \sum_{|C_x|>1} |C_x|.$$

This equality is the **class equation** of G ("class" is short for "conjugacy class"); the sum has one term for each nontrivial conjugacy class.

Proof. Since the conjugacy classes form a partition of G, we have $|G| = \sum |C_x|$, where the sum has one term for each conjugacy class. Separating trivial classes from nontrivial classes yields the class equation:

$$|G| = \sum_{|C_x|=1} |C_x| + \sum_{|C_x|>1} |C_x| = |Z(G)| + \sum_{|C_x|>1} |C_x|. \qquad \square$$

3.1.d. *p***-Groups.** The class equation yields properties of groups of order p^n, where p is prime.

Proposition 3.1.7. *Every group of order $p^n > 1$ (with p prime) has a nontrivial center.*

Proof. Let G have order $p^n > 1$. By Lagrange's Theorem, the index of any subgroup of G divides the order of G and is therefore a power of p. By Corollary 3.1.3, the order of every conjugacy class of G is a power of p. In particular, p divides every $|C_x| > 1$. Then p divides $\sum_{|C_x|>1} |C_x|$, and p divides $|G|$; by the class equation, p divides $|Z(G)|$. Therefore $|Z(G)| \neq 1$. $\qquad \square$

Lemma 3.1.8. *If $G/Z(G)$ is cyclic, then G is abelian.*

This is fun to prove and is left as an exercise.

Proposition 3.1.9. *Every group of order p^2 (with p prime) is abelian.*

Proof. Let $|G| = p^2$. By Proposition 3.1.7, $|Z(G)| = p$ or $|Z(G)| = p^2$. If $|Z(G)| = p^2$, then $G = Z(G)$ is abelian. If $|Z(G)| = p$, then $|G/Z(G)| = |G|/|Z(G)| = p$, $G/Z(G)$ is cyclic, and G is abelian by Lemma 3.1.8. (This implies $Z(G) = G$, so that the case $|Z(G)| = p$ does not actually happen.) $\qquad \square$

3.1.e. Normalizers. Conjugation also applies to subgroups, leading to another basic construction (normalizers).

Let G be a group. For each $g \in G$, $\alpha_g : x \longmapsto gxg^{-1}$ is an automorphism of G; the image gHg^{-1} of a subgroup H of G under α_g is also a subgroup of G. Hence G acts on the set of all its subgroups by inner automorphisms: $g \cdot H = gHg^{-1}$.

Under this action, $H \equiv H'$ if and only if $H = gH'g^{-1}$ for some $g \in G$. Two such subgroups are called **conjugate**, and the orbits are **conjugacy classes** of subgroups. We denote the conjugacy class of $H \leqslant G$ by C_H. By Lemma 3.1.1, conjugacy is an equivalence relation.

The **normalizer** of a subgroup H is the stabilizer of H under this action:

$$N(H) = N_G(H) = \{ a \in G \, ; \, aHa^{-1} = H \} \, .$$

The following results are straightforward:

Proposition 3.1.10. *The normalizer $N_G(H)$ of $H \leqslant G$ is the largest subgroup of G such that $H \lhd N_G(H)$.*

Proposition 3.1.11. *The number of conjugates of a subgroup equals the index of its normalizer.*

Exercises

1. Show that for every group homomorphism $\varphi : G \longrightarrow S_X$ there is a group action of G on X, defined by $g \cdot x = \varphi(g)(x)$. Show that this defines a one-to-one correspondence between groups actions of G on X and homomorphisms of G into S_X.

2. Give a direct proof that the centralizer of an element is a subgroup.

3. Show that the centralizer of x is the largest subgroup H such that $x \in Z(H)$.

4. Let G be a group of order n. Prove that an element of order k has at most n/k conjugates in G.

5. The **centralizer** of a subgroup $H \leqslant G$ is $C_G(H) = \bigcap_{h \in H} C_G(h) = \{ x \in G \, ; \, xh = hx$ for all $h \in H \}$. Show that $C_G(H)$ is a subgroup of G.

6. Show that $C_G(H) \lhd N_G(H)$ and that $N_G(H)/C_G(H)$ can be embedded into $\mathrm{Aut}(H)$.

7. Prove that the center of a group G is a characteristic subgroup of G.

8. Prove that every subgroup of $Z(G)$ is a normal subgroup of G.

9. Prove that every characteristic subgroup of a normal subgroup of a group G is a normal subgroup of G, and that every characteristic subgroup of a characteristic subgroup of G is a characteristic subgroup of G.

10. Let N be a characteristic subgroup of G and K/N be a characteristic subgroup of G/N, with $K \supseteq N$. Prove that K is a characteristic subgroup of G.

11. List the conjugacy classes of D_4.

12. List the conjugacy classes of the quaternion group Q.

13. Find the center of Q.

14. Find the center of D_n.

15. Prove that the inner automorphisms of a group G form a group which is isomorphic to $G/Z(G)$.

16. Prove Proposition 3.1.9.

17. Prove Lemma 3.1.8.

3.2. PERMUTATIONS

This section gives basic properties of finite symmetric groups and their elements. The results illustrate ideas introduced in Section 3.1. Recall (Example 1.2.2) that the symmetric group S_n is the group of all permutations of the set $I_n = \{1, 2, \ldots, n\}$; S_n has order $n!$.

We write permutations as left operators. An individual permutation σ can be specified by a table of values, arranged as a $2 \times n$ matrix in which the top row is $1\ 2\ \ldots\ n$ and the bottom row is $\sigma 1\ \sigma 2\ \ldots\ \sigma n$. For example,

$$\sigma = \begin{pmatrix} 1\ 2\ 3\ 4 \\ 4\ 2\ 1\ 3 \end{pmatrix}$$

is the permutation $\sigma \in S_4$ such that $\sigma 1 = 4$, $\sigma 2 = 2$, $\sigma 3 = 1$, and $\sigma 4 = 3$.

3.2.a. Cycles. When $2 \leqslant k \leqslant n$, a k-**cycle** is a permutation γ for which there exists distinct numbers $a_1, a_2, \ldots, a_k \leqslant n$ such that $\gamma a_1 = a_2$, $\gamma a_2 = a_3$, $\ldots$, $\gamma a_{k-1} = a_k$ (generally, $\gamma a_i = a_{i+1}$ for all $i < k$), $\gamma a_k = a_1$, and $\gamma x = x$ for all $x \neq a_1, \ldots, a_k$; then γ is denoted by $(a_1\ a_2\ \ldots\ a_k)$. Note that

$$(a_1\ a_2\ \ldots\ a_k) = (a_2\ a_3\ \ldots\ a_k\ a_1) = (a_3\ \ldots\ a_k\ a_1\ a_2) = \cdots = (a_k\ a_1\ a_2\ \ldots\ a_{k-1}).$$

For example,

$$\sigma = \begin{pmatrix} 1\ 2\ 3\ 4 \\ 4\ 2\ 1\ 3 \end{pmatrix}$$

is a 3-cycle: $\sigma = (1\ 4\ 3)$. In general, $\gamma = (a_1\ a_2\ \ldots\ a_k)$ satisfies $\gamma^k = 1$; since $\gamma^i a_1 = a_{i+1} \neq a_1$ when $0 \leqslant i < k$, it follows that a k-cycle γ has order k. A **cycle** is a k-cycle γ for some $2 \leqslant k \leqslant n$.

3.2.b. Transpositions. A **transposition** is a 2-cycle $\tau = (a\ b) = (b\ a)$ (where $a \neq b$). Every transposition has order 2.

Proposition 3.2.1. *Every permutation in S_n is a product of transpositions. Thus S_n is generated by all transpositions.*

Proof. This is trivial if $n = 1$. For $n > 1$ we proceed by induction on n.

Let $\sigma \in S_n$. If $\sigma n = n$, then $\sigma(\{1, \ldots, n-1\}) = \{1, \ldots, n-1\}$; by the induction hypothesis, the restriction ρ of σ to $\{1, \ldots, n-1\}$ is a product of transpositions; therefore σ is a product of transpositions.

If $\sigma n = a < n$, let $\tau = (a\ n)$. Then $\rho = \tau\sigma$ satisfies $\rho n = n$. By the above, ρ is a product of transpositions; therefore $\sigma = \tau\rho$ is a product of transpositions. $\square$

An unusual uniqueness property is associated with Proposition 3.2.1.

Proposition 3.2.2. *If*

$$\sigma = \tau_1\tau_2\ldots\tau_r = \upsilon_1\upsilon_2\ldots\upsilon_s$$

is a product of transpositions $\tau_1,\ldots,\tau_r$ *and* $\upsilon_1,\ldots,\upsilon_s$, *then* $r \equiv s$ (mod 2). *Thus a product of an even number of transpositions cannot equal a product of an odd number of transpositions.*

Proof. Recall that $\mathbb{Z}[X_1,\ldots,X_n]$ is the set of all polynomials $f = f(X_1,\ldots,X_n)$ with integer coefficients and n indeterminates (variables) $X_1,\ldots,X_n$. Define a group action of S_n on $\mathbb{Z}[X_1,\ldots,X_n]$ by

$$(\sigma \bullet f)(X_1,\ldots,X_n) = f(X_{\sigma 1},\ldots,X_{\sigma n}).$$

It is immediate that $1 \bullet f = f$ and that $\sigma \bullet (\tau \bullet f) = f$ for all $\sigma,\tau \in S_n$, as required. Furthermore the action of σ distributes products of polynomials: $\sigma \bullet (f_1 f_2 \ldots f_k) = (\sigma \bullet f_1)(\sigma \bullet f_2)\ldots(\sigma \bullet f_k)$ for all $f_1,\ldots,f_k \in \mathbb{Z}[X_1,\ldots,X_n]$.

In what follows,

$$f(X_1,\ldots,X_n) = \prod_{i,j\in I_n, i<j} (X_i - X_j).$$

We show that $\tau \bullet f = -f$ for every transposition τ. Let $\tau = (a\ b)$. Since $(a\ b) = (b\ a)$ we may assume $a < b$. Then

$$\tau \bullet (X_i - X_j)$$

$$= \begin{cases} X_b - X_a = -(X_a - X_b) & \text{if } i = a \text{ and } j = b, & (3.1) \\ X_b - X_j & \text{if } i = a \text{ and } b < j, & (3.2) \\ X_b - X_j = -(X_j - X_b) & \text{if } i = a \text{ and } i < j < b, & (3.3) \\ X_i - X_a & \text{if } i < a \text{ and } j = b, & (3.4) \\ X_i - X_a = -(X_a - X_i) & \text{if } a < i < j \text{ and } j = b, & (3.5) \\ X_i - X_j & \text{if } i,j \neq a,b. & (3.6) \end{cases}$$

Now $\tau \bullet f = \prod_{i,j\in I_n, i<j} \tau \bullet (X_i - X_j)$. Inspecting the six cases above shows that every term of $f = \prod_{i,j\in I_n, i<j} (X_i - X_j)$ appears exactly once in $\tau \bullet f$, though perhaps with a minus sign. Hence $\tau \bullet f = \pm f$. We now count minus signs. In the above, case (3.1) provides one minus sign; case (3.3) provides one minus sign for each $a < j < b$; and case (3.5) provides one minus sign for each $a < i < b$, as many as case (3.3). This yields an odd number of minus signs. Hence $\tau \bullet f = -f$.

If now $\sigma = \tau_1 \tau_2 \ldots \tau_r = \upsilon_1 \upsilon_2 \ldots \upsilon_s$ is a product of transpositions $\tau_1, \ldots, \tau_r$ and $\upsilon_1, \ldots, \upsilon_s$, then

$$\sigma \bullet f = \tau_1 \bullet (\tau_2 \bullet (\ldots (\tau_r \bullet f)) \ldots)) = (-1)^r f$$
$$= \upsilon_1 \bullet (\upsilon_2 \bullet (\ldots (\upsilon_s \bullet f)) \ldots)) = (-1)^s f;$$

this implies that $(-1)^r = (-1)^s$.   □

3.2.c. The Alternating Group. A permutation is **even** in case it is a product of an even number of transpositions and **odd** in case it is a product of an odd number of transpositions. By Proposition 3.2.2, an even permutation is not the product of an odd number of transpositions, and vice versa.

We see that a product of two even permutations is even, a product of two odd permutations is even, but products of one even permutation and one odd permutation (in either order) are odd. This resembles the multiplication on the cyclic group $\{+1, -1\} \subseteq \mathbb{Z}$, where $(+1)(+1) = (-1)(-1) = +1$ and $(+1)(-1) = (-1)(+1) = -1$. In fact, a homomorphism $\varepsilon : S_n \longrightarrow \{1, -1\}$ is defined by $\varepsilon(\sigma) = +1$ if σ is even, $\varepsilon(\sigma) = -1$ if σ is odd. The number $\varepsilon(\sigma)$ is the **sign** of σ and is variously denoted by ε_σ, sgn σ, or $(-1)^\sigma$; we use sgn σ.

The **alternating group** A_n is the group of all even permutations; A_n is the kernel of the sign homomorphism and is therefore a normal subgroup of S_n.

Proposition 3.2.3. *A permutation is even if and only if it is a product of 3-cycles. Thus A_n is generated by all 3-cycles.*

Proof. Inspection shows that $(a\ b\ c) = (a\ b)(c\ b)$; hence 3-cycles are even and products of 3-cycles are even.

To prove the converse, it suffices to show that every product $(a\ b)(c\ d)$ of two transpositions is a product of 3-cycles. If $\{a,b\} = \{c,d\}$, then $(a\ b) = (c\ d)$ and $(a\ b)(c\ d) = 1$ is an empty product of 3-cycles. If $\{a,b\} \cap \{c,d\}$ has one element, then we may assume that $\{a,b\} \cap \{c,d\} = \{b\} = \{d\}$ [since $(a\ b) = (b\ a)$ and $(c\ d) = (d\ c)$], and then $(a\ b)(c\ d) = (a\ b)(c\ b) = (a\ b\ c)$ is a 3-cycle. If $\{a,b\} \cap \{c,d\} = \varnothing$, then $(a\ b)(c\ d) = (a\ b)(c\ b)(b\ c)(d\ c) = (a\ b\ c)(b\ c\ d)$ is a product of two 3-cycles.   □

3.2.d. Cycle Decomposition. Suitable group actions provide insight into the structure of individual permutations. This requires the following definitions. The **support** of a permutation σ is the set

$$S(\sigma) = \{x \in I_n; \sigma x \neq x\}.$$

For instance, the support of a k-cycle $\gamma = (a_1\ a_2\ \ldots\ a_k)$ is $\{a_1, a_2, \ldots, a_k\}$; thus the order of a cycle is the order (the number of elements) of its support. In general, the complement of the support is the set $\{x \in I_n; \sigma x = x\}$ of **fixed**

points of σ. Two permutations σ and τ are **disjoint** in case their supports are disjoint $(S(\sigma) \cap S(\tau) = \emptyset)$. Permutations $(\sigma_i)_{i \in I}$ are **pairwise disjoint** in case their supports are pairwise disjoint $(S(\sigma_i) \cap S(\sigma_j) = \emptyset$ whenever $i \neq j)$.

Lemma 3.2.4. *Disjoint permutations commute.*

Proof. Note that $t \in S(\sigma)$ implies $\sigma t \in S(\sigma)$ ($\sigma t \neq t$ implies $\sigma \sigma t \neq \sigma t$).

Assume that σ and τ are disjoint. If $x \in S(\sigma)$, then $x \notin S(\tau)$, $\sigma \tau x = \sigma x$, and $\tau \sigma x = \sigma x$, since $\sigma x \in S(\sigma)$ implies $\sigma x \notin S(\tau)$. Exchanging σ and τ yields $\tau \sigma x = \tau x = \sigma \tau x$ for all $x \in S(\tau)$. If finally $x \notin S(\sigma) \cup S(\tau)$, then $\sigma \tau x = x = \tau \sigma x$. In every case $\sigma \tau x = \tau \sigma x$. □

Proposition 3.2.5. *Every permutation is a product of pairwise disjoint cycles; moreover this decomposition is unique, up to the order of the terms.*

Proof. Let $\sigma \in S_n$. Define an action of $\mathbb{Z}$ on $I_n = \{1, 2, \ldots, n\}$ by $i \cdot x = \sigma^i x$; note that $0 \cdot x = x$ and that $i \cdot (j \cdot x) = (i + j) \cdot x$, as required. This yields a partition of I_n into orbits. [The same orbits would obtain if we let $\langle \sigma \rangle \subseteq S_n$ act on I_n by evaluation.] If the orbit of x is trivial, then $\sigma x = 1 \cdot x = x$.

Let O be a nontrivial orbit. We show that the restriction of σ to O is a cycle. Take any $x \in O$ and let $S = S_{\mathbb{Z}}(x)$ be the stabilizer of x. Since O is finite, it follows from Proposition 3.1.2 that $S \neq 0$; hence $S = k\mathbb{Z}$ for some $k > 0$. Then S has index k in $\mathbb{Z}$; by Proposition 3.1.2, O has k elements.

We show that $O = \{\sigma^i x; 0 \leqslant i < k\}$. First, $x, \sigma x, \ldots, \sigma^{k-1} \in O$. Also $x, \sigma x, \ldots, \sigma^{k-1}$ are all distinct: if $\sigma^i x = \sigma^j x$, then $\sigma^{i-j} x = x$, $i - j \in S$, and k divides $i - j$, which if $0 \leqslant i, j < k$ implies $i = j$. Therefore the k elements of O are $x, \sigma x, \ldots, \sigma^{k-1} x$.

We have $\sigma(\sigma^{k-1} x) = \sigma^k x = k \cdot x = x$. Hence σ and the k-cycle

$$\gamma_O = (x \ \sigma x \ \sigma^2 x \ \ldots \ \sigma^{k-1} x)$$

have the same restriction to O. We see that O is the support of γ_O; hence cycles arising from different orbits are disjoint. Let $\mathscr{O}$ be the set of all nontrivial orbits. The cycles $(\gamma_O)_{O \in \mathscr{O}}$ are pairwise disjoint and we see that σ is the product $\sigma = \prod_{O \in \mathscr{O}} \gamma_O$ in which by Lemma 3.2.4 the terms can be written in any order.

Conversely, assume that σ is a product $\sigma = \prod_{i \in I} \gamma_i$ of pairwise disjoint cycles (which as above can be multiplied in any order). Let S_i be the support of γ_i. By the hypothesis, the sets S_i are nontrivial and pairwise disjoint. When $x \notin \bigcup_{i \in I} S_i$, we have $\gamma_i x = x$ for all i and $\sigma x = x$; hence the orbit of x (under the action of G as above) is trivial. When $x \in S_i$, we have $x \notin S_j$ and $\gamma_j x = x$ for all $j \neq i$, and $\sigma x = \gamma_i x$. Since γ_i is a cycle, the orbit of x is the support S_i of γ_i. This shows that the sets S_i coincide with the nontrivial orbits. Furthermore $\gamma_i = \gamma_O$ when $S_i = O$, since σ and γ_i have the same restriction to S_i. Hence $\prod_{O \in \mathscr{O}} \gamma_O$ and $\prod_{i \in I} \gamma_i$ have the same terms. □

The proof of Proposition 3.2.5 provides an efficient algorithm for writing permutations as products of pairwise disjoint cycles. For example, let

$$\sigma = \begin{pmatrix} 1\ 2\ 3\ 4\ 5\ 6\ 7\ 8\ 9 \\ 7\ 2\ 8\ 1\ 4\ 3\ 5\ 6\ 9 \end{pmatrix}.$$

Repeated applications of σ yield the orbits $\sigma 1 = 7$, $\sigma 7 = 5$, $\sigma 5 = 4$, $\sigma 4 = 1$; $\sigma 2 = 2$; $\sigma 3 = 8$, $\sigma 8 = 6$, $\sigma 6 = 3$; $\sigma 9 = 9$. Thus the orbits are $\{1,7,5,4\}$, $\{2\}$, $\{3,8,6\}$, and $\{9\}$. The restrictions of σ to the nontrivial orbits are the cycles $(1\ 7\ 5\ 4)$ and $(3\ 8\ 6)$; hence

$$\sigma = (1\ 7\ 5\ 4)(3\ 8\ 6).$$

3.2.e. Conjugacy Classes. The **cycle structure** of a permutation σ consists of the orders of the cycles in the decomposition of σ into a product of pairwise disjoint cycles; hence σ and $\tau \in S_n$ **have the same cycle structure** if their decompositions contain the same number of k-cycles for every $2 \leqslant k \leqslant n$.

It is readily verified that a k-cycle is even if k is odd and odd if k is even. Hence the sign of a permutation is quickly found from its cycle structure. For example, $\sigma = (1\ 7\ 5\ 4)(3\ 8\ 6)$ is odd, since $(1\ 7\ 5\ 4)$ is odd and $(3\ 8\ 6)$ is even.

Proposition 3.2.6. *Two permutations are conjugate in S_n if and only if they have the same cycle structure.*

Proof. Let $\alpha \in S_n$ and $\gamma = (a_1\ a_2\ \ldots\ a_k)$ be a k-cycle. We show that

$$\alpha\ (a_1\ a_2\ \ldots\ a_k)\ \alpha^{-1} = (\alpha a_1\ \alpha a_2\ \ldots\ \alpha a_n).$$

Indeed $(\alpha\gamma\alpha^{-1})(\alpha a_i) = \alpha\gamma a_i = \alpha a_{i+1}$ for all $i < k$, with $(\alpha\gamma\alpha^{-1})(\alpha a_k) = \alpha a_1$. Also $x \neq \alpha a_1, \ldots, \alpha a_k$ implies $\alpha^{-1} x \neq a_1, \ldots, a_k$, $\gamma\alpha^{-1} x = \alpha^{-1} x$, and $\alpha\gamma\alpha^{-1} x = x$. Thus $\alpha\gamma\alpha^{-1}$ is a k-cycle, with support $S(\alpha\gamma\alpha^{-1}) = \alpha S(\gamma)$.

If now $\sigma = \gamma_1\gamma_2\ldots\gamma_r$ is a product of cycles with pairwise disjoint supports $S(\gamma_1), \ldots, S(\gamma_r)$, then

$$\alpha\gamma\alpha^{-1} = (\alpha\gamma_1\alpha^{-1})(\alpha\gamma_2\alpha^{-1})\ldots(\alpha\gamma_r\alpha^{-1})$$

is a product of cycles of same orders, whose supports $\alpha S(\gamma_1), \ldots, \alpha S(\gamma_r)$ are pairwise disjoint (since α is injective). Thus σ and $\alpha\sigma\alpha^{-1}$ have the same cycle structure.

Conversely, assume that σ and τ have the same cycle structure. Write $\sigma = \gamma_1\gamma_2\ldots\gamma_r$ as the product of a sequence of pairwise disjoint cycles. Then $\tau = \delta_1\delta_2\ldots\delta_r$ is a product a pairwise disjoint cycles of the same orders; by permuting the terms if necessary, we can arrange that γ_i and δ_i have the same order for all i.

Let $\gamma_i = (a_1\, a_2\, \ldots\, a_k)$ and $\delta_i = (b_1\, b_2\, \ldots\, b_\ell)$. Since γ_i and δ_i have the same order, $k = \ell$, and there is a bijection $\alpha_i : S(\gamma_i) \longrightarrow S(\delta_i)$ such that $\alpha_i a_t = b_t$ for all t. Then $\alpha\gamma_i\alpha^{-1} = \delta_i$ holds for any permutation α which extends α_i. Since $S(\gamma_1),\ldots,S(\gamma_r)$ are pairwise disjoint and similarly for $S(\delta_1),\ldots,S(\delta_r)$, the bijections α_i can be patched together into a bijection α' of $S(\sigma) = \bigcup_{1 \leqslant i \leqslant r} S(\gamma_i)$ onto $S(\tau) = \bigcup_{1 \leqslant i \leqslant r} S(\delta_i)$ whose restriction to $S(\gamma_i)$ is α_i. Since $|S(\sigma)| = |S(\tau)|$, σ and τ have the same number of fixed points and there is a bijection α'' of the fixed points of σ onto the fixed points of τ. Patching α' and α'' together yields a permutation α whose restriction to $S(\sigma)$ is α'. Then the restriction of α to $S(\gamma_i)$ is α_i. Hence $\alpha\gamma_i\alpha^{-1} = \delta_i$ for all i; therefore $\alpha\sigma\alpha^{-1} = \tau$. $\square$

Proposition 3.2.6 determines the conjugacy classes of S_n. The equation

$$\alpha\,(a_1\, a_2\, \ldots\, a_k)\,\alpha^{-1} = (\alpha a_1\, \alpha a_2\, \ldots\, \alpha a_n)$$

in the proof can be used to determine centralizers.

Exercises

1. Show that S_n is generated by $(1\ 2),(2\ 3),\ldots,(n-1\ n)$.
2. Show that S_n is generated by the n-cycle $(1\ 2\ \ldots\ n)$ and transposition $(1\ 2)$.
3. Prove that a k-cycle is even if k is odd and odd if k is even.
4. Write

$$\sigma = \begin{pmatrix} 1\ 2\ 3\ 4\ 5\ 6\ 7\ 8 \\ 7\ 5\ 6\ 4\ 2\ 8\ 3\ 1 \end{pmatrix}$$

   as a product of pairwise disjoint cycles. Is σ even or odd?
5. Find the order of the permutation σ in the previous exercise.
6. Find the centralizer of the permutation σ in the previous exercise.
7. Write

$$\tau = \begin{pmatrix} 1\ 2\ 3\ 4\ 5\ 6\ 7\ 8 \\ 8\ 4\ 7\ 2\ 1\ 6\ 3\ 5 \end{pmatrix}$$

   as a product of pairwise disjoint cycles. Is τ even or odd?
8. Find the order of the permutation τ in the previous exercise.
9. Find the centralizer of the permutation τ in the previous exercise.
10. Let $\sigma = \gamma_1\gamma_2\ldots\gamma_r$ be a nonempty product of pairwise disjoint cycles of orders $k_1,k_2,\ldots,k_r$. Prove that the order of σ is the least common multiple of $k_1,k_2,\ldots,k_r$.
11. Prove that $Z(S_n) = 1$ if $n \geqslant 3$.
12. Show that there are $(k-1)!$ k-cycles with support $\{a_1,\ldots,a_k\}$.
13. Find the orders of all the conjugacy classes of S_4.
14. Find the orders of all the conjugacy classes of A_4. (*Warning:* Even permutations which are conjugate in S_4 are not necessarily conjugate in A_4.)
15. Show that A_4 has a normal subgroup that is isomorphic to V_4.

16. Show that $A_4 \cong \langle a,b \, ; \, a^3 = 1, \, b^2 = 1, \, aba = ba^2b \rangle$.

17. Show that $S_4 \cong \langle a,b \, ; \, a^4 = 1, \, b^2 = 1, \, (ba)^3 = 1 \rangle$.

18. Find the orders of all the conjugacy classes of S_5.

19. Find the orders of all the conjugacy classes of A_5.

20. Use the results of the previous exercise to prove that A_5 has no proper nontrivial normal subgroup.

21. How does Proposition 3.2.5 generalize to infinite symmetric groups?

3.3. THE SYLOW THEOREMS

When p is a prime number, a p-**subgroup** is a subgroup of order p^k for some k; a **Sylow p-subgroup** of a finite group G of order n is a subgroup S of order p^k such that $p^k \mid n$ and $p^{k+1} \nmid n$.

The Sylow Theorems are another basic tool of group theory. They state that Sylow p-subgroups exist, and give some of their properties. The Schur-Zassenhaus Theorem provides a first application.

3.3.a. Existence of Sylow Subgroups

Lemma 3.3.1. *Let A be a finite abelian group. If p divides $|A|$, then A contains an element of order p.*

Proof. By Theorem 2.2.6, A is a direct sum of cyclic q-groups (for various primes q). Since p divides $|A|$, one of these groups is a nontrivial p-group. Hence A contains an element a of order $p^k > 1$; then $a^{p^{k-1}} \in A$ has order p. $\square$

Lemma 3.3.1 does not require the full strength of Theorem 2.2.6 and can be proved by induction, as follows. If $|A| = p$, then A is cyclic and every $a \in A$, $a \neq 1$ has order p. Otherwise, take any $b \in A$, $b \neq 1$. If the order of b is a multiple kp of p, then $b^k \in A$ has order p. Otherwise, p does not divide the order of $B = \langle b \rangle \subseteq A$. Hence p divides $|A/B| = |A|/|B|$. By the induction hypothesis, A/B contains an element aB of order p. Now the order of aB in A/B divides the order of a in A (since $a^r = 1$ in A implies $a^r B = 1$ in A/B). Hence p divides the order of a, and as above A contains an element of order p.

The first Sylow Theorem is a partial converse to Lagrange's Theorem. When G is a group of order n, the order of every subgroup of G divides n, but it is not true in general that every divisor of n is the order of a subgroup of G. For example, A_4, which has order 12, contains no subgroup of order 6 (see the exercises). Theorem 3.3.2 states that a divisor of n is the order of a subgroup of G if that divisor is a power of a prime.

Theorem 3.3.2. *Let G be a finite group and p be a prime number. If p^k divides $|G|$, then G contains a subgroup of order p^k; in particular, G contains a Sylow p-subgroup.*

Proof. Let G be a finite group, $|G| = n$, and $p^k \mid n$. The result is trivial if $n = 1$, or more generally if $p^k = 1$. When $n > 1$ we proceed by induction on n. We may assume $p^k > 1$. Let $Z = Z(G)$ be the center of G. There are two cases.

Case 1. p divides $|Z|$. By Lemma 3.3.1, Z contains an element a of order p. Then Z contains all powers of a and the subgroup $N = \langle a \rangle$ of G is normal. Now $|G/N| = n/p < |G|$. By the induction hypothesis, G/N contains a subgroup S of order p^{k-1}. By Proposition 1.5.6, $S = T/N$, where $N \leqslant T \leqslant G$; T has order p^k, since $|T|/|N| = |S| = p^{k-1}$.

Case 2. p does not divide $|Z|$. However, p divides $|G|$. If p also divided the order of every nontrivial conjugacy class of G, then the class equation

$$|G| = |Z(G)| + \sum_{|C_x| > 1} |C_x|$$

would imply that p divides $|Z|$. Therefore G contains a nontrivial conjugacy class C_x whose order is not divisible by p. The centralizer $C(x)$ of x is not all of G (since $|C_x| > 1$), and p does not divide the index $[G : C(x)] = |C_x|$. Now a power of p which divides $|G| = |C(x)|[G : C(x)]$ must divide $|C(x)|$; thus $|G|$ and $|C(x)|$ are divisible by the same powers of p. By the induction hypothesis, $C(x) \subsetneqq G$ contains a subgroup of order p^k.   □

Corollary 3.3.3 (Cauchy's Theorem). *Let G be a finite group and p be a prime number. If p divides $|G|$, then G contains an element of order p.*

Corollary 3.3.4. *Let p be a prime number. The order of a finite group G is a power of p if and only if the order of every element of G is a power of p.*

Recall that a p-**group** is a group in which the order of every element is a power of p (where p is prime). By Corollary 3.3.4, a finite group G is a p-group if and only if the order of G is a power of p.

3.3.b. Properties of Sylow Subgroups

Lemma 3.3.5. *Let S be a Sylow p-subgroup of G. If $S \trianglelefteq G$, then S contains every p-subgroup of G; in particular, S is the only Sylow p-subgroup of G.*

Proof. Let T be a p-subgroup of G. If $S \trianglelefteq G$, then ST is a subgroup of G and $|ST| = |S||T|/|S \cap T|$ is a power of p. But $S \leqslant ST$; by the choice of S, $ST = S$, and $T \subseteq S$.   □

Theorem 3.3.6. *The number of Sylow p-subgroups of a finite group G divides $|G|$ and is congruent to 1 modulo p.*

Theorem 3.3.7. *All Sylow p-subgroups of a finite group are conjugate.*

Proof. The two theorems are proved together. Let G be a finite group and $\mathcal{S}$ be the set of all Sylow p-subgroups of G. When $a \in G$ and $S \in \mathcal{S}$, aSa^{-1} is a subgroup of G (since $\alpha_a : x \longmapsto axa^{-1}$ is an automorphism of G); in fact $aSa^{-1} \in \mathcal{S}$, since $|aSa^{-1}| = |S|$. Thus G acts on $\mathcal{S}$ by inner automorphisms.

Take any $S \in \mathcal{S}$. By the above, S acts on $\mathcal{S}$ by inner automorphisms. Since $aSa^{-1} = S$ for all $a \in S$, $\{S\}$ is an orbit. If $\{T\}$ is an orbit, then $aTa^{-1} = T$ for all $a \in S$, $S \subseteq N_G(T)$, and S is a Sylow p-subgroup of $N_G(T) \subseteq G$. But T is the only Sylow p-subgroup of $N_G(T)$, by Lemma 3.3.5; hence $T = S$. Thus $\{S\}$ is the only trivial orbit. Now the order of a nontrivial orbit is the index of a subgroup of S (a stabilizer) and is a power of p. Hence p divides the order of every nontrivial orbit. Since the orbits form a partition of $\mathcal{S}$, it follows that $|\mathcal{S}| \equiv 1 \pmod{p}$.

Now suppose that there are two conjugacy classes $\mathcal{C}'$ and $\mathcal{C}'' \subseteq \mathcal{S}$ of Sylow p-subgroups. Choose $S' \in \mathcal{C}'$ and $S'' \in \mathcal{C}''$. Under the action of S' on $\mathcal{S}$, $\mathcal{C}'$ and $\mathcal{C}''$ are unions of orbits. Since $\{S'\}$ is the only trivial orbit, we have $|\mathcal{C}'| \equiv 1$ and $|\mathcal{C}''| \equiv 0 \pmod{p}$. But $\mathcal{C}'$ and $\mathcal{C}''$ are also unions of orbits under the action of S''; since $\{S''\}$ is the only trivial orbit, we obtain $|\mathcal{C}'| \equiv 0$ and $|\mathcal{C}''| \equiv 1 \pmod{p}$. This contradiction proves Theorem 3.3.7. This in turn implies that $|\mathcal{S}|$ is the index of a normalizer and divides $|G|$. $\quad\square$

It follows from Theorem 3.3.7 that a Sylow p-subgroup is normal if and only if it is the only Sylow p-subgroup.

The **Sylow Theorems** are Theorems 3.3.3, 3.3.6, and 3.3.7; the name is sometimes applied to some of the following results.

Proposition 3.3.8. *Every p-subgroup of a finite group G is contained in a Sylow p-subgroup of G.*

Proof. Let H be a p-subgroup of G. As above, let H act by inner automorphisms on the set $\mathcal{S}$ of all Sylow p-subgroups of G. Since $|\mathcal{S}| \equiv 1 \pmod{p}$, there is at least one trivial orbit $\{S\}$. Then $aSa^{-1} = S$ for all $a \in H$ and $H \subseteq N_G(S)$. But S is a Sylow p-subgroup of $N_G(S) \subseteq G$; by Lemma 3.3.5, $H \subseteq S$. $\square$

Proposition 3.3.9. *Every subgroup which contains the normalizer of a Sylow p-subgroup is its own normalizer.*

Proof. Let S be a Sylow p-subgroup of G and H be a subgroup of G which contains $N_G(S)$. Let $a \in N_G(H)$ so that $aHa^{-1} = H$. Then S and aSa^{-1} are Sylow p-subgroups of H. By Theorem 3.3.7, $haSa^{-1}h^{-1} = S$ for some $h \in H$; thus $ha \in N_G(S) \subseteq H$, and $a \in H$. Hence $N_G(H) = H$. $\quad\square$

Proposition 3.3.10. *Let H be a p-subgroup of a finite group G which is not a Sylow p-subgroup of G. Then $H \underset{\neq}{<} N_G(H)$.*

Proof. Since H is not a Sylow p-subgroup of G, p divides $[G : H]$. If p does not divide $[G : N_G(H)]$, then $H \neq N_G(H)$. Now assume that p divides

$[G : N_G(H)]$. Let $\mathscr{C}$ be the conjugacy class of H. Then p divides $|\mathscr{C}| = [G : N_G(H)]$. Let H act on $\mathscr{C}$ by inner automorphisms. Then $\{H\}$ is a trivial orbit; since p divides $|\mathscr{C}|$, the number of trivial orbits is divisible by p. Therefore there are at least p trivial orbits. When $\{K\} \neq \{H\}$ is a trivial orbit, then $H \subseteq N_G(K)$ and $N_G(K) \neq K$; since H and K are conjugate, this implies $N_G(H) \neq H$.   □

Corollary 3.3.11. *In a p-group, every subgroup of G of index p is normal.*

3.3.c. The Schur-Zassenhaus Theorem. As an application of the Sylow theorems, we prove:

Theorem 3.3.12 (Schur-Zassenhaus). *Let the finite group G contain a normal subgroup N of index m such that $|N|$ and m are relatively prime. Then G contains a subgroup H of order m.*

Note that $N \cap H = 1$ (since $|N \cap H|$ must divide both $|N|$ and $|H|$); hence $G = NH$ and G is a semidirect product of N and H, by Proposition 2.3.2.

Proof. We saw (Theorem 2.4.8) that the result is true when N is abelian. In general, we proceed by induction on $|N|$. There is nothing to prove if $|N| = 1$. If $|N| > 1$, there is a prime p which divides $|N|$. Then p divides $|G|$, and the Sylow p-subgroups of G are nontrivial. Let S be a Sylow p-subgroup of G. Then the order of $SN/N \cong S/S \cap N$ is a power of p, and divides $|G/N| = m$; since $|N|$ and m are relatively prime, it follows that $SN/N = 1$ and $S \subseteq N$. Thus N contains every Sylow p-subgroup of G.

By Theorem 3.3.7, G has $[G : N_G(S)]$ Sylow p-subgroups, and N has $[N : N_N(S)]$ Sylow p-subgroups. Therefore $[G : N_G(S)] = [N : N_N(S)]$. This implies $|G|/|N_G(S)| = |N|/|N_N(S)|$ and

$$[N_G(S) : N_N(S)] = |N_G(S)|/|N_N(S)| = |G|/|N| = m.$$

Also $N_N(S) = N \cap N_G(S) \lhd N_G(S)$ (since $N \lhd G$) and $N_N(S)/S \lhd N_G(S)/S$, with $[N_G(S)/S : N_N(S)/S] = [N_G(S) : N_N(S)] = m$ and $|N_N(S)/S|$ relatively prime to m (since $|N_N(S)/S|$ divides $|N|$). Since S is nontrivial, the induction hypothesis yields a subgroup K/N of $N_N(S)/S$ of order m, where $S \subseteq K \subseteq N_N(S)$.

Let Z be the center of S, which is nontrivial by Proposition 3.1.7. Then $Z \lhd K$ and $S/Z \lhd K/Z$, with $[K/Z : S/Z] = [K : S] = m$ and $|S/Z|$ relatively prime to m (since $|S/Z|$ is a power of p). By the induction hypothesis, K/Z contains a subgroup L/Z of order m, where $Z \subseteq L \subseteq K$. We now have $Z \lhd L$, with $[L : Z] = m$ and $|Z|$ relatively prime to m (since $|Z|$ is a power of p). By the abelian case, $L \subseteq G$ contains a subgroup of order m.   □

It is known that in Theorem 3.3.12 any two subgroups of order m are conjugate; we will prove this in Section 3.7 when S is solvable (Theorem 3.7.15).

Exercises

1. Show that A_4 contains no subgroup of order 6.

2. Let $|G| = n$. Prove that 12 is the smallest value of n such that a divisor of n need not be the order of a subgroup of G. (Show that when $n \leqslant 11$, every divisor of n is the order of a subgroup of G.)

3. Let G be a finite abelian group. Prove that every divisor of $|G|$ is the order of a subgroup of G.

4. Find the Sylow subgroups of S_4.

5. Find the Sylow subgroups of S_5.

6. Show that every group G of order 18 has a normal subgroup $N \neq 1, G$.

7. Show that every group G of order 30 has a normal subgroup $N \neq 1, G$.

8. Show that every group G of order 56 has a normal subgroup $N \neq 1, G$.

9. Show that every group G of order 24 has a normal subgroup $N \neq 1, G$.

10. Let $p > q$ be prime numbers such that q divides $p - 1$. Prove that there are, up to isomorphism, exactly two groups of order pq.

3.4. SMALL GROUPS

Using the Sylow Theorems we now construct all groups of order up to 15.

3.4.a. Previous Results. Let p denote a prime number. We know that groups of order p are cyclic, and that groups of order p^2 are abelian (Proposition 3.1.9) and determined by Theorem 2.2.6.

Proposition 3.4.1. *Let p be a prime number. A group of order $2p$ is either cyclic or dihedral.*

Proof. By the Sylow Theorems (by Theorem 3.3.2), a group G of order $2p$ contains a subgroup P of order p and a subgroup Q of order 2. Since p and 2 are prime, $P \cap Q = 1$. Hence $|PQ| = |P||Q|/|P \cap Q| = |G|$ and $G = PQ$. Finally $P \triangleleft G$, since P has index 2. Thus G is a semidirect product of the cyclic groups P and Q, and is cyclic or dihedral by Corollary 2.3.5. □

Proposition 3.4.2. *Let $p > q$ be prime numbers. If q does not divide $p - 1$, then every group of order pq is cyclic.*

Proof. As in the previous proof, a group G of order pq contains a subgroup P of order p and a subgroup Q of order q (by Theorem 3.3.2). Since p and q are primes, $P \cap Q = 1$; hence $|PQ| = |P||Q|/|P \cap Q| = |G|$ and $G = PQ$. By Theorem 3.3.6, the number of subgroups of order p must divide pq and be $\equiv 1$ modulo p; since $p > q$, there is only one subgroup of order p, and $P \triangleleft G$. Similarly the number of subgroups of order q must divide pq and be $\equiv 1$ modulo q; it cannot be p, since q does not divide $p - 1$; hence there is only one subgroup of order q, and $Q \lhd G$. Thus G is a direct product of the cyclic groups P and Q, and is cyclic by Corollary 2.2.3. $\square$

We now know all groups of order 1, 2, 3, $4 = 2^2$, 5, $6 = 2 \cdot 3$, 7, $9 = 3^2$, $10 = 2 \cdot 5$, 11, 13, $14 = 2 \cdot 7$, and $15 = 5 \cdot 3$. This leaves groups of order 8 or 12.

3.4.b. Groups of Order 8. Up to isomorphism, there are three abelian groups of order 8: $\mathbb{Z}_2 \oplus \mathbb{Z}_2 \oplus \mathbb{Z}_2$, $\mathbb{Z}_2 \oplus \mathbb{Z}_4$, and $\mathbb{Z}_8$. Nonabelian examples include $D_4 \cong \langle a,b \, ; \, a^4 = b^2 = 1, \ bab^{-1} = a^{-1} \rangle$ and $Q \cong \langle a,b \, ; \, a^4 = 1, \ b^2 = a^2, \ bab^{-1} = a^{-1} \rangle$ (Example 2.6.5). Exercises showed that D_4 and Q are not isomorphic.

Proposition 3.4.3. *A group of order* 8 *is abelian or isomorphic to* D_4 *or* Q.

Proof. Let G be a nonabelian group of order 8. Not every element $a \neq 1$ of G has order 2 (otherwise, $xy = x^{-1}y^{-1} = (yx)^{-1} = yx$ for all $x,y \in G$). No $a \in G$ has order 8 (otherwise, G is cyclic). Hence G contains an element $a \neq 1$ of order 4. The subgroup $A = \langle a \rangle$ of G has index 2; hence $A \lhd G$. Let $b \notin A$. Then G is generated by a and b (since $A \subsetneqq \langle a,b \rangle \subseteq G$).

We have $b^2 \in A$ (since $bA \in G/A$ has order 2), with $b^2 \neq a, a^3$ (otherwise, b has order 8); hence $b^2 = 1$ or $b^2 = a^2$. Also $bab^{-1} \in A$ has order 4, with $bab^{-1} \neq a$ (otherwise, $ba = ab$ and $G = \langle a,b \rangle$ is abelian); hence $bab^{-1} = a^3 = a^{-1}$. This shows that the defining relations of D_4 (if $b^2 = 1$) or Q (if $b^2 = a^2$) hold in G. Therefore G is a homomorphic image of D_4 or Q; since $|G| = 8$, it follows that G is isomorphic to D_4 or to Q. (These presentations also follow from Hölder's Theorem, Proposition 2.6.7.) $\square$

3.4.c. Groups of Order 12. Up to isomorphism, there are two abelian groups of order 12: $\mathbb{Z}_2 \oplus \mathbb{Z}_2 \oplus \mathbb{Z}_3$ and $\mathbb{Z}_4 \oplus \mathbb{Z}_3 \cong \mathbb{Z}_{12}$. Nonabelian examples include A_4, $D_6 \cong \langle a,b \, ; \, a^6 = b^2 = 1, \ bab^{-1} = a^{-1} \rangle$, and $T \cong \langle a,b \, ; \, a^6 = 1, \ b^2 = a^3, \ bab^{-1} = a^{-1} \rangle$ (Example 2.6.6), which the exercises showed are not isomorphic to each other.

Proposition 3.4.4. *A group of order* 12 *is abelian or isomorphic to* A_4, D_6, *or* T.

Proof. Let G be a nonabelian group of order 12. By the Sylow Theorems (by Theorem 3.3.2), G contains a subgroup P of order 3. The group G acts on the four element set of left cosets of P by left multiplication: $g \cdot aP = g(aP)$. This provides a homomorphism $G \longrightarrow S_4$ with kernel $K \subseteq P$.

If $K = 1$, then G is isomorphic to a subgroup H of S_4 of order 12. We show that $H = A_4$. Let γ be a 3-cycle. Since H has index 2, two of $1, \gamma, \gamma^2$ must lie in the same left coset of H. Hence $\gamma \in H$, or $\gamma^2 \in H$ and $\gamma = \gamma^4 \in H$. Thus H contains all 3-cycles. By Proposition 3.2.3, $H = A_4$.

If $K \neq 1$, then $K = P$, and $P \triangleleft G$. By the Sylow Theorems (by Theorem 3.3.7), P is the only subgroup of order 3. Hence G has only two elements c and c^2 of order 3. Therefore c has only one or two conjugates and its centralizer $C_G(c)$ has order 6 or 12. By the Sylow Theorems (by Theorem 3.3.2), $C_G(c)$ contains an element d of order 2; since $c \in C_G(c)$, $cd = dc$.

Let $a = cd = dc$. Then a has order 6, $A = \langle a \rangle$ has index 2, and $A \triangleleft G$. Hence $G = \langle a, b \rangle$ for any $b \notin A$. Next $bab^{-1} \in A$ has order 6, with $bab^{-1} \neq a$ (otherwise, $ba = ab$ and $G = \langle a, b \rangle$ is abelian); hence $bab^{-1} = a^5 = a^{-1}$. Also $b^2 \in A$ and b^2 commutes with b; since $ba = a^5 b$ and $ba^2 = a^4 b$, it follows that $b^2 = 1$ or $b^2 = a^3$. Thus the defining relations of D_6 (if $b^2 = 1$) or T (if $b^2 = a^3$) hold in G. Therefore G is a homomorphic image of D_6 or T; since $|G| = 12$, it follows that G is isomorphic to D_6 or to T. (Again these presentations also follow from Hölder's Theorem, Proposition 2.6.7.) $\square$

3.4.d. Summary. Propositions 3.4.1 through 3.4.4 and previous results yield all groups of orders 1 to 15 (up to isomorphism):

> Order 1: 1;
> Order 2: $\mathbb{Z}_2$;
> Order 3: $\mathbb{Z}_3$;
> Order 4: $\mathbb{Z}_2 \oplus \mathbb{Z}_2 \cong V_4$, $\mathbb{Z}_3$;
> Order 5: $\mathbb{Z}_5$;
> Order 6: $\mathbb{Z}_2 \oplus \mathbb{Z}_3 \cong \mathbb{Z}_6$, $D_3 \cong S_3$;
> Order 7: $\mathbb{Z}_7$;
> Order 8: $\mathbb{Z}_2 \oplus \mathbb{Z}_2 \oplus \mathbb{Z}_2$, $\mathbb{Z}_2 \oplus \mathbb{Z}_4$, $\mathbb{Z}_8$, D_4, Q;
> Order 9: $\mathbb{Z}_3 \oplus \mathbb{Z}_3$, $\mathbb{Z}_9$;
> Order 10: $\mathbb{Z}_2 \oplus \mathbb{Z}_5 \cong \mathbb{Z}_{10}$, D_5;
> Order 11: $\mathbb{Z}_{11}$;
> Order 12: $\mathbb{Z}_2 \oplus \mathbb{Z}_2 \oplus \mathbb{Z}_3$, $\mathbb{Z}_4 \oplus \mathbb{Z}_3 \cong \mathbb{Z}_{12}$, A_4, D_6, T;
> Order 13: $\mathbb{Z}_{13}$;
> Order 14: $\mathbb{Z}_2 \oplus \mathbb{Z}_7 \cong \mathbb{Z}_{14}$, D_7;
> Order 15: $\mathbb{Z}_3 \oplus \mathbb{Z}_5 \cong \mathbb{Z}_{15}$.

Exercises

1. Find all groups of order 18. (Specify nonabelian groups by presentations.)

2. Find all groups of order 21.

3. Find all groups of order 28.

4. Find all groups of order 30.

3.5. COMPOSITION SERIES

Composition series analyze a group into simpler components, from which the group can be recovered by successive group extensions.

3.5.a. Normal Series. A **normal series** of a group G is a finite nonempty sequence

$$S : 1 = S_0 \lhd S_1 \lhd \cdots \lhd S_m = G$$

of subgroups of G (the **terms** of S) such that $S_0 = 1$, $S_m = G$, and $S_{i-1} \lhd S_i$ for all $i > 0$. The positive integer m is the **length** of the series; the quotient groups S_i/S_{i-1} $(i = 1, \ldots, m)$ are its **factors**. Normal series are often called **subnormal** series, which indicates that they may have terms which are not normal subgroups of G (see the exercises).

Two normal series $S : 1 = S_0 \lhd S_1 \lhd \cdots \lhd S_m = G$ and $T : 1 = T_0 \lhd T_1 \lhd \cdots \lhd T_n = G$ are **equivalent** in case they have the same factors, up to isomorphism and order of appearance: that is, in case $m = n$ and there exists a permutation σ of $1, 2, \ldots, m$ such that $T_i/T_{i-1} \cong S_{\sigma i}/S_{\sigma i-1}$ for all $i > 0$; equivalently, in case $m = n$ and there exists a permutation τ of $0, 1, \ldots, m-1$ such that $T_{i+1}/T_i \cong S_{\tau i+1}/S_{\tau i}$ for all $i < m$.

3.5.b. Refinement. A normal series T **refines** a normal series S (or is a **refinement** of S) in case every term of S is a term of T. In other words, the refinements of S are obtained by squeezing intermediate groups between the terms of S, to obtain a "finer" analysis of G.

Lemma 3.5.1 (Schreier's Theorem). *Any two normal series of a group G have equivalent refinements.*

Proof. Let $S : 1 = S_0 \lhd S_1 \lhd \cdots \lhd S_m = G$ and $T : 1 = T_0 \lhd T_1 \lhd \cdots \lhd T_n = G$ be two normal series of G. For all $0 \leqslant i < m$, $0 \leqslant j < n$, let

$$U_{ni+j} = S_i(S_{i+1} \cap T_j) \qquad \text{and} \qquad V_{mj+i} = T_j(T_{j+1} \cap S_i).$$

Also let $U_{mn} = V_{mn} = G$. Note that every integer $0 \leqslant k < mn$ can be written uniquely in the form $k = ni + j$ with $0 \leqslant i < m$, $0 \leqslant j < n$, and can be written uniquely in the form $k = mj' + i'$ with $0 \leqslant i' < m$, $0 \leqslant j' < n$.

Since $S_0 = T_0 = 1$, we have $U_{ni} = S_i$ and $V_{mj} = T_j$ for all $i < m$, $j < n$; in particular, $U_0 = V_0 = 1$. Also $U_{mn} = G = S_m$ and $V_{mn} = G = T_n$. In addition

$$S_i(S_{i+1} \cap T_n) = S_i S_{i+1} = S_{i+1} = U_{ni+n} \qquad \text{and}$$

$$T_j(T_{j+1} \cap S_m) = T_j T_{j+1} = T_{j+1} = V_{mj+m}$$

for all $i < m$, $j < n$. The next lemma (with $A = S_i$, $B = S_{i+1}$, $C = T_j$, $D = T_{j+1}$) implies that U_{ni+j} and V_{mj+i} are subgroups of G, $U_{ni+j} \lhd U_{ni+j+1}$, $V_{mj+i} \lhd V_{mj+i+1}$, and

$$U_{ni+j+1}/U_{ni+j} \cong V_{mj+i+1}/V_{mj+i}$$

for all $i < m$, $j < n$. Therefore U and V are equivalent normal series, which refine S and T; the permutation τ of $0, 1, \ldots, mn - 1$ which sends $mj + i$ to $ni + j$ satisfies $V_k/V_{k-1} \cong U_{\tau k+1}/U_{\tau k}$ for all $k < mn$. $\square$

Lemma 3.5.2 (Zassenhaus). *Assume* $A \trianglelefteq B \leqslant G$ *and* $C \trianglelefteq D \leqslant G$. *Then* $A(B \cap C)$, $A(B \cap D)$, $C(D \cap A)$, $C(D \cap B)$ *are subgroups of* G, $A(B \cap C) \trianglelefteq A(B \cap D)$, $C(D \cap A) \trianglelefteq C(D \cap B)$, *and*

$$(A(B \cap D))/(A(B \cap C)) \cong (C(D \cap B))/(C(D \cap A)).$$

Proof. First $A(B \cap C)$, $A(B \cap D)$ are subgroups of B, since $A \trianglelefteq B$. Also $B \cap C \trianglelefteq B \cap D$, since $C \trianglelefteq D$, and therefore $A(B \cap C) \trianglelefteq A(B \cap D)$. Let $S = B \cap D$, $T = A(B \cap C)$, and $U = A(B \cap D)$. Then $S \leqslant U$ and $T \trianglelefteq U$; also

$$ST = TS = A(B \cap C)(B \cap D) = A(B \cap D) = U.$$

We show that $S \cap T = (A \cap D)(B \cap C)$. If $s \in S \cap T$, then $s = at \in B \cap D$ for some $a \in A$, $t \in B \cap C$; this implies $t \in D$, $a = st^{-1} \in D$, and $s = at \in (A \cap D)$ $(B \cap C)$. Conversely, $A \cap D \subseteq S \cap T$, $B \cap C \subseteq S \cap T$, so that $(A \cap D)(B \cap C) \subseteq S \cap T$.

By the second isomorphism theorem, $S \cap T \trianglelefteq S$ and $ST/T \cong S/S \cap T$. Hence $(A \cap D)(B \cap C) \trianglelefteq B \cap D$ and

$$(A(B \cap D))/(A(B \cap C)) \cong (B \cap D)/(A \cap D)(B \cap C).$$

Exchanging A, B and C, D yields

$$(C(D \cap B))/(C(D \cap A)) \cong (B \cap D)/(A \cap D)(B \cap C).$$

This proves Lemma 3.5.2, and completes the proof of Lemma 3.5.1. $\square$

3.5.c. Composition Series. In a normal series $S : 1 = S_0 \trianglelefteq S_1 \trianglelefteq \cdots \trianglelefteq S_m = G$, each term S_i is a group extension of the previous term S_{i-1} by the factor S_i/S_{i-1}. Hence a normal series of a group G shows that G can be constructed by a sequence of group extensions from the factors of the series, whose structure is simpler. The factors of S are as simple as possible in structure when S has as many terms as possible.

A **composition series** is a normal series $S : 1 = S_0 \trianglelefteq S_1 \trianglelefteq \cdots \trianglelefteq S_m = G$ such that, for each $i > 0$, $S_{i-1} \underset{\neq}{\trianglelefteq} S_i$ and there is no $S_{i-1} \underset{\neq}{\trianglelefteq} N \underset{\neq}{\trianglelefteq} S_i$.

A group G is **simple** in case $G \neq 1$ and G contains no normal subgroup $1 \underset{\neq}{\trianglelefteq} N \underset{\neq}{\trianglelefteq} G$; equivalently, in case $1 \trianglelefteq G$ is the only normal series of G with nontrivial factors. Since the normal subgroups of S_i/S_{i-1} correspond to normal subgroups $S_{i-1} \lhd N \lhd S_i$, we have:

Proposition 3.5.3. *A normal series is a composition series if and only if all its factors are simple.*

Not every group has a composition series (see the exercises). However:

Proposition 3.5.4. *Every finite group has a composition series.*

Proof. When G has order n, a strictly increasing normal series $S : 1 = S_0 \lhdneq S_1 \lhdneq \cdots \lhdneq S_m = G$ has length $m \leqslant n$. Hence G has a strictly increasing normal series $S : 1 = S_0 \lhdneq S_1 \lhdneq \cdots \lhdneq S_m = G$ of maximal length; then S is a composition series. $\qquad\square$

3.5.d. The Jordan-Hölder Theorem. The Jordan-Hölder Theorem is a uniqueness property for composition series.

Theorem 3.5.5 (Jordan-Hölder). *Any two composition series of a group G are equivalent.*

Proof. Let $S : 1 = S_0 \lhd S_1 \lhd \cdots \lhd S_m = G$ be a composition series and T be a normal series which refines S. If $S_{i-1} = T_k \lhd T_{k+1} \lhd \cdots \lhd T_\ell = S_i$, then $\{T_k, T_{k+1}, \ldots, T_\ell\} = \{S_{i-1}, S_i\}$ (otherwise, there would exist $S_{i-1} \lhdneq T_j \lhdneq S_i$). This shows that the nontrivial factors of T are the factors of S.

Now let $S : 1 = S_0 \lhd S_1 \lhd \cdots \lhd S_m = G$ and $T : 1 = T_0 \lhd T_1 \lhd \cdots \lhd T_n = G$ be two composition series. By Schreier's Theorem (Lemma 3.5.1), S and T have equivalent refinements U and V. By the above, the nontrivial factors of U are the factors of S, and the nontrivial factors of V are the factors of T. Since U and V are equivalent, they have the same number of nontrivial factors; therefore S and T have the same number of factors, that is, $m = n$. Also there is a permutation σ such that $V_k/V_{k-1} \cong U_{\sigma k}/U_{\sigma k-1}$ for all $k > 0$. Then V_k/V_{k-1} is nontrivial if and only if $U_{\sigma k}/U_{\sigma k-1}$ is nontrivial; therefore σ induces a permutation τ of $1, 2, \ldots, n$ such that $T_i/T_{i-1} \cong S_{\sigma i}/S_{\sigma i-1}$ for all $i > 0$. $\qquad\square$

When a group G has a composition series, the **simple factors** of G are the factors of any composition series S of G; up to isomorphism, the simple factors of G do not depend on the choice of S.

Corollary 3.5.6. *When G has a composition series, every strictly ascending normal series of G can be refined to a composition series.*

The proof is an exercise.

3.5.e. Simplicity of A_n. All simple groups have now been determined. This **Classification Theorem** is one of the great achievements of twentieth-century mathematics. Its proof is enormous and was completed only recently (ca. 1985). In this book we only present the three most accessible families of simple groups.

By Lagrange's Theorem, a cyclic group C of prime order has no subgroup $1 \underset{\neq}{\leqslant} H \underset{\neq}{\leqslant} C$ and is therefore simple. Using the structure of finite abelian groups (Theorem 2.2.6), the reader will prove:

Proposition 3.5.7. *An abelian group is simple if and only if it is cyclic of prime order.*

Nonabelian simple groups are found by constructing composition series of sufficiently large finite groups. Dihedral groups are unsuitable (see the exercises), but symmetric groups yield a normal series $1 \trianglelefteq A_n \trianglelefteq S_n$ which we show is a composition series if $n \geqslant 5$. General linear groups are considered in the next section, and provide additional simple groups.

Proposition 3.5.8. A_n *is simple for all* $n \geqslant 5$.

Proof. First let $n = 5$. The reader will verify as an exercise that A_5 has five conjugacy classes, which consist of:

  1 identity element,
 15 products of two disjoint transpositions,
 20 3-cycles,
 12 5-cycles, and
 12 more 5-cycles.

A normal subgroup N of A_5 is the union of $\{1\}$ and other conjugacy classes. The orders of such unions are 1, 16, 21, 13, 25, 28, and over 30, none of which is a proper divisor of $|A_5| = 60$. Therefore A_5 has no normal subgroup $N \neq 1, A_5$.

For $n > 5$ we proceed by induction on n. Let $N \neq 1$, $N \trianglelefteq A_n$.

We show that N is **transitive**: that is, for each $i, j \leqslant n$, there exists $\sigma \in N$ such that $\sigma i = j$ (so that there is only one orbit under the action of N on $I_n = \{1, 2, \ldots, n\}$). Since $N \neq 1$, we have $\sigma a \neq a$ for some $\sigma \in N$ and $a \in I_n$. For any $i \neq a$ we can rig an even permutation α such that $\alpha a = a$ and $\alpha \sigma a = i$; then $i = \alpha \sigma \alpha^{-1} a$, with $\alpha \sigma \alpha^{-1} \in N$. For any $i, j \in I_n$ we now have $i = \mu a$ and $j = \nu a$ for some $\mu, \nu \in N$, whence $\nu \mu^{-1} i = j$ with $\nu \mu^{-1} \in N$.

We show that N contains a permutation $\sigma \neq 1$ with a fixed point ($\sigma a = a$ for some $a \in I_n$). Since $N \neq 1$, there exists $\nu \in N$ such that $\nu i = j \neq i$. Then $\nu j \neq j$. Let $k \neq i, j, \nu j$. If $\nu k = k$, we are done. Otherwise, $j, \nu j, k, \nu k$ are all distinct. Since $n \geqslant 6$, we can concoct an even permutation α such that $\alpha j = i$, $\alpha(\nu j) = j$, $\alpha k = k$, and $\alpha(\nu k) \neq i, j, k, \nu k$. Then $\mu = \alpha \nu \alpha^{-1} \in N$, $\mu i = \alpha \nu j = j = \nu i$, and $\mu k = \alpha \nu k \neq \nu k$. Then $\sigma = \mu^{-1} \nu \in N$ satisfies $\sigma i = i$ and $\sigma \neq 1$.

Let a be a fixed point of some $\sigma \neq 1$ in N ($\sigma a = a$). Let $B = \{\alpha \in A_n;$ $\alpha a = a\}$. Then $N \cap B \neq 1$. Also $B \cong A_{n-1}$ is simple by the induction hypothesis. Therefore $B \subseteq N$. But N is transitive; for each $\alpha \in A_n$ there exists $\nu \in N$ such that $\nu a = \alpha a$. Then $\nu^{-1}\alpha \in B \subseteq N$, which implies $\alpha \in N$. Thus $N = A_n$.

$\square$

3.5.f. Groups with Operators. The Jordan-Hölder Theorem extends to groups with operators.

Let S be a set [of operators]. An S-**normal series** of an S-group G is a finite nonempty sequence

$$T : 1 = T_0 \trianglelefteq_S T_1 \trianglelefteq_S \cdots \trianglelefteq_S T_m = G$$

of S-subgroups of G such that $T_0 = 1$, $T_m = G$, and $T_{i-1} \trianglelefteq_S T_i$ for all $i > 0$. The positive integer m is the length of the series; the quotient S-groups T_i/T_{i-1} ($i = 1, \ldots, s$) are its factors. Two S-normal series are equivalent in case they have the same factors, up to S-isomorphism and order of appearance.

An S-**composition series** of an S-group is an S-normal series $T : 1 = T_0 \trianglelefteq_S T_1$ $\trianglelefteq_S \cdots \trianglelefteq_S T_m = G$ such that, for each $i > 0$, $T_{i-1} \neq T_i$ and there is no $T_{i-1} \trianglelefteq_S N$ $\trianglelefteq_S T_i$ with $N \neq T_{i-1}, T_i$.

An S-group G is S-**simple** in case $G \neq 1$ and G has no normal S-subgroup $N \neq 1, G$. An S-normal series is an S-composition series if and only if all its factors are S-simple.

We saw in Section 1.7 that the Isomorphism Theorems extend to S-groups. So do their consequences, including the Zassenhaus Lemma. The proofs of the following results carry to S-groups with little or no modification.

Lemma 3.5.9 (Schreier's Theorem). *Any two S-normal series of an S-group have equivalent refinements.*

Theorem 3.5.10 (Jordan-Hölder). *Any two S-composition series of an S-group are equivalent.*

When $S = \varnothing$, Theorem 3.5.10 reduces to the Jordan-Hölder Theorem for groups (Theorem 3.5.5). But Theorem 3.5.10 has other consequences.

Let V be an abelian group on which a field K acts so that V is a vector space. Then a K-subgroup is a vector subspace; every K-subgroup is normal; V is K-simple if and only if V has dimension 1. Hence V has a K-composition series if and only if V has a finite basis; then Theorem 3.5.10 implies that all bases of V have the same number of elements. The details of this argument are an exercise.

Theorem 3.5.10 also yields new properties of groups that are similar to Theorem 3.5.5 but not consequences of it.

Any group G acts on itself by inner automorphisms ($^s x = sxs^{-1}$ for all $s, x \in G$). Under this action a G-subgroup is a normal subgroup of G, and a

G-normal series is a finite sequence

$$T : 1 = T_0 \lhd T_1 \lhd \cdots \lhd T_m = G$$

of normal subgroups of *G* (equivalently, a normal series in which every term is a normal subgroup of *G*). A **principal series** of *G* is a *G*-composition series; equivalently, a principal series is a finite maximal chain of normal subgroups of *G*.

Every finite group has a principal series. In general, a group *G* has a principal series if and only if it satisfies the ascending chain condition and the descending chain condition on normal subgroups (see the exercises); equivalently, if and only if every chain of normal subgroups of *G* is finite (Proposition B.2.3). Groups with this property are also said to have **finite length**.

Corollary 3.5.11. *Any two principal series of a group are equivalent.*

The **principal factors** of a group of finite length are the factors of its principal series. Principal factors are not necessarily simple (see the exercises).

Any group *G* also has a group *A* = Aut(*G*) of automorphisms which acts on *G* by evaluation ($^\alpha x = \alpha(x)$ for all $\alpha \in A$ and $x \in G$). An *A*-subgroup of *G* is a characteristic subgroup. A **characteristic series** is an *A*-composition series; equivalently, a characteristic series is a finite maximal chain of characteristic subgroups. Every finite group has a characteristic series.

Corollary 3.5.12. *Any two characteristic series of a group are equivalent.*

When a group has a characteristic series, its **characteristic factors** are the factors of its characteristic series.

Exercises

1. Show that A_4 has a normal series in which not every term is a normal subgroup of A_4.
2. Prove that $\mathbb{Z}$ does not have a composition series.
3. Prove the following. Let $N \lhd G$. If *N* has a composition series and *G/N* has a composition series, then *G* has a composition series.
4. Prove the following. Let $N \lhd G$. If *G* has a composition series, then *N* and *G/N* have composition series. (Show that *G* has a composition series in which *N* is a term.)
5. Find all composition series of D_4.
6. Find all composition series of A_4.
7. Find all composition series of *Q*.
8. Find all composition series of D_5.
9. Prove that an abelian group has a composition series if and only if it is finite.
10. Prove Corollary 3.5.6.

11. Prove Proposition 3.5.7.

12. Let A be a finite abelian group of order n. Show that the simple factors of A are determined by n.

13. Show that the simple factors of D_n are all abelian.

14. Prove that $1 \lhd A_n \lhd S_n$ is the only composition series of S_n if $n \geqslant 5$.

15. Show that a group of order p^n has a composition series of length n.

16. Let G be a group of order n and m be the length of its composition series. Show that $m \leqslant \log_2 n$. Show that there are arbitrarily large values of n such that $m = \log_2 n$.

The following exercise is more like a small research project.

17. Show that there is no simple nonabelian group of order $n < 60$.

The following exercises concern groups with operators.

18. Let V be a vector space over a field K, viewed as an abelian K-group. Show that V has a K-composition series if and only if V has a finite basis. Use Theorem 3.5.10 to prove that any two finite bases of V have the same number of elements.

19. Show that a group has a principal series if and only if it satisfies the a.c.c. and the d.c.c. on normal subgroups.

20. Show that A_4 has a principal series which is not a composition series.

21. Find the principal factors of S_n.

22. Show that a group has a characteristic series if and only if it satisfies the a.c.c. and the d.c.c. on characteristic subgroups.

23. Show that the characteristic factors of S_n coincide with its principal factors.

3.6. THE GENERAL LINEAR GROUP

Let V be a vector space over a field K with finite dimension $n \geqslant 2$. Recall that the general linear group $GL(V)$ is the group of all invertible linear transformations of V into V; $GL(V)$ is isomorphic to the group $GL(n,K)$ of all invertible $n \times n$ matrices with coefficients in K.

In this section we determine the simple factors of $GL(V)$ in case a composition series exists (e.g., if K is finite). This yields new simple groups.

3.6.a. The Special Linear Group. The **special linear group** of V is the subgroup $SL(V)$ of all $T \in GL(V)$ whose determinant is 1; it is isomorphic to the group $SL(n,K)$ of all $n \times n$ matrices with coefficients in K whose determinant is 1. Since the determinant is a homomorphism of $GL(V)$ [or of $GL(n,K)$] onto the multiplicative group K^* of all nonzero elements of K, we have:

Proposition 3.6.1. $SL(V) \lhd GL(V)$, and $GL(V)/SL(V) \cong K^*$.

Proposition 3.6.1 shows that a simple factor of $GL(V)$ is either a simple factor of $SL(V)$ or a simple factor of K^*. The latter are abelian. We now turn to the simple factors of $SL(V)$.

3.6.b. Elementary Transformations. An $n \times n$ matrix E with coefficients in K is **elementary** in case $A = I + \alpha E_{ij}$, where I is the identity matrix, $\alpha \in K$, $\alpha \neq 0$, $i \neq j$, and E_{ij} is the matrix in which the (i,j) entry is 1 and all other entries are 0; thus E has ones on the diagonal and zeroes elsewhere, except for the (i,j) entry which equals α. A linear transformation is **elementary** in case its matrix in some basis is elementary. All elementary matrices and transformations have determinant 1. If the linear transformation A is elementary, with matrix $I + \alpha E_{ij}$ in the basis $b_1, \dots, b_n$, then A adds a multiple of b_i to each $x \in V$:

$$Ax = A(x_1 b_1 + \cdots + x_n b_n) = x_1 b_1 + \cdots + x_n b_n + \alpha x_j b_i = x + \alpha x_j b_i.$$

$$(3.7)$$

When $A : V \longrightarrow V$ is a linear transformation, we denote by

$$F(A) = \{ x \in V ; \, Ax = x \} = \mathrm{Ker}\,(A - 1)$$

the subspace of all fixed points of A.

Lemma 3.6.2. *The following conditions on a linear transformation A are equivalent: (i) A is elementary; (ii) $A \in SL(V)$ and $\dim F(A) = n - 1$; (iii) $A \in SL(V)$ and $\mathrm{Im}\,(A - 1)$ has dimension 1.*

Proof. Let A have matrix $E = I + \alpha E_{ij}$ in some basis $b_1, \dots, b_n$ of V. Then $\det A = 1$. Equation (3.7) shows that $Ax = x$ if and only if $x_j = 0$; hence $\dim F(A) = n - 1$. [Equation (3.7) also shows that $\{b_1\}$ is a basis of $\mathrm{Im}\,(A - 1)$.]

Conversely, assume that $A \in SL(V)$ and $\dim F(A) = \dim \mathrm{Ker}\,(A - 1) = n - 1$ [equivalently, $A \in SL(V)$ and $\dim \mathrm{Im}\,(A - 1) = 1$]. The subspace $F(A)$ has a basis $b_2, \dots, b_n$; V has a basis $b_1, b_2, \dots, b_n$ with $b_1 \notin F(A)$. Then $Ab_1 = \delta b_1 + v$ for some $\delta \in K$ and $v \in F(A)$. The matrix of A is triangular, with $\delta, 1, \dots, 1$ on the diagonal; hence $\delta = \det A = 1$, $Ab_1 = b_1 + v$, with $v \neq 0$, since $b_1 \notin F(A)$, and $F(A)$ has a basis in which $b_2 = v$. Then $Ab_1 = b_1 + b_2$ and $Ab_i = b_i$ for all $i \geqslant 2$; the matrix of A in the basis $b_1, \dots, b_n$ is $I + E_{12}$. $\square$

Proposition 3.6.3. *$SL(V)$ is generated by all elementary transformations.*

Proof. We show that $SL(n, K)$ is generated by all elementary matrices. When $E = I + \alpha E_{ij}$ is elementary, the product of E and any suitable matrix M is obtained by adding α times the j-th row of M to the i-th row of M. Gauss-Jordan reduction implies that every matrix M with determinant 1 can be transformed into the identity matrix by a sequence of such operations. Therefore M^{-1} is a product of elementary matrices. $\square$

Lemma 3.6.4. *The elementary transformations constitute a conjugacy class of $GL(V)$. If $n = \dim V \geqslant 3$, then any two elementary transformations are conjugate in $SL(V)$.*

This does not hold when $\dim V = 2$ (see the exercises).

Proof. Let A be elementary and $C \in GL(V)$. The matrix M of A in some basis of V is elementary. There is a basis of V in which the matrix of CAC^{-1} is also M; hence CAC^{-1} is elementary.

Let A and B be elementary. The proof of Lemma 3.6.2 shows that there exists a basis $b_1, \ldots, b_n$ of V in which the matrix of A is $I + E_{12}$, and a basis $c_1, \ldots, c_n$ of V in which the matrix of B is $I + E_{12}$. There exists an invertible linear transformation C such that $Cb_i = c_i$ for all i. Then $CAC^{-1} = B$. Thus A and B are conjugate in $GL(V)$. Let $\delta = \det C \neq 0$. Since $n \geqslant 3$, we have $Bc_n = c_n$ and $B(\delta^{-1}c_n) = \delta^{-1}c_n$; hence the matrix of B in the basis $c_1' = c_1, \ldots, c_{n-1}' = c_{n-1}$, $c_n' = \delta^{-1}c_n$ is still $I + E_{12}$. Let C' be the invertible linear transformation such that $C'b_i = c_i'$ for all i. Then $C'AC'^{-1} = B$ and $\det C' = \delta^{-1}\det C = 1$. $\square$

3.6.c. Centers

Proposition 3.6.5. *Let $C \in GL(V)$. The following are equivalent:*

(1) *C commutes with all elementary transformations;*

(2) *every subspace of V of dimension 1 is invariant under C;*

(3) *$C = \lambda 1_V$ for some $\lambda \in K^*$;*

(4) *$C \in Z(GL(V))$.*

Proof. (1) implies (2). When A has matrix $I + E_{ij}$ in a basis $b_1, \ldots, b_n$ of V, then $\mathrm{Im}\,(A - 1)$ is the subspace of V generated by b_i. Therefore every one-dimensional subspace S of V is the image $S = \mathrm{Im}\,(A - 1)$ of some elementary transformation A. Then (1) implies

$$CS = \mathrm{Im}\,C(A - 1) = \mathrm{Im}\,C(A - 1)C^{-1} = \mathrm{Im}\,(A - 1) = S.$$

(2) implies (3). Let $b_1, \ldots, b_n$ be a basis of V. By (2), $Cb_i = \lambda_i b_i$ for some $\lambda_i \in K$. Also $C(b_i + b_j) = \mu(b_i + b_j)$ for some $\mu \in K$, which implies that $\lambda_i = \mu = \lambda_j$. Hence $\lambda_1 = \lambda_2 = \cdots = \lambda_n$. Then $\lambda = \lambda_1 = \cdots = \lambda_n$ satisfies $Cb_i = \lambda b_i$ for all i and $Cx = \lambda x$ for all $x \in V$. Also $\lambda \neq 0$, since $\det C = \lambda^n \neq 0$.

(3) implies (4). In fact (3) implies $ACx = A(\lambda x) = \lambda Ax = CAx$ for every linear transformation A. That (4) implies (1) is clear. $\square$

Corollary 3.6.6. $Z(GL(V)) \cong K^*$; $Z(SL(V))$ *is isomorphic to the multiplicative group of all n-th roots of unity of K.*

The multiplicative group of all n-th roots of unity of K is cyclic by Proposition 2.2.4; hence Corollary 3.6.6 provides more simple factors of $SL(V)$ and $GL(V)$ that are cyclic of prime orders and depend only on K (in fact, depend only on the number of n-th roots of unity in K). The remaining simple factors are those of $SL(V)/Z(SL(V))$.

3.6.d. The Projective Special Linear Group. The **projective special linear group** of V is $PSL(V) = SL(V)/Z(SL(V))$. (This group originated in projective geometry.)

Theorem 3.6.7. $PSL(V)$ *is simple when* $\dim V \geqslant 3$.

Proof. We show that a normal subgroup $N \supsetneqq Z(SL(V))$ of $SL(V)$ must contain an elementary transformation. By Lemma 3.6.4, this implies $N = SL(V)$.

Since $N \supsetneqq Z(SL(V))$, there is by Proposition 3.6.5 some $A \in N$ which does not commute with some elementary transformation D. Then $B = ADA^{-1}D^{-1} \in N$. We have $F(ADA^{-1}) \cap F(D) \subseteq F(B)$. By Lemma 3.6.4, ADA^{-1} is elementary; hence $\dim F(ADA^{-1}) = \dim F(D) = n - 1$ and $\dim F(B) \geqslant n - 2$. Also $\dim F(B) \neq n$, since $B \neq 1$. Hence $\dim F(B)$ equals $n - 1$ or $n - 2$.

If $\dim F(B) = n - 1$, then $B \in N$ is elementary.

Assume that $\dim F(B) = \dim \mathrm{Ker}\,(B - 1) = n - 2$. Then $\mathrm{Im}\,(B - 1)$ has dimension 2 and B is not elementary. Since $\dim V \geqslant 3$, $\mathrm{Im}\,(B - 1)$ is contained in a subspace U of V of dimension $n - 1$. Then $BU \subseteq (B - 1)U + U \subseteq U$. For any $u \in U$, $u \neq 0$ and $v \in V \backslash U$, there is an elementary transformation E such that $F(E) = U$ and $Ev = v + u$. Then $\mathrm{Im}\,(E - 1)$ is the subspace generated by u. Also $BEB^{-1}x = x$ for all $x \in U$ and

$$BEB^{-1}(Bv) = Bv + Bu.$$

Then $C = BEB^{-1}E^{-1} \in N$; also $U \subseteq F(C)$, so that C is either the identity or elementary. We show that $BEB^{-1} \neq E$ for some u and v; then $C \neq 1$ and N contains the elementary transformation C.

Assume that $F(B) \subsetneqq U$. Let $v \in F(B) \backslash U$. Since $\dim F(B) < \dim U$ there also exists $u \in U \backslash F(B)$. Then $Bu \neq u$, $BEB^{-1}v = v + Bu \neq Ev$, and $BEB^{-1} \neq E$.

Assume that $F(B) \subseteq U$. Let B' be the restriction of B to U. Then $F(B') = F(B) \neq 0, U$. By Proposition 3.6.5, $B' \notin Z(GL(U))$, and there exists $u \in U$ such that Bu and u generate different subspaces. Then BEB^{-1} is elementary by Lemma 3.6.4, and $BEB^{-1}(Bv) = Bv + Bu$ shows that $\mathrm{Im}\,(BEB^{-1} - 1)$ is the subspace generated by Bu. Therefore $BEB^{-1} \neq E$. $\square$

We leave the following result as a substantial exercise:

Theorem 3.6.8. *$PSL(V)$ is simple when* $\dim V = 2$ *and K has at least four elements.*

On the other hand, $PSL(V)$ is not simple when $\dim V = 2$ and K has two or three elements (see below).

3.6.e. Orders. We now find $|GL(n,K)|$, $|SL(n,K)|$, and $|PSL(n,K)|$.

Proposition 3.6.9. *When* $|K| = q$,

$$|GL(n,K)| = (q^n - 1)(q^n - q)\ldots(q^n - q^{n-1})$$

and $|SL(n,K)| = |GL(n,K)|/(q-1)$.

Proof. $|SL(n,K)| = |GL(n,K)|/(q-1)$ follows from Proposition 3.6.1.

Let $b_1,\ldots,b_n$ be a basis of V. When $A \in GL(V)$, then $Ab_1,\ldots,Ab_n$ is a basis of V; in fact this defines a one-to-one correspondence between the bases of V and the elements of $GL(V)$. Hence $|GL(V)|$ is the number of bases of V. Now $|V| = q^n$. In a basis $b_1,\ldots,b_n$, b_1 is any nonzero element of V and there are $q^n - 1$ possible choices for b_1. If $b_1,\ldots,b_i$ have been chosen, then b_{i+1} is any element of V which is not a linear combination of $b_1,\ldots,b_i$, and there are $q^n - q^i$ possible choices for b_{i+1}. $\qquad\square$

By Corollary 3.6.6, $|PSL(n,K)| = |SL(n,K)|/r$, where r is the number of n-th roots of unity in K.

We will show in Chapter 7 that there is one finite field K of order q for each power $q = p^k$ of a prime p. The groups $GL(n,K)$, $SL(n,K)$, $PSL(n,K)$ are then denoted by $GL(n,q)$, $SL(n,q)$, $PSL(n,q)$. Theorems 3.6.7 and 3.6.8 provide finite simple groups $PSL(n,q)$ for every $q = p^k$ and $n \geqslant 3$ ($n \geqslant 2$ if $q \geqslant 5$).

Let $q = 2$. Then $K = \mathbb{Z}_2$ and K contains one square root of unity. We have $|PSL(2,2)| = |SL(2,2)| = |GL(2,2)| = (4-1)(4-2) = 6$. Hence $PSL(2,2)$ is not simple. On the other hand, $|PSL(3,2)| = |SL(3,2)| = |GL(3,2)| = (8-1)(8-2)(8-4) = 168$; $PSL(3,2)$ is simple by Theorem 3.6.7, but it is not an alternating group.

If $q = 3$, then $K = \mathbb{Z}_3$ and K contains two square roots of unity. We have $|GL(2,3)| = (9-1)(9-3) = 48$, $|SL(2,3)| = 48/2 = 24$, and $|PSL(2,3)| = 24/2 = 12$. Hence $PSL(2,3)$ is not simple.

If $q = 4$, then K has characteristic 2 and contains only one square root of unity. We have $|GL(2,4)| = (16-1)(16-4) = 180$ and $|PSL(2,4)| = |SL(2,4)| = 180/3 = 60$. By Theorem 3.6.8, $PSL(2,4)$ is simple; it can be shown that $PSL(2,4) \cong A_5$.

Exercises

1. Show that in $SL(2,K)$, every elementary matrix is conjugate to some $I + \alpha E_{12}$.
2. Prove that the elementary 2×2 matrices $I + \alpha E_{12}$ and $I + \beta E_{12}$ are conjugate in $SL(2,K)$ if and only if $\alpha = \beta\gamma^2$ for some $\gamma \in K^*$.
3. Amend the proof of Theorem 3.6.7 to show that $Z(GL(V)) \subsetneq N \lhd GL(V)$ implies $SL(V) \subseteq N$.
4. Prove Theorem 3.6.8.
5. Show that $PSL(2,2) \cong D_3$.
6. To which known group of order 12 is $PSL(2,3)$ isomorphic?
7. Draw a table showing $|PSL(n,q)|$ for all $2 \leqslant n, q \leqslant 5$.
8. Let $|K| = q = p^k$ and S be the set of all upper triangular matrices with 1's on the main diagonal. Show that S is a Sylow p-subgroup of $SL(n,K)$.

3.7. SOLVABLE GROUPS

Solvable groups are a large class of groups that are close enough to abelian groups to inherit a number of their properties, such as stronger versions of the Sylow Theorems.

3.7.a. Commutators. In a group G, the **commutator** of $a, b \in G$ is the element $aba^{-1}b^{-1}$ of G (sometimes denoted by $[a,b]$); we see that $ab = ba$ if and only if $aba^{-1}b^{-1} = 1$.

The **commutator subgroup** or **derived group** of G is the subgroup G' of G (sometimes denoted by $[G,G]$) generated by all the commutators. We see that G is abelian if and only if $G' = 1$.

Proposition 3.7.1. G' is a **fully invariant** subgroup of G: $\eta(G') \subseteq G'$ for every endomorphism η of G.

Proof. If $c = aba^{-1}b^{-1}$ is a commutator, then $c^{-1} = bab^{-1}a^{-1}$ is a commutator and $\eta(c) = \eta(a)\eta(b)\eta(a)^{-1}\eta(b)^{-1}$ is a commutator. Hence every element d of G' is a product $d = c_1 c_2 \ldots c_k$ of commutators, and $\eta(d) = \eta(c_1)\eta(c_2)\ldots\eta(c_k)$ $\in G'$. $\qquad\qquad\square$

Proposition 3.7.2. G' is the smallest normal subgroup N of G such that G/N is abelian.

Proof. $G' \lhd G$ by Proposition 3.7.1. For all $a, b \in G$ we have $aba^{-1}b^{-1} \in G'$ and $ab \in G'ba = baG'$; hence G/G' is abelian. If, conversely, $N \lhd G$ and G/N is abelian, then $Naba^{-1}b^{-1} = NaNbNa^{-1}Nb^{-1} = N$ and $aba^{-1}b^{-1} \in N$ for all $a, b \in G$; therefore $G' \subseteq N$. $\qquad\qquad\square$

3.7.b. Solvable Groups. The **commutator series** of G is the infinite sequence

$$G \rhd G' \rhd G'' \rhd \cdots \rhd G^{(r)} \rhd \cdots,$$

where $G^{(r)}$ is defined by induction by: $G^{(0)} = G$ and $G^{(r+1)} = (G^{(r)})'$. A group G is **solvable** in case $G^{(r)} = 1$ for some r; then the commutator series of G can be truncated at r and viewed as a normal series. Solvable groups are sometimes called **metabelian**.

Proposition 3.7.3. *A group G is solvable if and only if it has a normal series whose factors are abelian.*

Proof. If G is solvable, then its commutator series is a normal series whose factors are abelian. Conversely, let $S : 1 = S_0 \lhd S_1 \lhd \cdots \lhd S_m = G$ be a normal series whose factors are abelian. Then $G^{(0)} \subseteq S_m$. If $r < m$, then S_{m-r}/S_{m-r-1} is abelian, $S'_{m-r} \subseteq S_{m-r-1}$, and $G^{(r)} \subseteq S_{m-r}$ implies $G^{(r+1)} \subseteq S'_{m-r} \subseteq S_{m-r-1}$. Therefore $G^{(r)} \subseteq S_{m-r}$ for all $r \leqslant m$; in particular, $G^{(m)} = 1$ and G is solvable. $\square$

Proposition 3.7.4. *A group G with a composition series is solvable if and only if all its simple factors are abelian.*

This result is an exercise. So are the next three propositions, which state that the class of solvable groups is closed under subgroups, homomorphic images, and group extensions.

Proposition 3.7.5. *Every subgroup of a solvable group is solvable.*

Proposition 3.7.6. *If G is solvable and $N \lhd G$, then G/N is solvable.*

Proposition 3.7.7. *Let $N \lhd G$. If N and G/N are solvable, then G is solvable.*

3.7.c. Examples. Abelian groups are solvable, and there are many non-abelian examples. All groups of order $n < 60$ are solvable; dihedral groups are solvable by Proposition 3.7.3; on the other hand, S_n is solvable if and only if $n \leqslant 4$ (if $n \geqslant 5$, then A_n is a nonabelian prime factor). These properties are exercises. The widest class of examples is provided by one of the major results of group theory:

Theorem 3.7.8 (Feit and Thompson). *Every group of odd order is solvable.*

Equivalently, every simple group of odd order is abelian; this also follows from the Classification Theorem.

We now give easier examples.

Proposition 3.7.9. *All finite p-groups are solvable.*

Proof. We prove by induction on n that a group G of order p^n is solvable. If $n = 1$, then G is abelian, hence solvable. In general, $Z(G)$ is solvable; $G/Z(G)$ is solvable by the induction hypothesis (since $Z(G) \neq 1$ by Proposition 3.1.7); hence G is solvable by Proposition 3.7.7. $\qquad\qquad\qquad\qquad\qquad\qquad$ □

Proposition 3.7.10. *Let $p \neq q$ be prime numbers. Every group G of order $p^n q$ is solvable.*

Proof. Let S be a Sylow p-subgroup of G. If $S \lhd G$, then G/S is cyclic, since $|G/S| = q$ is prime, S is solvable by Proposition 3.7.9, and G is solvable by Proposition 3.7.7.

Now assume that S is not normal. Then $N_G(S) = S$, since $[G : S] = q$ is prime and $S \leqslant N_G(S) \lneq G$. Let $\mathscr{S}$ be the set of all Sylow p-subgroups of G. Since $\mathscr{S}$ is the conjugacy class of S, $|\mathscr{S}| = [G : N_G(S)] = q$.

If the Sylow p-subgroups of G are pairwise disjoint ($S \cap T = 1$ for all $S \neq T$ in $\mathscr{S}$), then G contains $1 + q(p^n - 1)$ elements whose order is a power of p. This leaves at most $q - 1$ elements of order q. Hence G has only one Sylow q-subgroup Q. Then $Q \lhd G$ and G is solvable, since Q is solvable and G/Q, which has p^n elements, is solvable by Proposition 3.7.9.

Now assume that $S \cap T \neq 1$ for some $S \neq T$ in $\mathscr{S}$. Among all the intersections of two distinct Sylow p-subgroups of G there is one, M, which has the greatest possible number of elements. This subgroup M has the following properties.

Lemma 3.7.11. *Let M be an intersection of two distinct Sylow p-subgroups of G with the greatest possible number of elements. Then $H = N_G(M)$ has more than one Sylow p-subgroup; M is the intersection of all the Sylow p-subgroups of H; each Sylow p-subgroup of H is contained in a unique Sylow p-subgroup of G.*

Proof. We have $M \subsetneq S$ for some Sylow p-subgroup S of G. Proposition 3.3.10, applied to $M \subsetneq S$, yields $M \subsetneq N_S(M) = H \cap S$. Since $N_S(M) \subseteq S$ is a p-group, $N_S(M)$ is contained in a Sylow p-subgroup P of H, which is in turn contained in a Sylow p-subgroup T of G. Then $M \subsetneq N_S(M) \subseteq S \cap T$; by the choice of M, this implies $S = T$, so that $P \subseteq H \cap S = N_S(M)$ and $N_S(M) = P$ is a Sylow p-subgroup of H.

This shows that M is contained in a Sylow p-subgroup P of H. Since $M \lhd H = N_G(M)$, M is contained in every Sylow p-subgroup Q of H: indeed $Q = xPx^{-1}$ for some $x \in H$, so that $N \subseteq P$ implies $N = xNx^{-1} \subseteq Q$.

On the other hand, $M = S \cap T$ for some Sylow p-subgroups $S \neq T$ of G. Then $M \subseteq N_S(M) \cap N_T(S) \subseteq S \cap T$ shows that $M = N_S(M) \cap N_T(M)$; since $M \neq N_S(M), N_T(M)$, M is the intersection of two distinct Sylow p-subgroups of H. In particular, H has more than one Sylow p-subgroup. Also M is now the intersection of all the Sylow p-subgroups of H.

Finally let P be a Sylow p-subgroup of H. The p-group P is contained in a Sylow p-subgroup S of G. If P is contained in two distinct Sylow p-subgroups S and T of G, then $M \subsetneq P \subseteq S \cap T$, contradicting the choice of M. $\qquad\square$

We now resume the proof of Proposition 3.7.10. The number of Sylow p-subgroups of $H = N_G(M)$ divides $p^n q = |G|$ but is not divisible by p. Hence H has q Sylow p-subgroups. Now each Sylow p-subgroup of H is contained in a unique Sylow p-subgroup of G; since G also has q Sylow p-subgroups, M is contained in every Sylow p-subgroup of G. Therefore M is the intersection of all the Sylow p-subgroups of G; in particular, $M \trianglelefteq G$ (and $H = G$). Then M is also the intersection of any two distinct Sylow p-subgroups S, T of G (by the choice of M, we can't have $M \subsetneq S \cap T$).

The Sylow p-subgroups of G/M are pairwise disjoint (since M is the intersection of any two distinct Sylow p-subgroups of G). Hence G/M, whose order is $p^k q$ for some k, is solvable. Since the p-group M is solvable, it follows that G is solvable. $\qquad\square$

In Chapter 13 we will prove that every group of order $p^m q^n$ is solvable.

3.7.d. Metacyclic Groups. A group G is **metacyclic** in case G' and G/G' are cyclic.

Proposition 3.7.12. *A finite group G is metacyclic if and only if $G \cong \langle a, b\,;$ $a^m = b^n = 1,\ bab^{-1} = a^r\rangle$, where $mn = |G|$, $r^n \equiv 1$ (mod m), and $\gcd(r-1, m) = 1$.*

Proof. Let $G_{m,n,r,t} = \langle a, b\,; a^m = 1,\ b^n = a^t,\ bab^{-1} = a^r \rangle$. If G is metacyclic, then $G \cong G_{m,n,r,t}$ for some $m, n, r, t \geqslant 0$ with $m = |G'|$, $mn = |G|$, $r^n \equiv 1$, and $rt \equiv t$ (mod m), by Hölder's Theorem (Proposition 2.6.7).

We show that every commutator of $G_{m,n,r,t}$ is a power of $bab^{-1}a^{-1} = a^{r-1}$. Recall that $bab^{-1} = a^r$ implies $b^j a^i = a^{ir^j} b^j$, so that every element of $G_{m,n,r,t}$ can be written in the form $a^i b^j$. Hence

$$a^i b^j a^k b^\ell b^{-j} a^{-i} b^{-\ell} a^{-k} = a^i a^{kr^j} b^j b^\ell b^{-j} a^{-i} b^{-\ell} a^{-k} = a^{i+kr^j} b^\ell a^{-i} b^{-\ell} a^{-k}$$

$$= a^{i+kr^j} a^{-ir^\ell} a^{-k} = a^{k(r^j-1)-i(r^\ell-1)},$$

which is a power of a^{r-1} since $r-1$ divides $k(r^j-1) - i(r^\ell - 1)$. Hence $G'_{m,n,r,t} = \langle a^{r-1} \rangle$.

If $G \cong G_{m,n,r,t}$ is metacyclic, with $|G'| = m$, then $\langle a^{r-1} \rangle = \langle a \rangle$ and $\gcd(r-1, m) = 1$; since $rt - t \equiv 0$ (mod m), this implies $m \mid t$ and $b^n = 1$. If, conversely, $G \cong G_{m,n,r,0}$ with $r^n \equiv 1$ (mod m) and $\gcd(r-1, m) = 1$, then $G' \cong \langle a \rangle$ and $G/G' \cong \langle b \rangle$ are cyclic. $\qquad\square$

3.7.e. Hall Subgroups. The next three theorems are stronger versions of the Sylow Theorems, due to P. Hall, which hold in solvable groups.

Lemma 3.7.13. *A nontrivial finite solvable group G contains a nontrivial abelian normal p-subgroup for some prime number p.*

Proof. Let r be the smallest integer such that $G^{(r)} = 1$. Then $A = G^{(r-1)}$ is a nontrivial abelian group. There is a prime p which divides $|A|$; then $A(p) = \{ a \in A; \; a^{p^k} = 1 \text{ for some } k \geqslant 1 \}$ is a subgroup of A. By Corollary 3.3.4, $A(p)$ is a p-subgroup of A. Moreover $A(p)$ is a fully invariant subgroup of A: if η is an endomorphism of A, and $a^{p^k} = 1$, then $(\eta(a))^{p^k} = 1$. Now $A = G^{(r-1)}$ is a fully invariant subgroup of G; hence $A(p)$ is a fully invariant subgroup of G, in particular $A(p) \trianglelefteq G$. $\qquad\square$

Theorem 3.7.14. *Let G be a finite solvable group of order mn, where m and n are relatively prime. Then G contains a subgroup of order m.*

Proof. This holds if G is a p-group (e.g., if $|G| < 6$), and we proceed by induction on $|G|$. By Lemma 3.7.13, G contains a normal abelian p-subgroup $N \neq 1$. Let $|N| = p^k$. Then $p^k \mid mn$; since $\gcd(m,n) = 1$, either $p^k \mid m$ or $p^k \mid n$.

If $p^k \mid m$, then $|G/N| = (m/p^k)n$, with $\gcd(m/p^k, n) = 1$. By the induction hypothesis, G/N contains a subgroup of order m/p^k, which we can write in the form H/N where $N \subseteq H \subseteq G$. Then $|H| = m$.

If $p^k \mid n$, then $|G/N| = m(n/p^k)$ with $\gcd(m, n/p^k) = 1$. By the induction hypothesis, G/N contains a subgroup of order m, which we can write in the form K/N where $N \subseteq K \subseteq G$. Then $N \trianglelefteq K$ and K/N have relatively prime orders; by Schur's Theorem, $K \subseteq G$ contains a subgroup of order $|K/N| = m$. [The "abelian" case of Schur's Theorem (Theorem 2.4.8) suffices since N is abelian.] $\qquad\square$

The subgroups of order m in Theorem 3.7.14 are the **Hall subgroups** of G.

Theorem 3.7.15. *Let G be a finite solvable group of order mn, where m and n are relatively prime. All subgroups of G of order m are conjugate.*

Proof. Let $H, K \leqslant G$ have order m. If G is a p-group (e.g., if $|G| < 6$), then $H = K$. In general, we proceed by induction on $|G|$. As above, G contains a normal abelian p-subgroup $N \neq 1$, and $|N| = p^k$ divides either m of n.

Assume that $p^k \mid m$. Then $p \nmid n$. We have $|NH| = |H|(|N|/|N \cap H|) = mp^h$ for some h; then $mp^h \mid mn$, $p^h \mid n$, $p^h = 1$, $|NH| = m = |H|$, and $N \subseteq H$. Similarly $N \subseteq K$. By the induction hypothesis, H/N and K/N are conjugate in G/N: $H/N = (Nx)(K/N)(Nx)^{-1}$ for some $Nx \in G/N$. Since $N \subseteq H, K$, it follows that H and K are conjugate in G:

$$H = \bigcup_{h \in H} Nh = \bigcup_{k \in K} (Nx)(Nk)(Nx)^{-1}$$

$$= \bigcup_{k \in K} N(xkx^{-1}) = N(xKx^{-1}) = xKx^{-1},$$

since $N = xNx^{-1} \subseteq xKx^{-1}$.

Now assume that $p^k \mid n$. Then $N \cap H = N \cap K = 1$, $|NH| = p^k m$, and NH/N, $NK/N \leqslant G/N$ have order m. By the induction hypothesis, NH/N and NK/N are conjugate in G/N. As above, it follows that NH and NK are conjugate in G: $NH = xNKx^{-1}$ for some $x \in G$. Then H and xKx^{-1} are subgroups of NH order m. If $p^k < n$, it follows from the induction hypothesis that H and xKx^{-1} are conjugate in NH; hence H and K are conjugate in G. If $p^k = n$, then H and K are conjugate in G by Proposition 2.4.9. $\qquad\square$

Theorem 3.7.16. *Let G be a finite solvable group of order mn, where m and n are relatively prime. For every divisor ℓ of m, every subgroup of G of order ℓ is contained in a subgroup of order m.*

Proof. This holds if G is a p-group (e.g., if $|G| < 6$), by the Sylow theorems. In general we proceed by induction on $|G|$. Let H be a subgroup of G whose order ℓ divides m. As before, G contains a normal abelian p-subgroup $N \neq 1$, and $|N| = p^k$ divides either m of n.

Assume that $p^k \mid m$. Then $|NH/N| = |H|/|H \cap N|$ divides m, is relatively prime to n, and divides $|G/N| = (m/p^k)n$. Hence $|NH/N|$ divides m/p^k. By the induction hypothesis, NH/N is contained in a subgroup of G/N of order m/p^k, which has the form K/N, where $N \subseteq K$. Then K has order m and $H \subseteq NH \subseteq K$.

Assume that $p^k \mid n$. Then $H \cap N = 1$ and $|NH/N| = \ell$. By the induction hypothesis, NH/N is contained in a subgroup of G/N of order m, which has the form K/N, where $N \subseteq K$. Then $|K| = mp^k$ and $NH \subseteq K$. If $p^k < n$, then H is contained in a subgroup of $K \subseteq G$ of order m, by the induction hypothesis.

Assume that $p^k = n$. Let K be a subgroup of G of order m. Then $K \cap N = 1$, $|NK| = p^k m = mn$, and $NK = G$. Hence $|K \cap NH| = |K||NH|/|KNH| = \ell$. Then H and $L = K \cap NH$ are subgroups of NH of order ℓ. By Theorem 3.7.15, H and L are conjugate: $H = xLx^{-1}$ for some $x \in NH$. Hence H is contained in the subgroup xKx^{-1} of G, which has order m. $\qquad\square$

Exercises

1. Prove that a fully invariant subgroup of a fully invariant subgroup of a group G is a fully invariant subgroup of G.
2. Prove that $G^{(r)}$ is a fully invariant subgroup of G for all r.
3. Prove Proposition 3.7.4.
4. Prove Proposition 3.7.5.
5. Prove Proposition 3.7.6.
6. Prove Proposition 3.7.7.
7. Prove that all dihedral groups are solvable.
8. Find the commutator series of S_n.
9. Prove that S_n is solvable if and only if $n \leqslant 4$.
10. Show that every group of order less than 60 is solvable.

3.8. NILPOTENT GROUPS

Nilpotent groups form a smaller class than solvable groups, and have even more striking properties.

3.8.a. Central Series. A normal series $1 = S_0 \lhd S_1 \lhd \cdots \lhd S_m = G$ is **central** in case $S_i \lhd G$ and $S_{i+1}/S_i \subseteq Z(G/S_i)$ for all $i < m$. A normal series which is central (also called a **central series**) has abelian factors, but not conversely (see the exercises).

To give a general example of central series, define $Z_r = Z_r(G) \lhd G$ by induction as follows: $Z_0 = 1$; since $Z_r \lhd G$ and $Z(G/Z_r) \lhd G/Z_r$, there is a normal subgroup $Z_r \subseteq Z_{r+1} \lhd G$ such that

$$Z_{r+1}/Z_r = Z(G/Z_r).$$

In particular, $Z_1(G) = Z(G)$. The **ascending central series** of G is the infinite sequence

$$1 = Z_0 \lhd Z_1 \lhd \cdots \lhd Z_r \lhd Z_{r+1} \lhd \cdots.$$

If this series terminates (with $Z_r = G$ for some r), truncating it at r yields a normal series which is central.

Proposition 3.8.1. $Z_r(G)$ *is a characteristic subgroup of* G.

Proof. We show that when N is a characteristic subgroup of G and K/N is a characteristic subgroup of G/N, with $N \leqslant K$, then K is a characteristic subgroup of G. Since the center is a characteristic subgroup, it follows by induction on r that Z_r is a characteristic subgroup.

Let α be an automorphism of G. Since $\alpha(N) = N$, α induces an automorphism β of G/N, such that $\beta(Nx) = \alpha(Nx) = N\alpha(x)$ for all $x \in G$. Since $\beta(K/N) = K/N$, $x \in K$ implies $N\alpha(x) \in K/N$ and $\alpha(x) \in NK = K$. $\qquad\square$

The **descending central series** is the infinite sequence

$$G = G^0 \rhd G^1 \rhd G^2 \rhd \cdots \rhd G^r \rhd \cdots,$$

where $G^0 = G$ and G^{r+1} is the subgroup of G (sometimes denoted by $[G, G^r]$) generated by all commutators $aba^{-1}b^{-1}$ with $a \in G$ and $b \in G^r$. In particular, $G^1 = G'$. If this series terminates (if $G^r = 1$ for some r), then truncating it at r yields a normal series which is central:

Proposition 3.8.2. G^r *is a fully invariant subgroup of* G. *Furthermore* $G^r/G^{r+1} \subseteq Z(G/G^{r+1})$.

Proof. That G^r is fully invariant is proved by induction on r. Let η be an endomorphism of G. If G^r is fully invariant, then, for all $a \in G$ and $b \in G^r$,

$\eta(aba^{-1}b^{-1})$ is the commutator of $\eta(a) \in G$ and $\eta(b) \in G^r$ and is in G^{r+1}. Any $x \in G^r$ is a product of such commutators and their inverses; hence $\eta(x) \in G^{r+1}$.

Every $G^{r+1}b \in G^r/G^{r+1}$ commutes with all $G^{r+1}a \in G/G^{r+1}$, since $G^{r+1}ab = G^{r+1}ba$ for all $b \in G^r$. $\qquad\square$

3.8.b. Nilpotent Groups

Proposition 3.8.3. *The following conditions on a group G are equivalent:*

(1) *G has a [finite] central series;*

(2) *the ascending central series of G terminates ($Z_r = G$ for some r);*

(3) *the descending central series of G terminates ($G^r = 1$ for some r).*

A group is **nilpotent** in case it satisfies any of these equivalent conditions. All nilpotent groups are solvable, but not conversely; for instance, S_3 is solvable but not nilpotent (see the exercises).

Proof. We saw that (2) and (3) imply (1). Conversely, assume that G has a central series $S : 1 = S_0 \lhd S_1 \lhd \cdots \lhd S_m = G$. We show by induction that $G^{m-i} \subseteq S_i \subseteq Z_i$ for all $0 \leqslant i \leqslant m$; this implies $Z_m = G$ and $G^m = 1$.

We have $G^{m-m} = G = S_m$. Assume that $G^{m-i-1} \subseteq S_{i+1}$, where $i < m$. Let $a \in G$ and $b \in G^{m-i-1} \subseteq S_{i+1}$. Since $S_{i+1}/S_i \subseteq Z(G/S_i)$, we have $S_i ab = S_i ba$ and $aba^{-1}b^{-1} \in S_i$; thus S_i contains every generator of G^{m-i} and $G^{m-i} \subseteq S_i$.

We also have $S_0 = 1 = Z_0$. Assume that $S_i \subseteq Z_i$, where $i < m$. Then $G/Z_i \cong (G/S_i)/(Z_i/G_i)$, and there is a surjective homomorphism $\pi : G/S_i \longrightarrow G/Z_i$ with kernel Z_i/S_i, namely $S_i x \longmapsto Z_i x$. Since π is surjective,

$$S_{i+1}/Z_i = \pi(S_{i+1}/S_i) \subseteq \pi(Z(G/S_i)) \subseteq Z(G/Z_i) = Z_{i+1}/Z_i.$$

Hence $S_{i+1} \subseteq Z_{i+1}$. $\qquad\square$

Let G be nilpotent. When $Z_r = G$, the proof of Proposition 3.8.3 shows that $G^{r-i} \subseteq Z_i$ for all $0 \leqslant i \leqslant r$; hence $G^r = 1$. Similarly $G^r = 1$ implies $Z_r = G$. The smallest integer r such that $Z_r = G$ (such that $G^r = 1$) is the nilpotency **class** of G.

The following results state that the class of nilpotent groups is closed under subgroups, homomorphic images, and certain group extensions. The proofs are left as exercises.

Proposition 3.8.4. *Every subgroup of a nilpotent group is nilpotent.*

Proposition 3.8.5. *If G is nilpotent and $N \lhd G$, then G/N is nilpotent.*

Proposition 3.8.6. *Let $N \leqslant Z(G)$. If N and G/N are nilpotent, then G is nilpotent.*

Proposition 3.8.7. *If A and B are nilpotent, then $A \oplus B$ is nilpotent.*

3.8.c. Examples. All abelian groups are nilpotent.

Proposition 3.8.8. *Every finite p-group is nilpotent.*

Proof. We prove by induction on $|G|$ that a p-group G is nilpotent. We know (Proposition 3.1.7) that $Z(G) \neq 1$. Then $G/Z(G)$ is nilpotent by the induction hypothesis; $Z(G)$ is abelian, hence nilpotent; by Proposition 3.8.6, G is nilpotent. $\square$

Theorem 3.8.9. *For a finite group G the following conditions are equivalent:*

(1) *G is nilpotent;*

(2) *every Sylow subgroup of G is normal;*

(3) *G is a direct sum of p-groups.*

Proof. (1) implies (2). Let G be nilpotent. We show that $Z_i \subseteq H \leqslant G$ implies that $Z_{i+1} \subseteq N_G(H)$. Let $x \in Z_{i+1}$ and $h \in H$. Since $Z_i x \in Z(G/Z_i)$, we have $Z_i xh = Z_i hx$, $xhx^{-1}h^{-1} \in Z_i \subseteq H$, and $xhx^{-1} \in H$. Hence $x \in N_G(H)$.

Let S be a Sylow p-subgroup of G. Then $N_G(S) = N_G(N_G(S))$, by Proposition 3.3.9. Now $Z_0 = 1 \subseteq N_G(S)$. By the above, $Z_i \subseteq N_G(S)$ for all i. Since G is nilpotent, it follows that $N_G(S) = G$.

(2) implies (3). By (2) there is only one Sylow p-subgroup for each prime divisor p of $|G|$. Let $n = |G| = p_1^{t_1} p_2^{t_2} \ldots p_k^{t_k}$ be a product of powers of k distinct primes $p_1, \ldots, p_k$. Let S_i be the Sylow p_i-subgroup of G. We show by induction on j that $S_1 S_2 \ldots S_j = S_1 \oplus S_2 \oplus \cdots \oplus S_j$ for all $j \leqslant k$. This implies that $S_1 S_2 \ldots S_k = S_1 \oplus S_2 \oplus \cdots \oplus S_k$ has n elements, so that $G = S_1 \oplus S_2 \oplus \cdots \oplus S_k$.

If $j < k$, then $A = S_1 S_2 \ldots S_j$ and $B = S_{j+1}$ have the following properties: $A = S_1 \oplus S_2 \oplus \cdots \oplus S_j$, by the induction hypothesis; $A \cap B = 1$, since $|A|$ and $|B|$ are relatively prime; $A \trianglelefteq G$ and $B \trianglelefteq G$, by (2). Hence $AB = A \oplus B$ and $S_1 S_2 \ldots S_{j+1} = S_1 \oplus S_2 \oplus \cdots \oplus S_{j+1}$.

(3) implies (1) by Propositions 3.8.7 and 3.8.8. $\square$

Corollary 3.8.10. *When G is a finite nilpotent group, every divisor of $|G|$ is the order of a subgroup of G.*

The proof is an exercise. Corollary 3.8.10 does not extend to solvable groups; we saw in the exercises that the solvable group A_4 of order 12 does not contain a subgroup of order 6.

Exercises

1. Give an example of a normal series with abelian factors which is not a central series.

2. Find the ascending central series of S_n.
3. Find the descending central series of S_n.
4. Prove Proposition 3.8.4.
5. Prove Proposition 3.8.5.
6. Prove Proposition 3.8.6.
7. Prove Proposition 3.8.7.
8. Give an example of a group G with a normal subgroup N such that N and G/N are nilpotent but G is not nilpotent.
9. A **maximal** subgroup of a finite group G is a subgroup $M \underset{\neq}{<} G$ such that there exists no subgroup $M \underset{\neq}{<} H \underset{\neq}{<} G$. Prove that the following conditions on a finite group G are equivalent: (1) G is nilpotent; (2) every maximal subgroup of G is normal; (3) every maximal subgroup of G contains G'.
10. Prove Corollary 3.8.10.

4

RINGS

The roots of ring theory lie in the work of Dedekind and Hilbert. In the 1870s Dedekind showed that the factorization properties of integers extend to algebraic integers if elements are replaced by ideals; he also coined the words "ring," "ideal," and "field." Some 20 years later Hilbert laid the foundations of algebraic geometry, showing in particular that algebraic sets (over $\mathbb{C}$) are determined by ideals of polynomial rings. Some of these results will be found in Chapter 11.

The modern theory of rings and fields is built upon these earlier results and those of Galois (explained in Chapter 7). It was developed in the 1920s by Artin and Noether. With this work, of remarkable generality and depth, and the publication in 1930 of van der Waerden's *Moderne Algebra*, abstract algebra became one of the main branches of mathematics.

My advisor, Paul Dubreil, was a student of Emmy Noether. I am inordinately proud of this connection to the old masters.

This chapter contains general properties of rings and polynomials. Except in Section 4.1, a "ring" is an associative ring with an identity element.

Required results: basic properties of groups (Chapter 1). Zorn's Lemma (Theorem B.3.2) is used occasionally.

4.1. RINGS

This section contains the definition and basic properties of rings and of rings with an identity element.

4.1.a. Definition. A **ring** $(R, +, \cdot)$ is a set R together with two binary operations, an addition and a multiplication, such that:

(1) $(R, +)$ is an abelian group (the **additive group** of R);
(2) the multiplication is associative $(x(yz) = (xy)z)$;
(3) the multiplication is **distributive**: $(x + y)z = xz + yz$ and $z(x + y) = zx + zy$ for all $x, y, z \in R$.

(Rings as defined above are also called **associative rings**; a **nonassociative ring** only has properties (1) and (3).) In all later sections we also require:

(4) there exists an identity element (an element 1 such that $1x = x = x1$ for all $x \in R$).

When there is an identity element (for the multiplication), it is unique (see the exercises); we denote it by 1. The identity element is also called a **unity**; a ring with an identity element is also called a **ring with identity** or a **ring with unity**.

The **zero element** of a ring is the identity element 0 for its addition ($0 + x = x = x + 0$ for all $x \in R$).

4.1.b. Examples. Examples of rings with identity include $\mathbb{Z}$, the ring $\mathbb{Z}_n$ of integers modulo n, and all the other number systems in Appendix A.

Proposition 4.1.1. *For every abelian group A, the set* End(A) *of all endomorphisms of A is a ring with identity.*

Proof. A is denoted additively. The addition on End(A) is pointwise addition:

$$(\eta + \zeta)x = \eta x + \zeta x$$

for all $\eta, \zeta \in \mathrm{End}(A)$ and $x \in A$; commutativity in A ensures that $\eta + \zeta$ is an endomorphism. The multiplication on End(A) is composition of mappings: $(\eta \zeta)(x) = \eta(\zeta x)$. The proofs of (1), (2), and (3) are left as an exercise. $\square$

Every ring can be embedded in some End(A) (see the exercises).

Similarly let V be a vector space over a field K; the endomorphisms of V (the linear transformations $V \longrightarrow V$) constitute a ring $\mathrm{End}_K(V)$. A similar example is the ring $M_n(K)$ of all $n \times n$ matrices with coefficients in K.

More examples will be found in the exercises, and another major class of examples (polynomials) occupies Section 4.3.

4.1.c. Properties. Axioms (1), (2), and (3) imply further properties which will be used without comment.

(3') general distributivity: $(\sum_i x_i)(\sum_j y_j) = \sum_{i,j} x_i y_j$ for all $x_1, \ldots, x_m, y_1, \ldots, y_n \in R$ (see the exercises).

Infinite sums can be defined in a ring as follows (without using topology). The sum $\sum_{i \in I} x_i$ of a family $(x_i)_{i \in I}$ of elements of R is defined if $x_i = 0$ for almost all i (if the set $\{i \in I ; x_i \neq 0\}$ is finite); then $\sum_{i \in I} x_i$ is the finite sum $\sum_{i \in I, x_i \neq 0} x_i$ (the order of the terms does not matter, since the addition is commutative). By (3'):

(3'') $(\sum_{i \in I} x_i)(\sum_{j \in J} y_j) = \sum_{(i,j) \in I \times J} x_i y_j$ whenever $x_i = 0$ for almost all $i \in I$ and $y_j = 0$ for almost all $j \in J$.

(5) $x(y - z) = xy - xz$ and $(y - z)x = yx - zx$ for all $x, y, z \in R$. Indeed, recall that $y - z = y + (-z)$, so that $(y - z) + z = y$. By (3), $xy = x((y - z) + z) = x(y - z) + xz$. In the group $(R, +)$ this implies $x(y - z) = xy - xz$. Similarly $(y - z)x = yx - zx$.

(6) $x0 = 0 = 0x$ for all $x \in R$. By (5), $x0 = x(0 - 0) = x0 - x0 = 0$. Similarly $0x = 0$.

In the group $(R, +)$ the additive powers of an element x are written as integer multiples: $nx = x + x + \cdots + x$ if $n > 0$, $0x = 0$, and $(-n)x = -(nx)$ for all $n > 0$. (The notation x^n is reserved for multiplicative powers.) If R has an identity element, then $nx = (n1)x$. In general, integer multiples have additive properties: $(m + n)x = mx + nx$, $(mn)x = m(nx)$, $m(x + y) = mx + my$ for all $m, n \in \mathbb{Z}$ and $x, y \in R$, and a multiplicative property which is left as an exercise:

(7) $(nx)y = x(ny) = n(xy)$ for all $x, y \in R$ and $n \in \mathbb{Z}$.

A ring is **commutative** when its multiplication is commutative. Except for the quaternion algebra $\mathbf{Q}$, all the number systems are commutative, but $M_n(K)$ is not commutative if $n > 1$. Commutativity yields additional properties:

(8) if $xy = yx$, then $(xy)^n = x^n y^n$ for all $n \in \mathbb{N}$;

(9) (the **Binomial Theorem**) if $xy = yx$, then

$$(x + y)^n = \sum_{i+j=n} \binom{n}{i} x^i y^j$$

for all $n \in \mathbb{N}$, where

$$\binom{n}{i} = \frac{n!}{i!(n-i)!}.$$

Zero divisors in a ring are elements $a, b \neq 0$ such that $ab = 0$; then a is a **left zero divisor** and b is a **right zero divisor**. These elements occur in $\mathbb{Z}_n$ if n is not prime, and in $M_n(K)$ if $n > 1$. They determine the extent to which the multiplication is cancellative: by (5),

(10) $xy = xz$ implies $y = z$ if and only if x is not zero and is not a left zero divisor; $yx = zx$ implies $y = z$ if and only if x is not zero and is not a right zero divisor.

A **unit** or **invertible element** of a ring R with an identity element is an element $u \in R$ such that $uv = vu = 1$ for some $v \in R$; then v is unique, and is the **inverse** of u, denoted by u^{-1}. The units of R form a group under multiplication, the **group of units** $U(R)$ of R (see the exercises).

4.1.d. Homomorphisms. A **homomorphism** of rings of a ring R into a ring S is a mapping $\varphi : R \longrightarrow S$ which preserves both operations:

$$\varphi(x + y) = \varphi(x) + \varphi(y) \qquad \text{and} \qquad \varphi(xy) = \varphi(x)\varphi(y)$$

for all $x, y \in R$. When R and S have identity elements, a homomorphism of R into S is usually required to be a **homomorphism of rings with identity**, which also preserves the identity element:

$$\varphi(1) = 1.$$

A homomorphism of rings preserves the zero element ($\varphi(0) = 0$), integer multiples ($\varphi(nx) = n\,\varphi(x)$) and all sums, differences, products, and powers.

The identity mapping 1_R on a ring R is a ring homomorphism. The composition of two ring homomorphisms $\varphi : R \longrightarrow S$, $\psi : S \longrightarrow T$ is a ring homomorphism $\psi \circ \varphi : R \longrightarrow T$.

An **isomorphism** of a ring R onto a ring S is a bijective homomorphism of R onto S. Then the inverse bijection is an isomorphism of S onto R, and the rings R and S are **isomorphic**. As we did with groups, we treat isomorphic rings as if they were the same "abstract" ring.

4.1.e. Adjunction of an Identity. All our examples of rings are rings with identity. The following straightforward result shows that every ring can be embedded into a ring with identity.

Proposition 4.1.2. *For any ring R, $R^1 = R \times \mathbb{Z}$, with operations:*

$$(x,m) + (y,n) = (x + y, m + n), \qquad (x,m)(y,n) = (xy + nx + my, mn),$$

is a ring with identity. Furthermore $\iota : x \longmapsto (x,0)$ is an injective homomorphism of R into R^1.

Some properties are lost in this embedding (see the exercises), but in most situations an identity element may be assumed; we will do so in all later sections.

The ring R^1 in Proposition 4.1.1 has a universal property:

Proposition 4.1.3. *Every homomorphism φ of R into a ring with identity S factors uniquely through $\iota : R \longrightarrow R^1$ (there exists a homomorphism of rings with identity $\psi : R^1 \longrightarrow S$ unique such that $\varphi = \psi \circ \iota$).*

Proof. In R^1, $(x,m) = (x,0) + n(0,1)$, and $(0,1)$ is the identity element. If $\psi(1) = \psi(0,1) = 1$ and $\psi \circ \iota = \varphi$, then $\psi(x,m) = \varphi(x) + n1$; this shows that ψ is unique. Conversely, the mapping $\psi : (x,m) \longmapsto \varphi(x) + n1$ is readily shown to be a homomorphism with the required properties. $\qquad\square$

Exercises

1. In the definition of a ring with identity element, show that it is not necessary to require that the addition is commutative. [Let $(R, +, \cdot)$ be a set with two operations such that $(R, +)$ is a group and (2), (3), and (4) hold. Prove that $(R, +)$ is necessarily abelian.]

2. Prove that any two identity elements of a ring must be equal.

3. In a ring R with an identity element, $1 = 0$ if and only if $R = \{0\}$.

4. Prove (3'): in a ring R, $(\sum_{i \in I} x_i)(\sum_j y_j) = \sum_{i,j} x_i y_j$ holds for all $x_1, \ldots, x_m, y_1, \ldots, y_n \in R$.

5. Prove (7): in a ring R, $(nx)y = x(ny) = n(xy)$ holds for all $x, y \in R$ and $n \in \mathbb{Z}$.

6. Prove that in a ring R, $(mx)(ny) = (mn)(xy)$ holds for all $x, y \in R$ and $m, n \in \mathbb{Z}$.

7. Show that the set of all units of a ring with identity is a group under multiplication.

8. Show that $u \in R$ is a unit of R if and only if $xu = uy = 1$ for some $x, y \in R$; then $x = y = u^{-1}$.

9. Give addition and multiplication tables for $\mathbb{Z}_4$ and $\mathbb{Z}_5$.

10. Show that $\mathbb{Z}_n$ has zero divisors if and only if n is not a prime number.

11. Show that $\overline{a} \in \mathbb{Z}_n$ is a unit of $\mathbb{Z}_n$ if and only if $(a, n) = 1$.

12. Prove the following properties of integers: $a^{\phi(n)} \equiv 1 \pmod{n}$ whenever $(a, n) = 1$, where ϕ is Euler's ϕ function; if p is prime, then $a^p \equiv a$ for all a. (Use the previous exercise.)

13. Show that $\text{End}_K(V)$ is a ring for every vector space V over a field K.

14. Prove Proposition 4.1.2.

15. A ring R is **regular** (also called a **von Neumann regular**) in case there is for each $a \in R$ some $x \in R$ such that $axa = a$. Prove that R^1 is not regular.

16. Prove that every ring R with identity can be embedded into the ring $\text{End}(R, +)$ [hence every ring can be embedded into $\text{End}(A)$ for some abelian group A].

17. A **gaussian integer** is a complex number $a + ib$ in which a and b are integers. Show that gaussian integers constitute a ring with identity. What are the units?

18. Show that the complex numbers of the form $a + bi\sqrt{2}$, in which a and b are integers, constitute a ring with identity. What are the units?

4.2. HOMOMORPHISMS

As we saw in Chapter 1, when $\varphi : G \longrightarrow H$ is a homomorphism of groups, then $\text{Im } \varphi$ is a subgroup of H, $\text{Ker } \varphi$ is a normal subgroup of G, and φ can be factored into the projection $G \longrightarrow G/\text{Ker } \varphi$, the inclusion $\text{Im } \varphi \longrightarrow H$, and an isomorphism $G/\text{Ker } \varphi \longrightarrow \text{Im } \varphi$. We now extend these properties to rings (which from this point on are assumed to be rings with identity).

4.2.a. Subrings. A **subring** of a ring R is a ring S such that $S \subseteq R$ and the inclusion mapping $S \longrightarrow R$ is a homomorphism (of rings with identity). This relationship is sometimes denoted by $S \leqslant R$.

A subring of R is completely determined by its underlying set; the latter is also called a subring. The following straightforward result is often used as a definition of subrings:

Proposition 4.2.1. *A subset S of a ring R is [the underlying set of] a subring of R if and only if S is a subgroup of $(R, +)$, is closed under multiplication $(x, y \in S$ implies $xy \in S)$, and contains the identity element of R.*

In particular, a ring and its subrings have the same identity element. Subrings are often defined without this requirement; then a subring is called **unitary** in case it contains the identity element of the ring.

When $\varphi : R \longrightarrow S$ is a homomorphism of rings, the image

$$\mathrm{Im}\, \varphi = \varphi(R) = \{ \varphi(r) \, ; \, r \in R \}$$

of φ is a subring of S.

4.2.b. Quotient Rings. The kernel

$$K = \mathrm{Ker}\, \varphi = \varphi^{-1}(0) = \{ r \in R \, ; \, \varphi(r) = 0 \}$$

of a ring homomorphism $\varphi : R \longrightarrow S$ is a subgroup of $(R, +)$ and is closed under multiplication, but does not usually contain the identity element of R (since $\varphi(1) = 1 \neq 0$ unless $S = \{0\}$.) However, K has additional properties. If $x \in \mathrm{Ker}\, \varphi$, then $\varphi(xy) = \varphi(x)\varphi(y) = 0\varphi(y) = 0$, and $xy \in \mathrm{Ker}\, \varphi$ for all $y \in R$ (not just for all $y \in \mathrm{Ker}\, \varphi$). Similarly $x \in \mathrm{Ker}\, \varphi$ implies $yx \in \mathrm{Ker}\, \varphi$ for all $y \in R$.

An **ideal** of a ring R is a subgroup I of $(R, +)$ such that $x \in I$ implies $xy \in I$ and $yx \in I$ for all $y \in R$. This relationship is sometimes denoted by $I \lhd R$, or by $I \unlhd R$; we don't use this notation. A **proper** ideal also satisfies $I \neq R$; this condition is frequently added to the definition of ideals. It is traditional to denote ideals by German letters; we use this notation when R is commutative.

Proposition 4.2.2. *When I is an ideal of a ring R, the cosets of I are the elements of a ring R/I, whose operations are well defined by $(x + I) + (y + I) = (x + y) + I$ and $(x + I)(y + I) = xy + I$. The projection $\pi : x \longmapsto x + I$ is a ring homomorphism of R onto R/I, and $\mathrm{Ker}\, \pi = I$.*

Proof. First I is a normal subgroup of $(R, +)$, since $(R, +)$ is abelian. Hence R/I is already an abelian group, whose addition is well defined by $(x + I) + (y + I) = (x + y) + I$; the projection $\pi : x \longmapsto x + I$ is a homomorphism of abelian groups.

Let $u = x + a \in x + I$ and $v = y + b \in y + I$, with $a, b \in I$. Then $xb + ay + ab \in I$, since I is an ideal, and $uv = (x + a)(y + b) = xy + xb + ay + ab \in xy + I$.

This shows that $\{uv;\ u \in x + I,\ v \in y + I\}$ is contained in a single coset of I (namely, $xy + I$). Therefore there is a mapping of $R/I \times R/I$ into R/I which sends each pair $(x + I, y + I)$ of cosets of I onto the coset of I which contains $\{uv;\ u \in x + I,\ v \in y + I\}$. This mapping is the multiplication on R/I. We showed that $(x + I)(y + I) = xy + I$. In particular, $1 + I$ is the identity element of R/I. Then π is a homomorphism of rings. $\square$

The **quotient ring** (also called **factor ring**) of a ring R by an ideal I is the ring R/I in Proposition 4.2.2. For example, $R/0 \cong R$, and $R/R \cong 0$ is the **trivial** ring. The quotient rings of $\mathbb{Z}$ were found in Section A.3 to be $\mathbb{Z}/0 \cong \mathbb{Z}$ and all rings $\mathbb{Z}_n = \mathbb{Z}/(n)$ of integers modulo n ($n > 0$).

The ideals of R/I are determined by the following result, to be proved by our eager reader. A similar result constructs the subrings of R/I.

Proposition 4.2.3. *Let I be an ideal of a ring R. When $J \supseteq I$ is an ideal of R, then J/I is an ideal of R/I. Moreover every ideal of R/I can be written in this form for some unique ideal $J \supseteq I$.*

4.2.c. The Homomorphism Theorem. The main properties of homomorphisms in Section 1.6 extend readily to rings.

Theorem 4.2.4. *Let $\varphi : R \longrightarrow S$ and $\psi : R \longrightarrow T$ be ring homomorphisms. If φ is injective, then ψ factors through φ ($\psi = \chi \circ \varphi$ for some ring homomorphism $\chi : S \longrightarrow T$) if and only if $\operatorname{Im} \psi \subseteq \operatorname{Im} \varphi$, and then ψ factors uniquely through φ (χ is unique).*

Theorem 4.2.5. *Let $\varphi : R \longrightarrow S$ and $\psi : R \longrightarrow T$ be ring homomorphisms. If φ is surjective, then ψ factors through φ ($\psi = \chi \circ \varphi$ for some ring homomorphism $\chi : S \longrightarrow T$) if and only if $\operatorname{Ker} \varphi \subseteq \operatorname{Ker} \psi$, and then ψ factors uniquely through φ (χ is unique).*

Theorem 4.2.6 (Homomorphism Theorem). *When $\varphi : R \longrightarrow S$ is a homomorphism of rings:*

(1) $\operatorname{Im} \varphi$ *is a subring of S;*
(2) $K = \operatorname{Ker} \varphi$ *is an ideal of R;*
(3) $\varphi(x) = \varphi(y) \iff x + K = y + K$;
(4) $\operatorname{Im} \varphi \cong R/\operatorname{Ker} \varphi$; *in fact there exists an isomorphism $\theta : R/\operatorname{Ker} \varphi \longrightarrow \operatorname{Im} \varphi$ such that the following diagram commutes:*

$$
\begin{array}{ccc}
R & \xrightarrow{\ \varphi\ } & S \\
\pi \downarrow & & \uparrow \iota \\
R/\operatorname{Ker} \varphi & \xrightarrow[\ \theta\ ]{} & \operatorname{Im} \varphi
\end{array}
$$

where $\pi : R \longrightarrow R/\operatorname{Ker} \varphi$ is the projection and $\iota : \operatorname{Im} \varphi \longrightarrow S$ is the inclusion homomorphism.

The Isomorphism Theorems also extend to rings (see the exercises).

Proof. We prove Theorem 4.2.6. We saw that $\operatorname{Im} \varphi$ is a subring of S and that $K = \operatorname{Ker} \varphi$ is an ideal of R. Then Theorem 1.6.6 yields an isomorphism θ of abelian groups such that the above diagram commutes. Hence $\theta(x + K) = \varphi(x)$ for all $x \in R$. Therefore $\theta(1 + K) = \varphi(1) = 1$,

$$\theta((x + K)(y + K)) = \theta(xy + K) = \varphi(xy) = \varphi(x)\varphi(y) = \theta(x + K)\theta(y + K),$$

and θ is a ring isomorphism. $\qquad\qquad\Box$

A **homomorphic image** of a ring R is a ring S for which there exists a surjective homomorphism $R \longrightarrow S$. It follows from the Homomorphism Theorem that this happens if and only if S is isomorphic to the quotient of R by some ideal.

As another application of the Homomorphism Theorem, we show:

Proposition 4.2.7. *For each ring R [with identity] there is a unique homomorphism of rings of $\mathbb{Z}$ into R. Its image is the smallest subring of R and consists of all the integer multiples of the identity element; it is isomorphic either to $\mathbb{Z}$ or to $\mathbb{Z}_n$ for some $n > 0$.*

Proof. If $\varphi : \mathbb{Z} \longrightarrow R$ is a homomorphism and $\varphi(1) = 1$, then $\varphi(n) = n1$ for all n; hence φ is unique. Conversely, properties of integer multiplication in Section 4.1 show that the mapping $\varphi : n \longmapsto n1$ is a ring homomorphism of $\mathbb{Z}$ into R. Hence the image of φ, which consists of all integer multiples of 1, is a subring of R and is isomorphic to a quotient ring of $\mathbb{Z}$. $\qquad\Box$

In a ring R, an integer multiple $n1$ of 1 is an **integer of** R and is often denoted by n (even though it may happen that $n1 = 0$ with $n \neq 0$). Note that $n1 = 0$ if and only if $nx = 0$ for all $x \in R$. The **characteristic** of R is defined as follows. If $n1 \neq 0$ for all $n \neq 0$ [equivalently, if 1 has infinite order in $(R, +)$], then the characteristic of R is 0; otherwise, the characteristic of R is the positive integer n such that $n1 = 0$ [equivalently, n is the order of 1 in $(R, +)$, in case 1 has finite order].

4.2.d. Properties of Ideals. The following results are straighforward.

Proposition 4.2.8. *Every intersection of ideals of a ring R is an ideal of R.*

It follows from Proposition 4.2.7 that there exists for every subset S of a ring R a smallest ideal containing S; this is the ideal of R **generated by** S; it is often denoted by (S).

Proposition 4.2.9. *The ideal* (S) *of a ring* R *generated by a subset* S *of* R *is the set of all sums of elements of* R *of the form* xsy, *where* $s \in S$ *and* $x, y \in R$. *If* R *is commutative, then* (S) *is the set of all linear combinations of elements of* S *with coefficients in* R.

A **principal ideal** is an ideal $I = (a)$ which is generated by a single element a. If R is commutative, then $(a) = Ra$ by Proposition 4.1.7. By Proposition 1.4.4 all ideals of $\mathbb{Z}$ are principal; not all rings have this property (see the exercises).

A union of ideals is not in general an ideal. However:

Proposition 4.2.10. *The union of a nonempty chain of ideals of a ring* R *is an ideal of* R. *The union of a nonempty chain of proper ideals of a ring* R *[with identity element] is a proper ideal of* R.

A **maximal** ideal is a maximal proper ideal (an ideal $M \neq R$ such that $M \subsetneq I$ implies $I = R$ when I is an ideal).

Corollary 4.2.11. *Every proper ideal of a ring* R *is contained in a maximal ideal of* R.

Proof. Let $I \neq R$ be a proper ideal of a ring R. Let $\mathscr{S}$ be the set of all proper ideals of R which contain I. Then $\mathscr{S} \neq \emptyset$. Furthermore $\mathscr{S}$, ordered by inclusion, is inductive by Proposition 4.2.10. By Zorn's Lemma, $\mathscr{S}$ has a maximal element, which is a maximal ideal of R. $\qquad\square$

The **sum** of a family $(J_i)_{i \in I}$ of ideals of a ring is

$$\sum\nolimits_{i \in I} J_i = \{ \sum\nolimits_{i \in I} a_i \,;\, a_i \in J_i, \, a_i = 0 \text{ for almost all } i \}.$$

Proposition 4.2.12. *Every sum of ideals of a ring* R *is an ideal of* R.

By Proposition 4.2.12, $\sum_{i \in I} J_i$ is the smallest ideal which contains $\bigcup_{i \in I} J_i$.

Exercises

1. An element x of a ring R is **nilpotent** in case $x^n = 0$ for some n. Show that the nilpotent elements of a commutative ring form an ideal of R.
2. Let R be a ring and I be an ideal of R. Show that $I \neq R$ if and only if $1 \notin I$.
3. Prove Proposition 4.2.1.
4. Let R be a ring without identity element. Prove that every ideal of R is an ideal of R^1 [if $x \in R$ is identified with $\iota(x) \in R^1$].
5. Let $I \subseteq J$ be ideals of a ring R. Show that $(R/I)/(J/I) \cong R/J$.
6. Let S be a subring and I be an ideal of a ring R. Show that $S + I$ is a subring of R, I is an ideal of $S + I$, $S \cap I$ is an ideal of S, and $(S + I)/I \cong S/(S \cap I)$.

7. Prove Proposition 4.2.3.
8. Prove Theorem 4.2.5.
9. Prove Proposition 4.2.8.
10. Prove Proposition 4.2.9.
11. Prove Proposition 4.2.10.
12. Prove Proposition 4.2.12.
13. Prove the following: in a ring R without an identity element, the ideal (S) generated by a subset S is the set of all finite sums of elements of the form s, xs, sy, or xsy, where $s \in S$ and $x, y \in R$.
14. In a commutative ring R, show that the principal ideal $(a) = Ra$ is the set $\{xa; x \in R\}$ of all multiples of a.
15. Show that polynomials in two variables x and y with real coefficients form a ring with identity R, and that polynomials whose constant coefficient is 0 form an ideal I of R. Prove that I is not a principal ideal.
16. Find all ideals of $\mathbb{Z}_n$.

4.3. POLYNOMIALS

This section contains basic properties of polynomials in one indeterminate, polynomials in several indeterminates, and power series, with coefficients in a ring R with identity. We do not assume that R is commutative, but the results will be used only in this case.

4.3.a. Definition. Intuitively, a polynomial with one indeterminate X and coefficients in R is a linear combination $f(X) = a_n X^n + \ldots + a_1 X + a_0$ of powers of X with coefficients $a_0, a_1, \ldots, a_n \in R$.

The formal definition of polynomials does not use the undefined quantity X but takes advantage of the fact that a polynomial is determined by its coefficients. Formally, a **polynomial** with one indeterminate and coefficients in a ring R is an infinite sequence

$$a = (a_0, a_1, \ldots, a_n, \ldots)$$

of elements of R such that $a_n = 0$ for almost all n (such that the set $\{n \in \mathbb{N} \cup \{0\}; a_n \neq 0\}$ is finite); equivalently, such that there exists $n \geq 0$ such that $a_i = 0$ for all $i \geq n$. Addition is componentwise:

$$a + b = c, \qquad \text{where} \quad c_n = a_n + b_n \quad \text{for all } n,$$

and multiplication is defined by

$$ab = c, \qquad \text{where} \quad c_n = \sum_{i+j=n} a_i b_j \quad \text{for all } n.$$

Proposition 4.3.1. *When R is a ring, polynomials with one indeterminate and coefficients in R form a ring $R[X]$. If R is commutative, then $R[X]$ is commutative.*

Proof. First, $R[X]$ is an abelian group under addition; we recognize that $(R[X], +) \cong \bigoplus_{n \geqslant 0} R$.

We show that the multiplication on $R[X]$ is well defined. Let $a, b \in R[X]$. There exist $m, n \geqslant 0$ such that $a_i = 0$ for all $i \geqslant m$ and $b_j = 0$ for all $j \geqslant n$. If $k \geqslant m + n$, then $i + j = k$ implies $i \geqslant m$ or $j \geqslant n$, so that $c_k = \sum_{i+j=k} a_i b_j = 0$. Thus $c = ab \in R[X]$.

The multiplication on $R[X]$ inherits distributivity (and commutativity, if R is commutative) from the multiplication on R, and $1 = (1, 0, \ldots, 0, \ldots)$ is an identity element. Associativity is left as an exercise.   □

An injective homomorphism $\iota : R \longrightarrow R[X]$ is defined by

$$\iota(r) = (r, 0, \ldots, 0, \ldots).$$

It is common practice to identify $r \in R$ and $\iota(r) \in R[X]$; then R is a subring of $R[X]$.

The **indeterminate** X is

$$X = (0, 1, 0, \ldots, 0, \ldots).$$

(If the indeterminate is denoted by some other letter, say Y, then $R[X]$ is denoted by $R[Y]$.) It is common practice to denote $a \in R[X]$ by $a(X)$. We note that the indeterminate X commutes with every constant $r \in R$ (actually, with $\iota(r)$), with $rX = Xr = (0, r, 0, \ldots, 0, \ldots)$.

Now that X is defined, polynomials can be written in familiar form.

Proposition 4.3.2. $a(X) = a_0 + a_1 X + \cdots + a_n X^n$, *for every polynomial $a \in R[X]$ such that $a_i = 0$ for all $i \geqslant n$.*

Proof. Induction shows that $X^n = (0, \ldots, 0, 1, 0, \ldots, 0, \ldots)$ has coefficients $c_i = \delta_{in}$ ($c_n = 1$, $c_i = 0$, for all $i \neq n$). Hence $a_n X^n = (0, \ldots, 0, a_n, 0, \ldots, 0, \ldots)$ has coefficients $c_i = a_n \delta_{in}$. Then $a = a_0 + a_1 X + \cdots + a_n X^n$.   □

The operations on $R[X]$ were defined so that polynomials in the form $a_n X^n + \cdots + a_1 X + a_0$ can be added and multiplied in the usual way.

4.3.b. Degree. The **degree** of a polynomial $a \neq 0$ is the largest integer $n = \deg a$ such that $a_n \neq 0$; then a_n is the **leading coefficient** of a and $a_n X^n$ is the **leading term** of a. The degree of 0 is sometimes left undefined or is variously defined as $-1 \in \mathbb{Z}$ or as $-\infty$, or as any quantity such that $\deg 0 < \deg a$ for all $a \neq 0$.

A polynomial has degree at most n if and only if it can be written in the form $a_n X^n + \cdots + a_1 X + a_0$.

Proposition 4.3.3. *For all $a, b \neq 0$ in $R[X]$:*

(1) $\deg(a + b) \leqslant \max(\deg a, \deg b)$;
(2) *if $\deg a \neq \deg b$, then $\deg(a + b) = \max(\deg a, \deg b)$;*
(3) $\deg(ab) \leqslant \deg a + \deg b$;
(4) *if R has no zero divisors, then $\deg(ab) = \deg a + \deg b$.*

The proof is an exercise. Note that $\deg(a + b) < \max(\deg a, \deg b)$ happens (for instance, when $b = -a$). If R has zero divisors, then $\deg(ab) < \deg a + \deg b$ also happens (since $uv = 0$ in R implies $(uX^m)(vX^n) = 0$ in $R[X]$). Part (4) implies:

Corollary 4.3.4. *If R has no zero divisors, then $a(X)$ is a unit of $R[X]$ if and only if a is constant ($\deg a = 0$) and $a(0) = a_0 (= a)$ is a unit of R.*

4.3.c. Order. The **order** of a polynomial $a \neq 0$ is the smallest integer $n = \operatorname{ord} a$ such that $a_n \neq 0$; the order of 0 is ∞, so that $\operatorname{ord} 0 > \operatorname{ord} a$ for all $a \neq 0$. A polynomial has order at least n if and only if it is a multiple of X^n. The reader will happily prove:

Proposition 4.3.5. *For all $a, b \neq 0$ in $R[X]$:*

(1) $\operatorname{ord}(a + b) \geqslant \min(\operatorname{ord} a, \operatorname{ord} b)$;
(2) *if $\operatorname{ord} a \neq \operatorname{ord} b$, then $\operatorname{ord}(a + b) = \min(\operatorname{ord} a, \operatorname{ord} b)$;*
(3) $\operatorname{ord}(ab) \geqslant \operatorname{ord} a + \operatorname{ord} b$;
(4) *if R has no zero divisors, then $\operatorname{ord}(ab) = \operatorname{ord} a + \operatorname{ord} b$.*

4.3.d. Universal Property. The universal property of $R[X]$ constructs ring homomorphisms $\psi : R[X] \longrightarrow S$; necessarily, $\varphi = \psi \circ \iota : R \longrightarrow S$ is a homomorphism, and, in S, $\psi(X)$ commutes with every $\varphi(r)$ (since, in $R[X]$, X commutes with every $\iota(r)$). Conversely:

Proposition 4.3.6. *Let R and S be rings, $\varphi : R \longrightarrow S$ be a homomorphism, and $s \in S$ commute with $\varphi(r)$ for every $r \in R$. There is a homomorphism $\psi : R[X] \longrightarrow S$ unique such that $\psi \circ \iota = \varphi$ and $\psi(X) = s$; namely*

$$\psi(a_0 + a_1 X + \cdots + a_n X^n) = \varphi(a_0) + \varphi(a_1) s + \cdots + \varphi(a_n) s^n.$$

Moreover $\operatorname{Im} \psi$ is the smallest subring of S which contains $\operatorname{Im} \varphi$ and s.

If S is commutative, then $s \in S$ can be arbitrary.

Proof. A mapping $\psi : R[X] \longrightarrow S$ is defined by

$$\psi(a_0 + a_1 X + \cdots + a_n X^n) = \varphi(a_0) + \varphi(a_1)s + \cdots + \varphi(a_n)s^n.$$

It is immediate that ψ is a homomorphism. By Proposition 4.3.2, ψ is the only homomorphism such that $\psi \circ \iota = \varphi$ and $\psi(X) = s$. Hence Im ψ is a subring of S which contains Im φ and s. Conversely, every such subring of S contains all elements of the form $\varphi(a_0) + \varphi(a_1)s + \cdots + \varphi(a_n)s^n$ and contains Im ψ. □

For any ring homomorphism $\varphi : R \longrightarrow S$, Proposition 4.3.6 can be applied to $\iota \circ \varphi : R \longrightarrow S[X]$ and $X \in S[X]$ and yields a homomorphism $\psi : R[X] \longrightarrow S[X]$:

$$\psi(a_0 + a_1 X + \cdots + a_n X^n) = \varphi(a_0) + \varphi(a_1)X + \cdots + \varphi(a_n)X^n.$$

Thus every ring homomorphism $\varphi : R \longrightarrow S$ induces a homomorphism $\psi : R[X] \longrightarrow S[X]$. We usually denote $\psi(a)$ by $^\varphi a$.

4.3.e. Evaluation. The **value** of a polynomial $a(X) \in R[X]$ at any $x \in R$ is

$$a(x) = a_n x^n + \cdots + a_1 x + a_0.$$

This defines for each polynomial $a(X) \in R[X]$ a **polynomial function** $x \longmapsto a(x)$, which is a mapping of R into R.

If R is commutative, then Proposition 4.3.6, applied to $\varphi = 1_R$ and $x \in R$, shows that the **evaluation mapping** $\varepsilon_x : a \longmapsto a(x)$ is the homomorphism $\varepsilon_x : R[X] \longrightarrow R$ unique such that $\varepsilon_x(\iota(r)) = r$ for all $r \in R$ and $\varepsilon_x(X) = x$. More generally, this holds whenever $x \in R$ is **central**, that is, $xr = rx$ for all $x \in R$. In particular, the evaluation mapping ε_x is then an **evaluation homomorphism**.

The homomorphism ψ in Proposition 4.3.6 is an instance of evaluation mapping: the formula

$$\psi(a_0 + a_1 X + \cdots + a_n X^n) = \varphi(a_0) + \varphi(a_1)s + \cdots + \varphi(a_n)s^n$$

shows that $\psi(a) = (^\varphi a)(s)$.

4.3.f. Substitution. For any $a(X) = a_n X^n + \cdots + a_1 X + a_0 \in R[X]$ and $b(X) \in R[X]$, $a(b(X)) \in R[X]$ (also denoted by $a(b)$ or by $a \circ b$) is defined by

$$a(b(X)) = a_n b(X)^n + \cdots + a_1 b(X) + a_0.$$

This defines another operation on $R[X]$, **substitution**.

Substitution has good properties only if R is commutative. Then Proposition 4.3.6, applied to $\varphi = 1_R$ and $b(X) \in R[X]$ yields a unique homomorphism $\sigma_b :$

$R[X] \longrightarrow R[X]$ such that $\sigma_b(\iota(r)) = r$ for all $r \in R$ and $\sigma_b(X) = b(X)$; namely

$$\sigma_b(a_n X^n + \cdots + a_1 X + a_0) = a_n b(X)^n + \cdots + a_1 b(X) + a_0 = a(b(X)).$$

We call σ_b a **substitution homomorphism**. Then $(a(b))(x) = a(b(x))$ for all $x \in R$, and the reader will show that substitution is associative $(a \circ (b \circ c) = (a \circ b) \circ c$ for all $a,b,c \in R[X])$.

4.3.g. Polynomial Division. Polynomial or long division of a polynomial $a(X)$ by a polynomial $b(X) \neq 0$ requires repeated division by the leading coefficient b_n of $b(X)$. This is practicable only if every element of R is divisible by b_n; for instance, if b_n is a unit or if $b(X)$ is **monic** (its leading coefficient b_n is 1).

Proposition 4.3.7. *Let $b(X) \in R[X]$ be a nonzero polynomial whose leading coefficient is a unit of R. For every polynomial $a(X) \in R[X]$ there exists polynomials $q(X), r(X) \in R[X]$ such that*

$$a(X) = b(X)q(X) + r(X) \qquad and \qquad \deg r(X) < \deg b(X).$$

Furthermore $q(X)$ and $r(X)$ are unique.

Proof. Existence is proved by induction on $\deg a$. If $\deg a < \deg b$, then $q = 0$ and $r = a$ serve. Now let $\deg a = m \geqslant n = \deg b$. Then $a_m b_n^{-1} X^{m-n} b(X)$ has degree m and its leading term is $a_m b_n^{-1} X^{m-n} b_n X^n = a_m X^m$, which equals the leading term of a. Hence $a(X) - a_m b_n^{-1} X^{m-n} b(X)$ has degree less than m. By the induction hypothesis, $a(X) - a_m b_n^{-1} X^{m-n} b(X) = b(X)q(X) + r(X)$, with $\deg r < \deg b$; hence $a(X) = b(X)(a_m b_n^{-1} X^{m-n} + q(X)) + r(X)$.

To show that q and r are unique, we note that $\deg(bq) = \deg b + \deg q$ for all $q \neq 0$, since the leading coefficient of b is a unit. Assume that $a = bq_1 + r_1 = bq_2 + r_2$, where $q_1, q_2, r_1, r_2 \in R[X]$ and $\deg r_1, \deg r_2 < \deg b$. Then $b(q_1 - q_2) = r_2 - r_1$ has degree less than $\deg b$. If $q_1 - q_2 \neq 0$, then $\deg b(q_1 - q_2) = \deg b + \deg(q_1 - q_2) \geqslant \deg b$, a contradiction; therefore $q_1 = q_2$, which implies $r_1 = r_2$. $\square$

Corollary 4.3.8. *Let $t \in R$ and $a(X) \in R[X]$. If R is commutative, then $a(X)$ is a multiple of $X - t$ if and only if $a(t) = 0$.*

Proof. By Proposition 4.3.7, $a(X) = (X - t)q(X) + r$, where r is constant $(\deg r < 1)$. Since q and r are unique, $a(X)$ is a multiple of $X - t$ if and only if $r = 0$. Since evaluation at t is a homomorphism, $r = a(t)$. $\square$

4.3.h. Polynomials in Several Indeterminates. Intuitively, a polynomial with commuting indeterminates $(X_i)_{i \in I}$ and coefficients in R is a linear combination of monomials $X_{i_1}^{m_1} X_{i_2}^{m_2} \ldots X_{i_k}^{m_k}$, with coefficients in R.

The **exponent** m of a monomial $X_{i_1}^{m_1} X_{i_2}^{m_2} \ldots X_{i_k}^{m_k}$ is a family $m = (m_i)_{i \in I}$ of nonnegative integers such that $m_i = 0$ for almost all $i \in I$; m_i is the exponent of X_i. We denote by E_I the set of all such exponents. We recognize E_I as the free commutative monoid on the set I from Section 2.5.

The **degree** of an exponent $m = (m_i)_{i \in I} \in E_I$ is $\deg m = \sum_{i \in I} m_i$. We see that 0 (the exponent m with $m_i = 0$ for all i) has degree 0; all other exponents have positive degree.

Exponents are added componentwise: $(m_i)_{i \in I} + (n_i)_{i \in I} = (m_i + n_i)_{i \in I}$. We see that $\deg (m + n) = \deg m + \deg n$.

Formally a **polynomial with commuting indeterminates** indexed by I and coefficients in R is a family $a = (a_m)_{m \in E_I}$ of elements of R such that $a_m = 0$ for almost all m (such that the set $\{ m \in E_I ; a_m \neq 0 \}$ is finite). Addition is componentwise:

$$a + b = c, \qquad \text{where} \quad c_m = a_m + b_m \quad \text{for all } m,$$

and multiplication is defined by

$$ab = c, \qquad \text{where} \quad c_m = \sum_{j+k=m} a_j b_k \quad \text{for all } m,$$

using the addition on E_I.

Proposition 4.3.9. *When R is a ring, polynomials with indeterminates indexed by I and coefficients in R form a ring $R[(X_i)_{i \in I}]$. If R is commutative, then $R[(X_i)_{i \in I}]$ is commutative.*

$R[(X_i)_{i \in I}]$ is often denoted by $R[X]$; we don't use this notation unless $(X_i)_{i \in I}$ has only one element X.

Proof. First, $R[(X_i)_{i \in I}]$ is an abelian group under addition, which we recognize as the direct sum $\bigoplus_{m \in E_I} R$.

We show that the multiplication on $R[(X_i)_{i \in I}]$ is well defined. Let $a, b \in R[(X_i)_{i \in I}]$. Then $A = \{ j \in E_I ; a_j \neq 0 \}$ and $B = \{ k \in E_I ; b_k \neq 0 \}$ are finite; hence $A + B = \{ j + k ; j \in A, k \in B \}$ is finite. If $m \notin A + B$, then $j + k = m$ implies $j \notin A$ or $k \notin B$, and $a_j b_k = 0$; hence $c_m = 0$. Thus $c_m = 0$ for almost all m and $c = ab \in R[(X_i)_{i \in I}]$.

The multiplication on $R[(X_i)_{i \in I}]$ inherits distributivity (and commutativity, if R is commutative) from the multiplication on R, and 1 (whose coefficients are $c_m = 1$ if $m = 0$, $c_m = 0$ otherwise) is an identity element. Associativity is an exercise. $\qquad\square$

It is customary to identify $r \in R$ and the **constant** polynomial $(r_m)_{m \in E_I}$ such that $r_0 = r$ and $r_m = 0$ for all $m \neq 0$. The i-th indeterminate X_i is the polynomial $(c_m)_{m \in E_I}$ with $c_m = 1$ if $m_i = 1$, $m_j = 0$ for all $j \neq i$, and $c_m = 0$ otherwise. We see that the indeterminates commute with the constants and with each other.

In addition we see that X_i^t is the polynomial $(c_m)_{m \in E_I}$ with $c_m = 1$ if $m_i = t$, $m_j = 0$ for all $j \neq i$, and $c_m = 0$ otherwise. If now $k \in E_I$ is any exponent and $\{i \in I ; k_i \neq 0\} = i_1, \ldots, i_\ell$, then

$$X^k = X_{i_1}^{k_{i_1}} X_{i_2}^{k_{i_2}} \ldots X_{i_\ell}^{k_{i_\ell}}$$

is the polynomial $(c_m)_{m \in E_I}$ with $c_m = 1$ if $m = k$ and $c_m = 0$ for all $m \neq k$. In this product the terms can be written in any order, since the indeterminates commute with each other. Polynomials can now be written in the usual form:

Proposition 4.3.10. $a = \sum_{m \in E_I} a_m X^m$, for every $a \in R[(X_i)_{i \in I}]$.

The sum $\sum_{m \in E_I} a_m X^m$ exists since $a_m = 0$ for almost all m.

When $I = 1, 2, \ldots, k$ is finite, $R[(X_i)_{i \in I}]$ is also denoted by $R[X_1, X_2, \ldots, X_k]$ (or by $R[X,Y]$ or $R[X,Y,Z]$ if $k = 2$ or $k = 3$). Each $a \in R[(X_i)_{i \in I}]$ is commonly denoted by $a((X_i)_{i \in I})$ (by $(a(X_1, \ldots, X_k)$ if $I = \{1, 2, \ldots, k\})$.

The following result is often used as a definition of $R[X_1, \ldots, X_k]$.

Proposition 4.3.11. $R[X_1, \ldots, X_k, X_{k+1}] \cong (R[X_1, \ldots, X_k])[X_{k+1}]$.

The proof is by induction on k and an exercise.

The degree and order of a polynomial with several indeterminates are defined in the obvious way. Let $a((X_i)_{i \in I}) = \sum_{m \in E_I} a_m X^m \in R[(X_i)_{i \in I}]$. If $a \neq 0$, the **degree** of a is $\deg a = \max(\deg m ; a_m \neq 0)$; if $a = 0$, then $\deg a$ is, say, $-\infty$. If $a \neq 0$, the **order** of a is $\operatorname{ord} a = \min(\deg m ; a_m \neq 0)$; if $a = 0$, then $\operatorname{ord} a = \infty$. Propositions 4.3.3 and 4.3.5 extend to $R[(X_i)_{i \in I}]$ (see the exercises).

We also leave the universal property of $R[(X_i)_{i \in I}]$ (which implies Proposition 4.3.11) as an exercise:

Proposition 4.3.12. Let R and S be rings, $\varphi : R \longrightarrow S$ be a homomorphism, and $(s_i)_{i \in I}$ be any family of elements of S which commute with each other and with every $\varphi(r)$. There is a homomorphism $\psi : R[(X_i)_{i \in I}] \longrightarrow S$ unique such that $\psi \circ \iota = \varphi$ and $\psi(X_i) = s_i$ for all i; namely

$$\psi(a) = \psi \left(\sum_{m \in E_I} a_m X^m \right) = \sum_{m \in E_I} \varphi(a_m) s^m,$$

where s^m denotes $s_{i_1}^{m_{i_1}} s_{i_2}^{m_{i_2}} \ldots s_{i_\ell}^{m_{i_\ell}}$ when $\{i \in I ; m_i \neq 0\} = \{i_1, \ldots, i_\ell\}$. Moreover $\operatorname{Im} \psi$ is the smallest subring of S which contains $\operatorname{Im} \varphi$ and all s_i.

If S is commutative, then s_i can be chosen arbitrarily in S.

Proposition 4.3.12 provides a free commutative ring with a universal property similar to that of free groups. Given a set I, call a commutative ring R, together with a mapping $\iota : I \longrightarrow R$, **free** on I in case for every mapping φ of I into a commutative ring S there is a homomorphism $\psi : R \longrightarrow S$ of rings unique such that $\psi \circ \iota = \varphi$. By Proposition 4.3.12, $\mathbb{Z}[(X_i)_{i \in I}]$ is free on the set I (see the exercises).

Proposition 4.3.12 provides a number of constructions, the details of which are left as exercises.

Every homomorphism $\varphi : R \longrightarrow S$ of rings extends to a homomorphism $R[(X_i)_{i \in I}] \longrightarrow S[(X_i)_{i \in I}]$ of polynomial rings.

Polynomials in several indeterminates can be evaluated at elements of R; if R is commutative, this provides evaluation homomorphisms.

When R is commutative, Proposition 4.3.12 can also be applied to $\varphi = 1_R$ and any family $(b_i)_{i \in I}$ of elements of another polynomial ring $R[(Y_j)_{j \in J}]$, yielding a homomorphism $\sigma : R[(X_i)_{i \in I}] \longrightarrow R[(Y_j)_{j \in J}]$ unique such that $\sigma(\iota(r)) = r$ for all $r \in R$ and $\sigma(X_i) = b_i$ for all i; σ is a **substitution** homomorphism, since $\sigma(a)$ is obtained from a by the substitutions $X_i = b_i((Y_j)_{j \in J})$.

4.3.i. Formal Power Series. Power series are one of several generalizations of polynomial rings, some of which will be seen in the exercises.

A **formal power series** with one indeterminate and coefficients in R is an infinite sequence

$$a = (a_0, a_1, \ldots, a_n, \ldots)$$

of elements of R; $a = (a_0, a_1, \ldots, a_n, \ldots)$ is normally denoted by $a(X) = \sum_{n \geqslant 0} a_n X^n$. (Formal power series with several indeterminates are left as exercises.) Addition is componentwise:

$$a + b = c, \qquad \text{where} \quad c_n = a_n + b_n \quad \text{for all } n,$$

and multiplication is defined by

$$ab = c, \qquad \text{where} \quad c_n = \sum_{i+j=n} a_i b_j \quad \text{for all } n.$$

Proposition 4.3.13. *When R is a ring, formal power series with one indeterminate and coefficients in R form a ring $R[[X]]$. If R is commutative, then $R[[X]]$ is commutative.*

We have $(R[[X]], +) \cong \prod_{n \geqslant 0} R$, and $R[[X]]$ is an abelian group under addition. The multiplication on $R[[X]]$ inherits commutativity, distributivity, and associativity from the multiplication on R, as in Proposition 4.3.1.

Formal power series have no degree. The **order** of a formal power series $a \neq 0$ is the smallest integer $n = \text{ord}\, a$ such that $a_n \neq 0$; the order of 0 is ∞, so that $\text{ord}\, 0 > \text{ord}\, a$ for all $a \neq 0$. A power series has order at least n if and only if it is a multiple of X^n. The following result is proved like Proposition 4.3.5.

Proposition 4.3.14. *For all $a,b \neq 0$ in $R[[X]]$:*

(1) $\operatorname{ord}(a+b) \geqslant \min(\operatorname{ord}a, \operatorname{ord}b)$;

(2) *if* $\operatorname{ord}a \neq \operatorname{ord}b$, *then* $\operatorname{ord}(a+b) = \min(\operatorname{ord}a, \operatorname{ord}b)$;

(3) $\operatorname{ord}(ab) \geqslant \operatorname{ord}a + \operatorname{ord}b$;

(4) *if R has no zero divisors, then* $\operatorname{ord}(ab) = \operatorname{ord}a + \operatorname{ord}b$.

In the absence of a topology on R, we cannot define the sum of a series of elements of R, or evaluate a formal power series at an element of R. However, certain series can be added in $R[[X]]$ in a purely algebraic fashion (but for which the exercises give a topological interpretation).

Let $t_0(X), t_1(X), \ldots, t_k(X), \ldots$ be a sequence of formal power series, with $t_k(X) = (t_{k,0}, \ldots, t_{k,n}, \ldots) = t_{k,0} + \cdots + t_{k,n}X^n + \cdots$. We say that the sequence $t_0(X), \ldots, T_k(X), \ldots$ is **addible** in $R[[X]]$ in case, for every $n \geqslant 0$, $\operatorname{ord}t_k(X) \geqslant n$ for almost all k. If for instance $\operatorname{ord}t_k(X) \geqslant k$ for all k, then $t_0(X), \ldots, T_k(X), \ldots$ is addible.

Let $t_0(X), \ldots, t_k(X), \ldots$ be addible. For each $n \geqslant 0$, $t_{k,n} = 0$ holds for almost all k; hence the sum $\sum_{k \geqslant 0} t_{k,n}$ has only finitely many nonzero terms and can be calculated in R. The **sum** $S(X) = \sum_{k \geqslant 0} t_k(X) \in R[[X]]$ is the power series with coefficients $s_n = \sum_{k \geqslant 0} t_{k,n}$.

We give some applications of infinite sums in $R[[X]]$. First,

$$a(X) = a_0 + a_1 X + \cdots + a_n X^n + \cdots$$

holds for every formal power series $a \in R[[X]]$ (the sequence $a_0, \ldots, a_n X^n, \ldots$ in the right-hand side is addible, since $\operatorname{ord}a_n X^n \geqslant n$).

Proposition 4.3.15. *If R is commutative, then $a(X) = \sum_a {}_n X^n$ is a unit of $R[[X]]$ if and only if a_0 is a unit of R.*

Proof. If $ab = 1$ in $R[[X]]$, then $a_0 b_0 = 1$ in R and a_0 is a unit. We now prove the converse.

First we show that a is a unit in case $a_0 = 1$. Then $a = 1 - u$, where $u = 1 - a$ has order at least 1. We show that

$$b = 1 + u + u^2 + \cdots + u^n + \cdots$$

is an inverse of a. First $\operatorname{ord}u^n \geqslant n$ by Proposition 4.3.14, so that the sequence $1, u, \ldots, u^n, \ldots$ is addible. Let

$$b_n = 1 + u + u^2 + \cdots + u^n$$

be the partial sum. Then $b - b_n = u^{n+1} + \cdots$ has order $\operatorname{ord}(b - b_n) \geqslant n + 1$. Hence $\operatorname{ord}(ab - ab_n) > n$. On the other hand,

$$ab_n = (1 - u)(1 + u + \cdots + u^n) = 1 - u^{n+1}$$

and $\mathrm{ord}\,(ab_n - 1) = \mathrm{ord}\,u^{n+1} > n$. Hence $\mathrm{ord}\,(ab - 1) > n$, for all $n > 0$, which implies $ab - 1 = 0$. Similarly $ba = 1$.

If now a_0 is any unit, then $a_0^{-1}a(X)$, which has constant coefficient 1, has an inverse $b(X)$, and $a_0 b(X)$ is an inverse of $a(X)$. □

Substitution of power series is defined for any $a(X), b(X) \in R[[X]]$ such that $\mathrm{ord}\,b \geqslant 1$: then $\mathrm{ord}\,a_n b^n \geqslant n$, and there is a sum

$$a(b) = a_0 + a_1 b(X) + \cdots + a_n b(X)^n + \cdots$$

which is obtained from $a(X)$ by the substitution $X = b$. For example, the inverse $1 + u + u^2 + \cdots + u^n + \cdots$ of $1 - u$ is obtained by substituting $X = u$ in the series $1 + X + X^2 + \cdots + X^n + \cdots$. Substitution of power series shares the main properties of substitution of polynomials (see the exercises).

Exercises

1. Verify that the multiplication on $R[X]$ is associative.
2. Prove that the universal property of $R[X]$ characterizes $R[X]$ uniquely up to isomorphism.
3. Show that substitution in $R[X]$ is associative when R is commutative.
4. Let R have no zero divisors. Show that $a \longmapsto a \circ b$ is a homomorphism for all $b \in R[X]$ if and only if R is commutative.
5. Prove Proposition 4.3.4.
6. Prove Proposition 4.3.6.
7. Prove the following. Let $b \in R[X]$ be a nonconstant polynomial whose leading coefficient is a unit of R. For every polynomial $a \in R[X]$ of degree at most m there exists polynomials $c_0, c_1, \ldots, c_m \in R[X]$ such that $\deg c_i < \deg b$ for all i and

$$a = c_0 + c_1 b + \cdots + c_m b^m.$$

   Furthermore $c_0, \ldots, c_m$ are unique.
8. Prove Proposition 4.3.11.
9. Prove Proposition 4.3.12.
10. Define evaluation mappings $R[(X_i)_{i \in I}] \longrightarrow R$, which are homomorphisms if R is commutative.
11. Show that $\mathbb{Z}[(X_i)_{i \in I}]$ is a free commutative ring [with identity] on the set I.
12. Extend Proposition 4.3.3 to $R[(X_i)_{i \in I}]$.
13. Extend Proposition 4.3.5 to $R[(X_i)_{i \in I}]$.
14. Prove Proposition 4.3.13.
15. Prove Proposition 4.3.14.
16. Let R be commutative and $\mathfrak{m}$ be a maximal ideal of R. Show that $\mathfrak{m} + (X)$ is a maximal ideal of $R[[X]]$.
17. Prove that $R[[X]]$ is a metric space, with $d(a,b) = 2^{-\mathrm{ord}\,(a-b)}$ if $a \neq b$, $d(a,b) = 0$ if $a = b$. Show that the operations on $R[[X]]$ are continuous.

18. Show that the metric space $R[[X]]$ is the completion of $R[X]$.

19. Show that $\sum_{n \geqslant 0} a_n X^n$ is the sum of a series in the metric space $R[[X]]$.

20. If R is commutative, show that $(a+b)(c) = a(c) + b(c)$ and $(ab)(c) = a(c)b(c)$ for all $a,b,c \in R[[X]]$ with $\operatorname{ord} c \geqslant 1$ (so that $\sigma_c : a \longmapsto a(c)$ is an endomorphism of $R[[X]]$).

21. If R is commutative, show that $a \circ (b \circ c) = (a \circ b) \circ c$ for all $a,b,c \in R[[X]]$ with $\operatorname{ord} b,c \geqslant 1$.

22. Let R be a ring [with identity]. Define a ring of polynomials $R[X,Y]$ with two indeterminates X and Y which commute with the elements of R but not with each other. State and prove the appropriate universal property.

23. Define a ring of polynomials $R[(X_i)_{i \in I}]$ with noncommuting indeterminates $(X_i)_{i \in I}$. State and prove the appropriate universal property.

24. Define a ring $R[[(X_i)_{i \in I}]]$ of formal power series with commuting indeterminates $(X_i)_{i \in I}$.

4.4. DOMAINS AND FIELDS

This section contains basic properties of domains, including fields of fractions.

4.4.a. Definition. A **domain** or **integral domain** is a commutative ring [with identity] $R \neq 0$ such that $R \backslash \{0\}$ is closed under multiplication (if $a,b \neq 0$, then $ab \neq 0$); equivalently, a domain is a commutative ring which has no zero divisors. (The requirement that R be commutative is often omitted from this definition.) In a domain, $ab = ac$ implies $b = c$ whenever $a \neq 0$.

A **field** is a commutative ring K such that $K \backslash \{0\}$ is a group under multiplication; equivalently $K \neq 0$ and every nonzero element is a unit. Fields are domains; $\mathbb{Z}$ is a domain but not a field.

When R is a domain, $\deg(fg) = \deg f + \deg g$ holds for all nonzero polynomials $f,g \in R[(X_i)_{i \in I}]$; hence $R[(X_i)_{i \in I}]$ is a domain. (So is $R[[X]]$.) Thus $\mathbb{Z}[(X_i)_{i \in I}]$ is a domain; so is $K[(X_i)_{i \in I}]$ for every field K.

Proposition 4.4.1. *The characteristic of a domain is 0 or a prime number.*

Proof. Let R have characteristic $n \neq 0$ which is not a prime. Then $n = k\ell$ for some $0 < k, \ell < n$. In R we have $(k1)(\ell 1) = n1 = 0$ and $k1, \ell 1 \neq 0$, so R is not a domain. □

For instance $\mathbb{Z}_n$ (which has characteristic n) is a domain if and only if n is a prime, in which case $\mathbb{Z}_n$ is a field.

4.4.b. Prime and Maximal Ideals. A **prime** ideal of a commutative ring R is an ideal $\mathfrak{p} \neq R$ such that $ab \in \mathfrak{p}$ implies $a \in \mathfrak{p}$ or $b \in \mathfrak{p}$. For example, the principal ideal $(n) = \mathbb{Z}n$ of $\mathbb{Z}$ (with $n > 0$) is prime if and only if the number n is prime. The following result is an exercise:

Proposition 4.4.2. *Let R be a commutative ring and $\mathfrak{a}$ be an ideal of R. Then $R/\mathfrak{a}$ is a domain if and only if $\mathfrak{a}$ is a prime ideal.*

Recall that a **maximal** ideal of a ring R is a maximal proper ideal $\mathfrak{m}$ ($\mathfrak{m} \subsetneq R$, and there exists no ideal $\mathfrak{m} \subsetneq \mathfrak{a} \subsetneq R$).

Proposition 4.4.3. *Let R be a commutative ring and $\mathfrak{a}$ be an ideal of R. Then $R/\mathfrak{a}$ is a field if and only if $\mathfrak{a}$ is a maximal ideal.*

Proof. Let $\mathfrak{a}$ be an ideal which is not maximal. If $\mathfrak{a} = R$, then $R/\mathfrak{a} = 0$ is not a field (since a group may not be empty). If $\mathfrak{a} \neq R$, then there is an ideal $\mathfrak{a} \subsetneq \mathfrak{b} \subsetneq R$. By Proposition 4.2.6, $\mathfrak{b}/\mathfrak{a}$ is an ideal of $R/\mathfrak{a}$ and $0 \neq \mathfrak{b}/\mathfrak{a} \neq R/\mathfrak{a}$. In particular, the identity element $1 + \mathfrak{a}$ of $R/\mathfrak{a}$ is not in $\mathfrak{b}/\mathfrak{a}$ (otherwise, $\mathfrak{b}/\mathfrak{a} = R/\mathfrak{a}$). There exists $0 \neq b + \mathfrak{a} \in \mathfrak{b}/\mathfrak{a}$; $b + \mathfrak{a}$ is not a unit of $R/\mathfrak{a}$, since all its multiples lie in $\mathfrak{b}/\mathfrak{a}$. Hence $R/\mathfrak{a}$ is not a field.

Now let $\mathfrak{m}$ be a maximal ideal of R. Then $R/\mathfrak{m} \neq 0$. Let $0 \neq a + \mathfrak{m} \in R/\mathfrak{m}$. Then $a \notin \mathfrak{m}$ and $\mathfrak{m} \subsetneq Ra + \mathfrak{m} \subseteq R$. Since $\mathfrak{m}$ is maximal, this implies $Ra + \mathfrak{m} = R$; hence $1 = ra + m$ for some $r \in R$ and $m \in \mathfrak{m}$. Then $(r + \mathfrak{m})(a + \mathfrak{m}) = 1 + \mathfrak{m}$ in $R/\mathfrak{m}$, which shows that $a + \mathfrak{m}$ is a unit. Thus $R/\mathfrak{m}$ is a field. $\square$

Corollary 4.4.4. *In a commutative ring, every maximal ideal is prime.*

4.4.c. Fields of Fractions. The construction of the field of fractions of a domain generalizes the construction of $\mathbb{Q}$ in Section A.4. It will in turn be generalized in Chapter 11.

Let R be a domain. Define an equivalence relation $\equiv$ on $R \times (R \backslash \{0\})$ by

$$(a,b) \equiv (c,d) \iff ad = bc.$$

The equivalence class of $(a,b) \in R \times (R \backslash \{0\})$ is a **fraction** and is denoted by a/b or by $\frac{a}{b}$. The **field of fractions** $Q = Q(R)$ of R (also called the **quotient field** of R) is the set $Q = (R \times (R \backslash \{0\}))/\equiv$ of all fractions, with operations given by:

$$(a/b) + (c/d) = (ad + bc)/bc, \qquad (a/b)(c/d) = ac/bd.$$

It is straightforward that $\equiv$ is an equivalence relation, that the operations above are well defined, and that Q is a field with zero element $0/1$ and identity element $1/1$; if $a/b \neq 0$, then $(a/b)^{-1} = b/a$.

Proposition 4.4.5. *Let R be a domain. Then $Q(R)$ is a field and $\iota : a \longmapsto a/1$ is an injective homomorphism.*

If, for example, $R = \mathbb{Z}$, then $Q(R) \cong \mathbb{Q}$. In general, $Q(R)$ has a universal property:

Proposition 4.4.6. *Let R be a domain. Every injective homomorphism φ: $R \longrightarrow F$ of R into a field F factors uniquely through $\iota: R \longrightarrow Q(R)$: there is a homomorphism $\psi: Q(R) \longrightarrow F$ unique such that the following diagram commutes*

ψ *is given by* $\psi(a/s) = \varphi(a)\varphi(s)^{-1}$ *and is injective.*

Proof. When $b \neq 0$, the inverse of $\iota(b) = b/1$ in $Q = Q(R)$ is $1/b$. Hence $a/b = (a/1)(1/b) = \iota(a)\iota(b)^{-1}$. If $a/b = c/d$, then $ad = bc$, $\varphi(a)\varphi(d) = \varphi(b)\varphi(c)$, and $\varphi(a)\varphi(b)^{-1} = \varphi(c)\varphi(d)^{-1}$. Hence there exists a mapping $\psi: Q \longrightarrow F$ such that $\psi(a/b) = \varphi(a)\varphi(b)^{-1}$ for all $a/b \in Q$. It is immediate that ψ is a homomorphism and that $\psi \circ \iota = \varphi$.

Conversely, let $\psi: Q \longrightarrow F$ be a homomorphism (of rings with identity). Then ψ is injective, since Q is a field: if $q \neq 0$ in Q, then $\psi(q)\psi(q^{-1}) = \psi(qq^{-1}) = 1$ and $\psi(q) \neq 0$; we see that $\psi(q)^{-1} = \psi(q^{-1})$. If $\psi \circ \iota = \varphi$, then necessarily $\psi(a/b) = \varphi(a)\varphi(b)^{-1}$. Hence ψ is unique. $\square$

It is customary to identify $x \in R$ and $\iota(x) \in Q(R)$. Then R is a subring of the field $Q(R)$. By Proposition 4.4.6, $Q(R)$ is (up to isomorphism) the smallest such field. The field of fractions of R can also be characterized as follows:

Proposition 4.4.7. *Let K be a field and R be a subring of K. Then K is [isomorphic to] the field of fractions of R if and only if every element of K can be written in the form ab^{-1} with $a, b \in R$, $b \neq 0$.*

Proof. This condition is necessary. Conversely, let R and K be as in the statement. Then R is a domain and has a field of fractions Q. By Proposition 4.4.6 there is a homomorphism $\theta: Q \longrightarrow K$ such that $\theta \circ \iota$ is the inclusion homomorphism $R \longrightarrow K$, namely $\theta(a/b) = ab^{-1}$. Since Q is a field, θ is injective: if $0 \neq q \in Q$, then $\theta(q)\theta(q^{-1}) = \theta(1) = 1$ shows that $\theta(q) \neq 0$. Every element of K can be written in the form $ab^{-1} = \theta(a/b)$, so θ is surjective. $\square$

4.4.d. Rational Fractions. When K is a field, the polynomial ring $K[(X_i)_{i \in I}]$ (in one or several indeterminates) is a domain. The field of fractions of $K[(X_i)_{i \in I}]$ is the field of **rational fractions** $K((X_i)_{i \in I})$ with coefficients in K (with the same indeterminates $(X_i)_{i \in I}$ as $K[(X_i)_{i \in I}]$).

In the construction of $K((X_i)_{i \in I})$ the field K could be replaced by any domain; but the same fields would obtain (see the exercises).

If $(X_i)_{i \in I}$ is a finite sequence $X_1, \ldots, X_n$, then $K((X_i)_{i \in I})$ is denoted by $K(X_1, \ldots, X_n)$. In this case $K(X_1, \ldots, X_n)$ could be defined by induction:

Proposition 4.4.8. $K(X_1, \ldots, X_{n+1}) \cong K(X_1, \ldots, X_n)(X_{n+1})$.

Proof. This follows from Proposition 4.4.7. Indeed $F = K(X_1, \ldots, X_{n+1})$ is a field; $R = K(X_1, \ldots, X_n)[X_{n+1}]$ is [isomorphic to] a subring of F (which consists of all fractions f/g whose denominator g does not contain X_{n+1}), and every element of F can be written in the form $f/g = fg^{-1}$ with $f, g \in K[X_1, \ldots, X_{n+1}] \subseteq R$. $\qquad\square$

4.4.e. Formal Laurent Series. Finally we construct the field of fractions of $K[[X]]$, when K is a field.

A **formal Laurent series** with one indeterminate and coefficients in R is a family $a = (a_n)_{n \in \mathbb{Z}}$ such that $a_n = 0$ for almost all negative values of n (there is no restriction on a_n when $n > 0$); $a = (a_n)_{n \in \mathbb{Z}}$ is normally denoted by $a(X) = \sum_{n \in \mathbb{Z}} a_n X^n$. Addition is componentwise:

$$a + b = c, \qquad \text{where} \quad c_n = a_n + b_n \quad \text{for all } n,$$

and multiplication is defined by

$$ab = c, \qquad \text{where} \quad c_n = \sum_{i+j=n} a_i b_j \quad \text{for all } n.$$

Proposition 4.4.9. *When R is a commutative ring, formal Laurent series with one indeterminate and coefficients in R form a commutative ring $R((X))$.*

The proof is similar to the proof of Proposition 4.3.1 and is left as an exercise. We see that $R[[X]]$ is a subring of $R((X))$.

The **order** of a formal Laurent series $a \neq 0$ is the least integer $n = \operatorname{ord} a \in \mathbb{Z}$ such that $a_n \neq 0$; the order of 0 is ∞, so that $\operatorname{ord} 0 > \operatorname{ord} a$ for all $a \neq 0$. A Laurent series has order at least n if and only if it is a multiple of X^n. The following result is proved like Proposition 4.3.6.

Proposition 4.4.10. *For all $a, b \neq 0$ in $R((X))$:*

(1) $\operatorname{ord}(a + b) \geqslant \min(\operatorname{ord} a, \operatorname{ord} b)$;

(2) *if* $\operatorname{ord} a \neq \operatorname{ord} b$, *then* $\operatorname{ord}(a + b) = \min(\operatorname{ord} a, \operatorname{ord} b)$;

(3) $\operatorname{ord}(ab) \geqslant \operatorname{ord} a + \operatorname{ord} b$;

(4) *if R has no zero divisors, then* $\operatorname{ord}(ab) = \operatorname{ord} a + \operatorname{ord} b$.

Certain series can be added in $R((X))$ in a purely algebraic fashion. A sequence $t_0(X), t_1(X), \ldots, t_k(X), \ldots$ of formal Laurent series is **addible** in $R((X))$ in case, for every $n \geqslant 0$, $\operatorname{ord} t_k \geqslant n$ for almost all k. If, for instance, $\operatorname{ord} t_k \geqslant k$ for all k, then $t_0, \ldots, t_k, \ldots$ is addible. A **sum** $s(X) = \sum_{k \geqslant 0} t_k(X) \in R((X))$ is then defined as in Section 4.4.3. For instance, a Laurent series of order $n \neq \infty$ can be written as a sum $a(X) = \sum_{k \geqslant n} a_k X^k$, with $a_n \neq 0$.

Proposition 4.4.11. *When K is a field, $K((X))$ is [isomorphic to] the field of fractions of $K[[X]]$.*

Proof. By Proposition 4.3.15, a power series b is a unit of $K[[X]]$ if and only if $b_0 \neq 0$. It follows that $K((X))$ is a field: if $a(X) \neq 0$ in $K((X))$, then $a(X)$ has order $n \neq \infty$ and can be written in the form $a(X) = X^n b(X)$, where $b(X) = a_n + a_{n+1} X + \cdots + a_{n+k} X^k + \cdots$; hence $a(X)$ has an inverse in $K((X))$, namely $X^{-n} c(X)$, where $c(X)$ is an inverse of $b(X)$ in $K[[X]]$. Either $a(X) \in K[[X]]$ (if $n \geqslant 0$) or $a(X) = b(X)(X^{-n})^{-1}$ with $b(X), X^{-n} \in K[[X]]$ (if $n < 0$); hence $K((X))$ is the field of fractions of $K[[X]]$, by Proposition 4.4.7. $\quad\square$

Exercises

1. Prove Proposition 4.4.2.
2. Give a direct proof of Corollary 4.4.4.
3. Prove Proposition 4.4.5.
4. Let R be a domain and Q be its field of fractions. Show that the field of fractions of $R[X]$ is $Q(X)$.
5. Prove Proposition 4.4.9.
6. Prove that a Laurent series a of order n is a unit of $R((X))$ if and only if a_n is a unit of R.

5

FACTORIZATION IN RINGS

This chapter extends to some commutative rings the prime factorization property of integers, that every positive integer is a product of positive powers of primes. This is carried out with elements (Sections 5.1 and 5.4), then, more abstractly, with ideals (Sections 5.6 and 5.7). The results yield a number of properties of polynomial rings and introduce some important classes of rings. Commutative rings are studied in more depth in Chapter 11.

Required results: basic properties of integers (Section A.3); chain conditions (Section B.2); Zorn's Lemma (Theorem B.3.2); the results in Chapter 4.

All rings in this chapter are commutative rings with identity.

5.1. PRINCIPAL IDEAL DOMAINS

Principal ideal domains share all the arithmetic properties of $\mathbb{Z}$, yet are general enough to include the polynomial ring $K[X]$ when K is a field. This extends properties of integers to polynomials.

5.1.a. Definition. A **principal ideal domain** or PID is a domain R (a commutative ring with identity and no zero divisors) in which every ideal is principal. Since R is commutative with identity, the principal ideal generated by $a \in R$ is the set $(a) = Ra = \{ xa ; x \in R \}$ of all multiples of a.

5.1.b. Examples. We saw in Section 1.4 that $\mathbb{Z}$ is a PID.

Proposition 5.1.1. *When K is a field, the polynomial ring $K[X]$ is a principal ideal domain.*

Proof. It suffices to show that every nonzero ideal $\mathfrak{A}$ of $K[X]$ is principal. This follows from polynomial division (Proposition 5.3.6). Since $\mathfrak{A} \neq 0$ there exists a polynomial $0 \neq b \in \mathfrak{A}$ of smallest degree. We show that $\mathfrak{A} = (b)$. Already $(b) \subseteq \mathfrak{A}$. Conversely, let $a \in \mathfrak{A}$. Then $a = bq + r$ for some $q, r \in K[X]$, where $\deg r < \deg b$. Then $r = a - bq \in \mathfrak{A}$. If $r \neq 0$, then $\deg r < \deg b$ contradicts the choice of b; therefore $r = 0$, and $a = bq \in (b)$. Thus $\mathfrak{A} \subseteq (b)$. $\qquad\square$

On the other hand, polynomial rings with more than one indeterminate are not principal (see the exercises).

5.1.c. Properties. We now extend to all PIDs the arithmetic properties of $\mathbb{Z}$ proved in Section A.3. First we note some general properties.

Lemma 5.1.2. *In a domain R, $Ra = Rb$ if and only if $a = ub$ for some unit u.*

Proof. When u is a unit, $Ru = R$ and $Rub = Rb$. Conversely, let $Ra = Rb$. Then $a = ub$, $b = va$ for some $u, v \in R$. If $a = 0$, then $b = 0$, and conversely, and then $a = 1b$. Otherwise, $uva = a \neq 0$ implies $uv = 1$, and u is a unit. $\qquad\square$

A domain R is partitioned into equivalence classes by the equivalence relation $Ra = Rb$; equivalent elements are often called **associates**, and we call the equivalence classes **associate classes** of R. Uniqueness in various results can be achieved by choosing what we call one **representative** element in each nonzero associate class. For instance, the units of $\mathbb{Z}$ are $+1$ and -1; hence every associate class of $\mathbb{Z}$ contains one positive integer, and positive integers can be used as representative elements of $\mathbb{Z}$.

Lemma 5.1.3. *The ideals of a PID satisfy the ascending chain condition.*

Proof. Let $\mathfrak{a}_1 \subseteq \mathfrak{a}_2 \subseteq \cdots \subseteq \mathfrak{a}_n \subseteq \cdots$ be an ascending sequence of ideals of a PID. The union $\mathfrak{a} = \bigcup_{n \geqslant 0} \mathfrak{a}_n$ is an ideal; hence $\mathfrak{a} = (b)$ for some b. Then $b \in \mathfrak{a}_m$ for some m; hence $(b) \subseteq \mathfrak{a}_m \subseteq \mathfrak{a}_n \subseteq \mathfrak{a} = (b)$ and $\mathfrak{a}_m = \mathfrak{a}_n$ for all $n \geqslant m$. $\qquad\square$

An element p of a domain R is **irreducible** in case p is not 0 or a unit, and $p = ab$ implies that a or b is a unit of R (so that $Rp = Ra$ or $Rp = Rb$). If p is irreducible and u is a unit, then up is irreducible. An element p is **prime** in case Rp is a nonzero prime ideal of R; equivalently, in case p is not zero or a unit, and $ab \in Rp$ implies $a \in Rp$ or $b \in Rp$ (if $p \mid ab$, then $p \mid a$ or $p \mid b$). For example, the irreducible elements of $\mathbb{Z}$ are the primes and their opposites, and coincide with the prime elements of $\mathbb{Z}$. In general:

Lemma 5.1.4. *Let R be a PID. The following conditions on $p \in R$ are equivalent:*

(a) *p is irreducible;*

(b) *p is prime (Rp is a nonzero prime ideal);*

(c) *Rp is a maximal ideal.*

Proof. (b) implies (a). If p is prime and $p = ab$, then, say, $a \in Rp$; since $p \in Ra$, we have $Rp = Ra$, $p = au$ for some unit u, and $b = u$ is a unit.

(a) implies (c). Assume that Rp is contained in an ideal $\mathfrak{a}$ of R. Since R is a PID, $\mathfrak{a} = Ra$ for some $a \in R$. Then $p = ab$ for some $a \in R$. If p is irreducible, then either a is a unit and $\mathfrak{a} = R$ or b is a unit and $\mathfrak{a} = Rp$.

(c) implies (b) since maximal ideals are prime (Corollary 4.4.4). □

5.1.d. Unique Factorization

Lemma 5.1.5. *In a PID, every element, other than 0 and units, is a product of irreducible elements.*

Proof. Assume that this is false. Then there exists a principal ideal Ra generated by an element a which is not 0, not a unit, and not a product of [finitely many] irreducible elements. By the a.c.c. there exists a principal ideal Rm which is maximal with these properties. In particular, m is not irreducible. Since m is not 0 or a unit, we have $m = bc$, where b and c are not units. By Lemma 5.1.2, $Rm \subsetneq Rb$ and $Rm \subsetneq Rc$. By the maximality of Rm, b and c are products of irreducible elements; then so is $m = bc$. □

Theorem 5.1.6. *In a principal ideal domain, every element, other than 0 and units, is a nonempty product of irreducible elements. Furthermore if two products of irreducible elements*

$$p_1 p_2 \cdots p_m = q_1 q_2 \cdots q_n$$

are equal, then $m = n$ and the terms can be indexed so that $Rp_i = Rq_i$ for all i.

Equivalently, every nonzero element can be written uniquely (up to the order of the terms) as a product

$$u \, p_1^{a_1} \, p_2^{a_2} \cdots p_n^{a_n}$$

of a unit u and of positive powers of distinct representative irreducible elements.

Proof. By Lemma 5.1.5, it suffices to prove uniqueness in the first part of the statement. We proceed by induction on $\max(m,n)$. The result is clear if $m = n = 1$. Now assume that, say, $m > 1$. By Lemma 5.1.4, $q_i \in Rp_1$ for some i; then $Rp_1 = Rq_i$, since p_1 and q_i are irreducible. The q's can be indexed so that $Rq_1 = Rp_1$. By Lemma 5.1.2, $q_1 = up_1$ for some unit u.

We now have $p_2 \cdots p_m = uq_2 \cdots q_n$. If $n = 1$, then $p_2 \cdots p_m = u$ implies that $p_2, \ldots, p_n$ are units, a contradiction. Therefore $n > 1$. Then $q_2' = uq_2$ is irreducible, $Rq_2' = Rq_2$, and $p_2 \cdots p_m = q_2' \cdots q_n$. By the induction hypothesis, $m - 1 = n - 1$ and $q_2', \ldots, q_n$ can be indexed so that $Rp_i = Rq_i$ for all $i > 1$. □

5.1.e. l.c.m. and g.c.d. In a PID, the least common multiples and greatest common divisors can be obtained either from Theorem 5.1.6, or from ideals as in Section A.3. We begin with the latter approach.

In a domain R, an element a **divides** an element b in case $b = ar$ for some $r \in R$, that is, in case b is a multiple of a. This relationship is denoted by $a \mid b$; it is equivalent to $b \in Ra$ and to $Rb \subseteq Ra$. For example, 1 divides every element of R, but only units divide 1; 0 divides only 0, but every element a of R divides 0 (this only means $0 = 0a$ and does not make a a zero divisor).

Proposition 5.1.7. *Let R be a principal ideal domain and $a,b \in R$. There exists $m \in R$ such that:*

(i) $a \mid m$ and $b \mid m$;

(ii) *if $a \mid c$ and $b \mid c$, then $m \mid c$.*

Furthermore Rm is unique.

Proof. (i) and (ii) state that $Rm \subseteq Ra \cap Rb$, and that $Rc \subseteq Ra \cap Rb$ implies $Rc \subseteq Rm$. Now the ideal $Ra \cap Rb$ has properties (1) and (2). Therefore $Ra \cap Rb = Rn$ for some $n \in R$. Then (i) holds if and only if $Rm \subseteq Ra \cap Rb$, and (ii) holds if and only if $Rc \subseteq Ra \cap Rb$ implies $Rc \subseteq Rm$. Therefore (i) and (ii) hold if and only if $Rm = Ra \cap Rb = Rn$. □

The element m in Proposition 5.1.7 is the **least common multiple** or **l.c.m.** of a and b, and is usually denoted by $m = [a,b]$; we prefer $m = \mathrm{lcm}\,(a,b)$. If $a = 0$ or $b = 0$, then $m = 0$; otherwise, $m \neq 0$ (since $m \mid ab \neq 0$). In general, the l.c.m. can be required to be a representative element and is then unique.

A similar argument shows that every family $(a_i)_{i \in I}$ of nonzero elements has a l.c.m. $m = \mathrm{lcm}\,((a_i)_{i \in I})$.

Proposition 5.1.8. *Let R be a principal ideal domain and $a,b \in R$. There exists $d \in R$ such that:*

(i) $d \mid a$ and $d \mid b$;

(ii) *if $c \mid a$ and $c \mid b$, then $c \mid d$.*

Furthermore Rd is unique, and there exist $u,v \in R$ such that $d = ua + vb$.

Proof. Parts (i) and (ii) state that $Ra \subseteq Rd$, $Rb \subseteq Rd$, and $Ra \subseteq Rc$, $Rb \subseteq Rc$ implies $Rd \subseteq Rc$. Now

$$\mathfrak{d} = Ra + Rb = \{\, au + bv\,;\, u,v \in R \,\}$$

is an ideal of R. Hence $\mathfrak{d} = Rn$ for some $n \in R$. Then (i) holds if and only if $Rn \subseteq Rd$; (ii) holds if and only if $Rn \subseteq Rc$ implies $Rd \subseteq Rc$. Hence (i) and (ii) are equivalent to $Rd = Rn = \mathfrak{d}$. □

The element d in Proposition 5.1.8 is a **greatest common divisor** or **g.c.d.** of a and b, and is usually denoted by $d = (a,b)$ (often (a,b) also denotes the ideal $(a) + (b) = (d)$ generated by a and b); we prefer the notation $d = \gcd(a,b)$. If $a = 0$, then $d = b$ serves. In general, the g.c.d. can be required to be a representative element and is then unique.

A similar argument shows that every nonempty family $(a_i)_{i \in I}$ has a g.c.d. $d = \gcd((a_i)_{i \in I})$, which can be written in the form $d = \sum_{i \in I} u_i a_i$ (with $u_i = 0$ for almost all i).

Two nonzero elements a and b of a PID R are **relatively prime** in case 1 is a g.c.d. of a and b; equivalently, any $\gcd(a,b)$ is a unit; equivalently, $Ra + Rb = R$; equivalently, $ua + vb = 1$ for some $u, v \in R$. This relationship is usually written $(a,b) = 1$; we prefer $\gcd(a,b) = 1$. The following results are exercises:

Proposition 5.1.9. *In a PID, if* $\gcd(a,b) = 1$ *and* $\gcd(a,c) = 1$*, then* $\gcd(a,bc) = 1$*; if* $a \mid bc$ *and* $\gcd(a,b) = 1$*, then* $a \mid c$*.*

The l.c.m. and g.c.d. of two elements a and b can also be obtained from Theorem 5.1.6. Write $a = u\,p_1^{a_1} p_2^{a_2} \ldots p_k^{a_k}$ and $b = v\,q_1^{b_1} q_2^{b_2} \ldots q_\ell^{b_\ell}$ as products of a unit and of positive powers of distinct representative irreducible elements. The sequences $p_1, \ldots, p_k$ and $q_1, \ldots, q_\ell$ can be merged into one sequence so that a and b are products of a unit and of nonnegative powers of the same distinct representative irreducible elements: $a = u\,p_1^{a_1} p_2^{a_2} \ldots p_r^{a_r}$ and $b = v\,p_1^{b_1} p_2^{b_2} \ldots p_r^{b_r}$. The following result is an exercise:

Proposition 5.1.10. *In a PID, let* $a = u\,p_1^{a_1} p_2^{a_2} \ldots p_r^{a_r}$ *and* $b = v\,p_1^{b_1} p_2^{b_2} \ldots p_r^{b_r}$ *be products of a unit and of nonnegative powers of distinct representative irreducible elements. Then:*

(i) $a \mid b$ *if and only if* $a_i \leqslant b_i$ *for all* i*;*

(ii) $c = p_1^{c_1} p_2^{c_2} \ldots p_r^{c_r}$*, where* $c_i = \max(a_i, b_i)$*, is a l.c.m. of* a *and* b*;*

(iii) $d = p_1^{d_1} p_2^{d_2} \ldots p_r^{d_r}$*, where* $d_i = \min(a_i, b_i)$*, is a g.c.d. of* a *and* b*;*

(iv) $\operatorname{lcm}(a,b) \gcd(a,b) = tab$*, where* t *is a unit.*

5.1.f. Irreducible Polynomials. In the rest of this section K is a field and R is the ring $K[X]$ of polynomials in one indeterminate, which is a PID by Proposition 5.1.1.

A nonzero polynomial is **monic** in case its leading coefficient is 1. By Corollary 4.3.5, the units of $K[X]$ are the nonzero constants; therefore monic polynomials can be chosen as representative elements in $K[X]$.

Corollary 5.1.11. *Let K be a field. In $K[X]$, every nonzero polynomial can be written uniquely (up to the order of the terms) as a product of a constant and of positive powers of distinct monic irreducible polynomials.*

Which polynomials are irreducible depends on the field K. There are a few general results. A **root** of a polynomial $f(X) \in K[X]$ is an element a of K such that $f(a) = 0$. The following result is an exercise:

Proposition 5.1.12. *Let K be a field. In $K[X]$:*

(1) *every polynomial of degree 1 is irreducible;*

(2) *an irreducible polynomial of degree > 1 has no root in K;*

(3) *a polynomial of degree 2 or 3 is irreducible if and only if it has no root in K.*

Proposition 5.1.13. *For any field K, $K[X]$ contains infinitely many monic irreducible polynomials.*

Proof. (Euclid). We show that no finite sequence $p_1, \ldots, p_n$ of irreducible polynomials contains every monic irreducible polynomial. By Corollary 5.1.11, $f = 1 + p_1 p_2 \ldots p_n$, which is not constant, is a multiple of a monic irreducible polynomial p. Then $p = p_i$ would imply that $1 = f - p_1 \ldots p_n$ is a multiple of p_i, which is impossible; therefore $p \neq p_1, \ldots, p_n$. □

Let $K = \mathbb{C}$. By the Fundamental Theorem of Algebra, every nonconstant polynomial in $\mathbb{C}[X]$ has a root in $\mathbb{C}$. A field with this property is called **algebraically closed**. By Proposition 5.1.12, a polynomial is irreducible in $\mathbb{C}[X]$ if and only if it has degree 1.

Proposition 5.1.14. *In $\mathbb{R}[X]$, a polynomial is irreducible if and only if it either has degree 1, or has degree 2 and no root in $\mathbb{R}$.*

Proof. These polynomials are irreducible by Proposition 5.1.12. For the converse, take any nonconstant $f(X) \in \mathbb{R}[X]$. By Corollary 5.1.11, f can be factored in $\mathbb{C}[X]$ into polynomials of degree 1:

$$f(X) = a_n (X - r_1)(X - r_2) \ldots (X - r_n);$$

there, $n = \deg f$, a_n is the leading coefficient of f, and $r_1, \ldots, r_n \in \mathbb{C}$ are the (not necessarily distinct) roots of f in $\mathbb{C}$. Since A has real coefficients, applying complex conjugation yields another factorization:

$$f(X) = a_n (X - \bar{r}_1)(X - \bar{r}_2) \ldots (X - \bar{r}_n).$$

Since f has only one such factorization, this implies $\{r_1, \ldots, r_n\} = \{\bar{r}_1, \ldots, \bar{r}_n\}$. Therefore the roots of f in $\mathbb{C}$ can be classified into real roots and pairs of nonreal complex conjugate roots. Now $(X - z)(X - \bar{z}) = X^2 - (z + \bar{z})X + z\bar{z} \in \mathbb{R}[X]$, with no root in $\mathbb{R}$ if $z \notin \mathbb{R}$. Hence $f(X)$ is a product of polynomials of the form $X - r$ with $r \in \mathbb{R}$, and $(X - z)(X - \bar{z}) \in \mathbb{R}[X]$ with $z \in \mathbb{C} \backslash \mathbb{R}$; if f is irreducible in $\mathbb{R}[X]$, then f is in this form and has either degree 1, or degree 2 and no root in $\mathbb{R}$.   □

The case $K = \mathbb{Q}$ will be studied in Section 5.4.

Let $K = \mathbb{Z}_p$, where p is a prime number. In $\mathbb{Z}_p[X]$ there are only finitely many polynomials of degree $\leqslant n$. By Proposition 5.1.13, $\mathbb{Z}_p[X]$ has irreducible polynomials of arbitrarily high degree. The simplest of these are readily computed if p is small. As an example we find irreducible polynomials in $\mathbb{Z}_2[X]$ of degree up to 4.

Each coefficient of $f(X) \in \mathbb{Z}_2[X]$ is 0 or 1. Hence f has a root in $\mathbb{Z}_2$ if and only if either $a_0 = 0$ or f has an even number of nonzero terms.

By Proposition 5.1.12, X and $X + 1$ are irreducible, and the irreducible polynomials of degree 2 or 3 are $X^2 + X + 1$, $X^3 + X + 1$, and $X^3 + X^2 + 1$ (all others have a root in $\mathbb{Z}_2$).

The polynomials of degree 4 with no root are $X^4 + X + 1$, $X^4 + X^2 + 1$, $X^4 + X^3 + 1$, and $X^4 + X^3 + X^2 + X + 1$. If such a polynomial is not irreducible, then it is a product of two irreducible polynomials of degree 2; that is, it equals $(X^2 + X + 1)^2 = X^4 + X^2 + 1$. The remaining polynomials, $X^4 + X + 1$, $X^4 + X^3 + 1$, and $X^4 + X^3 + X^2 + X + 1$, are therefore irreducible.

Exercises

1. Show that a polynomial ring $R[(X_i)_{i \in I}]$ with more than one indeterminate is not a PID.
2. Show that the ring $R = \mathbb{Z}[i]$ of **Gauss integers** (= complex numbers of the form $a + bi$ with $a, b \in \mathbb{Z}$) is a PID. (Prove the following division property: for all $s, t \in R$, $t \neq 0$, there exist $q, r \in R$ such that $s = tq + r$ and $|r| < |t|$.)
3. Show that every family $(a_i)_{i \in I}$ of elements of a PID has a l.c.m. (which may be 0).
4. Show that every nonempty family $(a_i)_{i \in I}$ of elements of a PID has a g.c.d., which can be written in the form $d = \sum_{i \in I} u_i a_i$ with $u_i = 0$ for almost all i.
5. Prove that, in a PID, $\gcd(a, b) = \gcd(a, c) = 1$ implies $\gcd(a, bc) = 1$.
6. Prove that, in a PID, if $\gcd(a, b) = 1$ and $a \mid bc$, then $a \mid c$.
7. Prove Proposition 5.1.10.
8. Prove Proposition 5.1.12 (use Corollary 4.3.8).
9. Find all irreducible polynomials in $\mathbb{Z}_2[X]$ of degree 5.
10. Find all monic irreducible polynomials in $\mathbb{Z}_3[X]$ of degree up to 3. (Readers who are blessed with long winter evenings can try degree 4.)

5.2. POLYNOMIALS

In this section K is an algebraically closed field, so that $K[X]$ is a principal ideal domain with known irreducible elements. This yields various classical properties of polynomials in $K[X]$: relations between roots and coefficients, resultant, discriminant, and the solution of equations of degree $\leqslant 4$, all of which can be used in Chapter 7.

In fact these are some of the oldest properties in algebra. Algebra owes its name, and its existence as a separate field, to the ninth-century treatise *Al-Djabr wa'l muqabala* by al-Khwarizmi, the first systematic exposition of quadratic equations. The title means something like "the balancing" or "fitting" of "related quantities"; this may express the relationship between the two sides of quadratic equations, which Al-Khwarizmi wrote in the form $x^2 = ax + b$, $x^2 + b = ax$, and so on, with $a, b > 0$. Equations of degrees 3 and 4 were solved in the sixteenth century.

5.2.a. Elementary Symmetric Polynomials. Since K is algebraically closed, every nonconstant polynomial $f(X) \in K[X]$ can be factored uniquely into polynomials of degree 1:

$$f(X) = a_n (X - r_1) (X - r_2) \ldots (X - r_n);$$

(by Corollary 5.1.11); there, $n = \deg a$, a_n is the leading coefficient of f, and $r_1, \ldots, r_n \in K$ are the roots of f in K; $r_1, \ldots, r_n$ are not necessarily distinct.

The product $(X - r_1)(X - r_2) \ldots (X - r_n)$ can be expanded as follows. By distributivity, $(X - r_1)(X - r_2) \ldots (X - r_n)$ is a sum of terms of the form $t_1 t_2 \ldots t_n$ where, for each i, t_i is a term of $X - r_i$ (either $t_i = X$ or $t_i = -r_i$). These terms can be classified by the number k of times that $t_i = -r_i$. The result is

$$(X - r_1)(X - r_2) \ldots (X - r_n) = \sum_{0 \leqslant k \leqslant n} (-1)^k \, \mathfrak{s}_k \, X^{n-k},$$

where

$$\mathfrak{s}_k = \mathfrak{s}_k(r_1, \ldots, r_n) = \sum_{1 \leqslant i_1 < i_2 < \cdots < i_k \leqslant n} r_{i_1} r_{i_2} \ldots r_{i_k}.$$

For instance, $\mathfrak{s}_0 = 1$, $\mathfrak{s}_1 = r_1 + \cdots + r_n$, $\mathfrak{s}_2 = \sum_{1 \leqslant i < j \leqslant n} r_i r_j$, and $\mathfrak{s}_n = r_1 r_2 \ldots r_n$. In general, $\mathfrak{s}_k$ is the k-th **elementary symmetric function** of $r_1, \ldots, r_n$; it is symmetric in the sense that $\mathfrak{s}_k(r_{\sigma 1}, \ldots, r_{\sigma n}) = \mathfrak{s}_k(r_1, \ldots, r_n)$ for every permutation σ of $\{1, 2, \ldots, n\}$ (since $(X - r_{\sigma 1}) \ldots (X - r_{\sigma n}) = (X - r_1) \ldots (X - r_n)$). The polynomials

$$\mathfrak{s}_k = \mathfrak{s}_k^{(n)} = \sum_{1 \leqslant i_1 < i_2 < \cdots < i_k \leqslant n} X_{i_1} X_{i_2} \ldots X_{i_k} \in \mathbb{Z}[X_1, \ldots, X_n]$$

are the **elementary symmetric polynomials**. We have proved:

Proposition 5.2.1. *When K is an algebraically closed field, the coefficients $a_0, a_1, \ldots, a_n$ and roots $r_1, r_2, \ldots, r_n$ of a polynomial $f(X) \in K[X]$ of degree n are* *related by*

$$a_{n-k} = (-1)^k a_n \, \mathfrak{s}_k(r_1, \ldots, r_n),$$

where $\mathfrak{s}_k$ is the k-th elementary symmetric polynomial.

5.2.b. Derivatives. By Corollary 5.4.9, a nonconstant polynomial $f(X) \in K[X]$ can also be factored in the form

$$f(X) = a_n \, (X - r_1)^{m_1} \, (X - r_2)^{m_2} \ldots (X - r_k)^{m_k}$$

where $r_1, \ldots, r_k$ are the *distinct* roots of f. The positive integer m_i is the **order of multiplicity** of the root r_i. (It is the number of times that $X - r_i$ appears in the previous product $a = a_n(X - r_1)(X - r_2) \ldots (X - r_n)$ of not necessarily distinct irreducible polynomials.) A root is **simple** in case its order of multiplicity is 1 and **multiple** in case its order of multiplicity is greater than 1.

The order of multiplicity of a root can be tested (as we would in Calculus) by means of derivatives. The formal **derivative** of a polynomial $f(X) = a_n X^n + \cdots + a_i X^i + \cdots + a_1 X + a_0$ is defined (in any polynomial ring $R[X]$) by:

$$f'(X) = n a_n X^{n-1} + \cdots + i a_{i-1} X^{i-1} + \cdots + a_1.$$

This is a purely algebraic definition and has no interpretation as a limit. The derivative has algebraic properties whose proofs make interesting exercises:

Lemma 5.2.2. *For all polynomials $f, g \in R[X]$:*

(1) $(f + g)' = f' + g'$;
(2) $(fg)' = f'g + fg'$;
(3) $(f^k)' = k f^{k-1} f'$.

Since f' is a polynomial, f has **higher derivatives** $f'' = (f')', \ldots, f^{(k)} = (f^{(k-1)})', \ldots$. If $\deg f = n$, then $f^{(n)} = n! a_n$ and $f^{(k)} = 0$ for all $k > n$.

Proposition 5.2.3. *Let K be an algebraically closed field and $f(X) \in K[X]$ be a nonconstant polynomial with a root r with order of multiplicity m. Then $f^{(k)}(r) = 0$ for all $k < m$; if K has characteristic 0 or $p > m$, then $f^{(m)}(r) \neq 0$.*

Proof. We have $f(X) = (X - r)^m g(X)$, where $g(r) \neq 0$. Hence $f'(X) = m(X - r)^{m-1} g(X) + (X - r)^m g'(X) = (X - r)^{m-1} h(X)$, where $h(X) = mg(X) + (X - r)g'(X)$ and $h(r) = mg(r)$. If $m > 1$, then $f'(r) = 0$. By induction, $f^{(k)}(X) = (X - r)^{m-k} \ell(X)$, where $\ell(r) = m(m-1) \ldots (m-k+1)g(r)$, whenever $k \leqslant m$: if indeed $k < n$, then

$$f^{(k+1)}(X) = (m-k)(X-r)^{m-k-1}\ell(X) + (X-r)^{m-k}\ell'(X)$$
$$= (X-r)^{m-k-1}m(X),$$

where $m(X) = (m-k)\ell(X) + (X-r)\ell'(X)$ and $m(r) = m(m-1)\ldots(m-k+1)$ $(m-k)g(r)$. Therefore $f^{(k)}(r) = 0$ if $k < m$, and $f^{(m)}(r) = m!g(r)$; if K has characteristic 0 or characteristic $p > m$, then $p \nmid m!$ and $f^{(m)}(r) = m!g(r) \neq 0$. $\square$

5.2.c. The Resultant. The resultant detects when two polynomials have a common root, using only their coefficients. Let

$$f(X) = a_m (X-r_1)\ldots(X-r_m) \qquad \text{and} \qquad g(X) = b_n (X-s_1)\ldots(X-s_n)$$

be polynomials of degrees m and n ($a_m, b_n \neq 0$) with roots $r_1,\ldots,r_m, s_1,\ldots,s_n$ in K. The **resultant** of f and g is

$$R(f,g) = a_m^n b_n^m \prod_{1 \leqslant i \leqslant m, 1 \leqslant j \leqslant n} (r_i - s_j)$$
$$= a_m^n \prod_{1 \leqslant i \leqslant m} g(r_i) = (-1)^{mn} b_n^m \prod_{1 \leqslant j \leqslant n} f(s_j).$$

We see that $R(f,g) = 0$ if and only if f and g have a common root.

Proposition 5.2.4. *Let K be an algebraically closed field and $f(X) = a_m X^m + \cdots + a_0$, $g(X) = b_n X^n + \cdots + b_0 \in K[X]$. If $a_m, b_n \neq 0$, then*

$$R(f,g) = \begin{vmatrix} a_m & \cdots & \cdots & a_0 & & & \\ & \ddots & & & \ddots & & \\ & & a_m & \cdots & \cdots & a_0 \\ b_n & \cdots & \cdots & b_0 & & & \\ & \ddots & & & \ddots & & \\ & & b_n & \cdots & \cdots & b_0 \end{vmatrix}.$$

In this determinant each the first n rows is the row of coefficients of f, extended to size $m + n$ with zeros to the right and left, and the remaining m rows are constructed similarly from g. With Proposition 5.2.4 we can detect when two polynomials have a common root using only their coefficients.

Proof. Let D be the determinant above. To prove that $R(f,g) = D$, we regard D as a polynomial in $a_m,\ldots,a_0, b_n,\ldots,b_0$ with coefficients in $\mathbb{Z}$. Let

$F = A_m X^m + \cdots + A_0$, $G = B_n X^n + \cdots + B_0$, and

$$P = \begin{vmatrix} A_m & \cdots & & \cdots & A_0 & & & \\ & \ddots & & & & \ddots & & \\ & & A_m & \cdots & & \cdots & A_0 & \\ B_n & \cdots & & \cdots & B_0 & & & \\ & \ddots & & & & \ddots & & \\ & & B_n & \cdots & & \cdots & B_0 & \end{vmatrix} \in \mathbb{Z}[A_m, \ldots, A_0, B_n, \ldots, B_0, X].$$

Then $f(X) = F(a_m, \ldots, a_0, X)$, $g(X) = G(b_n, \ldots, b_0, X)$, and $D = P(a_m, \ldots, a_0, b_n, \ldots, b_0)$. A homomorphism

$$\Phi : \mathbb{Z}[A_m, \ldots, A_0, B_n, \ldots, B_0, X] \longrightarrow \mathbb{Z}[A_m, B_n, R_1, \ldots, R_m, S_1, \ldots, S_n, X]$$

is obtained from the substitutions $A_k = (-1)^k A_m \mathfrak{s}_k^{(m)}(R_1, \ldots, R_m)$ $(1 \leqslant k < m)$ and $B_k = (-1)^k B_n \mathfrak{s}_k^{(n)}(S_1, \ldots, S_n)$ $(1 \leqslant k < n)$, using the elementary symmetric polynomial with m or n indeterminates. By Proposition 5.2.1,

$$\Phi(F) = A_m (X - R_1) \ldots (X - R_m) \qquad \text{and}$$

$$\Phi(G) = B_n (X - S_1) \ldots (X - S_n).$$

Similarly $D = Q(a_m, b_n, r_1, \ldots, r_m, s_1, \ldots, s_n)$, where $Q = \Phi(P)$.

We show that Q is homogeneous of degree $m + n + mn$ (every monomial in Q has degree $m + n + mn$). The entry C_{ij} of P in row i and column j is either 0, or A_{m+i-j} (if $i \leqslant n$), or B_{i-j} (if $i \geqslant n$). Hence

$$P = \prod_{\sigma \in S_{m+n}} \text{sgn}\, \sigma \; C_{1,\sigma 1} \ldots C_{m+n, \sigma(m+n)}$$

$$= \prod_{\sigma \in S_{m+n}} \text{sgn}\, \sigma \prod_{1 \leqslant i \leqslant n} A_{m+i-\sigma i} \prod_{n < i \leqslant m+n} B_{i-\sigma i}.$$

When $(-1)^k A_m \mathfrak{s}_k^{(m)}(R_1, \ldots, R_m)$ and $(-1)^k B_n \mathfrak{s}_k^{(n)}(S_1, \ldots, S_n)$, which are homogeneous of degree $k + 1$, are substituted for A_k and B_k, the typical term

$$\prod_{1 \leqslant i \leqslant n} A_{m+i-\sigma i} \prod_{n < i \leqslant m+n} B_{i-\sigma i}$$

of P becomes a homogeneous polynomial of degree

$$\sum_{1 \leqslant i \leqslant n} m + i - \sigma i + 1 + \sum_{n < i \leqslant m+n} i - \sigma i + 1 = mn + n + m,$$

since $\sum_{1 \leqslant i \leqslant m+n} i = \sum_{1 \leqslant i \leqslant m+n} \sigma i$. Hence $Q = \Phi(P)$ is homogeneous of degree $m + n + mn$.

We show that $Q = A_m^n B_n^m \prod_{1 \leqslant i \leqslant m, 1 \leqslant j \leqslant n} (R_i - S_j)$.

Lemma 5.2.5. *Assume that* $a \in \mathbb{Z}[X_1,\ldots,X_n]$ *becomes* 0 *when* $X_j \neq X_i$ *is substituted for* X_i. *Then* $X_i - X_j$ *divides* a.

Proof. This is proved like Corollary 5.3.8. We may assume $i = 1$ and identify $\mathbb{Z}[X_1,\ldots,X_n]$ with $R[X_1]$, where $R = \mathbb{Z}[X_2,\ldots,X_n]$. In $R[X_1]$ we can divide a by $X_1 - X_j$: $a = (X_1 - X_j)q + r$, where $q,r \in R[X_1]$ and $\deg r < 1$, so that $r \in R$. Substituting $X_j \in R$ for X_1 yields $r = 0$. $\square$

We now resume the proof of Proposition 5.2.4. When $Q = \Phi(P)$ is written as a determinant, the first n rows are multiples of A_m and the remaining m rows are multiples of B_n. Therefore Q is divisible by $A_m^n B_n^m$. Next consider the equations

$$X^{n-1}\Phi(F) = A_m X^{m+n-1} + \cdots + A_0 X^{n-1} = 0$$

$$X^{n-2}\Phi(F) = A_m X^{m+n-2} + \cdots + A_0 X^{n-2} = 0$$

$$\vdots$$

$$\Phi(F) = A_m X^m + \cdots + A_0 = 0$$

$$X^{m-1}\Phi(G) = B_n X^{m+n-1} + \cdots + B_0 X^{m-1} = 0$$

$$X^{m-2}\Phi(G) = B_n X^{m+n-2} + \cdots + B_0 X^{m-2} = 0$$

$$\vdots$$

$$\Phi(G) = B_n X^n + \cdots + B_0 = 0.$$

Substituting $R_i = S_j$ in the above provides a homogeneous system of linear equations with a nonzero solution $S_j^{m+n-1},\ldots,1$ (since $\Phi(F)(R_i) = \Phi(G)(S_j) = 0$). The determinant of this system, which is obtained from Q by the same substitution, is 0. By Lemma 5.2.5, $R_i - S_j$ divides Q. Therefore Q is divisible by $R = A_m^n B_n^m \prod_{1 \leqslant i \leqslant m, 1 \leqslant j \leqslant n} (R_i - S_j)$. Since R is homogeneous with the same degree $m + n + mn$ as Q, we have $Q = tR$ for some integer t. Now P has a term $A_m^n B_0^m$; hence Q has one term $A_m^n B_n^m (-1)^m (S_1 S_2 \ldots S_n)^m$. Since R also has one such term, it follows that $Q = R$. Substituting $a_m, b_n, r_1, \ldots, r_m, s_1, \ldots, s_n$ then yields $D = a_m^n b_n^m \prod_{1 \leqslant i \leqslant m, 1 \leqslant j \leqslant n} (r_i - s_j)$. $\square$

Corollary 5.2.6. *Let* K *be an algebraically closed field. Two polynomials* $f(X), g(X) \in K[X]$ *have a common root in* K *if and only if* $R(f,g) = 0$.

5.2.d. The Discriminant. The discriminant detects when a polynomial has a multiple root, using only its coefficients. Discriminants also turn up in the solution of equations of degree $\leqslant 4$.

When $f(X) \in K[X]$ has degree $n \geqslant 2$ and roots $r_1,\ldots,r_n$ in K, the **discriminant** of f is

$$D(f) = a_n^{2n-2} \prod_{1 \leqslant i < j \leqslant n} (r_i - r_j)^2,$$

where a_n is the coefficient of X^n in f. Obviously:

Proposition 5.2.7. *Let K be an algebraically closed field. A polynomial $f(X) \in K[X]$ of degree at least 2 has a multiple root in K if and only if $D(f) = 0$.*

The next result computes $D(f)$ from the coefficients of f.

Proposition 5.2.8. *Let K be an algebraically closed field. When $f(X) \in K[X]$ has degree $n \geqslant 2$ and leading coefficient a_n,*

$$R(f, f') = (-1)^{n(n-1)/2} \, a_n \, D(f).$$

Proof. We have $f(X) = a_n \prod_{1 \leqslant j \leqslant n} (X - r_j)$. Hence

$$f'(X) = \sum_{1 \leqslant i \leqslant n} a_n \prod_{1 \leqslant j \leqslant n, j \neq i} (X - r_j),$$

$f'(r_i) = a_n \prod_{1 \leqslant j \leqslant n, j \neq i} (r_i - r_j)$, and Proposition 5.2.4 yields

$$
\begin{aligned}
R(f, f') &= a_n^{n-1} \prod_{1 \leqslant i \leqslant n} a'(r_i) \\
&= a_n^{n-1} \prod_{1 \leqslant i \leqslant n} a_n \prod_{1 \leqslant j \leqslant n, j \neq i} (r_i - r_j) \\
&= a_n^{n-1} a_n^n \prod_{1 \leqslant i, j \leqslant n, j \neq i} (r_i - r_j) \\
&= (-1)^{n(n-1)/2} \, a_n^{2n-1} \prod_{1 \leqslant i < j \leqslant n} (r_i - r_j)^2. \qquad \square
\end{aligned}
$$

The discriminant can also be expressed in terms of the **Vandermonde determinant**

$$
V(r_1, \ldots, r_n) = \begin{vmatrix}
1 & 1 & \cdots & 1 \\
r_1 & r_2 & \cdots & r_n \\
r_1^2 & r_2^2 & \cdots & r_n^2 \\
\vdots & \vdots & \ddots & \vdots \\
r_1^{n-1} & r_2^{n-1} & \cdots & r_n^{n-1}
\end{vmatrix}
$$

of the first n powers of $r_1, \ldots, r_n$. The reader will enjoy using Lemma 5.2.5 to prove:

Proposition 5.2.9. $V(r_1, \ldots, r_n) = \prod_{i > j} (r_i - r_j)$.

If now f has leading coefficient a_n and roots $r_1, \ldots, r_n$, then

$$D(f) = a_n^{2n-2} \, V(r_1, \ldots, r_n)^2.$$

5.2.e. Quadratic Equations. The discriminant of a quadratic polynomial $q(X) = aX^2 + bX + c$ (with $a \neq 0$) is found from

$$R(q,q') = \begin{vmatrix} a & b & c \\ 2a & b & 0 \\ 0 & 2a & b \end{vmatrix} = 4a^2c - ab^2.$$

By Proposition 5.2.8, $D = D(q) = b^2 - 4ac$.

If $q(X)$ has roots r and s then $D = a^2(r - s)^2$. For example, let $K = \mathbb{C}$ and $q \in \mathbb{R}[X]$. Then either $r,s \in \mathbb{R}$ and $D \geqslant 0$ or $r,s \in \mathbb{C}\backslash\mathbb{R}$, $s = \bar{r}$, $r - s$ is a real multiple of i, and $D < 0$. In this case the sign of the discriminant indicates whether the roots of q are real.

When K does not have characteristic 2, the reader will verify that the general quadratic equation $ax^2 + bx + c = 0$ (with $a \neq 0$) has solutions

$$\frac{-b \pm \sqrt{b^2 - 4ac}}{2a},$$

where $\pm\sqrt{b^2 - 4ac}$ are the square roots of the discriminant in K.

5.2.f. Equations of Degree 3. If K does not have characteristic 3, a polynomial $f(X) = aX^3 + bX^2 + cX + d$ of degree 3 can be put in the form $a(X^3 + pX + q)$ by substituting $X - b/3a$ for X. We see that p and q are rational functions of a,b,c,d; for instance it is readily verified that $p = (3ac - b^2)/3a^2$. The discriminant of $X^3 + pX + q$ is found from the resultant

$$\begin{vmatrix} 1 & 0 & p & q & 0 \\ 0 & 1 & 0 & p & q \\ 3 & 0 & p & 0 & 0 \\ 0 & 3 & 0 & p & 0 \\ 0 & 0 & 3 & 0 & p \end{vmatrix} = 4p^3 + 27q^2$$

By Proposition 5.2.8, $D = D(X^3 + pX + q) = -4p^3 - 27q^2$.

If the roots of $X^3 + pX + q$ are r,s,t, then $D = (r - s)^2(r - t)^2(s - t)^2$. For example, let $K = \mathbb{C}$ and $p,q \in \mathbb{R}$. Then either $r,s,t \in \mathbb{R}$ and $D \geqslant 0$ or, say, $t \in \mathbb{R}$ and $r,s \in \mathbb{C}\backslash\mathbb{R}$, $s = \bar{r}$, and $D < 0$; hence the sign of the discriminant again indicates whether all roots of $f(X)$ are real.

The third-degree equation $x^3 + px + q = 0$ can be solved by Cardano's sixteenth-century method when K does not have characteristic 2 or 3. With $x = u + v$, the equation becomes

$$(u + v)^3 + p(u + v) + q = u^3 + v^3 + (3uv + p)(u + v) + q = 0.$$

If we can arrange that $3uv = -p$, then u^3 and v^3 satisfy $u^3v^3 = -p^3/27$ and $u^3 + v^3 = -q$; hence $u' = u^3$ and $v' = v^3$ are the roots

$$u' = \frac{-q + \sqrt{q^2 + 4p^3/27}}{2}, \qquad v' = \frac{-q - \sqrt{q^2 + 4p^3/27}}{2},$$

of the **resolvant polynomial** $X^2 + qX - p^3/27$. Note that the resolvant polynomial has discriminant $(1/27)(4p^3 + 27q^2)$. The roots x_1, x_2, x_3 of $X^3 + pX + q$ are all

$$x = u + v = \sqrt[3]{\frac{-q + \sqrt{q^2 + 4p^3/27}}{2}} + \sqrt[3]{\frac{-q - \sqrt{q^2 + 4p^3/27}}{2}}$$

in which u and v are cube roots of u' and v' such that $uv = -p/3$.

The roots r_1, r_2, r_3 of the original polynomial $aX^3 + bX^2 + cX + d$ are $r_i = x_i - b/3a$.

5.2.g. Equations of Degree 4. The roots r_1, r_2, r_3, r_4 of the fourth degree equation

$$f(x) = ax^4 + bx^3 + cx^2 + dx = e = 0 \qquad (5.1)$$

can be found by the following method when K does not have characteristic 2. Substituting $X - b/4a$ for X puts $f(X) = aX^4 + bX^3 + cX^2 + dX + e$ in the form

$$g(X) = 1/a \; f(X - b/4a) = X^4 + pX^2 + qX + r.$$

We see that p, q, and r are rational functions of a, b, c, d, e; for instance it is readily verified that $p = (8ac - 3b^2)/8a^2$.

Let x_1, x_2, x_3, x_4 be the roots of the equation

$$x^4 + px^2 + qx + r = 0. \qquad (5.2)$$

Then $x_i = r_i + b/4a$, and $x_1 + x_2 + x_3 + x_4 = 0$, by Proposition 5.2.1. Let

$$u = (x_1 + x_2)(x_3 + x_4) = -(x_1 + x_2)^2$$

$$v = (x_1 + x_3)(x_2 + x_4) = -(x_1 + x_3)^2$$

$$w = (x_1 + x_4)(x_2 + x_3) = -(x_1 + x_4)^2;$$

equivalently,

$$u = (r_1 + r_2 + b/2a)(r_3 + r_4 + b/2a) = -(r_1 + r_2 + b/2a)^2,$$

$$v = (r_1 + r_3 + b/2a)(r_2 + r_4 + b/2a) = -(r_1 + r_3 + b/2a)^2,$$

$$w = (r_1 + r_4 + b/2a)(r_2 + r_3 + b/2a) = -(r_1 + r_4 + b/2a)^2.$$

To solve (5.1) and (5.2), we construct a polynomial of degree 3 whose roots are u, v, and w. Its coefficients are found by computing $u + v + w$, $uv + uw + vw$, and uvw from the elementary symmetric functions $\sum_{i<j} x_i x_j = p$, $\sum_{i<j<k} x_i x_j x_k = -q$, and $x_1 x_2 x_3 x_4 = r$. First,

$$u + v + w = 2 \sum_{i<j} x_i x_j = 2p.$$

Next, $uv + uw + vw = \sum_{i<j} x_i^2 x_j^2 + 3\sum_{j<k,\, j,k \neq i} x_i^2 x_j x_k + 6x_1 x_2 x_3 x_4$. Since

$$0 = \left(\sum_i x_i \right) \left(\sum_{i<j<k} x_i x_j x_k \right) = \sum_{j<k,\, j,k \neq i} x_i^2 x_j x_k + 4x_1 x_2 x_3 x_4 \qquad \text{and}$$

$$p^2 = \left(\sum_{i<j} x_i x_j \right)^2 = \sum_{i<j} x_i^2 x_j^2 + 2 \sum_{j<k,\, j,k \neq i} x_i^2 x_j x_k + 6x_1 x_2 x_3 x_4,$$

we have $\sum_{j<k,\, j,k \neq i} x_i^2 x_j x_k = -4r$ and

$$uv + uw + vw = p^2 - 4r.$$

Finally,

$$uvw = \sum_{i,j,k} x_i^3 x_j^2 x_k + 2 \sum_{i;j<k<\ell} x_i^3 x_j x_k x_\ell$$

$$+ 2 \sum_{i<j<k} x_i^2 x_j^2 x_k^2 + 4 \sum_{i<j;k<\ell} x_i^2 x_j^2 x_k x_\ell,$$

where it is understood that i, j, k, ℓ are all distinct in the summations. Now

$$pr = \left(\sum_{i<j} x_i x_j \right)(x_1 x_2 x_3 x_4) = \sum_{i<j;k<\ell} x_i^2 x_j^2 x_k x_\ell,$$

$$q^2 = \left(\sum_{i<j<k} x_i x_j x_k \right)^2 = \sum_{i;j<k<\ell} x_i^3 x_j x_k x_\ell + 2 \sum_{i<j;k<\ell} x_i^2 x_j^2 x_k x_\ell,$$

$$0 = \left(\sum_i x_i \right)^2 (x_1 x_2 x_3 x_4) = \sum_{i;j<k<\ell} x_i^3 x_j x_k x_\ell + 2\sum_{i<j;k<\ell} x_i^2 x_j^2 x_k x_\ell,$$

$$0 = \left(\sum_i x_i \right) \left(\sum_{i<j} x_i x_j \right) \left(\sum_{i<j<k} x_i x_j x_k \right)$$

$$= \sum_{i,j,k} x_i^3 x_j^2 x_k + 3 \sum_{i;j<k<\ell} x_i^3 x_j x_k x_\ell$$

$$+ 3 \sum_{i<j<k} x_i^2 x_j^2 x_k^2 + 8 \sum_{i<j;k<\ell} x_i^2 x_j^2 x_k x_\ell,$$

and it readily follows that

$$uvw = -q^2.$$

Hence u, v, w are the solutions of the equation

$$x^3 - 2px^2 + (p^2 - 4r)x + q^2 = 0. \tag{5.3}$$

The polynomial $r(X) = X^3 - 2pX^2 + (p^2 - 4r)X + q^2 \in K[X]$ whose roots are u, v, w is the **resolvant polynomial** of A. Since

$$u - v = (x_1 - x_4)(x_3 - x_2),$$
$$u - w = (x_1 - x_3)(x_4 - x_2),$$
$$v - w = (x_1 - x_2)(x_4 - x_3),$$

$r(X)$ and $g(X)$ have the same discriminant. The discriminant of f is $D(f) = a^6 D(f/a) = a^6 D(g) = a^6 D(r)$.

To solve equation (5.2), first solve equation (5.3) to determine u, v, and w. (Cardano's method can be used if K does not have characteristic 3.) Then $x_1 + x_2 = u'$ is a square root of $-u$, $x_1 + x_3 = v'$ is a square root of $-v$, and $x_1 + x_4 = w'$ is a square root of $-w$. The square roots must be chosen so that

$$u'v'w' = (x_1 + x_2)(x_1 + x_3)(x_1 + x_4) = x_1^2 \sum_i x_i + \sum_{i<j<k} x_i x_j x_k = -q.$$

Then $u' + v' + w' = 3x_1 + x_2 + x_3 + x_4 = 2x_1$, $u' - v' - w' = x_2 - x_1 - x_3 - x_4 = 2x_2$, and similarly $-u' + v' - w' = 2x_3$, $-u' - v' + w' = 2x_4$. Thus the solutions of (5.2) are

$$x_1 = \tfrac{1}{2}(u' + v' + w'),$$
$$x_2 = \tfrac{1}{2}(u' - v' - w'),$$
$$x_3 = \tfrac{1}{2}(-u' + v' - w'),$$
$$x_4 = \tfrac{1}{2}(-u' - v' + w')$$

(where $u'v'w' = -q$). The solutions of (5.1) are

$$r_1 = -\frac{b}{4a} + \frac{1}{2}(u' + v' + w'),$$
$$r_2 = -\frac{b}{4a} + \frac{1}{2}(u' - v' - w'),$$
$$r_3 = -\frac{b}{4a} + \frac{1}{2}(-u' + v' - w'),$$
$$r_4 = -\frac{b}{4a} + \frac{1}{2}(-u' - v' + w').$$

Exercises

1. Prove Lemma 5.2.2.
2. Define partial derivatives in $R[X,Y]$ and prove their main properties.
3. Prove Proposition 5.2.9.
4. Verify that $D(X^3 + pX + q) = -4p^3 - 27q^2$.
5. Solve the quadratic equation $aX^2 + bX + c = 0$ (where $a \neq 0$ and K is algebraically closed and does not have characteristic 2).
6. Let $K = \mathbb{C}$, $p,q \in \mathbb{R}$, and $D = -4p^3 - 27q^2$. Show that Cardano's method yields real roots if $D \geqslant 0$ and one real root, two complex conjugate roots if $D < 0$.
7. Find $D(X^4 + pX^2 + qX + r)$.

5.3. RATIONAL FRACTIONS

Another application of principal ideal domains is the additive decomposition of rational fractions into partial fractions.

5.3.a. Reduced Form. In what follows K is a field. A rational fraction (in one indeterminate) is an element of the field of fractions $K(X)$ of the principal ideal domain $K[X]$, where K is a field.

Lemma 5.3.1. *Every rational fraction can be written uniquely in the form* f/g *with* $f,g \in K[X]$, g *monic, and* $\gcd(f,g) = 1$.

Proof. By definition, every rational fraction has the form a/b where $a,b \in K[X]$. Let $d = \gcd(a,b)$, $a = fd$, and $b = gd$. Since $\gcd(fd,gd) = d\gcd(f,g)$ we have $\gcd(f,g) = 1$. Furthermore $cf/cg = f/g$ for every $c \in K$, $c \neq 0$; hence we can arrange that g is monic.

Assume that $f/g = a/b$ with $\gcd(f,g) = \gcd(a,b) = 1$ and g,b monic. Then $bf = ag$. By Proposition 5.1.9, $b \mid g$ and $g \mid b$; hence $b = g$, and then $a = f$. □

A rational fraction f/g is **reduced** in case g is monic and $\gcd(f,g) = 1$. (Actually this is a property of the polynomials f and g.)

5.3.b. Partial Fractions. A **partial fraction** is a rational fraction f/q^r where $r \geqslant 1$, q is a monic irreducible polynomial, and $\deg f < \deg q$. This implies $\gcd(f,q) = 1$, so that a partial fraction is reduced. The main result of this section is:

Theorem 5.3.2. *Every rational fraction can be written uniquely as the sum of a polynomial and of partial fractions with distinct denominators.*

5.3.c. Proof of Main Result. The proof of Theorem 5.3.2 proceeds in three steps.

For the first step of the proof, we call a rational fraction f/g **polynomial-free** in case $\deg f < \deg g$. It is immediate that sums and products of polynomial-free fractions are polynomial free.

Lemma 5.3.3. *Every rational fraction can be written uniquely as the sum of a polynomial and a polynomial-free fraction.*

Proof. Given f/g, with g monic and $\gcd(f,g) = 1$, polynomial division yields $f = gp + r$, where $\deg r < \deg g$. Then $f/g = p + r/g$, with r/g polynomial-free and $\gcd(r,g) = \gcd(f,g) = 1$.

Conversely, let p, r, and h be polynomials with h monic, $\gcd(r,h) = 1$ and $\deg r < \deg h$. Then $p + r/h = (hp + r)/h$, with h monic and $\gcd(hp + r,h) = \gcd(r,h) = 1$. If $f/g = p + r/h$, then $g = h$ and $f = hp + r$ by Lemma 5.3.1, and the uniqueness of p and r/g follows from the uniqueness of p and r in the polynomial division. $\square$

For the second step of the proof, we use the following lemma:

Lemma 5.3.4. *When* $\deg f < \deg gh$ *and* $\gcd(g,h) = 1$, *there exist unique polynomials* u,v *such that* $f/gh = u/g + v/h$, $\deg u < \deg g$, *and* $\deg v < \deg h$. *If* $\gcd(f,gh) = 1$, *then* $\gcd(u,g) = \gcd(v,h) = 1$.

Proof. Since $\gcd(g,h) = 1$, there exist polynomials p,q such that $hp + gq = 1$ and polynomials s,t such that $hs + gt = f$. Polynomial division yields $s = gy + u$, $t = hz + v$, where $\deg u < \deg g$ and $\deg v < \deg h$. Then $f = hu + ghy + gv + ghz$. If $y + z \neq 0$, then

$$\deg(hu + gv) < \deg g + \deg h \leqslant \deg gh(y + z)$$

and $\deg f < \deg(hu + ghy + gv + ghz)$; therefore $y + z = 0$ and $hu + gv = f$. Then $f/gh = u/g + v/h$. Any polynomial which divides u and g, or v and h, also divides f and gh; hence $\gcd(f,gh) = 1$ implies $\gcd(u,g) = \gcd(v,h) = 1$.

Assume that $f/gh = u/g + v/h = u'/g + v'/h$, where $\deg u, \deg u' < \deg g$ and $\deg v, \deg v' < \deg h$. Then $hu + gv = hu' + bv' = f$ and $h(u - u') = g(v' - v)$. Since $\gcd(g,h) = 1$ this implies $g \mid u - u'$ and $h \mid v' - v$. Since $\deg(u - u') < \deg g$, $\deg(v' - v) < \deg h$, it follows that $u - u' = v' - v = 0$. $\square$

Lemma 5.3.5. *When* $\deg f < \deg g$ *and* $\gcd(f,g) = 1$, *there exist unique polynomials* $h_1,\ldots,h_k,q_1,\ldots,q_k$ *such that* $q_1,\ldots,q_k$ *are distinct monic irreducible polynomials;* $\deg h_i < \deg q_i^{r_i}$ *and* $q_i \nmid h_i$ *for all* i; *and*

$$\frac{f}{g} = \frac{h_1}{q_1^{r_1}} + \frac{h_2}{q_2^{r_2}} + \cdots + \frac{h_k}{q_k^{r_k}} \tag{5.4}$$

Proof. By Corollary 5.1.11, g can be written uniquely as a product $g = q_1^{r_1} q_2^{r_2} \ldots q_k^{r_k}$ of positive powers of distinct, monic irreducible polynomials. (The constant is missing since g is monic.) The existence of $h_1, \ldots, h_k$ is proved by induction on k using Lemma 5.3.4. The result is true if $k = 1$. Assume that it is true for $k - 1 \geqslant 1$, and let $g = q_1^{r_1} q_2^{r_2} \ldots q_k^{r_k}$, $\deg f < \deg g$, $\gcd(f, g) = 1$, and $g_1 = q_1^{r_1} q_2^{r_2} \ldots q_{k-1}^{r_{k-1}}$. Then $(g_1, q_k^{r_k}) = 1$, and Lemma 5.3.4 yields polynomials h and h_k unique such that $f/g = h/g_1 + h_k/q_k^{r_k}$, $\deg h < \deg g_1$, $\deg h_k < \deg q_k^{r_k}$, and $(h, g_1) = (h_k, q_k^{r_k}) = 1$. Since q_k is irreducible, this implies $q_k \nmid h_k$. The induction hypothesis then applies to h/g_1 and puts f/g in the required form.

Conversely, let $q_1, \ldots, q_k$ be distinct monic irreducible polynomials and $h_1, \ldots, h_k$ be polynomials such that $\deg h_i < \deg q_i^{r_i}$, $q_i \nmid h_i$ for all i and (5.4) holds. Let

$$\frac{h_1}{q_1^{r_1}} + \frac{h_2}{q_2^{r_2}} + \cdots + \frac{h_k}{q_k^{r_k}} = \frac{n}{d},$$

where $n = h_1 q_2^{r_2} \ldots q_k^{r_k} + \cdots + h_k q_1^{r_1} \ldots q_{k-1}^{r_{k-1}}$ and $d = q_1^{r_1} q_2^{r_2} \ldots q_k^{r_k}$. We see that $\deg n < \deg d$. Also $q_i \nmid n$, since q_i divides all the terms of n except for the term which starts with h_i. Hence $\gcd(n, d) = 1$, since an irreducible polynomial which divides d must be one of the q_i's and cannot divide n. By Lemma 5.3.1, $n = f$ and $d = g$; hence $g = q_1^{r_1} q_2^{r_2} \ldots q_k^{r_k}$ is (up to the order of the terms) the factorization of g into a irreducible polynomials. Hence $q_1, \ldots, q_k$ are unique. The uniqueness of $h_1, \ldots, h_k$ is clear if $k = 1$ and is proved by induction on k: if

$$\frac{f}{g} = \frac{h_1}{q_1^{r_1}} + \frac{h_2}{q_2^{r_2}} + \cdots + \frac{h_k}{q_k^{r_k}},$$

then $f/g = h/g_1 + h_k/q_k^{r_k}$, where $g_1 = q_1^{r_1} \ldots q_{k-1}^{r_{k-1}}$; by the uniqueness in Lemma 5.3.4, h_k must be as above, and the induction hypothesis yields the uniqueness of $h_1, \ldots, h_{k-1}$. $\square$

The last step of the proof writes a fraction h/q^r, with $\deg h < \deg q^r$, as a sum of partial fractions.

Lemma 5.3.6. *If $\deg h < \deg p^r$, there exist unique polynomials $f_1, \ldots, f_r$ such that $\deg f_i < \deg p$ for all i and*

$$\frac{h}{p^r} = \frac{f_1}{p} + \frac{f_2}{p^2} + \cdots + \frac{f_r}{p^r}. \tag{5.5}$$

Proof. The proof is by induction on r. The result is clear of $r = 1$. If $r \geqslant 2$, then (5.5) is equivalent to $h = f_1 p^{r-1} + \cdots + f_{r-1} p + f_r$. Division yields polynomials q, r unique such that $h = pq + r$ and $\deg r < \deg p$. We see that $r = f_r$, which yields the existence and uniqueness of f_r. Also $q = f_1 p^{r-2} + \cdots + f_{r-1}$ so that

$$\frac{q}{p^{r-1}} = \frac{f_1}{p} + \cdots + \frac{f_{r-1}}{p^{r-1}},$$

with $\deg q < \deg p^{r-1}$, and the induction hypothesis yields the existence and uniqueness of $f_1, \ldots, f_{r-1}$. $\qquad\qquad\qquad\qquad\qquad\qquad\qquad\qquad\qquad\qquad\square$

5.4. UNIQUE FACTORIZATION DOMAINS

Unique factorization domains share most of the arithmetic properties of principal ideal domains but include the polynomial rings $K[X_1, \ldots, X_n]$, where K is a field. This extends properties of integers to polynomials in several indeterminates.

5.4.a. Definition. A **unique factorization domain** or UFD is a domain R (a commutative ring with identity and no zero divisors) in which:

(1) every element (except for 0 and units) is a nonempty product of irreducible elements;

(2) if two products of irreducible elements $p_1 p_2 \ldots p_m = q_1 q_2 \ldots q_n$ are equal, then $m = n$ and the terms can be indexed so that $R p_i = R q_i$ for all i.

Recall that an element p is irreducible in case p is not 0 or a unit, and $p = xy$ implies that one of x, y is a unit.

As we saw in Section 5.1, a domain is a UFD if and only if every nonzero element can be written uniquely (up to the order of the terms) as the product of a unit u and of positive powers of distinct representative irreducible elements.

By Theorem 5.1.6, every PID is a UFD. Thus $\mathbb{Z}$ is a UFD; so is $K[X]$ for every field K. Theorem 5.4.4 below provides examples of UFDs which are not PIDs.

5.4.b. Properties. Consequences of Theorem 5.1.6 hold in every UFD. When $a = u p_1^{a_1} p_2^{a_2} \ldots p_k^{a_k}$ and $b = v q_1^{b_1} q_2^{b_2} \ldots q_\ell^{b_\ell} \in D$ are products of a unit and of positive powers of distinct representative irreducible elements, the sequences $p_1, \ldots, p_k$ and $q_1, \ldots, q_\ell$ can be merged so that a and b are products of a unit and of nonnegative powers of the same distinct representative irreducible elements: $a = u p_1^{a_1} p_2^{a_2} \ldots p_r^{a_r}$ and $b = v p_1^{b_1} p_2^{b_2} \ldots p_r^{b_r}$. As in Section 5.1 we obtain the following result, whose proof is an exercise:

Proposition 5.4.1. *In a UFD, let $a = u p_1^{a_1} p_2^{a_2} \ldots p_r^{a_r}$ and $b = v p_1^{b_1} p_2^{b_2} \ldots p_r^{b_r}$ be products of a unit and of nonnegative powers of distinct representative irreducible elements. Then:*

(1) $a \mid b$ *if and only if* $a_i \leqslant b_i$ *for all i;*

(2) *a and b have a l.c.m. $m = \mathrm{lcm}(a, b)$, namely $m = w p_1^{c_1} p_2^{c_2} \ldots p_r^{c_r}$, where $c_i = \max(a_i, b_i)$ and w is any unit;*

(3) *a and b have a g.c.d.* $d = \gcd(a,b)$, *namely* $d = z\,p_1^{d_1}p_2^{d_2}\ldots p_r^{d_r}$, *where* $d_i = \min(a_i,b_i)$ *and z is any unit;*

(4) $\mathrm{lcm}(a,b)\gcd(a,b) = tab$, *where t is a unit.*

The l.c.m. and g.c.d. can be required to be representative elements, or products of positive powers of representative irreducible elements (without a unit factor), and are then unique. On the other hand, the g.c.d. of a and b cannot necessarily be written in the form $ua + vb$ (see the exercises). The l.c.m. and g.c.d. also exist if $a = 0$: namely, $\mathrm{lcm}(0,b) = 0$ and $\gcd(0,b) = b$.

A similar argument shows that every finite sequence $a_1,\ldots,a_n$ of elements of a UFD has a l.c.m. $m = \mathrm{lcm}(a_1,\ldots,a_n)$ and a g.c.d. $d = \gcd(a_1,\ldots,a_n)$.

Two nonzero elements a and b of a UFD are **relatively prime** in case 1 is a g.c.d. of a and b; equivalently, any $\gcd(a,b)$ is a unit; equivalently, there is no irreducible element which divides both a and b. We denote this relationship by $\gcd(a,b) = 1$. If $\gcd(a,b) = \gcd(a,c) = 1$, then $\gcd(a,bc) = 1$ (see the exercises).

Lemma 5.4.2. *In a UFD, if* $\gcd(a,b) = 1$ *and* $a \mid bc$, *then* $a \mid c$.

Proof. We can write a, b, c in the form

$$a = u\,p_1^{a_1}\ldots p_r^{a_r}, \qquad b = v\,p_1^{b_1}\ldots p_r^{b_r}, \qquad c = w\,p_1^{c_1}\ldots p_r^{c_r},$$

where u,v,w are units and $p_1,\ldots,p_n$ are distinct irreducible elements. Then $bc = vw\,p_1^{b_1+c_1}\ldots p_r^{b_r+c_r}$. If $a_i > 0$, then $a_i \leqslant b_i + c_i$ (since $a \mid bc$), and $b_i = 0$ (since $\gcd(a,b) = 1$). Hence $a_i \leqslant c_i$ for all i and $a \mid c$. $\qquad\square$

Corollary 5.4.3. *In a UFD, irreducible elements are prime (if p is irreducible and* $p \mid ab$, *then* $p \mid a$ *or* $p \mid b$).

Proof. If $p \nmid a$, then $\gcd(p,a) = 1$ and $p \mid b$ by Lemma 5.4.2. $\qquad\square$

5.4.c. Polynomials. The main result of this section is:

Theorem 5.4.4. *When R is a unique factorization domain, then* $R[X]$ *is a unique factorization domain.*

Theorem 5.4.4 implies that $\mathbb{Z}[X_1,\ldots,X_n]$ and $K[X_1,\ldots,X_n]$ (where K is a field) are UFDs. This provides examples of UFDs which are not PIDs. Actually Theorem 5.4.4 holds for any number of indeterminates (see the exercises).

The proof, given below, uses the quotient field Q of R, whose elements are fractions a/b with $a,b \in R$, $b \neq 0$; if $a \neq 0$ it may be assumed that $\gcd(a,b) = 1$. In fact Theorem 5.4.4 follows from the relationship between $R[X]$ and the UFD $Q[X]$.

5.4.d. Primitive Polynomials. A polynomial $p(X) \in R[X]$ is **primitive** in case its coefficients are relatively prime; equivalently, in case no irreducible element of R divides every coefficient of $p(X)$.

Lemma 5.4.5. *Every nonzero polynomial $f(X) \in Q[X]$ can be written in the form $f(X) = (a/b)f^*(X)$ where $a, b \in R$ are relatively prime and $f^*(X) \in R[X]$ is primitive; furthermore a, b, f^* are unique up to multiplication by units of R.*

Proof. We have $f(X) = a_n/b_n X^n + \cdots + a_1/b_1 X + a_0/b_0$, where $b_0, \ldots, b_n \neq 0$. Then $b = \mathrm{lcm}(b_n, \ldots, b_0) \in R$ serves as a common denominator and $f(X)$ can be written in the form $f(X) = (1/b)(c_n X^n + \cdots + c_0)$ with $c_n, \ldots, c_0 \in R$. Factoring out $a = \gcd(c_n, \ldots, c_0) \in R$ yields $f(X) = (a/b)f^*(X)$ with f^* primitive.

Assume that $(a/b)g(X) = (c/d)h(X)$, where $\gcd(a, b) = \gcd(c, d) = 1$ in R and $g, h \in R[X]$ are primitive. Then $adg(X) = bch(X)$. Since g is primitive, the g.c.d. of the coefficients of g is a unit and the g.c.d. of the coefficients of $adg(X)$ is (up to multiplication by a unit) ad. Similarly the g.c.d. of the coefficients of $bch(X)$ is bc. Since $adg(X) = bch(X)$ we have $bc = adu$ for some unit u. Dividing by ad yields $g(X) = uh(X)$. Also a and c divide each other, by Lemma 5.4.2, so that $c = va$ for some unit v; similarly $d = wb$ for some unit w. $\square$

Lemma 5.4.6 (Gauss). *If f and $g \in R[X]$ are primitive, then fg is primitive.*

Proof. Let $f(X) = a_m X^m + \cdots + a_0$ and $g(X) = b_n X^n + \cdots + b_0$. Then $f(X) g(X) = c_{m+n} X^{m+n} + \cdots + c_0$, where $c_k = \sum_{i+j=k} a_i b_j$.

Let $p \in R$ be irreducible. Since f and g are primitive, $p \nmid a_i$ and $p \nmid b_j$ for some i, j. Let k, ℓ be the smallest such i, j so that $p \mid a_i$ for all $i < k$ and $p \mid b_j$ for all $j < \ell$, but $p \nmid a_k, b_\ell$. By Corollary 5.4.3, $p \nmid a_k b_\ell$. However, p divides every other term of the sum $c_{k+\ell} = \sum_{i+j=k+\ell} a_i b_j$, since $i < k$ implies $p \mid a_i$ and $i > k$ implies $j < \ell$ and $p \mid b_j$. Therefore $p \nmid c_{k+\ell}$. Thus no irreducible element divides all coefficients of fg. $\square$

Lemma 5.4.7. *Let R be a UFD with field of fractions Q and $p \in R[X]$ be primitive. Then p is irreducible in $R[X]$ if and only if p is irreducible in $Q[X]$.*

Proof. If p is irreducible in $Q[X]$, then p is irreducible in $R[X]$. Assume that p is not irreducible in $Q[X]$. Then $p = fg$ for some $f, g \in Q[X]$ with $\deg f, \deg g < \deg p$. By Lemma 5.4.5, $p(X) = tf^*(X)g^*(X)$ for some $t \in Q$ and some primitive $f^*, g^* \in R[X]$ with $\deg f^* = \deg f$, $\deg g^* = \deg g$. But f^*g^* is primitive by Lemma 5.4.6. By the uniqueness in Lemma 5.4.5, t is a unit of R. Hence p is not irreducible in $R[X]$. $\square$

Corollary 5.4.8. *Let $p(X) = tp^*(X) \in Q[X]$, where $t \in Q$, $t \neq 0$ and $p^* \in R[X]$ is primitive. Then p is irreducible in $Q[X]$ if and only if p^* is irreducible in $R[X]$.*

We can now prove Theorem 5.4.4. Let $f(X) \in R[X]$ and $d \in R$ be a g.c.d. of the coefficients of f. Then $f(X) = df^*(X)$, where f^* is primitive. In R, d is a product of irreducible elements. Also $f^*(X)$ is a product (in $Q[X]$) of irreducible elements $p_1(X),\ldots,p_n(X)$ of $Q[X]$. By Lemma 5.4.5, $p_i(X) = t_i p_i^*(X)$ for some $t_i \in Q$ and primitive $p_i^* \in R[X]$. By Lemma 5.4.7, p_i^* is irreducible in $R[X]$. Then $f^*(X) = tp_1^*(X)\ldots p_n^*(X)$, where $t \in Q$ and all $p_i^* \in R[X]$ are primitive and irreducible. Since f^* and $p_1^*\ldots p_n^*$ are primitive, t is a unit of R, and f^* is a product of irreducible polynomials.

In particular, an irreducible element of $R[X]$ is either an irreducible element of R or an irreducible primitive polynomial.

Uniqueness of the factorization into irreducible elements of $R[X]$ follows from uniqueness in R and in $Q[X]$. Assume that two products of irreducible elements of $R[X]$ are equal:

$$p_1 \cdots p_k p_{k+1} \cdots p_r = q_1 \cdots q_\ell q_{\ell+1} \cdots q_s.$$

We can arrange that $p_1,\ldots,p_k,q_1,\ldots,q_\ell$ are irreducible elements of R and $p_{k+1},\ldots,p_r,q_{\ell+1},\ldots,q_s$ are primitive irreducible polynomials.

Let $a = p_1 \ldots p_k$, $b = q_1 \ldots q_\ell \in R$, and $f = p_{k+1} \ldots p_r$, $g = q_{\ell+1} \ldots q_s \in R[X]$. By Lemma 5.4.6, f and g are primitive. Taking g.c.d.'s of coefficients in both sides of $af(X) = bg(X)$ then yields $a = ub$ for some unit u of R. Since R is a UFD, this implies $k = \ell$, and $q_1,\ldots,q_l$ can be indexed so that $Rp_i = Rq_i$ for all i (so that $p_i = u_i q_i$ for some unit u_i of R). We also have $uf(X) = g(X)$. By Lemma 5.4.7, $up_{k+1},p_{k+2},\ldots,p_r,q_{\ell+1},\ldots,q_s$ are irreducible in $Q[X]$. Since $Q[X]$ is a UFD, we have $r = s$, and $q_{\ell+1},\ldots,q_s$ can be indexed so that $p_j(X) = v_j q_j(X)$ for some unit v_j of $Q[X]$ (so that $(p_j) = (q_j)$ holds in $Q[X]$) for all $j \geqslant k$. Then v_j is a unit of R: indeed $v_j = c/d$ for some $c,d \in R$; since p_j and q_j are primitive and $dp_j(X) = cq_j(X)$, taking g.c.d.'s on both sides yields $c = vd$ for some unit v of R and $v_j = v$. Therefore $(p_j) = (q_j)$ holds in $R[X]$ for all $j \geqslant k$.   □

5.4.e. Eisenstein's Criterion. When R is a UFD, Corollary 5.4.8 shows that the irreducible polynomials of $Q[X]$ are determined by those of $R[X]$. For example, the irreducible polynomials of $\mathbb{Q}[X]$ are determined by those of $\mathbb{Z}[X]$.

Proposition 5.4.9 (Eisenstein's Criterion). *Let R be a UFD and $f(X) = a_n X^n + \cdots + a_0 \in R[X]$ be primitive. Assume that there exists an irreducible element p of R such that $p \mid a_i$ for all $i < n$, $p^2 \nmid a_0$, and $p \nmid a_n$. Then f is irreducible.*

Proof. Assume that $f = gh$, where $g(X) = b_r X^r + \cdots + b_0$ and $h(X) = c_s X^s + \cdots + c_0 \in R[X]$ are not constant. Then $a_k = \sum_{i+j=k} b_i c_j$ for all k. In particular, $a_0 = b_0 c_0$. Since $p \mid a_0$ we have, say, $p \mid b_0$, by Corollary 5.4.3. Then $p \nmid c_0$, since $p^2 \nmid a_0$. Also $p \nmid b_r$, since $p \nmid a_n = b_r c_s$. Therefore there exists a least $k > 0$ such that $p \nmid b_k$. Then $p \mid b_i$ for all $i < k$, and p divides every term of $a_k = \sum_{i+j=k} b_i c_j$, except for $b_k c_0$; therefore $p \nmid a_k$. Since $0 < k \leqslant r < n$ this contradicts the hypothesis. $\qquad\square$

For example, $3X^2 - 4X + 2$ is irreducible in $\mathbb{Z}[X]$ (hence also in $\mathbb{Q}[X]$); $Y^3 + 2X^2Y - 3X$ is irreducible in $\mathbb{Q}[X,Y] \cong \mathbb{Q}[X][Y]$.

Proposition 5.4.10. *Let $f(X) \in R[X]$ and $\mathfrak{a}$ be an ideal of R. If $f(X)$ is monic and the image of $f(X)$ in $(R/\mathfrak{a})[X]$ is irreducible, then $f(X)$ is irreducible.*

The proof is an exercise. For example, we saw that $X^3 + X + 1$ is irreducible in $\mathbb{Z}_2[X]$; therefore $X^3 + 4X^2 - 3X - 1$ is irreducible in $\mathbb{Z}[X]$ (and in $\mathbb{Q}[X]$).

Exercises

1. Prove Proposition 5.4.1.
2. Let K be a field. Show that the g.c.d. of X and Y in $K[X,Y]$ cannot be written in the form $uX + vY$ (with $u,v \in K[X,Y]$).
3. Show that every finite sequence $a_1,\ldots,a_n$ of elements of a UFD has a l.c.m. $m = [a_1,\ldots,a_n]$ and a g.c.d. $d = (a_1,\ldots,a_n)$.
4. Prove that in a UFD, $(a,b) = (a,c) = 1$ implies $(a,bc) = 1$.
5. Show that $\mathbb{Z}[X]$ is not a PID.
6. Show that the ring $\mathbb{Z}[i\sqrt{5}]$ of all complex numbers of the form $a + bi\sqrt{5}$ with $a,b \in \mathbb{Z}$ is not a UFD.
7. Let R be a UFD. Show that $R[(X_i)_{i \in I}]$ is a UFD for all $(X_i)_{i \in I}$. (Follow the proof of Theorem 5.4.4.)
8. Prove Proposition 5.4.10.

5.5. NOETHERIAN RINGS

Noetherian rings are a large class of commutative rings in which factorization can be carried out with ideals rather than elements. This section contains the definition and some examples, including (again) polynomial rings.

5.5.a. Definition. A commutative ring with identity R is **noetherian** in case its ideals satisfy the ascending chain condition (a.c.c.). In detail, the following conditions on R are equivalent by Proposition B.2.1:

(1) Every ascending sequence

$$\mathfrak{a}_1 \subseteq \mathfrak{a}_2 \subseteq \cdots \subseteq \mathfrak{a}_j \subseteq \mathfrak{a}_{j+1} \subseteq \cdots$$

of ideals of R terminates (there exists $n > 0$ such that $\mathfrak{a}_j = \mathfrak{a}_n$ for all $j \geqslant n$).

(2) There is no strictly ascending sequence of ideals of R

$$\mathfrak{a}_1 \subsetneqq \mathfrak{a}_2 \subsetneqq \cdots \subsetneqq \mathfrak{a}_j \subsetneqq \mathfrak{a}_{j+1} \subsetneqq \cdots .$$

(3) Every nonempty set S of ideals of R has a maximal element (an element $\mathfrak{m}$ of S such that there is no $\mathfrak{m} \subsetneqq \mathfrak{a} \in S$).

R is noetherian in case it satisfies all of these conditions. The next result gives a fourth equivalent condition.

Proposition 5.5.1. *A commutative ring with identity R is noetherian if and only if every ideal of R is finitely generated (as an ideal).*

Proof. Let R be noetherian and $\mathfrak{a}$ be an ideal of R. Let $\mathscr{S}$ be the set of all finitely generated ideals of R contained in $\mathfrak{a}$. Then $\mathscr{S} \neq \emptyset$ (e.g., $(0) \in \mathscr{S}$). By the a.c.c., $\mathscr{S}$ has a maximal element $\mathfrak{m}$. If $\mathfrak{m} \subsetneqq \mathfrak{a}$, then for any $a \in \mathfrak{a} \backslash \mathfrak{m}$ the ideal $\mathfrak{m} + Ra$ of R is contained in $\mathfrak{a}$ and is finitely generated (by a and the generators of $\mathfrak{m}$), contradicting the maximality of $\mathfrak{m}$. Therefore $\mathfrak{m} = \mathfrak{a}$ and $\mathfrak{a} \in \mathscr{S}$ is finitely generated.

Conversely, assume that every ideal of R is finitely generated. Let

$$\mathfrak{a}_1 \subseteq \mathfrak{a}_2 \subseteq \cdots \subseteq \mathfrak{a}_j \subseteq \mathfrak{a}_{j+1} \subseteq \cdots$$

be an ascending sequence of ideals of R. Then $\mathfrak{a} = \bigcup_{j \geqslant 1} \mathfrak{a}_j$ is an ideal of R and is finitely generated by, say, $x_1, \ldots, x_k$. Then $x_j \in \mathfrak{a}_{n_j}$ for some n_j. If $n \geqslant n_1, \ldots, n_k$, then $\mathfrak{a}_n \subseteq \mathfrak{a}$ contains $x_1, \ldots, x_k$; $\mathfrak{a}_n = \mathfrak{a}$; and $\mathfrak{a}_n = \mathfrak{a}_j = \mathfrak{a}$ for all $j \geqslant n$. □

Since R is commutative with identity, the ideal of R generated by $x_1, \ldots, x_k$ is $\mathfrak{a} = Rx_1 + \cdots + Rx_k$; every element of $\mathfrak{a}$ is a linear combination $r_1 x_1 + \cdots + r_k x_k$ of $x_1, \ldots, x_k$, and the generating set $\{x_1, \ldots, x_k\}$ is sometimes called a **basis** of $\mathfrak{a}$. (This does not mean that every element of $\mathfrak{a}$ can be written uniquely in the form $r_1 x_1 + \cdots + r_k x_k$.)

5.5.b. Examples. By Lemma 5.1.3, every PID is noetherian (this also follows from Proposition 5.5.1). Hence $\mathbb{Z}$ is noetherian, so is the polynomial ring $K[X]$ in one indeterminate for every field K. The next result imply that the UFDs $\mathbb{Z}[X_1, \ldots, X_n]$ and $K[X_1, \ldots, X_n]$ are noetherian.

Theorem 5.5.2 (Hilbert Basis Theorem). *Let R be a commutative ring with identity. If R is noetherian, then $R[X]$ is noetherian.*

Proof. Let $\mathfrak{A}$ be an ideal of $R[X]$. For each $n \geqslant 0$ let $\mathfrak{a}_n$ be the set of all $r \in R$ for which there exists a polynomial $f(X) \in \mathfrak{A}$ of degree $\leqslant n$ in which the coefficient of X^n is r. Then $\mathfrak{a}_n$ is an ideal of R (since $\mathfrak{A}$ is an ideal of $R[X]$), and $\mathfrak{a}_n \subseteq \mathfrak{a}_{n+1}$ (since $f(X) \in \mathfrak{A}$ implies $Xf(X) \in \mathfrak{A}$). The ascending sequence

$$\mathfrak{a}_1 \subseteq \mathfrak{a}_2 \subseteq \cdots \subseteq \mathfrak{a}_j \subseteq \mathfrak{a}_{j+1} \subseteq \cdots$$

of ideals of R terminates with some $\mathfrak{a}_m$ ($\mathfrak{a}_j = \mathfrak{a}_m$ for all $j \geqslant m$). Also $\mathfrak{a}_1, \ldots, \mathfrak{a}_m$ have finite generating sets $S_1, \ldots, S_m$. For each $s \in S_j$ let $g_s(X) \in \mathfrak{A}$ be a polynomial of degree $\leqslant j$ in which the coefficient of X^j is s. We show that $\mathfrak{A}$ coincides with the ideal $\mathfrak{B}$ generated by the finite set $\{ g_s ; s \in S_1 \cup \cdots \cup S_m \}$.

We have $\mathfrak{B} \subseteq \mathfrak{A}$, since $\mathfrak{A}$ contains all generators of $\mathfrak{B}$. We show by induction on $\deg f$ that every $f(X) \in \mathfrak{A}$ is in $\mathfrak{B}$. This already holds if $f = 0$. Now let $f(X) = a_n X^n + \cdots + a_0$ have degree $n \geqslant 0$. Then $a_n \in \mathfrak{a}_n$. If $n \leqslant m$, then

$$a_n = r_1 s_1 + \cdots + r_k s_k$$

for some $r_1, \ldots, r_k \in R$ and $s_1, \ldots, s_k \in S_n$. Hence $g(X) = r_1 g_{s_1}(X) + \cdots + r_k g_{s_k}(X)$ $\in \mathfrak{B}$ has degree n and leading coefficient a_n. Then $\deg(f - g) < n$, $f - g \in \mathfrak{B}$ by the induction hypothesis, and $f \in \mathfrak{B}$. If $n > m$, then $\mathfrak{a}_n = \mathfrak{a}_m$;

$$a_n = r_1 s_1 + \cdots + r_k s_k$$

for some $r_1, \ldots, r_k \in R$ and $s_1, \ldots, s_k \in S_m$; $g(X) = r_1 g_{s_1}(X) + \cdots + r_k g_{s_k}(X) \in \mathfrak{B}$ has degree m and leading coefficient a_n; and $X^{n-m} g(X) \in \mathfrak{B}$ has degree n and leading coefficient a_n. As above, $f(X) - X^{n-m} g(X) \in \mathfrak{B}$ by the induction hypothesis, and $f \in \mathfrak{B}$. $\square$

Theorem 5.5.3. *Let R be a commutative ring with identity. If R is noetherian, then $R[[X]]$ is noetherian.*

Proof. This is proved like Theorem 5.5.2, using orders instead of degrees. Let $\mathfrak{A}$ be an ideal of $R[[X]]$. For each $n \geqslant 0$ let $\mathfrak{a}_n$ be the set of all $r \in R$ for which there exists a power series $f(X) \in \mathfrak{A}$ of order $\geqslant n$ in which the coefficient of X^n is r. Then $\mathfrak{a}_n$ is an ideal of R (since $\mathfrak{A}$ is an ideal of $R[[X]]$), and $\mathfrak{a}_n \subseteq \mathfrak{a}_{n+1}$ (since $f(X) \in \mathfrak{A}$ implies $Xf(X) \in \mathfrak{A}$). By the a.c.c., the ascending sequence

$$\mathfrak{a}_0 \subseteq \mathfrak{a}_1 \subseteq \cdots \subseteq \mathfrak{a}_j \subseteq \mathfrak{a}_{j+1} \subseteq \cdots$$

of ideals of R terminates with some $\mathfrak{a}_m$ ($\mathfrak{a}_j = \mathfrak{a}_m$ for all $j \geqslant m$). Also $\mathfrak{a}_j$ has a finite generating set $s_{j1}, s_{j2}, \ldots, s_{jr}$. For each s_{jk} let $g_{jk}(X) \in \mathfrak{A}$ be a power series of order $\geqslant j$ in which the coefficient of X^j is s_{jk}. We show that $\mathfrak{A}$ is generated by the finite set $g_{01}, \ldots, g_{mt}$.

Let $f(X) \in \mathfrak{A}$. For each $n \leqslant m$ we show by induction on n that there exists $c_{01}, \ldots, c_{mt} \in R$ such that $f(X) - (c_{01} g_{01}(X) + \cdots + c_{mt} g_{mt}(X))$ has order at least $n + 1$. First, $f(X) = a_0 + a_1 X + \cdots$ with $a_0 \in \mathfrak{a}_0$. Since $\mathfrak{a}_0$ is generated by $s_{01}, \ldots, s_{0r}$, we have $a_0 = c_{01} s_{01} + \cdots + c_{0r} s_{0r}$ for some $c_{01}, \ldots, c_{0r} \in R$; then $f(X) - c_{01} g_{01}(X) - \cdots - c_{0r} g_{0r}(X)$ has constant term 0 and order at least 1.

Now assume that

$$h(X) = f(X) - (c_{01}g_{01}(X) + \cdots + c_{mt}g_{mt}(X)) = a'_n X^n + a'_{n+1}X^{n+1} + \cdots$$

has order at least n, where $0 < n < m$. Then $h \in \mathfrak{A}$, $a'_n \in \mathfrak{a}_n$, and

$$a'_n = r_{n1}s_{n1} + \cdots + r_{nr}s_{nr}$$

for some $r_{n1}, \ldots, r_{nr} \in R$, since $\mathfrak{a}_n$ is generated by $s_{n1}, \ldots, s_{nr}$. Hence $g(X) = r_{n1}g_{n1}(X) + \cdots + r_{nr}g_{nr}(X)$ has order at least n, and the coefficient of X^n in $g(X)$ is a'_n; hence $h(X) - g(X)$ has order at least $n + 1$.

We now have $c_{01}, \ldots, c_{mt} \in R$ such that

$$f_1(X) = f(X) - (c_{01}g_{01}(X) + \cdots + c_{m-1,s}g_{m-1,s}(X)$$
$$+ c_{m1}g_{m1}(X) + \ldots + c_{mt}g_{mt}(X))$$

has order at least $m + 1$. We complete the proof by constructing power series $d_{m1}(X), \ldots, d_{mt}(X)$ such that $f_1(X) = d_{m1}(X)g_{m1}(X) + \cdots + d_{mt}(X)g_{mt}(X)$.

The coefficients of X^{n-m} in $d_{m1}(X), \ldots, d_{mt}(X)$ are constructed by induction on $n \geq m$ as follows. Assume that $n > m$ and that we have constructed the coefficients of $X^0, X^1, \ldots, X^{n-m-1}$ in $d_{m1}(X), \ldots, d_{mt}(X)$, so that we have polynomials $_n d_{m1}(X), \ldots, _n d_{mt}(X)$ of degree less than $n - m$ such that

$$h(X) = f_1(X) - (_n d_{m1}(X)g_{m1}(X) + \cdots + _n d_{mt}(X)g_{mt}(X))$$
$$= a'_n X^n + a'_{n+1}X^{n+1} + \cdots$$

has order at least n. (This holds if $n = m + 1$.) Then $h(X) \in \mathfrak{A}$, $a'_n \in \mathfrak{a}_n = \mathfrak{a}_m$, and $a'_n = r_{m1}s_{m1} + \cdots + r_{mt}s_{mt}$ for some $r_{m1}, \ldots, r_{mt} \in R$. Hence

$$g(X) = r_{m1}X^{n-m}g_{m1}(X) + \cdots + r_{mt}X^{n-m}g_{mt}(X)$$

has order at least n, and the coefficient of X^m in g is a'_n. Therefore

$$h(X) - g(X) = f_1(X) - ((_n d_{m1}(X) + r_{m1}X^{n-m})g_{m1}(X) + \cdots$$
$$+ (_n d_{mt}(X) + r_{mt}X^{n-m})g_{mt}(X))$$

has order at least $n + 1$. The coefficients of X^{n-m} in $d_{m1}(X), \ldots, d_{mt}(X)$ are $r_{m1}, \ldots, r_{mt}$.

We now have power series $d_{m1}(X), \ldots, d_{mt}(X)$ such that

$$f_1(X) - (d_{m1}(X)g_{m1}(X) + \cdots + d_{mt}(X)g_{mt}(X))$$

has order at least n for all $n > m$; therefore

$$f_1(X) = d_{m1}(X)g_{m1}(X) + \cdots + d_{mt}(X)g_{mt}(X). \qquad \square$$

5.6. PRIMARY DECOMPOSITION

This section contains some general ideal theory, with which we prove a factorization theorem for noetherian rings, using ideals rather than elements.

5.6.a. Operations on Ideals. The sum and intersection of ideals were seen in Section 5.1.

The **product** of a finite nonempty sequence $\mathfrak{a}_1, \mathfrak{a}_2, \ldots, \mathfrak{a}_n$ of ideals of a ring is the ideal $\mathfrak{a}_1 \mathfrak{a}_2 \ldots \mathfrak{a}_n$ generated by their product $\{a_1 a_2 \ldots a_n; a_1 \in \mathfrak{a}_1, a_2 \in \mathfrak{a}_2, \ldots, a_n \in \mathfrak{a}_n\}$ as subsets. (The notation $\mathfrak{a}_1 \mathfrak{a}_2 \ldots \mathfrak{a}_n$ is traditional and should not be read as a product of subsets.) The product of the empty sequence of ideals of R is defined to be R. The product of ideals is associative (see below); diligent readers will make sure that $\mathfrak{a}_1 \mathfrak{a}_2 \ldots \mathfrak{a}_n$ coincides with the product $\mathfrak{a}_1 \mathfrak{a}_2 \ldots \mathfrak{a}_n$ constructed by induction from products of two ideals (as in Section 1.1).

Proposition 5.6.1. *The product of ideals has the following properties:*

(1) $\mathfrak{a}_1 \mathfrak{a}_2 \ldots \mathfrak{a}_n \subseteq \mathfrak{a}_1 \cap \mathfrak{a}_2 \cap \cdots \cap \mathfrak{a}_n$;

(2) $\mathfrak{a}(\mathfrak{b}\mathfrak{c}) = (\mathfrak{a}\mathfrak{b})\mathfrak{c}$;

(3) $\mathfrak{a} \sum_{i \in I} \mathfrak{b}_i = \sum_{i \in I} \mathfrak{a}\mathfrak{b}_i$, $(\sum_{i \in I} \mathfrak{b}_i)\mathfrak{a} = \sum_{i \in I} \mathfrak{b}_i\mathfrak{a}$.

Proof. We prove the first half of (3) and leave the remaining properties to the reader. By definition, $\mathfrak{p} = \mathfrak{a} \sum_{i \in I} \mathfrak{b}_i$ is the ideal generated by all products ab where $a \in \mathfrak{a}$ and $b \in \sum_{i \in I} \mathfrak{b}_i$, whereas $\mathfrak{s} = \sum_{i \in I} \mathfrak{a}\mathfrak{b}_i$ consists of all sums $\sum_{i \in I} c_i$ with $c_i \in \mathfrak{a}\mathfrak{b}_i$ and $c_i = 0$ for almost all i. Since $\mathfrak{b}_i \subseteq \sum_{i \in I} \mathfrak{b}_i$, we have $\mathfrak{a}\mathfrak{b}_i \subseteq \mathfrak{p}$ for all i; hence $\mathfrak{s} \subseteq \mathfrak{p}$. Conversely, let $a \in \mathfrak{a}$ and $b = \sum_{i \in I} b_i \in \sum_{i \in I} \mathfrak{b}_i$, where $b_i \in \mathfrak{b}_i$ and $b_i = 0$ for almost all i. Then $ab_i \in \mathfrak{a}\mathfrak{b}_i$ for all i, and $ab = \sum_{i \in I} ab_i \in \mathfrak{s}$. Hence $\mathfrak{s}$ contains every generator of $\mathfrak{p}$ and $\mathfrak{p} \subseteq \mathfrak{s}$. $\square$

If R is commutative, then the product of ideals is commutative and can be used to prove the following property:

Proposition 5.6.2 (Chinese Remainder Theorem). *Let $\mathfrak{a}_1, \ldots, \mathfrak{a}_n$ be ideals of a commutative ring R [with identity]. If $\mathfrak{a}_i + \mathfrak{a}_j = R$ whenever $i \neq j$, then the homomorphism $R \longrightarrow R/\mathfrak{a}_1 \times \cdots \times R/\mathfrak{a}_n$ induced by the projections $\pi_i : R \longrightarrow R/\mathfrak{a}_i$ is surjective.*

Proof. Let $\varphi : R \longrightarrow R/\mathfrak{a}_1 \times \cdots \times R/\mathfrak{a}_n$ be induced by the projections $\pi_i : R \longrightarrow R/\mathfrak{a}_i$. Let $\mathfrak{b}_j = \prod_{i \neq j} \mathfrak{a}_i$. If $\mathfrak{a}_j + \mathfrak{b}_j \subsetneqq R$, then $\mathfrak{a}_j + \mathfrak{b}_j \subseteq \mathfrak{m}$ for some prime ideal $\mathfrak{m}$ of R. Since $\mathfrak{b}_j \subseteq \mathfrak{m}$ and $\mathfrak{m}$ is a prime ideal, we have $\mathfrak{a}_k \subseteq \mathfrak{m}$ for some $k \neq j$, and then $\mathfrak{a}_j + \mathfrak{a}_k \subseteq \mathfrak{m}$, contradicting $\mathfrak{a}_j + \mathfrak{a}_k = R$. Therefore $\mathfrak{a}_j + \mathfrak{b}_j = R$. Hence $\pi_j(\mathfrak{b}_j) = R/\mathfrak{a}_j$. On the other hand, $\pi_i(\mathfrak{b}_j) = 0$ if $i \neq j$, since $\mathfrak{b}_j \subseteq \mathfrak{a}_i$. It follows that $\varphi(\mathfrak{b}_1 + \cdots + \mathfrak{b}_n)$ is all of $R/\mathfrak{a}_1 \times \cdots \times R/\mathfrak{a}_n$. $\square$

The conclusion of Proposition 5.6.2 can be stated as

$$R/(\mathfrak{a}_1 \cap \cdots \cap \mathfrak{a}_n) \cong R/\mathfrak{a}_1 \times \cdots \times R/\mathfrak{a}_n$$

(since Ker $\varphi = \mathfrak{a}_1 \cap \cdots \cap \mathfrak{a}_n$) or as follows: for each $x_1, \ldots, x_n \in R$, there exists $x \in R$ such that $x - x_i \in \mathfrak{a}_i$ for all i.

When R is commutative, the **quotient** of an ideal $\mathfrak{a}$ of R by a subset S of R (also called the **transporter** of S into $\mathfrak{a}$) is

$$\mathfrak{a} : S = \{ x \in R; \, xS \subseteq \mathfrak{a} \} = \{ x \in R; \, xs \in \mathfrak{a} \text{ for all } s \in S \}.$$

The following properties are straightforward:

Proposition 5.6.3. *When $S \subseteq R$ and $\mathfrak{a}, \mathfrak{b}, \mathfrak{c}, (\mathfrak{a}_i)_{i \in I}$ are ideals of the commutative ring R [with identity]:*

(1) $\mathfrak{a} : S$ *is an ideal and $\mathfrak{a} \subseteq \mathfrak{a} : S$;*

(2) $\mathfrak{a} : S = R$ *if and only if $S \subseteq \mathfrak{a}$;*

(3) $\mathfrak{c} \subseteq \mathfrak{a} : \mathfrak{b} \iff \mathfrak{b}\mathfrak{c} \subseteq \mathfrak{a}$;

(4) $(\bigcap_{i \in I} \mathfrak{a}_i) : S = \bigcap_{i \in I} (\mathfrak{a}_i : S)$;

(5) $(\mathfrak{a} : \mathfrak{b}) : \mathfrak{c} = \mathfrak{a} : \mathfrak{b}\mathfrak{c}$.

5.6.b. Radicals. From now on we assume that R is commutative. Recall that an ideal $\mathfrak{p} \neq R$ of R is **prime** in case $ab \in \mathfrak{p}$ implies $a \in \mathfrak{p}$ or $b \in \mathfrak{p}$; equivalently, in case $a, b \in R \backslash \mathfrak{p}$ implies $ab \in R \backslash \mathfrak{p}$; equivalently, in case $\mathfrak{a}\mathfrak{b} \subseteq \mathfrak{p}$ implies $\mathfrak{a} \subseteq \mathfrak{p}$ or $\mathfrak{b} \subseteq \mathfrak{p}$ (whenever $\mathfrak{a}$ and $\mathfrak{b}$ are ideals of R). For example, a (principal) ideal $\mathbb{Z}a$ of $\mathbb{Z}$ (with $a > 0$) is prime if and only if a is a prime number.

Proposition 5.6.4. *Let $\mathfrak{a}$ be an ideal of a commutative ring R. The intersection of all prime ideals of R containing $\mathfrak{a}$ is $\{ x \in R; \, x^n \in \mathfrak{a} \text{ for some } n > 0 \}$.*

Proof. Let $x \in R$ and $\mathfrak{r}$ be the intersection of all prime ideals which contain $\mathfrak{a}$. If $x \notin \mathfrak{r}$, then $x \notin \mathfrak{p}$ for some prime ideal $\mathfrak{p} \supseteq \mathfrak{a}$, $x^n \notin \mathfrak{p}$ for all $n > 0$, and $x^n \notin \mathfrak{a}$ for all $n > 0$.

Conversely, assume that $x^n \notin \mathfrak{a}$ for all $n > 0$. By Zorn's Lemma, there is an ideal $\mathfrak{m}$ of R which contains $\mathfrak{a}$, contains none of the x^n, and is maximal with these properties. We show that $\mathfrak{m}$ is prime; since $x \notin \mathfrak{m}$, it follows that $x \notin \mathfrak{r}$. Assume that $a, b \notin \mathfrak{m}$. By the choice of $\mathfrak{m}$, $\mathfrak{m} + Ra$ contains some x^k, and $\mathfrak{m} + Rb$ contains some x^ℓ; then $x^k = m_1 + ra$, $x^\ell = m_2 + sb$ yields $x^{k+\ell} = (m_1 + ra)(m_2 + sb) \in \mathfrak{m} + Rab$, $\mathfrak{m} + Rab \neq \mathfrak{m}$, and $ab \notin \mathfrak{m}$. $\square$

The **radical** of an ideal $\mathfrak{a}$ is the intersection $\mathrm{Rad}\,\mathfrak{a}$ of all prime ideals containing $\mathfrak{a}$; equivalently, the set

$$\mathrm{Rad}\,\mathfrak{a} = \{ x \in R; \, x^n \in \mathfrak{a} \text{ for some } n > 0 \}.$$

The radical of $\mathfrak{a}$ is sometimes denoted by $\sqrt{\mathfrak{a}}$.

The following result follows from Proposition 5.6.4.

Corollary 5.6.5. *For an ideal $\mathfrak{a}$ of a commutative ring R the following conditions are equivalent:*

(1) $\mathfrak{a}$ *is an intersection of prime ideals of R;*

(2) $x^n \in \mathfrak{a}$ *implies $x \in \mathfrak{a}$;*

(3) $\mathrm{Rad}\,\mathfrak{a} = \mathfrak{a}$.

An ideal is **semiprime** in case it satisfies the equivalent conditions in Corollary 5.6.5. For example, an ideal $\mathbb{Z}a$ of $\mathbb{Z}$ (with $a > 0$) is semiprime if and only if a is a product $a = p_1 \ldots p_r$ of distinct primes (see exercises).

Lemma 5.6.6. *If $\mathrm{Rad}\,\mathfrak{a}$ is finitely generated, then $(\mathrm{Rad}\,\mathfrak{a})^n \subseteq \mathfrak{a}$ for some $n > 0$.*

Proof. Assume that $\mathfrak{r} = \mathrm{Rad}\,\mathfrak{a} = Rc_1 + \cdots + Rc_k$. We have $c_i^{n_i} \in \mathfrak{a}$ for some $n_i > 0$. Let $n = n_1 + \cdots + n_k$. Every product of n elements of $\mathfrak{r}$ is a sum of integer multiples of products $c_1^{t_1} \ldots c_k^{t_k}$ with $t_1 + \cdots + t_k = n$. Now $t_1 + \cdots + t_k = n$ implies $t_i \geqslant n_i$ for some i and $c_1^{t_1} \ldots c_k^{t_k} \in \mathfrak{a}$. Hence $\mathfrak{a}$ contains all products of n elements of $\mathfrak{r}$, which generate $\mathfrak{r}^n$. $\square$

5.6.c. Primary Ideals. An ideal $\mathfrak{q} \neq R$ of a commutative ring R [with identity] is **primary** in case $ab \in \mathfrak{q}$ implies $a \in \mathfrak{q}$ or $b \in \mathrm{Rad}\,\mathfrak{q}$; equivalently, in case $\mathfrak{ab} \subseteq \mathfrak{q}$ implies $\mathfrak{a} \subseteq \mathfrak{q}$ or $\mathfrak{b} \subseteq \mathrm{Rad}\,\mathfrak{q}$ (whenever $\mathfrak{a}$ and $\mathfrak{b}$ are ideals of R). For example, an ideal $\mathbb{Z}a$ of $\mathbb{Z}$ (with $a > 0$) is primary if and only if a is a positive power of a prime number.

Proposition 5.6.7. *The radical of a primary ideal is a prime ideal.*

Proof. Let $\mathfrak{q}$ be primary and $\mathfrak{p} = \mathrm{Rad}\,\mathfrak{q}$. If $ab \in \mathfrak{p}$ and $a \notin \mathfrak{p}$, then $a^n b^n \in \mathfrak{q}$, $a^n \notin \mathfrak{q}$, $(b^n)^m \in \mathfrak{q}$, and $b \in \mathfrak{p}$. $\square$

An ideal is $\mathfrak{p}$-**primary** when it is primary with radical $\mathfrak{p}$.

Lemma 5.6.8. *The intersection of finitely many $\mathfrak{p}$-primary ideals is $\mathfrak{p}$-primary.*

Proof. We prove that the intersection of two $\mathfrak{p}$-primary ideals $\mathfrak{q}_1$ and $\mathfrak{q}_2$ is $\mathfrak{p}$-primary; then the result follows by induction. If $x \in \mathfrak{p}$, then $x^m \in \mathfrak{q}_1$, $x^n \in \mathfrak{q}_2$, and $x^{m+n} \in \mathfrak{q}_1 \cap \mathfrak{q}_2$; therefore $\mathrm{Rad}\,(\mathfrak{q}_1 \cap \mathfrak{q}_2) = \mathfrak{p}$. Then $ab \in \mathfrak{q}_1 \cap \mathfrak{q}_2$ and $b \notin \mathfrak{p}$ implies $a \in \mathfrak{q}_1 \cap \mathfrak{q}_2$, since $\mathfrak{q}_1$ and $\mathfrak{q}_2$ are $\mathfrak{p}$-primary. $\square$

5.6.d. Primary Decomposition. We now let R be noetherian.

An ideal $\mathfrak{i}$ is called **irreducible** (short for "intersection irreducible") in case $\mathfrak{i}$ is proper ($\mathfrak{i} \neq R$) and is not the intersection $\mathfrak{i} = \mathfrak{a} \cap \mathfrak{b}$ of ideals $\mathfrak{a}, \mathfrak{b} \neq \mathfrak{i}$.

Lemma 5.6.9. *In a noetherian ring, every proper ideal is the intersection of finitely many irreducible ideals.*

Proof. Assume that the result is false. Then the set $\mathcal{S}$ of all proper ideals which are not intersections of finitely many irreducible ideals is nonempty. Since R is noetherian, $\mathcal{S}$ has a maximal element $\mathfrak{m}$. Now $\mathfrak{m}$ is not irreducible (is not the intersection of a family of one irreducible ideal); hence $\mathfrak{m} = \mathfrak{a} \cap \mathfrak{b}$ for some ideals $\mathfrak{m} \subsetneqq \mathfrak{a}$, $\mathfrak{m} \subsetneqq \mathfrak{b}$. Since $\mathfrak{m}$ is maximal, $\mathfrak{a}$ and $\mathfrak{b}$ are intersections of finitely many irreducible ideals. But then so is $\mathfrak{m}$; this is the required contradiction. $\qquad\square$

Lemma 5.6.10. *In a noetherian ring, every irreducible ideal is primary.*

Proof. Let $\mathfrak{a}$ be irreducible; assume that $ab \in \mathfrak{a}$, $b \notin \operatorname{Rad}\mathfrak{a}$. Let $\mathfrak{a}_n = \mathfrak{a} : b^n$. We see that $a \in \mathfrak{a}_1$, $\mathfrak{a}_n$ is an ideal, and

$$\mathfrak{a} \subseteq \mathfrak{a}_1 \subseteq \mathfrak{a}_2 \subseteq \cdots \subseteq \mathfrak{a}_n \subseteq \mathfrak{a}_{n+1} \subseteq \cdots$$

(since $xb^n \in \mathfrak{a}$ implies $xb^{n+1} \in \mathfrak{a}$). In a noetherian ring, this ascending sequence terminates, and $\mathfrak{a}_{2n} = \mathfrak{a}_n$ for some n. Let $\mathfrak{b} = \mathfrak{a} + Rb^n$. If $x \in \mathfrak{a}_n \cap \mathfrak{b}$, then $x = t + yb^n$ for some $t \in \mathfrak{a}$, $xb^n = tb^n + yb^{2n} \in \mathfrak{a}$, $yb^{2n} \in \mathfrak{a}$, $y \in \mathfrak{a}_{2n} = \mathfrak{a}_n$, $yb^n \in \mathfrak{a}$, and $x \in \mathfrak{a}$. Thus $\mathfrak{a}_n \cap \mathfrak{b} = \mathfrak{a}$. Now $\mathfrak{b} \neq \mathfrak{a}$, since $b \notin \operatorname{Rad}\mathfrak{a}$. If $\mathfrak{a}$ is irreducible, then $\mathfrak{a}_n = \mathfrak{a}$, $\mathfrak{a}_1 = \mathfrak{a}$, and $a \in \mathfrak{a}$. $\qquad\square$

It follows from the last two lemmas that every ideal $\mathfrak{a}$ of a noetherian ring is the intersection $\mathfrak{a} = \mathfrak{q}_1 \cap \mathfrak{q}_2 \cap \cdots \cap \mathfrak{q}_r$ of finitely many primary ideals. The primary ideals $\mathfrak{q}_1, \ldots, \mathfrak{q}_r$ are a **primary decomposition** of $\mathfrak{a}$. In such an intersection it may be assumed that no term is **redundant** (such that $\mathfrak{q}_i \supseteq \bigcap_{j \neq i} \mathfrak{q}_j$, so that $\bigcap_j \mathfrak{q}_j = \bigcap_{j \neq i} \mathfrak{q}_j$), and that the radicals $\operatorname{Rad}\mathfrak{q}_1, \ldots, \operatorname{Rad}\mathfrak{q}_r$ are all distinct (by Lemma 5.6.8); we call $\mathfrak{q}_1, \ldots, \mathfrak{q}_r$ a **reduced** primary decomposition of I when $\mathfrak{q}_1, \ldots, \mathfrak{q}_r$ have these properties. We have proved:

Theorem 5.6.11. *Every ideal of a commutative noetherian ring has a reduced primary decomposition.*

For example, if $n = p_1^{r_1} \ldots p_k^{r_k}$ is a positive integer, written as a product of positive powers of distinct primes, then $\mathbb{Z}n = \mathbb{Z}p_1^{r_1} \cap \cdots \cap \mathbb{Z}p_k^{r_k}$ is a reduced primary decomposition of $\mathbb{Z}n$.

5.6.e. Uniqueness. Reduced primary decompositions have the following uniqueness property:

Theorem 5.6.12. *If*

$$\mathfrak{a} = \mathfrak{q}_1 \cap \mathfrak{q}_2 \cap \cdots \cap \mathfrak{q}_r = \mathfrak{q}_1' \cap \mathfrak{q}_2' \cap \cdots \cap \mathfrak{q}_s'$$

are two reduced primary decompositions of $\mathfrak{a}$, then $r = s$ and $\mathfrak{q}_1', \ldots, \mathfrak{q}_s'$ can be indexed so that $\operatorname{Rad}\mathfrak{q}_i = \operatorname{Rad}\mathfrak{q}_i'$ for all i.

Proof. We show that in a reduced primary decomposition $\mathfrak{a} = \mathfrak{q}_1 \cap \mathfrak{q}_2 \cap \cdots \cap \mathfrak{q}_r$, the distinct prime ideals $\mathfrak{p}_1 = \mathrm{Rad}\,\mathfrak{q}_1, \ldots, \mathfrak{p}_r = \mathrm{Rad}\,\mathfrak{q}_r$ coincide with the prime ideals of the form $\mathfrak{a} : x = \mathfrak{a} : \{x\}$, where $x \in R \backslash \mathfrak{a}$. These ideals are the **associated prime ideals** of $\mathfrak{a}$.

For any $1 \leqslant i \leqslant r$ let $\mathfrak{b} = \bigcap_{j \neq i} \mathfrak{q}_j$. Then $\mathfrak{a} = \mathfrak{b} \cap \mathfrak{q}_i \subsetneqq \mathfrak{b}$. By Lemma 5.6.6, $\mathfrak{p}_i^n \subseteq \mathfrak{q}_i$ for some n; then $\mathfrak{b}\mathfrak{p}_i^n \subseteq \mathfrak{b} \cap \mathfrak{q}_i = \mathfrak{a}$. Let n be minimal such that $\mathfrak{b}\mathfrak{p}_i^n \subseteq \mathfrak{a}$. Then $n \geqslant 1$ (since $\mathfrak{b} \not\subseteq \mathfrak{a}$), $\mathfrak{b}\mathfrak{p}_i^{n-1} \not\subseteq \mathfrak{a}$, and there exists $x \in \mathfrak{b}\mathfrak{p}_i^{n-1} \backslash \mathfrak{a}$. Then $x \in \mathfrak{b}$, $x \notin \mathfrak{q}_i$ (otherwise, $x \in \mathfrak{b} \cap \mathfrak{q}_i = \mathfrak{a}$), and $\mathfrak{a} : x \subseteq \mathfrak{q}_i : x \subseteq \mathfrak{p}_i$ (since $x \notin \mathfrak{q}_i$ and $\mathfrak{q}_i$ is $\mathfrak{p}_i$-primary). On the other hand, $x\mathfrak{p}_i \subseteq \mathfrak{b}\mathfrak{p}_i^n \subseteq \mathfrak{a}$ and $\mathfrak{p}_i \subseteq \mathfrak{a} : x$. Thus $\mathfrak{p}_i = \mathfrak{a} : x$ for some $x \notin \mathfrak{a}$.

Conversely, assume that $\mathfrak{p} = \mathfrak{a} : x$ is a prime ideal, where $x \notin \mathfrak{a}$. Then $x \notin \mathfrak{q}_i$ for some i. Let $\mathfrak{b} = \prod_{x \notin \mathfrak{q}_i} \mathfrak{q}_i$. Then $x\mathfrak{b} \subseteq \mathfrak{q}_i$ for all i (either $x \in \mathfrak{q}_i$ or $\mathfrak{b} \subseteq \mathfrak{q}_i$), and $\mathfrak{b} \subseteq \mathfrak{p}$. Since $\mathfrak{p}$ is prime, this implies $\mathfrak{q}_i \subseteq \mathfrak{p}$ for some i such that $x \notin \mathfrak{q}_i$. Then $\mathfrak{p}_i = \mathrm{Rad}\,\mathfrak{q}_i \subseteq \mathfrak{p}$. Conversely, $\mathfrak{a} : x \subseteq \mathfrak{q}_i : x \subseteq \mathfrak{p}_i$, since $x \notin \mathfrak{q}_i$ and $\mathfrak{q}_i$ is $\mathfrak{p}_i$-primary. Thus $\mathfrak{p} = \mathfrak{p}_i$. $\square$

In Theorem 5.6.12 the primary ideals themselves need not be unique. However, it can be proved that if

$$\mathfrak{q}_1 \cap \mathfrak{q}_2 \cap \cdots \cap \mathfrak{q}_r = \mathfrak{q}_1' \cap \mathfrak{q}_2' \cap \cdots \cap \mathfrak{q}_r'$$

are two reduced primary decompositions, with $\mathrm{Rad}\,\mathfrak{q}_i = \mathrm{Rad}\,\mathfrak{q}_i' = \mathfrak{p}_i$ for all i, then $\mathfrak{q}_i = \mathfrak{q}_i'$ if $\mathfrak{p}_i$ is minimal in $\{\mathfrak{p}_1, \ldots, \mathfrak{p}_r\}$. (This happens for all i is R is a PID.)

Exercises

1. In a commutative ring, a product of principal ideals is principal: namely $(a_1)(a_2) \cdots (a_n) = (a_1 a_2 \ldots a_n)$.

2. Prove that $\mathfrak{a}_1 \mathfrak{a}_2 \ldots \mathfrak{a}_n \subseteq \mathfrak{a}_1 \cap \mathfrak{a}_2 \cap \cdots \cap \mathfrak{a}_n$ for all ideals $\mathfrak{a}_1, \ldots, \mathfrak{a}_n$ of a ring.

3. When R is a PID, prove that $(Ra)(Rb) = Ra \cap Rb$ whenever $(a, b) = 1$.

4. Prove that $\mathfrak{a}(\mathfrak{b}\mathfrak{c}) = (\mathfrak{a}\mathfrak{b})\mathfrak{c}$ for all ideals $\mathfrak{a}, \mathfrak{b}, \mathfrak{c}$ of a ring.

5. Prove that $(\sum_{i \in I} \mathfrak{b}_i)\mathfrak{a} = \sum_{i \in I} \mathfrak{b}_i \mathfrak{a}$ for all ideals $\mathfrak{a}$ and $(\mathfrak{b}_i)_{i \in I}$ of a ring.

6. Let R be a PID and $Ra \neq 0, R$ be a principal ideal of R. When is Ra a prime ideal?

7. Let R be a PID and $Ra \neq 0, R$ be a principal ideal of R. Describe $\mathrm{Rad}\,Ra$ in terms of the prime factorization of a.

8. Prove that $\mathrm{Rad}(\mathfrak{a} \cap \mathfrak{b}) = (\mathrm{Rad}\,\mathfrak{a}) \cap (\mathrm{Rad}\,\mathfrak{b})$.

9. Let R be a PID and $Ra \neq 0, R$ be a principal ideal of R. When is Ra a semiprime ideal?

10. Let R be a PID and $Ra \neq 0, R$ be a principal ideal of R. When is Ra a primary ideal?

11. Let $\mathfrak{a}$ be an ideal whose radical $\mathrm{Rad}\,\mathfrak{a}$ is a maximal ideal. Prove that $\mathfrak{a}$ is primary.

12. In a noetherian ring, prove that $\mathrm{Rad}\,\mathfrak{a}$ is the intersection of the associated prime ideals of $\mathfrak{a}$.

5.7. DEDEKIND DOMAINS

Dedekind domains are a class of rings in which factorization can be carried out with ideals, using products rather than intersections. We will revisit these rings in Chapter 11.

5.7.a. Fractional Ideals. Before defining Dedekind domains, we extend properties of ideals to the following subsets. Let R be a domain, and Q be the field of fractions of R. A **fractional ideal** of R is a subset of Q of the form

$$\mathfrak{a}/c = \{a/c \in Q \,;\, a \in \mathfrak{a}\},$$

where $\mathfrak{a}$ is an ideal of R and $c \in R$, $c \neq 0$.

Every ideal $\mathfrak{a}$ of R is a fractional ideal (since $\mathfrak{a} = \mathfrak{a}/1$). Conversely, a fractional ideal $\mathfrak{a}/c$ contained in R is an ideal of R (then $\mathfrak{a}/c = \mathfrak{a} : c$).

Lemma 5.7.1. *When* $q_1, q_2, \ldots, q_r \in Q$,

$$Rq_1 + Rq_2 + \cdots + Rq_r \subseteq Q$$

is a fractional ideal of R. If R is noetherian, then every fractional ideal of R is of this form.

Proof. Let $q_i = a_i/c_i$, where $a_i, c_i \in R$, $c_i \neq 0$. Since R is a domain, $c = c_1 c_2 \ldots c_r \neq 0$. We see that $Rq_1 + \cdots + Rq_r = (Rb_1 + \cdots + Rb_r)/c$, where $b_1, \ldots, b_r \in R$. Hence $Rq_1 + \cdots + Rq_r$ is a fractional ideal. If R is noetherian, then every ideal of R is finitely generated, and every fractional ideal has the form $(Rb_1 + \cdots + Rb_r)/c = Rb_1/c + \cdots + Rb_r/c$ for some $b_1, \ldots, b_r \in R$. $\square$

A fractional ideal of the form $\mathfrak{A} = Rq_1 + Rq_2 + \cdots + Rq_r$ (where $q_1, \ldots, q_r \in Q$) is called **finitely generated**. (As we will see, this means that $\mathfrak{A}$ is finitely generated as an R-module.)

The operations on ideals of R extend to fractional ideals. The following result is an exercise:

Proposition 5.7.2. *When $\mathfrak{A}$ and $\mathfrak{B}$ are fractional ideals of a domain R, then:*

(1) $\mathfrak{A} \cap \mathfrak{B}$ *is a fractional ideal of R;*

(2) $\mathfrak{A} + \mathfrak{B} = \{a + b \,;\, a \in \mathfrak{A}, \, b \in \mathfrak{B}\}$ *is a fractional ideal of R;*

(3) *the set $\mathfrak{AB}$ of all finite sums $a_1 b_1 + \cdots + a_r b_r$, where $a_1, \ldots, a_r \in \mathfrak{A}$, $b_1, \ldots, b_r \in \mathfrak{B}$, is a fractional ideal of R;*

(4) *if $\mathfrak{B} \neq 0$, then $\mathfrak{A} : \mathfrak{B} = \{q \in Q \,;\, q\mathfrak{B} \subseteq \mathfrak{A}\}$ is a fractional ideal of R; in particular, $\mathfrak{A}' = \{q \in Q \,;\, q\mathfrak{A} \subseteq R\} = R : \mathfrak{A}$ is a fractional ideal of R when $\mathfrak{A} \neq 0$.*

A fractional ideal $\mathfrak{A}$ is **invertible** in case there exists a fractional ideal $\mathfrak{B}$ of R such that $\mathfrak{A}\mathfrak{B} = R$.

Lemma 5.7.3. (1) *A fractional ideal* $\mathfrak{A} \neq 0$ *of R is invertible if and only if* $\mathfrak{A}\mathfrak{A}' = R$.

(2) *Every nonzero principal ideal is invertible.*

(3) *Every invertible fractional ideal is finitely generated.*

Proof. (1) Assume that $\mathfrak{A}\mathfrak{B} = R$. Then $\mathfrak{B} \subseteq \mathfrak{A}'$, and conversely $\mathfrak{A}' = \mathfrak{A}'R = \mathfrak{A}'\mathfrak{A}\mathfrak{B} \subseteq R\mathfrak{B} = \mathfrak{B}$, so that $\mathfrak{B} = \mathfrak{A}'$.

(2) If $\mathfrak{A} = Ra \neq 0$ is a principal ideal of R, then $\mathfrak{A}' = R/a$ and $\mathfrak{A}\mathfrak{A}' = R$.

(3) Assume that $\mathfrak{A}\mathfrak{B} = R$. Then $1 = a_1 b_1 + \cdots + a_r b_r$ for some $a_1,\ldots,a_r \in \mathfrak{A}$, $b_1,\ldots,b_r \in \mathfrak{B}$. Hence $Ra_1 + \cdots + Ra_r \subseteq \mathfrak{A}$. Conversely, $a \in \mathfrak{A}$ implies $a = (ab_1)a_1 + \cdots + (ab_r)a_r$, with all $ab_i \in \mathfrak{A}\mathfrak{B} = R$, so $\mathfrak{A} = Ra_1 + \cdots + Ra_r$. $\quad\square$

When $\mathfrak{A} = \mathfrak{a}/c$ is a fractional ideal, (2) implies $\mathfrak{A} = \mathfrak{a}\mathfrak{c}'$, where $\mathfrak{c} = Rc$.

5.7.b. Dedekind Domains. A **Dedekind domain** is a domain which satisfies the following equivalent conditions:

Theorem 5.7.4. *For a domain R the following conditions are equivalent:*

(1) *Every nonzero ideal of R is a product of prime ideals of R.*

(2) *Every nonzero ideal of R can be written uniquely as a product of positive powers of distinct prime ideals.*

(3) *Every nonzero ideal of R is invertible (as a fractional ideal).*

(4) *Every nonzero fractional ideal of R is invertible.*

Then R is noetherian, and every prime ideal of R is a maximal ideal of R.

Condition (4) implies that the nonzero fractional ideals of R form an abelian group under multiplication of ideals; condition (2) implies that this is a free abelian group (on the set $\mathscr{P}$ of all nonzero prime ideals of R). For the proof we note:

Lemma 5.7.5. *Let* $\mathfrak{p}_1,\ldots,\mathfrak{p}_r,\mathfrak{q}_1,\ldots,\mathfrak{q}_s$ *be invertible prime ideals. If*

$$\mathfrak{a} = \mathfrak{p}_1\mathfrak{p}_2\cdots\mathfrak{p}_r = \mathfrak{q}_1\mathfrak{q}_2\cdots\mathfrak{q}_s,$$

then $r = s$ and (up to the order of the terms) $\mathfrak{p}_i = \mathfrak{q}_i$ for all i.

Proof. The proof is by induction on r. If $r = 0$, then $\mathfrak{a} = R$ and $s = 0$ (otherwise $\mathfrak{a} \subseteq \mathfrak{q}_1 \subsetneq R$). If $r > 0$, then let $\mathfrak{p}_r$, say, be minimal among $\{\mathfrak{p}_1,\ldots,\mathfrak{p}_r\}$. Then $\mathfrak{q}_1\ldots\mathfrak{q}_s \subseteq \mathfrak{p}_r$ implies $\mathfrak{q}_j \subseteq \mathfrak{p}_r$, and $\mathfrak{p}_1\ldots\mathfrak{p}_r \subseteq \mathfrak{q}_j$ implies $\mathfrak{p}_i \subseteq \mathfrak{q}_j \subseteq \mathfrak{p}_r$ for some i. By the choice of $\mathfrak{p}_r$ we have $\mathfrak{p}_i = \mathfrak{q}_j = \mathfrak{p}_r$, and $\mathfrak{q}_1,\ldots,\mathfrak{q}_s$ can be reindexed so that $q_s = \mathfrak{p}_r$. Then multiplication by $\mathfrak{p}'_r = q'_s$ yields

$$\mathfrak{p}_1 \mathfrak{p}_2 \dots \mathfrak{p}_{r-1} = q_1 q_2 \dots q_{s-1}$$

and the induction hypothesis yields $r = s$ and (after reindexing) $\mathfrak{p}_i = q_i$ for all i. $\qquad\square$

Proof of Theorem 5.7.4. (3) implies (4). Let $\mathfrak{A} = \mathfrak{a}/c$ be a fractional ideal. If $\mathfrak{a}$ in invertible, with $\mathfrak{a}\mathfrak{B} = R$, then $(\mathfrak{a}/c)(c\mathfrak{B}) = \mathfrak{a}\mathfrak{B} = R$ and $\mathfrak{A}$ is invertible.

(4) implies R noetherian: by Lemma 7.3, every ideal of R is finitely generated as a fractional ideal, hence is finitely generated as an ideal.

(4) implies (1). If (1) does not hold, then the set $\mathscr{S}$ of all ideals of R which are not products of prime ideals is not empty. Since R is noetherian by (4), $\mathscr{S}$ has a maximal element $\mathfrak{a}$. The ring R is an empty product of prime ideals; hence $\mathfrak{a} \neq R$, and $\mathfrak{a} \subseteq \mathfrak{p}$ for some prime (actually, maximal) ideal $\mathfrak{p}$ of R. Then $\mathfrak{a} = \mathfrak{a}\mathfrak{p}\mathfrak{p}' \subseteq \mathfrak{a}\mathfrak{p}' \subseteq \mathfrak{p}\mathfrak{p}' = R$ and $\mathfrak{a}\mathfrak{p}'$ is an ideal of R. Also $\mathfrak{a} \subsetneq \mathfrak{a}\mathfrak{p}'$ (otherwise, $\mathfrak{p} = \mathfrak{a}'\mathfrak{a}\mathfrak{p} = \mathfrak{a}'\mathfrak{a}\mathfrak{p}'\mathfrak{p} = R$). Hence $\mathfrak{a}\mathfrak{p}' \notin \mathscr{S}$ is a product of prime ideals of R. Then so is $\mathfrak{a} = \mathfrak{a}\mathfrak{p}'\mathfrak{p}$; this is the required contradiction.

(4) implies (2) by Lemma 7.5 (since (4) implies (1)). Obviously (2) implies (1).

(1) implies (3). We show that every prime ideal $\mathfrak{p}$ is invertible; this implies that every product of prime ideals is invertible. Let $a \in \mathfrak{p}$, $a \neq 0$. By (1), Ra is a product of prime ideals $Ra = \mathfrak{p}_1 \dots \mathfrak{p}_r$. Also Ra is invertible, by Lemma 5.7.3; hence $\mathfrak{p}_1, \dots, \mathfrak{p}_r$ are invertible. Now $\mathfrak{p}_1 \dots \mathfrak{p}_r = Ra \subseteq \mathfrak{p}$ implies $\mathfrak{p}_i \subseteq \mathfrak{p}$ for some i. By Lemma 5.7.6 below, every invertible prime ideal is a maximal ideal; it follows that $\mathfrak{p} = \mathfrak{p}_i$ is invertible (and that every prime ideal of R is maximal). $\qquad\square$

Lemma 5.7.6. *When* (1) *holds, every invertible prime ideal is a maximal ideal.*

Proof. Let $\mathfrak{p}$ be an invertible prime ideal. We want to show that $\mathfrak{p} + Ra = R$ for every $a \in R$, $a \notin \mathfrak{p}$. By (1), $\mathfrak{p} + Ra$ and $\mathfrak{p} + Ra^2$ are products of prime ideals:

$$\mathfrak{p} + Ra = \mathfrak{p}_1 \mathfrak{p}_2 \dots \mathfrak{p}_r, \qquad \mathfrak{p} + Ra^2 = q_1 q_2 \dots q_s.$$

Since $\mathfrak{p}$ is a prime ideal, $\overline{R} = R/\mathfrak{p}$ is a domain. The projection $x \longmapsto \overline{x}$ of R onto $\overline{R}$ induces a bijection between the ideals of R containing $\mathfrak{p}$ and the ideals of $\overline{R}$. We see that $\overline{\mathfrak{p} + Ra} = \overline{R}\overline{a}$ and $\overline{\mathfrak{p} + Ra^2} = \overline{R}\overline{a}^2$ are principal ideals of $\overline{R}$, and that

$$\overline{q}_1 \dots \overline{q}_s = \overline{R}\overline{a}^2 = (\overline{R}\overline{a})^2 = \overline{\mathfrak{p}}_1^2 \dots \overline{\mathfrak{p}}_r^2.$$

By Lemma 5.7.3, $\overline{R}\overline{a}$ and $\overline{R}\overline{a}^2$ are invertible. Therefore $\overline{\mathfrak{p}}_1, \dots, \overline{\mathfrak{p}}_r, \overline{q}_1, \dots, \overline{q}_s$ are invertible. Also $\overline{\mathfrak{p}}_1, \dots, \overline{\mathfrak{p}}_r, \overline{q}_1, \dots, \overline{q}_s$ are prime ideals of $\overline{R}$ (since $\mathfrak{p}_1, \dots, \mathfrak{p}_r, q_1, \dots, q_s$ are prime ideals of R and contain $\mathfrak{p}$). By Lemma 5.7.5, $\overline{q}_1 \dots \overline{q}_s = \overline{\mathfrak{p}}_1^2 \dots \overline{\mathfrak{p}}_r^2$

now implies that $s = 2r$ and that $\mathfrak{q}_1, \ldots, \mathfrak{q}_s$ can be indexed so that $\overline{\mathfrak{p}}_i = \overline{\mathfrak{q}}_{2i-1} = \overline{\mathfrak{q}}_{2i}$ for all i. Then $\mathfrak{p}_i = \mathfrak{q}_{2i-1} = \mathfrak{q}_{2i}$ for all i. Therefore $\mathfrak{p} + Ra^2 = (\mathfrak{p} + Ra)^2$.

Now $\mathfrak{p} \subseteq (\mathfrak{p} + Ra)^2 \subseteq \mathfrak{p}^2 + Ra$. In fact $\mathfrak{p} \subseteq \mathfrak{p}^2 + \mathfrak{p}a$, since $x + ya \in \mathfrak{p}$ with $x \in \mathfrak{p}^2$ implies $ya \in \mathfrak{p}$ and $y \in \mathfrak{p}$. Hence $\mathfrak{p} \subseteq \mathfrak{p}(\mathfrak{p} + Ra) \subseteq \mathfrak{p}$ and $\mathfrak{p} = \mathfrak{p}(\mathfrak{p} + Ra)$. But $\mathfrak{p}$ is invertible; hence $R = \mathfrak{p}'\mathfrak{p} = \mathfrak{p}'\mathfrak{p}(\mathfrak{p} + Ra) = \mathfrak{p} + Ra$. $\qquad\square$

5.7.c. Principal Ideals. Lemma 5.7.3 and condition (3) in Theorem 5.7.4 show that every PID is a Dedekind domain. Dedekind domains which are not PIDs will be constructed in Chapter 11. In the rest of this section we show that Dedekind domains are not much more general than PIDs: we prove that every ideal in a Dedekind domain is generated by at most two elements.

By Theorem 5.7.4, every nonzero ideal $\mathfrak{a}$ of a Dedekind domain R is a product of positive powers of distinct prime ideals. It is convenient to write $\mathfrak{a}$ in the form

$$\mathfrak{a} = \prod_{\mathfrak{p} \in \mathscr{P}} \mathfrak{p}^{e_\mathfrak{a}(\mathfrak{p})},$$

where $\mathscr{P}$ is the set of all nonzero prime ideals of R, each exponent $e_\mathfrak{a}(\mathfrak{p})$ is a nonnegative integer, and $e_\mathfrak{a}(\mathfrak{p}) = 0$ for almost all $\mathfrak{p} \in \mathscr{P}$. We note that $e_\mathfrak{a}(\mathfrak{p})$ is the largest integer k such that $\mathfrak{a} \subseteq \mathfrak{p}^k$: indeed $\mathfrak{a} = \prod_{\mathfrak{p} \in \mathscr{P}} \mathfrak{p}^{e_\mathfrak{a}(\mathfrak{p})} \subseteq \mathfrak{p}^{e_\mathfrak{a}(\mathfrak{p})}$, and $\mathfrak{a} \subseteq \mathfrak{p}^k$ implies $\mathfrak{b} = \mathfrak{a}(\mathfrak{p}^k)' \subseteq R$, $\mathfrak{a} = \mathfrak{b}\mathfrak{p}^k$, and $e_\mathfrak{a}(\mathfrak{p}) \geqslant k$.

Proposition 5.7.7. *Let $\mathfrak{a}$ be a nonzero ideal of a Dedekind domain R and $\mathfrak{p}_1, \ldots, \mathfrak{p}_n$ be nonzero prime ideals of R. There exists a principal ideal $\mathfrak{b}$ of R such that $e_\mathfrak{a}(\mathfrak{p}_i) = e_\mathfrak{b}(\mathfrak{p}_i)$ for all i.*

Proof. For each i let $\mathfrak{a}_i = \mathfrak{p}_i^{e_\mathfrak{a}(\mathfrak{p}_i)}$ and $\mathfrak{b}_i = \mathfrak{p}_i^{e_\mathfrak{a}(\mathfrak{p}_i)+1}$. Then $\mathfrak{a} \subseteq \mathfrak{a}_i$, $\mathfrak{a} \not\subseteq \mathfrak{b}_i$, and there exists $a_i \in \mathfrak{a} \backslash \mathfrak{b}_i$. If $\mathfrak{b}_i + \mathfrak{b}_j \neq R$, then $\mathfrak{b}_i + \mathfrak{b}_j \subseteq \mathfrak{q}$ for some prime ideal $\mathfrak{q}$, and $\mathfrak{b}_i$, $\mathfrak{b}_j \subseteq \mathfrak{q}$ implies $\mathfrak{p}_i$, $\mathfrak{p}_j \subseteq \mathfrak{q}$ and $\mathfrak{p}_i = \mathfrak{q} = \mathfrak{p}_j$ (since all prime ideals of R are maximal). Hence $\mathfrak{b}_i + \mathfrak{b}_j = R$ whenever $i \neq j$. By the Chinese Remainder Theorem (Proposition 5.6.2), there exists $b \in R$ such that $b + \mathfrak{b}_i = a_i + \mathfrak{b}_i$ for all i. For each $\mathfrak{p}_i$ we have $b \in \mathfrak{a}_i$ (since $a_i + \mathfrak{b}_i \subseteq \mathfrak{a}_i$) and $b \notin \mathfrak{b}_i$ (since $a_i \notin \mathfrak{b}_i$). Hence $e_\mathfrak{a}(\mathfrak{p}_i)$ is the largest integer k such that $Rb \subseteq \mathfrak{p}_i^k$, and $\mathfrak{b} = Rb$ satisfies $e_\mathfrak{b}(\mathfrak{p}_i) = e_\mathfrak{a}(\mathfrak{p}_i)$ for all i. $\qquad\square$

Corollary 5.7.8. *Every ideal of a Dedekind domain is generated by at most two elements.*

Proof. Let $\mathfrak{a}$ be a nonzero ideal of a Dedeking domain R; let $c \in \mathfrak{a}$, $c \neq 0$, and $\mathfrak{c} = Rc$. By Proposition 5.7.7, there exists a principal ideal $\mathfrak{b} = Rb$ such that $e_\mathfrak{a}(\mathfrak{p}) = e_\mathfrak{b}(\mathfrak{p})$ whenever $e_\mathfrak{c}(\mathfrak{p}) > 0$. We show that $\mathfrak{a} = \mathfrak{b} + \mathfrak{c} = Rb + Rc$.

If $e_\mathfrak{c}(\mathfrak{p}) = 0$, then $\mathfrak{c} \not\subseteq \mathfrak{p}$, $\mathfrak{a} \not\subseteq \mathfrak{p}$, $\mathfrak{b} + \mathfrak{c} \not\subseteq \mathfrak{p}$, and $e_\mathfrak{a}(\mathfrak{p}) = e_{\mathfrak{b}+\mathfrak{c}}(\mathfrak{p}) = 0$. Now let $e_\mathfrak{c}(\mathfrak{p}) > 0$ and $k = e_\mathfrak{a}(\mathfrak{p}) = e_\mathfrak{b}(\mathfrak{p})$. Then $\mathfrak{a} \subseteq \mathfrak{p}^k$, $\mathfrak{b} + \mathfrak{c} \subseteq \mathfrak{b} + \mathfrak{a} \subseteq \mathfrak{p}^k$, but $\mathfrak{b} + \mathfrak{c} \not\subseteq \mathfrak{p}^{k+1}$, since $\mathfrak{b} \not\subseteq \mathfrak{p}^{k+1}$. Hence $e_{\mathfrak{b}+\mathfrak{c}}(\mathfrak{p}) = k = e_\mathfrak{a}(\mathfrak{p})$. Thus $e_\mathfrak{a}(\mathfrak{p}) = e_{\mathfrak{b}+\mathfrak{c}}(\mathfrak{p})$ for all $\mathfrak{p}$; therefore $\mathfrak{a} = \mathfrak{b} + \mathfrak{c}$. $\qquad\square$

Actually we have shown a little more: given $\mathfrak{a} \neq 0$ there is for each $c \in \mathfrak{a}$, $c \neq 0$ some $b \in \mathfrak{a}$ such that $\mathfrak{a}$ is generated by c and b.

Exercises

1. Find all fractional ideals of $\mathbb{Z}$.
2. Find all fractional ideals of a PID.
3. If $p_1, \ldots, p_k$ are irreducible elements of a PID, and $(p_i) \neq (p_j)$ whenever $i \neq j$, then $(p_1^{r_1}) \ldots (p_k^{r_k}) = (p_1)^{r_1} \cap \cdots \cap (p_k)^{r_k}$.
4. Prove Proposition 5.7.2.
5. Prove directly that when every nonzero prime ideal of a domain R is invertible, then every nonzero prime ideal of R is maximal.
6. Let $\mathfrak{a}$ be a nonzero ideal of a Dedekind domain R. Show that every ideal of $R/\mathfrak{a}$ is principal.

6

FIELD EXTENSIONS

Fields are another major algebraic structure. To study fields, one looks at field extensions, which tell how a smaller field fits into a larger one. This chapter gives general properties of fields and field extensions, and some properties of transcendental extensions. Algebraic extensions are studied further in Chapter 7.

Required results: basic properties of rings, fields, and polynomials in Chapter 4 and Sections 5.1 and 5.4. Section 6.3 requires Zorn's Lemma (Theorem B.3.2). Properties of cardinal numbers (Corollary B.5.12) are used in Section 6.4.

6.1. FIELDS

This section contains a number of elementary properties of fields.

6.1.a. Homomorphisms. Recall that a field is a commutative ring K (necessarily a domain) such that $K^* = K \backslash \{0\}$ is a group under multiplication; equivalently, every nonzero element of K is a unit. $\mathbb{Q}, \mathbb{R}, \mathbb{C}$ are fields; $\mathbb{Z}_n$ is a field if and only if n is prime.

A **homomorphism** of fields of a field K into a field L is a homomorphism of rings with identity (a mapping $\varphi : K \longrightarrow L$ such that $\varphi(1) = 1$ and $\varphi(x + y) = \varphi(x) + \varphi(y)$, $\varphi(xy) = \varphi(x)\varphi(y)$ for all $x, y \in K$).

Proposition 6.1.1. *Every homomorphism of fields is injective.*

Proof. In the above, $x \neq 0$ in K implies $\varphi(x)\varphi(x^{-1}) = \varphi(xx^{-1}) = 1 \neq 0$ and $\varphi(x) \neq 0$; hence $\mathrm{Ker}\,\varphi = 0$. $\qquad\square$

Thus the basic relationship between two fields is (up to isomorphism) inclusion. This is why the study of field extensions constitutes so much of field theory.

Proposition 6.1.2. *Every field homomorphism* $\varphi : E \longrightarrow F$ *induces a homomorphism* $E[X] \longrightarrow F[X]$, *which sends* $f(X) = a_0 + \cdots + a_n X^n \in E[X]$ *onto* $(^\varphi f)(X) = \varphi a_0 + \cdots + (\varphi a_n) X^n \in F[X]$. *For all* $f \in E[X]$ *and* $c \in E$, $\varphi(f(c)) = (^\varphi f)(\varphi c)$.

The proof is an exercise.

A **subfield** of a field F is a field K such that $K \subseteq F$ and the inclusion mapping $K \longrightarrow F$ is a homomorphism of fields. The following result is clear:

Proposition 6.1.3. K *is [the underlying set of] a subfield of a field F if and only if K is an additive subgroup of F (equivalently, $0 \in K$, and $x - y \in K$ for all $x, y \in K$) and $K^* = K \backslash \{0\}$ is a multiplicative subgroup of F^* (equivalently, $1 \in K$, and $xy^{-1} \in K$ for all $x, y \in K$).*

For example, $\mathbb{Q}$ is a subfield of $\mathbb{R}$; $\mathbb{R}$ is a subfield of $\mathbb{C}$. The **Homomorphism Theorem** for fields states:

Proposition 6.1.4. *When* $\varphi : K \longrightarrow L$ *is a homomorphism of fields, then* $\operatorname{Im} \varphi$ *is a subfield of L and* $K \cong \operatorname{Im} \varphi$.

6.1.b. The Characteristic. By Proposition 4.2.7, a field K contains a smallest subring, which consists of all the integer multiples of 1 and is isomorphic either to $\mathbb{Z}$ or to $\mathbb{Z}_p$ for some $p > 0$. Recall that the characteristic of K is 0 if this subring is $\cong \mathbb{Z}$, and p if this subring is $\cong \mathbb{Z}_p$; then p is a prime by Proposition 4.4.1. Proposition 4.4.6, applied to the quotient field $Q(\mathbb{Z}) \cong \mathbb{Q}$, yields:

Proposition 6.1.5. *Every field K has a smallest subfield, which is isomorphic to $\mathbb{Q}$ if K has characteristic 0, and to $\mathbb{Z}_p$ if K has characteristic $p \neq 0$.*

In a field of characteristic 0, $nx = 0$ implies $n = 0$ or $x = 0$, since $n \neq 0$ implies $n1 \neq 0$ and $nx = (n1)x \neq 0$. In a field of characteristic $p \neq 0$, $px = 0$ holds for all x.

Proposition 6.1.6. *In a field of characteristic $p \neq 0$, $(x + y)^p = x^p + y^p$ holds for all x, y; hence $x^p = y^p$ implies $x = y$.*

Proof. This follows from the Binomial Theorem:

$$(x + y)^p = \sum_{0 \leqslant k \leqslant p} \binom{p}{k} x^k y^{p-k}.$$

If $0 < k < p$, then p divides the numerator of

$$\binom{p}{k} = \frac{p!}{k!\,(p-k)!}$$

but not its denominator; hence p divides $\binom{p}{k}$ and

$$\binom{p}{k} x^k y^{p-k} = 0$$

in K. Hence $(x+y)^p = x^p + y^p$. Thus $x \longmapsto x^p$ is a homomorphism and is injective by Proposition 6.1.1. $\square$

Fields of characteristic $p \neq 0$ in which $x \longmapsto x^p$ is bijective (not just injective) are called **perfect**; they will be studied further in Chapter 7.

6.1.c. Subfields. Subfields have a number of general properties. The following result is an exercise:

Proposition 6.1.7. *An intersection of subfields of a field F is a subfield of F.*

Likewise the union of a chain of subfields of F is a subfield of F. We prove a slightly more general property:

Proposition 6.1.8. *The union of a chain of fields is a field.*

Proof. A chain of fields is a family of fields $(K_i)_{i \in I}$ such that for every $i, j \in I$, one of K_i, K_j is a subfield of the other. In particular, all K_i have the same zero element and the same identity element. The operations on $K = \bigcup_{i \in I} K_i$ are defined as follows. Let $x, y \in K$. Since $(K_i)_{i \in I}$ is a chain we have $x, y \in K_i$ for some $i \in I$. If also $x, y \in K_j$, then (since $(K_i)_{i \in I}$ is a chain) $x + y$ is the same in K_i and K_j; hence $x + y$ does not depend on i. By definition, $x + y$ is the same in K and in K_i. The definition of xy is similar. Since each K_i is a field it is immediate that K is a field. $\square$

By Proposition 6.1.7, there is for each subset S of a field F a smallest subfield of F which contains S (namely, the intersection of all the subfields of F which contain S). This is the subfield of F **generated** by S.

The **composite** of a nonempty family $(K_i)_{i \in I}$ of subfields of a field F is the subfield of F generated by the union $\bigcup_{i \in I} K_i$. It is traditionally denoted by $\prod_{i \in I} K_i$ (by $K_1 K_2 \ldots K_n$ if $I = \{1, 2, \ldots, n\}$) and should not be confused with a product of subsets.

6.1.d. Subrings. For an easier description of the subfield generated by a subset, and of composites of subfields, we look at the generation of subrings of fields.

Proposition 6.1.9. *Let F be a field, K be a subfield of F and S be a subset of F.*

(1) *The subring $K[S]$ of F generated by $K \cup S$ is the set of all linear combinations with coefficients in K of products of powers of elements of S.*

(2) *The subfield $K(S)$ of F generated by $K \cup S$ is the set of all $ab^{-1} \in F$ with $a, b \in K[S]$, $b \neq 0$, and is isomorphic to the field of fractions of $K[S]$.*

Proof. (1) Let $(X_s)_{s \in S}$ be a family of indeterminates indexed by S (with $X_s \neq X_t$ if $s \neq t$). By the universal property of $K[(X_s)_{s \in S}]$ (Proposition 4.3.12), there is a unique ring homomorphism $\psi : K[(X_s)_{s \in S}] \longrightarrow F$ such that $\psi(x) = x$ for all $x \in K$ and $\psi(X_s) = s$ for all $s \in S$; ψ sends $\sum_m a_m X_{s_1}^{m_{s_1}} \ldots X_{s_t}^{m_{s_t}} \in K[(X_s)_{s \in S}]$ to $\sum_m a_m s_1^{m_{s_1}} \ldots s_t^{m_{s_t}} \in F$; Im ψ is the smallest subring of F which contains K and all $s \in S$. Thus $K[S] = \text{Im } \psi$, and Im ψ consists of all linear combinations with coefficients in K of products of powers of elements of S.

(2) By Proposition 4.4.6, $Q = \{ab^{-1} ; a, b \in K[S], b \neq 0\}$ is a subfield of F (which is isomorphic to the field of fractions of $K[S]$). We see that Q contains $K \cup S \subseteq K[S]$. Conversely, a subfield of F which contains $K \cup S$ must contain $K[S]$ and contain all $ab^{-1} \in Q$. Hence $Q = K(S)$. $\qquad\square$

The notation $K[S]$ is used because of the relationship between $K[S]$ and the polynomial ring $K[(X_s)_{s \in S}]$, but the two should not be confused. Similarly $K(S)$ should not be mistaken for a field of rational fractions. When $S = \{s_1, \ldots, s_n\}$ is finite, $K[S]$ is denoted by $K[s_1, \ldots, s_n]$, and $K(S)$ by $K(s_1, \ldots, s_n)$. By Proposition 6.1.9:

Corollary 6.1.10. *Let K be a subfield of F and $s_1, \ldots, s_n \in F$, $S \subseteq F$.*

(1) $x \in K[s_1, \ldots, s_n] \iff x = f(s_1, \ldots, s_n)$ *for some* $f \in K[X_1, \ldots, X_n]$;

(2) $x \in K(s_1, \ldots, s_n) \iff x = r(s_1, \ldots, s_n)$ *for some* $r \in K(X_1, \ldots, X_n)$;

(3) $x \in K[S] \iff x \in K[s_1, \ldots, s_n]$ *for some* $s_1, \ldots, s_n \in S$;

(4) $x \in K(S) \iff x \in K(s_1, \ldots, s_n)$ *for some* $s_1, \ldots, s_n \in S$.

When K is the smallest subfield of F, then the subfield of F generated by S is also generated by $K \cup S$, and is described by Proposition 6.1.9. Similarly:

Corollary 6.1.11. *Let $(K_i)_{i \in I}$ be subfields of a field F (where $I \neq \emptyset$). Then $x \in \prod_{i \in I} K_i$ if and only if $x = ab^{-1}$ for some $a, b \in R$, $b \neq 0$, where R is the set of all finite sums of finite products of elements of $\bigcup_{i \in I} K_i$.*

Proof. Let K be the smallest subfield of F. Then $\prod_{i \in I} K_i = K(\bigcup_{i \in I} K_i)$. In Proposition 6.1.9, a product of an element of K and powers of elements of $\bigcup_{i \in I} K_i$ is a product of elements of $\bigcup_{i \in I} K_i$, and conversely. $\qquad\square$

Corollary 6.1.12. *Let K and L be subfields of a field F. Then $x \in KL$ if and only if $x = ab^{-1}$ for some $a, b \in R$, $b \neq 0$, where R is the set of all finite sums of products of an element of K and an element of L.*

6.2. FIELD EXTENSIONS

A **field extension** or **extension** of a field K is a field E of which K is a subfield. We write this relationship as an inclusion $K \subseteq E$ when it is understood that E and K are fields.

A field extension of K can also be defined as a field F for which there is a homomorphism $\varphi : K \longrightarrow F$ (which is necessarily injective by Proposition 6.1.1). The two definitions are equivalent up to isomorphism. If K is a subfield of E and $E \cong F$, then there is a homomorphism of K into F. If, conversely, $\varphi : K \longrightarrow F$ is a homomorphism, then $\mathrm{Im}\, \varphi \cong K$ is a subfield of F; there is also a field $E \cong F$ of which K is a subfield (e.g., the disjoint union $K \cup (F \backslash \mathrm{Im}\, \varphi)$ with the appropriate operations).

6.2.a. Degree. The first property of any extension E of a field K is that it is a vector space over K: E is an abelian group under addition, and $K \subseteq E$ acts on E by multiplication in E.

The **degree** of an extension E of a field K, denoted by $[E : K]$, is the dimension of E, regarded as a vector space over K. For example, $[\mathbb{C} : \mathbb{R}] = 2$, and $[\mathbb{R} : \mathbb{Q}]$ is infinite (in fact $[\mathbb{R} : \mathbb{Q}] = |\mathbb{R}|$).

An extension $K \subseteq E$ is **finite** in case its degree is finite; we also say that E is **finite over** K. This does not mean that the set E is finite.

Proposition 6.2.1. $[F : K] = [F : E][E : K]$ *whenever* $K \subseteq E \subseteq F$.

Proof. With $K \subseteq E \subseteq F$, let $(\beta_j)_{j \in J}$ be a basis of F over E and $(\alpha_i)_{i \in I}$ be a basis of E over K. Every element of F can be written uniquely as a linear combination of the elements of $(\beta_j)_{j \in J}$ with coefficients in E, and every such coefficient can be written uniquely as a linear combination of the elements of $(\alpha_i)_{i \in I}$ with coefficients in K. Hence every element of F can be written uniquely as a linear combination of the elements of $(\alpha_i \beta_j)_{i \in I, j \in J}$ with coefficients in K: $(\alpha_i \beta_j)_{i \in I, j \in J}$ is a basis of E over K, and $[F : K] = |I \times J| = [F : E][E : K]$. $\square$

When $K \subseteq E, F$ are field extensions of K, a K-**homomorphism** is a homomorphism φ which is the identity on K ($\varphi(x) = x$ for all $x \in K$). Then φ is, in particular, a linear transformation of E into F (viewed as vector spaces over K). The induced homomorphism $E[X] \longrightarrow F[X]$ is also a K-homomorphism.

6.2.b. Simple Extensions. An extension E of a field K is **finitely generated** in case $E = K(s_1, \ldots, s_n)$ for some $s_1, \ldots, s_n \in E$ and **simple** in case $E = K(s)$ for some $s \in E$ (i.e., E is singly generated); then s is a **primitive element** of E. A simple extension E is not like a simple group and may contain intermediate fields $K \subsetneq F \subsetneq E$ (see the exercises).

Examples of simple extensions are readily constructed. The field of rational fractions $K(X)$ is a simple extension of K (generated by X). The next result is a basic construction of field theory, which we will soon show yields all finite simple extensions.

Proposition 6.2.2. *Let K be a field and $E = K[X]/(q)$, where $q(X) \in K[X]$ is an irreducible polynomial. Then E is a field extension of K and has a basis $1, \alpha, \alpha^2, \ldots, \alpha^{n-1}$ over K, where $n = \deg q$ and $q(\alpha) = 0$ in E. Hence $E = K[\alpha] = K(\alpha)$; E is a simple extension of K; and $[E : K] = \deg q$.*

Proof. By Lemma 5.1.4, (q) is a maximal ideal of $K[X]$; hence E is a field. There is a homomorphism $\varphi : K \longrightarrow E$ which sends $x \in K$ to $x + (q) \in K[X]/(q)$. In what follows we identify x and $\varphi(x)$ when $x \in K$.

Let $\alpha = X + (q) \in E$. The projection $K[X] \longrightarrow E$ sends X onto α and $x \in K$ onto $\varphi(x) = x$. Hence it coincides with the evaluation homomorphism $K[X] \longrightarrow E$, $f(X) \longmapsto f(\alpha)$; in other words, $f(X) + (q) = f(\alpha)$ for all $f(X) \in K[X]$. In particular, every element of E has the form $f(\alpha)$ for some $f(X) \in K[X]$. By Corollary 6.1.10, $E = K[\alpha]$. Hence $E = K(\alpha)$. Also $q(\alpha) = q(X) + (q) = 0$ in E.

Let $\deg q = n$. For each $f(X) \in K[X]$ we have $f = qb + r$ with $\deg r < n$, and $f(\alpha) = r(\alpha)$. Hence every element of E has the form $r(\alpha)$ where $\deg r < n$. Furthermore r is unique: if $r(\alpha) = s(\alpha)$, then $r + (q) = s + (q)$ and $q \mid r - s$; if also $\deg r, \deg s < \deg q$, then $r = s$. Thus every element of E can be written in the form $r(\alpha) = r_0 + r_1 \alpha + \cdots + r_{n-1} \alpha^{n-1}$ for some unique $r_0, \ldots, r_{n-1} \in K$. Thus $1, \alpha, \alpha^2, \ldots, \alpha^{n-1}$ is a basis of E over K. Then $[E : K] = n$. $\square$

Proposition 6.2.2 shows that every irreducible polynomial q with coefficients in K has a root α in an extension of K, which we can then view as constructed by adjoining α to K. For example, $q(X) = X^2 + 1 \in \mathbb{R}[X]$ is irreducible and $\mathbb{R}[X]/(q) \cong \mathbb{C}$ (see the exercises); thus Proposition 6.2.2 provides a rigorous way to construct $\mathbb{C}$ by adjoining to $\mathbb{R}$ an element i such that $i^2 = -1$.

The simple extension constructed in Proposition 6.2.2 has a universal property, which we will soon extend to all finite simple extensions:

Lemma 6.2.3. *Let $K(\alpha) = K[X]/(q)$ be as in Proposition 6.2.2, $\varphi : K \longrightarrow L$ be a field homomorphism, and β be a root of $^{\varphi}q$ in L. There exists a unique field homomorphism $\psi : K(\alpha) \longrightarrow L$ which extends φ and sends α onto β.*

Proof. Recall that $(^{\varphi}q)(X) = \varphi(a_n)X^n + \cdots + \varphi(a_1)X + \varphi(a_0)$ when $q = a_n X^n + \cdots + a_1 X + a_0$. By the universal property of $K[X]$ (Proposition 4.3.6) there is a homomorphism $\chi : K[X] \longrightarrow L$ such that $\chi(x) = \varphi(x)$ for all $x \in K$ and $\chi(X) = \beta$; namely $\chi(f(X)) = (^{\varphi}f)(\beta)$. Then $\chi(q(X)) = (^{\varphi}q)(\beta) = 0$; hence $(q) \subseteq \operatorname{Ker} \chi$ and χ factors through the projection $K[X] \longrightarrow K[X]/(q)$:

$$K \overset{\subseteq}{\longrightarrow} K[X] \longrightarrow K[X]/(q) = K(\alpha)$$

$$\varphi \searrow \quad \downarrow \chi \quad \nearrow \psi$$

$$L$$

This yields a homomorphism $\psi : K(\alpha) \longrightarrow L$ which extends φ and sends $\alpha = X + (q)$ onto $\chi(X) = \beta$. Since $1, \alpha, \ldots, \alpha^{n-1}$ is a basis of $K(\alpha)$ there is only one such homomorphism. $\square$

6.2.c. Algebraic and Transcendental Elements. We can now construct all simple extensions (up to K-isomorphisms).

Proposition 6.2.4. *Let $K \subseteq E$ be a field extension and $\alpha \in E$. Let $\varphi : K[X] \longrightarrow E$, $f(X) \longmapsto f(\alpha)$ be the evaluation homomorphism. Then either:*

(1) *φ is injective, in which case $K[\alpha] \cong K[X]$ and $K(\alpha) \cong K(X)$, or*

(2) *$\operatorname{Ker} \varphi$ is generated by a unique monic irreducible polynomial $q(X)$, in which case $[K(\alpha) : K] = \deg q$ and $K[\alpha] = K(\alpha) \cong K[X]/(q)$; the isomorphism sends α onto $X + (q)$.*

Proof. By Corollary 6.1.10, $K[\alpha] = \operatorname{Im} \varphi$. Hence (1) holds (when φ is injective). Now assume that φ is not injective. Then $\operatorname{Ker} \varphi$ is a prime ideal of $K[X]$ (since $K[\alpha]$ has no zero divisors). Since $K[X]$ is a principal ideal domain, $\operatorname{Ker} \varphi = (q)$ is generated by an irreducible polynomial q, which may be assumed to be monic and is then unique. Then $K[X]/(q) \cong K[\alpha]$; the isomorphism is induced by φ and sends $X + (q)$ onto α. By Proposition 6.2.2, $K[\alpha]$ is a field, so that $K(\alpha) = K[\alpha]$, and $[K(\alpha) : K] = \deg q$. $\square$

An element α of a field extension is **algebraic over K** in case there exists a nonzero polynomial $f(X) \in K[X]$ such that $f(\alpha) = 0$ and **transcendental over K** otherwise. For example, every complex number is algebraic over $\mathbb{R}$; $\sqrt[3]{2}$ and $1 + \sqrt{5} \in \mathbb{R}$ are algebraic over $\mathbb{Q}$. It has been proved by other methods that e and $\pi \in \mathbb{R}$ are transcendental over $\mathbb{Q}$; it can be shown that most real numbers are in fact transcendental over $\mathbb{Q}$ (see the exercises). In general, Proposition 6.2.4 yields characterizations of algebraic and transcendental elements:

Corollary 6.2.5. *Let E be an extension of a field K and $\alpha \in E$. The following conditions on α are equivalent:*

(1) *α is transcendental over K (if $f(X) \in K[X]$ and $f(\alpha) = 0$, then $f = 0$);*

(2) *$K(\alpha) \cong K(X)$;*

(3) *$[K(\alpha) : K]$ is infinite.*

Corollary 6.2.6. *Let E be an extension of a field K and $\alpha \in E$. The following conditions on α are equivalent:*

(1) *α is algebraic over K (there exists a polynomial $0 \neq f(X) \in K[X]$ such that $f(\alpha) = 0$);*

(2) *there exists a monic irreducible polynomial* $q \in K[X]$ *such that* $q(\alpha) = 0$;
(3) $[K(\alpha) : K]$ *is finite.*

In part (2), *q is unique;* $f(\alpha) = 0$ *if and only if* $q \mid f$; $K(\alpha) \cong K[X]/(q)$; *and* $[K(\alpha) : K] = \deg q.$

When α is algebraic over K, the unique monic irreducible polynomial $q(X) \in K[X]$ in part (2) is the **irreducible** or **minimal** polynomial **of** α, which we denote by $\mathrm{Irr}_K(\alpha)$. The **degree** of α over K is the degree of $\mathrm{Irr}_K(\alpha)$, and equals the degree of $K(\alpha)$ over K.

For example, every $a \in K$ is algebraic over K; $\mathrm{Irr}_K(a) = X - a$ and a has degree 1 over K. The complex number i is algebraic over $\mathbb{R}$; $\mathrm{Irr}_\mathbb{R}(i) = X^2 + 1$ and i has degree 2 over $\mathbb{R}$.

The last part of Proposition 6.2.4 now shows that the universal property of $K[X]/(q)$ in Lemma 6.2.3 extends to every simple algebraic extension:

Corollary 6.2.7. *Let* $K(\alpha)$ *be a simple algebraic extension of* K *and* $q = \mathrm{Irr}_K(\alpha)$. *Let* $\varphi : K \longrightarrow L$ *be a field homomorphism and* β *be a root of* $^\varphi q$ *in* L. *There exists a unique field homomorphism* $\psi : K(\alpha) \longrightarrow L$ *which extends* φ *and sends* α *onto* β. $\qquad\qquad\square$

6.2.d. Algebraic Extensions. An extension E of a field K is **algebraic** in case every element of E is algebraic over K; we also say that E is **algebraic over** K. For example $\mathbb{C}$ is an algebraic extension of $\mathbb{R}$, but $\mathbb{R}$ is not algebraic over $\mathbb{Q}$.

Algebraic extensions have a number of properties, whose proofs make wonderful exercises and are left to the reader.

Proposition 6.2.8. *Let* E *be an extension of a field* K.

(1) *Every finite extension is an algebraic extension.*
(2) *If* $E = K(s_1, \ldots, s_n)$ *and* $s_1, \ldots, s_n$ *are algebraic over* K, *then* E *is finite over* K *(hence* E *is algebraic over* K).
(3) *If* $E = K(S)$ *and every* $s \in S$ *is algebraic over* K, *then* E *is algebraic over* K.

Proposition 6.2.9. *Let* $K \subseteq E \subseteq F$ *be fields. If* F *is algebraic over* K, *then* E *is algebraic over* K *and* F *is algebraic over* E. *If, conversely,* E *is algebraic over* K *and* F *is algebraic over* E, *then* F *is algebraic over* K.

Proposition 6.2.10.

(1) *If* E *is algebraic over* K *and the composite* EF *exists, then* EF *is algebraic over* F.
(2) *A composite of algebraic extensions of* K *is an algebraic extension of* K.

Corollary 6.2.11. *If $K \subseteq E$ is a finite extension and the composite EF exists, then EF is a finite extension of F.*

Proof. Let $\alpha_1, \ldots, \alpha_n$ be a basis of E over K. Then $E = K(\alpha_1, \ldots, \alpha_n)$, and $\alpha_1, \ldots, \alpha_n$ are algebraic over K (since E is). Hence $EF = F(\alpha_1, \ldots, \alpha_n)$ is finite over F, by Proposition 6.2.8. $\square$

Corollary 6.2.12. *If α and β are algebraic over K, then $\alpha + \beta$, $\alpha - \beta$, $\alpha\beta$, and (if $\beta \neq 0$) $\alpha\beta^{-1}$ are algebraic over K.*

Proof. Let $\alpha, \beta \in E \supseteq K$ be algebraic over K. By Proposition 6.2.8, $K(\alpha, \beta) \subseteq E$ is algebraic over K. Hence so are $\alpha + \beta$, $\alpha - \beta$, $\alpha\beta$, $\alpha\beta^{-1} \in K(\alpha, \beta)$. $\square$

6.2.e. Transcendental Extensions. A field extension $K \subseteq E$ is **transcendental** in case it is not an algebraic extension (i.e., some element of E is transcendental over K) and **totally transcendental** in case every element of $E \backslash K$ is transcendental over K. We also say that E is transcendental, or totally transcendental, **over** K.

For example, $\mathbb{R}$ is transcendental, but not totally transcendental, over $\mathbb{Q}$. The next result shows that rational fraction fields are totally transcendental extensions:

Proposition 6.2.13. $K((X_i)_{i \in I})$ *is a totally transcendental extension of K for any field K.*

Proof. We need to show that if $\alpha \in K((X_i)_{i \in I})$ is algebraic over K, then $\alpha \in K$.

First we prove this when $|I| = 1$. Let $E = K(\chi)$, where χ is transcendental over K. Let $\alpha \in E$. Then $\alpha = f(\chi)/g(\chi) \in K(\chi)$, where $f, g \in K[X]$ and $g \neq 0$. If $\alpha \notin K$, then $\alpha g(X) \neq f(X)$, since $\alpha g(X) \notin K[X]$, and $f(X) - \alpha g(X) \neq 0$ in $K(\alpha)[X]$. Then $f(\chi) - \alpha g(\chi) = 0$ shows that $\chi \in E$ is algebraic over $K(\alpha)$. Hence $[K(\chi) : K(\alpha)]$ is finite; since $[K(\chi) : K]$ is infinite, it follows that $[K(\alpha) : K]$ is infinite, and α is transcendental over K. Thus $K(\chi)$ is totally transcendental over K whenever χ is transcendental over K. In particular, $K(X)$ is totally transcendental over K.

When $I = \{1, 2, \ldots, n\}$ is finite, we proceed by induction on n. Let $\alpha \in K(X_1, \ldots, X_n)$ be algebraic over K. Then $\alpha \in K(X_1, \ldots, X_{n-1})(X_n)$ is algebraic over $K(X_1, \ldots, X_{n-1})$. By the case $|I| = 1$, $\alpha \in K(X_1, \ldots, X_{n-1})$, and it follows from the induction hypothesis that $\alpha \in K$.

Now let I be arbitrary. Let $\alpha = f/g \in K((X_i)_{i \in I})$ be algebraic over K. Since the polynomials f and g have only finitely many nonzero coefficients, α belongs to $K((X_i)_{i \in J})$ for some finite subset J of I, and it follows from the finite case that $\alpha \in K$. $\square$

Every field extension is a totally transcendental extension of an algebraic extension. In detail:

Proposition 6.2.14. *Let $K \subseteq E$ be a field extension. Let A be the set of all elements of E that are algebraic over K. Then A is a subfield of E. Moreover A is algebraic over K and E is totally transcendental over A.*

Proof. A is a field by Corollary 6.2.12, and is algebraic over K. Let $\alpha \in E$ be algebraic over A. Then $A(\alpha)$ is finite over A, is algebraic over A, and is algebraic over K, by Propositions 6.2.8 and 6.2.9. Hence α is algebraic over K and $\alpha \in A$. Thus E is totally transcendental over A. $\qquad\square$

The study of transcendental extensions will be continued in Sections 6.4 and 6.5.

Exercises

1. Prove that $[\mathbb{R} : \mathbb{Q}] = \mathfrak{c} \ (= |\mathbb{R}|)$.
2. Prove that every complex number is algebraic over $\mathbb{R}$.
3. Prove directly that $1 + \sqrt{5}$ is algebraic over $\mathbb{Q}$.
4. Show that there are countably many irreducible polynomials in $\mathbb{Q}[X]$.
5. Let A be the set of all real numbers that are algebraic over $\mathbb{Q}$. Use the previous exercise to show that A is countable, so that $|R \backslash A| = \mathfrak{c}$.
6. Show that the simple extension $E = \mathbb{Q}(\sqrt[6]{2})$ of $\mathbb{Q}$ has intermediate fields $\mathbb{Q} \subsetneqq F \subsetneqq E$.
7. Let $E = K(s_1, \ldots, s_n)$, where $s_1, \ldots, s_n$ are algebraic over K. Show that E is a finite extension of K.
8. Let $E = K(S)$, where every $s \in S$ is algebraic over K. Show that E is algebraic over K.
9. Let E be algebraic over K and F be algebraic over E. Show that F is algebraic over K.
10. Let $K \subseteq E \subseteq F$ and F be algebraic over K. Show that E is algebraic over K and F is algebraic over E.
11. Let E be algebraic over K. Show that the composite EF is algebraic over F (if the composite EF exists).
12. Prove that a composite of algebraic extensions of K is an algebraic extension of K.
13. Prove that the union of a chain of algebraic extensions of K is an algebraic extension of K.

6.3. ALGEBRAIC CLOSURE

The **algebraic closure** of a field K is a field $\overline{K}$ which is algebraic over K and algebraically closed. We show that $\overline{K}$ exists and is (up to K-isomorphism) the largest algebraic extension of K.

6.3.a. Algebraically Closed Fields

Proposition 6.3.1. *For a field K the following conditions are equivalent:*

(1) *Every nonconstant polynomial $f(X) \in K[X]$ has a root in K ($f(r) = 0$ for some $r \in K$).*

(2) *Every irreducible polynomial in $K[X]$ has degree 1.*

(3) *The only algebraic extension of K is K itself.*

Proof. (1) implies (2). If $q(X) \in K[X]$ is irreducible and has a root r in K, then $X - r$ divides $q(X)$, and q is a constant multiple of $X - r$ and has degree 1.

(2) implies (3). Let α be algebraic over K. Then $q(X) = \mathrm{Irr}_K(\alpha)$ is monic and irreducible; if q has degree 1, then $q(X) = X - r$ for some $r \in K$, and $q(\alpha) = 0$ yields $\alpha = r \in K$.

(3) implies (2) by Proposition 6.2.2: if $E = K[X]/(q) = K$, then $\deg q = [E : K] = 1$.

(2) implies (1) since a nonconstant polynomial is a nonempty product of irreducible polynomials.   □

A field K is **algebraically closed** in case it satisfies any of these conditions. For example, the **Fundamental Theorem of Algebra** states that $\mathbb{C}$ is algebraically closed.

6.3.b. Existence. Let K be a field.

Lemma 6.3.2. *There exists an algebraic extension of K which contains a root of every nonconstant polynomial with coefficients in K.*

Proof (**Kempf**). Since every nonconstant polynomial has an irreducible factor, it suffices to construct an extension E of K in which every irreducible polynomial $q(X) \in K[X]$ has a root. Let $(q_i(X))_{i \in I}$ be the set of all irreducible polynomials $q_i(X) \in K[X]$, written as a family. For each $q_j(X) = a_n X^n + \cdots + a_1 X + a_0 \in K[X]$, there is a polynomial $q_j(X_j) = a_n X_j^n + \cdots + a_1 X_j + a_0 \in K[(X_i)_{i \in I}]$. Let $\mathfrak{A}$ be the ideal of $K[(X_i)_{i \in I}]$ generated by all $q_i(X_i)$.

We show that $\mathfrak{A} \neq K[(X_i)_{i \in I}]$. Otherwise, $1 = \sum_{j \in J} u_j((X_i)_{i \in I}) q_j(X_j)$ for some finite subset J of I and polynomials $u_j((X_i)_{i \in I}) \in K[(X_i)_{i \in I}]$ ($j \in J$). Since J is finite, applying Proposition 6.2.2 finitely many times yields an extension $E \supseteq K$ in which each q_j has a root α_j. By the universal property of $K[(X_i)_{i \in I}]$, there is a homomorphism $\varphi : K[(X_i)_{i \in I}] \longrightarrow E$ such that $\varphi(x) = x$ for all $x \in K$, $\varphi(X_j) = \alpha_j$ for all $j \in J$, and, say, $\varphi(X_i) = 0$ for all $i \in I \backslash J$. Then $\varphi(q_j(X_j)) = q_j(\alpha_j) = 0$ for all $j \in J$, and $1 = \varphi(1) = \sum_{j \in J} \varphi(u_j) \varphi(q_j(X_j)) = 0$, a contradiction.

Since $\mathfrak{A} \neq K[(X_i)_{i \in I}]$, $\mathfrak{A}$ is contained in a maximal ideal $\mathfrak{M}$ of $K[(X_i)_{i \in I}]$. The quotient ring $E = K[(X_i)_{i \in I}]/\mathfrak{M}$ is a field by Proposition 4.4.3. There is a homomorphism $\varphi : K \longrightarrow E$ which sends $x \in K$ to $x + \mathfrak{M} \in E$. In what follows we identify x and $\varphi(x)$ when $x \in K$.

Let $\alpha_i = X_i + \mathfrak{M} \in E$. The projection $K[(X_i)_{i \in I}] \longrightarrow E$ sends X_i onto α_i (for every $i \in I$) and every $x \in K$ to $\varphi(x) = x$. Hence it coincides with the evaluation homomorphism $K[(X_i)_{i \in I}] \longrightarrow E$, $f((X_i)_{i \in I}) \longmapsto f((\alpha_i)_{i \in I})$; in other words, $f((X_i)_{i \in I}) + \mathfrak{M} = f((\alpha_i)_{i \in I})$ for all $f((X_i)_{i \in I}) \in K[(X_i)_{i \in I}]$. In particular $q_i(\alpha_i) = q_i(X_i) + \mathfrak{M} = 0$ in E, and every q_i has a root in E. Also $E = K((\alpha_i)_{i \in I})$ is algebraic over K. $\qquad \square$

The exercises indicate alternate proofs of Lemma 6.3.2.

Theorem 6.3.3. *For every field K there exists an algebraic extension $\overline{K}$ of K which is algebraically closed.*

Proof. Construct a sequence of fields

$$K = E_0 \subseteq E_1 \subseteq \cdots \subseteq E_n \subseteq E_{n+1} \subseteq \cdots$$

by induction: given E_n, E_{n+1} is the algebraic extension of E_n constructed by Lemma 6.3.2, which contains a root of every nonconstant polynomial with coefficients in E_n. By induction, E_n is algebraic over K. Hence $\overline{K} = \bigcup_{n \geqslant 0} E_n$ is algebraic over K. Let $f(X) \in \overline{K}[X]$ be a nonconstant polynomial. Since f has only finitely many nonzero coefficients, we have $f \in E_n[X]$ for some n. Hence f has a root in $E_{n+1} \subseteq \overline{K}$. Thus $\overline{K}$ is algebraically closed. $\qquad \square$

6.3.c. Homomorphisms. The following result is a fundamental property of algebraic extensions and will imply the uniqueness of $\overline{K}$:

Theorem 6.3.4. *Let E be an algebraic extension of K. Every homomorphism of K into an algebraically closed field L can be extended to E.*

Proof. We use Corollary 6.2.7 and Zorn's Lemma. Let $\mathscr{S}$ be the set of all pairs (F, ψ), where $K \subseteq F \subseteq E$ and $\psi : F \longrightarrow L$ is a field homomorphism which extends φ. Then $\mathscr{S} \neq \varnothing$, since $(K, \varphi) \in \mathscr{S}$. Order $\mathscr{S}$ by $(F, \psi) \leqslant (G, \chi)$ if and only if F is a subfield of G and χ extends ψ. If $((F_i, \psi_i)_{i \in I}$ is a chain in $\mathscr{S}$, then $F = \bigcup_{i \in I} F_i$ is the union of a chain of subfields of E and is a subfield of E, and there is a homomorphism $\psi : F \longrightarrow L$ which extends all ψ_i (and is well defined by $\psi(x) = \psi_i(x)$ whenever $x \in F_i$). Thus $\mathscr{S}$ is inductive.

By Zorn's Lemma, $\mathscr{S}$ has a maximal element (M, μ). Assume $M \neq E$. Let $\alpha \in E \backslash M$ and $a = \mathrm{Irr}_M(\alpha)$. Since L is algebraically closed, $^\mu q$ has a root in L. By Corollary 6.2.7, μ can be extended to $M(\alpha)$, which contradicts the maximality of (M, μ). Therefore $M = E$. $\qquad \square$

By Theorem 6.3.4, every algebraic extension of K is (up to K-isomorphism) contained in every algebraically closed extension of K.

6.3.d. Uniqueness. Theorem 6.3.4 is not a universal property: in the theorem, a homomorphism $K \longrightarrow L$ may be extended to E in more than one way. Nevertheless, Theorem 6.3.4 implies the uniqueness of the algebraic closure:

Proposition 6.3.5. *All algebraically closed fields which are algebraic over K are K-isomorphic.*

Proof. Let L and M be algebraically closed fields which are algebraic over K. By Theorem 6.3.4, the inclusion $K \longrightarrow M$ extends to a K-homomorphism $\varphi : L \longrightarrow M$. Now φ is injective by Proposition 6.1.1. Also M is algebraic over Im $\varphi \supseteq K$, by Proposition 6.2.9; since Im $\varphi \cong L$ is algebraically closed, it follows that $M = \mathrm{Im}\,\varphi$, and φ is an isomorphism. $\square$

We denote by $\overline{K}$ any one of the algebraic closures of K. For example, $\mathbb{C} = \overline{\mathbb{R}}$: indeed $\mathbb{C}$ is algebraically closed, and is algebraic over $\mathbb{R}$ since $[\mathbb{C} : \mathbb{R}] = 2$. By Theorem 6.3.4:

Proposition 6.3.6. *Let E be an algebraic extension of K. Then $\overline{E}$ is an algebraic closure of K. For any algebraic closure $\overline{K}$ of K, E is K-isomorphic to an intermediate field $K \subseteq F \subseteq \overline{K}$.*

The algebraic closure of K can now be characterized in several ways:

(1) $\overline{K}$ is an algebraically closed, algebraic extension of K (by definition).
(2) $\overline{K}$ is a maximal algebraic extension of K: indeed $\overline{K}$ is algebraically closed if and only if there is no algebraic extension $\overline{K} \subsetneqq L$.
(3) $\overline{K}$ is (up to K-isomorphism) the largest algebraic extension of K.
(4) $\overline{K}$ is a minimal algebraically closed extension of K: if $L \subseteq \overline{K}$ is algebraically closed, then $\overline{K} = L$ since $\overline{K}$ is algebraic over $L \supseteq K$.
(5) $\overline{K}$ is (up to K-isomorphism) the smallest algebraically closed extension of K (by Theorem 6.3.4).

We also note the following property:

Proposition 3.7. *Every K-endomorphism of $\overline{K}$ is a K-automorphism.*

Proof. Let $\varphi : \overline{K} \longrightarrow \overline{K}$ be a K-endomorphism of $\overline{K}$. Then φ is injective, by Proposition 6.1.1. Also $\overline{K}$ is algebraic over Im $\varphi \supseteq K$; since Im $\varphi \cong \overline{K}$ is algebraically closed, it follows that $\overline{K} = \mathrm{Im}\,\varphi$ and that φ is an isomorphism. $\square$

6.3.e. Conjugates. A **conjugate** over K of an element α of $\overline{K}$ is the image $\sigma\alpha$ of α under a K-automorphism σ of $\overline{K}$.

For example, $\mathbb{C}$ has exactly two $\mathbb{R}$-automorphisms: indeed, if $\tau \in \mathrm{Aut}_{\mathbb{R}}(\mathbb{C})$, then $(\tau i)^2 + 1 = \tau(i^2 + 1) = 0$, and either $\tau i = i$, in which case τ is the identity on $\mathbb{C}$, or $\tau i = -i$, in which case $\tau(x + iy) = x - iy$ and τ is ordinary complex conjugation. Hence a complex number z has two conjugates over $\mathbb{R}$: itself and its ordinary conjugate $\overline{z}$.

More generally, we have:

Proposition 6.3.8. *Each $\alpha \in \overline{K}$ has finitely many conjugates over K, which are the roots of $\mathrm{Irr}_K(\alpha)$ in $\overline{K}$.*

Proof. Let $q(X) = \mathrm{Irr}_K(\alpha) \in K[X]$. If σ is a K-automorphism of $\overline{K}$, then $^\sigma q = q$, $q(\sigma\alpha) = (^\sigma q)(\sigma\alpha) = \sigma(q(\alpha)) = 0$, and $\sigma\alpha$ is a root of q. In particular, α has at most $\deg q$ conjugates.

Conversely, let $\beta \in \overline{K}$ be a root of q. By Corollary 6.2.7, there is a K-homomorphism $\varphi : K(\alpha) \longrightarrow \overline{K}$ such that $\varphi\alpha = \beta$. By Theorem 6.3.4, φ can be extended to a K-endomorphism σ of $\overline{K}$. By Proposition 6.3.7, σ is in fact a K-automorphism of $\overline{K}$. Hence $\beta = \sigma\alpha$ is a conjugate of α. $\quad\square$

Exercises

1. Show that the field of all complex numbers that are algebraic over $\mathbb{Q}$ is an algebraic closure of $\mathbb{Q}$.
2. Show that the algebraic closure of $\mathbb{Z}_p$ is countable.
3. Show that every algebraically closed field is infinite.
4. Show that $|K[X]| \leqslant \aleph_0 |K|$; conclude that an algebraic extension of a field K has at most $\aleph_0 |K|$ elements.
5. Use the previous exercise to construct a set $S \supseteq K$ such that every algebraic extension of K is isomorphic to an extension $F \subseteq S$. Then use Zorn's Lemma to give another proof of Lemma 6.3.2.
6. Give another proof of Lemma 6.3.2, using Proposition 6.2.2 and ordinal induction.
7. Find all conjugates (over $\mathbb{Q}$) of $\sqrt[3]{2} \in \overline{\mathbb{Q}}$.
8. Find all conjugates (over $\mathbb{Q}$) of $\sqrt{2} + \sqrt{3} \in \overline{\mathbb{Q}}$.

6.4. TRANSCENDENCE DEGREE

This section studies transcendental extensions, for which we prove a dimension property.

6.4.a. Algebraic Independence. We begin by extending Proposition 6.2.4 from one element to arbitrary subsets. Let E be a field extension of K and S be a subset of E. By the universal property of $K\left[(X_\alpha)_{\alpha \in S}\right]$, there exists a homomorphism $\varphi : K\left[(X_\alpha)_{\alpha \in S}\right] \longrightarrow E$ unique such that $\varphi(x) = x$ for all $x \in K$ and $\varphi(X_\alpha) = \alpha$ for all $\alpha \in S$; φ is the evaluation homomorphism which sends

$$f((X_\alpha)_{\alpha \in S}) = \sum_m a_m X_{\alpha_1}^{m_{\alpha_1}} \ldots X_{\alpha_t}^{m_{\alpha_t}} \in K\left[(X_\alpha)_{\alpha \in S}\right]$$

to its value

$$f(S) = \sum_m a_m \alpha_1^{m_{\alpha_1}} \ldots \alpha_t^{m_{\alpha_t}} \in E;$$

$\mathrm{Im}\,\varphi$ is the subring $K[S] \subseteq E$ generated by K and S.

The subset S of E is **algebraically independent over** K in case φ is injective; equivalently, in case 0 is the only polynomial $f((X_\alpha)_{\alpha \in S})$ such that $f(S) = 0$; equivalently, in case the elements of S satisfy no nontrivial polynomial equation with coefficients in K. Then $K[S] \cong K\left[(X_\alpha)_{\alpha \in S}\right]$ and $K(S) \cong K((X_\alpha)_{\alpha \in S})$. For example, $\{\alpha\}$ is algebraically independent over K if and only if α is transcendental over K.

Since a polynomial has only finitely many nonzero coefficients, a subset S is algebraically independent over K if and only if every finite subset of S is algebraically independent over K.

The union $S = \bigcup_{i \in I} S_i$ of a chain $(S_i)_{i \in I}$ of algebraically independent subsets is algebraically independent: indeed every finite subset of S is contained in some S_i and hence is algebraically independent.

Lemma 6.4.1. *Let E be a field extension of a field K. A subset S of E is algebraically independent over K if and only if no $\beta \in S$ is algebraic over $K(S \backslash \{\beta\}) \subseteq E$.*

Proof. Assume that $\beta \in S$ is algebraic over $K(S \backslash \{\beta\})$. Then $f(\beta) = 0$ for some $f(X_\beta) \in K(S \backslash \{\beta\})[X_\beta]$, $f \neq 0$. By Proposition 6.1.9, every coefficient of f has the form ab^{-1} with $a, b \in K[S \backslash \{\beta\}]$, $b \neq 0$; multiplying f by all the b's yields a polynomial $g(X_\beta) \in K[S \backslash \{\beta\}][X_\beta]$, $g \neq 0$ such that $g(\beta) = 0$. By Proposition 6.1.9, every coefficient of g has the form $f(S \backslash \{\beta\})$ for some $f \in K[(X_\alpha)_{\alpha \in S \backslash \{\beta\}}]$; hence there is a polynomial $h \in K\left[(X_\alpha)_{\alpha \in S}\right] \cong K[(X_\alpha)_{\alpha \in S \backslash \{\beta\}}][X_\beta]$, $h \neq 0$, such that $h(S) = 0$, and S is not algebraically independent over K.

Conversely, assume that S is not algebraically independent over K. Then $f(S) = 0$ for some $f((X_\alpha)_{\alpha \in S}) \in K\left[(X_\alpha)_{\alpha \in S}\right]$, $f \neq 0$. The polynomial f is not constant and some X_β actually appears in f. Then β satisfies a nontrivial algebraic equation $f(S) = 0$ with coefficients in $K[S \backslash \{\beta\}]$, and is algebraic over $K(S \backslash \{\beta\})$. □

Lemma 6.4.2. *Let E be a field extension of a field K and $S \subseteq E$. If S is algebraically independent over K and $\beta \in E$ is transcendental over $K(S)$, then $S \cup \{\beta\}$ is algebraically independent over K.*

Proof. We have $\beta \notin S$. Assume that $f((X_\alpha)_{\alpha \in S \cup \{\beta\}}) \in K[(X_\alpha)_{\alpha \in S \cup \{\beta\}}]$ and $f(S) = 0$. Then β does not appear in f. Otherwise, β would satisfy an algebraic equation with coefficients in $K(S)$ and would be algebraic over $K(S)$. Hence $f \in K\left[(X_\alpha)_{\alpha \in S}\right]$, $f(S) = 0$, and $f = 0$, since S is algebraically independent. □

6.4.b. Transcendence Bases

Lemma 6.4.3. *Let E be an extension of a field K. For a subset S of E the following properties are equivalent:*

(1) S is a maximal algebraically independent subset of E.

(2) *S is algebraically independent over K and E is algebraic over K(S).*

(3) *S is minimal such that E is algebraic over K(S).*

Proof. (1) implies (2) by Lemma 6.4.2.

(2) implies (1). Assume that S is algebraically independent over K and that E is algebraic over $K(S)$. Then every $\beta \notin S$ is algebraic over $K(S)$; by Lemma 6.4.1, $S \cup \{\beta\}$ is not algebraically independent over K.

(2) implies (3). Assume that S is algebraically independent over K and that E is algebraic over $K(S)$. If $T \subseteq S$ and E is algebraic over $K(T)$, then T is a maximal algebraically independent subset (since (2) implies (1)); hence $S = T$.

(3) implies (2). Assume that E is algebraic over $K(S)$. If S is not algebraically independent over K, then some $\beta \in S$ is algebraic over $K(S \backslash \{\beta\})$, by Lemma 6.4.1. Then $K(S)$ is algebraic over $K(S \backslash \{\beta\})$. By Proposition 6.2.9, E is algebraic over $K(S \backslash \{\beta\})$, and S is not minimal with this property. $\square$

A **transcendence base** of a field extension $K \subseteq E$ is a subset of E which satisfies the equivalent conditions in Lemma 6.4.3.

Proposition 6.4.4. *Let E be an extension of a field K and $S \subseteq T \subseteq E$ be subsets such that S is algebraically independent over K and E is algebraic over $K(T)$. Then E has a transcendence base $S \subseteq B \subseteq T$.*

Proof. By Zorn's Lemma, the set of all algebraically independent subsets U with $S \subseteq U \subseteq T$ has a maximal element M. If $\beta \in T \backslash M$, then $M \cup \{\beta\}$ is not algebraically independent and β is algebraic over $K(M)$ by Lemma 6.4.2. Hence every element of T is algebraic over $K(M)$. By Propositions 6.2.8 and 6.2.9, every element of $K(T)$ is algebraic over $K(M)$, and every element of E is algebraic over $K(M)$. Thus M is a transcendence base of E. $\square$

When B is a transcendence base of E over K, then E is algebraic over $K(B)$, and $K(B)$ is totally transcendental over K (by Proposition 6.2.13). Thus every extension is an algebraic extension of a totally transcendental extension. (By Proposition 6.2.14, every extension is also a totally transcendental extension of an algebraic extension.)

6.4.c. Transcendence Degree

Theorem 6.4.5. *Let E be an extension of a field K. All transcendence bases of E have the same number of elements.*

The proof of Theorem 6.4.5 is similar to the proof that all bases of a vector space have the same number of elements. First we establish an **exchange**

property:

Lemma 6.4.6. *Let B and C be transcendence bases of E. For each $\beta \in B$ there exists $\gamma \in C$, with either $\gamma = \beta$ or $\gamma \notin B$, such that $(B \setminus \{\beta\}) \cup \{\gamma\}$ is a transcendence base of E.*

Proof. The result is trivial if $\beta \in C$, and we may assume that $\beta \notin C$. Let $B' = B \setminus \{\beta\}$. If every element of C is algebraic over $K(B')$, then, by Propositions 6.2.8 and 6.2.9, $K(C)$ is algebraic over $K(B')$ and $\beta \in E$ is algebraic over $K(B')$, which contradicts Lemma 6.4.1. Therefore some $\gamma \in C$ is transcendental over $K(B')$. In particular, $\gamma \notin B'$; in fact, $\gamma \notin B$, since $\beta \notin C$. By Lemma 6.4.2, $B'' = B' \cup \{\gamma\} = (B \setminus \{\beta\}) \cup \{\gamma\}$ is algebraically independent over K.

If β is transcendental over $K(B'')$, then $B \cup \{\gamma\} = B'' \cup \{\beta\}$ is algebraically independent by Lemma 6.4.2, which contradicts Lemma 6.4.1 since γ is algebraic over $K(B)$. Therefore β is algebraic over $K(B'')$. Thus every element of B is algebraic over $K(B'')$; $K(B)$ is algebraic over $K(B'')$; and E is algebraic over $K(B'')$. $\qquad\square$

Proof of Theorem 6.4.5. Let $K \subseteq E$ have two transcendence bases B and C.

If $C = \{\gamma_1, \ldots, \gamma_n\}$ is finite, then Lemma 6.4.6 implies that B is finite and that $|B| \leqslant |C|$: if indeed B contains distinct elements $\beta_1, \ldots, \beta_n, \beta_{n+1}, \ldots$, then n applications of Lemma 6.4.6 show that $B' = (B \setminus \{\beta_1, \ldots, \beta_n\}) \cup \{\gamma_1, \ldots, \gamma_n\}$ is a transcendence base, and then Lemma 6.4.6 yields a contradiction if applied to B', $\beta_{n+1} \in B'$, and C. Exchanging B and C yields $|B| = |C|$.

Now assume that B is infinite. By the above, C is infinite. Each $\beta \in B$ is algebraic over $K(C)$ and satisfies a polynomial equation with coefficients in $K(C)$. Since a polynomial has only finitely many nonzero coefficients, there is a finite subset S_β of $K(C)$ such that β is algebraic over $K(S_\beta)$. By Corollary 6.1.10, part (4), there is a finite subset T_β of C such that β is algebraic over $K(T_\beta)$. Then every element of B is algebraic over $K(\bigcup_{\beta \in B} T_\beta)$; by Lemma 6.4.3, C is minimal with this property, and it follows that $C = \bigcup_{\beta \in B} T_\beta$. Hence $|C| \leqslant |\aleph_0||B|$; by Corollary B.5.12, $|C| \leqslant |B|$. Exchanging B and C yields $|B| = |C|$. $\qquad\square$

The **transcendence degree** $\mathrm{tr.d.}_K(E)$ of a field extension $K \subseteq E$ is the number of elements of the transcendence bases of E. For example, an algebraic extension has transcendence degree 0; $K(X)$ has transcendence degree 1 over K; in general, $K((X_i)_{i \in I})$ has transcendence degree $|I|$ over K. Conversely, an extension with transcendence degree $|I|$ is (up to K-isomorphism) an algebraic extension of $K((X_i)_{i \in I})$.

Proposition 6.4.7. *When $K \subseteq E \subseteq F$, then $\mathrm{tr.d.}_K(F) = \mathrm{tr.d.}_K(E) + \mathrm{tr.d.}_E(F)$.*

Proof. Let B be a transcendence base of E over K and C be a transcendence base of F over E. Then $B \subseteq E$ and C are disjoint, since no element of C is algebraic over E. We show that $B \cup C$ is a transcendence base of F over K.

Every element of E is algebraic over $K(B \cup C) \supseteq K(B)$, and so is every element of C. By Propositions 6.2.8 and 6.2.9, every element of $E(C)$ is algebraic over $K(B \cup C)$; every element of F is algebraic over $E(C)$, hence every element of F is algebraic over $K(B \cup C)$.

A finite subset of $B \cup C$ can be written as $B' \cup C'$, where $B' \subseteq B$ and $C' \subseteq C$. We show by induction on $|C'|$ that $B' \cup C'$ is algebraically independent over K. This is true if $C' = \emptyset$, since $B' \subseteq B$ is algebraically independent over K. Otherwise, let $\gamma \in C'$. If γ is algebraic over $K(B' \cup C' \setminus \{\gamma\})$, then γ is algebraic over $E(C' \setminus \{\gamma\}) \supseteq K(B' \cup C' \setminus \{\gamma\})$, which contradicts Lemma 6.4.1, since $C' \subseteq C$ is algebraically independent over E. Therefore γ is transcendental over $K(B' \cup C' \setminus \{\gamma\})$. By the induction hypothesis, $B' \cup C' \setminus \{\gamma\}$ is algebraically independent over K; hence $B' \cup C'$ is algebraically independent over K, by Lemma 6.4.2. Thus every finite subset of $B \cup C$ is algebraically independent over K. $\qquad\square$

6.5. LÜROTH'S THEOREM

In this section we prove Lüroth's Theorem, which states that the field of rational fractions $K(X)$ (in one indeterminate) has the following property: every field $K \subsetneq E \subseteq K(X)$ is isomorphic to $K(X)$.

Theorem 6.5.1 (Lüroth). *If $K \subsetneq E \subseteq K(X)$, then $E = K(\alpha)$ for some $\alpha \in E$.*

Lüroth's Theorem has the following generalization (not proved here): if $K \subsetneq E \subseteq K(X_1, \ldots, X_n)$ and tr.d.$_K E = 1$, then $E = K(\alpha)$ for some $\alpha \in E$. By Proposition 6.2.13, α is transcendental over K, so that $E \cong K(X)$. Intermediate fields $K \subseteq F \subseteq K(X_1, \ldots, X_n)$ have not been classified in general.

The proof of Lüroth's Theorem begins with polynomial lemmas.

Lemma 6.5.2. *Assume that*

$$u(Y)f(X) + v(Y)g(X) = k(X)\ell(X, Y),$$

where $f(X), g(X) \in K[X]$ are relatively prime, $u(Y), v(Y) \in K[Y]$ are relatively prime, u and v are not both constant, $k(X) \in K[X]$, and $\ell(X, Y) \in K[X, Y]$. Then k is constant.

Proof. We may assume that u is not constant. Then u has a root $\alpha \in \overline{K}$. We have $k(X)\ell(X, \alpha) = v(\alpha)g(X)$, with $v(\alpha) \neq 0$; otherwise, $\mathrm{Irr}_K(\alpha)$ would divide both u and v. Therefore $k(X)$ divides $g(X)$ in $\overline{K}[X]$. Since k and g have coefficients in K, polynomial division shows that $k(X)$ divides $g(X)$ in $K[X]$. Hence $k(X)$ divides $u(Y)f(X)$ (in $K[X, Y]$), and divides $f(X)$ (in $K[X]$). Therefore $k(X)$ divides $f(X)$ and $g(X)$, and must be constant. $\qquad\square$

Lemma 6.5.3. *If $f(X), g(X) \in K[X]$ are relatively prime and not both constant, then $f(X) - Yg(X)$ is irreducible in $K(Y)[X]$.*

Proof. We have $f(X) - Yg(X) \in K[X, Y]$. If there is a factorization $f(X) - Yg(X) = k(X, Y)\ell(X, Y)$ with $k, \ell \in K[X, Y]$, then Y does not appear in, say, $k(X, Y)$, and it follows from Lemma 6.5.2 that k is constant. Hence $f(X) - Yg(X)$ is irreducible in $K[X, Y] = K[Y][X]$. Viewed as a polynomial in X with coefficients in $K[Y]$, $f(X) - Yg(X)$ is irreducible and not constant, hence is primitive. By Lemma 5.4.7, $f(X) - Yg(X)$ is irreducible in $K(Y)[X]$. $\qquad\square$

To prove Theorem 6.5.1, let $K \subsetneq E \subseteq K(\chi)$, where χ is transcendental over K (for instance, $K(\chi) = K(X)$). Take any $\alpha \in E \backslash K$. Since $E \subseteq K(\chi)$, we have $\alpha = f(\chi)/g(\chi)$, where $f, g \in K[X]$, $g \neq 0$, and we may assume that $f(X)$ and $g(X)$ are relatively prime. By Lemma 6.5.3, $f(X) - Yg(X)$ is irreducible in $K(Y)[X]$, and its degree in X is $m = \max(\deg f, \deg g)$. But $K(\alpha) \cong K(Y)$, since α is transcendental over K by Proposition 6.2.13. Hence $h(X) = f(X) - \alpha g(X)$ is irreducible in $K(\alpha)[X]$. Since $f(\chi) - \alpha g(\chi) = 0$, it follows that $h = \text{Irr}_{K(\alpha)}(\chi)$. Hence $[K(\chi) : K(\alpha)] = \deg h = m$. Since $K(\alpha) \subseteq E$, it follows that $d = [K(\chi) : E]$ is finite, d divides m, and $[E : K(\alpha)] = m/d$.

We construct $\alpha \in E \backslash K$ such that $d = m$; this implies $E = K(\alpha)$. Let $q(X) = \text{Irr}_E(\chi) \in E(X)$. Then $\deg q = [K(\chi) : E] = d$. Also $q(X) \notin K(X)$, since χ is not algebraic over K. Hence $q(X)$ has a coefficient $\alpha \in E \backslash K$. As above, $\alpha = f(\chi)/g(\chi)$, with $f, g \in K[X]$, $g \neq 0$, and we may assume that $f(X)$ and $g(X)$ are relatively prime. Then q divides $f - \alpha g$ in $E(X)$; hence

$$f(X) - \alpha g(X) = q(X) r(X)$$

for some $r(X) \in E[X]$,

$$g(\chi)f(X) - f(\chi)g(X) = q(X)s(X)$$

for some $s(X) = g(\chi)r(X) \in K(\chi)[X]$, and

$$g(Y)f(X) - f(Y)g(X) = \overline{q}(X, Y)\overline{s}(X, Y)$$

in $K(Y)[X] \cong K(\chi)[X]$, where $q(X) = \overline{q}(X, \chi)$ and $s(X) = \overline{s}(X, \chi)$ in $K(\chi)[X]$. Since $K[Y]$ is a UFD, there is a factorization

$$g(Y)f(X) - f(Y)g(X) = q^*(X, Y)s^*(X, Y) \tag{6.1}$$

in $K[Y][X] \cong K[X, Y]$, with $q^*(X, Y) = h(Y)\overline{q}(X, Y)$ for some $h(Y) \in K(Y)$. Since $q(X) \in E[X]$ is monic, $h(Y)$ is the leading coefficient of $q^* \in K[Y][X]$, and $h(Y) \in K[Y]$. Also $\alpha = f(\chi)/g(\chi)$ is a coefficient of $q(X) \in K(\chi)[X]$; hence $h(Y)f(Y)/g(Y)$ is a coefficient of $q^* \in K[Y][X]$ and $k(Y) = h(Y)f(Y)/g(Y) \in K[Y]$. Since f and g are relatively prime, $g(Y)$ divides $h(Y)$. Hence $\deg h \geqslant \deg g$ and $\deg k \geqslant \deg f$.

Since both $h(Y)$ and $k(Y)$ appear as coefficients of $q^* \in K[Y][X]$, the degree in Y of $q^* \in K[X,Y]$ is at least $m = \max(\deg f, \deg g)$. On the other hand, the degree in Y of $q^*(X,Y)s^*(X,Y) = g(Y)f(X) - f(Y)g(X)$ is at most m. Hence the degree in Y of $q^*(X,Y)$ is m, and Y does not appear in $s^*(X,Y)$. By Lemma 6.5.2, $s^*(X,Y)$ is constant. Hence equation (6.1) implies $q^*(Y,X) = -q^*(X,Y)$, so that the degree in Y of q^* equals its degree in X, which is $\deg q = d$. Hence $m = d$. This implies $E = K(\alpha)$, and Theorem 6.5.1 is proved. $\square$

7

GALOIS THEORY

For centuries algebra was concerned only with polynomial equations. Algebra began with al-Khowarizmi's treatise on quadratic equations. The next advances were formulas for the solutions of equations of degree three and four, published by Cardano in his *Ars magna* (1545), and the development of modern notation around 1600. Finding similar formulas for equations of degree five turned out to be much more difficult. This became the outstanding problem in algebra; it was not solved until Ruffini in 1799 and Abel in 1824 proved that there are in fact no such formulas. (Abelian groups are named after the latter.)

It remained for Galois in 1830 to define what we now call the Galois group of an equation (viewed as a group of permutations of its roots) and to tie the solvability by radicals of an equation (of any degree) to the solvability of its Galois group. To Galois we also owe the use of the word "group" in its modern sense. These results were very difficult to express without our modern concepts. Modern Galois theory is largely the work of Artin and has become an essential tool of commutative algebra.

Galois Theory generally studies algebraic extensions of a given field K. One may assume that these extensions are contained in an algebraic closure $\overline{K}$ of K.

Extensions $K \subseteq E \subseteq \overline{K}$ are studied by means of K-homomorphisms of E into $\overline{K}$. This gives rise to two classes of extensions: in a **separable** extension, K-homomorphisms distinguish (separate) the elements of K from the elements of $E \backslash K$; in a **normal** extension the K-homomorphisms of E into $\overline{K}$ send E onto E and induce a group $\mathrm{Aut}_K(E)$ of automorphisms of E. When E is both normal and separable, the group $\mathrm{Aut}_K(E)$ determines the degree of E over K and all intermediate fields $K \subseteq F \subseteq E$: this is the Fundamental Theorem of Galois theory.

Topics in this chapter include normal extensions, separable extensions, the Fundamental Theorem, and its applications to solvability of equations by radicals, and to constructibility with straight edge and compass.

Required results: Chapter 6, Section 6.1 and especially Sections 6.2 and 6.3; Section 6.4 is used in Section 7.10. Properties of polynomials from Section 5.2 are used in various places: the derivative is used widely; the elementary symmetric polynomials are used in Sections 7.4 and 7.10; and more specialized properties of polynomials of degree 3 or 4 are used in Section 7.6. Section 7.10 requires basic properties of solvable groups from Section 3.7. Euler's φ function (Section 2.2) is helpful in Section 7.7.

7.1. NORMAL EXTENSIONS

This section contains basic properties of normal extensions and the construction of splitting fields, which throws light on these extensions. The results yield a construction of all finite fields.

7.1.a. Splitting Fields. A polynomial $f(X) \in K[X]$ **splits** in a field extension E of K in case it has a factorization $f(X) = a(X - \alpha_1)(X - \alpha_2)\ldots(X - \alpha_n)$ in $E[X]$; there $a \in K$ is the leading coefficient of f and $\alpha_1, \ldots, \alpha_n$ are the roots of f in E. For example, every $f(X) \in K[X]$ splits in any algebraic closure $\overline{K}$ of K.

A **splitting field** of a polynomial $f(X) \in K[X]$ is a field extension E of K such that f splits in E and E is generated by the roots of f in E. More generally, let $\mathscr{S} \subseteq K[X]$ be any set of polynomials. A **splitting field** of $\mathscr{S}$ (over K) is a field extension E of K such that every $f \in \mathscr{S}$ splits in E and E is generated by the roots in E of all $f \in \mathscr{S}$. By Proposition 6.2.8, E is algebraic over K.

Lemma 7.1.1. *Every set $\mathscr{S} \subseteq K[X]$ of polynomials has a splitting field $F \subseteq \overline{K}$.*

Proof. Let R be the set of all roots of all $f \in \mathscr{S}$ in the algebraic closure $\overline{K}$ of K; $K(R) \subseteq \overline{K}$ is a splitting field of $\mathscr{S}$. $\qquad\square$

Lemma 7.1.2. *Let E and F be splitting fields of $\mathscr{S} \subseteq K[X]$, with $K \subseteq F \subseteq \overline{K}$. Then $\varphi E = F$ for every K-homomorphism $\varphi : E \longrightarrow \overline{K}$.*

Proof. For each $f \in \mathscr{S}$ we have $f(X) = a(X - \alpha_1)\ldots(X - \alpha_n)$ in $E[X]$ and $f(X) = a(X - \beta_1)\ldots(X - \beta_n)$ in $F[X]$. Applying φ yields $^\varphi f = a(X - \varphi\alpha_1)\ldots(X - \varphi\alpha_n)$. Since factorization in $F[X]$ is unique, $\varphi\{\alpha_1, \ldots, \alpha_n\} = \{\beta_1, \ldots, \beta_n\}$. If R is the set of all roots of all $f \in \mathscr{S}$ in E, and S is the set of all roots of all $f \in \mathscr{S}$ in F, then $\varphi R = S$ and $\varphi E = \varphi(K(R)) = K(\varphi R) = K(S) = F$. $\qquad\square$

When $K \subseteq F \subseteq \overline{K}$ is a splitting field of $f \in K[X]$, the proof of Lemma 7.1.2 shows that every K-homomorphism $F \longrightarrow \overline{K}$ permutes the roots of f. This idea will be explored further in Section 7.6.

By Lemma 7.1.1, every set $\mathscr{S} \subseteq K[X]$ has a splitting field $F \subseteq \overline{K}$. Now any splitting field E of $\mathscr{S}$ is algebraic over K; by Theorem 6.3.4, the inclusion $K \longrightarrow \overline{K}$ can be extended to a K-homomorphism $E \longrightarrow \overline{K}$. Then Lemma

7.1.2 implies that E and F are K-isomorphic and we have proved:

Proposition 7.1.3. *Every set $\mathscr{S} \subseteq K[X]$ of polynomials has a splitting field, and all splitting fields of $\mathscr{S}$ are K-isomorphic.*

7.1.b. Normal Extensions

Proposition 7.1.4. *Let $\overline{K}$ be an algebraic closure of K and $K \subseteq E \subseteq \overline{K}$ be an algebraic extension of K. The following conditions are equivalent:*

(1) *E is a splitting field of a set of polynomials.*

(2) *$\varphi E = E$ for every K-homomorphism $\varphi : E \longrightarrow \overline{K}$.*

(3) *$\varphi E \subseteq E$ for every K-homomorphism $\varphi : E \longrightarrow \overline{K}$.*

(4) *$\sigma E = E$ for every K-automorphism of $\overline{K}$.*

(5) *$\sigma E \subseteq E$ for every K-automorphism of $\overline{K}$.*

(6) *Every irreducible polynomial $q \in K[X]$ which has a root in E splits in E.*

Proof. (1) implies (2) by Lemma 7.1.2, and (2) implies (3).

By Theorem 6.3.4, every K-homomorphism $\varphi : E \longrightarrow \overline{K}$ extends to a K-endomorphism σ of $\overline{K}$, which by Proposition 6.3.7 is a K-automorphism of $\overline{K}$; then $\varphi = \sigma_{|E}$. If, conversely, σ is a K-automorphism of $\overline{K}$, then $\varphi = \sigma_{|E}$ is a K-homomorphism of E into $\overline{K}$. Therefore (2) is equivalent to (4), and (3) is equivalent to (5).

(3) implies (6). Let $q(X) \in K[X]$ be irreducible with a root $\alpha \in E$. We may assume that q is monic. Then $q = \mathrm{Irr}_K(\alpha)$. If $\beta \in \overline{K}$ is a root of q, then Corollary 6.2.7 provides a K-homomorphism $\psi : K(\alpha) \longrightarrow \overline{K}$ such that $\psi\alpha = \beta$. By Theorem 6.3.4, ψ extends to a K-homomorphism $E \longrightarrow \overline{K}$. If (3) holds, then $\beta = \psi\alpha \in E$. Thus E contains every root of q in $\overline{K}$; therefore q splits in E.

(6) implies (1). For each $\alpha \in E$ let $q_\alpha = \mathrm{Irr}_K(\alpha)$. Then q_α has a root α in E; by (6), q_α splits in E, and E contains all the roots of q_α in $\overline{K}$. Hence E is generated by (in fact, consists of) the roots of all polynomials q_α. Thus E is a splitting field of $\{ q_\alpha \, ; \, \alpha \in E \}$. $\qquad\square$

A **normal** extension of a field K is an algebraic extension $K \subseteq E \subseteq \overline{K}$ of K which satisfies the equivalent conditions in Proposition 7.1.4.

Normal extensions can be characterized by their conjugates. Recall that a conjugate (over K) of an element α of $\overline{K}$ is the image $\sigma\alpha$ of α under a K-automorphism σ of $\overline{K}$.

A **conjugate** (over K) of an extension $K \subseteq E \subseteq \overline{K}$ is the image σE of E under a K-automorphism σ of $\overline{K}$. When $E \subseteq \overline{K}$, then φE is a conjugate of E for every K-homomorphism $\varphi : E \longrightarrow \overline{K}$: by Theorem 6.3.4, φ can be extended to a K-endomorphism σ of $\overline{K}$, which by Proposition 6.3.7 is a K-automorphism of $\overline{K}$. Hence Proposition 7.1.4 yields:

Proposition 7.1.5. *Let $K \subseteq E \subseteq \overline{K}$ be an algebraic extension of K. The following conditions are equivalent:*

(1) *E is a normal extension of K.*

(2) *E contains all conjugates of all elements of E.*

(3) *E has only one conjugate.*

7.1.c. Properties. Normal extensions have the following properties.

Proposition 7.1.6. *The following holds:*

(1) *If F is a normal extension of K and $K \subseteq E \subseteq F$, then F is a normal extension of E.*

(2) *If E is a normal extension of K and the composite EF exists, then EF is a normal extension of KF.*

(3) *If $(E_i)_{i \in I}$ is a family of normal extensions of K, then the composite $\prod_{i \in I} E_i$ and intersection $\bigcap_{i \in I} E_i$ are normal extensions of K.*

The proofs of these properties make wonderful exercises and are left to the reader.

It is not true that when E is a normal extension of K and F is a normal extension of E, then F is a normal extension of K. In Section 7.4 we will see how to construct a counterexample.

Proposition 7.1.6 implies that for each algebraic extension $K \subseteq E \subseteq \overline{K}$, there is a smallest normal extension which contains E (namely, the intersection of all normal extensions which contain E). It can be constructed as follows:

Proposition 7.1.7. *Let $K \subseteq E \subseteq \overline{K}$ be an algebraic extension of K. The smallest normal extension $F \subseteq \overline{K}$ of K which contains E is the composite of all the conjugates of E.*

Proof. A normal extension which contains E must contain all conjugates of E and must contain their composite F. Conversely, F contains E and is a normal extension, since every K-automorphism of $\overline{K}$ sends a conjugate of E onto a conjugate of E and therefore sends F into F. □

7.1.d. Finite Fields. With splitting fields we can determine all finite fields.

A finite field F has characteristic $p \neq 0$ and is a finite extension of $\mathbb{Z}_p$ for some prime p. Hence F has order $|F| = p^n$, where $n = [F : \mathbb{Z}_p]$. The group $F^* = F \backslash \{0\}$ is cyclic, by Proposition 2.2.4, and has order $p^n - 1$; therefore $x^{p^n - 1} = 1$ for all $x \in F^*$ and $x^{p^n} = x$ for all $x \in F$.

Theorem 7.1.8. *Let p be a prime number and $n > 0$ be an integer. Up to isomorphism, there is exactly one field F of order p^n; F is a splitting field of $X^{p^n} - X \in \mathbb{Z}_p[X]$ and its elements are the roots of $X^{p^n} - X$.*

Proof. If there is a field of order p^n, then we saw that $x^{p^n} - x = 0$ for all $x \in F$, so that the p^n elements of F are roots of $f(X) = X^{p^n} - X$. Since f has at most p^n roots, it follows that F consists of all the roots of f. In particular, F is a spitting field of f and is unique up to isomorphism.

Conversely, let F be a splitting field of $f(X) = X^{p^n} - X \in \mathbb{Z}_p[X]$. Then F has characteristic p. Let $\alpha, \beta \in F$ be roots of f. By Proposition 6.1.6, $(\alpha - \beta)^{p^n} = \alpha^{p^n} - \beta^{p^n} = \alpha - \beta$ and $(\alpha\beta^{-1})^{p^n} = \alpha^{p^n}(\beta^{p^n})^{-1} = \alpha\beta^{-1}$. Hence the roots of f form a subfield of F. Since F is generated by the roots of f, F consists of all the roots of f. Also the derivative of f is $f' = -1$; by Proposition 5.2.3, f has no multiple roots. Therefore $|F| = p^n$. $\square$

We denote by $GF(q)$ the finite field (also called a **Galois field**) of order $q \; (= p^n)$. Properties of these fields make fine exercises.

Exercises

1. Let F be a normal extension of K and $K \subseteq E \subseteq F$. Prove that F is a normal extension of E.

2. Let E be a normal extension of K, and let the composite EF exist. Prove that EF is a normal extension of KF.

3. Let E and F be normal extensions of K. Prove that EF and $E \cap F$ are normal extensions of K.

4. Draw the addition and multiplication tables of $GF(4)$.

5. Draw the addition and multiplication tables of $GF(8)$.

6. Let K have characteristic $p \neq 0$. Show that K has a subfield of order p^n if and only if $X^{p^n} - X$ splits in K, and then K has only one subfield of order p^n.

7. Show that a field of order p^n has a subfield of order p^m if and only if m divides n.

8. Let K be a field and L, M be subfields of K of orders p^m and p^n respectively. Show that $L \cap M$ has order p^d, where $d = (m,n)$.

7.2. SEPARABLE EXTENSIONS

A polynomial $f \in K[X]$ is separable when it has no multiple root in $\overline{K}$. An algebraic extension $K \subseteq E$ is separable when all its elements have separable irreducible polynomials. We will see in Section 7.4 that a normal extension $K \subseteq E \subseteq \overline{K}$ is separable when it has enough K-homomorphisms $E \longrightarrow \overline{K}$ to separate the elements of K from the elements of $E \backslash K$. First we show that the number of K-homomorphisms $E \longrightarrow \overline{K}$ is directly related to polynomial separability.

7.2.a. The Separability Degree. The **separability degree** of an algebraic extension E of K is the number of K-homomorphisms of E into an algebraic closure $\overline{K}$ of K. By Proposition 6.3.5, the separability degree does not depend on the choice of $\overline{K}$; it is denoted by $[E:K]_s$.

Proposition 7.2.1. *When $E = K(\alpha)$ is a simple algebraic extension, then $[E:K]_s$ is the number of distinct roots of $\mathrm{Irr}_K(\alpha)$ in $\overline{K}$.*

Proof. Let $q(X) = \mathrm{Irr}_K(\alpha)$. If $\varphi : E \longrightarrow \overline{K}$ is a K-homomorphism, then $({}^{\varphi}q)(\varphi\alpha) = \varphi(q(\alpha)) = 0$ and $\varphi\alpha$ is a root of q. Conversely, for each root $\beta \in \overline{K}$ of q, there is by Lemma 6.2.3 (applied to the inclusion homomorphism $K \longrightarrow \overline{K}$) exactly one K-homomorphism $E \longrightarrow \overline{K}$ such that $\varphi\alpha = \beta$. □

Proposition 7.2.2. $[F:K]_s = [F:E]_s [E:K]_s$ *when* $K \subseteq E \subseteq F$.

Proof. We may assume that $E \subseteq \overline{K}$. Let φ be a K-homomorphism of E into $\overline{K}$. By Theorem 6.3.4 and Corollary 6.3.7, there is a K-automorphism σ of $\overline{K}$ which extends φ.

If ψ is an E-homomorphism of F into $\overline{K}$, then $\sigma \circ \psi$ is a K-homomorphism of F into $\overline{K}$ which extends φ. If, conversely, χ is a K-homomorphism of F into $\overline{K}$ which extends φ, then $\sigma^{-1} \circ \chi$ is an E-homomorphism of F into $\overline{K}$. This provides a bijection between the E-homomorphisms $F \longrightarrow \overline{K}$ and the K-homomorphisms $F \longrightarrow \overline{K}$ which extend φ. Therefore there are $[F:E]_s$ K-homomorphisms $F \longrightarrow \overline{K}$ which extend φ.

The K-homomorphisms of F into $\overline{K}$ can be classified by their restrictions to E, which are K-homomorphisms of E into $\overline{K}$. There are $[E:K]_s$ K-homomorphisms of E into $\overline{K}$, each of which is the restriction of $[F:E]_s$ K-homomorphisms of F into $\overline{K}$. This yields $[F:E]_s [E:K]_s$ K-homomorphisms of F into $\overline{K}$. □

Proposition 7.2.3. *When $K \subseteq E$ is a finite extension,* $[E:K]_s \leqslant [E:K]$.

Proof. E is finitely generated: $E = K(\alpha_1,\ldots,\alpha_k)$ for some $\alpha_1,\ldots,\alpha_k \in E$. The proof is by induction on k.

If $k = 1$, then $E = K(\alpha)$ is simple. Let $q = \mathrm{Irr}_K(\alpha)$. Then $[E:K] = \deg q$, whereas $[E:K]_s$ is the number of distinct roots of q in $\overline{K}$, by Proposition 7.2.1, and cannot exceed the degree or q. Hence $[E:K]_s \leqslant [E:K]$.

If $k > 1$, then $E = F(\alpha_k)$, where $F = K(\alpha_1, \ldots, \alpha_{k-1})$. By Proposition 7.2.2 and the induction hypothesis, $[E : K]_s = [E : F]_s [F : K]_s \leqslant [E : F][F : K] = [E : K]$. $\qquad\square$

Corollary 7.2.4. *A finite algebraic extension has only finitely many conjugates, and is contained in a finite normal extension.*

Proof. This follows from Propositions 7.2.3 and 7.1.7. $\qquad\square$

7.2.b. Irreducible Polynomials. A polynomial $f(X) \in K[X]$ is **separable** in case f has no multiple roots in $\overline{K}[X]$. Proposition 7.2.1 and the proof of Proposition 7.2.3 show that $[E : K]_s$ is determined by $[E : K]$ and by the extent to which irreducible polynomials of elements of E fail to be separable.

Lemma 7.2.5. *Let $q(X) \in K[X]$ be an irreducible polynomial.*

(1) *If K has characteristic 0, then q is separable.*
(2) *If K has characteristic $p \neq 0$, then all roots of q in $\overline{K}$ have the same order of multiplicity which is a power p^m of p, and there exists a separable irreducible polynomial $s(X) \in K[X]$ such that $q(X) = s(X^{p^m})$.*

Proof. Let $\alpha \in \overline{K}$ be a root of q. Then α is algebraic over K and $\mathrm{Irr}_K(\alpha) = q$. If α is a multiple root of q, then $q'(\alpha) = 0$ (Proposition 5.2.3), q divides q', and $q' = 0$ (since $\deg q' < \deg q$). This cannot happen if K has characteristic 0, which proves (1).

Now let K have characteristic $p \neq 0$. If $q(X) = a_n X^n + \cdots + a_1 X + a_0$ has a multiple root, then $q'(X) = n a_n X^{n-1} + \cdots + a_1 = 0$. This can happen only if $a_i = 0$ whenever $p \nmid i$, so that $q(X)$ contains only powers of X^p. Thus $q(X) = r(X^p)$ for some polynomial $r(X) \in K[X]$; r is irreducible, since a nontrivial factorization of r would yield a nontrivial factorization of q, and $\deg r < \deg q$. If r is not separable, then $r(X) = t(X^p) = q(X^{p^2})$ for some irreducible polynomial $t \in K[X]$, with $\deg t < \deg r$. This continues until $q(X) = s(X^{p^m})$ for some separable irreducible polynomial s.

We now have $s(X) = \prod_{i \in I} (X - \beta_i)$, where $\beta_1, \ldots, \beta_k$ are the distinct roots of s in $\overline{K}$. Since $\overline{K}$ is algebraically closed, we have $\beta_i = \alpha_i^{p^m}$ for some $\alpha_i \in \overline{K}$. By Proposition 6.1.6, $\alpha_1, \ldots, \alpha_k$ are distinct and

$$q(X) = \prod_{i \in I} (X^{p^m} - \alpha_i^{p^m}) = \prod_{i \in I} (X - \alpha_i)^{p^m}.$$

Hence every root of Q in $\overline{K}$ has multiplicity p^m. $\qquad\square$

Corollary 7.2.6. *Let E be a finite extension of K.*

(1) *If K has characteristic 0, then $[E : K]_s = [E : K]$.*
(2) *If K has characteristic $p \neq 0$, then $[E : K] = p^m [E : K]_s$ for some $m \geqslant 0$.*

Proof. This is proved like Proposition 7.2.3. E is finitely generated, say $E = K(\alpha_1, \ldots, \alpha_k)$; the proof is by induction on k. If $k = 1$, then $E = K(\alpha_1)$ is simple, and the result follows from Proposition 7.2.1 and Lemma 7.2.5. If $k > 1$, then $E = F(\alpha_k)$, where let $F = K(\alpha_1, \ldots, \alpha_{k-1})$. By Proposition 7.2.2, $[E : K]_s = [F(\alpha_k) : F]_s [F : K]_s$, $[E : K] = [F(\alpha_k) : F][F : K]$, and the result follows from the case $k = 1$ and the induction hypothesis. $\square$

The **inseparability degree** of a finite extension $K \subseteq E$ is the quotient $[E : K]_i = [E : K]/[E : K]_s$. By Proposition 7.2.2, $[F : E]_i [E : K]_i = [F : K]_i$ whenever $K \subseteq E \subseteq F$ are finite extensions.

7.2.c. Separable Extensions. Let E be an algebraic extension of K. An element α of E is **separable over K** in case $\mathrm{Irr}_K(\alpha)$ is separable; E is a **separable** extension of K (or, is **separable over K**) in case every element of E is separable over K.

Proposition 7.2.7. *If K has characteristic 0, then every algebraic extension of K is separable.*

Proof. This follows from Lemma 7.2.5. $\square$

Proposition 7.2.8. *For a finite extension E of K the following are equivalent:*

(1) *E is a separable extension of K (every element of E is separable over K).*

(2) *E is generated by separable elements.*

(3) *$[E : K]_s = [E : K]$ (there are $[E : K]$ K-homomorphisms $E \longrightarrow \overline{K}$).*

Proof. (1) implies (2) is clear.

(2) implies (3). This is proved by induction like Corollary 7.2.6. If (2) holds, then $E = K(\alpha_1, \ldots, \alpha_k)$ is generated by finitely many elements $\alpha_1, \ldots, \alpha_k$ which are separable over K. If $k = 1$, then $[K(\alpha_1) : K]_s = [K(\alpha_1) : K]$ by Proposition 7.2.1, since α_1 and $\mathrm{Irr}_K(\alpha_1)$ are separable. If $k > 1$, then $E = F(\alpha_k)$, where $F = K(\alpha_1, \ldots, \alpha_{k-1})$; by Proposition 7.2.2,

$$[E : K]_s = [F(\alpha_k) : F]_s [F : K]_s = [F(\alpha_k) : F][F : K] = [E : K],$$

by the case $k = 1$ and the induction hypothesis.

(3) implies (1). Let $\alpha \in E$. By Corollary 7.2.6 we have $[K(\alpha) : K]_s \leqslant [K(\alpha) : K]$ and $[E : K(\alpha)]_s \leqslant [E : K(\alpha)]$. Also

$$[E : K(\alpha)]_s [K(\alpha) : K]_s = [E : K]_s = [E : K] = [E : K(\alpha)][K(\alpha) : K].$$

Therefore $[K(\alpha) : K]_s = [K(\alpha) : K] = \deg \mathrm{Irr}_K(\alpha)$, which shows that $\mathrm{Irr}_K(\alpha)$ and α are separable over K. $\square$

7.2.d. Properties. Our indefatigable reader will prove:

Proposition 7.2.9. *The following holds:*

(1) *If $E = K(S)$ is algebraic over K and every element of S is separable over K, then E is separable over K.*

(2) *If $K \subseteq E \subseteq F$, then F is separable over K if and only if F is separable over E and E is separable over K.*

(3) *If E is separable over K and EF exists, then EF is separable over KF.*

Theorem 7.2.10 (Primitive Element). *Every finite separable extension is simple.*

Proof. Let E be a finite separable extension of K. If K is finite, then E is finite, $E \backslash \{0\}$ is a cyclic group under multiplication by Proposition 2.2.4, and E is singly generated as an extension.

Now assume that K is infinite. We show that every finite separable extension $E = K(\alpha, \beta)$ of K with two generators is simple; it follows by induction on k that every finite separable extension $K(\alpha_1, \ldots, \alpha_k)$ of K is simple. Let $[E : K] = n$. There are n K-homomorphisms $\varphi_1, \ldots, \varphi_n$ of E into $\overline{K}$. Let

$$f(X) = \prod_{i \neq j} (\varphi_i \alpha + (\varphi_i \beta) X - \varphi_j \alpha - (\varphi_j \beta) X) \in \overline{K}[X].$$

Since K is infinite, we cannot have $f(t) = 0$ for all $t \in K$. Hence $f(t) \neq 0$ for some $t \in K$. Then $\varphi_1(\alpha + t\beta), \ldots, \varphi_n(\alpha + t\beta)$ are all distinct. Therefore there are at least n K-homomorphisms of $K(\alpha + t\beta)$ into $\overline{K}$. This implies $[K(\alpha + t\beta) : K] \geqslant n$. But $K(\alpha + t\beta) \subseteq E$; therefore $E = K(\alpha + t\beta)$. (The "primitive element" in Theorem 7.2.10 is the single generator of E.) $\square$

Corollary 7.2.11. *Let E be separable over K. If every element of E has degree at most n over K, then E has degree at most n over K.*

Proof. Let $\alpha \in E$ have maximal degree $m \leqslant n$. If $K(\alpha) \neq E$, then $K(\alpha, \beta) \supsetneq K(\alpha)$ for some $\beta \in E$, and Theorem 7.2.10 implies $K(\alpha, \beta) = K(\gamma)$ where $\gamma \in E$ has degree greater than m, a contradiction. $\square$

Exercises

1. Let $E = K(S)$ be algebraic over K. If every element of S is separable over K, prove that E is separable over K.
2. If $K \subseteq E \subseteq F$, and F is separable over K, show that F is separable over E and E is separable over K.
3. If F is separable over E and E is separable over K, show that F is separable over K.
4. If E is separable over K and EF exists, prove that EF is separable over KF.
5. Show that $\mathrm{Aut}(GF(p^n))$ is cyclic of order n.

7.3. PURELY INSEPARABLE EXTENSIONS

A purely inseparable extension of a field K is an algebraic extension in which only the elements of K are separable over K. This section completes Section 7.2 and gives basic properties and examples of purely inseparable extensions, with applications to perfect fields. The results can be used later but are not necessary for what follows.

7.3.a. Definition. Let K be a field of characteristic $p \neq 0$. Since $\overline{K}$ is algebraically closed, each $a \in K$ has a p-th root $\alpha \in \overline{K}$ (such that $\alpha^p = a$), which is unique by Proposition 6.1.6. Hence each $a \in K$ has a unique p^m-th root in $\overline{K}$. Proposition 6.1.6 also implies that

$$K^{1/p} = \{\alpha \in \overline{K}; \alpha^p \in K\} \qquad \text{and}$$

$$K^{1/p^\infty} = \{\alpha \in \overline{K}; \alpha^{p^k} \in K \text{ for some } k \geqslant 0\}$$

are subfields of $\overline{K}$ containing K.

Proposition 7.3.1. *Let E be an algebraic extension of a field K of characteristic $p \neq 0$. The following conditions are equivalent:*

(1) *No element of $E \backslash K$ is separable over K.*

(2) *For each $\alpha \in E$, $\operatorname{Irr}_K(\alpha) = X^{p^m} - a$ for some $m \geqslant 0$ and $a \in K$.*

(3) *Each $\alpha \in E$ has a power α^{p^m} in K.*

(4) *There is a K-homomorphism of E into K^{1/p^∞}.*

(5) *There is only one K-homomorphism of E into $\overline{K}$.*

Proof. (1) implies (2). Let $\alpha \in E$ and $q = \operatorname{Irr}_K(\alpha)$. By Lemma 7.2.5, $q(X) = s(X^{p^m})$ for some separable irreducible polynomial $s(X) \in K[X]$ and $m \geqslant 0$. Then $s(\alpha^{p^m}) = q(\alpha) = 0$, $s = \operatorname{Irr}_K(\alpha^{p^m})$, $\alpha^{p^m} \in E$ is separable over K, $a = \alpha^{p^m} \in K$ by (1), $s(X) = X - a$, and $q(X) = X^{p^m} - a$.

Obviously (2) implies (3).

(3) implies (4). By Theorem 6.3.4 there is a K-homomorphism $\varphi : E \longrightarrow \overline{K}$. For each $\alpha \in E$ we have $\alpha^{p^m} \in K$ for some $m \geqslant 0$, $(\varphi\alpha)^{p^m} = \varphi(\alpha^{p^m}) \in K$, and $\varphi\alpha \in K^{1/p^\infty}$.

(4) implies (5). By Proposition 6.1.6, p-th roots are unique in $\overline{K}$; hence p^m-th roots are unique in $\overline{K}$. Now let $\varphi : E \longrightarrow K^{1/p^\infty}$ be a K-homomorphism. For each $\alpha \in E$ we have $(\varphi\alpha)^{p^m} \in K$ for some $m \geqslant 0$; hence $\varphi(\alpha^{p^m}) \in K$ and $\alpha^{p^m} \in K$. Then any K-homomorphism ψ of E into $\overline{K}$ must send α onto the p^m-th root of α^{p^m} in $\overline{K}$. This uniquely determines $\psi\alpha$: if $\alpha^{p^m}, \alpha^{p^{m+k}} \in K$ and $\psi\alpha$ is the p^m-th root of α^{p^m} in $\overline{K}$, then $\psi\alpha$ is also the p^{m+k}-th root of $\alpha^{p^{m+k}}$ in $\overline{K}$, and therefore does not depend on the choice of m (as long as $\alpha^{p^m} \in K$).

(5) implies (1). Let $\alpha \in E$ be separable over K and $q = \mathrm{Irr}_K(\alpha)$, $n = \deg q$. By Proposition 7.2.1, there are n K-homomorphisms of $K(\alpha)$ into $\overline{K}$, which extend to at least n K-homomorphisms of E into $\overline{K}$. By (5), $n = 1$ and $\alpha \in K$. $\square$

Let K be a field of characteristic $p \neq 0$ and E be an extension of K. An algebraic element α of E is **purely inseparable** over K in case $\mathrm{Irr}_K(\alpha) = X^{p^m} - a$ for some $m \geqslant 0$ and $a \in K$; equivalently, $\alpha^{p^m} \in K$ for some $m \geqslant 0$. A **purely inseparable** extension of K is an algebraic extension which satisfies the equivalent conditions in Proposition 7.3.1; equivalently, in which every element is purely inseparable over K.

Proposition 7.3.1 shows that, up to K-isomorphism, K^{1/p^∞} is the largest purely inseparable extension of K.

7.3.b. Properties. The following properties are exercises:

Proposition 7.3.2. *The following holds:*

(1) *If $E = K(S)$ is algebraic over K and every element of S is purely inseparable over K, then E is purely inseparable over K.*

(2) *If $K \subseteq E \subseteq F$, then F is purely inseparable over K if and only if F is purely inseparable over E and E is purely inseparable over K.*

(3) *If E is purely inseparable over K and EF exists, then EF is purely inseparable over KF.*

Proposition 7.3.3. *Let E be an algebraic extension of K. Then*

$$S = \{\alpha \in E ; \alpha \text{ is separable over } K\}$$

is a subfield of E; S is separable over K and E is purely inseparable over S.

Proof. This follows from Proposition 7.2.9. Let $\alpha, \beta \in S$. Then $K(\alpha, \beta)$ is separable over K, and $\alpha - \beta, \alpha\beta^{-1} \in K(\alpha, \beta)$ are separable over K. Thus S is a subfield of E. Clearly S is a separable extension of K. If $\gamma \in E$ is separable over S, then γ is separable over K and $\gamma \in S$, so E is purely inseparable over S. $\square$

Proposition 7.3.4. *Let E be a normal extension of K, G be the group of all K-automorphisms of E, and*

$$F = \{\alpha \in E ; \sigma\alpha = \alpha \text{ for all } \sigma \in G\}.$$

Then E is separable over F, and F is purely inseparable over K.

Proof. We may assume that $E \subseteq \overline{K}$. We have $K \subseteq F$, and F is a subfield of E, since every $\sigma \in G$ is an automorphism. Let $\alpha \in F$. Every K-homomorphism

φ of $K(\alpha)$ into $\overline{K}$ extends to a K-automorphism ψ of E into $\overline{K}$; by Proposition 7.1.4, $\psi E = E$, and $\psi|_E$ is a K-automorphism of E. By the choice of α this implies $\varphi\alpha = \psi\alpha = \alpha$. Hence $[K(\alpha):K]_s = 1$ and α is purely inseparable over K. Thus F is purely inseparable over K.

Let $\alpha \in E$. By Proposition 6.3.8, α has finitely many distinct conjugates $\alpha_1 = \alpha$, $\alpha_2,\ldots,\alpha_r$ over K, which all belong to E since E is normal. Every $\sigma \in G$ can be extended to a K-automorphism of $\overline{K}$; hence σ permutes $\alpha_1,\ldots,\alpha_r$, and the polynomial $f(X) = (X-\alpha_1)\ldots(X-\alpha_r) \in E[X]$ has the following properties: ${}^\sigma f = f$ for every $\sigma \in G$ so that $f \in F[X]$, f is separable, and $f(\alpha) = 0$. Therefore $\mathrm{Irr}_F(\alpha)$ divides f and α is separable over F. Thus E is separable over F. $\square$

Corollary 7.3.5. *Let K have characteristic $p \neq 0$ and $\alpha \in \overline{K}$. Then $\alpha \in K^{1/p^\infty}$ if and only if $\sigma\alpha = \alpha$ for every K-automorphism σ of $\overline{K}$.*

This makes a nice exercise.

7.3.c. Perfect Fields. A field K is **perfect** in case either K has characteristic 0, or K has characteristic $p \neq 0$ and every element of K has a p-th root in K.

Proposition 7.3.6. *Finite fields and algebraically closed fields are perfect.*

Proof. Let F be a finite field, of characteristic $p \neq 0$. By Proposition 6.1.6, $\pi : x \longmapsto x^p$ is an endomorphism of F. Hence π is injective; since F is finite, π is bijective. $\square$

Proposition 7.3.7. *For a field K of characteristic p, the following conditions are equivalent:*

(1) *K is perfect.*
(2) *Every algebraic extension of K is separable.*
(3) *$K^{1/p} = K$.*
(4) *$K^{1/p^\infty} = K$.*

Proof. (1), (3), and (4) are equivalent, since p-th roots (and p^k-th roots) are unique in $\overline{K}$; (2) implies (4), since K^{1/p^∞} is a purely inseparable extension of K.

(4) implies (2). By Proposition 7.3.1, (4) implies that the only purely inseparable extension of K is K itself. By Proposition 7.1.7 every algebraic extension of K is contained in a normal extension. By Proposition 7.3.4, every normal extension of K is a separable extension of a purely inseparable extension of K, and is separable over K by (4). Therefore every algebraic extension of K is separable over K. $\square$

Corollary 7.3.8. *Every algebraic extension of a perfect field is perfect.*

The proof is an exercise.

Exercises

1. If $E = K(S)$ is algebraic over K and every element of S is purely inseparable over K, prove that E is purely inseparable over K.
2. If $K \subseteq E \subseteq F$, prove that F is purely inseparable over K if and only if F is purely inseparable over E and E is purely inseparable over K.
3. If E is purely inseparable over K and EF exists, prove that EF is purely inseparable over KF.
4. Let K have characteristic $p \neq 0$ and $\alpha \in \overline{K}$. Show that $\alpha \in K^{1/p^\infty}$ if and only if $\sigma\alpha = \alpha$ for every K-automorphism σ of $\overline{K}$.
5. Let K have characteristic $p \neq 0$. Show that $K(X)$ is not perfect.
6. Prove that every algebraic extension of a perfect field is perfect.

7.4. GALOIS EXTENSIONS

A **Galois extension** of a field K is a normal and separable extension E of K. This definition ensures that the K-homomorphisms of E into $\overline{K}$ are K-automorphisms of E, and constitute a group, the **Galois group** of E over K, which we denote by $\mathrm{Gal}(E/K)$. (This does not mean there is a "quotient" E/K.)

The main result of this section (the Fundamental Theorem of Galois theory) provides a bijection between intermediate fields $K \subseteq F \subseteq E$ and subgroups of $\mathrm{Gal}(E/K)$, when E is a finite Galois extension.

7.4.a. Fixed Fields. When E is a field and G is a group of automorphisms of E, the **fixed field** of G is the subfield

$$\mathrm{Fix}_E(G) = \{\, \alpha \in E \,;\, \sigma\alpha = \alpha \text{ for all } \sigma \in G \,\}$$

of E of all elements of E which are fixed under every $\sigma \in G$.

Proposition 7.4.1 (Artin). *Let E be a field, G be a finite group of automorphisms of E, and $K = \mathrm{Fix}_E(G)$. Then E is a Galois extension of K, $[E : K] = |G|$, and $\mathrm{Gal}(E/K) = G$.*

Proof. First we show that E is separable over K. (This also follows from Proposition 7.3.4.) The group G is a group of permutations of E and acts on E by evaluation. For each $\alpha \in E$, let $\alpha_1 = \alpha$, $\alpha_2,\ldots,\alpha_r$ be the distinct elements of the orbit of α under G. Then $r \leqslant |G|$ and every $\sigma \in G$ permutes $\alpha_1,\ldots,\alpha_r$. Hence the polynomial $f_\alpha(X) = (X - \alpha_1)\ldots(X - \alpha_r) \in E[X]$ has the following properties: $^\sigma f_\alpha = f_\alpha$ for every $\sigma \in G$, so that $f_\alpha \in K[X]$; f_α is separable; and $f_\alpha(\alpha) = 0$. Therefore α is algebraic over K and $\mathrm{Irr}_K(\alpha)$ divides f_α. Hence α

is separable over K and has degree at most $r \leqslant |G|$. Thus E is algebraic over K. By Corollary 7.2.11, E is separable over K and has degree $[E : K] \leqslant |G|$. Also E is a splitting field of $(f_\alpha)_{\alpha \in E}$ and is a normal extension of K.

Let $\overline{K} \supseteq E$ (for instance we may let $\overline{K} = \overline{E}$, by Proposition 6.3.6). Since E is a normal extension of K, a K-homomorphism of E into $\overline{K}$ is a K-automorphism of E, and conversely. Since E is separable over K, the number of K-automorphisms of E is $|\mathrm{Gal}(E/K)| = [E : K] \leqslant |G|$. But G is a subgroup of $\mathrm{Gal}(E/K)$. Therefore $\mathrm{Gal}(E/K) = G$ and $[E : K] = |G|$. $\qquad\square$

Proposition 7.4.2. *If E is a Galois extension of K, then the fixed field of* $\mathrm{Gal}(E/K)$ *is K.*

Proof. This follows from Proposition 7.3.4, but we give a direct proof. We may assume that $E \subseteq \overline{K}$. Let $G = \mathrm{Gal}(E/K)$. Then $K \subseteq \mathrm{Fix}_E(G)$. Let $\alpha \in \mathrm{Fix}_E(G)$. Every K-homomorphism φ of $K(\alpha)$ into $\overline{K}$ extends to a K-automorphism ψ of E into $\overline{K}$; by Proposition 7.1.4, $\psi E = E$, and ψ is a K-automorphism of E. By the choice of α this implies $\varphi \alpha = \psi \alpha = \alpha$. Thus $[K(\alpha) : K]_s = 1$. Since $K(\alpha) \subseteq E$ is separable over K, this implies $[K(\alpha) : K] = 1$ and $\alpha \in K$. $\qquad\square$

7.4.b. Fundamental Theorem

Theorem 7.4.3 (Fundamental Theorem of Galois Theory). *Let E be a finite Galois extension of K.*

If F is a subfield of E which contains K, then E is a finite Galois extension of F and F is the fixed field of $\mathrm{Gal}(E/F)$.

If H is a subgroup of $\mathrm{Gal}(E/K)$, *then the fixed field $F = \mathrm{Fix}_E(H)$ of H is a subfield of E which contains K and $H = \mathrm{Gal}(E/F)$.*

This defines an order reversing, one-to-one correspondence between subgroups of $\mathrm{Gal}(E/K)$ *and intermediate fields $K \subseteq F \subseteq E$.*

Proof. If H is a subgroup of $\mathrm{Gal}(E/K)$, then $F = \mathrm{Fix}_E(H)$ is a field, $K \subseteq F \subseteq E$, and $\mathrm{Gal}(E/F) = H$ by Proposition 7.4.1. If, conversely, $K \subseteq F \subseteq E$, then E is a finite Galois extension of F, by Propositions 7.2.9, 7.1.6, and 7.2.9, and the fixed field of $\mathrm{Gal}(E/F)$ is F, by Proposition 7.4.2. Thus $H \longmapsto \mathrm{Fix}_E(H)$ and $F \longmapsto \mathrm{Gal}(E/F)$ are mutually inverse bijections.

Furthermore $H \subseteq H'$ implies $\mathrm{Fix}_E(H) \supseteq \mathrm{Fix}_E(H')$, and $K \subseteq F \subseteq F' \subseteq E$ implies $\mathrm{Gal}(E/F) \supseteq \mathrm{Gal}(E/F')$. $\qquad\square$

7.4.c. Consequences

Corollary 7.4.4. *Let E be a finite Galois extension of K. Let $K \subseteq F_i \subseteq E$ ($i = 1,2,3$) and $H_i = \mathrm{Gal}(E/F_i)$ (so that $F_i = \mathrm{Fix}_E(H_i)$).*

(1) $F_1 = F_2 F_3$ *if and only if $H_1 = H_2 \cap H_3$.*
(2) $F_1 = F_2 \cap F_3$ *if and only if $H_1 = \langle H_2 \cup H_3 \rangle$.*

(3) F_1 and F_2 are conjugate if and only if H_1 and H_2 are conjugate in $\mathrm{Gal}(E/K)$.

Proof. (1) Let $K \subseteq F \subseteq E$ and $H = \mathrm{Gal}(E/F)$ (so that $F = \mathrm{Fix}_E(H)$). Then $F_i \subseteq F$ implies $H_i = \mathrm{Gal}(E/F_i) \supseteq \mathrm{Gal}(E/F) = H$; conversely, $H_i \supseteq H$ implies $F_i = \mathrm{Fix}_E(H_i) \subseteq \mathrm{Fix}_E(H) = F$. Hence $F_2 F_3 \subseteq F$ is equivalent to $F_2, F_3 \subseteq F$; $H_2, H_3 \supseteq H$; and $H_2 \cap H_3 \supseteq H$; and F_1 is the smallest such field F if and only if H_1 is the largest such subgroup H.

(2) Similarly $F \subseteq F_i$ if and only if $H \supseteq H_i$; hence $F_1 = F_2 \cap F_3$ if and only if $H_1 = \langle H_2 \cup H_3 \rangle$ (= the subgroup of $\mathrm{Gal}(E/K)$ generated by $H_2 \cup H_3$).

(3) Note that F_1 and F_2 are conjugate if and only if $F_2 = \tau F_1$ for some $\tau \in \mathrm{Gal}(E/K)$: indeed every $\tau \in \mathrm{Gal}(E/K)$ can be extended to a K-automorphism of $\overline{K}$ so that τF_1 is a conjugate of F_1; if, conversely, $F_2 = \sigma F_1$ is a conjugate of F_1, where σ is a K-automorphism of $\overline{K}$, then $\sigma E = E$, since E is normal and $F_2 = \tau F_1$, where $\tau = \sigma|_E \in \mathrm{Gal}(E/K)$.

Let $K \subseteq F \subseteq E$ and $\sigma, \tau \in \mathrm{Gal}(E/K)$. Then $\sigma \in \mathrm{Gal}(E/\tau F)$ is equivalent to $\sigma \tau \alpha = \tau \alpha$ for all $\alpha \in F$; $\tau^{-1} \sigma \tau \alpha = \alpha$ for all $\alpha \in F$; $\tau^{-1} \sigma \tau \in \mathrm{Gal}(E/F)$; and $\sigma \in \tau \mathrm{Gal}(E/F) \tau^{-1}$. Thus $\mathrm{Gal}(E/\tau F) = \tau \mathrm{Gal}(E/F) \tau^{-1}$. $\square$

Corollary 7.4.5. *Let E be a finite Galois extension of K and $K \subseteq F \subseteq E$. Then F is a normal extension of K if and only if $\mathrm{Gal}(E/F)$ is a normal subgroup of $\mathrm{Gal}(E/K)$, and then $\mathrm{Gal}(E/K)/\mathrm{Gal}(E/F) \cong \mathrm{Gal}(F/K)$.*

Proof. The first part of the statement follows from Corollary 7.4.4. Now assume that F is a normal extension of K. By Propositions 7.2.9 and 6.2.9, F is a Galois extension of K. When $\sigma \in \mathrm{Gal}(E/K)$, then $\sigma F = F$ by Proposition 7.1.4, and σ induces a K-automorphism of F. This provides a homomorphism $\Phi : \mathrm{Gal}(E/K) \longrightarrow \mathrm{Gal}(F/K)$ which assigns to $\sigma \in \mathrm{Gal}(E/K)$ its restriction $\Phi(\sigma) = \sigma|_F$ to F. We see that $\mathrm{Ker}\, \Phi = \mathrm{Gal}(E/F)$. $\square$

Proposition 7.4.6. *Let E be a finite Galois extension of K and F be a field extension of K such that EF is defined. Then EF is a finite Galois extension of F, E is a finite Galois extension of $E \cap F$, and $\mathrm{Gal}(EF/F) \cong \mathrm{Gal}(E/E \cap F)$.*

Proof. By Proposition 6.2.10 and Propositions 7.1.6 and 7.2.9, EF is a Galois extension of F and E is a Galois extension of $E \cap F$. Also E is finite over $E \cap F \supseteq K$.

Let σ be an F-automorphism of EF. Since E is a normal extension of K, we have $\sigma E = E$ and σ induces a K-automorphism of E. Let Φ be the homomorphism of $\mathrm{Gal}(EF/F)$ into $\mathrm{Gal}(E/K)$ which assigns to each $\sigma \in \mathrm{Gal}(EF/F)$ its restriction $\Phi(\sigma) = \sigma|_E$ to E. Since EF is generated by $E \cup F$, an automorphism of EF is uniquely determined by its restrictions to E and F; hence Φ is injective. We show that $\mathrm{Im}\, \Phi = \mathrm{Gal}(E/E \cap F)$.

If $\alpha \in E \cap F$, then α is fixed under $\sigma|_E$ for every $\sigma \in \mathrm{Gal}(EF/F)$. If, conversely, $\alpha \in E$ is fixed under $\sigma|_E$ for every $\sigma \in \mathrm{Gal}(EF/F)$, then α is fixed under every $\sigma \in \mathrm{Gal}(EF/F)$ so that $\alpha \in F$ by Proposition 7.4.2 and $\alpha \in E \cap F$.

Hence $\text{Fix}_E(\text{Im }\Phi) = E \cap F$, and Theorem 7.4.3 yields $\text{Im }\Phi = \text{Gal}(E/E \cap F)$. Thus $\text{Gal}(EF/F) \cong \text{Gal}(E/E \cap F)$. In particular, $\text{Gal}(EF/F)$ is finite, and EF is a finite extension of F, by Proposition 7.4.1. $\qquad\square$

7.4.d. Symmetric Rational Functions. We show that every finite group is a Galois group.

Let $K(X_1,\ldots,X_n)$ be the field of rational functions in n indeterminates over a field K. Recall that the **elementary symmetric polynomials** $\mathfrak{s}_1,\ldots,\mathfrak{s}_n \in K[X_1,\ldots,X_n]$ are

$$\mathfrak{s}_k(X_1,\ldots,X_k) = \sum\nolimits_{1 \leqslant i_1 < i_2 < \cdots < i_k \leqslant n} X_{i_1} X_{i_2} \ldots X_{i_k}.$$

For example, $\mathfrak{s}_1(X_1,\ldots,X_n) = X_1 + \cdots + X_n$ and $\mathfrak{s}_n(X_1,\ldots,X_n) = X_1 \ldots X_n$. By Proposition 5.2.1,

$$(X - X_1)(X - X_2)\ldots(X - X_n)$$

$$= X^n - \mathfrak{s}_1 X^{n-1} + \cdots + (-1)^k \mathfrak{s}_k X^{n-k} + \cdots + (-1)^n \mathfrak{s}_n.$$

Proposition 7.4.7. $K(X_1,\ldots,X_n)$ *is a Galois extension of* $K(\mathfrak{s}_1,\ldots,\mathfrak{s}_n)$, *whose Galois group is isomorphic to the symmetric group* S_n.

Proof. Let $E = K(X_1,\ldots,X_n)$ and $S = K(\mathfrak{s}_1,\ldots,\mathfrak{s}_n) \subseteq E$. For each permutation σ of $\{1,2,\ldots,n\}$, there is a K-automorphism $\overline{\sigma}$ of E given for all $R \in K(X_1,\ldots,X_n)$ by

$$(\overline{\sigma}R)(X_1,\ldots,X_n) = R(X_{\sigma 1},\ldots,X_{\sigma n}).$$

The group G of all $\overline{\sigma}$ is a finite group of automorphisms of E and is isomorphic to S_n. Let $F = \text{Fix}_E(G)$. By Proposition 7.4.1, E is a Galois extension of F, and $\text{Gal}(E/F) = G \cong S_n$. In particular, $[E : F] = n!$.

Since $\mathfrak{s}_k$ is symmetric ($\mathfrak{s}_k(X_{\sigma 1},\ldots,X_{\sigma n}) = \mathfrak{s}_k(X_1,\ldots,X_n)$ for all $\sigma \in S_n$), we have $\mathfrak{s}_1,\ldots,\mathfrak{s}_n \in F$ and $S = K(\mathfrak{s}_1,\ldots,\mathfrak{s}_n) \subseteq F$. Hence $[E : S] \geqslant n!$. Now E is the splitting field of the separable polynomial $(X - X_1)\ldots(X - X_n) \in S[X]$. Since any S-automorphism of E must permute the roots $X_1,\ldots,X_n$, and is uniquely determined by its values on $X_1,\ldots,X_n$, there are at most $n!$ S-automorphisms of E and $[E : S] \leqslant n!$. Therefore $S = F$. $\qquad\square$

Corollary 7.4.8. *The elementary symmetric polynomials are algebraically independent over* K.

Proof. We just saw that $K(X_1,\ldots,X_n)$ is algebraic over $K(\mathfrak{s}_1,\ldots,\mathfrak{s}_n)$. By Proposition 6.4.4, $K(X_1,\ldots,X_n)$ has a transcendence base $B \subseteq \{\mathfrak{s}_1,\ldots,\mathfrak{s}_n\}$. In fact $B = \{\mathfrak{s}_1,\ldots,\mathfrak{s}_n\}$, since B has n elements by Theorem 6.4.5. In particular, $\mathfrak{s}_1,\ldots,\mathfrak{s}_n$ are algebraically independent over K. $\qquad\square$

A rational fraction $r(X_1,\ldots,X_n) \in K(X_1,\ldots,X_n)$ is **symmetric** in case $r(X_{\sigma 1},\ldots,X_{\sigma n}) = r(X_1,\ldots,X_n)$ for all $\sigma \in S_n$. The proof of Proposition 7.4.7 shows that r is symmetric if and only if $r \in K(\mathfrak{s}_1,\ldots,\mathfrak{s}_n)$. Thus:

Corollary 7.4.9. *A rational function of n indeterminates $X_1,\ldots,X_n$ is symmetric if and only if it is a rational function of the elementary symmetric polynomials $\mathfrak{s}_1(X_1,\ldots,X_n),\ldots,\mathfrak{s}_n(X_1,\ldots,X_n)$.*

Corollary 7.4.10. *Every finite group is isomorphic to a Galois group.*

Proof. Every group G of finite order n acts on itself by left multiplication and is thereby isomorphic to a subgroup of S_n. Hence G is isomorphic to a subgroup H of $\mathrm{Gal}(E/S)$, where $E = R(X_1,\ldots,X_n)$ is the Galois extension of $S = K(\mathfrak{s}_1,\ldots,\mathfrak{s}_n)$ in Proposition 7.4.7. If $F = \mathrm{Fix}_E(H)$ is the fixed field of H, then E is a Galois extension of H and $\mathrm{Gal}(E/F) = H \cong G$, by Theorem 7.4.3. $\square$

Exercises

1. Let K be a finite field of characteristic p. Show that K is a Galois extension of $\mathbb{Z}_p$ and that $\mathrm{Gal}(K/\mathbb{Z}_p)$ is cyclic.
2. Let K be a field of characteristic 0 and $\varepsilon \in \overline{K}$ be a root of unity ($\varepsilon^n = 1$ for some $n > 0$). Show that $K(\varepsilon)$ is Galois over K and that $\mathrm{Gal}(K(\varepsilon)/K)$ is abelian.
3. Use Corollary 7.4.10 to construct normal extensions $K \subseteq E$ and $E \subseteq F$ such that $K \subseteq F$ is not normal.

7.5. INFINITE GALOIS EXTENSIONS

Theorem 7.4.3 can be extended to any infinite Galois extension E of K if $\mathrm{Gal}(E/K)$ is provided with a suitable topology. This result is due to Krull.

7.5.a. The Finite Topology. The **finite topology** on $\mathrm{Gal}(E/K)$ has a basis which consists of all sets

$$\{\tau \in \mathrm{Gal}(E/K); \ \tau_{|S} = \sigma_{|S}\},$$

where $\sigma \in \mathrm{Gal}(E/K)$ and S is a finite subset of E.

Before extending Theorem 7.4.3, we show that this finite topology on $\mathrm{Gal}(E/K)$ is determined by certain subgroups of finite index.

Lemma 7.5.1. *The basis of the finite topology on $\mathrm{Gal}(E/K)$ consists of all left cosets of Galois groups $\mathrm{Gal}(E/F) \leqslant \mathrm{Gal}(E/K)$ of E over finite extensions $K \subseteq F \subseteq E$ of K.*

Proof. When S is a finite subset of E, then $K(S) \in E$ is a finite extension of K, E is Galois over $K(S)$, and $\tau_{|S} = \sigma_{|S}$ is equivalent to $\tau_{|K(S)} = \sigma_{|K(S)}$, to $\sigma^{-1}\tau \in \mathrm{Gal}\,(E/K(S))$, and to $\tau \in \sigma\,\mathrm{Gal}\,(E/K(S))$. If, conversely, $K \subseteq F \subseteq E$ is finite over K, then $F = K(S)$ for some finite set, and, as above, $\tau \in \sigma\,\mathrm{Gal}\,(E/F)$ if and only if $\tau_{|S} = \sigma_{|S}$. □

Lemma 7.5.2. *Let $K \subseteq E$ be a Galois extension and $K \subseteq F \subseteq E$ be an intermediate field.*

(1) $[\mathrm{Gal}\,(E/K) : \mathrm{Gal}\,(E/F)] = [F : K]$.

(2) *F is normal over K if and only if $\mathrm{Gal}\,(E/F)$ is normal in $\mathrm{Gal}\,(E/K)$.*

Proof. (1) Every K-homomorphism of F into $\overline{K}$ is the restriction to F of a K-homomorphism $\sigma \in \mathrm{Gal}\,(E/K)$ of E into $\overline{K}$. Now σ and $\tau \in \mathrm{Gal}\,(E/K)$ have the same restriction to F if and only if $\tau^{-1}\sigma \in \mathrm{Gal}\,(E/F)$, if and only if σ and τ are in the same left coset of $\mathrm{Gal}\,(E/F)$. Hence there are $[\mathrm{Gal}\,(E/K) : \mathrm{Gal}\,(E/F)]$ K-homomorphisms of F into $\overline{K}$, and $[F : K] = [\mathrm{Gal}\,(E/K) : \mathrm{Gal}\,(E/F)]$.

(2) If F is normal over K, then (as in the proof of Corollary 7.4.5) F is Galois over K and restriction to F provides a homomorphism $\mathrm{Gal}\,(E/K) \longrightarrow \mathrm{Gal}\,(F/K)$ whose kernel is $\mathrm{Gal}\,(E/F)$; hence $\mathrm{Gal}\,(E/F)$ is a normal subgroup of $\mathrm{Gal}\,(E/K)$.

Conversely, assume that $\mathrm{Gal}\,(E/F)$ is a normal subgroup of $\mathrm{Gal}\,(E/K)$. We may assume that $E \subseteq \overline{K}$. Let $\varphi : F \longrightarrow \overline{K}$ be a K-homomorphism. Then φ can be extended to $\sigma : E \longrightarrow \overline{K}$; since E is normal, $\sigma \in \mathrm{Gal}\,(E/K)$. Let $\alpha \in F$. For every $\tau \in \mathrm{Gal}\,(E/F)$, we have $\sigma^{-1}\tau\sigma \in \mathrm{Gal}\,(E/F)$, $\sigma^{-1}\tau\sigma\alpha = \alpha$, and $\tau\sigma\alpha = \sigma\alpha$. Hence $\varphi\alpha = \sigma\alpha \in F$ by Proposition 7.4.2. Thus F is normal over K. □

Let $\mathscr{F}$ denote the set of all Galois groups $\mathrm{Gal}\,(E/F) \subseteq \mathrm{Gal}\,(E/K)$ of finite extensions $K \subseteq F \subseteq E$ of K.

Lemma 7.5.3. *When $K \subseteq E$ is a Galois extension:*

(1) *every $H \in \mathscr{F}$ has finite index in $\mathrm{Gal}\,(E/K)$;*

(2) $\bigcap_{H \in \mathscr{F}} H = 1$;

(3) *$\mathscr{F}$ is closed under finite intersections;*

(4) *every subgroup $H \in \mathscr{F}$ contains a normal subgroup $N \in \mathscr{F}$.*

Proof. (1) follows from Lemma 7.5.2, part (1).

(2) Assume that $\sigma \in H$ for every $H \in \mathscr{F}$. If $\alpha \in E$, then $K(\alpha)$ is finite over K, $\sigma \in \mathrm{Gal}\,(E/K(\alpha)) \in \mathscr{F}$, and $\sigma\alpha = \alpha$. Thus $\sigma = 1$.

(3) If $H_1 = \mathrm{Gal}\,(E/F_1)$ and $H_2 = \mathrm{Gal}\,(E/F_2) \in \mathscr{F}$, where F_1 and F_2 are finite over K. Then $F = F_1 F_2$ (which is finitely generated over K) is finite over K. Now $\sigma\alpha = \alpha$ for all $\alpha \in F_1 F_2$ if and only if $\sigma\alpha = \alpha$ for all $\alpha \in F_1$ and all $\alpha \in F_2$; hence $\mathrm{Gal}\,(E/F) = H_1 \cap H_2$.

(4) Let $H = \mathrm{Gal}(E/F)$, where F is finite over K. By Corollary 7.2.4, F is contained in a finite normal extension $F \subseteq N \subseteq E$. Then $\mathrm{Gal}(E/N) \in \mathscr{F}$ is normal by Lemma 7.5.2 and $\mathrm{Gal}(E/N) \subseteq \mathrm{Gal}(E/F)$. $\square$

Lemma 7.5.3 implies that every Galois group has the following property: the identity is the intersection of subgroups of finite index. Thus not every group can be a Galois group. However, we saw in Section 7.4 that every finite group is a Galois group.

Proposition 7.5.4. *When E is a Galois extension of K, let:*

$\mathscr{N}$ be the set of all cosets of normal subgroups $N \in \mathscr{F}$,

$\mathscr{L}$ be the set of all left cosets of subgroups $H \in \mathscr{F}$,

$\mathscr{R}$ be the set of all right cosets of subgroups $H \in \mathscr{F}$.

Then $\mathscr{N}$, $\mathscr{L}$, and $\mathscr{R}$ are bases for the finite topology on $\mathrm{Gal}(E/K)$.

Proof. For $\mathscr{L}$ this follows from Lemma 7.5.1. If $A, B \in \mathscr{N}$ and $\sigma \in A \cap B$, then $A = M\sigma$, $B = N\sigma$, where $M, N \in \mathscr{F}$ are normal, and $A \cap B = M\sigma \cap N\sigma = (M \cap N)\sigma \in \mathscr{N}$ by Lemma 7.5.3. Thus $\mathscr{N}$ is a basis for a topology on $\mathrm{Gal}(E/K)$; and similarly for $\mathscr{R}$.

When $\sigma \in A \in \mathscr{L}$, then $\sigma \in N \subseteq A$ for some $N \in \mathscr{N}$ by Lemma 7.5.3. Conversely, $\sigma \in N \in \mathscr{N}$ implies $\sigma \in N \in \mathscr{L}$, since $\mathscr{N} \subseteq \mathscr{L}$. Therefore $\mathscr{L}$ and $\mathscr{N}$ are bases of the same topology on $\mathrm{Gal}(E/K)$. Similarly, $\mathscr{R}$ and $\mathscr{N}$ are bases of the same topology on $\mathrm{Gal}(E/K)$. This topology is the finite topology by Lemma 7.5.1. $\square$

Thus an open set in the finite topology on $G = \mathrm{Gal}(E/K)$ is a union of members of $\mathscr{N}$ (equivalently, a union of members of $\mathscr{L}$ or $\mathscr{R}$). Every subgroup $H \in \mathscr{F}$ has finite index by Lemma 7.5.3 and is both open ($H \in \mathscr{L}$) and closed ($G \backslash H$ is a finite union of left cosets of H). It is readily verified that the operations on G are continuous. If G is finite, then $\{1\} = \mathrm{Gal}(E/E)$ is open, and the finite topology on G is the discrete topology.

Proposition 7.5.5. *When E is a Galois extension of G, then $\mathrm{Gal}(E/K)$ (with the finite topology) is compact Hausdorff and totally disconnected.*

Proof. Compactness follows from Tychonoff's Theorem, but we give a direct proof. Let $G = \mathrm{Gal}(E/K)$. If $\sigma, \tau \in G$ and $\sigma \neq \tau$, then $\sigma^{-1}\tau \notin H$ for some $H \in \mathscr{F}$ (since $\bigcap_{H \in \mathscr{F}} H = 1$ by Lemma 7.5.3), so that $\tau \notin \sigma H$. Then τH, σH are disjoint neighborhoods of σ and τ; this shows that G is Hausdorff. Also τH and $G \backslash \tau H$ (which is a union of left cosets of H) are both open; since $\tau \in \tau H$ and $\sigma \in G \backslash \tau H$, this shows that G is totally disconnected.

To prove that G is compact, we show that every ultrafilter $\mathscr{U}$ on G converges to some $\sigma \in G$. To construct σ, we show that a mapping $\sigma : E \longrightarrow E$ is well defined by $\sigma\alpha = \tau\alpha$ whenever $F \subseteq E$ is finite over K, $\alpha \in F$, $H = \mathrm{Gal}(E/F)$, and $\tau H \in \mathscr{U}$. Each $\alpha \in E$ belongs to a finite extension $F \subseteq E$ of K (e.g., to $K(\alpha)$). Then $H = \mathrm{Gal}(E/F) \in \mathscr{F}$ has finite index and $G \in \mathscr{U}$ is the union of finitely many left cosets of H; since $\mathscr{U}$ is an ultrafilter, there is some left coset $\tau H \in \mathscr{U}$ and a value $\tau\alpha \in E$ of $\sigma\alpha$. Now let $F, F' \subseteq E$ be finite over K, $\alpha \in F \cap F'$, $H = \mathrm{Gal}(E/F)$, $H' = \mathrm{Gal}(E/F')$, and $\tau H, \tau'H' \in \mathscr{U}$. Then $\tau H \cap \tau'H' \in \mathscr{U}$ is not empty and contains some υ; then $\upsilon^{-1}\tau \in \mathrm{Gal}(E/F)$ and $\upsilon^{-1}\tau' \in \mathrm{Gal}(E/F')$, and $\tau\alpha = \upsilon\alpha = \tau'\alpha$, since $\alpha \in F \cap F'$. Hence σ is well defined.

If $\alpha, \beta \in E$, then $F = K(\alpha, \beta)$ is a finite extension of K, $H = \mathrm{Gal}(E/F) \in \mathscr{F}$, some $\tau H \in \mathscr{U}$, and $\sigma\alpha = \tau\alpha$, $\sigma\beta = \tau\beta$. Hence $\sigma(\alpha + \beta) = \sigma\alpha + \sigma\beta$ and $\sigma(\alpha\beta) = (\sigma\alpha)(\sigma\beta)$. Thus σ is a K-endomorphism of E. Then $\sigma E = E$ by Proposition 7.1.4 and $\sigma \in G$.

Let $H = \mathrm{Gal}(E/F) \in \mathscr{F}$ (where $F \subseteq E$ is finite over K). As above $\tau H \in \mathscr{U}$ for some $\tau \in G$. Then $\sigma\alpha = \tau\alpha$ for all $\alpha \in F$. Hence $\tau^{-1}\sigma \in \mathrm{Gal}(E/F) = H$ and $\sigma H = \tau H \in \mathscr{U}$. It follows that $\mathscr{U}$ contains every neighborhood of σ, and so converges to σ. □

7.5.b. Krull's Theorem

Proposition 7.5.6. *Let E be a Galois extension of K, H be a subgroup of $\mathrm{Gal}(E/K)$, and $F = \mathrm{Fix}_E(H)$ be the fixed field of H. Then E is a Galois extension of F and $\mathrm{Gal}(E/F)$ is the closure of H in the finite topology on $\mathrm{Gal}(E/K)$.*

Proof. First $\tau \in \overline{H}$ if and only if τU intersects H for all $U \in \mathscr{F}$.

Let $\tau \in \overline{H}$ and $\alpha \in F$. Then $F = K(\alpha)$ is finite over K and $U = \mathrm{Gal}(E/F) \in \mathscr{F}$. Hence τU contains some $\eta \in H$. Then $\eta = \tau\upsilon$ for some $\upsilon \in U$; since $\eta\alpha = \upsilon\alpha = \alpha$, it follows that $\tau\alpha = \alpha$. This shows $\overline{H} \subseteq \mathrm{Gal}(E/F)$.

Conversely, let $\tau \in \mathrm{Gal}(E/F)$. Let $U = \mathrm{Gal}(E/L) \in \mathscr{F}$, where $L \subseteq E$ is finite over K. By Theorem 7.2.10, $L = K(\alpha)$ for some $\alpha \in E$. By Proposition 7.1.7, there is a field $N \subseteq E$ which contains $F(\alpha)$ and is finite normal over F. In particular, N is a finite Galois extension of F. Restriction to N yields a homomorphism $\Phi : \mathrm{Gal}(E/F) \longrightarrow \mathrm{Gal}(N/F)$, $\Phi(\sigma) = \sigma|_N$. Now $F = \mathrm{Fix}_E(H)$ is the fixed field of H; hence the fixed field $\mathrm{Fix}_N(\Phi(H))$ of $\Phi(H)$ is F. By Theorem 7.4.3, $\mathrm{Gal}(N/F) = \Phi(H)$. Hence $\Phi(\tau) = \Phi(\eta)$ for some $\eta \in H$. Then $\tau|_N = \eta|_N$, $\tau|_L = \eta|_L$, and $\tau^{-1}\eta \in \mathrm{Gal}(E/L) = U$. Hence $\eta \in \tau U \cap H$. This shows $\tau \in \overline{H}$. □

Using Proposition 7.5.6 instead of Proposition 7.4.1, the proof of Theorem 7.4.3 yields:

Theorem 7.5.7 (Krull). *Let E be a Galois extension of K.*

If F is a subfield of E which contains K, then E is a Galois extension of F and F is the fixed field of $\mathrm{Gal}(E/F)$.

If H is a closed subgroup of $\mathrm{Gal}(E/K)$ in the finite topology, then the fixed field of H is a subfield of E which contains K and $H = \mathrm{Gal}(E/F)$.

This defines an order reversing, one-to-one correspondance between closed subgroups of $\mathrm{Gal}(E/K)$ *and intermediate fields* $K \subseteq F \subseteq E$.

If E is finite over K, then $\mathrm{Gal}(E/K)$ is finite, its finite topology is the discrete topology, every subgroup is closed, and Theorem 7.5.7 reduces to Theorem 7.4.3.

With Theorem 7.5.7, Corollaries 7.4.4 and 7.4.5 extend to arbitrary Galois extensions (if subgroups are replaced by closed subgroups); see exercises.

7.5.c. An Example. We conclude with McCarthy's example of a Galois group which contains uncountably many subgroups of finite index, of which only countably many are closed. This shows that a Galois group may have comparatively few closed subgroups and that subgroups of finite index need not be closed in the Krull topology.

Let $K = \mathbb{Q}$ and E be the composite of all extensions $\mathbb{Q}(\alpha) \subseteq \mathbb{C}$ of $\mathbb{Q}$ in which $\alpha^2 \in \mathbb{Q}$; equivalently, $E = \mathbb{Q}(R)$, where

$$R = \{ \sqrt{p};\ p \text{ is a prime number or } p = -1 \}.$$

Since $\mathrm{Irr}_{\mathbb{Q}}(\sqrt{p}) = X^2 - p$ has degree 2, each $\sqrt{p} \in R$ has only two conjugates in E, $\sqrt{p}$ and $-\sqrt{p}$, and we have $\sigma(\sqrt{p}) = \sqrt{p}$ or $\sigma(\sqrt{p}) = -\sqrt{p}$ for every automorphism $\sigma \in G = \mathrm{Gal}(E/\mathbb{Q})$. For each subset S of R, there is an automorphism $\sigma \in G$ such that $\sigma(\sqrt{p}) = -\sqrt{p}$ for all $\sqrt{p} \in S$ and $\sigma(\sqrt{p}) = \sqrt{p}$ for all $\sqrt{p} \in T \backslash S$. Hence $|G| \geqslant 2^{\aleph_0} > \aleph_0$. Also $\sigma^2 = 1$ for every $\sigma \in G$ (since $\sigma(\sigma(\sqrt{p})) = \sqrt{p}$ for every $\sqrt{p} \in R$). Therefore G is abelian and is a vector space over $\mathbb{Z}_2$. Let B be a basis of G over $\mathbb{Z}_2$. Then B is uncountable; otherwise, G would be countable. For each $\beta \in B$, $B \backslash \{\beta\}$ generates a subgroup of G of index 2. Therefore $G = \mathrm{Gal}(E/\mathbb{Q})$ contains uncountably many subgroups of finite index.

On the other hand, E is countable, since it is algebraic over $\mathbb{Q}$. If $F \subseteq E$ is finite over $\mathbb{Q}$, then $F = \mathbb{Q}(\alpha)$ for some $\alpha \in E$ by Proposition 7.2.7. Therefore there are only countably many subfields $F \subseteq E$ which are finite over K. By Theorem 7.5.7 (and Lemma 7.5.2), $\mathrm{Gal}(E/\mathbb{Q})$ has only countably many closed subgroups of finite index.

Exercises

1. Show that $\mathrm{Gal}(E/K)$ is closed in the finite topology on E^E.

2. Show that the multiplication $(\sigma, \tau) \longmapsto \sigma\tau$ and inverse $\sigma \longmapsto \sigma^{-1}$ are continuous in the finite topology on $\mathrm{Gal}(E/K)$.

3. Let G be a group. Show that the intersection of finitely many subgroups of G of finite index is a subgroup of finite index.

4. Let G be a group. Show that every subgroup of G of finite index contains a normal subgroup of G of finite index.

5. Let G be a group. Show that the identity is the intersection of subgroups of G of finite index if and and only if G can be embedded into (= is isomorphic to a subgroup of) a direct product of finite groups. (Such groups are called **profinite**.)

6. Extend Corollary 7.4.4 to arbitrary Galois extensions.

7. Extend Corollary 7.4.5 to arbitrary Galois extensions.

7.6. POLYNOMIALS

In this section we compute the Galois group of the splitting field of a polynomial of small degree. More examples of Galois groups will be seen in later sections.

When E is a splitting field of $f(X) \in K[X]$, the K-automorphisms of E permute the roots of f (Proposition 7.6.1). The resulting permutations indicate how the roots of f can be constructed from K. This idea will be explored further in Sections 7.9 and 7.10.

7.6.a. Permutations. The **Galois group** of a polynomial $f(X) \in K[X]$ over the field K is the group $\mathrm{Gal}(f/K) = \mathrm{Aut}_K(E)$ of K-automorphisms of a splitting field E of f. If E is separable over K (equivalently, if f is separable), then by the fundamental Theorem of Galois Theory (Theorem 7.4.3), the intermediate fields $K \subseteq L \subseteq E$ are determined by the subgroups of the finite group $\mathrm{Gal}(f/K)$.

In general, the fixed field F of $\mathrm{Gal}(f/K)$ is, by Proposition 7.3.4, a purely inseparable extension of K, and E is a finite Galois extension of F, whose Galois group is $\mathrm{Gal}(f/K)$. By Theorem 7.2.10, E is a simple extension of F. Hence E is the splitting field of a separable irreducible polynomial $q(X) \in F[X]$, and $\mathrm{Gal}(f/K) = \mathrm{Gal}(q/F)$ is the Galois group of a separable irreducible polynomial.

Proposition 7.6.1. *Let $f(X) \in K[X]$ be a polynomial with n distinct roots.*

(1) *Every $\tau \in \mathrm{Gal}(f/K)$ permutes the roots of f; hence $\mathrm{Gal}(f/K)$ is isomorphic to a subgroup G of S_n.*

(2) *If f is separable and irreducible, then n divides $|G|$ and G is a transitive subgroup of S_n.*

Proof. (1) Let E be a splitting field of f and $\alpha_1, \ldots, \alpha_n$ be the distinct roots of A in E. Let τ be a K-automorphism of E. Then $f(\alpha) = 0$ implies $f(\tau\alpha) = ({}^\tau f)(\tau\alpha) = \tau(f(\alpha)) = 0$. Hence τ permutes the roots of f. The restriction of τ to $\{\alpha_1, \ldots, \alpha_n\}$ induces a permutation $\sigma \in S_n$ such that $\tau\alpha_i = \alpha_{\sigma i}$ for all i. Since E is generated by $\alpha_1, \ldots, \alpha_n$, σ is uniquely determined by τ. This provides an injective homomorphism $\mathrm{Gal}(f/K) \longrightarrow S_n$.

(2) Let f be separable and irreducible. Then f has degree n and $f = \mathrm{Irr}_K(\alpha_i)$ (for any i). Hence $[K(\alpha_i):K] = n$. By Theorem 7.4.3, $\mathrm{Gal}(E/K)$ has a sub-

group $\mathrm{Gal}(E/K(\alpha_i))$ of index n; hence n divides $|G|$. For any i, j there is by Corollary 6.2.7 a K-automorphism τ of E such that $\tau\alpha_i = \alpha_j$; hence G is transitive (for any i, j there is some $\sigma \in G$ such that $\sigma i = j$). $\qquad\square$

When $f(X) \in K[X]$ is separable and irreducible, Proposition 7.6.1 implies the following. If $\deg f = 2$, then $\mathrm{Gal}(f/K) \cong S_2 \cong \mathbb{Z}_2$. If $\deg f = 3$, then either $\mathrm{Gal}(f/K) \cong S_3$ or $\mathrm{Gal}(f/K) \cong A_3 \cong \mathbb{Z}_3$.

7.6.b. Example. Let $f(X) = X^3 - 2 \in \mathbb{Q}[X]$; f is irreducible by Eisenstein's criterion (Proposition 5.4.9) and is separable since $\mathbb{Q}$ has characteristic 0. Let $E \subseteq \mathbb{C}$ be the splitting field of f.

The complex roots of f are $\rho = \sqrt[3]{2} \in \mathbb{R}$, $j\rho$, and $j^2\rho$, where

$$j = -\frac{1}{2} + i\frac{\sqrt{3}}{2} \in \mathbb{C}$$

is a primitive cube root of unity. Hence $E = \mathbb{Q}(\rho, j\rho, j^2\rho) = \mathbb{Q}(\rho, j)$, and E has an intermediate field $\mathbb{Q}(\rho) \subseteq \mathbb{R}$. We see that $[\mathbb{Q}(\rho) : \mathbb{Q}] = 3$ and $[E : \mathbb{Q}(\rho)] = 2$. Hence $[E : \mathbb{Q}] = 6$ and $G = \mathrm{Gal}(A/\mathbb{Q}) \cong S_3$.

We know that $S_3 \cong D_3$ is generated by the 3-cycle $(1\ 2\ 3)$ and the transposition $(2\ 3)$. Hence G is generated by γ and τ, where

$$\gamma\rho = j\rho, \qquad \gamma(j\rho) = j^2\rho, \qquad \gamma(j^2\rho) = \rho; \qquad \gamma j = j,$$

$$\tau\rho = \rho, \qquad \tau(j\rho) = j^2\rho, \qquad \tau(j^2\rho) = j\rho; \qquad \tau j = j^2,$$

and $G = \{1, \gamma, \gamma^2, \tau, \gamma\tau, \gamma^2\tau\}$. The subgroups of G are 1, $\{1, \tau\}$, $\{1, \gamma\tau\}$, $\{1, \gamma^2\tau\}$, $\{1, \gamma, \gamma^2\}$, and G. Therefore there are four fields $\mathbb{Q} \subsetneqq F \subsetneqq E$. The fixed field $F(\{1, \tau\})$ of $\{1, \tau\}$ has degree 3 over $\mathbb{Q}$, contains ρ (since $\tau\rho = \rho$), and contains $\mathbb{Q}(\rho)$, which also has degree 3; therefore $F(\{1, \tau\}) = \mathbb{Q}(\rho)$. Similarly $F(\{1, \gamma\tau\}) = \mathbb{Q}(j^2\rho)$ (since $\gamma\tau(j^2\rho) = j^2\rho$); $F(\{1, \gamma^2\tau\}) = \mathbb{Q}(j\rho)$ (since $\gamma^2\tau(j\rho) = j\rho$); and $F(\{1, \gamma, \gamma^2\}) = \mathbb{Q}(j)$ (since $\gamma j = j$), which is normal over $\mathbb{Q}$, since $\{1, \gamma, \gamma^2\} \lhd G$. Thus the fields $\mathbb{Q} \subsetneqq F \subsetneqq E$ are $\mathbb{Q}(\rho)$, $\mathbb{Q}(j^2\rho)$, $\mathbb{Q}(j\rho)$, and $\mathbb{Q}(j)$.

7.6.c. Polynomials of Degree 3. Let $f(X) \in K[X]$ be a separable polynomial of degree $n \geqslant 2$. Recall (Section 5.2) that the discriminant of f is

$$D = D(f) = a^{2n-2} \prod_{1 \leqslant i < j \leqslant n} (\alpha_i - \alpha_j)^2,$$

where a is the leading coefficient of A and $\alpha_1, \ldots, \alpha_n$ are the roots of f in the splitting field E of A (which we may assume is contained in $\overline{K}$). By Proposition 5.2.8, $D \in K$; $D \neq 0$, since f is separable.

In E, D is the square of

$$d = a^{n-1} \prod_{1 \le i < j \le n} (\alpha_i - \alpha_j).$$

If $\sigma \in \mathrm{Gal}(A/K)$ induces an even permutation of $\{1, 2, \ldots, n\}$, then $\sigma d = d$; otherwise, $\sigma d = -d$. If $d \in K$ (if D has a square root in K), then $\mathrm{Gal}(A/K)$ cannot induce odd permutations and is isomorphic to a subgroup of A_n. If $d \notin K$ (if D does not have a square root in K), then d is not in the fixed field of $\mathrm{Gal}(A/K)$, and $\mathrm{Gal}(A/K)$ must induce at least one odd permutation.

If $n = 3$, then $\mathrm{Gal}(f/K) \cong A_3$ or $\mathrm{Gal}(f/K) \cong S_3$, and we obtain:

Proposition 7.6.2. *Let $f(X) \in K[X]$ be a separable irreducible polynomial of degree 3. If $D(f)$ has a square root in K, then $\mathrm{Gal}(f/K) \cong A_3$; otherwise, $\mathrm{Gal}(f/K) \cong S_3$.*

We saw that the discriminant of $X^3 + pX + q$ is $-4p^3 - 27q^2$. For example, $f(X) = X^3 - 2 \in \mathbb{Q}[X]$ has discriminant $-27 \times 2^2 = -108$; -108 does not have a square root in $\mathbb{Q}$; therefore $\mathrm{Gal}(f/\mathbb{Q}) \cong S_3$.

Proposition 7.6.2 shows that the Galois group of f reflects the construction of the roots $\alpha_1, \alpha_2, \alpha_3$ of f. Cardano's formula shows that in general, $\alpha_1, \alpha_2, \alpha_3$ can be reached by first adjoining a square root of $D(f)$ to K (if necessary), and then adjoining cube roots of elements of $K(\sqrt{D(f)})$. If, for example, $f(X) = X^3 - 2 \in \mathbb{Q}[X]$, then the roots of f can be reached by adjoining first $\sqrt{-108}$ (equivalently, $\sqrt{-3}$), then $\sqrt[3]{2}$; this corresponds to the structure $\mathbb{Q} \subset \mathbb{Q}(j) \subset E$ of the splitting field, and to the structure $S_3 \supset A_3 \supset 1$ of the Galois group.

7.6.d. Polynomials of Degree 4. Let $f(X) = aX^4 + bX^3 + cX^2 + dX + e \in K[X]$ be a separable irreducible polynomial of degree 4 with splitting field E, which we may assume is contained in $\overline{K}$, and roots where $\alpha_1, \alpha_2, \alpha_3, \alpha_4 \in \overline{K}$. By Proposition 7.6.1, $\mathrm{Gal}(f/K)$ is isomorphic to a transitive subgroup G of S_4 of order 4, 8, 12, or 24.

When K does not have characteristic 2, we saw in Section 5.2 that $f(X)$ has a resolvant polynomial $r(X)$ of degree 3, which can be used to solve the equation $f(x) = 0$ in $\overline{K}$; namely $r(X) = (X - u)(X - v)(X - w)$, where

$$u = \left(\alpha_1 + \alpha_2 + \frac{b}{2a}\right)\left(\alpha_3 + \alpha_4 + \frac{b}{2a}\right),$$

$$v = \left(\alpha_1 + \alpha_3 + \frac{b}{2a}\right)\left(\alpha_2 + \alpha_4 + \frac{b}{2a}\right),$$

$$w = \left(\alpha_1 + \alpha_4 + \frac{b}{2a}\right)\left(\alpha_2 + \alpha_3 + \frac{b}{2a}\right),$$

We saw that $r(X) \in K[X]$, that $D(f) = a^6 D(r)$, in particular, r is separable and u, v, w are all distinct; and that $\alpha_1, \alpha_2, \alpha_3, \alpha_4 \in K(u', v', w')$, where u', v', w' are square roots of $-u, -v, -w$ such that $u'v'w'$ is a certain element of K.

Let $F = K(u, v, w) \subseteq \overline{K}$ be the splitting field of r. By the above, $E \subseteq F(u', v')$, and $[E : F] \leqslant 4$. On the other hand, F is Galois over K, and $[F : K] = |\mathrm{Gal}(F/K)| = |\mathrm{Gal}(r/K)|$ divides 6.

Proposition 7.6.3. *Let $f(X) \in K[X]$ be a separable irreducible polynomial of degree* 4, *where K does not have characteristic* 2; *let F be a splitting field of its resolvant.*

(1) $[F : K]$ *divides* 6.

(2) *If $[F : K] = 6$, then $\mathrm{Gal}(f/K) \cong S_4$.*

(3) *If $[F : K] = 3$, then $\mathrm{Gal}(f/K) \cong A_4$.*

(4) *If $[F : K] = 2$, then $\mathrm{Gal}(f/K) \cong D_4$ if f is irreducible over F, otherwise $\mathrm{Gal}(f/K) \cong \mathbb{Z}_4$.*

(5) *If $[F : K] = 1$, then $\mathrm{Gal}(f/K) \cong V_4$.*

Proof. As before, each $\tau \in \mathrm{Gal}(f/K)$ induces a permutation σ of $\{1, 2, 3, 4\}$ (such that $\tau\alpha_i = \alpha_{\sigma i}$), and $\mathrm{Gal}(f/K)$ is isomorphic to a subgroup G of S_4.

Let $V = \{1, (1\ 2)(3\ 4), (1\ 3)(2\ 4), (1\ 4)(2\ 3)\}$; V is a normal subgroup of S_4 and is isomorphic to V_4. The centralizer C of $(1\ 2)(3\ 4) \in V$ consists of all permutations σ such that either $\sigma\{1, 2\} = \{1, 2\}$, $\sigma\{3, 4\} = \{3, 4\}$, or $\sigma\{1, 2\} = \{3, 4\}$, $\sigma\{3, 4\} = \{1, 2\}$. Since $(1\ 2)(3\ 4)$ has three conjugates in S_4, C has order 8, and consists of $(1\ 2)$, $(3\ 4)$, $(1\ 3\ 2\ 4)$, $(1\ 4\ 3\ 2)$, and the elements of V, all of which commute with $(1\ 2)(3\ 4)$. As a group, $C \cong D_4$; also C is one of the three Sylow 2-subgroups of S_4. The centralizers of $(1\ 3)(2\ 4)$ and $(1\ 4)(2\ 3)$ are the remaining Sylow 2-subgroups and consist of similar permutations. We see that a permutation commutes with all the elements of V if and only if it belongs to V.

Each $\tau \in \mathrm{Gal}(f/K)$ also permutes u, v, and w. Let $\tau \in \mathrm{Gal}(f/K)$ induce $\sigma \in S_4$. If $\sigma \in V$, then the definition of u, v, w shows that $\tau u = u$, $\tau v = v$, and $\tau w = w$. If, conversely, u, v, and w are fixed under τ, then $\tau u = u \neq v, w$, and we must have $\sigma\{1, 2\} = \{1, 2\}$, $\sigma\{3, 4\} = \{3, 4\}$ or $\sigma\{1, 2\} = \{3, 4\}$, $\sigma\{3, 4\} = \{1, 2\}$; hence σ is in the centralizer of $(1\ 2)(3\ 4)$; similarly σ is in the centralizers of $(1\ 3)(2\ 4)$ and $(1\ 4)(2\ 3)$, and $\sigma \in V$. Therefore $\tau \in \mathrm{Gal}(E/F)$ if and only if $\sigma \in V$, and $\mathrm{Gal}(E/F) \cong G \cap V$. By Corollary 7.4.5, $\mathrm{Gal}(F/K) \cong G/G \cap V$.

We now turn to $G \cong \mathrm{Gal}(E/K)$ which, by Proposition 7.6.1, is a transitive subgroup of S_4 of order 4, 8, 12, or 24. We saw that $[E : F] \leqslant 4$ and $[F : K] \leqslant 6$. If $|G| = 24$, then $\mathrm{Gal}(f/K) \cong S_4$, $[E : K] = 24$, and $[F : K] = 6$.

If $|G| = 12$, then $G = A_4$, since A_4 is the only subgroup of S_4 of order 12. Hence $\mathrm{Gal}(f/K) \cong A_4$, $V \subseteq G$, $\mathrm{Gal}(F/K) = G/V$, and $[F : K] = 3$.

If $|G| = 8$, then G is one of the three Sylow 2-subgroups of S_4, and $\mathrm{Gal}(f/K) \cong D_4$; again $V \subseteq G$ and $[F : K] = 2$. Since V is transitive and $V \subseteq G$, there exists for any $i \neq j$ some $\tau \in \mathrm{Gal}(f/K)$ such that $\tau \alpha_i = \alpha_j$ and τ induces some $\sigma \in V$. Then $\tau \in \mathrm{Gal}(E/F)$, τ permutes the roots of $\mathrm{Irr}_F(\alpha_i)$ by Proposition 7.6.1, α_j is a root of $\mathrm{Irr}_F(\alpha_i)$, and $\mathrm{Irr}_F(\alpha_j) = \mathrm{Irr}_F(\alpha_i)$. Thus $\alpha_1, \alpha_2, \alpha_3, \alpha_4$ are roots of $\mathrm{Irr}_F(\alpha_1)$. This implies that f and $\mathrm{Irr}_F(\alpha_1)$ are proportional, and f remains irreducible when viewed as an element of $F[X]$.

If $|G| = 4$, then either $G = V$ or G is cyclic generated by a 4-cycle. If $G = V$, then $\mathrm{Gal}(f/K) \cong V_4$, $\mathrm{Gal}(F/K) = 1$, and $[F:K] = 1$. If $G \cong \mathbb{Z}_4$, then $|G \cap V| = 2$ and $[F:K] = 2$. In this case $G \cap V \cong \mathrm{Gal}(E/F)$ is not transitive, and Proposition 7.6.1 shows that f is not irreducible over F. □

Additional techniques for finding Galois groups will be seen in Section 11.6.

Exercises

1–10. Find the Galois groups of the following polynomials:

1. $X^3 - X - 1$, over $\mathbb{Q}$. Also find all intermediate fields of the splitting field.

2. $X^3 - X - 1$, over $\mathbb{Q}(\sqrt{-23})$.

3. $X^3 - 10$, over $\mathbb{Q}$. Also find all intermediate fields of the splitting field.

4. $X^3 - 10$, over $\mathbb{Q}(\sqrt{2})$.

5. $X^3 - 10$, over $\mathbb{Q}(\sqrt{-3})$.

6. $(X^2 - 2)(X^2 - 3)$, over $\mathbb{Q}$. Also find all intermediate fields of the splitting field.

7. $X^4 - 3$, over $\mathbb{Q}$. Also find all intermediate fields of the splitting field.

8. $X^4 - 3$, over $\mathbb{Q}(\sqrt{3})$.

9. $X^4 - 3$, over $\mathbb{Q}(\sqrt{-3})$.

10. $X^4 + X + 3$, over $\mathbb{Q}$.

7.7. CYCLOTOMY

This section provides a fine example of abstract algebra and number theory coming to each other's aid. We start with basic properties of cyclotomic fields and polynomials. These are used to prove a theorem of Dirichlet, that for any $n > 0$ there are infinitely many primes $p \equiv 1 \pmod{n}$. Then we can show that every finite abelian group is the Galois group of some extension of $\mathbb{Q}$. The results are not used later, except for part a, and for Proposition 7.7.7 which is used in Section 7.11.

7.7.a. Roots of Unity. An n-th **root of unity** in a field K is an element $\varepsilon \in K$ such that $\varepsilon^n = 1$. The n-th roots of unity form a multiplicative subgroup of K, which by Proposition 2.2.4 is cyclic. A n-th root of unity ε is a **primitive** n-th root of unity in case it has order n ($\varepsilon^k \neq 1$ if $0 < k < n$).

A **root of unity** in a field K is a n-th root of unity for some $n > 0$. If n is the order of ε (if n is the smallest integer $k > 0$ such that $\varepsilon^k = 1$), then ε is a primitive n-th root of unity. If K has characteristic $p \neq 0$, then $p \nmid n$, since $\varepsilon^{pk} = (\varepsilon^k)^p = 1$ implies $\varepsilon^k = 1$.

7.7.b. Cyclotomic Polynomials. If $K = \mathbb{C}$, then the n-th roots of unity in $\mathbb{C}$ are the complex numbers $\varepsilon_k = \cos 2k\pi/n + i \sin 2k\pi/n$; ε_k is primitive if and only if $\gcd(k,n) = 1$. Thus $\mathbb{C}$ has $\phi(n)$ primitive n-th roots of unity, where ϕ is Euler's ϕ function.

The n-th **cyclotomic polynomial** is

$$\Phi_n(X) = \prod_{\varepsilon \text{ primitive}} (X - \varepsilon) \in \mathbb{C}[X],$$

where the product runs over all primitive n-th roots of unity in $\mathbb{C}$. We see that Φ_n is separable and has degree $\phi(n)$.

We prove some basic properties of Φ_n.

Lemma 7.7.1. $X^n - 1 = \prod_{d|n} \Phi_d(X)$.

Proof. We have $X^n - 1 = \prod_\varepsilon (X - \varepsilon)$, where the product runs over all n-th roots of unity. Each n-th root of unity ε has an order d which is a divisor of n, and is then a primitive d-th root of unity. Classifying all terms $X - \varepsilon$ by the order of ε yields $X^n - 1 = \prod_{d|n} (\prod_{\varepsilon \text{ has order } d} (X - \varepsilon)) = \prod_{d|n} \Phi_d(X)$. □

The n-th roots of unity are arranged in a circle, and the cyclotomic polynomials derive their name (greek for "circle cutting") from the partition $X^n - 1 = \prod_{d|n} \Phi_d(X)$ of the n-th roots of unity into sets of roots of equal order. The formula $X^n - 1 = \prod_{d|n} \Phi_d(X)$ implies the property $n = \sum_{d|n} \phi(d)$ of ϕ.

By Lemma 7.7.1, $\Phi_n(X) = (X^n - 1)/\prod_{d|n, \, d<n} \Phi_d(X)$; this permits the recursive computation of $\Phi_n(X)$. By definition, $\Phi_1(X) = X - 1$; hence

$$\Phi_2(X) = (X^2 - 1)/(X - 1) = X + 1;$$

$$\Phi_3(X) = (X^3 - 1)/(X - 1) = X^2 + X + 1;$$

$$\Phi_6(X) = (X^6 - 1)/(X - 1)(X + 1)(X^2 + X + 1) = X^2 - X + 1;$$

and so forth. If p is prime, then

$$\Phi_p(X) = (X^p - 1)/(X - 1) = X^{p-1} + X^{p-2} + \cdots + X + 1.$$

In general:

Proposition 7.7.2. Φ_n is monic and has integer coefficients.

Proof. This holds if $n = 1$. For $n > 1$ the proof is by induction on n: $\Phi_n(X) = (X^n - 1)/\prod_{d|n, \, d<n} \Phi_d(X)$ is obtained by polynomial division of $X^n - 1 \in \mathbb{Z}[X]$ by the the monic polynomial $\prod_{d|n, \, d<n} \Phi_d(X) \in \mathbb{Z}[X]$. □

Proposition 7.7.3. Φ_n *is irreducible in* $\mathbb{Q}[X]$.

Proposition 7.7.3 implies $\Phi_n = \mathrm{Irr}_\mathbb{Q}(\varepsilon)$ when ε is a primitive n-th root of unity.

Proof. Assume that Φ_n is not irreducible in $\mathbb{Q}[X]$. Let $q(X) \in \mathbb{Z}[X]$ be an irreducible factor of Φ_n. Then $\Phi_n(X) = q(X)r(X)$ with $r(X) \in \mathbb{Z}[X]$ and $\deg q$, $\deg r > 1$. Since Φ_n is monic, the leading coefficients of q and r are ± 1, and we may arrange that q and r are monic.

Let ε and ζ be complex roots of q and r, respectively. Since q is irreducible in $\mathbb{Q}[X]$, $q = \mathrm{Irr}_\mathbb{Q}(\varepsilon)$. Also ε and ζ are primitive n-th roots of unity; hence $\zeta = \varepsilon^k$ for some $k > 0$. Choose ε and ζ so that k is as small as possible. Let p be a prime divisor of k. Then p does not divide n (otherwise, ζ would not have order n) and ε^p is a primitive n-th root of unity and a root of Φ_n. Also $(\varepsilon^p)^{k/p} = \zeta$; by the choice of ε and ζ, ε^p is not a root of q. Hence ε^p is a root of r. Since k is as small as possible, we have $k = p$. Furthermore $q(X)$ divides $r(X^p)$ in $\mathbb{Q}[X]$, since $q = \mathrm{Irr}_\mathbb{Q}(\varepsilon)$ and ε is a root of $r(X^p)$. Thus $r(X^p) = q(X)s(X)$ for some $s(X) \in \mathbb{Q}[X]$; since q is monic, polynomial division in $\mathbb{Z}[X]$ shows that $s \in \mathbb{Z}[X]$.

The projection $m \longmapsto \overline{m}$ of $\mathbb{Z}$ onto the field $\mathbb{Z}_p$ induces a homomorphism $f(X) \longmapsto \overline{f}(X)$ of $\mathbb{Z}[X]$ onto $\mathbb{Z}_p[X]$; if $f(X) = a_m X^m + \cdots + a_0$, then $\overline{f}(X) = \overline{a}_m X^m + \cdots + \overline{a}_0$. We have $\deg \overline{q}$, $\deg \overline{r} > 1$, since q and r are monic. Let $\overline{r}(X) = X^m + \overline{r}_{m-1} X^{m-1} + \cdots + \overline{r}_0$; since $\mathbb{Z}_p$ has characteristic p, we have

$$\overline{r}^p(X) = X^{mp} + \overline{r}_{m-1}^p X^{(m-1)p} + \cdots + \overline{r}_0^p = \overline{r(X^p)}.$$

Hence $\overline{q}$ divides $\overline{r}^p$ and $\overline{q}, \overline{r}$ have a common irreducible factor $\overline{h} \in \mathbb{Z}_p[X]$. Now $qr = \Phi_n$ divides $X^n - 1$ in $\mathbb{Z}[X]$; hence $\overline{q}\overline{r}$ divide $X^n - \overline{1}$ in $\mathbb{Z}_p[X]$ and $\overline{h}^2$ divides $X^n - \overline{1}$. Hence $X^n - \overline{1} \in \mathbb{Z}_p[X]$ is not separable. However, the derivative of $X^n - \overline{1}$ is $nX^{n-1} \neq 0$, since p does not divide n, and $X^n - \overline{1}$ cannot have multiple roots in an algebraic closure of $\mathbb{Z}_p$. This is the required contradiction. $\square$

7.7.c. Dirichlet's Theorem.

Dirichlet's Theorem states that for any relatively prime integers m and n, there are infinitely many primes $p \equiv m \pmod{n}$. Cyclotomic polynomials can be used to prove the following particular case:

Theorem 7.7.4 (Dirichlet). *For each positive integer n there are infinitely many prime numbers $p \equiv 1 \pmod{n}$.*

Theorem 7.7.4 follows from two lemmas.

Lemma 7.7.5. *Let $m, n \geqslant 2$ be integers. Then $|\Phi_n(m)| > m - 1$.*

Proof. $\Phi_n(m)$ is the product of $\phi(n)$ complex numbers of the form $m - \varepsilon$, where $|\varepsilon| = 1$, $\varepsilon \neq 1$. We have $|m - \varepsilon| > |m - 1| \geqslant 1$, and it follows that $|\Phi_n(m)| > (m - 1)^{\phi(n)} \geqslant m - 1$. $\qquad\qquad\square$

Lemma 7.7.6. *Let $m, n \geqslant 1$ be integers and p be a prime divisor of $\Phi_n(m)$. Then $p \nmid m$ and either $p \equiv 0$ or $p \equiv 1 \pmod{n}$.*

Proof. By Lemma 7.7.1, $\Phi_n(m)$ divides $m^n - 1$. Therefore p divides $m^n - 1$ and does not divide m. Let k be the order of $\overline{m}$ in the multiplicative group $\mathbb{Z}_p^*$ of nonzero elements of the field $\mathbb{Z}_p$; in particular, $m^k \equiv 1 \pmod{p}$. By Lagrange's Theorem, $k \mid p - 1$. Also $k \mid n$, since $\overline{m}^n = \overline{1}$. Let $n = k\ell$.

If $\ell = 1$, then $n = k \mid p - 1$ and $p \equiv 1 \pmod{n}$. Now assume that $\ell > 1$. Since $k \mid n$, every divisor of k is a divisor of n, and Lemma 7.7.1 yields

$$(X^k)^\ell - 1 = X^n - 1 = \Phi_n(X) \prod\nolimits_{d \mid n, \ d < n} \Phi_d(X)$$

$$= \Phi_n(X) \, (X^k - 1) \, f(X),$$

where $f(X)$ is a product of cyclotomic polynomials. Hence

$$\Phi_n(X) f(X) = ((X^k)^\ell - 1)/(X^k - 1)$$

$$= (X^k)^{\ell-1} + (X^k)^{\ell-2} + \cdots + X^k + 1,$$

so that $\Phi_n(m)$, and p, divide $m^{k\ell-k} + m^{k\ell-2k} + \cdots + m^k + 1$. Since $m^k \equiv 1 \pmod{p}$ this yields $\ell \equiv 0 \pmod{p}$. Hence $p \mid n = k\ell$. $\qquad\square$

Proof of Theorem 7.7.4. We may assume $n > 1$. For each $k \geqslant 1$ we have $\Phi_{kn}(kn) > 1$ by Lemma 7.7.5, and $\Phi_{kn}(kn)$ has a prime divisor p. By Lemma 7.7.6, $p \nmid kn$ and hence $p \equiv 1 \pmod{kn}$. In particular, $p > kn$ and $p \equiv 1 \pmod{n}$. Thus there are arbitrarily large primes $p \equiv 1 \pmod{n}$. $\qquad\square$

7.7.d. Cyclotomic Fields. The n-th **cyclotomic field** is $\mathbb{Q}_n = \mathbb{Q}(\varepsilon) \subseteq \mathbb{C}$, where ε is a primitive n-th root of unity.

Recall that an automorphism of the additive group $\mathbb{Z}_n$ has the form $\overline{x} \longmapsto \overline{k}\overline{x}$, where $\overline{k}$ is a unit of the ring $\mathbb{Z}_n$ ($\gcd(k, n) = 1$); hence the group of automorphisms $\mathrm{Aut}(\mathbb{Z}_n)$ of the cyclic group $\mathbb{Z}_n$ is isomorphic to the group of units U_n of the ring $\mathbb{Z}_n$. If p is a prime, then $\mathbb{Z}_p$ is a field and U_p is cyclic of order $p - 1$.

Proposition 7.7.7. $\mathbb{Q}_n$ *is a Galois extension of* $\mathbb{Q}$; $[\mathbb{Q}_n : \mathbb{Q}] = \phi(n)$ *and* $\mathrm{Gal}(\mathbb{Q}_n/\mathbb{Q}) \cong U_n$.

Proof. Let ε be a primitive n-th root of unity. We see that $\mathbb{Q}_n = \mathbb{Q}(\varepsilon)$ is a splitting field of $X^n - 1$ and is therefore Galois over $\mathbb{Q}$. By Proposition 7.7.3, $\mathrm{Irr}_{\mathbb{Q}}(\varepsilon) = \Phi_n$; hence $[\mathbb{Q}_n : \mathbb{Q}] = \deg \Phi_n = \phi(n)$.

Let $C \cong \mathbb{Z}_n$ be the multiplicative group of all n-th roots of unity ($=$ all powers of ε). Every $\sigma \in \mathrm{Gal}(\mathbb{Q}_n/\mathbb{Q})$ preserves products and permutes the roots of $X^n - 1$ (by Proposition 7.6.1); hence $\sigma|_C$ is an automorphism of C, and restriction to C provides a homomorphism $\Psi : \mathrm{Gal}(\mathbb{Q}_n/\mathbb{Q}) \longrightarrow \mathrm{Aut}(C) \cong U_n$. If $\sigma|_C$ is the identity, then $\sigma\varepsilon = \varepsilon$ and σ is the identity on $\mathbb{Q}_n$; hence Ψ is injective. Ψ is surjective since $|\mathrm{Gal}(\mathbb{Q}_n/\mathbb{Q})| = \phi(n) = |U_n|$. $\square$

Additional properties of $\mathbb{Q}_n$ are given in the exercises.

Proposition 7.7.8. *Every finite abelian group is the Galois group of some Galois extension of $\mathbb{Q}$.*

Proof. A finite abelian group G is a direct sum $G = C_1 \oplus C_2 \oplus \cdots \oplus C_k$ of cyclic groups $C_1, C_2, \ldots, C_k$ of orders $n_1, n_2, \ldots, n_k$. By Dirichlet's Theorem 7.7.4, there exists distinct primes $p_1, p_2, \ldots, p_k$ such that $p_i \equiv 1 \pmod{n_i}$ for all i. Let $n = p_1 p_2 \cdots p_k$. Then $\mathbb{Q}_n$ is a Galois extension of $\mathbb{Q}$ and $\mathrm{Gal}(\mathbb{Q}_n/\mathbb{Q}) \cong U_n$.

If $\gcd(r,s) = 1$, then $\mathbb{Z}_{rs} \cong \mathbb{Z}_r \oplus \mathbb{Z}_s$; since $(u,v) \in \mathbb{Z}_r \oplus \mathbb{Z}_s$ is a unit if and only if u and v are units, we have $U_{rs} \cong U_r \oplus U_s$. Therefore $\mathrm{Gal}(\mathbb{Q}_n/\mathbb{Q}) \cong U_{p_1} \oplus \cdots \oplus U_{p_k}$. Now U_{p_i} is cyclic of order $p_i - 1$; since n_i divides $p_i - 1$, U_{p_i} contains a subgroup H_i of index n_i. Then $U_{p_i}/H_i \cong C_i$, $(U_{p_1} \oplus \cdots \oplus U_{p_k})/(H_1 \oplus \cdots \oplus H_k) \cong G$, and $\mathrm{Gal}(\mathbb{Q}_n/\mathbb{Q})$ contains a subgroup H such that $\mathrm{Gal}(\mathbb{Q}_n/\mathbb{Q})/H \cong G$. Its fixed field $F = \mathrm{Fix}_{\mathbb{Q}_n}(H)$ is Galois over $\mathbb{Q}$ and $\mathrm{Gal}(F/\mathbb{Q}) \cong \mathrm{Gal}(\mathbb{Q}_n/\mathbb{Q})/H \cong G$. $\square$

Exercises

1. Find Φ_8, Φ_{12}, Φ_{18}.
2. Show that $\Phi_n(0) = \pm 1$, with $\Phi_n(0) = 1$ if $n > 1$ is odd.
3. Prove that $\Phi_{2n}(X) = \Phi_n(-X)$ whenever $n > 1$ is odd.
4. Let p be a prime. Show that $\Phi_{np}(X) = \Phi_n(X^p)$ if $p \mid n$; otherwise, $\Phi_{np}(X) = \Phi_n(X^p)/\Phi_n(X)$.
5. Let n be divisible by p^2 for some prime p. Prove that the sum of all the primitive n-th roots of unity is 0 in $\mathbb{C}$.
6. Show that $\mathbb{Q}_m \mathbb{Q}_n = \mathbb{Q}_{\mathrm{lcm}(m,n)}$.
7. Show that $\mathbb{Q}_m \cap \mathbb{Q}_n = \mathbb{Q}_{\gcd(m,n)}$. (Use the previous exercise and Proposition 7.4.6.)
8. Show that $\mathrm{Gal}(\mathbb{Q}_n/\mathbb{Q})$ need not be cyclic.

7.8. THE NORM AND TRACE

This section contains basic properties of the norm, trace, and K-homomorphisms of a finite extension E of K, which will be used in the next section.

7.8.a. Linear Independence of Homomorphisms

Proposition 7.8.1. *Let* E, F *be extensions of* K. *Distinct* K-*homomorphisms of* E *into* F *are linearly independent over* F.

Proof. Assume that there is an equality

$$\gamma_1 \sigma_1 + \cdots + \gamma_n \sigma_n = 0, \tag{7.1}$$

in which $\sigma_1, \ldots, \sigma_n : E \longrightarrow F$ are distinct and $\gamma_1, \ldots, \gamma_n \in F$ are not all 0. Choose (7.1) so that n is as small as possible. Then $\gamma_1, \ldots, \gamma_n \neq 0$ and $n \geqslant 2$, since $\sigma_1 \neq 0$. Since $\sigma_1 \neq \sigma_2$, there exists $\alpha \in E$ such that $\sigma_1 \alpha \neq \sigma_2 \alpha$. Then

$$\gamma_1 (\sigma_1 \alpha)(\sigma_1 \beta) + \cdots + \gamma_n (\sigma_n \alpha)(\sigma_n \beta) = \gamma_1 \sigma_1 (\alpha \beta) + \cdots + \gamma_n \sigma_n (\alpha \beta) = 0$$

for all $\beta \in E$, and

$$(\gamma_1 \sigma_1 \alpha) \sigma_1 + \cdots + (\gamma_n \sigma_n \alpha) \sigma_n = 0. \tag{7.2}$$

Multiplying (7.1) by $\sigma_1 \alpha$ and subtracting from (7.2) cancels the first term and yields

$$\gamma_2 (\sigma_2 \alpha - \sigma_1 \alpha) \sigma_2 + \cdots + \gamma_n (\sigma_n \alpha - \sigma_1 \alpha) \sigma_n = 0.$$

This is an equality of length less than n, in which the first coefficient is not 0, contradicting the choice of n. $\qquad\square$

7.8.b. The Norm and Trace.

Let V be a finite-dimensional vector space, and let $T : V \longrightarrow V$ be a linear transformation. Recall that the determinant and trace of T can be computed from its matrix M in any basis: the determinant of T is the determinant of M, and the trace of T is the sum of the diagonal entries of M. If

$$c(X) = \det (T - XI) = (-1)^n X^n + (-1)^{n-1} c_{n-1} X^{n-1} + \cdots + c_0$$

is the characteristic polynomial of T (and also of its matrix in any basis), then c_0 is the determinant of T and c_{n-1} is the trace of T.

We now let E be a finite extension of K. For each $\alpha \in E$, $T_\alpha : E \longrightarrow E$, $T_\alpha \gamma = \alpha \gamma$, is a K-linear transformation of E. The **norm** and **trace** of α, which we denote by $N_K^E(\alpha)$ and $\mathrm{Tr}_K^E(\alpha)$, or by $N(\alpha)$ and $\mathrm{Tr}(\alpha)$, are the determinant and trace of T_α:

$$N_K^E(\alpha) = \det T_\alpha, \qquad \mathrm{Tr}_K^E(\alpha) = \mathrm{tr} T_\alpha.$$

$N_K^E(\alpha)$ and $\mathrm{Tr}_K^E(\alpha)$ are elements of K.

For example, $\mathbb{C}$ is a finite extension of $\mathbb{R}$. When $z = a + bi \in \mathbb{C}$, then $T_z(x + iy) = (ax - by) + i(bx + ay)$. In the basis $\{1, i\}$, the matrix of T_z is

$$M = \begin{pmatrix} a & -b \\ b & a \end{pmatrix}.$$

Hence the norm of z is $\det M = a^2 + b^2 = z\overline{z}$, and the trace of z is $\operatorname{tr} M = 2a = z + \overline{z}$.

We show that the norm and trace of $\alpha \in E$ can be computed from the K-homomorphisms of E into $\overline{K}$, and from the conjugates of α.

Lemma 7.8.2. *Let E be a finite extension of K. Let $\alpha \in E$ and $q(X) = \operatorname{Irr}_K(\alpha)$. Then*

$$\det(T_\alpha - XI) = (-1)^n q^k(X),$$

where $n = [E : K]$ and $k = [E : K(\alpha)]$.

Proof. We have $T_{\alpha+\beta} = T_\alpha + T_\beta$, $T_{\alpha\beta} = T_\alpha T_\beta$ for all $\alpha, \beta \in E$. Hence $f(T_\beta) = T_{f(\beta)}$ for every $\beta \in E$ and $f(X) \in K[X]$. In particular, $q(T_\alpha) = T_{q(\alpha)} = 0$. (This implies that $q(X)$ is the minimal polynomial of T_α.)

Let $c(X) = \det(T_\alpha - XI)$ be the characteristic polynomial of T_α. In $K[X]$, $c(X)$ is the product of its leading coefficient $(-1)^n$ and of monic irreducible polynomials $r_1(X), \ldots, r_k(X)$. If $\lambda \in \overline{K}$ is a root of $r_j(X)$, then $c(\lambda) = 0$, λ is an eigenvalue of T_α, $T_\alpha \gamma = \lambda \gamma$ for some $\gamma \neq 0$, $T_\alpha^i \gamma = \lambda^i \gamma$ for all $i \geqslant 0$, $q(T_\alpha)\gamma = q(\lambda)\gamma$ for some $\gamma \neq 0$, and $q(\lambda) = 0$. Therefore $\operatorname{Irr}_K(\lambda) = r_j = q$. Hence $c(X) = (-1)^n q^k(X)$. The exponent k is found by comparing $\deg c = [E : K]$ and $\deg q = [K(\alpha) : K]$. $\square$

Proposition 7.8.3. *Let E be a finite extension of K of degree n. Let $\alpha \in E$, $q = \operatorname{Irr}_K(\alpha)$, $\alpha_1, \ldots, \alpha_r$ be the distinct roots of q in $\overline{K}$, and $\sigma_1, \ldots, \sigma_s$ be the distinct K-homomorphisms of E into $\overline{K}$. Then $r, s \mid n$, $n/s = [E : K]_i$, and*

$$N_K^E(\alpha) = (\alpha_1 \ldots \alpha_r)^{n/r} = ((\sigma_1 \alpha) \ldots (\sigma_s \alpha))^{n/s}$$

$$\operatorname{Tr}_K^E(\alpha) = \left(\frac{n}{r}\right)(\alpha_1 + \cdots + \alpha_r) = \left(\frac{n}{s}\right)(\sigma_1 \alpha + \cdots + \sigma_s \alpha)$$

Proof. By Lemma 7.2.5, all roots of q have the same order of multiplicity m. In the above, $rm = \deg q = [K(\alpha) : K]$ and $s = [E : K]_s$. Then

$$q(X) = (X - \alpha_1)^m \ldots (X - \alpha_r)^m$$

$$= X^{rm} - m(\alpha_1 + \cdots + \alpha_r)X^{rm-1} + \cdots + (-1)^{rm}(\alpha_1 \ldots \alpha_r)^m.$$

By Lemma 7.8.2, $c(X) = \det(T_\alpha - XI) = (-1)^n q^k(X)$, where $k = [E : K(\alpha)]$ $= n/rm$. The coefficient of X^{n-1} in $c(X)$ is $(-1)^{n-1} km(\alpha_1 + \cdots + \alpha_r)$; this yields the trace of α. The norm of α is the constant coefficient of $c(X)$, $(\alpha_1 \ldots \alpha_r)^{km}$.

By Proposition 7.2.1, there are r K-homomorphisms $\tau_1, \ldots, \tau_r$ of $K(\alpha)$ into $\overline{K}$, which can be numbered so that $\tau_i \alpha = \alpha_i$. Each τ_i can be extended to E in $t = [E : K(\alpha)]_s$ different ways. Therefore $s = tr$, $\sigma_1 \alpha + \cdots + \sigma_s \alpha = t(\alpha_1 + \cdots + \alpha_r)$, and $(\sigma_1 \alpha) \ldots (\sigma_s \alpha) = (\alpha_1 \ldots \alpha_r)^t$. $\qquad\square$

Proposition 7.8.3 is often used as definition of the norm and trace. The following consequences are clear:

Corollary 7.8.4. *If* $\alpha \in K$, *then* $N_K^E(\alpha) = \alpha^n$ *and* $\mathrm{Tr}_K^E = n\alpha$, *where* $n = [E : K]$. *If* E *is not separable over* K, *then* $\mathrm{Tr}_K^E(\alpha) = 0$ *for all* $\alpha \in E$. *If* E *is Galois over* K, *then*

$$N_K^E(\alpha) = \prod_{\sigma \in \mathrm{Gal}(E/K)} \sigma\alpha \qquad and \qquad \mathrm{Tr}_K^E(\alpha) = \sum_{\sigma \in \mathrm{Gal}(E/K)} \sigma\alpha.$$

7.8.c. Properties

Proposition 7.8.5. *When* E *is a finite extension of* K,

$$N_K^E(\alpha\beta) = N_K^E(\alpha) N_K^E(\beta) \qquad and \qquad \mathrm{Tr}_F^E(\alpha + \beta) = \mathrm{Tr}_K^E(\alpha) + \mathrm{Tr}_K^E(\beta)$$

for all $\alpha, \beta \in E$.

Proof. We have $T_{\alpha\beta} = T_\alpha T_\beta$ and $T_{\alpha+\beta} = T_\alpha + T_\beta$; hence $\det T_{\alpha\beta} = \det T_\alpha \det T_\beta$ and $\mathrm{tr}\, T_{\alpha+\beta} = \mathrm{tr}\, T_\alpha + \mathrm{tr}\, T_\beta$. (The result also follows from Proposition 7.8.3, since $\sigma_1, \ldots, \sigma_s$ are homomorphisms.) $\qquad\square$

Proposition 7.8.6. *If* $K \subseteq E \subseteq F$ *are finite extensions of* K, *then*

$$N_K^E(N_E^F(\alpha)) = N_K^F(\alpha) \qquad and \qquad \mathrm{Tr}_K^E(\mathrm{Tr}_E^F(\alpha)) = \mathrm{Tr}_K^F(\alpha)$$

for all $\alpha \in F$.

Proof. We may assume that $F \subseteq \overline{K}$. Let $n = [E : K]$, $\sigma_1, \ldots, \sigma_s$ be the distinct K-homomorphisms of E into $\overline{K}$, $m = [F : E]$, and $\tau_1, \ldots, \tau_t$ be the distinct E-homomorphisms of F into $\overline{K}$. Then $n/s = [E : K]_i$, $m/t = [F : E]_i$, and there are K-homomorphisms $\sigma_1', \ldots, \sigma_s'$ of F into $\overline{K}$ which extend $\sigma, \ldots, \sigma_s$. If $\tau : F \longrightarrow \overline{K}$ is a K-homomorphism, then $\tau|_E = \sigma_i$ for some i, $\sigma_i'^{-1}\tau : F \longrightarrow \overline{K}$ is an F-homomorphism, and $\tau = \sigma_i'\tau_j$ for some j. Thus the K-homomorphisms of F into $\overline{K}$ are the st distinct maps $\sigma_i'\tau_j$. Also $[F : K] = [F : E][E : K] = mn$

and $[F:K]_i = [F:E]_i [E:K]_i = mn/st$. By Proposition 7.8.3,

$$N_K^F(\alpha) = (\textstyle\prod_{i,j} \sigma_i' \tau_j \alpha)^{mn/st} = (\textstyle\prod_i \sigma_i' (\textstyle\prod_j \tau_j \alpha)^{m/t})^{n/s}$$
$$= (\textstyle\prod_i \sigma_i' N_E^F(\alpha))^{n/s} = (\textstyle\prod_i \sigma_i N_E^F(\alpha))^{n/s} = N_K^E(N_E^F(\alpha)),$$
$$\mathrm{Tr}_K^F(\alpha) = \left(\frac{mn}{st}\right) \textstyle\sum_{i,j} \sigma_i' \tau_j \alpha = \left(\frac{n}{s}\right) \textstyle\sum_i \sigma_i' \left(\frac{m}{t}\right) \textstyle\sum_j \tau_j \alpha$$
$$= \left(\frac{n}{s}\right) \textstyle\sum_i \sigma_i' \mathrm{Tr}_{F/E}(\alpha) = \left(\frac{n}{s}\right) \textstyle\sum_i \sigma_i \mathrm{Tr}_{F/E}(\alpha)$$
$$= \mathrm{Tr}_K^E(\mathrm{Tr}_E^F(\alpha)),$$

since $N_E^F(\alpha), \mathrm{Tr}_E^F(\alpha) \in E$. $\qquad\qquad\square$

7.9. CYCLIC EXTENSIONS

A **cyclic** extension is a Galois extension whose Galois group is cyclic. In this section we construct all cyclic extensions of K, when K has all its roots of unity. The results will be used in Section 7.10.

7.9.a. Hilbert's Theorem 90. First we prove:

Lemma 7.9.1 (Hilbert's Theorem 90, multiplicative form). *Let E be a cyclic extension of K, τ be a generator of $\mathrm{Gal}(E/K)$, and $\alpha \in E$. Then $N_K^E(\alpha) = 1$ if and only if $\alpha = \tau\gamma/\gamma$ for some $\gamma \in E$, $\gamma \neq 0$.*

Proof. When $\gamma \in E$, then $N(\tau\gamma/\gamma) = 1$: indeed Corollary 7.8.4 yields

$$N(\tau\gamma) = \textstyle\prod_{\sigma \in \mathrm{Gal}(E/K)} \sigma\tau\gamma = \textstyle\prod_{\sigma \in \mathrm{Gal}(E/K)} \sigma\gamma = N(\gamma).$$

Conversely, let $n = [E:K]$ so that $\mathrm{Gal}(E/K) = \{1, \tau, \tau^2, \ldots, \tau^{n-1}\}$. By Proposition 7.8.1,

$$1 + \alpha\tau + \alpha(\tau\alpha)\tau^2 + \cdots + \alpha(\tau\alpha)\ldots(\tau^{n-2}\alpha)\tau^{n-1} \neq 0,$$

and there exists $\beta \in E$ such that

$$\delta = \beta + \alpha(\tau\beta) + \alpha(\tau\alpha)(\tau^2\beta) + \cdots + \alpha(\tau\alpha)\ldots(\tau^{n-2}\alpha)(\tau^{n-1}\beta) \neq 0.$$

If $N(\alpha) = \alpha(\tau\alpha)(\tau^2\alpha)\ldots(\tau^{n-1}\alpha) = 1$, then $\alpha(\tau\delta) = \delta$, and $\alpha = \tau\gamma/\gamma$, where $\gamma = \delta^{-1} \neq 0$. $\qquad\square$

Lemma 7.9.2 (Hilbert's Theorem 90, additive form). *Let E be a cyclic extension of K, τ be a generator of $\mathrm{Gal}(E/K)$, and $\alpha \in E$. Then $\mathrm{Tr}_K^E(\alpha) = 0$ if and only if $\alpha = \tau\gamma - \gamma$ for some $\gamma \in E$.*

Proof. When $\gamma \in E$, then $\mathrm{Tr}(\tau\gamma - \gamma) = 0$: indeed Corollary 7.8.4 yields

$$\mathrm{Tr}(\tau\gamma) = \sum_{\sigma \in \mathrm{Gal}(E/K)} \sigma\tau\gamma = \sum_{\sigma \in \mathrm{Gal}(E/K)} \sigma\gamma = \mathrm{Tr}(\gamma).$$

Conversely, let $n = [E : K]$, so that $\mathrm{Gal}(E/K) = \{1, \tau, \tau^2, \ldots, \tau^{n-1}\}$, $\tau^n = 1$. By Proposition 7.8.1, $\mathrm{Tr}(\beta) \neq 0$ for some $\beta \in E$. Let

$$\delta = \alpha(\tau\beta) + (\alpha + \tau\alpha)(\tau^2\beta) + \cdots + (\alpha + \tau\alpha + \cdots + \tau^{n-2}\alpha)(\tau^{n-1}\beta).$$

If $\mathrm{Tr}(\alpha) = \alpha + \tau\alpha + \cdots + \tau^{n-1}\alpha = 0$, then

$$\tau\delta = (\tau\alpha)(\tau^2\beta) + \cdots + (\tau\alpha + \cdots + \tau^{n-2}\alpha)(\tau^{n-1}\beta) - \alpha\beta,$$

$$\delta - \tau\delta = \alpha(\tau\beta) + \alpha(\tau^2\beta) + \cdots + \alpha(\tau^{n-1})\beta + \alpha\beta = \alpha\,\mathrm{Tr}(\beta),$$

and $\alpha = \tau\gamma - \gamma$, where $\gamma = -\delta/\mathrm{Tr}(\beta)$. $\qquad\square$

7.9.b. The General Case. We now turn to cyclic extensions.

Proposition 7.9.3. *Let $n > 0$ and K be a field whose characteristic is either 0 or not a divisor of n, and which contains a primitive n-th root of unity.*

(1) *If E is a cyclic extension of K of degree n, then $E = K(\alpha)$ where $\alpha^n \in K$.*

(2) *If $E = K(\alpha)$ with $\alpha^n \in K$, then E is a cyclic extension of K whose degree k divides n, and $\alpha^k \in K$.*

Proof. Let $\varepsilon \in K$ be a primitive n-th root of unity.

(1) Let E be a cyclic extension of K of degree n and τ be a generator of $\mathrm{Gal}(E/K)$. Then $N(\varepsilon) = \varepsilon^n = 1$. By Lemma 7.9.1, $\tau\alpha = \varepsilon\alpha$ for some $\alpha \in E$, $\alpha \neq 0$. Then $\tau\alpha^n = (\tau\alpha)^n = \alpha^n$. Since τ generates $\mathrm{Gal}(E/K)$, α^n is fixed under every K-automorphism of E, and $\alpha^n \in K$. Also the conjugates of α are $\alpha, \varepsilon\alpha, \varepsilon^2\alpha, \ldots, \varepsilon^{n-1}\alpha \in K(\alpha)$. Since $K(\alpha)$ contains n conjugates of α, there are at least n K-homomorphisms of $K(\alpha)$ into $\overline{K}$ and $[K(\alpha) : K] \geqslant n$. But $K(\alpha) \subseteq E$; therefore $E = K(\alpha)$.

(2) Now let $E = K(\alpha)$, where $\alpha^n = a \in K$, and we may assume that $\alpha \neq 0$. Then α is a root of $X^n - a \in K[X]$, and the roots of $X^n - a$ are $\alpha, \varepsilon\alpha, \varepsilon^2\alpha, \ldots, \varepsilon^{n-1}\alpha \in E$. Thus $X^n - a$ is separable and E is its splitting field, and is Galois over K.

If $\sigma \in \mathrm{Gal}(E/K)$, then $\sigma\alpha$ is a root of $X^n - a$ and $\sigma\alpha = \varepsilon^i\alpha$ for some i. Also $\sigma\alpha = \tau\alpha$ implies $\sigma = \tau$, since $E = K(\alpha)$. Hence there is an injective homomorphism $\sigma \longmapsto \varepsilon^i$ of $\text{Gal}(E/F)$ into the multiplicative group of all n-th roots of unity. Since the latter is cyclic of order n, it follows that $\text{Gal}(E/K)$ is cyclic and that $k = |\text{Gal}(E/K)|$ divides n. If τ generates $\text{Gal}(E/K)$, and $\tau\alpha = \varepsilon^j\alpha$, then ε^j has order k and $\tau(\alpha^k) = (\tau\alpha)^k = \alpha^k$, which implies that $\alpha^k \in K$. $\qquad\square$

A primitive n-th root of unity is readily adjoined to K if necessary:

Proposition 7.9.4. *Let $\varepsilon \in \overline{K}$ be a root of unity. Then ε is a primitive n-th root of unity for some $n > 0$; if K has characteristic $p \neq 0$, then $p \nmid n$. Also $E = K(\varepsilon)$ is a Galois extension of K of degree n, and $\text{Gal}(E/K)$ is abelian.*

Proof. First, ε is a primitive n-th root of unity, where n is the order of ε (the smallest $k > 0$ such that $\varepsilon^k = 1$). Let K have characteristic $p \neq 0$. Let $\ell = n/p^k$, where p^k is the largest power of p which divides n. Then $p \nmid \ell$. Also $(\varepsilon^\ell)^{p^k} = 1$; since p-th roots are unique in K, this implies $\varepsilon^\ell = 1$. Hence $n = \ell$ is not divisible by p.

We see that $E = K(\varepsilon) \subseteq \overline{K}$ contains $1, \varepsilon, \ldots, \varepsilon^{n-1}$; hence E is the splitting field of the separable polynomial $X^n - 1 \in K[X]$ and is Galois over K.

Let $\sigma, \tau \in \text{Gal}(E/K)$. Then $\sigma\varepsilon$ is a root of $X^n - 1$; hence $\sigma\varepsilon = \varepsilon^i$ for some i. Similarly $\tau\varepsilon = \varepsilon^j$ for some j. Then $\sigma\tau\varepsilon = \sigma(\varepsilon^j) = \varepsilon^{ij} = \tau\sigma\varepsilon$. Since $E = K(\varepsilon)$, it follows that $\sigma\tau = \tau\sigma$. $\qquad\square$

In Proposition 7.9.4, $\text{Gal}(E/K)$ need not be cyclic (see exercises for Section 7.7).

7.9.c. Characteristic p. If K has characteristic $p \neq 0$, then there is no primitive n-th root of unity when p divides n, nor can one be adjoined to K. Furthermore $K(\alpha)$ is not separable when $\alpha^n \in K$. Nevertheless, a result very similar to Proposition 7.9.3 holds in this case.

Lemma 7.9.5. *Let K be a field of characteristic $p \neq 0$ and $a \in K$. If α is a root of $X^p - X - a$ in some extension of K, then the roots of $X^p - X - a$ are $\alpha, \alpha + 1, \ldots, \alpha + p - 1$. Hence either $X^p - X - a$ splits in K or $X^p - X - a$ is irreducible in $K[X]$.*

Proof. First $\overline{k}^p = \overline{k}$ holds for all $\overline{k} \in \mathbb{Z}_p$, since $\mathbb{Z}_p^*$ is a group of order $p - 1$. Hence $i^p = i$ holds for $i = 0, 1, \ldots, p - 1$ in any field of characteristic p. If α is a root of $X^p - X - a \in K[X]$, this implies $(\alpha + i)^p - (\alpha + i) = (\alpha^p + i^p) - (\alpha + i) = \alpha^p - \alpha$, so that $\alpha, \alpha + 1, \ldots, \alpha + p - 1$ are roots of $X^p - X - a$. If $X^p - X - a$ has a root in K, then all its roots are in K.

Assume that $X^p - X - a$ has no root in K, and let $f(X) \in K[X]$ be a nonconstant monic polynomial which divides $X^p - X - a$. In $\overline{K}$, $X^p - X - a$ has a factorization

$$X^p - X - a = \prod_{i=0,1,\ldots,p-1} (X - \alpha - i).$$

Hence f has a factorization

$$f(X) = \prod_{i \in S} (X - \alpha - i),$$

where $S \subseteq \{0, 1, \ldots, p - 1\}$. If $|S| = s$, then f has degree s and the coefficient of X^{s-1} in $f(X) \in K[X]$ is the sum of s terms $-(\alpha + i)$ and equals $-s\alpha + j \in K$ for some $j = 0, 1, \ldots, p - 1$. If $s \neq 0$ in K, this implies $\alpha \in K$; therefore $s = 0$ in K and f has degree p. Thus $X^p - X - a$ is irreducible.   □

Proposition 7.9.6 (Artin-Schreier). *Let K be a field of characteristic $p \neq 0$.*

(1) *If E is an extension of K of degree p, then $E = K(\alpha)$ where $\alpha^p - \alpha \in K$.*

(2) *If $E = K(\alpha)$ with $\alpha^p - \alpha \in K$, $\alpha \notin K$, then E is a cyclic extension of K of degree p.*

Proof. (1) An extension E of degree p is cyclic. Let τ be a generator of $\mathrm{Gal}(E/K)$. We have $\mathrm{Tr}(1) = p1 = 0$; by Lemma 7.9.2, $\tau\alpha = \alpha + 1$ for some $\alpha \in E$. Then $\tau^i\alpha = \alpha + i$ for all $i = 0, 1, \ldots, p - 1$, α has at least p conjugates, and $[K(\alpha) : K] \geqslant p$. Since $K(\alpha) \subseteq E$ this implies $E = K(\alpha)$. Also

$$\tau(\alpha^p - \alpha) = (\tau\alpha)^p - \tau\alpha = (\alpha + 1)^p - (\alpha + 1) = \alpha^p - \alpha.$$

Therefore $\alpha^p - \alpha$ is fixed under all τ^i, and $\alpha^p - \alpha \in K$.

(2) Conversely, let $E = K(\alpha)$, where $\alpha^p - \alpha = a \in K$, $\alpha \notin K$. By Lemma 7.9.5, $q(X) = X^p - X - a \in K[X]$ is irreducible, and its roots are α, $\alpha + 1, \ldots,$ $\alpha + p - 1$. Hence $q = \mathrm{Irr}_K(\alpha)$, $[E : K] = p$, and E, which is the splitting field of the separable polynomial q, is Galois over K. Furthermore there is a K-automorphism τ of E such that $\tau\alpha = \alpha + 1$. Then $\alpha, \tau\alpha, \ldots, \tau^{p-1}\alpha = \alpha, \alpha + 1, \ldots, \alpha + p - 1$ are all distinct; hence τ has order at least p and generates $\mathrm{Gal}(E/K)$.   □

7.10. SOLVABILITY BY RADICALS

We saw in Section 5.2 that the roots of a polynomial f of degree $\leqslant 4$ can generally be calculated from its coefficients by means of arithmetic operations and radicals (n-th roots). Restating this property in terms of the splitting field of f leads to a criterion for solvability by radicals and to Abel's Theorem, which states that the general equation of degree n is not solvable by radicals when $n \geqslant 5$.

7.10.a. Definition. We say that $\alpha \in \overline{K}$ is **radical** over K in case either $\alpha^n \in K$, where the characteristic of K does not divide n, or $\alpha^p - \alpha \in K$, if K has characteristic $p \neq 0$. (Separability considerations, and the results of Section 7.9, suggest $\alpha^p - \alpha \in K$ rather than $\alpha^p \in K$ when K has characteristic p.) A

radical extension of K is a simple extension $E = K(\alpha)$ where α is radical over K.

For example, $\mathbb{C} = \mathbb{R}(i)$ is a radical extension of $\mathbb{R}$.

An extension E of K is **solvable by radicals** in case there exist radical extensions

$$K = F_0 \subseteq F_1 \subseteq F_2 \subseteq \cdots \subseteq F_r$$

such that $E \subseteq F_r$ (so that every element of E can be reached by successive adjunctions of radical elements). A polynomial with coefficients in K is **solvable by radicals** over K in case its splitting field is solvable by radicals over K.

Thus we saw in Section 5.2 that the roots of $aX^2 + bX + c \in K[X]$ in $\overline{K}$ are

$$\frac{-b \pm \sqrt{b^2 - 4ac}}{2a}$$

(when K does not have characteristic 2). This formula shows that the splitting field of $aX^2 + bX + c$ is the radical extension $E = K(\sqrt{b^2 - 4ac})$ of K.

For a polynomial $X^3 + pX + q \in K[X]$ of degree 3 (where K does not have characteristic 2 or 3), Cardano's method puts the roots in the form $u' + v'$, where u' and v' are cube roots of $u, v \in F = K(\sqrt{q^2 + 4p^3/27})$. Hence the splitting field of $X^3 + pX + q$ is contained in $F(\sqrt[3]{u}, \sqrt[3]{v})$ and is solvable by radicals over K. Similar considerations apply to polynomials of degree 4 (see the exercises).

7.10.b. Properties. By Propositions 7.9.4, 7.9.3, and 7.9.6, radical extensions are separable. Therefore:

Proposition 7.10.1. *If E is solvable by radicals over K, then E is separable over K.*

Proposition 7.10.2.

(1) *If E is solvable by radicals over K and $K \subseteq F \subseteq E$, then F is solvable by radicals over K and E is solvable by radicals over F.*

(2) *If E is solvable by radicals over K and F is solvable by radicals over E, then F is solvable by radicals over K.*

(3) *If E is solvable by radicals over K and EF exists, then EF is solvable by radicals over F.*

The proofs make fine exercises for our tireless reader.

Theorem 7.10.3. *An extension E of a field K is solvable by radicals if and only if E is contained in a finite Galois extension of K whose Galois group is solvable.*

In fact solvable groups are named for this property.

Proof. The proof requires careful adjunction of roots of unity to K, so that we can use the results in Section 7.9. Let E be solvable by radicals over K. We may assume that $E \subseteq \overline{K}$ and that there are finite radical extensions

$$K = E_0 \subseteq E_1 \subseteq E_2 \subseteq \cdots \subseteq E_r = E,$$

with $E_{i+1} = E_i(\alpha_i)$, where α_i is radical over E_i.

Let $n = [E : K]$. If K has characteristic $p \neq 0$, let $m = n/p^k$, where p^k is the largest power of p which divides n; if K has characteristic 0, let $m = n$. Let $F = K(\varepsilon)$, where $\varepsilon \in \overline{K}$ is a primitive m-th root of unity. Then F contains a primitive ℓ-th root of unity $\varepsilon^{m/\ell}$ for each divisor ℓ of m, that is, for each divisor ℓ of n which is not divisible by the characteristic of K.

By Proposition 7.10.2, EF is solvable by radicals over F, and hence over K. In fact there are finite extensions

$$K \subseteq F = F_0 \subseteq F_1 \subseteq F_2 \subseteq \cdots \subseteq F_r = EF$$

such that each F_{i+1} is radical over F_i. We may let $F_{i+1} = F_i(\alpha_i)$, with α_i as above. Then $[F_{i+1} : F_i]$ divides $[E_{i+1} : E_i]$ and $[E : K]$. By the choice of F, F contains a primitive ℓ-th root of unity whenever ℓ divides $[E : K]$ and is not divisible by the characteristic of K. If K has characteristic $p \neq 0$, then, by Propositions 7.9.3 and 7.9.6, $[F_{i+1} : F_i]$ is either p or not divisible by p.

The smallest normal extension N of K which contains EF is the composite of the conjugates $\sigma(EF)$ of EF. There are finitely many K-homomorphisms $\sigma : EF \longrightarrow \overline{K}$, and each conjugate $\sigma(EF)$ is K-isomorphic to EF and is solvable by radicals over K. Then $N \subseteq \overline{K}$ is a Galois extension of K. We have $E \subseteq N$ and show that $\mathrm{Gal}(N/K)$ is solvable.

There are finite extensions

$$K \subseteq F = N_0 \subseteq N_1 \subseteq N_2 \subseteq \cdots \subseteq N_s = N$$

such that each N_{j+1} is radical over N_j, which may be constructed so that $N_{j+1} = N_j(\sigma\alpha_i)$ for some i and $\sigma : EF \longrightarrow \overline{K}$, with $\sigma E_i \subseteq N_j$. Then either $\sigma\alpha_i^{n_j} \in N_j$ where n_j divides $[E_{i+1} : E_i]$, so that the characteristic of K does not divide n_j and F contains a primitive n_j-th root of unity, or $(\sigma\alpha_i)^p - \sigma\alpha_i \in N_j$, where $p \neq 0$ is the characteristic of K. By Proposition 7.9.4, F is Galois over K, and $\mathrm{Gal}(F/K)$ is abelian. By Propositions 7.9.3 and 7.9.6, N_{i+1} is Galois over N_i, and $\mathrm{Gal}(N_{i+1}/N_i)$ is cyclic. We now have a normal series

$$1 = \mathrm{Gal}(N/N_s) \lhd \cdots \lhd \mathrm{Gal}(N/N_1) \lhd \mathrm{Gal}(N/N_0) \lhd \mathrm{Gal}(N/K)$$

whose factors (by Corollary 7.4.5) are isomorphic to the abelian groups

$$\mathrm{Gal}(K/F), \mathrm{Gal}(N_1/N_0), \ldots, \mathrm{Gal}(N_s/N_{s-1}).$$

Thus $\mathrm{Gal}(N/K)$ is solvable.

For the converse it suffices to prove that a finite Galois extension $E \subseteq \overline{K}$ of K with a solvable Galois group is solvable by radicals.

Again we first need to adjoin suitable roots of unity to K. Let $n = [E : K]$. If K has characteristic $p \neq 0$, let $m = n/p^k$, where p^k is the largest power of p which divides n; if K has characteristic 0, let $m = n$. Let $F = K(\varepsilon)$, where $\varepsilon \in \overline{K}$ is a primitive m-th root of unity. As before, $F \subseteq \overline{K}$ contains a primitive ℓ-th root of unity for each divisor ℓ of n which is not divisible by the characteristic of K. By Proposition 7.4.6, EF is a finite Galois extension of F, and $\mathrm{Gal}(EF/F) \cong \mathrm{Gal}(E/E \cap F)$ is a subgroup of $\mathrm{Gal}(E/K)$. Hence $[EF : F]$ divides $[E : K]$ and $\mathrm{Gal}(EF/F)$ is solvable. We show that EF is solvable by radicals over F; then $E \subseteq EF$ is solvable by radicals over K by Proposition 7.10.2.

Since $\mathrm{Gal}(EF/F)$ is solvable, it has a normal series

$$1 = H_0 \lhd H_1 \lhd \cdots \lhd H_{r-1} \lhd H_r = \mathrm{Gal}(EF/F)$$

whose factors H_{i+1}/H_i are cyclic of prime order. Hence there are finite extensions

$$F = F_r \subseteq F_{r-1} \subseteq \cdots \subseteq F_1 \subseteq F_0 = EF,$$

where F_i is the fixed field of H_i, so that F_i is a Galois extension of F_{i+1} and $\mathrm{Gal}(F_i/F_{i+1}) \cong H_{i+1}/H_i$ is cyclic of prime order q_i. Since q_i divides $[EF : F]$ and $[E : K]$, F contains a primitive q_i-th root of unity when q_i is not the characteristic of K. Hence Propositions 7.9.3 and 7.9.6 apply: every F_i is a radical extension of F_{i+1}, and EF is solvable by radicals over F. $\square$

7.10.c. Abel's Theorem. The **general polynomial** $f(X)$ of degree n over a field K is

$$f(X) = A_n X^n + A_{n-1} X^{n-1} + \cdots + A_0 \in K(A_0, \ldots, A_n)[X],$$

where $A_0, \ldots, A_n$ are indeterminates.

The general polynomial of degree 2 is solvable by radicals when K does not have characteristic 2. Indeed the formula

$$X_1, X_2 = \frac{-B \pm \sqrt{B^2 - 4AC}}{2A}$$

shows that the roots X_1, X_2 of the quadratic equation $AX^2 + BX + C = 0$ lie in an extension of $K(A, B, C)$ which is solvable by radicals. Similar results in

Section 5.2 show that the general polynomials of degree 3 and 4 are solvable by radicals if K does not have characteristic 2 or 3. Thus the solvability by radicals of the general polynomial of degree n expresses the idea that there is a *formula* which calculates the solutions of every equation of degree n (from its coefficients, using arithmetic operations and k-th roots).

We now find the Galois group of the general polynomial f of degree n. For this, we show that f can also be defined by its roots. Let

$$g(X) = A(X - X_1)(X - X_2)\ldots(X - X_n) \in K(A, X_1, \ldots, X_n)[X],$$

where $A, X_1, \ldots, X_n$ are indeterminates. Since

$$(X - X_1)(X - X_2)\ldots(X - X_n)$$
$$= X^n - \mathfrak{s}_1 X^{n-1} + \cdots + (-1)^k \mathfrak{s}_k X^{n-k} + \cdots + (-1)^n \mathfrak{s}_n,$$

the coefficients of $g(X) = a_n X^n + \cdots + a_1 X + a_0$ are $a_n = A$ and

$$a_k = (-1)^k A \mathfrak{s}_{n-k}(X_1, \ldots, X_n)$$

when $k < n$. Thus the coefficients of g lie in the field $K(A)(\mathfrak{s}_1, \ldots, \mathfrak{s}_n) \subseteq K(A)(X_1, \ldots, X_n)$ of symmetric rational fractions (with coefficients in $K(A)$). The splitting field of g is $K(A)(X_1, \ldots, X_n)$; by Proposition 7.4.7, the Galois group of g is isomorphic to the symmetric group S_n.

By Corollary 7.4.8, $a_0, \ldots, a_{n-1}$ are algebraically independent over $K(A)$. Hence there is an isomorphism $K(a_0, \ldots, a_n) \cong K(A_0, \ldots, A_n)$ which sends $g(X)$ to $f(X)$. Therefore f and g have isomorphic Galois groups, and:

Proposition 7.10.4. *The Galois group of the general polynomial of degree n is isomorphic to S_n.*

We saw in Chapter 3 that S_n is solvable if and only if $n \leqslant 4$ (since $A_n \lhd S_n$ is a nonabelian simple group when $n \geqslant 5$). Therefore we have proved:

Theorem 7.10.5 (Abel). *The general polynomial of degree n is solvable by radicals if and only if $n \leqslant 4$.*

Thus there is no formula which calculates the solutions of every equation $a_n x^n + \cdots + a_0 = 0$ of degree $n \geqslant 5$ from its coefficients (using arithmetic operations and k-th roots). This result holds in all characteristics.

Exercises

1. Let E be solvable by radicals over K and $K \subseteq F \subseteq E$; prove that F is solvable by radicals over K and that E is solvable by radicals over F.

2. Let E be solvable by radicals over K and F be solvable by radicals over E; prove that F is solvable by radicals over K.

3. Let E be solvable by radicals over K. If EF exists, prove that EF is solvable by radicals over KF.

4. When K does not have characteristic 2 or 3, show that every polynomial of degree 4 in $K[X]$ is solvable by radicals over K.

7.11. GEOMETRIC CONSTRUCTIONS

In this section we prove the impossibility of certain classical constructions in euclidean geometry. We show that trisection of angles, duplication of cubes, and quadrature of circles are not possible in general using straight edge and compass alone.

7.11.a. Constructibility. Constructibility by straight edge and compass refers to ancient procedures in plane euclidean geometry by which precise constructions of points and other figures are obtained using only a straight edge ruler and a compass to draw straight lines and circles.

Basic constructions by straight edge and compass include the following:

(1) Given two points A and B and a point C on a straight line L, one can construct a point D on L such that $CD = AB$.

(2) Given three points A, B, C, the point D such that $ABCD$ is a parallelogram can be constructed with a compass from the distances $CD = AB$ and $BD = AC$.

(3) The parallel to any straight line AB through any point C can be constructed by (2).

(4) The perpendicular bisector of a line segment AB can be constructed from the intersections of two circles of equal radius centered at A and B.

(5) The perpendicular to any straight line AB through any point C can be constructed by (4) and (3).

Given two points P and Q, points which are **constructible** (short for: constructible by straight edge and compass) from P and Q are defined recursively as follows, along with **constructible** straight lines and circles:

(a) P and Q are constructible.

(b) If two points $A \neq B$ are constructible, then the straight line AB, and the circle through A with center B, are constructible.

(c) If two lines (straight lines or circles) are constructible, then their intersections are constructible.

(d) Every constructible point or line can be obtained from P and Q by finitely many applications of (b) and (c).

For instance, if the point C and the straight line L are constructible, then the parallel to L through C is constructible by (3), and the perpendicular to L through C is constructible by (5).

7.11.b. Properties. The algebra of constructibility becomes clearer if a cartesian system of coordinates is used to represent every point in the euclidean plane by a pair of real numbers, or by a single complex number. This can be achieved so that P and Q are represented by 0 and 1. A complex number is **constructible** (from 0 and 1) in case the corresponding point is constructible from P and Q.

Lemma 7.11.1.

(1) *The set of all constructible complex numbers is a field.*

(2) *If z^2 is constructible, then z is constructible.*

(3) *If $z = x + iy$, with $x, y \in \mathbb{R}$, then z is constructible if and only if x and y are constructible.*

Proof. (1) We already know that 0 and 1 are constructible; we call the corresponding points O and I.

Let $a, b \in \mathbb{C}$ be constructible and A, B be the corresponding points.

The sum $a + b$ corresponds to the fourth point C of the parallelogram $OABC$ and therefore is constructible. The difference $a - b$ is constructed similarly from the parallelogram $ABOC$.

If $b \neq 0$, the product ab and quotient a/b correspond to points C and D such that the triangles OIA and OBC are similar, and the triangles OID and OBA are similar. To construct, say, D, we construct a point B' on the straight line OI such that $OB' = OB$; then, using the compass, a point A' such that the triangles OBA and $OB'A'$ are equal. Then D is the intersection of OA' and the parallel to $A'B'$ through I. The construction of C is similar. Thus ab and a/b are constructible.

(2) Let z be one of the complex square roots of b. Then z lies on the perpendicular bisector of BB' (where B' is as above). To construct $|z| = \sqrt{|b|}$, construct B'' on the straight line OI such that $IB'' = OB = |b|$; let D be at the intersection of the circle with diameter OB'' and the perpendicular to OB'' through I. Then $ID = \sqrt{|b|}$. Thus z is constructible, from the intersection of the circle of center O and radius $\sqrt{|b|}$ with the perpendicular bisector of BB'.

(3) The x- and y-axis are constructible. If a point A is constructible, then its perpendicular projections B and C onto the axes are constructible; if conversely B and C are constructible, then the perpendiculars to the axes through B and C are constructible and their intersection A is constructible. Hence $x + iy \in \mathbb{C}$ is constructible if and only if x and iy are constructible; y and iy can be constructed from each other with the circle of radius $|y|$ and center 0. $\square$

7.11.c. Main Result. With analytic geometry, intersections of straight lines and circles can be found by solving linear or quadratic equations. Hence constructible complex numbers are algebraic over $\mathbb{Q}$. More precisely:

Theorem 7.11.2. *A complex number is constructible if and only if it is contained in a finite normal extension $F \subseteq \mathbb{C}$ of $\mathbb{Q}$ whose degree is a power of 2.*

Proof. Call a complex number **2-constructible** in case it is contained in a finite normal extension $F \subseteq \mathbb{C}$ of $\mathbb{Q}$ whose degree is a power of 2. (It follows from Section 7.10 that 2-constructible numbers are those which can be reached from $\mathbb{Q}$ by successive adjunctions of square roots.)

First we show that 2-constructible numbers have properties (1), (2), and (3) in Lemma 7.11.1.

Lemma 7.11.3.

(0) *When $z_1, \ldots, z_n$ are 2-constructible, there exists a finite normal extension $F \subseteq \mathbb{C}$ of $\mathbb{Q}$ which contains $z_1, \ldots, z_n$ and whose degree is a power of 2.*

(1) *The set of all 2-constructible complex numbers is a field.*

(2) *If z^2 is 2-constructible, then z is 2-constructible.*

(3) *If $z = x + iy$, with $x, y \in \mathbb{R}$, then z is 2-constructible if and only if x and y are 2-constructible.*

Proof. (0) Let a and b be 2-constructible so that there exists finite normal extensions $E, F \subseteq \mathbb{C}$ of $\mathbb{Q}$ such that $a \in E$ and $b \in F$, whose degrees are powers of 2. By Proposition 7.1.6, the composite EF is a normal extension of $\mathbb{Q}$, which contains a and b, and $[EF : \mathbb{Q}]$ is a power of 2 by Proposition 7.4.6. Thus (0) holds if $n = 2$. The general case follows by induction.

(1) follows from (0).

(2) Let z^2 be 2-constructible so that $z^2 \in E$ for some normal extension $E \subseteq \mathbb{C}$ of $\mathbb{Q}$ whose degree is a power of 2. If $z \in E$, then z is 2-constructible. Otherwise, $[E(z) : E] = 2$ by Proposition 7.9.3. Let F be the composite of all the conjugates of $E(z)$ in $\mathbb{C}$. Then F is a normal extension of $\mathbb{Q}$. Also F is obtained from E by successive adjunctions of the finitely many conjugates of z; hence $[F : E]$ is a power of 2.

(3) By (2), i is 2-constructible. Let $z = x + iy$ be 2-constructible. By (0), z and i are contained in a finite normal extension $F \subseteq \mathbb{C}$ of $\mathbb{Q}$ whose degree is a power of 2. Let $\sigma : \mathbb{C} \longrightarrow \mathbb{C}$ be complex conjugation. Since F is a normal extension of $\mathbb{Q}$, we have $\sigma F \subseteq F$ and $x - iy = \overline{z} \in F$. Therefore $x = (z + \overline{z})/2 \in F$ and $y = (\overline{z} - z)/2i \in F$ are 2-constructible.

If, conversely, $x, y \in \mathbb{R}$ are 2-constructible, then x, y, and i are contained in a finite normal extension $F \subseteq \mathbb{C}$ of $\mathbb{Q}$ whose degree is a power of 2, and so is $z = x + iy$.   □

We now complete the proof of Theorem 7.11.2. To prove that a constructible complex number z is 2-constructible, we use induction on the number of times that part (c) of the definition is used in the construction of z. We already know that 0 and 1 are 2-constructible.

Let the point P which corresponds to z be constructed as the intersection of two lines (straight lines or circles) obtained, using part (b) of the definition, from simpler points A, B, C, D which by the induction hypothesis correspond to 2-constructible complex numbers. By Lemma 7.11.3 there is a finite normal extension $F \subseteq \mathbb{C}$ of $\mathbb{Q}$ which contains all the coordinates of A, B, C, D and whose degree is a power of 2. Then the equation

$$(x_B - x_A)(y - y_A) = (y_B - y_A)(x - x_A)$$

of the straight line AB has coefficients in F; so does the equation

$$(x - x_B)^2 + (y - y_B)^2 = (x_A - x_B)^2 + (y_A - y_B)^2$$

of the circle through A with center B. Now P is the intersection of two such straight lines or circles. The intersection of two straight lines with coefficients in F has coordinates in F. The intersection of two circles

$$x^2 + y^2 + a_1 x + b_1 y + c_1 = 0 = x^2 + y^2 + a_2 x + b_2 y + c_2$$

with coefficients in F is also the intersection of either circle with the straight line $(a_1 - a_2)x + (b_1 - b_2)y + (c_1 - c_2) = 0$, which has coefficients in F. The intersection of a straight line $y = ax + b$ and circle $x^2 + y^2 + cx + dy + f = 0$ with coefficients in F is determined by a quadratic equation

$$x^2 + (ax + b)^2 + cx + d(ax + b) + f = 0 \tag{E}$$

whose coefficients and discriminant δ are in F, hence are 2-constructible. By Lemma 7.11.3, $\sqrt{\delta}$ is 2-constructible; hence the solutions of (E) are 2-constructible, and the corresponding intersection points have 2-constructible coordinates. In either case our point P is 2-constructible.

Now we show that conversely, a 2-constructible complex number z is constructible. There is a finite normal extension $F \subseteq \mathbb{C}$ of $\mathbb{Q}$ which contains z, whose degree is a power of 2. By Propositions 3.7.6 and 3.7.3, the Galois group of F has a normal series

$$1 = G_0 \lhd G_1 \lhd \cdots \lhd G_n = \mathrm{Gal}(F/\mathbb{Q})$$

whose factors have order 2. Therefore there exists intermediate fields

$$\mathbb{Q} = F_0 \subseteq F_1 \subseteq F_2 \subseteq \cdots \subseteq F_n = F$$

such that $[F_{i+1} : F_i] = 2$ for all i. By Proposition 7.9.3, $F_{i+1} = F_i(c_i)$, where $c_i^2 \in F_i$. If every element of F_i is constructible, then Lemma 7.11.1 shows that c_i is constructible, so that every element of F_{i+1} is constructible. By induction, every element of F_n is constructible, including z. $\qquad \Box$

7.11.d. Examples. With Theorem 7.11.2 we can show that three classical problems in euclidean geometry, the trisection of angles, the duplication of cubes, and the quadrature of circles, have no solution.

Corollary 7.11.4. *There is no construction by straight edge and compass that can trisect arbitrary angles.*

Proof. This construction would divide any angle into three equal parts. We show that a $\pi/3$ angle cannot be so divided using only straight edge and compass. Otherwise, the complex number $\varepsilon = e^{i\pi/9}$ would be constructible. But ε is a primitive 18-th root of unity, and we saw (Proposition 7.7.7) that $\mathbb{Q}(\varepsilon) = \mathbb{Q}_{18}$ is Galois over $\mathbb{Q}$ of degree $\phi(18) = 6$. Therefore there cannot be a finite normal extension of $\mathbb{Q}$ which contains ε and whose degree is a power of 2. $\qquad \Box$

In the same spirit, eager readers will show:

Corollary 7.11.5. *There is no construction by straight edge and compass that can duplicate arbitrary cubes (construct a cube whose volume is twice the volume of a given cube).*

Corollary 7.11.6. *There is no construction by straight edge and compass that can square arbitrary circles (construct a square whose area is that of a given circle).*

7.11.e. Regular Polygons. A **Fermat prime** is a prime of the form $2^{2^k} + 1$. It is known that $2^{2^0} + 1 = 3$, $2^{2^1} + 1 = 5$, $2^{2^2} + 1 = 17$, $2^{2^3} + 1 = 257$, and $2^{2^4} + 1 = 65537$ are prime. No other Fermat primes have been discovered.

Lemma 7.11.7. *If $2^k + 1$ is prime, then k is a power of 2.*

Proof. If k is not a power of 2, then $k = ij$, where i is a power of 2 and j is odd. Since j is odd,

$$X^j + 1 = (X + 1)(X^{j-1} - X^{j-2} + \cdots - X + 1).$$

Substituting $X = 2^i$ yields a factorization of $2^k + 1$. $\qquad \Box$

Proposition 7.11.8 (Gauss). *A regular polygon with n sides can be constructed by straight edge and compass from its radius if and only if n is the product of a power of 2 and distinct Fermat primes.*

Proof. Let ε_n be a primitive n-th root of unity. By Proposition 7.7.7, $\mathbb{Q}_n = \mathbb{Q}(\varepsilon_n)$ is a Galois extension of $\mathbb{Q}$, with degree $\phi(n)$. A regular polygon with n sides can be constructed by straight edge and compass from its radius (from its center and one vertex) if and only if ε_n is constructible. Now ε_n is constructible if and only if $[\mathbb{Q}(\varepsilon_n) : \mathbb{Q}]$ is a power of 2, if and only if $\phi(n)$ is a power of 2.

Let $n = 2^k p_1^{k_1} \ldots p_r^{k_r}$, where $p_1, \ldots, p_r$ are distinct odd primes. By Corollary 2.2.12,

$$\phi(n) = 2^{k-1} \, p_1^{k_1 - 1}(p_1 - 1) \ldots p_r^{k_r - 1}(p_r - 1)$$

We see that $\phi(n)$ is a power of 2 if and only if $k_1 = \cdots = k_r = 1$ and $p_1, \ldots, p_r$ are primes of the form $2^k + 1$, that is, by Lemma 7.11.7, Fermat primes. $\quad\square$

Exercises

1. Give a construction by straight edge and compass that duplicates arbitrary squares (construct a square whose area is twice the area of a given square).

2. Prove that there is no construction by straight edge and compass that can duplicate arbitrary cubes (construct a cube whose volume is twice the volume of a given cube).

3. Prove that there is no construction by straight edge and compass that can square arbitrary circles (construct a square whose area is that of a given circle).

8

ORDERED FIELDS

This chapter studies how compatible order relations affect fields. The main results, from a 1926 paper by Artin and Schreier, give new abstract views of real and complex numbers, and a more algebraic proof that $\mathbb{C}$ is algebraically closed.

Required results: Chapter 7, Sections 7.1, 7.2, and 7.4 (Galois theory) and Proposition 7.9.3 and 7.9.6 (cyclic extensions).

8.1. ORDERED FIELDS

This section contains the first properties of ordered fields and the characterization of $\mathbb{R}$ as the largest archimedean ordered field.

8.1.a. Definition. An **ordered field** (short for **totally ordered field**) is a field F together with a total order relation $\leqslant$ such that:

(1) $x \leqslant y$ implies $x + z \leqslant y + z$;
(2) if $z \geqslant 0$, then $x \leqslant y$ implies $xz \leqslant yz$.

$\mathbb{R}$ and $\mathbb{Q}$ are ordered fields; other examples will be found in the exercises. We note some consequences of the axioms, whose proofs are exercises:

(3) F has characteristic 0.
(4) $x > 0 \iff -x < 0$.
(5) If $z \leqslant 0$, then $x \leqslant y$ implies $xz \geqslant yz$.
(6) $x^2 > 0$ for all $x \neq 0$.
(7) $x > y > 0$ implies $0 < y^{-1} < x^{-1}$.

Proposition 8.1.1. *A field F is ordered, and $P \subseteq F$ is its set of positive elements, if and only if P is closed under addition and multiplication and F is the disjoint union $F = P \cup \{0\} \cup (-P)$, where $-P = \{-x; x \in P\}$.*

Proof. Let F be ordered, $P = \{x \in F ; x > 0\}$ be its set of positive elements, and $P' = \{x \in F ; x < 0\}$. Then P is closed under addition and multiplication and F is the disjoint union $F = P \cup \{0\} \cup P'$. By (4), $P' = -P$.

Conversely assume that F is a disjoint union $F = P \cup \{0\} \cup (-P)$, with P closed under addition and multiplication. Let $x < y$ in F if and only if $y - x \in P$. It is immediate that $\leqslant$ is a total order relation under which F is an ordered field. Note that $P = \{x \in F ; x > 0\}$.   □

In an ordered field F the absolute value $|x|$ of $x \in F$ is defined in the usual way: $|x| = x$ if $x \geqslant 0$, $|x| = -x \geqslant 0$ if $x \leqslant 0$. This absolute value has values in F and is similar but not related to the valuations and absolute values in the next chapter (unless $F \subseteq \mathbb{R}$). We see that $|xy| = |x||y|$ and $|x + y| \leqslant |x| + |y|$ for all $x, y \in F$. The absolute value on F is used to supplement our beloved algebra with an occasional argument from analysis.

8.1.b. Orderability

Proposition 8.1.2. *A field F can be ordered if and only if -1 is not the sum of squares of elements of F.*

Proof. In an ordered field, (6) shows that $1 = 1^2 > 0$ and that sums of squares are nonnegative; then $-1 < 0$ by (4), and -1 is not a sum of squares.

Conversely, let F be a field and S be the set of all nonempty sums of nonzero squares in F. Assume that $-1 \notin S$. Then $0 \notin S$, since $x_1^2 + \cdots + x_n^2 = 0$ (with $x_1, \ldots, x_n \neq 0$) would imply $n > 1$, $-x_1^2 = x_2^2 + \cdots + x_n^2$, and $-1 = (x_2/x_1)^2 + \cdots + (x_n/x_1)^2 \in S$. Also S is closed under addition, and is closed under multiplication since $(\sum_i x_i^2)(\sum_j y_j^2) = \sum_{i,j} (x_i y_j)^2$. In fact S is a multiplicative subgroup of $F^* = F \backslash \{0\}$, since $1 \in S$ and $x = \sum_i x_i^2 \in S$ implies $x^{-1} = x/x^2 = \sum_i (x_i/x)^2 \in S$.

By Zorn's Lemma, there is a subset M of F which contains S, is closed under addition, is a multiplicative subgroup of F^*, and is maximal with these properties. Then $0 \notin M$, $0 \notin -M$, and $M \cap (-M) = \emptyset$ (since M is closed under addition, $a \in M \cap (-M)$ would imply $0 = a - a \in M$). We show that $F = M \cup \{0\} \cup (-M)$; then F can be ordered by Proposition 8.1.1.

Let $a \in F$. Assume that $a \neq 0$, $-a \notin M$. If $x, y \in M$, then $x + ay \neq 0$ (since $a \neq -x/y$). If $x, y \in M \cup \{0\}$, then $x + ay = 0$ implies $x = y = 0$. Let

$$T = \{x + ay ; x, y \in M \cup \{0\} \text{ and } x, y \text{ are not both } 0\}.$$

Then $0 \notin T$; T contains a, contains M and $S \subseteq M$, and is closed under addition. If $u + av$, $x + ay \in T$ (with $u, v, x, y \in M \cup \{0\}$), then $(u + av)(x + ay) \neq 0$ and $(u + av)(x + ay) = (ux + a^2 vy) + a(uy + vx)$ with $ux + a^2 vy$, $uy + vx \in M \cup \{0\}$ (since $a^2 \in M$); thus T is closed under multiplication. Also $1 \in T$ and $t = x + ay \in T$ implies that $(x + ay)^{-1} = (x + ay)/t^2 = (x/t^2) + a(y/t^2) \in T$ (since $t^2 \in M$). Thus T is a multiplicative subgroup of F^*. Since M is maximal, we must have $T = M$; hence $a \in M$.   □

It follows from Theorem 8.1.2 that $\mathbb{C}$ cannot be ordered. Orderable fields are also called **formally real** fields and are studied further in the next section.

8.1.c. Archimedean Fields. An element x of an ordered field F is **infinite** in case $x > n$ for every integer $n = n1$ of F. An ordered field is **archimedean** in case it has no infinite element; equivalently, if for each $x \in F$ there is an integer $n > x$; equivalently, if for each $x \in F$, $x > 0$ there is an integer $n > 0$ such that $1/n < x$. For example, $\mathbb{Q}$ and $\mathbb{R}$ are archimedean. The exercises show that not every ordered field is archimedean.

Our next result characterizes $\mathbb{R}$ as the largest archimedean ordered field.

Theorem 8.1.3. *An ordered field is archimedean if and only if it is isomorphic (as an ordered field) to a subfield of $\mathbb{R}$.*

Proof. If $F \subseteq \mathbb{R}$, then an infinite element of F would be infinite in $\mathbb{R}$; therefore F is archimedean.

Conversely, let F be archimedean. Since F has characteristic 0, we may assume that $\mathbb{Q}$ is a subfield of F. Since $1 > 0$ in F, positive integers are positive in F, and so are positive rationals, so that $\mathbb{Q}$ is an ordered subfield of F (when $q, r \in \mathbb{Q}$, then $q < r$ in $\mathbb{Q}$ implies $q < r$ in F; since $\mathbb{Q}$ and F are totally ordered, $q < r$ in $\mathbb{Q}$ if and only if $q < r$ in F). Moreover $\mathbb{Q}$ is "dense" in F:

Lemma 8.1.4. *If $x < y$ in F, then $x < r < y$ for some $r \in \mathbb{Q}$.*

Proof. Since F is archimedean, there exist integers k, ℓ and $m > 0$ such that $-x < -k$ (then $x > k$), $y < \ell$, and $y - x > 1/m$. Then $k < x < y < \ell$ and $k + i/m \leqslant x$ holds if $i = 0$ but not if $i \geqslant m(\ell - k)$. Hence there is a greatest integer i such that $k + i/m \leqslant x$. Then

$$x < k + \frac{i+1}{m} \leqslant x + \frac{1}{m} < y. \qquad \square$$

To embed F into $\mathbb{R}$, we use the following properties of $\mathbb{R}$: every Cauchy sequence of rational numbers has a limit on $\mathbb{R}$; two Cauchy sequences $a = (a_1, \ldots, a_n, \ldots)$ and $b = (b_1, \ldots, b_n, \ldots)$ of rational numbers have the same limit in $\mathbb{R}$ if and only if for every $\varepsilon \in \mathbb{Q}$, $\varepsilon > 0$ there exists $N \in \mathbb{N}$ such that $|a_n - b_n| < \varepsilon$ for all $n \geqslant N$. (These properties are clear from the construction of $\mathbb{R}$ as the metric space completion of $\mathbb{Q}$; see Section A.5.)

We show that every $x \in F$ is the limit (in F) of a Cauchy sequence of rational numbers. By Lemma 8.1.4 there is for each $n \in \mathbb{N}$ a rational number a_n such that $x - 1/n < a_n < x + 1/n$. Then $|a_n - a_m| < 2/n$ for all $m \geqslant n$, and $a = (a_1, \ldots, a_n, \ldots)$ is a Cauchy sequence. Moreover $x = \lim_F a_n$ is the limit of $(a_1, \ldots, a_n, \ldots)$ in F: for every $\varepsilon \in \mathbb{Q}$, $\varepsilon > 0$ there exists $N \in \mathbb{N}$ such that $|a_n - x| < \varepsilon$ for all $n \geqslant N$; by Lemma 8.1.4, this holds for every positive $\varepsilon \in F$.

Let $a = (a_1,\ldots,a_n,\ldots)$ and $b = (b_1,\ldots,b_n,\ldots)$ be Cauchy sequences of rational numbers such that $x = \lim_F a_n = \lim_F b_n$. For every $\varepsilon \in \mathbb{Q}$, $\varepsilon > 0$ there exists $N \in \mathbb{N}$ such that $x - \varepsilon < a_n, b_n < x + \varepsilon$ for all $n \geqslant N$; then $|a_n - b_n| < \varepsilon$ (in $\mathbb{Q}$) for all $n \geqslant N$. Hence a and b have the same limit $\lim_{\mathbb{R}} a_n = \lim_{\mathbb{R}} b_n$ in $\mathbb{R}$. Therefore there is a mapping ρ of F into $\mathbb{R}$ such that $\rho(x) = \lim_{\mathbb{R}} a_n$ whenever $a = (a_1,\ldots,a_n,\ldots)$ is a Cauchy sequence of rational numbers and $\lim_F a_n = x$.

It is immediate that $x = \lim_F a_n$ and $y = \lim_F b_n$ imply $x + y = \lim_F(a_n + b_n)$ and $xy = \lim_F(a_n b_n)$; therefore $\rho(x + y) = \rho(x) + \rho(y)$ and $\rho(xy) = \rho(x)\rho(y)$, and ρ is a homomorphism. In particular, ρ is injective. If $x > 0$ in F, then $x > q > 0$ for some $q \in \mathbb{Q}$ and there is a Cauchy sequence $a = (a_1,\ldots,a_n,\ldots)$ of rational numbers with $x = \lim_F a_n$ and $a_n \geqslant q$ for all n; hence $\rho(x) = \lim_{\mathbb{R}} a_n > 0$. Thus $x < y$ in F implies $\rho(x) < \rho(y)$ in $\mathbb{R}$; since $\mathbb{R}$ and F are totally ordered, $x < y$ in F if and only if $\rho(x) < \rho(y)$ in $\mathbb{R}$. $\qquad\square$

Exercises

1. In an ordered field, prove that $x > 0$ if and only if $-x < 0$.
2. In an ordered field, prove that $x \leqslant y$ implies $xz \geqslant yz$ when $z \leqslant 0$.
3. In an ordered field, prove that $x^2 > 0$ for all $x \neq 0$.
4. In an ordered field, prove that $x > y > 0$ implies $0 < y^{-1} < x^{-1}$.
5. Let K be an ordered field. Show that $K(X)$ is an ordered field, with $f/g > 0$ in $K(X)$ if and only if $a/b > 0$, where a and b are the leading coefficients of f and $g \in K[X]$. Show that $K(X)$ is not archimedean.
6. Show that $\mathbb{Q}$ can be ordered in only one way.
7. Show that $\mathbb{R}$ can be ordered in only one way.
8. Let F be an ordered field (e.g., $\mathbb{Q}$). Let $a = (a_1,\ldots,a_n,\ldots)$ and $b = (b_1,\ldots,b_n,\ldots)$ be Cauchy sequences in F. Show that $a + b = (a_1 + b_1,\ldots,a_n + b_n,\ldots)$ and $ab = (a_1 b_1,\ldots,a_n b_n,\ldots)$ are Cauchy sequences.

8.2. REAL CLOSED FIELDS

This section studies formally real fields and real closed fields, which are generalizations of $\mathbb{R}$.

8.2.a. Formally Real Fields. A field F is **formally real** in case -1 is not a sum of squares in F. By Proposition 8.1.2, F is formally real if and only if it can be ordered. In particular, a formally real field has characteristic 0. All subfields of a formally real field are formally real.

Lemma 8.2.1. *Let F be a formally real field and $E = F(\alpha) \neq F$ be a simple extension of F.*

(1) *If α^2 is a positive element of F, then E is formally real.*

(2) *If $[E : F]$ is odd, then E is formally real.*

Proof. (1) Assume $\alpha^2 = a \in F$. Then every element of E has the form $x + y\alpha$ for some unique $x, y \in F$. If E is not formally real, then $-1 = \sum_i (x_i + y_i\alpha)^2$ for some $x_i, y_i \in F$, which implies $\sum_i x_i^2 + ay_i^2 = -1$ and $a < 0$.

(2) is proved by induction on $n = [E : F]$. The result holds (for all E and F) if $n = 1$. Now assume that $n > 1$. Let $\alpha \in F \backslash E$. Then $[F(\alpha) : F]$ and $[E : F(\alpha)]$ divide $[E : F] = n$ and are odd; if $F \subsetneqq F(\alpha) \subsetneqq E$, then E is formally real by the induction hypothesis. Now let $E = F(\alpha)$; let $q(X) = \mathrm{Irr}_F(\alpha)$, which has degree n. Every element of E has the form $f(\alpha)$ for some unique polynomial $f(X) \in F[X]$ of degree less than n.

If E is not formally real, then $-1 = \sum_i f_i(\alpha)^2$, where $f_i(X) \in F[X]$, $\deg f_i < n$. Hence $1 + \sum_i f_i(X)^2 = q(X)g(X)$ for some $g(X) \in F[X]$. In F, the leading coefficient of f_i^2 is a square and is positive; therefore the degree of $1 + \sum_i f_i^2$ is even and less than $2n$. Hence the degree of g is odd and less than n. Then one of the irreducible factors $p(X)$ of $g(X)$ must have odd degree $k < n$. Now the polynomial p has a root β in some simple extension $F(\beta)$ of F, with $p(X) = \mathrm{Irr}_F(\beta)$ and $[F(\beta) : F] = k$. Then $1 + \sum_i f_i(\beta)^2 = q(\beta)g(\beta) = 0$ and $F(\beta)$ is not formally real, which contradicts the induction hypothesis. $\qquad\square$

8.2.b. Real Closed Fields. A field R is **real closed** in case it is formally real, and there is no formally real algebraic extension $E \supsetneqq R$. For example, $\mathbb{R}$ is real closed: up to isomorphism, $\mathbb{C}$ is the only algebraic extension $E \supsetneqq \mathbb{R}$, and $\mathbb{C}$ is not formally real.

Proposition 8.2.2. *A real closed field R has the following properties:*

(1) *Every positive element of R is a square in R.*

(2) *Every polynomial of odd degree in $R[X]$ has a root in R.*

(3) *R can be ordered in only one way.*

Proof. (3) follows from (1) and Proposition 8.1.1. Lemma 8.2.1 yields (1) and (2) as follows: If a positive element a of F is not a square in F, then $X^2 - a \in F[X]$ is irreducible, and there is an algebraic extension $R(\alpha) \supsetneqq R$ in which $\alpha^2 = a$, which is formally real by Lemma 8.2.1. Similarly, if $f(X) \in R[X]$ has odd degree and no root in R, then f has an irreducible factor q of odd degree with no root in R, and there is an algebraic extension $R(\alpha) \supsetneqq R$ in which $\mathrm{Irr}_R(\alpha) = q$; then $[R(\alpha) : R]$ is odd and $R(\alpha)$ is formally real by Lemma 8.2.1. In either case R cannot be real closed. $\qquad\square$

Theorem 8.2.3. *A formally real field R is real closed if and only if:*

(1) *every positive element of R is a square in R; and*

(2) *every polynomial of odd degree in $R[X]$ has a root in R.*

Proof. These conditions are necessary by Proposition 8.2.2. For the converse we first prove the following generalization of the Fundamental Theorem of Algebra:

Lemma 8.2.4. *Let R be an ordered field with properties (1) and (2) in Theorem 8.2.3. Let $C = R(i)$, where $i^2 = -1$. Then C is algebraically closed.*

Proof. Note that every element a of R is a square in C: if $a > 0$, then a is a square in R; if $a < 0$, then $-a$ is a square in R, and $-a = b^2$ yields $a = (ib)^2$.

Next we show that every element $a + ib$ of C is a square in C. Since $b/2 \in R$ is a square in C, it suffices to show that $c + 2i$ is a square in C for every $c (= 2a/b) \in R$: $c + 2i = (x + iy)^2 = (x^2 - y^2) + 2ixy$, equivalently $x^2 - y^2 = c$ and $xy = 1$, for some $x, y \in R$. With $y = x^{-1}$, the equation $x^2 - y^2 = c$ reads $x^4 - cx^2 - 1 = 0$. Now the equation $t^2 - ct - 1 = 0$ has two solutions t_1, t_2 in R by (1), since its discriminant $c^2 + 4$ is positive. Furthermore t_1, say, is positive, since $t_1 t_2 = -1 < 0$. Hence $t_1 = x^2$ for some $x \in R$, and then $c + 2i = (x + i/x)^2$.

Now take any $\alpha \in \overline{C}$. Then α and its conjugates generate a finite Galois extension E of C, which is also a finite Galois extension of R. Let $G = \text{Gal}(E/R)$. Then $|G| = [E : R] = 2[E : C]$ is even.

Let S be a Sylow 2-subgroup of G and $F = \text{Fix}_E(S)$ be its fixed field. Then $[F : R] = [G : S]$ is odd. If $\alpha \in F \backslash R$, then the irreducible polynomial $\text{Irr}_R(\alpha)$ has odd degree, since $[R(\alpha) : R]$ divides $[F : R]$, and cannot have a root in R, contradicting (2). Therefore $F = R$ and $G = S$ is a 2-group.

If $\alpha \notin C$, then $\text{Gal}(E/C)$ is a nontrivial 2-group and contains a subgroup H of index 2. Its fixed field $F = \text{Fix}_E(H)$ has $[F : C] = 2$. If $\beta \in F \backslash C$, then $\text{Irr}_C(\beta)$ has degree 2 and cannot have a root in C. But we showed that every element of C has a square root in C; therefore every quadratic polynomial with coefficients in C has a root in C. This contradiction shows that there is no $\alpha \in \overline{C} \backslash C$. □

We can now complete the proof of Theorem 8.2.3. Let R be a formally real field with properties (1) and (2). By Lemma 8.2.4, $C = R(i)$ is the algebraic closure $\overline{R}$ of R. If $E \supsetneq R$ is a proper algebraic extension of R, then (up to isomorphism) $E = C$; hence -1 is a square in E and E is not formally real. Thus R is real closed. □

Corollary 8.2.5. *If R is real closed, then $[\overline{R} : R] = 2$.*

We will show in Section 8.4 that this property characterizes real closed fields.

Corollary 8.2.6. *If R is real closed, then $q(X) \in R[X]$ is irreducible if and only if q either has degree 1 or is a quadratic polynomial without roots in R.*

Proof. This is proved like the corresponding result for $\mathbb{R}$, using the R-automorphism $i \longmapsto -i$ of $\overline{R} = R(i)$. The details are an exercise. □

Corollary 8.2.7. *Let A be the field of all real numbers that are algebraic over* $\mathbb{Q}$. *Then A is real closed.*

Proof. A is formally real, and has properties (1) and (2) in Theorem 8.2.3, since any real real number which is algebraic over A is contained in A. □

8.2.c. Real Closure. A **real closure** of an ordered field F is a real closed field R which is algebraic over F and whose unique order relation induces the given order relation on F.

Theorem 8.2.8. *Every ordered field has a real closure.*

Proof. Let F be an ordered field. Let E be the subfield of $\overline{F}$ generated by the square roots of the positive elements of F. By part (1) of Lemma 8.2.1, $F(\alpha_1,\ldots,\alpha_n)$ is formally real for any such square roots $\alpha_1,\ldots,\alpha_n$. Hence E is formally real: if $-1 = \sum_i \beta_i^2$ holds in E, then $-1 = \sum_i \beta_i^2$ holds in some $F(\alpha_1,\ldots,\alpha_n)$, which we just saw is impossible.

By Zorn's Lemma there is a subfield R of $\overline{F}$ which contains E, is formally real, and is maximal with these properties. Then a proper algebraic extension of R is (up to isomorphism) contained in $\overline{F}$ and cannot be formally real by the maximality of R. Thus R is real closed. Also $R \subseteq \overline{F}$ is algebraic over F. Finally every positive element of F is a square in $E \subseteq R$ and is positive in R; hence every negative element of F is negative in R, and it follows that the order relation on F is induced by the unique order relation on R. □

In Section 8.3 we prove that the real closure of F is unique up to isomorphism.

Exercises

1. Let F be formally real. Show that $F(X)$ is formally real.
2. Let K and L be real closed fields. Show that every field homomorphism of K into L is order preserving.
3. Let R be a real closed field and $f(X) \in R[X]$. If $f(a)f(b) < 0$ for some $a < b$ in R, prove that f has a root $a < r < b$ in R. (Use Corollary 8.2.6.)

8.3. STURM'S THEOREM

Sturm's Theorem provides an algorithm which counts the number of roots of separable polynomials. This is used to prove the uniqueness of real closures.

8.3.a. Roots. First we prove two lemmas that locate the roots of polynomials.

Lemma 8.3.1. *Let R be a real closed field and* $f(X) \in R[X]$. *If* $f(a)f(b) < 0$ *for some* $a < b$ *in R, then* f *has a root* $a < r < b$ *in R.*

The proof uses Corollary 8.2.6 and is an exercise.

Lemma 8.3.2. *Let* $f(X) = X^n + a_{n-1}X^{n-1} + \cdots + a_0 \in F[X]$, *where* F *is an ordered field. Then* $|r| \leqslant 1 + |a_{n-1}| + \cdots + |a_0|$ *holds for every root* r *of* f *in* F.

Proof. $|r|$ is the absolute value in F. The result is obvious if $|r| \leqslant 1$. If $|r| > 1$, then dividing $f(r) = 0$ by r^{n-1} yields $r = -a_{n-1} - a_{n-2}/r - \cdots - a_0/r^{n-1}$ and $|r| \leqslant |a_{n-1}| + \cdots + |a_0|$. □

8.3.b. Sturm Sequences. Let R be a real closed field and $f(X) \in R[X]$. A **Sturm sequence of** f **over** an interval $[a,b]$ (where $a < b$ in R) is a sequence $f_0 = f, f_1 = f', f_2, \ldots, f_k$ of polynomials $f_i(X) \in R[X]$ such that:

(a) $f_i(a), f_i(b) \neq 0$ for all i;

(b) if $f_i(r) = 0$ with $a < r < b$ and $0 < i < k$, then $f_{i-1}(r)f_{i+1}(r) < 0$;

(c) f_k has no root $a \leqslant r \leqslant b$.

(The original definition is slightly different; see the exercises.) The **standard sequence** of f has $f_0 = f$, $f_1 = f'$, and f_i defined by induction for $i \geqslant 2$ by

$$f_{i-1} = f_i g_{i-1} - f_{i+1},$$

where $\deg f_{i+1} < \deg f_i$; the sequence ends with the first k such that $f_{k+1} = 0$.

Proposition 8.3.3. *Let* R *be a real closed field and* $f(X) \in R[X]$ *be a non-constant separable polynomial. The standard sequence of* f *is a Sturm sequence of* f *over some interval* $[a,b]$.

Proof. The definition of the standard sequence $f_0, \ldots, f_k$ shows that f_{i-1}, f_i and f_i, f_{i+1} have the same g.c.d. This g.c.d. is f_k, since f_k divides f_{k-1}. Since f is separable, f and f' are relatively prime in $\overline{R}[X]$, and f_k is a nonzero constant. Hence (c) holds (for any interval $[a,b]$). Consecutive polynomials f_{i-1} and f_i are relatively prime and cannot have a common root in R; hence $f_i(r) = 0$ implies $f_{i-1}(r) = f_i(r)g_{i-1}(r) - f_{i+1}(r) = -f_{i+1}(r)$ and $f_{i-1}(r)f_{i+1}(r) < 0$, proving (b) (for any interval $[a,b]$). Finally, Lemma 8.3.2 provides an interval $[a,b]$ such that all the roots of $f_0, f_1, \ldots, f_k$ in R satisfy $a < x < b$. Then (a) holds. □

8.3.c. Sturm's Theorem. In an ordered field, the **number of sign changes** of a sequence $x_0, \ldots, x_k$ of nonzero elements is the number of indices $i > 0$ such that $x_i x_{i-1} < 0$. If $S : f_0, \ldots, f_k$ is a Sturm sequence, and $f_i(x) \neq 0$ for all i, then $V_S(x)$ denotes the number of sign changes of the sequence $f_0(x), \ldots, f_k(x)$.

Theorem 8.3.4 (Sturm). *Let R be a real closed field, and let S be a Sturm sequence of $f(X) \in R[X]$ over an interval $[a,b]$. The number of roots of f in $[a,b]$ is equal to $V_S(a) - V_S(b)$.*

Proof. Let $a < r_1 < r_2 < \cdots < r_\ell < b$ be the distinct roots of $f_0, \ldots, f_k$ in $[a,b]$. By Lemma 8.3.1, $V_S(x)$ is constant in each interval $r_j < x < r_{j+1}$, and in the intervals $[a, r_1)$ and $(r_\ell, b]$. To prove Sturm's Theorem, we show that crossing a root of $f_0 = f$ decreases this constant by 1. In detail, let $r_{j-1} < x < r_j < y < r_{j+1}$ (with $a \leqslant x$ if $j = 0$ and $y \leqslant b$ if $j = \ell$). We show that $V_S(x) = V_S(y) + 1$ if r_j is a root of f; otherwise, $V_S(x) = V_S(y)$.

If r_j is not a root of f_i, then $f_i(x)$ and $f_i(y)$ have the same sign by Lemma 8.3.1; hence differences between $V_S(x)$ and $V_S(y)$ arise at indices i such that $f_i(r_j) = 0$. Let r_j be a root of f_i, where $i \geqslant 1$. Then $i < k$. By (b), $f_{i-1}(r_j)$ and $f_{i+1}(r_j)$ have opposite signs; also $f_{i-1}(r_j), f_{i+1}(r_j) \neq 0$, so that, by Lemma 8.3.1, $f_{i-1}(x)$, $f_{i-1}(r_j)$, and $f_{i-1}(y)$ all have the same sign, and $f_{i+1}(x)$, $f_{i+1}(r_j)$, and $f_{i+1}(y)$ all have the opposite sign. Therefore the sequences

$$f_{i-1}(x), \; f_i(x), \; f_{i+1}(x) \qquad \text{and} \qquad f_{i-1}(y), \; f_i(y), \; f_{i+1}(y)$$

both have one sign change. It follows that $f_1(x), \ldots, f_k(x)$ and $f_1(y), \ldots, f_k(y)$ have the same number of sign changes; if r_j is a root of f_1 (hence not a root of f), then $f_0(x), f_1(x), \ldots, f_k(x)$ and $f_0(y), f_1(y), \ldots, f_k(y)$ have the same number of sign changes, and $V_S(x) = V_S(y)$.

If r_j is not a root of f, then $f_0(x)$ and $f_0(y)$ have the same sign by Lemma 8.3.1. We may assume that r_j is not a root of f_1, so that $f_1(x)$ and $f_1(y)$ have the same sign. Then the sequences $f_0(x), f_1(x)$ and $f_0(y), f_1(y)$ have the same number of sign changes. Therefore $V_S(x) = V_S(y)$.

If r_j is a root of f, then $f(X) = (X - r_j)g(X)$, where $g(r_j) \neq 0$ (since f is separable) and g has no root in $[x,y]$. Then $f'(X) = g(X) + (X - r_j)g'(X)$ also has no root in $[x,y]$. By Lemma 8.3.1, $g(x)$, $g(r_j)$, and $g(y)$ all have the same sign, and $f'(x)$, $f'(r_j) = g(r_j)$, and $f'(y)$ all have the same sign. Since $x < r_j < y$, it follows that $f(x)f'(x) < 0 < f(y)f'(y)$. Hence the sequence $f_0(x), f_1(x)$ has one sign change, whereas the sequence $f_0(y), f_1(y)$ has none. Therefore $V_S(x) = V_S(y) + 1$. $\qquad \square$

8.3.d. The Real Closure. Using Sturm's Theorem, we can prove:

Lemma 8.3.5. *Let R be a real closure of F and $F \subseteq E \subseteq R$ be a finite extension. Every order preserving homomorphism of F into a real closed field R' extends to an order preserving homomorphism of E into R'.*

Proof. By Theorem 7.2.10, $E = F(\alpha)$. Then $q = \mathrm{Irr}_F(\alpha)$ is separable and has a root α in R. Let $\alpha_1 < \cdots < \alpha_m$ be the distinct roots of q in R. The definition of the standard sequence $S : q_0, \ldots, q_k$ of q shows that $q_i \in F[X]$. Lemma 8.3.2 yields $a, b \in F$ such that all the roots (in R) of $q_0, \ldots, q_k$ lie in the interval (a,b).

Let $\varphi : F \longrightarrow R'$ be an order preserving homomorphism. The standard sequence of $^\varphi q$ in R' is $T : {}^\varphi q_1, \ldots, {}^\varphi q_k$. Since φ is order preserving, all the roots of $^\varphi q_1, \ldots, {}^\varphi q_k$ in R' lie in the interval $(\varphi a, \varphi b)$, and $V_T(\varphi a) - V_T(\varphi b) = V_S(a) - V_S(b)$. By Sturm's Theorem, $^\varphi q$ has $m > 0$ roots $\beta_1 < \cdots < \beta_m$ in R'. Therefore φ extends to a homomorphism $\psi : E \longrightarrow R'$.

We now construct an order-preserving homomorphism $E \longrightarrow R'$ which extends φ. Since R is real closed, we have $\alpha_{i+1} - \alpha_i = \gamma_i^2$ for some $\gamma_i \in R$. By the above, there is a homomorphism $\psi : F(\alpha_1, \ldots, \alpha_m, \gamma_1, \ldots, \gamma_{m-1}) \longrightarrow R'$ which extends φ. Then $\psi\alpha_1, \ldots, \psi\alpha_m$ are roots of $^\varphi q$ in R'. Also $\psi\alpha_{i+1} - \psi\alpha_i = \psi\gamma_i^2 > 0$. Hence $\psi\alpha_1 < \cdots < \psi\alpha_m$, and it follows that $\psi\alpha_i = \beta_i$ for all i. We show that the restriction of ψ to $F(\alpha_1, \ldots, \alpha_m) \supseteq E$ is order preserving. Let $\delta \in F(\alpha_1, \ldots, \alpha_m)$ be positive. Then $\delta = \gamma^2$ for some $\gamma \in R$. By the above there is a homomorphism $\chi : F(\alpha_1, \ldots, \alpha_m, \gamma_1, \ldots, \gamma_{m-1}, \gamma) \longrightarrow R'$ which extends φ. As before, this implies $\chi\alpha_i = \beta_i$ for all i. Therefore ψ and χ have the same restriction to $F(\alpha_1, \ldots, \alpha_m)$. Hence $\psi\delta = \chi\delta = \chi\gamma^2$ is positive in R'. $\square$

Proposition 8.3.6. *Let R be a real closure of F. Every order-preserving homomorphism of F into a real closed field R' extends to an order-preserving homomorphism of R into R'.*

Proof. Let $\mathscr{S}$ be the set of all pairs (E, ψ) where $F \subseteq E \subseteq R$ and $\psi : E \longrightarrow R'$ is an order-preserving F-homomorphism. Let $(E, \psi) \leqslant (K, \chi)$ in $\mathscr{S}$ if and only if $E \subseteq K$ and $\psi = \chi_{|K}$. By Zorn's Lemma, $\mathscr{S}$ has a maximal element (M, μ). If $M \neq R$, then Lemma 8.3.5 μ can be extended to any finite extension of M and M is not maximal. Hence $M = R$, and there is an order-preserving homomorphism $R \longrightarrow R'$ which extends φ. $\square$

Theorem 8.3.7. *Any two real closures of an ordered field F are F-isomorphic (as ordered fields).*

Proof. Let R and R' be real closures of F. By Proposition 8.3.6, the inclusion $F \subseteq R'$ extends to an order-preserving F-homomorphism $\xi : R \longrightarrow R'$; ξ is an isomorphism, since R' is algebraic over $\text{Im } \xi$. $\square$

Exercises

1. Use Sturm's Theorem to show that $X^3 - 7X + 7 \in \mathbb{R}[X]$ has two real roots between 1 and 2.

2. Use Sturm's Theorem to determine the number of real roots of $X^4 + 2X^2 - 3X - 1 \in \mathbb{R}[X]$.

3. Find the standard sequence of $X^3 + pX + q$.

4. Let R be real closed and $f(X) \in R[X]$ be a separable polynomial with a root $r \in R$. Show that there exists $a < r < b$ in R such that $f(x)f'(x) < 0$ and $f(y)f'(y) > 0$ for all $a < x < r < y < b$. [The original definition of Sturm sequences requires this property of f_0 and f_1 instead of requiring $f_1 = f'$.]

8.4. THE ARTIN-SCHREIER THEOREM

When R is a real closed field, then $[\overline{R} : R] = 2$ (Corollary 8.2.5), and irreducible polynomials in $R[X]$ have degree at most 2 (Corollary 8.2.6). Either property characterizes real closed fields; in fact:

Theorem 8.4.1 (Artin-Schreier). *For a field K the following properties are equivalent:*

(1) $[\overline{K} : K]$ *is finite.*

(2) *There is an upper bound for the degrees of irreducible polynomials in $K[X]$.*

(3) K *is real closed.*

Proof. (3) implies (2) by Corollary 8.2.6.
(2) implies (1). First we prove:

Lemma 8.4.2. *If there is a upper bound for the degrees of irreducible polynomials in $K[X]$, then K is perfect.*

Proof. This holds if K has characteristic 0. Assume that K has characteristic $p \neq 0$ and that $a \in K$ is not a p-th power in K. To prove the lemma, we show that $f(X) = X^{p^k} - a \in K[X]$ is irreducible for all $k > 0$.

In $K[X]$, f is a product $f = q_1 \ldots q_r$ of monic irreducible polynomials. Let $\alpha \in \overline{K}$ be a root of f. Then $\alpha^{p^k} = a$, $f(X) = X^{p^k} = \alpha^{p^k} = (X - \alpha)^{p^k}$, and each q_i is a power of $X - \alpha$: $q_i(X) = (X - \alpha)^{t_i}$. If $t = \min(t_1, \ldots, t_r)$, then $q = (X - \alpha)^t$ is irreducible and divides $q_1, \ldots, q_r$; it follows that $q_1 = \ldots = q_r = q$ and that $f = q^r$. In particular, $p^k = rt$, and r is a power of p. On the other hand, $(-\alpha)^{rt} = -a$. Since p is either odd or equal to 2, this implies $\alpha^{tr} = a$. But $\alpha^t \in K$, since $q \in K[X]$, and $(\alpha^t)^r = a$ is not a p-th power in K. Hence p does not divide r. Therefore $r = 1$, and f is irreducible. $\qquad\square$

With Lemma 8.4.2 we can show that (2) implies (1) in Theorem 8.4.1. If every irreducible polynomial in $K[X]$ has degree at most n, then K is perfect by the lemma. Hence $\overline{K}$ is separable over K. Also every element of $\overline{K}$ has degree at most n over K. Hence $[\overline{K} : K] \leqslant n$, by Corollary 7.2.11.
(1) implies (3). Again we start with a lemma:

Lemma 8.4.3. *Assume that $[\overline{K} : K]$ is finite. Then K is perfect. Also either $\overline{K} = K(i)$, where $i \notin K$, $i^2 = -1$, or there exists a field $K \subseteq F \subseteq \overline{K}$ such that $\overline{K}$ is Galois over F of prime degree and -1 is a square in F.*

Proof. Let $n = [\overline{K} : K]$. Every $\alpha \in \overline{K}$ has degree at most n, every irreducible polynomial in $K[X]$ has degree at most n, and K is perfect by Lemma 8.4.2. Hence $\overline{K}$ is Galois over K.

Let $i \in \overline{K}$ be a root of $X^2 + 1 \in K[X]$. If $K(i) \subsetneq \overline{K}$, then $\overline{K}$ is Galois over $K(i)$. Its Galois group $\mathrm{Gal}(\overline{K}/K(i))$ has a subgroup H of prime order. Let $F = \mathrm{Fix}_{\overline{K}}(H)$ be the fixed field of H. Then $\overline{K}$ is Galois over F of prime degree, since $\mathrm{Gal}(\overline{K}/F) = H$ has prime order, and $i \in K(i) \subseteq F$. $\square$

With Lemma 8.4.3 we can show that (1) implies (3) in Theorem 8.4.1.

Assume that $\overline{K} = K(i)$. Then every element of $\overline{K}$ can be written in the form $x + iy$ for some unique $x, y \in K$. Also $\overline{K}$ has a K-automorphism $x + iy \longmapsto x - iy$. It follows that the sum of two squares in K is a square in K: for any $x, y \in K$, we have $x + iy = u^2$ for some $u \in \overline{K}$, so that $x^2 + y^2 = (x + iy)\overline{(x + iy)} = u^2\overline{u}^2 = (u\overline{u})^2$ is a square in K. Hence a sum of squares in K is a square. Since $-1 = i^2$ is not a square in K, -1 is not a sum of squares and K is formally real. On the other hand, the only algebraic extension $E \supsetneq K$ of K is (up to isomorphism) $\overline{K}$, which is not formally real. Hence K is real closed.

If $\overline{K} \neq K(i)$, then by Lemma 8.4.3 there is a field $K \subseteq F \subseteq \overline{K}$ such that $\overline{K}$ is Galois over F of prime degree and -1 is a square in F. We let F be a field which contains a square root of -1 and prove that a Galois extension $F \subseteq E \subseteq \overline{F}$ of prime degree p cannot be algebraically closed. Therefore the case $\overline{K} \neq K(i)$ is vacuous.

First assume that F has characteristic p. By Proposition 7.9.6, $E = F(\alpha)$, where $a = \alpha^p - \alpha \in F$, $\alpha \notin F$. Then $\mathrm{Irr}_F(\alpha) = X^p - X - a$ and $1, \alpha, \ldots, \alpha^{p-1}$ is a basis of E over F. Let $\beta = c_0 + c_1\alpha + \cdots + c_{p-1}\alpha^{p-1} \in E$. Then

$$\beta^p = c_0^p + c_1^p\alpha^p + \cdots + c_{p-1}^p\alpha^{(p-1)p} = c_0^p + c_1^p(\alpha + a) + \cdots + c_{p-1}^p(\alpha + a)^{p-1}.$$

If $\beta^p - \beta = a\alpha^{p-1}$, then $c_{p-1}^p - c_{p-1} = a$ (since $1, \alpha, \ldots, \alpha^{p-1}$ is a basis) and the irreducible polynomial $X^p - X - a \in F[X]$ has a root c_{p-1} in F, a contradiction. Therefore $X^p - X - a\alpha^{p-1} \in E[X]$ has no root in E, and E is not algebraically closed.

Next assume that F does not have characteristic p. We may assume that E contains a primitive p-th root ε of unity (otherwise E is not algebraically closed). Then ε is a root of $(X^p - 1)/(X - 1) \in F[X]$ and $[F(\varepsilon) : F] < p$. Since $[F(\varepsilon) : F]$ divides $[E : F] = p$, it follows that $[F(\varepsilon) : F] = 1$ and $\varepsilon \in F$. Thus F contains a primitive p-th root of unity. By Proposition 7.9.3, $E = F(\alpha)$, where $\alpha^p \in F$. We show that $X^p - \alpha \in E[X]$ has no root in E, so that E is not algebraically closed.

Assume that $\beta^p = \alpha$ for some $\beta \in E$. Let $\sigma \in \mathrm{Gal}(E/F)$. Then $\beta^{p^2} = \alpha^p \in K$ and $(\sigma\beta)^{p^2} = \beta^{p^2}$. Hence $\zeta = (\sigma\beta)/\beta \in E$ satisfies $\zeta^{p^2} = 1$, $\zeta^p \in F$ (since ζ^p is a p-th root of unity), and $(\sigma\zeta)^p = \zeta^p$. Then $\varepsilon = (\sigma\zeta)/\zeta$ satisfies $\varepsilon^p = 1$, and $\varepsilon \in F$. We now have $\sigma\beta = \zeta\beta$, $\sigma^2\beta = (\sigma\zeta)(\sigma\beta) = (\varepsilon\zeta)(\zeta\beta) = \varepsilon\zeta^2\beta$; by induction, $\sigma^k\beta = \varepsilon^{k(k-1)/2}\zeta^k\beta$ for all $k \geqslant 0$. Hence $\beta = \sigma^p\beta = \varepsilon^{p(p-1)/2}\zeta^p\beta$ and $\varepsilon^{p(p-1)/2}\zeta^p = 1$.

If p is odd, then p divides $p(p-1)/2$ and $\varepsilon^{p(p-1)/2} = 1$. This also holds if $p = 2$: then $\zeta^4 = 1$ and $\zeta^2 = \pm 1$; since -1 is a square in F, it follows that $\zeta \in F$ and $\varepsilon = (\sigma\zeta)/\zeta = 1$. Hence $\zeta^p = \varepsilon^{p(p-1)/2}\zeta^p = 1$ and $\sigma\alpha = \sigma\beta^p = \zeta^p\beta^p = \alpha$. Since $E = F(\alpha)$ this implies $\sigma = 1$ and we have proved that $\mathrm{Gal}\,(E/F) = 1$, a contradiction. Therefore there is no $\beta \in E$ such that $\beta^p = \alpha$. $\qquad\square$

Exercises

1. Let F be a field of characteristic $p \neq 0$. Assume that $X^p - X - a \in F[X]$ has no root in F. Prove that $X^p - X - a$ is irreducible.

2. As in the proof of Theorem 8.4.1 (case $\overline{K} \neq K(i)$), let p be a prime number, F be a field which does not have characteristic p and contains a primitive p-th root of unity and a square root of -1, and $F \subseteq E = F(\alpha)$ be a Galois extension of degree p, where $\alpha^p = a \in F$. Show that $X^p - a \in F[X]$ is irreducible.

3. Let C be an algebraically closed field. Show that $\mathrm{Aut}(C)$ has no finite subgroup of order greater than 2.

9

FIELDS WITH VALUATIONS

Absolute values and valuations are generalizations of the absolute values on $\mathbb{R}$ and $\mathbb{C}$. Their study gives further insight into $\mathbb{R}$ and $\mathbb{C}$, and into finite extensions.

Required results: Chapter 6 and Sections 7.2, 7.4, 7.8, and 7.9 of Chapter 7. All rings in this chapter are commutative rings with an identity element.

9.1. ABSOLUTE VALUES

9.1.a. Definition. An **absolute value** on a field F is a mapping $v : F \longrightarrow \mathbb{R}$ such that for all $x, y \in F$:

   (i) $v(x) \geqslant 0$, with $v(x) = 0$ if and only if $x = 0$;

   (ii) $v(xy) = v(x)v(y)$;

   (iii) $v(x + y) \leqslant v(x) + v(y)$.

Often $v(x)$ is denoted by $|x|$ or by $|x|_v$. Absolute values are also called **real valuations** or **real-valued** valuations.

The axioms imply $v(1) = 1$; $v(x^{-1}) = 1/v(x)$ whenever $x \neq 0$; $v(-1) = 1$; $v(-x) = v(x)$ for all x; and $v(n) = v(1 + 1 + \cdots + 1) \leqslant v(1) + v(1) + \cdots + v(1) = n$ for all $n > 1$ ($n \in \mathbb{Z}$ also denotes $n1 \in F$).

Condition (iii) also implies the inequality

$$|v(x) - v(y)| \leqslant v(x - y)$$

for all $x, y \in F$; the proof is an exercise.

9.1.b. Examples. Obvious examples of absolute values are the usual absolute values on $\mathbb{Q}$ and $\mathbb{R}$ and the usual absolute value or modulus on $\mathbb{C}$.

On every field F there is a **trivial** absolute value $t(x) = 1$ for all $x \neq 0$.

For less trivial examples let $F = K(X)$ be the field of rational fractions in one indeterminate over a field K. An absolute value v_∞ on $K(X)$ is defined by

$$v_\infty(f(X)/g(X)) = \begin{cases} e^{\deg f - \deg g} & \text{if } f \neq 0, \\ 0 & \text{if } f = 0, \end{cases}$$

The reader will verify that v_∞ is well defined and that (ii) and (iii) hold. Similarly an absolute value v_0 on $K(X)$ is well defined by

$$v_0(f(X)/g(X)) = \begin{cases} e^{\operatorname{ord} g - \operatorname{ord} f} & \text{if } f \neq 0, \\ 0 & \text{if } f = 0; \end{cases}$$

equivalently, $v_0(f/g) = e^{-m}$ if $m \in \mathbb{Z}$, $f(X)/g(X) = X^m h(X)/k(X)$, and X does not divide h or k ($h(0), k(0) \neq 0$). Substitution yields an absolute value v_a on $K(X)$ for each $a \in K$, defined by

$$v_a(f(X)/g(X)) = v_0(f(X - a)/g(X - a)).$$

All of these absolute values are trivial on K ($v(x) = 1$ for all $x \in K$, $x \neq 0$) and satisfy a stronger version of (iii), namely

(iii*) $v(x + y) \leqslant \max(v(x), v(y))$.

Property (iii*) implies

(iii**) if $v(x) \neq v(y)$, then $v(x + y) = \max(v(x), v(y))$.

Indeed, if, say, $v(x) < v(y)$, then (iii*) yields $v(x + y) \leqslant v(y)$ and $v(y) = v((x + y) + (-x)) \leqslant \max(v(x + y), v(x)) = v(x + y)$ (since $v(y) \not\leqslant v(x)$).

9.1.c. Archimedean Absolute Values. An absolute value v on a field F is **archimedean** in case there is no $x \in F$ such that $v(n) \leqslant v(x)$ for all $n \in \mathbb{N}$.

Proposition 9.1.1. *For an absolute value v on a field F the following properties are equivalent:*

(1) v *is nonarchimedean.*

(2) $v(n) \leqslant 1$ *for every integer n.*

(3) (iii*) *holds:* $v(x + y) \leqslant \max(v(x), v(y))$ *for all $x, y \in F$.*

Proof. (1) implies (2). If v is not archimedean, there exists $a \in F^*$ such that $v(n) \leqslant v(a)$ for every positive integer n. Then the set $\{v(n); n > 0\}$ is bounded above. This implies $v(n) \leqslant 1$ for all $n > 0$ (hence for all $n \in \mathbb{Z}$): otherwise, the set $\{v(n), v(n^2), \ldots, v(n^k), \ldots\}$ has no upper bound.

(2) implies (3). (2) implies

$$v(x+y)^n = v\left(\sum_{0 \leqslant k \leqslant n} \binom{n}{k} x^k y^{n-k}\right)$$

$$\leqslant \sum_{0 \leqslant k \leqslant n} v(x)^k \, v(y)^{n-k} \leqslant (n+1)(\max(v(x),v(y))^n .$$

Hence $v(x+y) \leqslant (n+1)^{1/n} \max(v(x),v(y))$ for all $n > 0$. Hence $v(x+y) \leqslant \max(v(x),v(y))$, since $\lim_{n \to \infty} (n+1)^{1/n} = 1$.

(3) implies (1). (iii*) implies $v(n1) \leqslant v(1)$ for all n.   □

For example, the usual absolute values on $\mathbb{R}$ and $\mathbb{Q}$ are archimedean; Proposition 9.1.1 shows that the absolute values v_∞ and v_a on $K(X)$ are nonarchimedean. The trivial absolute value on any field is nonarchimedean.

Corollary 9.1.2. *On a field of characteristic $p \neq 0$, every absolute value is nonarchimedean.*

Proof. In such a field F, $n^p = n$ for every integer n, and $n^{p-1} = 1$ if $n \neq 0$ in F (if $p \nmid n$ in $\mathbb{Z}$); hence $v(n) = 0$ or $v(n) = 1$.   □

9.1.d. The Value Group. By (ii), an absolute value $v : F \longrightarrow \mathbb{R}$ induces a multiplicative homomorphism of $F^* = F\backslash\{0\}$ into the multiplicative group $\mathbb{P}$ of all positive real numbers. The **value group** of v is the multiplicative group $G_v = v(F^*) = (\mathrm{Im}\ v)\backslash\{0\}$. It is a subgroup of $\mathbb{P}$.

Proposition 9.1.3. *A nontrivial subgroup of $\mathbb{P}$ is either cyclic or dense in $\mathbb{P}$.*

Proof. Let $1 \lneqq G \leqslant \mathbb{P}$ and m be the greatest lower bound of $\{g \in G ; g > 1\}$. We show that G is cyclic if $m > 1$ and dense in $\mathbb{P}$ if $m = 1$.

Assume that $m > 1$. If $m \notin G$, then m is the limit of a sequence $g_1 > \cdots > g_n > g_{n+1} > \cdots$ of elements of G. Since $m^2 > m$, we have $m < g_{n+1} < g_n < m^2$ for some n. Then $g = g_n/g_{n+1} \in G$ and $1 < g < m$, contradicting the definition of m. Therefore $m \in G$. For each $g \in G$ we now have $m^k \leqslant g < m^{k+1}$ for some $k \in \mathbb{Z}$, $h = g/m^k \in G$, $1 \leqslant h < m$, $h = 1$, and $g = m^k$. Thus $G = \langle m \rangle$ is cyclic.

Now assume that $m = 1$. We prove that every real number $r > 1$ is the limit of a sequence of elements of G. We have

$$r = e^t = \lim_{u \to \infty} \left(1 + \frac{t}{u}\right)^u ,$$

where $t = \ln r > 0$. Now $m = 1$ is the limit of a decreasing sequence $g_1, \ldots, g_n, \ldots$ of elements of G. We have $g_n = 1 + t/u_n$ for some $u_n > 0$ and $u_n \leqslant k_n < u_n + 1$ for some integer $k_n \geqslant 0$. Then $\lim_{n \to \infty} u_n = \infty$, since $\lim_{n \to \infty} g_n = 1$. Also

$$\left(1 + \frac{t}{u_n}\right)^{u_n} \leqslant \left(1 + \frac{t}{u_n}\right)^{k_n} < \left(1 + \frac{t}{u_n}\right)^{u_n} \left(1 + \frac{t}{u_n}\right) .$$

Hence

$$\lim_{n\to\infty} g_n^{k_n} = \lim_{n\to\infty} \left(1 + \frac{t}{u_n}\right)^{k_n} = \lim_{n\to\infty} \left(1 + \frac{t}{u_n}\right)^{u_n} = \lim_{u\to\infty} \left(1 + \frac{t}{u}\right)^{u} = r . \qquad \square$$

By Proposition 9.1.3, the value group of a nontrivial absolute value is either dense or cyclic. An absolute value v is **discrete** in case its value group G_v is cyclic. For example, the absolute values v_∞, v_0 on $K(X)$ are discrete; the usual absolute value on $\mathbb{Q}$ is not discrete.

9.1.e. Equivalence. Every absolute value v on a field F induces a distance function d on F, namely

$$d(x,y) = v(x - y).$$

Indeed (i) and (iii) show that $d(x,y) \geqslant 0$, $d(x,y) = 0$ if and only if $x = y$, and $d(x,z) \leqslant d(x,y) + d(y,z)$. This makes F a metric space and, in particular, a topological space. It is immediate that the operations on F and the mapping v are continuous (see the exercises).

Proposition 9.1.4. *Let v and w be absolute values on a field F, with value groups G_v and G_w. The following properties are equivalent:*

(1) *v and w induce the same topology on F.*

(2) *$v(a) < 1$ if and only if $w(a) < 1$.*

(3) *There exists $k > 0$ such that $w(x) = v(x)^k$ for all $x \in F$.*

(4) *There exists an order preserving isomorphism $\theta : G_v \longrightarrow G_w$ such that $w(a) = \theta(v(a))$ for all $a \neq 0$.*

Proof. (1) implies (2). If $v(a) < 1$, then $\lim_{n\to\infty} v(a)^n = 0$ in $\mathbb{R}$ and $\lim_{n\to\infty} a^n = 0$ in F for the topology induced by v. Since w induces the same topology, $U = \{x \in F ; w(x) < 1\}$ is a neighborhood of 0 and $a^n \in U$ for some n. Then $w(a^n) < 1$ and $w(a) < 1$. Exchanging v and w yields the converse implication.

(2) implies (3). If (2) holds, then $v(a) > 1$ if and only if $w(a) > 1$, and $v(a) = 1$ if and only if $w(a) = 1$. Hence v is trivial if and only if w is trivial, and then (3) holds. Otherwise, $v(a) > 1$ and $w(a) > 1$ for some $a \in F$, and $w(a) = v(a)^k$ for some real number $k > 0$.

We show that $w(x) = v(x)^k$ for every $x \in F^*$. If $v(x) = 1$, then $w(x) = 1 = v(x)^k$. Otherwise, $v(x) = v(a)^t$ for some $t \in \mathbb{R}$. If p/q is a rational number and $p/q < t$, then $v(a)^{p/q} < v(x)$, $v(a^p) < v(x^q)$, $v(a^p/x^q) < 1$, $w(a^p/x^q) < 1$, $w(a^p) < w(x^q)$, and $w(a)^{p/q} < w(x)$. Similarly $p/q > t$ implies $w(a)^{p/q} > w(x)$. Therefore $w(a)^t = w(x)$. Hence $w(x) = w(a)^t = v(a)^{kt} = v(x)^k$.

(3) implies (4). Let $\theta(g) = g^k$.

(4) implies (1). If (4) holds, then the metric spaces on F defined by v and w have the same open disks and therefore have the same open sets.   □

Two absolute values are **equivalent** in case they satisfy the equivalent conditions in Proposition 9.1.4. Equivalent absolute values are also called **dependent**. When v and w are equivalent, v is trivial if and only if w is trivial; v is discrete if and only if w is discrete; by Proposition 9.1.1, v is nonarchimedean if and only if w is nonarchimedean.

9.1.f. Absolute Values on $\mathbb{Q}$. In addition to the usual absolute value $v_\infty(x) = |x|$ on $\mathbb{Q}$, a discrete, nonarchimedean absolute value v_p on $\mathbb{Q}$ is defined for each prime number p by

$$v_p(m/n) = \begin{cases} 0 & \text{if } m/n = 0, \\ p^{-k} & \text{if } k \in \mathbb{Z}, \ m/n = p^k t/u, \text{ and } p \text{ does not divide } t, u; \end{cases}$$

v_p is well defined, since every nonzero rational number can be written uniquely in the form $p^k t/u$ with $k \in \mathbb{Z}$ and $p \nmid t, u$. Properties (ii) and (iii*) are exercises.

We now prove that this constructs essentially all absolute values on $\mathbb{Q}$.

Theorem 9.1.5. *Every nontrivial absolute value on $\mathbb{Q}$ is equivalent either to v_∞ or to v_p for some unique prime number p.*

Proof. Let v be an absolute value on $\mathbb{Q}$.

Assume that v is nonarchimedean. By Proposition 9.1.1, $v(n) \leqslant 1$ for every $n > 1$. If $v(n) = 1$ for all $n > 1$, then $v(x) = 1$ for every $x \in \mathbb{Q}^*$; since we assumed that v is not trivial, we have $v(n) < 1$ for some $n > 1$. Then

$$I = \{\, n \in \mathbb{Z}\,;\, v(n) < 1 \,\}$$

is a nonzero, proper ideal of $\mathbb{Z}$. Moreover I is a prime ideal: if $m, n \notin I$, then $v(m) = v(n) = 1$, $v(mn) = 1$, and $mn \notin I$. Therefore I is generated by a prime number p. Then $v(n) = 1$ if and only if $p \nmid n$. Let $c = 1/v(p) > 1$. If $m/n = p^k t/u$, where $p \nmid t, u$, then $v(m/n) = v(p)^k v(t)/v(u) = c^{-k}$; thus v is equivalent to v_p. There p is unique, since v_p and v_q are not equivalent when $p \neq q$.

Now assume that v is archimedean. We show that $v(n) > 1$ for all $n > 1$. For any $m, n > 1$ there exists an integer $k \geqslant 0$ such that $n^k \leqslant m < n^{k+1}$; repeated division by n yields integers $0 \leqslant r_0, \ldots, r_k < n$ such that $m = r_0 + r_1 n + \cdots + r_k n^k$. Then $v(r_i) \leqslant r_i < n$ and $v(m) \leqslant n(k+1)u^k$, where $u = \max(1, v(n))$. Also $n^k \leqslant m$ yields $k \leqslant \log_n m = \ln m / \ln n$ and

$$v(m) \leqslant n(\log_n m + 1)\, u^{\log_n m}.$$

Replacing m by m^r yields $v(m)^r = v(m^r) \leqslant n(r\log_n m + 1)\, u^{r\log_n m}$ and

$$v(m) \leqslant (nr\log_n m + n)^{1/r}\, u^{\log_n m}$$

for all $r > 0$. Since $\lim_{r\to\infty}(nr\log_n m + n)^{1/r} = 1$, it follows that

$$v(m) \leqslant u^{\log_n m} = \max\,(1, v(n))^{\log_n m}.$$

Since v is archimedean we have $v(m) > 1$ for some $m > 1$ by Proposition 9.1.1. Then the inequality $v(m) \leqslant \max\,(1, v(n))^{\log_n m}$ shows that $v(n) > 1$ for all $n > 1$. Then $v(m) \leqslant v(n)^{\log_n m} = v(n)^{\ln m/\ln n}$, $v(m)^{1/\ln m} \leqslant v(n)^{1/\ln n}$ for all $m, n > 1$, $v(m)^{1/\ln m} = v(n)^{1/\ln n}$ for all $m, n > 1$, and $\ln v(m)/\ln m = \ln v(n)/\ln n$ for all $m, n > 1$. Therefore $c = \log_n v(n) = \ln v(n)/\ln n > 0$ does not depend on $n > 1$. Then $v(n) = n^c$ for all $n > 1$, and it follows that $v(x) = |x|^c$ for every $x \in \mathbb{Q}$. Thus v is equivalent to the usual absolute value v_∞. □

9.1.g. Absolute Values on $K(X)$. Our next result is somewhat similar to Theorem 9.1.5. For every irreducible polynomial $q(X) \in K[X]$, a discrete, non-archimedean absolute value v_q on $K(X)$ is well defined by

$$v_q(f/g) = \begin{cases} 0 & \text{if } f/g = 0, \\ e^{-k} & \text{if } k \in \mathbb{Z},\ f/g = q^k t/u,\ \text{and } q \text{ does not divide } t, u. \end{cases}$$

Indeed every rational fraction $f/g \neq 0$ can be written uniquely in the form $f/g = q^k t/u$, where $k \in \mathbb{Z}$ and $q \nmid t, u$. We see that v_q is trivial on K ($v(a) = 1$ for all $a \in K$). For instance, $v_a = v_{X-a}$ for each $a \in K$.

Theorem 9.1.6. *Every nontrivial absolute value on $K(X)$ which is trivial on K is equivalent to v_∞ or to v_q for some unique monic irreducible polynomial q.*

Proof. Let v be an absolute value on $K(X)$ which is trivial on K but not trivial. By Proposition 9.1.1, v is nonarchimedean.

First assume that $v(X) \leqslant 1$. Since v is trivial on F, (iii*) yields $v(a_0 + a_1 X + \cdots + a_n X^n) \leqslant 1$ for every $f(X) = a_0 + a_1 X + \cdots + a_n X^n \in K[X]$. If $v(f) = 1$ for every $f(X) \in K[X]$, $f \neq 0$, then $v(f/g) = 1$ for every $f/g \in K(X)$, $f/g \neq 0$; since we assumed that v is not trivial, we have $v(f) < 1$ for some $f \in K[X]$. Then

$$\mathfrak{A} = \{\, f(X) \in K[X]\,;\ v(f) < 1 \,\}$$

is a nonzero, proper ideal of $K[X]$. Moreover $\mathfrak{A}$ is a prime ideal: if $f, g \notin \mathfrak{A}$, then $v(f) = v(g) = 1$, $v(fg) = 1$, and $fg \notin \mathfrak{A}$. Therefore $\mathfrak{A}$ is generated by a monic irreducible polynomial q. Then $v(f) = 1$ if and only if q does not divide f. Let $c = 1/v(q) > 1$. If $f = q^k t/u$, where q does not divide t, u, then $v(f) = $

$v(q)^k v(t)/v(u) = c^{-k}$; thus v is equivalent to v_q. Moreover q is unique: if v_q and v_r are equivalent, then $v_q(r) < 1$ and $v_r(q) < 1$, $q \mid r$ and $r \mid q$, and $q = r$.

Now assume that $v(X) = c > 1$. If $a_n \neq 0$, then $v(a_n X^n) = c^n$ and (iii**) yields $v(a_0 + a_1 X + \cdots + a_n X^n) = c^n$, since $v(a_i X^i) < v(a_n X^n)$ for all $i < n$. Thus $v(f) = c^{\deg f}$ and $v(f/g) = c^{\deg f - \deg g}$ whenever $f \neq 0$, and v is equivalent to v_∞. $\square$

Exercises

1. Prove that $|v(x) - v(y)| \leqslant v(x - y)$ for every absolute value v.

2. Define absolute values on a domain. Prove that every absolute value on a domain can be extended uniquely to an absolute value on its field of fractions.

3. Verify that v_∞ is an absolute value on $K(X)$.

4. Verify that v_0 is an absolute value on $K(X)$.

5. Verify that v_p is an absolute value on $\mathbb{Q}$ for every prime p.

6. Verify that v_q is an absolute value on $K(X)$ for every monic irreducible polynomial q.

7. Extend Proposition 9.1.1 to any domain.

8. Let $v : F \longrightarrow \mathbb{R}$ be an absolute value on a field F. Prove that v^k (defined by: $v^k(a) = v(a)^k$ for all a) is an absolute value on F for all $0 < k < 1$ (for all $k > 0$, if v is nonarchimedean).

9. Prove that every absolute value is continuous in its own topology.

10. Prove that the operations on a field are continuous in the topology induced by any absolute value.

11. Prove that $|x| \left(\prod_{p \text{ prime}} v_p(x) \right) = 1$ for all $x \in \mathbb{Q}$, $x \neq 0$.

12. Amend the definition of v_q so that the previous exercise can be extended to $K(X)$.

9.2. COMPLETION

The completion of a field with an absolute value is its completion as a metric space. The construction of $\mathbb{R}$ from $\mathbb{Q}$ by Cauchy sequences is an example.

9.2.a. Cauchy Sequences. A field F with an absolute value $v : F \longrightarrow \mathbb{R}$ is **complete** (with respect to v) in case it is complete as a metric space. For example, $\mathbb{R}$ and $\mathbb{C}$ are complete with respect to the ordinary absolute values.

In detail, $\ell \in F$ is the limit $\ell = \lim_{n \to \infty} a_n$ of a sequence $a = (a_n)_{n>0}$ of elements of F (with respect to v) in case there exists for every real number $\varepsilon > 0$ some integer K such that $v(a_n - \ell) < \varepsilon$ for all $n \geqslant K$. We also say that the sequence a converges to ℓ.

A **Cauchy sequence** is a sequence $a = (a_n)_{n>0}$ of elements of F such that there exists for every real number $\varepsilon > 0$ some integer K such that $v(a_m - a_n) < \varepsilon$ for all $m, n \geqslant K$. Every sequence which converges (to some $\ell \in F$) has this property. The field F is complete (with respect to v) in case, conversely, every Cauchy sequence converges.

9.2.b. Construction

Theorem 9.2.1. *Let F be a field with an absolute value $v : F \longrightarrow \mathbb{R}$. There exists a field extension $\widehat{F}_v = \widehat{F}$ of F and an absolute value $\hat{v} : \widehat{F} \longrightarrow \mathbb{R}$ which extends v, such that $\widehat{F}$ is complete (with respect to $\hat{v}$) and F is dense in $\widehat{F}$.*

Proof. Let C be the set of all Cauchy sequences of elements of F. The sum and product of two Cauchy sequences are defined componentwise:

$$(a_n)_{n>0} + (b_n)_{n>0} = (a_n + b_n)_{n>0},$$
$$(a_n)_{n>0} \, (b_n)_{n>0} = (a_n b_n)_{n>0},$$

and are readily seen to be Cauchy sequences. Hence C is a commutative ring; the identity element of C is the constant sequence 1 (the sequence $1 = (1_n)_{n>0}$ with $1_n = 1$ for all n).

A sequence a is **null** in case $\lim a_n = 0$; equivalently, in case $\lim v(a_n) = 0$. It is immediate that null sequences constitute an ideal N of C. We show that $\widehat{F} = C/N$ has all the required properties.

First we prove that $\widehat{F}$ is a field. Let $a + N \neq 0$ in $\widehat{F}$ so that $a \in C \backslash N$. Then $\lim v(a_n) > 0$, and there exists $\alpha > 0$ and $K > 0$ such that $v(a_n) \geqslant \alpha$ for all $n \geqslant K$. In particular, $a_n \neq 0$ for all $n \geqslant K$. Let

$$b_n = \begin{cases} 1/a_n & \text{if} \quad a_n \neq 0, \\ 1 & \text{if} \quad a_n = 0. \end{cases}$$

Then $b = (b_n)_{n>0}$ is a Cauchy sequence, since

$$v(b_m - b_n) = v\left(\frac{1}{a_m} - \frac{1}{a_n}\right) = v\left(\frac{a_n - a_m}{a_m a_n}\right) \leqslant \frac{v(a_n - a_m)}{\alpha^2}$$

for all $m, n \geqslant K$. Moreover $a_n b_n = 1$ for all $n \geqslant K$, so that $ab - 1 \in N$. Thus $a + N$ has an inverse $b + N$ in $\widehat{F}$.

The constant sequence homomorphism $\iota : F \longrightarrow C$ assign to each $x \in F$ the constant sequence $\iota(x) = (x_n)_{n>0}$ in which $x_n = x$ for all n, and induces a homomorphism $x \longmapsto \iota(x) + N$ of F into $\widehat{F}$. In what follows we identify $x \in F$ and $\iota(x) + N \in \widehat{F}$; then $\widehat{F}$ is a field extension of F.

When a is a Cauchy sequence in F, then $(v(a_n))_{n>0}$ is a Cauchy sequence in $\mathbb{R}$, since $|v(a_m) - v(a_n)| \leqslant v(a_m - a_n)$; hence $(v(a_n))_{n>0}$ converges in $\mathbb{R}$. Moreover $a - b \in N$ implies $\lim v(a_n - b_n) = 0$, and the inequality $|v(a_n) - v(b_m)| \leqslant v(a_n - b_n)$ then shows $\lim v(a_n) = \lim v(b_n)$. Hence a mapping $\hat{v} : \widehat{F} \longrightarrow \mathbb{R}$ is well defined by

$$\hat{v}(a + N) = \lim_{n \to \infty} v(a_n).$$

If $x \in F$ and $x_n = x$ is the constant sequence, then $\hat{v}(x) = \lim v(x_n) = v(x)$; thus $\hat{v}$ extends v. Since $\hat{v}$ extends v, $a + N = \lim a_n$ hold in $\widehat{F}$ for every Cauchy sequence $a = (a_n)_{n>0}$. Therefore F is dense in $\widehat{F}$.

Let $(b_n)_{n>0}$ be a Cauchy sequence in $\widehat{F}$. For each $n > 0$ there is some $a_n \in F$ such that $\hat{v}(b_n - a_n) < e^{-n}$. Then $a = (a_n)_{n>0}$ is a Cauchy sequence in F and $a + N = \lim_{n \to \infty} a_n = \lim_{n \to \infty} b_n$ in $\widehat{F}$. Hence $\widehat{F}$ is complete. $\square$

Proposition 9.2.2. *When v is a nonarchimedean absolute value on a field F, v and $\hat{v}$ have the same value group.*

Proof. $\hat{v}$ is nonarchimedean by Proposition 9.1.1. Let $\alpha \in \widehat{F}^*$. Since F is dense in $\widehat{F}$, there exists $x \in F$ such that $\hat{v}(x - \alpha) < \hat{v}(\alpha)$. Then $\hat{v}(\alpha) = \hat{v}(x) = v(x)$. $\square$

9.2.c. Universal Property. We show that the field $\widehat{F}$ in Theorem 9.2.1 is unique up to isomorphism. This follows from its universal property.

Proposition 9.2.3. *Let F be a field with an absolute value v. Let $\widehat{F}$ be a field extension of F with an absolute value $\hat{v}$ which extends v, such that $\widehat{F}$ is complete (with respect to $\hat{v}$) and F is dense in $\widehat{F}$.*

Let K be a complete field with respect to an absolute value w. Every homomorphism $F \longrightarrow K$ which preserves absolute values can be extended uniquely to a homomorphism $\widehat{F} \longrightarrow K$ which preserves absolute values.

$\widehat{F}$ is unique, up to isomorphisms which preserve absolute values.

Proof. Let $\varphi : F \longrightarrow K$ preserve absolute values. Then φ is continuous. Since F is dense in $\widehat{F}$, φ extends uniquely to a continuous mapping $\psi : \widehat{F} \longrightarrow K$, as follows. Every $\alpha \in \widehat{F}$ is the limit $\alpha = \lim a_n$ of a sequence of elements of F. Then $(a_n)_{n>0}$ is a Cauchy sequence in F, $(\varphi(a_n))_{n>0}$ is a Cauchy sequence in K, and there is a limit $\ell = \lim \varphi(a_n)$ in K. If $\lim a_n = \lim b_n$, then $\lim(a_n - b_n) = 0$, $\lim(\varphi(a_n) - \varphi(b_n)) = 0$, and $\lim \varphi(a_n) = \lim \varphi(b_n)$. Hence a mapping $\psi : \widehat{F} \longrightarrow K$ is well defined by $\psi(\alpha) = \lim \varphi(a_n)$ whenever $\alpha = \lim a_n$.

It is immediate that ψ extends φ and is a homomorphism. Since $\hat{v}$ is continuous (see the exercises), $\alpha = \lim a_n$ implies $\hat{v}(\alpha) = \lim \hat{v}(a_n) = \lim v(a_n)$. Therefore ψ preserves absolute values, since φ does. Conversely, any such homomorphism χ must be continuous and $\chi(\alpha) = \lim \varphi(a_n)$ must hold whenever $\alpha = \lim a_n$. Therefore ψ is unique.

The universal property of $\widehat{F}$ implies its uniqueness in the usual fashion; the details are left as an exercise. $\square$

The field $\widehat{F}_v = \widehat{F}$, which exists by Theorem 9.2.1 and is unique (up to isomorphism) by Proposition 9.2.3, is the **completion** of F with respect to v.

9.2.d. Examples. The obvious example of completion is $\mathbb{R}$, which is the completion of $\mathbb{Q}$ with respect to the ordinary absolute value; this follows from Proposition 9.2.3 (regardless of how $\mathbb{R}$ is constructed) since $\mathbb{R}$ is complete and $\mathbb{Q}$ is dense in $\mathbb{R}$.

The completion of $\mathbb{Q}$ with respect to v_p is described in some detail below by a construction which will be generalized in Section 9.4. The completion of $K(X)$ with respect to v_0 is $K((X))$ (see the exercises).

9.2.e. p-Adic Numbers. For each prime number p, we denote by $\mathbb{Q}_p$ the completion of $\mathbb{Q}$ with respect to v_p; the elements of $\mathbb{Q}_p$ are p-**adic numbers**. We denote by $\mathbb{J}_p$ the ring $\{x \in \mathbb{Q}_p; \hat{v}_p(x) \leqslant 1\}$; the elements of $\mathbb{J}_p$ are p-**adic integers**.

To describe $\mathbb{J}_p$, we note that for every $k > 0$, every integer $x \in \mathbb{Z}$ can be written uniquely in the form

$$x = x_0 + x_1 p + \cdots + x_k p^k,$$

where $x_0, \ldots, x_k \in \mathbb{Z}$ and $0 \leqslant x_i < p$ for all $i < k$. (This is proved below but is readily shown by repeated division.) We show that every p-adic integer x can be written uniquely in the form

$$x = x_0 + x_1 p + \cdots + x_k p^k + \cdots,$$

where $x_k \in \mathbb{Z}$ and $0 \leqslant x_k < p$ for every k.

A p-adic integer is the limit of a Cauchy sequence $(a_n)_{n>0}$ in which we may assume that $v_p(a_n) \leqslant 1$ for all n. Let

$$R = \{x \in \mathbb{Q}; v_p(x) \leqslant 1\} = \{m/n \in \mathbb{Q}; p \nmid n\}.$$

We see that R is a subring of $\mathbb{Q}$ and that

$$Rp = \{x \in \mathbb{Q}; v_p(x) < 1\} = \{m/n \in \mathbb{Q}; p \mid m \text{ and } p \nmid n\}$$

is an ideal of R. We see that $v_p(x) \leqslant p^{-k}$ if and only if $x \in Rp^k$.

We claim that every $x = m/n \in R$ can be written uniquely in the form $x = x_0 + pr$ with $x_0 \in \mathbb{Z}$, $0 \leqslant x_0 < p$, and $r \in R$. If indeed $p \nmid n$, then $un + vp = 1$ for some integers u and v, $mu = pq + x_0$ for some integers q, x_0 with $0 \leqslant x_0 < p$, and

$$m/n = (mun + mvp)/n = mu + (mv/n)p = x_0 + (q + (mv/n))p = x_0 + pr,$$

where $r = q + (mv/n) \in R$. Then x_0 and r are unique, since $y - z \notin Rp$ when $0 \leqslant y, z < p$, $y \neq z$.

By induction on n, it follows from the claim that, for every $k \geqslant 0$, every $x \in R$ can be written uniquely in the form

$$x = x_0 + x_1 p + \cdots + x_k p^k + p^{k+1} r$$

for every k, where $x_0, \ldots, x_k \in \mathbb{Z}$, $0 \leqslant x_i < p$ for all $i \leqslant k$, and $r \in R$. In particular, this holds for every $x \in \mathbb{Z}$. The uniqueness shows that $x_0, \ldots, x_k$ do not change if k is increased. We call $x_0, \ldots, x_k, \ldots$ the *coefficients* of x.

Now let $(a_n)_{n>0}$ be a Cauchy sequence of elements of R. For every $k > 0$ there is some $K > 0$ such that $v_p(a_m - a_n) < p^{-k}$ for all $m, n \geqslant K$. Then $a_m - a_n \in Rp^{k+1}$ and a_m, a_n have the same first k coefficients $x_0, \ldots, x_k$. This constructs a series $x_0 + x_1 p + \cdots + x_k p^k + \cdots$ whose k-th partial sum $s_k = x_0 + x_1 p + \cdots + x_k p^k$ satisfies $a_m - s_k \in Rp^{k+1}$ for all $m > K$. We see that $\lim_{n \to \infty} a_n = \lim_{k \to \infty} s_k$. Therefore every p-adic integer x can be written in the form

$$x = x_0 + x_1 p + \cdots + x_k p^k + \cdots,$$

where $x_k \in \mathbb{Z}$ and $0 \leqslant x_k < p$ for every k. Conversely, every such series converges in $\mathbb{J}_p$. To prove uniqueness, assume that $x_0 + x_1 p + \cdots + x_k p^k + \cdots = y_0 + y_1 p + \cdots + y_k p^k + \cdots$ (where $0 \leqslant x_k, y_k < p$ for all k). Then $x_0 - y_0 = (y_1 - x_1)p + \cdots$, $v_p(x_0 - y_0) < 1/p$, p divides $x_0 - y_0$, and $x_0 = y_0$. Then $x_1 + \cdots + x_k p^{k-1} + \cdots = y_1 + \cdots + y_k p^{k-1} + \cdots$ and $x_1 = y_1$; by induction, $x_k = y_k$ for all k. We call $x_0, \ldots, x_k, \ldots$ the *coefficients* of $x = x_0 + x_1 p + \cdots + x_k p^k + \cdots \in \mathbb{J}_p$.

The reader will verify that x is a unit in $\mathbb{J}_p$ if and only if $x_0 \neq 0$; that $\mathbb{Q}_p$ is the field of fractions of $\mathbb{J}_p$; and that every fraction $x/y \in \mathbb{Q}_p$ can be simplified so that its denominator is a power of p. Hence p-adic numbers can be described as sums of Laurent series in p:

Proposition 9.2.4. *Every nonzero p-adic number can be written uniquely in the form*

$$x = x_m p^m + x_{m+1} p^{m+1} + \cdots + x_n p^n + \cdots,$$

where $m \in \mathbb{Z}$, $x_n \in \mathbb{Z}$ and $0 \leqslant x_n < p$ for all $n \geqslant m$, and $x_m \neq 0$. Conversely, every such series converges to a p-adic number. Then $\hat{v}_p(x) = p^{-m}$. In particular, x is a p-adic integer if and only if $m \geqslant 0$.

Exercises

1. Show that the sum and product of two Cauchy sequences are Cauchy sequences.
2. Let F be a field which is complete with respect to a nonarchimedean absolute value. Prove that a series $\sum_{n \geqslant 0} a_n$ converges in F if and only if $\lim_{n \to \infty} a_n = 0$.

3. Let F be a field with an absolute value v. Let $\hat{F}$ and K be field extensions of F with absolute values $\hat{v}$ and w which extend v such that $\hat{F}$ and K are complete (with respect to $\hat{v}$ and w) and F is dense in $\hat{F}$ and K. Prove that there is an F-isomorphism $\theta : \hat{F} \longrightarrow K$ which preserves absolute values $(\theta(\hat{v}(\alpha)) = w(\theta(\alpha))$ for all $\alpha \in \hat{F})$.

4. Prove directly that every integer $z \in \mathbb{Z}$ can be written uniquely in the form $z = r_0 + r_1 p + \cdots + r_k p^k$ where $r_0, \ldots, r_k \in \mathbb{Z}$ and $0 \leqslant r_i < p$ for all $i < k$.

5. Let $R = \{x \in \mathbb{Q} ; v_p(x) \leqslant 1\}$. Show that $R/Rp \cong \mathbb{Z}_p$.

6. Let $x = x_0 + x_1 p + \cdots + x_k p^k + \cdots \in \mathbb{J}_p$ (where $0 \leqslant x_k < p$ for all k). Show that x is a unit in $\mathbb{J}_p$ if and only if $x_0 \neq 0$.

7. Prove that $\mathbb{Q}_p$ is the field of fractions of $\mathbb{J}_p$.

8. Prove that every nonzero p-adic number can be written uniquely in the form $x = x_m p^m + x_{m+1} p^{m+1} + \cdots + x_n p^n + \cdots$, with $m \in \mathbb{Z}$, $x_n \in \mathbb{Z}$ and $0 \leqslant x_n < p$ for all $n \geqslant m$, and $x_m \neq 0$.

9. Show that every domain D with a absolute value has a completion $\hat{D}$ which is a domain and has a suitable universal property.

10. Let D be a domain. Show by an example that $Q(\hat{D})$ and $\widehat{Q(D)}$ need not be isomorphic.

11. Let K be a field. Show that $K[[X]]$ is the completion of $K[X]$ with respect to v_0.

12. Let K be a field. Show that $K((X))$ is the completion of $K(X)$ with respect to v_0.

9.3. EXTENDING ABSOLUTE VALUES

The main results in this chapter concern the extension problem for absolute values: given an absolute value v on a field K and a finite field extension E of K, can v be extended to an absolute value on E, and, if so, what can we say about these extensions?

In this section we solve the extension problem for archimedean absolute values; the solution follows from general properties of complete fields, from which we can determine all complete archimedean fields. The nonarchimedean case is considered in Section 9.5.

9.3.a. Normed Vector Spaces. Let K be a field with an absolute value v and V be a vector space over K. A **norm** on V is a mapping $x \longmapsto \|x\|$ of V into $\mathbb{R}$ such that

   (i) $\|x\| \geqslant 0$, with $\|x\| = 0$ if and only if $x = 0$;

   (ii) $\|ay\| = v(a)\|y\|$; and

   (iii) $\|x + y\| \leqslant \|x\| + \|y\|$,

for all $x, y \in V$.

Our interest in normed vector spaces comes from the following example. If E is a field extension of K, and w is an absolute value on E which extends v, then E is a vector space over K and w is a norm on E.

In general, a norm on V induces a distance function $\|x - y\|$ on V and therefore a topology on V.

Lemma 9.3.1. *Let K be a field which is complete with respect to an absolute value v and V be a finite-dimensional normed vector space over K.*

(1) *In any basis of V, a sequence is a Cauchy sequence in V if and only if its sequence of i-th coordinates is a Cauchy sequence in K for every i.*

(2) *In any basis of V, a sequence converges in V if and only if its sequence of i-th coordinates converges in K for every i.*

(3) *In any basis of V, the i-th coordinate function is continuous for every i.*

(4) *V is complete.*

Proof. Let $e_1, \ldots, e_n$ be a basis of V. Let $(x_k)_{k>0}$ be a sequence of elements of V and $x_k = \sum_{1 \leqslant i \leqslant n} x_{ki} e_i$, where $x_{ki} \in K$. The inequality

$$\|\textstyle\sum_{1 \leqslant i \leqslant n} x_{ki} e_i\| \leqslant \sum_{1 \leqslant i \leqslant n} v(x_{ki}) \|e_i\|$$

shows that if the sequence $(x_{ki})_{k>0}$ is Cauchy in K (or converges in K) for every $i \leqslant n$, then $(x_k)_{k>0}$ is Cauchy in V (or converges in V). The converses holds if $n = 1$; for $n > 1$ we proceed by induction on n, beginning with (1).

Assume that $n > 1$, and let $(x_k)_{k>0}$ be a Cauchy sequence in V. Assume that, say, $(x_{kn})_{k>0}$ is not a Cauchy sequence. Then there exists $\varepsilon > 0$ such that $v(x_{in} - x_{jn}) \geqslant \varepsilon$ does not hold for all $i, j > N$ regardless of the choice of N. For every $k > 0$ there exists $i_k, j_k > k$ such that $v(x_{i_k n} - x_{j_k n}) \geqslant \varepsilon$, in particular $x_{i_k n} - x_{j_k n} \neq 0$. Let

$$y_k = (x_{i_k n} - x_{j_k n})^{-1} (x_{i_k} - x_{j_k}).$$

Then $\lim_{k \to \infty} y_k = 0$, since $v((x_{i_k n} - x_{j_k n})^{-1}) \leqslant 1/\varepsilon$ and $(x_k)_{k>0}$ is Cauchy; hence $(y_k)_{k>0}$ is a Cauchy sequence in V. Let $y_k = \sum_{1 \leqslant i \leqslant n} y_{ki} e_i$, where $y_{ki} \in K$. The definition of y_k shows that $y_{kn} = 1$ for all k. By the induction hypothesis (applied to $(y_k - e_n)_{k>0}$) $(y_{ki})_{k>0}$ is a Cauchy sequence in K for every $i < n$. Since K is complete, $y_i = \lim_{k \to \infty} y_{ki}$ exists for all $i \leqslant n$. Then

$$0 = \lim_{k \to \infty} y_k = \lim_{k \to \infty} \textstyle\sum_{1 \leqslant i \leqslant n} y_{ki} e_i = \sum_{1 \leqslant i \leqslant n} y_i e_i.$$

Since $y_n = 1$ this shows that $e_1, \ldots, e_n$ are not linearly independent, which is the required contradiction. This proves (1).

With the same notation, if $\lim_{k \to \infty} x_k = x$ holds in V, then $(x_k)_{k>0}$ is a Cauchy sequence, each $(x_{ki})_{k>0}$ is a Cauchy sequence by (1), and each $(x_{ki})_{k>0}$ has a limit in K since K is complete. This proves (2). If similarly $(x_k)_{k>0}$ is a Cauchy sequence in V, then each $(x_{ki})_{k>0}$ has a limit x_i in K and then $\sum_{1 \leqslant i \leqslant n} x_i e_i = \lim_{k \to \infty} x_k$. This proves (3) and also shows that the i-th coordinate function preserves limits of sequences, which implies (4).   □

By (3), all norms on V induce the same topology on V (see the exercises).

9.3.b. Uniqueness. We now turn to extensions.

Theorem 9.3.2. *Let K be a complete field with respect to a nontrivial absolute value v and E be a finite extension of K, of degree n. Assume that there is an absolute value w on E which extends v. Then E is complete with respect to w; moreover w is unique and is given by*

$$w(\alpha) = v(N_K^E(\alpha))^{1/n}.$$

Proof. E is complete by Lemma 9.3.1. Assume that $w(\alpha^n) \neq v(N_K^E(\alpha))$ for some $\alpha \in E$. Then $\alpha \neq 0$. Since $w(\alpha^n) > v(N(\alpha))$ implies $w((\alpha^{-1})^n) < v(N(\alpha^{-1}))$, we may assume that $w(\alpha^n) < v(N(\alpha))$. Then $\beta = \alpha^n N(\alpha)^{-1}$ satisfies $w(\beta) < 1$ and $N(\beta) = 1$. Hence $\lim_{k\to\infty} \beta^k = 0$. On the other hand, the definition of $N(\beta)$ shows that in any basis of E over K, $N(\beta)$ is a polynomial function of the coordinates of β. By Lemma 9.3.1, N is continuous, and $\lim_{k\to\infty} N(\beta^k) = 0$. This contradicts $N(\beta) = 1$. Therefore $w(\alpha^n) = v(N(\alpha))$ for all $\alpha \in E$. $\qquad\square$

9.3.c. Existence. The formula $w(\alpha) = v(N_K^E(\alpha))^{1/n}$ Theorem 9.3.2 actually yields an absolute value on E if $n = 2$.

Lemma 9.3.3. *Let K be a complete field with respect to an absolute value v and $b, c \in K$. If $v(b)^2 > 4v(c)$, then the polynomial $q(X) = X^2 - bX + c$ has a root in K.*

Proof. We use an iterative construction to produce a root of q. First $x \in K$ is a nonzero root of q if and only if $x = b - (c/x)$. Now $v(x) \geqslant \frac{1}{2}v(b) > 0$ implies $x \neq 0$ and

$$v(b - (c/x)) \geqslant v(b) - (v(c)/v(x)) \geqslant v(b) - 2(v(c)/v(b)) \geqslant v(b) - \tfrac{1}{2}v(b) = \tfrac{1}{2}v(b)$$

since $v(c) < \frac{1}{4}v(b)^2$. Therefore we can define a sequence $x_1, \ldots, x_n, \ldots$ of elements of K by

$$x_1 = \tfrac{1}{2}b, \qquad x_{n+1} = b - (c/x_n);$$

by the above, $v(x_n) \geqslant \frac{1}{2}v(b) > 0$ and $x_n \neq 0$.

We show that $(x_n)_{n>0}$ is a Cauchy sequence. Let $r = 4v(c)/v(b)^2 < 1$. We have

$$v(x_{n+2} - x_{n+1}) = v(c)\, v\left(\frac{1}{x_{n+1}} - \frac{1}{x_n}\right) = \frac{v(c)\, v(x_n - x_{n+1})}{v(x_n)v(x_{n+1})}$$

$$\leqslant \frac{4v(c)}{v(b)^2}\, (v(x_{n+1} - x_n)) = r\,(v(x_{n+1} - x_n)).$$

Therefore $v(x_{n+2} - x_{n+1}) \leqslant r^n v(x_2 - x_1)$, and $(x_n)_{n>0}$ is a Cauchy sequence. Since K is complete, $(x_n)_{n>0}$ has a limit x in K. Then $x = b - (c/x)$ and x is a root of q. □

Proposition 9.3.4. *Let K be a field which is complete with respect to a non-trivial absolute value v and does not have characteristic 2. Let E be a field extension of K of degree 2. Then $w(\alpha) = v(N_K^E(\alpha))^{1/2}$ is an absolute value on E which extends v.*

Proof. E is a Galois extension of K (e.g. by Proposition 7.9.3) and has a nontrivial K-automorphism $\alpha \longmapsto \overline{\alpha}$. Then $N(\alpha) = \alpha \overline{\alpha}$.

Let $w(\alpha) = v(N(\alpha))^{1/2}$. If $\alpha \in K$, then $N(\alpha) = \alpha^2$ and $w(\alpha) = v(\alpha^2)^{1/2} = v(\alpha)$. Properties (i) and (ii) hold, since $N(\alpha) = 0$ if and only if $\alpha = 0$, and $N(\alpha\beta) = N(\alpha)N(\beta)$ for all $\alpha, \beta \in E$. Also $w(\overline{\alpha}) = w(\alpha)$, since $N(\overline{\alpha}) = N(\alpha)$.

It remains to show (iii): $w(\alpha + \beta) \leqslant w(\alpha) + w(\beta)$. We first consider the case where $\beta = \overline{\alpha}$. Then (iii) reads $v(\alpha + \overline{\alpha}) \leqslant 2w(\alpha)$ and $v(\alpha + \overline{\alpha})^2 \leqslant 4v(\alpha\overline{\alpha})$. Let $b = \alpha + \overline{\alpha}$ and $c = \alpha\overline{\alpha}$; we want to show that $v(b)^2 \leqslant 4v(c)$. If $v(b)^2 > 4v(c)$, then by Lemma 9.3.3 the polynomial $q(X) = X^2 - bX + c = (X - \alpha)(X - \overline{\alpha})$ has a root in K; hence $\alpha \in K$, but then $b = 2\alpha$ and $v(b)^2 = (v(2\alpha))^2 = 4v(\alpha^2) = 4v(c)$. Thus (iii) holds when $\beta = \overline{\alpha}$.

Now let $\beta = 1$. Then (iii) reads $w(\alpha + 1) \leqslant w(\alpha) + 1$ and is equivalent to $v(N(\alpha + 1)) \leqslant v(N(\alpha)) + 2(v(N(\alpha)))^{1/2} + 1$. But

$$v(N(\alpha + 1)) = v((\alpha + 1)(\overline{\alpha} + 1)) = v(\alpha\overline{\alpha} + \alpha + \overline{\alpha} + 1)$$

$$\leqslant v(N(\alpha)) + v(\alpha + \overline{\alpha}) + 1 \leqslant v(N(\alpha)) + 2(v(N(\alpha)))^{1/2} + 1$$

by the case $\beta = \overline{\alpha}$. Thus (iii) holds when $\beta = 1$. Then (ii) yields $w(\alpha + \beta) \leqslant w(\alpha) + w(\beta)$ for all α and β. □

9.3.d. The Archimedean Case. We can now prove the following characterization of $\mathbb{R}$ and $\mathbb{C}$:

Theorem 9.3.5 (Ostrowski). *Up to isomorphisms which preserve absolute values, $\mathbb{R}$ and $\mathbb{C}$ are the only fields that are complete with respect to an archimedean absolute value.*

Proof. Let F be complete with respect to an archimedean absolute value v. By Corollary 9.1.2, F has characteristic 0. Up to isomorphism, $\mathbb{Q} \subseteq F$; by Theorem 9.1.5, $v_{|\mathbb{Q}}$ is equivalent to the usual absolute value on $\mathbb{Q}$. By Proposition 9.2.3, there is an isomorphism which preserves absolute values of $\mathbb{R} = \hat{\mathbb{Q}}$ onto a subfield of F. Hence we may assume that $\mathbb{R} \subseteq F$ and that $v_{|\mathbb{R}}$ is the usual absolute value.

If F contains an element i such that $i^2 = -1$, then $\mathbb{C} = \mathbb{R}(i) \subseteq F$ and the uniqueness in Theorem 9.3.2 implies that $v_{|\mathbb{C}}$ is the usual absolute value, since both induce the usual absolute value on $\mathbb{R}$. If, on the other hand, F contains no element i such that $i^2 = -1$, then $E = F(i)$ contains i, v extends to an absolute value w on E by Proposition 9.3.4, and E is complete by Theorem 9.3.2. Then $\mathbb{C} = \mathbb{R}(i) \subseteq E$, and as above, $w_{|\mathbb{C}}$ is the usual absolute value. If we can prove that $E = \mathbb{C}$, then it will follow that $F = \mathbb{R}$. In other words, we may now assume that $\mathbb{C} \subseteq F$ and that $v_{|\mathbb{C}}$ is the usual absolute value.

Assume that F contains some $\alpha \notin \mathbb{C}$. Let

$$r = \text{g.l.b.} \ \{v(z - \alpha); z \in \mathbb{C}\}$$

The inequality $|v(x - \alpha) - v(y - \alpha)| \leqslant v(x - y) = |x - y|$ for all $x, y \in \mathbb{C}$ shows that $f(z) = v(z - \alpha)$ is a continuous mapping of $\mathbb{C}$ into $\mathbb{R}$. Hence

$$D = \{z \in \mathbb{C}; \ v(z - \alpha) \leqslant r + 1\}$$

is a closed nonempty subset of $\mathbb{C}$. Moreover D is bounded, since

$$|x - y| = v((x - \alpha) - (y - \alpha)) \leqslant v(x - \alpha) + v(y - \alpha) \leqslant 2(r + 1)$$

for all $x, y \in D$. Hence D is compact, and f attains a minimum value $f(z_0)$ on D. Since

$$r = \text{g.l.b.} \ \{v(z - \alpha); z \in D\},$$

we have $f(z_0) = r$. Thus there exists $z \in \mathbb{C}$ such that $f(z) = r$.

Now let

$$C = \{z \in \mathbb{C}; \ v(z - \alpha) = r\}.$$

C is closed, since f is continuous, bounded since $C \subseteq D$, and we just showed that $C \neq \varnothing$. We now show that C is open. Since no such subset of $\mathbb{C}$ can exist, this is the required contradiction.

To show that C is open we let $x \in C$, $y \in \mathbb{C}$, and $|x - y| < r$, and we show that $y \in C$. Then $\beta = \alpha - x$ and $z = y - x$ satisfy $v(\beta) = r$ and $|z| < r$, and we want to show that $v(z - \beta) = r$. For any $n > 0$ let ω be a primitive n-th root of unity. Then

$$\beta^n - z^n = (\beta - z)(\beta - \omega z) \ldots (\beta - \omega^{n-1} z).$$

Now

$$v(\beta - \omega^i z) = v(\alpha - x - \omega^i z) \geqslant r$$

for all i. Hence

$$v(\beta - z) r^{n-1} \leqslant v((\beta - z)(\beta - \omega z) \ldots (\beta - \omega^{n-1} z))$$
$$= v(\beta^n - z^n) \leqslant v(\beta)^n + |z|^n = r^n + |z|^n$$

and $v(\beta - z) \leqslant r + (|z|^n/r^{n-1})$. Since this holds for all n and $|z| < r$, it follows that $v(\beta - z) \leqslant r$. But $v(\beta - z) = v(\alpha - x - z) \geqslant r$. Therefore $v(z - \beta) = r$. □

It follows from Theorem 9.3.5 that every field with an archimedean absolute value is, up to isomorphism, a subfield of $\mathbb{C}$ with the usual absolute value.

The usual absolute value on a subfield of $\mathbb{C}$ can always be extended to a larger subfield of $\mathbb{C}$. Thus Theorem 9.3.5 solves the extension problem for archimedean absolute values. The nonarchimedean case will be tackled in Section 9.5.

Exercise

1. Let K be a field which is complete with respect to an absolute value v, and let V be a finite-dimensional vector space over K. Prove that all norms on V induce the same topology on V.

9.4. VALUATIONS

Valuations are a generalization of nonarchimedean absolute values. They are more flexible than absolute values; we will see in the next section that they extend more readily to field extensions.

9.4.a. Ordered Abelian Groups. A **partially ordered group** is a group G with a partial order relation $\leqslant$ which is compatible with the operation on $G : x \leqslant y$ implies $xz \leqslant yz$ and $zx \leqslant zy$.

An **ordered abelian group** is a totally ordered abelian group. In an ordered abelian group, $x < y$ implies $xz < yz$ and $x^{-1} > y^{-1}$ (since $x^{-1} < y^{-1}$ would imply $1 = xx^{-1} < yx^{-1} < yy^{-1} = 1$). Ordered abelian groups are torsion free (if e.g. $x > 1$, then $x^n > 1$ for all $n > 0$). They are often denoted additively, but with valuations we prefer the multiplicative notation.

For example, the multiplicative group $\mathbb{P}$ of all positive real numbers is an ordered abelian group; so is every subgroup of $\mathbb{P}$. The direct product $\mathbb{P}^n$ of $n > 0$ copies of $\mathbb{P}$ is an ordered abelian group under the **lexicographic** order, in which $(x_1, \ldots, x_n) < (y_1, \ldots, y_n)$ if and only if there exists $1 \leqslant k \leqslant n$ such that $x_i = y_i$ for all $i < k$ and $x_k < y_k$.

Two ordered abelian groups G and H are **isomorphic** in case there exists an order-preserving isomorphism $\theta : G \longrightarrow H$. Note that the inverse isomorphism θ^{-1} is also order preserving: if $h < h'$, then $\theta^{-1}(h) < \theta^{-1}(h')$, since G is totally ordered and $\theta^{-1}(h) \geqslant \theta^{-1}(h')$ would imply $h \geqslant h'$.

9.4.b. Valuations. Let G be an ordered abelian group. Let $G \cup \{0\}$ be obtained from G by adjoining an element $0 \notin G$ such that $0 < g$ and $0g = g0 = 0$ for all $g \in G$.

A **valuation** on a field F with values in G is a mapping $v : F \longrightarrow G \cup \{0\}$ such that

(i) $v(x) = 0$ if and only if $x = 0$;

(ii) $v(xy) = v(x)v(y)$; and

(iii*) $v(x + y) \leqslant \max(v(x), v(y))$,

for all $x, y \in F$. (Since $G \cup \{0\}$ is totally ordered, we have $v(x) \leqslant v(y)$ or $v(y) \leqslant v(x)$, so that $\max(v(x), v(y))$ exists.) For example, a nonarchimedean absolute value is a valuation with values in $\mathbb{P}$.

In general, $v(1) = 1$; $v(1/x) = 1/v(x)$ for all $x \neq 0$; $v(-1) = 1$; $v(-x) = v(x)$ for all x; $v(n) = v(1 + 1 + \cdots + 1) \leqslant v(1) = 1$ for all $n > 1$; and, as before,

(iii**) if $v(x) \neq v(y)$, then $v(x + y) = \max(v(x), v(y))$.

By (ii), a valuation $v : F \longrightarrow G \cup \{0\}$ induces a multiplicative homomorphism of $F^* = F \backslash \{0\}$ into G. The **value group** of v is the subgroup $G_v = v(F^*) = (\operatorname{Im} v) \backslash \{0\}$ of G. There would be no loss of generality in assuming from the beginning that G_v is all of G.

Two valuations $v : F \longrightarrow G \cup \{0\}$ and $w : F \longrightarrow H \cup \{0\}$ on F are **equivalent** in case there exists an order preserving isomorphism $\theta : G_v \longrightarrow G_w$ such that $w(a) = \theta(v(a))$ for all $a \in F^*$.

9.4.c. Examples. Every nonarchimedean absolute value is a valuation. In fact a valuation is [equivalent to] a necessarily nonarchimedean absolute value if and only if its value group is [isomorphic to] a subgroup of $\mathbb{P}$.

We show by another example that valuations are more general than nonarchimedean absolute values. Let $F = K(X_1, \ldots, X_n)$ be the field of rational fractions in n indeterminates over any field K. Define a valuation $v_0 : F \longrightarrow \mathbb{P}^n \cup \{0\}$ as follows. Any $f/g \in F^*$ can be written uniquely in the form

$$f(X_1, \ldots, X_n)/g(X_1, \ldots, X_n) = X_1^{m_1} \ldots X_n^{m_n} \, h(X_1, \ldots, X_n)/k(X_1, \ldots, X_n),$$

where $m_1, \ldots, m_n \in \mathbb{Z}$ and $h(0, \ldots, 0), k(0, \ldots, 0) \neq 0$; let

$$v_0(f/g) = (e^{-m_1}, \ldots, e^{-m_n}) \in \mathbb{P}^n.$$

Also let $v_0(0) = 0$. Properties (ii) and (iii*) are readily verified. However, the value group $G \subseteq \mathbb{P}^n$ of v_0 cannot be embedded in $\mathbb{P}$ if $n > 1$. Indeed $\mathbb{P}$ is archimedean in the sense that for every $a, b \in \mathbb{P}$, $a, b > 1$, there exists some n such that $a^n > b$. Every subgroup of $\mathbb{P}$ is therefore archimedean. But G is not archimedean when $n > 1$: if $a = (1, e, 1, \ldots, 1) = v_0(1/X_2)$ and $b = (e, 1, \ldots, 1) = v_0(1/X_1)$, then $a, b \in G$ and $a, b > 1 = (1, \ldots, 1)$, but $a^n = (1, e^n, 1, \ldots, 1) < b$ for all n. Therefore G is not isomorphic (as an ordered abelian group) to a subgroup of $\mathbb{P}$.

9.4.d. Valuation Rings. We now give alternate definitions of valuations. Let $v : F \longrightarrow G \cup \{0\}$ be a valuation. The **valuation ring** of v is the set

$$\mathfrak{o}_v = \{ x \in F \,;\, v(x) \leqslant 1 \}.$$

Proposition 9.4.1. *Let* $v : F \longrightarrow G \cup \{0\}$ *be a valuation and* $\mathfrak{o} = \mathfrak{o}_v$ *be the valuation ring of* v.

(1) $\mathfrak{o}$ *is a subring of* F *and* F *is the field of fractions of* $\mathfrak{o}$; *in fact for every* $x \in F^*$ *we have* $x \in \mathfrak{o}$ *or* $x^{-1} \in \mathfrak{o}$.

(2) *The group of units of* $\mathfrak{o}_v$ *is* $\mathfrak{u}_v = \{ x \in F \,;\, v(x) = 1 \}$.

(3) $\mathfrak{o}$ *has exactly one maximal ideal* $\mathfrak{m}_v = \{ x \in F \,;\, v(x) < 1 \}$, *which consists of all the elements of* $\mathfrak{o}$ *which are not units in* $\mathfrak{o}$.

(4) *The ideals of* $\mathfrak{o}$ *form a chain.*

Proof. (1) is immediate. If x is a unit in $\mathfrak{o}$, then $v(x) \leqslant 1$, $v(x^{-1}) \leqslant 1$, and $v(x) = 1$. If, conversely, $v(x) = 1$, then $v(x^{-1}) = 1$ in F, $x^{-1} \in \mathfrak{o}$, and x is a unit in $\mathfrak{o}$. Now $\mathfrak{m} = \mathfrak{m}_v = \{ x \in F\bullet ;\, v(x) < 1 \}$ is an ideal of $\mathfrak{o}$; $\mathfrak{m}$ is a maximal ideal since every element of $\mathfrak{o} \backslash \mathfrak{m}$ is a unit. By (4), $\mathfrak{m}$ is the only maximal ideal of $\mathfrak{o}$.

We now prove (4). Let $\mathfrak{a}$ and $\mathfrak{b}$ be nonzero ideals of $\mathfrak{o}$. Assume that $\mathfrak{a} \not\subseteq \mathfrak{b}$. Let $a \in \mathfrak{a} \backslash \mathfrak{b}$. If $v(x) \leqslant v(a)$, then $v(xa^{-1}) \leqslant 1$ in F, $xa^{-1} \in \mathfrak{o}$, and $x = (xa^{-1})a \in \mathfrak{a}$. Let $b \in \mathfrak{b}$, $b \neq 0$. If $v(a) \leqslant v(b)$, then similarly $ab^{-1} \in \mathfrak{o}$ and $a = (ab^{-1})b \in \mathfrak{b}$, a contradiction; therefore $v(a) > v(b)$ and $b \in \mathfrak{a}$. Thus $\mathfrak{b} \subseteq \mathfrak{a}$. $\qquad\square$

Valuation rings can be characterized abstractly as follows:

Proposition 9.4.2. *Let* $\mathfrak{o}$ *be a domain,* $\mathfrak{u}$ *be its group of units, and* $F = Q(\mathfrak{o})$ *be its field of fractions. The following conditions are equivalent:*

(1) $\mathfrak{o}$ *is the valuation ring of some valuation on* F.

(2) *The ideals of* $\mathfrak{o}$ *form a chain.*

(3) *For every* $x \in F^*$ *we have* $x \in \mathfrak{o}$ *or* $x^{-1} \in \mathfrak{o}$.

Then $G = F^*/\mathfrak{u}$ *is an ordered abelian group and the projection* $F \longrightarrow G \cup \{0\}$ *is a valuation on* F *whose valuation ring is* $\mathfrak{o}$.

Proof. (1) implies (2) by Proposition 9.4.1.

(2) implies (3). Let $x = a/b \in F^*$ (with $a,b \in \mathfrak{o}$, $a,b \neq 0$). If $\mathfrak{o}a \subseteq \mathfrak{o}b$, then $a = bt$ for some $t \in \mathfrak{o}$ and $x = a/b = t \in \mathfrak{o}$. Therefore $x \notin \mathfrak{o}$ implies $\mathfrak{o}a \not\subseteq \mathfrak{o}b$, $\mathfrak{o}b \subseteq \mathfrak{o}a$ by (2), and $x^{-1} = b/a \in \mathfrak{o}$.

(3) implies (1). The group of units $\mathfrak{u}$ of $\mathfrak{o}$ is a multiplicative subgroup of F^*. Let $G = F^*/\mathfrak{u}$. An order relation $\leqslant$ on G is well defined by

$$x\mathfrak{u} \leqslant y\mathfrak{u} \qquad \text{if and only if} \quad xy^{-1} \in \mathfrak{o}.$$

If indeed $x\mathfrak{u} = x'\mathfrak{u}$ and $y\mathfrak{u} = y'\mathfrak{u}$, then $xy^{-1} \in x'y'^{-1}\mathfrak{u}$, and $xy^{-1} \in \mathfrak{o}$ if and only if $x'y'^{-1} \in \mathfrak{o}$. We see that $\leqslant$ is reflexive and transitive; if $xy^{-1} \in \mathfrak{o}$ and $yx^{-1} =$

$(xy^{-1})^{-1} \in \mathfrak{o}$, then $xy^{-1} \in \mathfrak{u}$ and $x\mathfrak{u} = y\mathfrak{u}$. Moreover $x\mathfrak{u} \leqslant y\mathfrak{u}$ implies $(x\mathfrak{u})(z\mathfrak{u}) \leqslant (y\mathfrak{u})(z\mathfrak{u})$, and it follows from (3) that G is an ordered abelian group.

A valuation $v = v_\mathfrak{o} : F \longrightarrow G \cup \{0\}$ is now defined by

$$v_\mathfrak{o}(x) = \begin{cases} x\mathfrak{u} & \text{if } x \neq 0, \\ 0 & \text{if } x = 0. \end{cases}$$

Indeed (i) and (ii) are immediate; (iii*) holds if x, y, or $x + y$ is 0; if $x, y, x + y \neq 0$ and, say, $v(x) \leqslant v(y)$, then $xy^{-1} \in \mathfrak{o}$, $(x + y)y^{-1} = xy^{-1} + 1 \in \mathfrak{o}$, and $v(x + y) \leqslant v(y)$. We see that $v(x) \leqslant 1$ (= $1\mathfrak{u}$) if and only if $x \in \mathfrak{o}$; thus $\mathfrak{o} = \mathfrak{o}_v$. $\qquad\square$

In the above, $v_\mathfrak{o}$ is the valuation **induced** by $\mathfrak{o}$.

A **valuation ring** (or a **valuation domain**) is a domain which satisfies the equivalent conditions in Proposition 9.4.2. A valuation ring **of** a field F is a valuation ring whose fraction field is F; equivalently, a subring $\mathfrak{o}$ of F such that $x \in \mathfrak{o}$ or $x^{-1} \in \mathfrak{o}$ holds for every $x \in F^*$. A valuation is completely determined (up to equivalence) by its valuation ring:

Proposition 9.4.3. *Let v be a valuation on a field F and $\mathfrak{o} = \mathfrak{o}_v$ be its valuation ring. Then v is equivalent to $v_\mathfrak{o}$. In particular, two valuations are equivalent if and only if they have the same valuation ring.*

Proof. v and $v_\mathfrak{o}$ induce surjective multiplicative homomorphisms

where $\mathfrak{u}$ is the group of units of $\mathfrak{o}$. We see that $\operatorname{Ker} v = \mathfrak{u} = \operatorname{Ker} v_\mathfrak{o}$. Therefore there is a multiplicative isomorphism $\theta : F^*/\mathfrak{u} \longrightarrow G_v$ such that $v = \theta \circ v_\mathfrak{o}$; θ is order preserving since $v(x) \leqslant v(y)$ is equivalent to $v(xy^{-1}) \leqslant 1$, $xy^{-1} \in \mathfrak{o}$, $x\mathfrak{u} \leqslant y\mathfrak{u}$, and $v_\mathfrak{o}(x) \leqslant v_\mathfrak{o}(y)$. $\qquad\square$

9.4.e. The Residue Class Field. Let v be a valuation on a field F, $\mathfrak{o}$ be its valuation ring, and $\mathfrak{m}$ be the maximal ideal of $\mathfrak{o}$. Then $\overline{F}_v = \mathfrak{o}/\mathfrak{m}$ is a field and the projection $\pi = \pi_v : \mathfrak{o} \longrightarrow \overline{F}_v$ is a homomorphism; the coset $\pi_v(x) = x + \mathfrak{m}$ is the **residue class** of x and $\overline{F}_v = \mathfrak{o}/\mathfrak{m}$ is the **residue class field** of v. For example, the residue class field of the valuation v_p on $\mathbb{Q}$ is $\cong \mathbb{Z}_p$ (see the exercises).

Proposition 9.4.4. *Let v be a nonarchimedean absolute value on a field F. The residue class fields of v and $\hat{v}$ are isomorphic.*

Proof. Let $\mathfrak{o}$ be the valuation ring of v, $\mathfrak{m}$ be its maximal ideal, $\hat{\mathfrak{o}}$ be the valuation ring of $\hat{v}$, and $\hat{\mathfrak{m}}$ be its maximal ideal. Since $\hat{v}$ extends v, we have

$\hat{o} \cap F = o$ and $\hat{m} \cap F = m$. Hence there is a homomorphism $\theta : o/m \longrightarrow \hat{o}/\hat{m}$ such that the following square commutes:

where the horizontal maps are projections. Then θ is injective: since $\hat{m} \cap F = m$, $\theta(\pi_o(x)) = 0$ implies $\pi_{\hat{o}}(x) = 0$, $x \in \hat{m}$, $x \in m$, and $\pi_o(x) = 0$. Also there is for each $\alpha \in \hat{o} \backslash \hat{m}$ some $x \in F$ such that $\hat{v}(x - \alpha) < \hat{v}(\alpha) = 1$. By (iii**), $v(x) = \hat{v}(x) = \hat{v}(\alpha) = 1$. Hence $x \in o$, $x - \alpha \in \hat{m}$, and $\pi_{\hat{o}}(\alpha) = \pi_{\hat{o}}(x) = \theta(\pi_o(x))$. Therefore θ is surjective.   □

9.4.f. Discrete Valuations. A valuation is **discrete** in case its value group is cyclic (isomorphic to $\mathbb{Z}$). Since $\mathbb{P}$ contains cyclic groups, every discrete valuation is equivalent to a (discrete) nonarchimedean absolute value.

Let v be a discrete valuation on a field F and $o = o_v$, $m = m_v$, $u = u_v$. Then G_v is cyclic and has a single generator $v(p)$; since $v(p)^{-1}$ also generates G_v, we may assume that $v(p) < 1$. Then $p \in m$. If $x \in F^*$, then $v(x) = v(p)^k$ for some $k \in \mathbb{Z}$ so that $u = x/p^k \in u$ and $x = up^k$; if, conversely, $x = up^k$, with $u \in u$, then $v(x) = v(p)^k$, so that $k = \log_{v(p)} v(x)$, and then $u = x/p^k$. This proves:

Proposition 9.4.5. *Let v be a discrete valuation on a field F and $v(p) < 1$ generate G_v. Every $x \in F^*$ can be written uniquely in the form $x = up^k$ for some $u \in u_v$ and $k \in \mathbb{Z}$.*

Using Proposition 9.4.5, the reader will show that valuation rings of discrete valuations (called **discrete** valuation rings) can be characterized as follows:

Proposition 9.4.6. *Let o be a domain and $F = Q(o)$ be its field of fractions. Then o is the valuation ring of a discrete valuation on F if and only if o is a principal ideal domain with a unique prime ideal $\mathfrak{p} \neq 0$.*

In Proposition 9.4.5 the element p generates the unique prime ideal $\mathfrak{p} = m_v$ and is, up to multiplication by units, the only prime element of o_v.

$\mathbb{J}_p$ is a discrete valuation ring. In fact the description of $\mathbb{J}_p$ in Section 9.2 extends to all discrete valuation rings:

Proposition 9.4.7. *Let v be a discrete valuation on a field F and $v(p) < 1$ generate G_v. Let $\mathfrak{r}$ be a set of representatives of the residue classes of o_v. Every $x \in F^*$ can be written uniquely in the form*

$$x = r_k p^k + r_{k+1} p^{k+1} + \cdots + r_n p^n + \cdots,$$

where $k \in \mathbb{Z}$ and $r_k, r_{k+1}, \ldots \in \mathfrak{r}$, $r_k \notin m_v$; then $v(x) = v(p)^k$.

Proof. By definition, every residue class $x + \mathfrak{m} \subseteq \mathfrak{o}$ contains exactly one element of $\mathfrak{r}$. Let $x \in F^*$. As before we have $v(x) = v(p)^k = v(p^k)$ for some unique $k \in \mathbb{Z}$ and $x = up^k$ for some unique $u \in \mathfrak{o} \setminus \mathfrak{m}$. Then $u \in r_k + \mathfrak{m}$ for some unique $r_k \in \mathfrak{r}$ and $r_k \notin \mathfrak{m}$. Since $\mathfrak{m}$ is generated by p, we have $u = r_k + x_{k+1}p$ and $x = r_k p^k + x_{k+1} p^{k+1}$ for some unique $x_{k+1} \in \mathfrak{o}$. Then $x_{k+1} = r_{k+1} + x_{k+2}p$ and $x = r_k p^k + r_{k+1} p^{k+1} + x_{k+2} p^{k+2}$ for some unique $r_{k+1} \in \mathfrak{r}$ and $x_{k+2} \in \mathfrak{o}$. Continuing in this fashion yields the unique series expansion $x = r_k p^k + r_{k+1} p^{k+1} + \cdots + r_n p^n + \cdots$, which converges to x since $v(r_n p^n + \cdots) \leqslant v(p)^n$ for all n. $\square$

For example, let $F = K(X)$ and $v = v_0$. Then $v(X) = 1/e$ generates G_v. By Proposition 9.4.7, every rational fraction $f/g \in K(X)$ can be expanded into a Laurent series (which converges to f/g in the topology induced by v_0).

Since a discrete valuation v on a field F is equivalent to a nonarchimedean absolute value, Section 9.2 yields a completion $\widehat{F}_v$. Our next result generalizes Proposition 9.2.4.

Proposition 9.4.8. *Let v be a discrete valuation on a field F and $v(p) < 1$ generate G_v. Let $\mathfrak{r}$ be a set of representatives of the residue classes of $\mathfrak{o}_v$. Every $x \in \widehat{F}_v$ can be written uniquely in the form*

$$x = r_k p^k + r_{k+1} p^{k+1} + \cdots + r_n p^n + \cdots ,$$

where $k \in \mathbb{Z}$ and $r_k, r_{k+1}, \ldots \in \mathfrak{r}$, $r_k \notin \mathfrak{m}_v$. Conversely, every such series converges to an element of $\widehat{F}_v$.

Proof. By Proposition 9.4.4, $\mathfrak{r}$ is also a set of representatives of the residue classes of $\hat{\mathfrak{o}}_v$. Moreover $\hat{v}(p) = v(p)$ is a generator of $G_{\hat{v}} = G_v$. Therefore the first part of the statement follows from Proposition 9.4.7. The second part is clear since $\widehat{F}_v$ is complete. $\square$

Exercises

1. Verify that $\mathbb{P}^n$ is an ordered abelian group under the lexicographic order.
2. Verify that every valuation v has property (iii**): if $v(x) \neq v(y)$, then $v(x + y) = \max(v(x), v(y))$.
3. Verify that v_0 is a valuation on $K(X_1, \ldots, X_n)$.
4. Find the residue class field of the valuation v_p on $\mathbb{Q}$.
5. Let K be a field and $q(X) \in K[X]$ be an irreducible polynomial. Find the residue class field of the valuation v_q on $K(X)$.
6. A **place** on a field F with values in a field Q is a mapping $\pi : F \longrightarrow Q \cup \{\infty\}$ such that

    $\mathfrak{o}_\pi = \{x \in F \, ; \, \pi(x) \neq \infty\}$ is a subring of F and $\pi_{|\mathfrak{o}_\pi}$ is a homomorphism;

    if $\pi(x) = \infty$, then $x^{-1} \in \mathfrak{o}_\pi$ and $\pi(x^{-1}) = 0$.

Show that every valuation v on a field F gives rise to a place $\pi_v : F \longrightarrow \overline{F} \cup \{\infty\}$, where $\overline{F}$ is the residue class field and $\mathfrak{o}_\pi = \mathfrak{o}_v$.

7. Let π be a place on a field F with values in a field Q. Show that $\{\pi(x) \in Q; \pi(x) \neq \infty\}$ is a subfield of Q.

8. Define equivalence of places. Show that every place π is equivalent to the place π_v of the valuation v determined by the valuation ring $\mathfrak{o}_\pi$.

9. Let $\mathfrak{o}$ be a domain and $F = Q(\mathfrak{o})$ be the field of fractions of $\mathfrak{o}$. Show that $\mathfrak{o}$ is the valuation ring of a discrete valuation on F if and only if $\mathfrak{o}$ is a principal ideal domain with a unique prime ideal $\mathfrak{p} \neq 0$.

10. Prove that a valuation ring is discrete if and only if it is noetherian.

11. Prove that every ordered abelian group G is the value group of a valuation.

9.5. EXTENDING VALUATIONS

Any valuation on a field K can be extended to every field extension of K (Theorem 9.5.2). With this result we attack the extension problem for nonarchimedean absolute values and obtain further properties of finite field extensions.

9.5.a. Homomorphisms. First we prove:

Lemma 9.5.1. *Let R be a subring of a field K. Every homomorphism φ of R into an algebraically closed field L can be extended to a valuation ring of K.*

Proof. Let $\mathscr{S}$ be the set of all pairs (S, ψ) where $S \supseteq R$ is a subring of K and $\psi : S \longrightarrow L$ is a homomorphism which extends φ. Let $(S, \psi) \leqslant (T, \chi)$ if and only if $S \subseteq T$ and $\psi = \chi_{|S}$. It is immediate that $\mathscr{S}$ is an inductive partially ordered set. By Zorn's Lemma, $\mathscr{S}$ has a maximal element (V, ω). Then ω cannot be extended to any subring $V \subsetneqq S \subseteq K$. We show that V is a valuation ring of K. It suffices to prove that R is a valuation ring of K if (R, φ) is maximal in $\mathscr{S}$ (if φ cannot be extended to any subring $S \supsetneqq R$ of K).

First $\operatorname{Im} \varphi \subseteq L$ is a domain and $\mathfrak{m} = \operatorname{Ker} \varphi$ is a prime ideal of R. We claim that every $a \in R \backslash \mathfrak{m}$ is a unit of R. Indeed

$$S = \{ x/a^n \in K ; x \in R, \ k \geqslant 0 \}$$

is a subring of K, which contains R since $a^0 = 1$. Moreover $x/a^m = y/a^n$ implies $a^n x = a^m y$, $\varphi(a)^n \varphi(x) = \varphi(a)^m \varphi(y)$, and $\varphi(x)/\varphi(a)^m = \varphi(y)/\varphi(a)^n$, since $\varphi(a) \neq 0$. Hence a mapping $\psi : S \longrightarrow L$ is well defined by $\psi(x/a^n) = \varphi(x)/\varphi(a)^n$. It is immediate that ψ is a homomorphism which extends φ. Since R is maximal, we have $S = R$ and $a^{-1} = 1/a \in R$. Thus every $a \in R \backslash \mathfrak{m}$ is a unit of R. Hence $\mathfrak{m}$ is a maximal ideal of R. Therefore $F = \operatorname{Im} \varphi \cong R/\mathfrak{m}$ is a field.

Let $c \in K \backslash R$ and $S = R[c]$ be the subring of K generated by R and c. We show that $\mathfrak{m}S = S$. Recall that $S = R[c]$ is the image of the evaluation homomorphism $\varepsilon_c : f \longmapsto f(c)$ of $R[X]$ into K, whose kernel

$$\mathfrak{A} = \{ f(X) \in R[X]; \ f(c) = 0 \}$$

is an ideal of $R[X]$. Now $\varphi : R \longrightarrow F$ extends to a surjective homomorphism $\overline{\varphi} : R[X] \longrightarrow F[X]$; if $f(X) = a_n X^n + \cdots + a_0$, then $\overline{\varphi}(f) = {}^\varphi f = \varphi(a_n) X^n + \cdots + \varphi(a_0)$. Since $\overline{\varphi}$ is surjective, $\mathfrak{B} = \overline{\varphi}(\mathfrak{A})$ is an ideal of $F[X]$.

Assume that $\mathfrak{B} \neq F[X]$. Then $\mathfrak{B}$ consists of all the multiples of some polynomial $g(X) \in F[X]$ which is not a nonzero constant. Since L is algebraically closed, g has a root $\gamma \in L$. Let $\varepsilon_\gamma : h \longmapsto h(\gamma)$ be the evaluation homomorphism of $F[X]$ into L. Then $\varepsilon_\gamma(\overline{\varphi}(f)) = 0$ for all $f \in \mathfrak{A}$, since $\overline{\varphi}(f) \in \mathfrak{B}$ is a multiple of g. Thus $\mathfrak{A} = \mathrm{Ker}\, \varepsilon_c \subseteq \mathrm{Ker}\, (\varepsilon_\gamma \circ \overline{\varphi})$ and $\varepsilon_\gamma \circ \overline{\varphi}$ factors through ε_c:

$$\begin{array}{ccc} R[X] & \xrightarrow{\ \overline{\varphi}\ } & F[X] \\ {\scriptstyle \varepsilon_c} \downarrow & & \downarrow {\scriptstyle \varepsilon_\gamma} \\ S & \dashrightarrow & L \\ & {\scriptstyle \psi} & \end{array}$$

This yields a homomorphism $\psi : S \longrightarrow L$ which extends φ. Since we assumed $c \notin R$, this contradicts the maximality of R. Therefore $\mathfrak{B} = F[X]$. Then $1 = \overline{\varphi}(f)$ for some $f = a_n X^n + \cdots + a_0 \in \mathfrak{A}$. Hence $\varphi(a_0) = 1$ and $\varphi(a_i) = 0$ for all $i > 0$, so that $a_0 - 1 \in \mathfrak{m}$ and $a_i \in \mathfrak{m}$ for all $i > 0$. On the other hand, $a_n c^n + \cdots + a_1 c + a_0 = 0$, since $f \in \mathfrak{A}$. Hence $1 = 1 - a_n c^n - \cdots - a_1 c - a_0 \in \mathfrak{m}S$ and $\mathfrak{m}S = S$.

We can now show that R is a valuation ring. Let $c \in K$, $c \neq 0$. Assume that $c \notin R$ and $c^{-1} \notin R$. By the above, $\mathfrak{m}R[c] = R[c]$ and $\mathfrak{m}R[c^{-1}] = R[c^{-1}]$. Then

$$m_k c^k + \cdots + m_1 c + m_0 = 1 = n_\ell c^{-\ell} + \cdots + n_1 c^{-1} + n_0 \qquad (9.1)$$

for some $m_0, \ldots, m_k, n_0, \ldots, n_\ell \in \mathfrak{m}$. We have $k, \ell \geqslant 1$ (since $1 \notin \mathfrak{m}$) and may assume that $k \geqslant \ell$ and that $k + \ell$ is minimal such that (9.1) holds for some $m_0, \ldots, m_k, n_0, \ldots, n_\ell \in \mathfrak{m}$. Then $1 - n_0$ is a unit of R (since $1 - n_0 \notin \mathfrak{m}$) and $(1 - n_0) c^\ell = n_\ell + n_{\ell-1} c + \cdots + n_1 c^{\ell-1}$ yields

$$c^k = (1 - n_0)^{-1} (n_\ell c^{k-\ell} + n_{\ell-1} c^{k-\ell+1} + \cdots + n_1 c^{k-1}).$$

Substituting for c^k in $m_k c^k + \cdots + m_1 c + m_0 = 1$ then lowers k by 1, which contradicts the minimality of $k + \ell$. $\qquad \square$

Theorem 9.5.2. *Let K be a subfield of E. Every valuation on K can be extended to a valuation on E.*

Proof. Let $v : K \longrightarrow G \cup \{0\}$ be a valuation on K. Let $\mathfrak{o}$ be the valuation ring of v, $\mathfrak{m}$ be its maximal ideal, $\mathfrak{u} = \mathfrak{o} \backslash \mathfrak{m}$ be its group of units, and $F = \mathfrak{o}/\mathfrak{m}$ be the residue class field. Let L be the algebraic closure of F. By Lemma 9.5.1, the homomorphism $\pi : \mathfrak{o} \longrightarrow \mathfrak{o}/\mathfrak{m} = F \subseteq L$ extends to a homomorphism $\psi :$ $\mathfrak{O} \longrightarrow L$, where $\mathfrak{O}$ is a valuation ring of E and $\mathfrak{o} \subseteq \mathfrak{O}$. Let $\mathfrak{M}$ be the maximal ideal of $\mathfrak{O}$ and $\mathfrak{U}$ be its group of units.

We show that $\mathfrak{O} \cap K = \mathfrak{o}$, $\mathfrak{M} \cap K = \mathfrak{m}$, and $\mathfrak{U} \cap K = \mathfrak{u}$. We already have $\mathfrak{o} \subseteq$ $\mathfrak{O} \cap K$. If $x \in \mathfrak{m}$, then $\pi(x) = 0$, $\psi(x) = 0$, x is not a unit of $\mathfrak{O}$, and $x \in \mathfrak{M}$. If now $x \in K$ and $x \notin \mathfrak{o}$, then $x^{-1} \in \mathfrak{m}$, $x^{-1} \in \mathfrak{M}$, and $x \notin \mathfrak{O}$; hence $\mathfrak{O} \cap K = \mathfrak{o}$. Similarly, if $x \in \mathfrak{o}$ and $x \notin \mathfrak{m}$, then x is a unit of $\mathfrak{o}$, x is a unit of $\mathfrak{O}$, and $x \notin \mathfrak{M}$; hence $\mathfrak{M} \cap \mathfrak{o} = \mathfrak{m} = \mathfrak{M} \cap K$. Then $\mathfrak{U} \cap K = \mathfrak{u}$.

The inclusion $K^* \longrightarrow E^*$ induces a homomorphism

$\mu : K^*/\mathfrak{u} \longrightarrow E^*/\mathfrak{U}$, which is injective since $\mathfrak{U} \cap K = \mathfrak{u}$. Moreover $\mu(x\mathfrak{u}) \leqslant$ $\mu(y\mathfrak{u})$ in $E^*/\mathfrak{U}$ is equivalent to $x\mathfrak{u} \leqslant y\mathfrak{u}$ in $K^*/\mathfrak{u}$, since both are equivalent to $xy^{-1} \in \mathfrak{O} \cap K = \mathfrak{o}$. Therefore, up to the isomorphism $K^*/\mathfrak{u} \cong \mathrm{Im}\, \mu$, the valuation $v_{\mathfrak{O}}$ on E extends the valuation $v_{\mathfrak{o}}$ on K. But $v_{\mathfrak{o}}$ is equivalent to the given valuation v. Therefore v can be extended to E. $\qquad\square$

9.5.b. The Ramification Index. We use the following notation when E is an extension of K: v is a valuation on K, with valuation ring $\mathfrak{o}$, residue class field $\overline{K} = \mathfrak{o}/\mathfrak{m}$, and value group G_v; and w is a valuation on E which extends v, with valuation ring $\mathfrak{O}$, residue class field $\overline{E} = \mathfrak{O}/\mathfrak{M}$, and value group G_w. Since w extends v, we have $\mathfrak{O} \cap K = \mathfrak{o}$ and $\mathfrak{M} \cap K = \mathfrak{m}$.

Proposition 9.5.3. *When E is an extension of K and w extends v, then $[G_w : G_v] \leqslant [E : K]$.*

Proof. Let $(\alpha_i)_{i \in I}$ be a linearly dependent family of nonzero elements of E. We have $\sum_{i \in I} c_i \alpha_i = 0$ for some $c_i \in K$ which are almost all zero but not all zero. Then $m = \max_{i \in I} w(c_i \alpha_i) > 0$. We have $m = w(c_j \alpha_j)$ for some $j \in I$. If $w(c_i \alpha_i) < m$ for all $i \neq j$, then $w(\sum_{i \in I} c_i \alpha_i) = w(c_j \alpha_j) > 0$ by (iii**), contradicting $\sum_{i \in I} c_i \alpha_i = 0$; therefore $w(c_i \alpha_i) = w(c_j \alpha_j)$ for some $i \neq j$ and $w(\alpha_i) G_v = w(\alpha_j) G_v$. This shows that if the cosets $w(\alpha_i) G_v$ are all distinct, then the family $(\alpha_i)_{i \in I}$ is linearly independent over K. $\qquad\square$

The index $e(E : K) = [G_w : G_v]$ is the **ramification index** of E over K relative to w. (Actually, it is the ramification index of w.) By Proposition 9.5.3, $e(E : K)$ is finite if $K \subseteq E$ is finite.

We can now prove:

Theorem 9.5.4. *Let E be a finite extension of K. Every nonarchimedean absolute value on K can be extended to a nonarchimedean absolute value on E.*

Proof. A nonarchimedean absolute value on K is a valuation v such that $G_v \subseteq \mathbb{P}$. By Theorem 9.5.2, v can be extended to a valuation w on any finite extension E of K. Let $e = e(E : K)$ be the ramification index. Then $g \longmapsto g^e$ is a homomorphism of G_w into $\mathbb{P}$, which is injective, since the ordered abelian group G_w is torsion free. Now each $r \in \mathbb{P}$ has a positive e-th root in $\mathbb{P}$; this yields an injective endomorphism $r \longmapsto r^{1/e}$ of $\mathbb{P}$. Composition yields an injective homomorphism $\theta : g \longmapsto (g^e)^{1/e}$ of G_w into $\mathbb{P}$ which is the identity on G_v. Then $\theta \circ w$ is a nonarchimedean absolute value on E which extends v. $\qquad \square$

Using the same homomorphism $g \longmapsto g^e$ the reader will show:

Corollary 9.5.5. *Let E be a finite extension of K and w extend v. If v is discrete, then w is discrete.*

9.5.c. The Residue Class Degree. Let E be an extension of K. Since $\mathfrak{O} \cap K = \mathfrak{o}$, $\mathfrak{M} \cap K = \mathfrak{m}$, and $\mathfrak{M} \cap \mathfrak{o} = \mathfrak{m}$, there is a homomorphism $\varphi : \overline{K} \longrightarrow \overline{E}$ such that the following square commutes:

$$
\begin{array}{ccc}
\mathfrak{o} & \longrightarrow & \overline{K} = \mathfrak{o}/\mathfrak{m} \\
\cap \downarrow & & \vdots \downarrow \varphi \\
\mathfrak{O} & \longrightarrow & \overline{E} = \mathfrak{O}/\mathfrak{M}
\end{array}
$$

where the horizontal maps are projections. Hence $\overline{E}$ is an extension of $\overline{K}$. It is convenient to identify the residue classes $\overline{x} = x + \mathfrak{m} \in \overline{K}$ and $\overline{x} = x + \mathfrak{M} \in \overline{E}$ for each $x \in K$, so that $\overline{K}$ is a subfield of $\overline{E}$. The **residue class degree** of E over K relative to w is the degree $f(E : K) = [\overline{E} : \overline{K}]$.

Proposition 9.5.6. *When E is an extension of K and w extends v, then $e(E : K)\, f(E : K) \leqslant [E : K]$.*

Proposition 9.5.6 is a consequence of the following Lemma.

Lemma 9.5.7. *Let E be an extension of K. Let $(\alpha_i)_{i \in I}$ be a family of nonzero elements of E such that the cosets $w(\alpha_i)G_v$ are all different. Let $(\beta_j)_{j \in J}$ be elements of $\mathfrak{O}$ whose residue classes $(\overline{\beta}_j)_{j \in J}$ are linearly independent over $\overline{K}$. Then the products $(\alpha_i \beta_j)_{i \in I, j \in J}$ are linearly independent over K.*

Proof. Assume that $\sum_{i \in I, j \in J} c_{ij}\alpha_i\beta_j = 0$, where $c_{ij} \in K$, $c_{ij} = 0$ for almost all i, j, and $c_{ij} \neq 0$ for some i, j.

For each i let $\gamma_i = \sum_{j \in J} c_{ij}\beta_j \in E$ and $c_{it} = \max_{j \in J} w(c_{ij})$. If $c_{ij} \neq 0$ for some j, then $c_{it} \neq 0$ and $c_{ij}/c_{it} \in \mathfrak{o}$ for all j; since $(\overline{\beta}_j)_{j \in J}$ are linearly independent over $\mathfrak{o}/\mathfrak{m}$, we have

$$\sum_{j \in J} \overline{(c_{ij}/c_{it})}\,\overline{\beta}_j \neq 0 \quad \text{in} \quad \mathfrak{O}/\mathfrak{M},$$

$\sum_{j \in J}(c_{ij}/c_{it})\beta_j \notin \mathfrak{M}$, $w(\sum_{j \in J}(c_{ij}/c_{it})\beta_j) = 1$, and $w(\sum_{j \in J} c_{ij}\beta_j) = w(c_{it})$. Thus either $w(\gamma_i) = 0$ or $w(\gamma_i) \in G_v$.

As in the proof of Proposition 9.5.3, let $m = \max_{i \in I}(w(\gamma_i\alpha_i))$. Then $m > 0$ and $m = w(\gamma_k\alpha_k)$ for some $k \in I$. If $w(\gamma_i\alpha_i) < m$ for all $i \neq k$, then $w(\sum_{i \in I}\gamma_i\alpha_i) = w(\gamma_k\alpha_k) > 0$ by (iii**), contradicting $\sum_{i \in I}\gamma_i\alpha_i = 0$; therefore $w(\gamma_i\alpha_i) = w(\gamma_k\alpha_k) > 0$ for some $i \neq k$. Then $w(\alpha_i)G_v = w(\alpha_k)G_v$, which contradicts the hypothesis. $\square$

Proposition 9.5.8. *If K is complete for a discrete absolute value v and E is a finite extension of K, then $e(E:K)\,f(E:K) = [E:K]$.*

Proof. The notation is as before. Let $v(p) < 1$ generate G_v and $w(\rho) < 1$ generate G_w (which is cyclic by Corollary 9.5.5). Since $e = e(E:K)$ is finite, we have $w(\rho)^e = v(p)$ and $\rho^e = up$ for some $u \in \mathfrak{U}$. Then the cosets $G_v, \rho G_v, \ldots, \rho^{e-1}G_v$ are all distinct. Let $f = f(E:K)$ and $\beta_1, \ldots, \beta_f$ be elements of $\mathfrak{O}$ whose residue classes $\overline{\beta}_1, \ldots, \overline{\beta}_f$ constitute a basis of $\overline{E}$ over $\overline{K}$. By Lemma 9.5.7, the products $\rho^i\beta_j$ ($0 \leqslant i < e$, $1 \leqslant j \leqslant f$) are linearly independent over K.

Let $\mathfrak{r}$ be a set of representatives of the residue classes of $\mathfrak{o}$ and

$$\mathfrak{R} = \{r_1\beta_1 + \cdots + r_f\beta_f; r_1, \ldots, r_f \in \mathfrak{r}\}.$$

Since $\overline{\beta}_1, \ldots, \overline{\beta}_f$ is a basis of $\overline{E}$ over $\overline{K}$, $\mathfrak{R}$ is a set of representatives of the residue classes of $\mathfrak{O}$. Let $\alpha \in E^*$. As in the proof of Proposition 9.4.7, $w(\alpha) = w(\rho)^k$ for some $k \in \mathbb{Z}$, $k = e\ell + i$ for some $\ell \in \mathbb{Z}$, $0 \leqslant i < e$, $w(\alpha) = w(\rho^i p^\ell)$, and $\alpha = u\rho^i p^\ell$ for some $u \in \mathfrak{U}$; $u = \sum_{1 \leqslant j \leqslant f} r_{\ell i j}\beta_j + \mu$ for some $r_{\ell i j} \in \mathfrak{r}$ and $\mu \in \mathfrak{M}$,

$$\alpha = \sum_{1 \leqslant j \leqslant f} r_{\ell i j} p^\ell \rho^i \beta_j + \alpha',$$

where $\alpha' = \rho^i p^\ell \mu$, $w(\alpha') < w(\rho^i p^\ell) = w(\rho)^k$, and $w(\alpha') \leqslant w(\rho)^{k+1}$. Continuing in this fashion, we obtain a series expansion

$$\alpha = \sum_{m \geqslant \ell, 0 \leqslant i < e, 1 \leqslant j \leqslant f} r_{mij} p^m \rho^i \beta_j.$$

Since K is complete each series $\sum_{m \geqslant \ell} r_{mij} p^m$ converges in K, and we see that the elements $\rho^i\beta_j$ ($0 \leqslant i < e$, $1 \leqslant j \leqslant f$) span E and hence constitute a basis. $\square$

9.5.d. Finite Extensions

Theorem 9.5.9. *Let v be a nonarchimedean absolute value on a field K, $\hat{K} = \hat{K}_v$ be the completion of K, $E = K(\alpha)$ be a finite extension of K, and $q(X) = \mathrm{Irr}_K(\alpha)$. Let $q_1(X), \ldots, q_r(X)$ be the distinct monic irreducible factors of $q(X)$ in $\hat{K}[X]$. There are exactly r absolute values $w_1, \ldots, w_r$ on E which extend v; moreover $\hat{E}_{w_i} \cong \hat{K}[X]/(q_i)$, $[\hat{E}_{w_i} : \hat{K}] = \deg q_i$, and $\sum_i [\hat{E}_{w_i} : \hat{K}] \leqslant [E : K]$.*

If E is separable over K, then $[E : K] = \sum_i [\hat{E}_{w_i} : \hat{K}]$.

Recall that every finite separable extension is simple (Theorem 7.2.10).

Proof. Let $\hat{E}_i = \hat{K}[X]/(q_i)$. Then $\hat{E}_i$ is a finite extension of $\hat{K}$ and $\hat{E}_i = \hat{K}(\alpha_i)$, where $\mathrm{Irr}_{\hat{K}}(\alpha_i) = q_i$; in fact $[\hat{E}_i : \hat{K}] = \deg q_i$ and $\sum_i [\hat{E}_i : \hat{K}] = \sum_i \deg q_i \leqslant \deg q = [E : K]$. If E is separable over K, then q is separable, $q = q_1 \ldots q_r$, and $[E : K] = \sum_i [\hat{E}_i : \hat{K}]$.

By Theorem 9.5.4 there is an absolute value $\hat{v}_i$ on $\hat{E}_i$ which extends $\hat{v}$. Since $\hat{K}$ is complete, $\hat{v}_i$ is unique and $\hat{E}_i$ is complete, by Theorem 9.3.2. Now $(q) \subseteq (q_i)$ in $\hat{K}[X]$. Hence the inclusion $K[X] \subseteq \hat{K}[X]$ induces a K-homomorphism $\varphi_i : E \cong K[X]/(q) \longrightarrow \hat{K}[X]/(q_i) \cong \hat{E}_i$ such that the following square commutes:

$$\begin{array}{ccc} K[X] & \longrightarrow & E \\ {\scriptstyle \subseteq} \downarrow & & \downarrow {\scriptstyle \varphi_i} \\ \hat{K}[X] & \longrightarrow & \hat{E}_i \end{array}$$

where the horizontal maps are projections. In particular, $\varphi_i(\alpha) = \alpha_i$. Hence $\hat{E}_i$ is a field extension of E and $\hat{v}_i$ induces an absolute value w_i on E which extends v.

Conversely, let w be an absolute value on E which extends v. The completion $\hat{E}_w$ contains $\alpha \in E$ and $\hat{K}$, and there is an evaluation homomorphism $\varepsilon_\alpha : f(X) \longmapsto f(\alpha)$ of $\hat{K}[X]$ into $\hat{E}_w$. Since α is algebraic over $\hat{K} \supseteq K$, $\mathrm{Im}\, \varepsilon_\alpha = \hat{K}[\alpha]$ is a field and a finite extension of $\hat{K}$, and is complete by Theorem 9.3.2. Therefore $\hat{E}_w = \hat{K}[\alpha]$ and ε_α is surjective. Hence $\mathrm{Ker}\, \varepsilon_\alpha$ is a maximal ideal of $\hat{K}[X]$, and is generated by a monic irreducible polynomial $r(X) \in \hat{K}[X]$, which divides q since $\varepsilon_\alpha(q) = q(\alpha) = 0$. Therefore $r = q_i$ for some i, and there is a $\hat{K}$-isomorphism $\hat{E}_w \cong \hat{K}[X]/(q_i) = \hat{E}_i$. Up to this isomorphism, $\hat{w}$ and $\hat{v}_i$ induce the same absolute value $\hat{v}$ on $\hat{K}$, coincide by Theorem 9.3.2, and induce the same absolute value on E. Thus $w = w_i$.

In particular, there is a $\hat{K}$-isomorphism $\hat{E}_{w_i} \cong \hat{E}_i = \hat{K}[X]/(q_i)$ which takes α to α_i; in particular, $[\hat{E}_{w_i} : \hat{K}] = [\hat{E}_i : \hat{K}] = \deg q_i$.

If $w_i = w_j$, then $\hat{E}_{w_i} = \hat{E}_{w_j}$, and there is a $\hat{K}$-isomorphism of $\hat{E}_i$ onto $\hat{E}_j$ which takes α_i onto α_j. Hence $q_i = \mathrm{Irr}(\alpha_i) = \mathrm{Irr}(\alpha_j) = q_j$. Therefore $w_1, \ldots, w_r$

are distinct. If E is separable, then q is separable, $q = q_1 \ldots q_r$, and $[E : K] = \deg q = \sum_i \deg q_i = \sum_i [\hat{E}_{w_i} : \hat{K}]$. $\square$

9.5.e. Discrete Valuations. Theorem 9.5.9 can be improved if v is discrete. Let $e_i = e(E : K)$ and $f_i = f(E : K)$ be the ramification index and residue class degree of E relative to w_i. Then $[\hat{E}_{w_i} : \hat{K}] = e(\hat{E}_{w_i} : \hat{K}) f(\hat{E}_{w_i} : \hat{K})$ by Proposition 9.5.8, $e(\hat{E}_{w_i} : \hat{K}) = e_i$ by Proposition 9.2.2, and $f(\hat{E}_{w_i} : \hat{K}) = f_i$ by Proposition 9.4.4. Hence:

Theorem 9.5.10. *Let v be a discrete absolute value on a field K, $E = K(\alpha)$ be a finite extension of K of degree n, $w_1, \ldots, w_r$ be the distinct absolute values on E which extend v, and e_i, f_i be the ramification index and residue class degree of E relative to w_i. Then $\sum_i e_i f_i \leq n$; if E is separable over K, then $\sum_i e_i f_i = n$.*

Further improvements can be obtained if E is a Galois extension of K. Then it can be shown that w_i and w_j are conjugate for all i, j ($w_i = w_j \circ \sigma$ for some $\sigma \in \mathrm{Gal}(E/K)$). Therefore $e_i = e_j = e$ and $f_i = f_j = f$ for all i, j, and we obtain:

Theorem 9.5.11. *Let v be a discrete absolute value on a field K, E be a finite Galois extension of K of degree n, and $w_1, \ldots, w_r$ be the distinct absolute values on E which extend v. Then E has the same ramification index e and residue class degree f relative to all w_i, and $efr = n$.*

Exercises

1. Let E be a finite extension of K, v be a valuation on K, and w be a valuation on E which extends v. Prove that if v is discrete, then w is discrete.

2. Let E be an extension of K and w, w' be valuations on E which extend the same valuation v on K. If w and w' are conjugate over K (if $w' = w \circ \sigma$ for some K-automorphism σ of E), show that $e(E : K)$ and $f(E : K)$ are the same for w and w'.

10

MODULES

The French algebraist Albert Châtelet called abelian groups "modules." An abelian group on which a ring R acts thus became an R-module. Modules (in this modern sense) appear in van der Waerden much as in Sections 10.7 and 10.8 below; they became important with the development of homological algebra in the 1940s and 1950s.

This chapter contains basic properties of modules, submodules, homomorphisms, and free modules; the structure theorem for finitely generated modules over PIDs; and applications to linear algebra. Chapters 11, 12, and 13 give further applications to groups and to rings. The study of modules in general is continued in Chapter 14.

Required results: basic properties of groups (Chapters 1 and 2) and rings (Chapter 4), chain conditions (Section B.2), and the results on PIDs in Section 5.1. The Jordan-Hölder and Krull-Schmidt Theorems for modules are derived from the corresponding results for groups with operators (Sections 1.7 and 2.9). Properties of cardinal numbers (specifically, Corollary B.5.12) are used in the proof of Proposition 10.5.7; basic properties of cyclotomic polynomials from Section 7.7 are used in the proof of Theorem 10.6.7.

Except in Section 10.1, all rings in this chapter have an identity element, and all modules are unital.

10.1. MODULES

Groups act on sets. Rings act on abelian groups; the resulting structures are modules. Later examples and applications will show the flexibility and usefulness of this new concept.

10.1.a. Left R-Modules. Let R be a ring, which does not necessarily have an identity element. A **left R-module** is an abelian group M together with a left action $R \times M \longrightarrow M$, $(r,x) \longmapsto rx$ of R on M (the **R-module structure** on M) such that:

(1) (mixed associativity) $r(sx) = (rs)x$;

(2) (distributivity) $r(x + y) = rx + ry$, $(r + s)x = rx + sx$

hold for all $r, s \in R$, $x, y \in M$. If R has an identity element 1 (a unity), an R-module M is **unital** (or **unitary**) in case

(3) (unital) $1x = x$

for all $x \in M$. The existence of a left R-module structure on M is often indicated by the notation ${}_R M$.

10.1.b. Examples. The definition of modules is very similar to the definition of vector spaces. In fact, when K is a field, a unital K-module is a vector space over K, and conversely. But modules are much more general and complex than vector spaces; indeed:

every abelian group is a unital $\mathbb{Z}$-module (with $nx = x + x + \cdots + x$ if $n \in \mathbb{Z}$ is positive);

every ring R is a left R-module; the ring operations provide the addition and action of R. This module is denoted by ${}_R R$ to distinguish it from the ring R. If R has an identity element, then ${}_R R$ is a unital R-module.

10.1.c. Properties. We note some elementary consequences of the axioms. Let M be a left R-module. Let $r, s \in R$ and $x, y \in M$. By (1), products such as rsx may be written without parentheses. Furthermore:

(4) $r0 = 0 = 0x$: by (2), $r0 = r(0 + 0) = r0 + r0$; hence $r0 = 0$. Similarly, $0x = (0 + 0)x = 0x + 0x$ yields $0x = 0$.

(5) $r(x - y) = rx - ry$, $(r - s)x = rx - sx$: by (2), $r(x - y) + ry = r(x - y + y) = rx$; the second equality is similar.

In a left R-module M we can form finite **linear combinations** $r_1 x_1 + \cdots + r_n x_n$ of elements $x_1, \ldots, x_n$ of M with coefficients $r_1, \ldots, r_n$ in R.

Infinite sums and infinite linear combinations are defined in a left R-module M as follows. In the absence of a topology on M, the sum $\sum_{i \in I} m_i$ of infinitely many elements m_i of M is defined if and only if $m_i = 0$ for almost all i (i.e., the set $\{i \in I ;\ m_i \neq 0\}$ is finite); then $\sum_{i \in I} m_i = \sum_{i \in I,\, m_i \neq 0} m_i$. For instance, an infinite linear combination $\sum_{i \in I} r_i x_i$ of arbitrary elements x_i of M with coefficients $r_i \in R$ is defined if $r_i = 0$ for almost all i; then $\sum_{i \in I} r_i x_i = \sum_{i \in I,\, r_i \neq 0} r_i x_i$.

10.1.d. Equivalent Definition. Module structures on an abelian group M can also be defined as ring homomorphisms. If A is an abelian group, $\mathrm{End}(A)$ is the ring of endomorphisms of A, written on the left; this is also denoted by $\mathrm{End}_{\mathbb{Z}}(A)$.

Proposition 10.1.1. *Let M be an abelian group.*

There is a one-to-one correspondence between left R-module structures $R \times M \longrightarrow M$ on M and ring homomorphisms $R \longrightarrow \text{End}_{\mathbb{Z}}(M)$.

There is a one-to-one correspondence between unital left R-module structures on M and homomorphisms $R \longrightarrow \text{End}(M)$ of rings with identity elements.

Proof. Let M be a left R-module. For each $r \in R$, the mapping $\varphi_r : M \longrightarrow M$, $\varphi_r(x) = rx$, is an endomorphism of the additive group M by (2). Furthermore $\varphi_r \circ \varphi_s = \varphi_{rs}$, since $\varphi_r(\varphi_s(x)) = r(sx) = (rs)x = \varphi_{rs}(x)$ by (1), and $\varphi_r + \varphi_s = \varphi_{r+s}$, since $\varphi_r(x) + \varphi_s(x) = rx + sx = (r+s)x = \varphi_{r+s}(x)$ by (2). Hence $\varphi : r \longmapsto \varphi_r$ is a ring homomorphism of R into $\text{End}(M)$. If R has an identity element and M is unital, then $\varphi(1) = 1_M$ is the identity on M.

If, conversely, $\varphi : R \longrightarrow \text{End}(M)$ is a ring homomorphism, then R acts on M by $rx = \varphi_r(x)$, and the calculations above show that (1) and (2) hold; so does (3), if φ is a homomorphism of rings with identity elements ($\varphi(1) = 1$). $\square$

By Proposition 4.1.3, every ring homomorphism $R \longrightarrow \text{End}(M)$ induces a unique homomorphism $R^1 \longrightarrow \text{End}(M)$ of rings with identity elements. Hence it may be assumed in the general study of left R-modules that R has an identity element and that modules are unital; we will do so in all subsequent sections.

The kernel of $R \longrightarrow \text{End}(M)$ consists of all $r \in R$ such that $rM = 0$ ($rx = 0$ for all $x \in M$); it is an ideal of R called the **annihilator** of M, which we denote by $\text{Ann}(M)$.

A left R-module M is **faithful** in case the homomorphism $R \longrightarrow \text{End}(M)$ is injective; equivalently, in case $\text{Ann}(M) = 0$. (Then $R \longrightarrow \text{End}(M)$ is a faithful (= injective) representation of R by endomorphisms of M.)

10.1.e. Right R-Modules. A **right R-module** is an abelian group M together with a right action $R \times M \longrightarrow M$, $(r, x) \longmapsto xr$ of R on M such that:

(1*) $(xr)s = x(rs)$;

(2*) $x(r + s) = xr + xs$, $(x + y)r = xr + yr$

hold for all $r, s \in R$, $x, y \in M$. If R has an identity element, then M is a unital right R-module in case

(3*) $x1 = x$

also holds for all $x \in M$. The existence of a right R-module structure on M is often indicated by the notation M_R.

The following construction reduces right modules to left modules. When R is a ring, the **opposite** ring R^{op} has the same underlying abelian group as R and the opposite multiplication: if rs is the product of r and s in R, then the product $r * s$ of r and s in R^{op} is $r * s = sr$. It is an easy exercise to verify that R^{op} is a ring. The ring R is commutative if and only if $R^{\text{op}} = R$.

Proposition 10.1.2. *Every [unital] right R-module is a [unital] left R^{op}-module, and conversely.*

Proof. Let M be a right R-module. Define a left action of R^{op} on M by $rx = xr$ for all $r \in R$ and $x \in M$. Then (2) follows from (2*); (1) follows from (1*) since $r(sx) = (xs)r = x(sr) = (sr)x = (r*s)x$. If M is a unital right R-module, then M is a unital left R^{op}-module. The converse is similar. $\square$

By Proposition 10.1.2, a right R-module structure on an abelian group M can be defined as a ring homomorphism $R^{op} \longrightarrow \text{End}_{\mathbb{Z}}(M)$. Alternately, we can write the endomorphisms of M on the right, which reverses the multiplication on $\text{End}(M)$: if $x(\eta\zeta) = (x\eta)\zeta$, then $\eta\zeta = \zeta \circ \eta$; the resulting endomorphism ring is $\text{End}_{\mathbb{Z}}^{op}(M)$. Then a right R-module structure on M is a ring homomorphism $R \longrightarrow \text{End}_{\mathbb{Z}}^{op}(M)$.

Corollary 10.1.3. *When R is commutative, then every left R-module is also a right R-module, and vice versa.*

When R is commutative we often refer to left R-modules or right R-modules as just R-modules.

10.1.f. Bimodules. Let R and S be rings. A **left R-, right S-bimodule**, or, simply, **R-S-bimodule**, is a left R-module M which is also a right S-module so that

(6) $r(xs) = (rx)s$

for all $r \in R$, $x \in M$, and $s \in S$.
For example:

Proposition 10.1.4. *R is a left R-, right R-bimodule. Every left R-module is an R-$\mathbb{Z}$-bimodule. If R is commutative, then every left R-module is an R-R-bimodule.*

An R-S-bimodule structure on an abelian group M can be viewed as a pair of ring homomorphisms $R \longrightarrow \text{End}_{\mathbb{Z}}(M)$, $S^{op} \longrightarrow \text{End}_R(M)$; or as a pair of ring homomorphisms $S \longrightarrow \text{End}_{\mathbb{Z}}^{op}(M)$, $R \longrightarrow \text{End}_S(M)$ (see the exercises). ($\text{End}_R(M)$ is the set of all $\eta \in \text{End}_{\mathbb{Z}}(M)$ such that $\eta(rx) = r\eta(x)$ for all $r \in R$, $x \in M$.)

10.1.g. Homomorphisms. A **homomorphism** $\varphi : M \longrightarrow N$ of left R-modules (also called an **R-homomorphism**) is a homomorphism of abelian groups which preserves the action of R; equivalently, a mapping $\varphi : M \longrightarrow N$ such that

$$\varphi(x + y) = \varphi(x) + \varphi(y) \qquad \text{and} \qquad \varphi(rx) = r\varphi(x)$$

for all $x, y \in M$ and $r \in R$; equivalently, a mapping $\varphi : M \longrightarrow N$ such that

$$\varphi(rx + sy) = r\varphi(x) + s\varphi(y)$$

for all $x, y \in M$ and $r, s \in R$. Homomorphisms of right modules and homomorphisms of bimodules are defined similarly.

A module homomorphism $\varphi : M \longrightarrow N$ preserves all linear combinations:

$$\varphi(\textstyle\sum_{i \in I} r_i x_i) = \sum_{i \in I} r_i \varphi(x_i)$$

for all $x_i \in M$ and $r_i \in R$ such that $r_i = 0$ for almost all i.

An **endomorphism** of a left R-module M is a homomorphism of M into M. A module homomorphism which is injective is a **monomorphim** or is **monic**; a module homomorphism which is surjective is an **epimorphim** or is **epic**. An **isomorphism** of left R-modules is a bijective homomorphism; then the inverse bijection is also a module isomorphism. Homomorphisms of vector spaces are also called **linear transformations**.

The identity mapping 1_M on a module M is a module homomorphism. Module homomorphisms can be composed: if $\varphi : M \longrightarrow N$ and $\psi : N \longrightarrow P$ are homomorphisms of left R-modules, then so is $\psi \circ \varphi : M \longrightarrow P$.

Module homomorphisms can also be added pointwise: if $\varphi, \psi : M \longrightarrow N$ are homomorphisms of left R-modules, then $\varphi + \psi : M \longrightarrow N$ is defined by

$$(\varphi + \psi)(x) = \varphi(x) + \psi(x)$$

for all $x \in M$ and is a homomorphism, since the addition on N is commutative.

Module homomorphisms will be studied in more detail in Section 10.3.

Exercises

1. Prove that a linear combination of linear combinations of elements $(x_i)_{i \in I}$ of a left R-module is a linear combination of $(x_i)_{i \in I}$.

2. Let A be an abelian group. When does there exist a unital $\mathbb{Z}_n$-module structure on A?

3. Let A be an abelian group. When does there exist a unital $\mathbb{Q}$-module structure on A?

4. Let A be an abelian group. Prove that there is only one unital left $\mathbb{Z}$-module structure on A.

5. Let A be an abelian group. Prove that there is at most one unital left $\mathbb{Q}$-module structure on A.

6. Let M be a [unital] left R-module and $\varphi : S \longrightarrow R$ be a homomorphism of rings [with identity elements]. Make M into a [unital] left S-module.

7. Let M be a [unital] left R-module and I be an ideal of R contained in the annihilator of M. Make M into a [unital] left R/I-module.

8. Let V be an abelian group and K be a field. Show that a unital $K[X]$-module structure on V is determined by a unital K-module structure on V and a linear transformation $T : V \longrightarrow V$; in fact this constructs a one-to-one correspondence.

9. Let V be a vector space over a field K. How can V become a $K[X]$-$K[X]$-bimodule?

10. Let M be an abelian group. Show that there is a one-to-one correspondence between R-S-bimodule structures on M and pairs of ring homomorphisms $R \longrightarrow \mathrm{End}_{\mathbb{Z}}(M)$, $S^{\mathrm{op}} \longrightarrow \mathrm{End}_R(M)$.

10.2. SUBMODULES

This section reviews basic properties of submodules. The results are stated for left R-modules; except as noted, similar definitions and results apply to right R-modules, and to R-S-bimodules.

In what follows R is a ring with identity element and all R-modules are unital.

10.2.a. Submodules. A **submodule** of a left R-module M is a left R-module A such that, as a set, $A \subseteq M$, and the inclusion mapping $A \longrightarrow M$ is a homomorphism. This relationship is often denoted by $A < M$ or by $A \leqslant M$. As with groups and rings, a submodule is completely determined by its underlying set. The following straightforward result is often taken as the definition of submodules.

Proposition 10.2.1. *Let M be a left R-module. Then A is a submodule of M if and only if A is a subgroup of M (e.g., $A \neq \varnothing$, and $a, b \in A$ implies $a - b \in A$) and $a \in A$ implies $ra \in A$ for all $r \in R$.*

Equivalently, $A \subseteq M$ is a submodule of M if and only if $A \neq \varnothing$, and $a, b \in A$ implies $ra + sb \in A$ for all $r, s \in R$.

For example, the submodules of a vector space are its subspaces; the submodules of an abelian group (viewed as a $\mathbb{Z}$-module) are its subgroups.

The submodules of $_R R$ are the **left ideals** of R. A **right ideal** of R is a left ideal of R^{op}; equivalently, a submodule of R_R; equivalently, an additive subgroup I of R such that $x \in I \Longrightarrow xr \in I$ for all $r \in R$. Thus an ideal of R, also called a **two-sided ideal**, is a left ideal of R which is also a right ideal. If R is commutative, then ideals, left ideals, and right ideals all coincide.

Submodules arise from module homomorphisms in two ways:

Proposition 10.2.2. *Let $\varphi : M \longrightarrow N$ be a homomorphism of left R-modules. Then*

$$\mathrm{Im}\, \varphi = \{\, \varphi(x) \in N \,;\, x \in M \,\}$$

is a submodule of N, and

$$\text{Ker } \varphi = \{ x \in M ; \varphi(x) = 0 \}$$

is a submodule of M.

10.2.b. Properties of Submodules. The following results are straightforward.

Proposition 10.2.3. *Let M be a left R-module. Every intersection of submodules of M is a submodule of M. The union of a nonempty chain of submodules of M is a submodule of M.*

It follows from Proposition 10.2.3 that there exists for every subset S of a module M a smallest submodule of M containing S; this is the submodule of M **generated** by S or **spanned** by S. The following result does not quite extend to bimodules (see the exercises):

Proposition 10.2.4. *Let M be a left R-module and S be a subset of M. The submodule of M generated by S is the set of all linear combinations of elements of S with coefficients in R.*

A **cyclic** submodule is a submodule which is generated by a single element. In a left R-module, the cyclic submodule generated by a is $Ra = \{ ra ; r \in R \}$, by Proposition 10.2.4. (Like Proposition 10.2.4, this result extends to right modules but not to bimodules.) More generally, the submodule generated by a finite subset $\{ a_1, \ldots, a_n \}$ is the set

$$Ra_1 + \cdots + Ra_n = \{ r_1 a_1 + \cdots + r_n a_n ; r_1, \ldots, r_n \in R \}$$

of all linear combinations of $a_1, \ldots, a_n$; such submodules are called **finitely generated**.

The **sum** of a family $(A_i)_{i \in I}$ of submodules is their sum as additive subgroups:

$$\sum_{i \in I} A_i = \{ \sum_{i \in I} a_i ; a_i \in A_i, a_i = 0 \text{ for almost all } i \}.$$

Proposition 10.2.5. *Every sum of submodules is a submodule.*

$\sum_{i \in I} A_i$ is the smallest submodule which contains the union $\bigcup_{i \in I} A_i$.

The **product** of a left ideal and a submodule generalizes the product of ideals in a ring (Section 5.6). If L is a left ideal of R and A is a submodule of a left R-module M, then the **product** of L and A is the set LA of all finite sums of elements of the form ℓx with $\ell \in L$ and $x \in A$; equivalently, the set of all linear combinations of elements of A with coefficients in L. Since L is a left ideal of R, LA is a submodule of M; LA is the submodule generated by the set product $\{ \ell a ; \ell \in L, a \in A \}$, with which it should not be confused.

The following properties are exercises:

Proposition 10.2.6. *Let L, L', $(L_i)_{i \in I}$ be left ideals of R and A, $(A_i)_{i \in I}$ be submodules of a left R-module M. Then:*

(1) $LA \subseteq A$;

(2) $L(L'A) = (LL')A$;

(3) $L \sum_{i \in I} A_i = \sum_{i \in I} LA_i$, $(\sum_{i \in I} L_i)A = \sum_{i \in I} L_iA$.

10.2.c. Quotient Modules. Because modules are built from abelian groups, any submodule can be used to construct a quotient module. Inspired by the proofs of Propositions 1.5.4, 1.5.6, and 1.7.3, our diligent reader will show:

Proposition 10.2.7. *Let M be a left R-module and N be a submodule of M. The cosets of N are the elements of a left R-module M/N, whose operations are well defined by $(x + N) + (y + N) = (x + y) + N$ and $r(x + N) = rx + N$. The projection $\pi : M \longrightarrow M/N$ is a module homomorphism of M onto M/N, and $\mathrm{Ker}\,\pi = N$.*

Proposition 10.2.8. *Let M be a left R-module and A be a submodule of M. Every submodule of M/A is of the form B/A for some unique submodule B of M which contains A.*

More generally let $\varphi : M \longrightarrow M'$ be a module homomorphism. If A is a submodule of M, then $\varphi(A)$ is a submodule of M' contained in $\mathrm{Im}\,\varphi$. If A' is a submodule of M', then $\varphi^{-1}(A')$ is a submodule of M which contains $\mathrm{Ker}\,\varphi$. This induces a one-to-one, order-preserving correspondence between the submodules of $\mathrm{Im}\,\varphi$ and the submodules of M which contain $\mathrm{Ker}\,\varphi$.

Exercises

1. Let M be a left R-module. Show that A is a submodule of M if and only if A is a subgroup of M, and $a \in A$ implies $ra \in A$ for all $r \in R$.

2. Let M be a left R-module. Show that $A \subseteq M$ is a submodule of M if and only if $A \neq \emptyset$, and $a,b \in A$ implies $ra + sb \in A$ for all $r,s \in R$.

3. Let $\varphi : M \longrightarrow N$ be a homomorphism of left R-modules. Prove that $\mathrm{Im}\,\varphi$ is a submodule of N.

4. Let $\varphi : M \longrightarrow N$ be a homomorphism of left R-modules. Prove that $\mathrm{Ker}\,\varphi$ is a submodule of M.

5. Let M be a left R-module. Prove that every intersection of submodules of M is a submodule of M.

6. Let M be a left R-module. Prove that the union of a nonempty chain of submodules of M is a submodule of M.

7. Let M be a left R-module and S be a subset of M. Show that the submodule of M generated by S is the set of all linear combinations of elements of S with coefficients in R.

8. Let M be an R-S-bimodule. Show that the submodule of M generated by a subset $X \subseteq M$ is the set of all finite sums of elements of M of the form rxs, with $r \in R$, $x \in X$, and $s \in S$.

9. Prove that every sum of submodules is a submodule.

10. Let L and L' be left ideals of R and A be a submodule of a left R-module M. Prove that $L(L'A) = (LL')A$.

11. Let L be a left ideal of R and $(A_i)_{i \in I}$ be submodules of a left R-module M. Prove that $L \sum_{i \in I} A_i = \sum_{i \in I} LA_i$.

12. Let $(L_i)_{i \in I}$ be left ideals of R and A be a submodule of a left R-module M. Prove that $(\sum_{i \in I} L_i)A = \sum_{i \in I} L_i A$.

13. A submodule of a R-S-bimodule can be multiplied on the left by a left ideal of R and on the right by a right ideal of S. Study this operation.

14. Prove Proposition 10.2.7.

15. Let M be a left R-module and N be a submodule of M. Prove that every submodule of M/N is of the form A/N for some unique submodule A of M which contains N.

10.3. HOMOMORPHISMS

Modules are groups with operators. Hence modules inherit all the properties of groups with operators, including the Jordan-Hölder Theorem and the Krull-Schmidt Theorem. This section contains basic properties up to the Jordan-Hölder Theorem (Krull-Schmidt is in Section 10.4) and includes a few properties that are specific to modules.

In what follows R is a ring with identity element and all R-modules are unital. All results are stated for left R-modules; except as noted, similar definitions and results apply to right R-modules, and to R-S-bimodules.

10.3.a. Factoring Homomorphisms. Theorems 1.6.2 and 1.6.4 provide two basic results which factor one module homomorphism through another.

Theorem 10.3.1. *Let M, M', and M'' be left R-modules and $\varphi : M' \longrightarrow M$, $\psi : M'' \longrightarrow M$ be homomorphisms. If φ is injective, then ψ factors through φ (i.e., $\psi = \varphi \circ \chi$ for some homomorphism $\chi : M'' \longrightarrow M'$) if and only if $\mathrm{Im}\, \psi \subseteq \mathrm{Im}\, \varphi$, and then ψ factors uniquely through φ (i.e., χ is unique):*

Proof. If $\operatorname{Im} \psi \subseteq \operatorname{Im} \varphi$, then Theorem 1.6.2 provides an abelian group homomorphism $\chi : M'' \longrightarrow M'$ such that $\psi = \varphi \circ \chi$; it suffices to prove that χ is a module homomorphism. If $x \in M''$, $r \in R$, then

$$\varphi(\chi(rx)) = \psi(rx) = r\,\psi(x) = r\,\varphi(\chi(x)) = \varphi(r\,\chi(x))$$

and $\chi(rx) = r\,\chi(x)$, since φ is injective. $\square$

Theorem 10.3.2. *Let M, M', and M'' be left R-modules and $\varphi : M \longrightarrow M'$, $\psi : M \longrightarrow M''$ be homomorphisms. If φ is surjective, then ψ factors through φ ($\psi = \chi \circ \varphi$ for some homomorphism $\chi : M' \longrightarrow M''$) if and only if $\operatorname{Ker} \varphi \subseteq \operatorname{Ker} \psi$; and then ψ factors uniquely through φ (χ is unique):*

Proof. If $\operatorname{Ker} \varphi \subseteq \operatorname{Ker} \psi$, then Theorem 1.6.4 provides an abelian group homomorphism $\chi : M' \longrightarrow M''$ such that $\psi = \chi \circ \varphi$; it suffices to prove that χ is a module homomorphism. Let $x' \in M'$. Since φ is surjective we have $x' = \varphi(x)$ for some $x \in M$. For all $r \in R$,

$$\chi(rx') = \chi(r\,\varphi(x)) = \chi(\varphi(rx)) = \psi(rx) = r\,\psi(x) = r\,\chi(\varphi(x)) = r\,\chi(x').\quad \square$$

Let $\varphi : M \longrightarrow N$ and $\psi : N \longrightarrow P$ be module homomorphisms. Then $\psi \circ \varphi = 0$ ($\psi(\varphi(x)) = 0$ for all $x \in M$) if and only if $\operatorname{Im} \varphi \subseteq \operatorname{Ker} \psi$; hence Theorems 10.3.1 and 10.3.2 yield:

Corollary 10.3.3. *Let $\varphi : M \longrightarrow N$ and $\psi : N \longrightarrow P$ be module homomorphisms. Then $\psi \circ \varphi = 0$ ($\psi(\varphi(x)) = 0$ for all $x \in M$) if and only if φ factors through the inclusion $\operatorname{Ker} \psi \longrightarrow N$.*

Corollary 10.3.4. *Let $\varphi : M \longrightarrow N$ and $\psi : N \longrightarrow P$ be module homomorphisms. Then $\psi \circ \varphi = 0$ ($\psi(\varphi(x)) = 0$ for all $x \in M$) if and only if ψ factors through the projection $N \longrightarrow N/\operatorname{Im} \varphi$.*

Due to the similarity between Corollaries 10.3.3 and 10.3.4, the quotient module $N/\operatorname{Im} \varphi$ is known as the **cokernel** of φ.

10.3.b. Isomorphisms

Theorem 10.3.5 (Homomorphism Theorem). *When $\varphi : M \longrightarrow N$ is a homomorphism of left R-modules:*

(1) Im φ is a submodule of N;
(2) $K = \text{Ker } \varphi$ is a submodule of M;
(3) $\varphi(x) = \varphi(y) \iff x + K = y + K$;
(4) Im $\varphi \cong M/\text{Ker } \varphi$; in fact there exists a module isomorphism θ of $M/\text{Ker } \varphi$ onto Im φ such that the following diagram commutes:

$$
\begin{array}{ccc}
M & \xrightarrow{\ \varphi\ } & N \\
\pi \downarrow & & \uparrow \iota \\
M/\text{Ker } \varphi & \xrightarrow{\ \theta\ } & \text{Im } \varphi
\end{array}
$$

(where $\pi : M \longrightarrow M/\text{Ker } \varphi$ is the projection and $\iota : \text{Im } \varphi \longrightarrow N$ is the inclusion homomorphism).

This is proved like Theorems 1.6.5 and 1.7.4.

As an application of Theorem 10.3.5 we construct all **cyclic** (singly generated) left R-modules. The **annihilator** of an element a of a left R-module M is

$$\text{Ann}(a) = \{r \in R;\ ra = 0\};$$

this is a left ideal of R. Note that $\text{Ann}(M) = \bigcap_{a \in M} \text{Ann}(a)$.

Proposition 10.3.6. *A left R-module M is cyclic if and only if it is isomorphic to R/L for some left ideal L of R; if $M = Ra$ then $M \cong R/\text{Ann}(a)$.*
If R is commutative, then $\text{Ann}(Ra) = \text{Ann}(a)$.

This result extends to right modules but not to bimodules.

Proof. Let $M = Ra$ be cyclic. There is a module homomorphism $\varphi : {}_R R \longrightarrow M$, defined by $\varphi(r) = ra$. We see that Im $\varphi = M$ and Ker $\varphi = \text{Ann}(a)$; by Theorem 10.3.5, $M \cong R/\text{Ann}(a)$. If, conversely, L is a left ideal of R, then R/L is generated (as a left R-module) by $1 + L$, and is cyclic; hence any module which is isomorphic to R/L is cyclic.

Let $s \in R$. If $s(ra) = 0$ for all $r \in R$, then $sa = 0$. The converse holds if R is commutative; then $\text{Ann}(Ra) = \text{Ann}(a)$. $\qquad \square$

Finally, the Isomorphism Theorems for groups with operators (Theorems 1.7.5 and 1.7.6) yield similar results for modules, which can also be proved directly (see the exercises).

Theorem 10.3.7 (First Isomorphism Theorem). *Let M be a left R-module and $A \subseteq B$ be submodules of M. Then A is a submodule of B, B/A is a submodule of M/A, and*

$$(M/A)/(B/A) \cong M/B\ ;$$

more precisely, there exists a module isomorphism $\theta : (M/A)/(B/A) \longrightarrow M/B$
such that the following diagram commutes:

$$M \xrightarrow{\pi} M/A \xrightarrow{\sigma} (M/A)/(B/A)$$
$$\searrow^{\rho} \qquad \swarrow_{\theta}$$
$$M/B$$

where π, ρ, σ *are the projections.*

Theorem 10.3.8 (Second Isomorphism Theorem). *Let M be a left R-module and A, B be submodules of M. Then* $A + B$ *and* $A \cap B$ *are submodules of M and*

$$(A + B)/B \cong A/A \cap B \;;$$

more precisely, there exists a module isomorphism $\theta : A/A \cap B \longrightarrow (A + B)/B$
such that the following diagram commutes:

$$
\begin{array}{ccc}
A & \xrightarrow{\iota} & A + B \\
{\scriptstyle \pi}\downarrow & & \downarrow{\scriptstyle \rho} \\
A/A \cap B & \xrightarrow[\theta]{} & A + B/B
\end{array}
$$

where π *and* ρ *are the projections and* ι *is the inclusion homomorphism.*

10.3.c. The Jordan-Hölder Theorem. A **composition series** of a module M is a finite, strictly ascending sequence of submodules

$$0 = A_0 \subsetneqq A_1 \subsetneqq \cdots \subsetneqq A_m = M$$

such that there is no submodule $A_{i-1} \subsetneqq B \subsetneqq A_i$. The **factors** of the series are the quotient modules A_i/A_{i-1} $(i > 0)$. Equivalently, a composition series is a maximal chain of submodules which is finite. Not every module has one; for example, the $\mathbb{Z}$-module $\mathbb{Z}$ does not have a composition series.

A module M is **simple** in case $M \neq 0$ and M has no submodule $A \neq 0, M$. Cyclic groups of prime order, and one-dimensional vector spaces, are simple modules. An ascending sequence of submodules $0 = A_0 \subseteq A_1 \subseteq \cdots \subseteq A_m = M$ is a composition series if and only if all its factors A_i/A_{i-1} are simple.

Two composition series

$$0 = A_0 \subsetneqq A_1 \subsetneqq \cdots \subsetneqq A_m = M \qquad \text{and} \qquad 0 = B_0 \subsetneqq B_1 \subsetneqq \cdots \subsetneqq B_n = M$$

are **equivalent** in case they have the same length and (up to isomorphisms) the same factors (in case $m = n$ and there is a permutation $\sigma \in S_n$ such that

$B_i/B_{i-1} \cong A_{\sigma i}/A_{\sigma(i-1)}$ for all $i > 0$). The general Jordan-Hölder Theorem (Theorem 3.5.10) and Schreier's Theorem (Lemma 3.5.9) imply:

Theorem 10.3.9 (Jordan-Hölder Theorem for Modules). *Any two composition series of a module are equivalent.*

Corollary 10.3.10. *When a left R-module M has a composition series of length n, every strictly ascending chain of submodules of M can be refined to a composition series and has length at most n.*

A module M has **finite length** in case it has a composition series; equivalently, in case it has a maximal chain of submodules which is finite. Then every maximal chain of submodules is finite by Corollary 10.3.10, and Proposition B.2.3 yields:

Proposition 10.3.11. *A module has finite length if and only if its submodules satisfy the ascending chain condition and the descending chain condition.*

When a module M has finite length, all maximal chains of submodules of M have the same length, by Theorem 10.3.9 and Corollary 10.3.10; the **length** of M is the length (the number of factors) of its composition series; equivalently, the length (the number of elements, minus one) of its maximal chains of submodules.

Exercises

1. Given a set S of operators, show that there exists a ring R such that abelian S-groups [with operators] coincide with R-modules.
2. Prove Corollary 10.3.3.
3. Prove Corollary 10.3.4.
4. Prove Theorem 10.3.5.

10.4. DIRECT SUMS AND PRODUCTS

Direct sums and products help construct modules from simpler modules, as we will see in Sections 10.7 and 12.1; their universal properties help build diagrams. This section contains basic properties, including some that are specific to modules.

In what follows R is a ring with identity element and all R-modules are unital. As before all results are stated for left R-modules; except as noted, similar definitions and results apply to right R-modules, and to R-S-bimodules.

10.4.a. Direct Products. The **direct product** of a family $(M_i)_{i \in I}$ of left R-modules is their cartesian product $\prod_{i \in I} M_i$ (the set of all families $(x_i)_{i \in I}$ such that $x_i \in M_i$ for all $i \in I$), with the pointwise operations:

$$(x_i)_{i \in I} + (y_i)_{i \in I} = (x_i + y_i)_{i \in I} , \qquad r(x_i)_{i \in I} = (rx_i)_{i \in I} .$$

It is immediate that $\prod_{i \in I} M_i$ a left R-module. If $I = \varnothing$, then $\prod_{i \in I} M_i = 0$. If $I = \{1\}$, then $\prod_{i \in I} M_i = M_1$. If $I = \{1, 2, \ldots, n\}$, then the direct product $\prod_{i \in I} M_i$ of $M_1, M_2, \ldots, M_n$ is also denoted by $M_1 \times M_2 \times \cdots \times M_n$, and $(x_i)_{i \in I}$ by $(x_1, x_2, \ldots, x_n)$.

The direct product $\prod_{i \in I} M_i$ comes with **projections** $\pi_i : \prod_{i \in I} M_i \longrightarrow M_i$; π_i assigns to each $(x_i)_{i \in I} \in \prod_{i \in I} M_i$ its i component x_i. The projections are epimorphisms; in fact the left R-module structure on $\prod_{i \in I} M_i$ is the only module structure such that the projections are module homomorphisms.

The direct product and its projections have a universal property:

Proposition 10.4.1. *Let M be a left R-module and $\varphi_i : M \longrightarrow M_i$ be a module homomorphism for each $i \in I$. There exists a module homomorphism $\varphi : M \longrightarrow \prod_{i \in I} M_i$ unique such that the following diagram commutes for every i:*

$$
\begin{array}{ccc}
M & \overset{\varphi}{\dashrightarrow} & \prod_{i \in I} M_i \\
& {\varphi_i}\searrow & \downarrow {\pi_i} \\
& & M_i
\end{array}
$$

The proof is an exercise. Proposition 10.4.1 implies that a family $(\varphi_i)_{i \in I}$ of module homomorphisms $\varphi_i : M_i \longrightarrow N_i$ induces a homomorphism $\varphi = \prod_{i \in I} \varphi_i : \prod_{i \in I} M_i \longrightarrow \prod_{i \in I} N_i$ unique such that every square

$$
\begin{array}{ccc}
\prod_{i \in I} M_i & \overset{\varphi}{\dashrightarrow} & \prod_{i \in I} N_i \\
\downarrow & & \downarrow \\
M_i & \underset{\varphi_i}{\longrightarrow} & N_i
\end{array}
$$

commutes (the vertical maps are projections); namely, $\varphi((x_i)_{i \in I}) = (\varphi_i(x_i))_{i \in I}$. (This can also be shown directly.) If $I = \{1, 2, \ldots, n\}$, $\prod_{i \in I} \varphi_i$ is also denoted by $\varphi_1 \times \varphi_2 \times \cdots \times \varphi_n$.

10.4.b. Direct Sums. The **direct sum** $\bigoplus_{i \in I} M_i$ (also called **external direct sum**) of a family $(M_i)_{i \in I}$ of left R-modules is the submodule

$$\bigoplus_{i \in I} M_i = \{ (x_i)_{i \in I} \in \prod_{i \in I} M_i ; \, x_i = 0 \text{ for almost all } i \}$$

of $\prod_{i \in I} M_i$. If $I = \{1, 2, \ldots, n\}$, then the direct sum $\bigoplus_{i \in I} M_i$ is also denoted by $M_1 \oplus M_2 \oplus \cdots \oplus M_n$, and it coincides with the direct product $M_1 \times M_2 \times \cdots \times M_n$.

The direct sum comes with **injections** $\iota_i : M_i \longrightarrow \bigoplus_{i \in I} M_i$; when $a \in M_i$, $\iota_i(a)$ has components

$$\iota_i(a)_j = \begin{cases} a & \text{if } j = i, \\ 0 & \text{if } j \neq i. \end{cases}$$

Lemma 10.4.2. *Direct sum injections are injective module homomorphisms. Moreover every $x \in \bigoplus_{i \in I} M_i$ can be written uniquely in the form $x = \sum_{i \in I} \iota_i(x_i)$, with $x_i \in M_i$ for all i and $x_i = 0$ for almost all i.*

Proof. Let $x = (x_i)_{i \in I} \in \bigoplus_{i \in I} M_i$. If J is a finite subset of I, then $y = \sum_{i \in J} \iota_i(x_i)$ has components $y_i = x_i$ if $i \in J$, $y_i = 0$ if $i \notin J$; this implies

$$\sum_{i \in I} \iota_i(x_i) = \sum_{i \in I, x_i \neq 0} \iota_i(x_i) = x.$$

If, conversely, $x = \sum_{i \in I} \iota_i(y_i)$, with $y_i \in M_i$ and $y_i = 0$ for almost all i, then $y = (y_i)_{i \in I} \in \bigoplus_{i \in I} M_i$, $y = \sum_{i \in I} \iota_i(y_i) = x$, and $y_i = x_i$ for all i. $\square$

The direct sum and its injections have the following universal property:

Proposition 10.4.3. *Let M be a left R-module and $\varphi_i : M_i \longrightarrow M$ be a module homomorphism for each $i \in I$. There exists a module homomorphism $\varphi : \bigoplus_{i \in I} M_i \longrightarrow M$ unique such that the following diagram commutes for every i:*

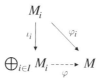

Proof. Let M be a module and $\varphi_i : M_i \longrightarrow M$ be module homomorphisms. Then $\sum_{i \in I} \varphi_i(x_i)$ is defined for every $x \in \bigoplus_{i \in I} G_i$. If $\varphi \circ \iota_i = \varphi_i$ for all i, then

$$\varphi(x) = \varphi(\sum_{i \in I} \iota_i(x_i)) = \sum_{i \in I} \varphi(\iota_i(x_i)) = \sum_{i \in I} \varphi_i(x_i)$$

for all $x = (x_i)_{i \in I} \in \bigoplus_{i \in I} M_i$ by Lemma 10.4.2. Therefore there is only one such homomorphism φ.

Conversely, a mapping $\varphi : \bigoplus_{i \in I} M_i \longrightarrow M$ is defined by

$$\varphi(x) = \sum_{i \in I} \varphi_i(x_i),$$

since $x_i = 0$ for almost all i. It is immediate that φ is a module homomorphism and that $\varphi \circ \iota_i = \varphi_i$ for all i. $\square$

The universal property in Proposition 10.4.3 implies that a family $(\varphi_i)_{i \in I}$ of homomorphisms $\varphi_i : M_i \longrightarrow N_i$ induces a homomorphism $\varphi = \bigoplus_{i \in I} \varphi_i :$ $\bigoplus_{i \in I} M_i \longrightarrow \bigoplus_{i \in I} N_i$ unique such that every square

$$
\begin{array}{ccc}
M_i & \xrightarrow{\;\varphi_i\;} & N_i \\
\downarrow & & \downarrow \\
\bigoplus_{i \in I} M_i & \dashrightarrow[\varphi] & \bigoplus_{i \in I} N_i
\end{array}
$$

commutes (the vertical maps are injections); namely $\varphi((x_i)_{i \in I}) = (\varphi_i(x_i))_{i \in I}$. (This can also be shown directly.) If $I = \{1, 2, \ldots, n\}$, $\bigoplus_{i \in I} \varphi_i$ is also denoted by $\varphi_1 \oplus \varphi_2 \oplus \cdots \oplus \varphi_n$.

Direct sums of modules have another characterization in terms of homomorphisms. To state this, we note that the **sum** of a possibly infinite family $(\varphi_i)_{i \in I} : M \longrightarrow N$ of module homomorphisms can be defined pointwise as follows: $\sum_{i \in I} \varphi_i$ is defined if and only if, for each $x \in M$, $\varphi_i(x) = 0$ for almost i; then $(\sum_{i \in I} \varphi_i)(x) = \sum_{i \in I} \varphi_i(x)$. It is readily verified that $\sum_{i \in I} \varphi_i$ is a module homomorphism.

Proposition 10.4.4. *When $(M_i)_{i \in I}$ is a family of left R-modules, $M \cong \bigoplus_{i \in I} M_i$ if and only if there exist homomorphisms $M_i \xrightarrow{\mu_i} M \xrightarrow{\rho_i} M_i$ such that*

(a) $\rho_i \circ \mu_i = 1_{M_i}$ *for all i;*

(b) $\rho_i \circ \mu_j = 0$ *whenever $i \neq j$;*

(c) *for each $x \in M$, $\rho_i(x) = 0$ for almost all i;*

(d) $\sum_{i \in I} \mu_i \circ \rho_i = 1_M$.

Proof. The injections $\iota_i : M_i \longrightarrow \bigoplus_{i \in I} M_i$ and projections $\pi_i : \bigoplus_{i \in I} M_i \longrightarrow M_i$ have properties (a), (b), (c), and (d), since $x = \sum_{i \in I} \iota_i(x_i)$ for all $x = (x_i)_{i \in I} \in \bigoplus_{i \in I} M_i$. If $\theta : \bigoplus_{i \in I} M_i \longrightarrow M$ is an isomorphism, then $\mu_i = \theta \circ \iota_i : M_i \longrightarrow M$ and $\rho_i = \pi_i \circ \theta^{-1} : M \longrightarrow M_i$ inherit (a), (b), (c), and (d) from ι_i and π_i; for example,

$$\theta(x) = \theta(\textstyle\sum_{i \in I} \iota_i(\pi_i(x))) = \sum_{i \in I} \theta(\iota_i(\pi_i(x))) = \sum_{i \in I} \mu_i(\rho_i(\theta(x)))$$

for all $x \in \bigoplus_{i \in I} M_i$.

Conversely, assume that μ_i and ρ_i have properties (a), (b), (c), and (d). By Proposition 10.4.3 there exists a homomorphism $\theta : \bigoplus_{i \in I} M_i \longrightarrow M$ such that $\theta \circ \iota_i = \mu_i$ for all i (where ι_i is the injection); if $x = (x_i)_{i \in I} \in \bigoplus_{i \in I} M_i$, then $\theta(x) = \sum_{i \in I} \mu_i(x_i)$. If $\theta(x) = 0$, then $\sum_{j \in I} \mu_j(x_j) = 0$ implies

$$0 = \rho_i(\textstyle\sum_{j \in I} \mu_j(x_j)) = \sum_{j \in I} \rho_i(\mu_j(x_j)) = x_i$$

for every i, by (a) and (b); hence $x = 0$, which shows that θ is injective. If $y \in M$, then $x = (\rho_i(y))_{i \in I} \in \bigoplus_{i \in I} M_i$ by (c), and (d) yields $y = \sum_{i \in I} \mu_i(\rho_i(y)) = \theta(x)$; thus θ is surjective. $\qquad\square$

Corollary 10.4.5. *A left R-module M is isomorphic to $A \oplus B$ if and only if there exist homomorphisms $A \underset{\pi}{\overset{\mu}{\rightleftarrows}} M \underset{\nu}{\overset{\rho}{\rightleftarrows}} B$ such that $\pi \circ \mu = 1_A$, $\rho \circ \nu = 1_B$, $\pi \circ \nu = 0$, $\rho \circ \mu = 0$, and $\mu \circ \pi + \nu \circ \rho = 1_M$.*

10.4.c. Internal Direct Sums. Direct sums can also be characterized in terms of submodules rather than homomorphisms.

Lemma 10.4.6. *When $(M_i)_{i \in I}$ is a family of left R-modules, $M \cong \bigoplus_{i \in I} M_i$ if and only if M contains submodules $A_i \cong M_i$ $(i \in I)$ such that every element of M can be written uniquely in the form $x = \sum_{i \in I} a_i$ with $a_i \in A_i$ for all i and $a_i = 0$ for almost all i.*

Proof. By Lemma 10.4.2, every element of $\bigoplus_{i \in I} M_i$ can be written uniquely in the form $x = \sum_{i \in I} t_i$ with $t_i \in \mathrm{Im}\, \iota_i$ for all i and $t_i = 0$ for almost all i. If $\theta : \bigoplus_{i \in I} M_i \longrightarrow M$ is an isomorphism, then $A_i = \theta(\mathrm{Im}\, \iota_i) \cong \mathrm{Im}\, \iota_i \cong M_i$ is a submodule of M and every element of M can be written uniquely in the form $\theta(x) = \sum_{i \in I} a_i$, with $a_i = \theta(t_i) \in A_i$ for all i and $a_i = 0$ for almost all i.

Conversely, assume that M contains submodules $A_i \cong M_i$ $(i \in I)$ such that every element of M can be written uniquely in the form $x = \sum_{i \in I} a_i$, with $a_i \in A_i$ for all i and $a_i = 0$ for almost all i. The inclusion homomorphisms $A_i \longrightarrow M$ induce a homomorphism $\theta : \bigoplus_{i \in I} A_i \longrightarrow M$ such that $\theta((a_i)_{i \in I}) = \sum_{i \in I} a_i$ for every $(a_i)_{i \in I} \in \bigoplus_{i \in I} A_i$. The hypothesis shows that θ is bijective. Thus $M \cong \bigoplus_{i \in I} A_i \cong \bigoplus_{i \in I} M_i$. $\qquad\square$

A module M is the **internal direct sum** of submodules $(A_i)_{i \in I}$ in case every element x of M can be written uniquely in the form $x = \sum_{i \in I} a_i$, with $a_i \in A_i$ for all i and $a_i = 0$ for almost all i. The notation $M = \bigoplus_{i \in I} A_i$ is also used to denote internal direct sums. Indeed every external direct sum is isomorphic to an internal direct sum, and conversely: by Lemma 10.4.2, an external direct sum $\bigoplus_{i \in I} M_i$ is the internal direct sum of its submodules $\mathrm{Im}\, \iota_i$; conversely, an internal direct sum $\bigoplus_{i \in I} A_i$ of submodules is isomorphic to the external direct sum $\bigoplus_{i \in I} A_i$, by Lemma 10.4.6.

Proposition 10.4.7. *A left R-module M is the internal direct sum of submodules $(A_i)_{i \in I}$ if and only if:*

(i) $A_i \cap (\sum_{j \neq i} A_j) = 0$ *for all i;*
(ii) $M = \sum_{i \in I} A_i$.

Proof. Assume that every element x of M can be written uniquely in the form $x = \sum_{i \in I} a_i$, with $a_i \in A_i$ for all i and $a_i = 0$ for almost all i. Then $M =$

$\sum_{i \in I} A_i$. If $a_i \in A_i \cap (\sum_{j \neq i} A_j)$, then $a_i = \sum_{j \neq i} a_j$ can be written in this form in two ways, and $a_i = 0$ (and $a_j = 0$ for all $j \neq i$). Thus (i) and (ii) hold.

Conversely, assume (i) and (ii). By (ii), every element x of M can be written in the form $x = \sum_{i \in I} a_i$, with $a_i \in A_i$ for all i and $a_i = 0$ for almost all i. If $\sum_{i \in I} a_i = \sum_{i \in I} b_i$ (with $b_i \in A_i$ for all i and $b_i = 0$ for almost all i), then, for each i,

$$a_i - b_i = \sum_{j \neq i} b_j - a_j \in A_i \cap (\sum_{j \neq i} A_j);$$

by (i), this implies $a_i - b_i = 0$. Thus $M = \bigoplus_{i \in I} A_i$. □

Proposition 10.4.7 can be improved by placing any total order relation on the set I: then condition (i) can be replaced by

(i′) $A_i \cap (\sum_{j < i} A_j) = 0$.

This makes a fine exercise. In particular:

Corollary 10.4.8. *A left R-module M is the internal direct sum of submodules $A_1, \ldots, A_n$ if and only if:*

(i′) $A_i \cap (A_1 + A_2 + \cdots + A_{i-1}) = 0$ *for all $i > 1$;*
(ii) $M = A_1 + A_2 + \cdots + A_n$.

Corollary 10.4.9. *A left R-module M is an internal direct sum $M = A \oplus B$ if and only if $A \cap B = 0$ and $A + B = M$.*

A submodule A of M is a **direct summand** of M in case $M = A \oplus B$ for some submodule B of M.

Corollary 10.4.10. *Let $\sigma : M \longrightarrow N$ and $\mu : N \longrightarrow M$ be homomorphisms of left R-modules. If $\sigma \circ \mu = 1_N$, then $M = \operatorname{Im} \mu \oplus \operatorname{Ker} \sigma$.*

Proof. First, $\operatorname{Im} \mu \cap \operatorname{Ker} \sigma = 0$, since $\mu(y) \in \operatorname{Im} \mu \cap \operatorname{Ker} \sigma$ implies $y = \sigma(\mu(y)) = 0$. Also $M = \operatorname{Im} \mu + \operatorname{Ker} \sigma$: for each $x \in M$, we have $\sigma(x) = \sigma(\mu(\sigma(x)))$, $x - \mu(\sigma(x)) \in \operatorname{Ker} \sigma$, and $x = \mu(\sigma(x)) + (x - \mu(\sigma(x))) \in \operatorname{Im} \mu + \operatorname{Ker} \sigma$. □

If $\sigma \circ \mu = 1$, then σ is an epimorphism (=is surjective) and μ is a monomorphism (= is injective). Conversely, an epimorphism $\sigma : M \longrightarrow N$ **splits** in case $\sigma \circ \mu = 1_N$ for some homomorphism $\mu : N \longrightarrow M$; a monomorphism $\mu : N \longrightarrow M$ **splits** in case $\sigma \circ \mu = 1_N$ for some homomorphism $\sigma : M \longrightarrow N$. (Then M *splits* into a direct sum, by Corollary 10.4.10, hence the terminology.)

10.4.d. The Krull-Schmidt Theorem. The Krull-Schmidt Theorem for modules is a corollary of the Krull-Schmidt Theorem for groups with operators (Theorem 2.9.2).

A left R-module M is **indecomposable** in case $M \neq 0$ and $M = A \oplus B$ implies $A = 0$ or $B = 0$. We saw in Section 2.2 that infinite cyclic groups and cyclic

p-groups are indecomposable $\mathbb{Z}$-modules. A vector space is indecomposable if and only if it has dimension 1.

By Corollary 10.3.11, a left R-module M has finite length (has a composition series) if and only if it satisfies the ascending chain condition and the descending chain condition on submodules (if and only if it has finite length as defined in Section 2.9). Proposition 2.9.1 yields:

Proposition 10.4.11. *A module of finite length is a direct sum of finitely many indecomposable modules.*

The Krull-Schmidt Theorem (2.9.2) implies that the decomposition in Proposition 10.4.11 is unique, up to isomorphism and the order of the terms.

Theorem 10.4.12 (Krull-Schmidt). *Let M be a left R-module of finite length. If*

$$M \cong A_1 \oplus A_2 \oplus \cdots \oplus A_m \cong B_1 \oplus B_2 \oplus \cdots \oplus B_n,$$

where $A_1, \ldots, A_m$ and $B_1, \ldots, B_n$ are indecomposable, then $m = n$ and $B_1, \ldots, B_n$ can be indexed so that $A_i \cong B_i$ for all $i \leqslant n$ and the **exchange property** *holds:*

$$M \cong A_1 \oplus \cdots \oplus A_k \oplus B_{k+1} \oplus \cdots \oplus B_n$$

for all $k < n$.

Exercises

1. Let A_i be a submodule of M_i for each $i \in I$. Show that $\bigoplus_{i \in I} A_i$ is a submodule of $\bigoplus_{i \in I} M_i$, and that $(\bigoplus_{i \in I} M_i)/(\bigoplus_{i \in I} A_i) \cong \bigoplus_{i \in I} M_i/A_i$.

2. Show that a submodule of $A \oplus B$ need not equal the direct sum $C \oplus D$ of a submodule C of A and a submodule D of B.

3. Prove the following associativity property of direct products: if $I = \bigcup_{j \in J} I_j$ is a partition of I, then $\prod_{i \in I} M_i \cong \prod_{j \in J} (\prod_{i \in I_j} M_i)$.

4. Prove the following associativity property of direct sums: if $I = \bigcup_{j \in J} I_j$ is a partition of I, then $\bigoplus_{i \in I} M_i \cong \bigoplus_{j \in J} (\bigoplus_{i \in I_j} M_i)$.

5. Prove that the direct product of a family of modules is characterized up to isomorphism by its universal property.

6. Prove that the direct sum of a family of modules is characterized up to isomorphism by its universal property.

7. Let I be a totally ordered set. Prove that a left R-module M is the internal direct sum of submodules $(A_i)_{i \in I}$ if and only if:

   (i) $A_i \cap (\sum_{j < i} A_j) = 0$ for all i;
   (ii) $M = \sum_{i \in I} A_i$.

8. Show that there is a one-to-one correspondence between homomorphisms φ: $\bigoplus_{i \in I} M_i \longrightarrow \prod_{j \in J} N_j$ and families $(\varphi_{i,j})_{i \in I, j \in J}$ of homomorphisms $\varphi_{i,j} : M_i \longrightarrow N_j$.

9. Show that $A \cong A'$, $B \cong B'$ implies $A \oplus B \cong A' \oplus B'$.

10. Show that $A \oplus B \cong A' \oplus B'$ does not imply $A \cong A'$ even when $B \cong B'$.

10.5. FREE MODULES

Of all modules, free modules are most like vector spaces. This section contains basic properties of free modules and module presentations. In what follows R is a ring with identity element and all R-modules are unital. Results are stated (mostly) for left R-modules; similar definitions and results apply to right R-modules but not to bimodules.

10.5.a. Definition. Let M be a left R-module and X be a subset of M. We saw (Proposition 10.2.4) that the submodule of M generated by X is the set of all linear combinations $\sum_{x \in X} r_x x$ with coefficients $r_x \in R$. But an element of M might be written in this form in many different ways. The simplest way to construct a module M generated by X is to have every element x of M written uniquely as a linear combination $x = \sum_{x \in X} r_x x$. Then M is free on X.

We give a more formal definition. A subset X of a left R-module M is **linearly independent** over R in case $\sum_{x \in X} r_x x = 0$, with $r_x \in R$ and $r_x = 0$ for almost all $x \in X$, implies $r_x = 0$ for all $x \in X$; that is, if no nontrivial linear combination $\sum_{x \in X} r_x x$ equals 0. Then $\sum_{x \in X} r_x x = \sum_{x \in X} s_x x$ implies $\sum_{x \in X} (r_x - s_x) x = 0$ and $r_x = s_x$ for all x. A family $(x_i)_{i \in I} \subseteq M$ is linearly independent if and only if $\sum_{i \in I} r_i x_i = 0$ (with $r_i \in R$ for all i and $r_i = 0$ for almost all i) implies $r_i = 0$ for all i; this implies $x_i \neq x_j$ whenever $i \neq j$.

A **basis** of a left R-module M is a linearly independent subset of M which generates (spans) M. Thus $X \subseteq M$ is a basis of M if and only if every element of M can be written uniquely as a linear combination $\sum_{x \in X} r_x x$ with coefficients $r_x \in R$ (and $r_x = 0$ for almost all x). Equivalently, $(e_i)_{i \in I} \subseteq M$ is a basis of M if and only if every element x of M can be written uniquely as a linear combination $x = \sum_{i \in I} x_i e_i$ with coefficients x_i in R (with $x_i = 0$ for almost all i); the coefficients $(x_i)_{i \in I}$ are the **coordinates** of x in the basis $(e_i)_{i \in I}$.

A left R-module is **free** in case it has a basis, and is then **free on** that basis. For example, all vector spaces are free (this is proved in the next section). Among $\mathbb{Z}$-modules, $\mathbb{Z}$ is free, but finite abelian groups are not free, and neither is $\mathbb{Q}$ (see the exercises). For any ring R, $_R R$ is free (with basis $\{1\}$).

10.5.b. Construction. Coordinates show that bases are related to direct sums. This leads to a construction of all free modules. Let $\bigoplus_{i \in I} {}_R R$ denote the direct sum of $|I|$ copies of $_R R$: $\bigoplus_{i \in I} {}_R R = \bigoplus_{i \in I} M_i$ where $M_i = {}_R R$ for all i.

Proposition 10.5.1. $\bigoplus_{i \in I} {}_R R$ *is a free left R-module; it has a basis* $(e_i)_{i \in I}$, *the* **standard basis**, *in which* e_i *has components* $(e_i)_i = 1$, $(e_i)_j = 0$ *for all* $j \neq i$.

A left R-module M has a basis $(e_i)_{i \in I}$ *if and if and only if it is isomorphic to* $\bigoplus_{i \in I} {}_R R$; *the isomorphism* $M \longrightarrow \bigoplus_{i \in I} {}_R R$ *assigns to each element x of M its coordinates in the basis* $(e_i)_{i \in I}$.

A left R-module is free if and only if it is isomorphic to $\bigoplus_{i \in I} {}_R R$ *for some I.*

Proof. Let $\iota_i : {}_R R \longrightarrow \bigoplus_{i \in I} {}_R R$ be the injection. By Lemma 10.4.2, every element of $\bigoplus_{i \in I} {}_R R$ can be written uniquely as a sum $\sum_{i \in I} \iota_i(r_i)$, with $r_i \in R$ and $r_i = 0$ for almost all i. Since $\iota_i(r_i) = r_i \iota_i(1)$, this shows that $(e_i)_{i \in I} = (\iota_i(1))_{i \in I}$ is a basis of $\bigoplus_{i \in I} {}_R R$. Thus $\bigoplus_{i \in I} {}_R R$ is free, and so is every left R-module $M \cong \bigoplus_{i \in I} {}_R R$.

Conversely, let M be a free left R-module and $(e_i)_{i \in I}$ be a basis of M. By definition of a basis, the mapping $\theta : (r_i)_{i \in I} \longmapsto \sum_{i \in I} r_i e_i$ of $\bigoplus_{i \in I} {}_R R$ into M is a bijection. It is a homomorphism, since $\sum_{i \in I} r_i e_i + \sum_{i \in I} s_i e_i = \sum_{i \in I} (r_i + s_i) e_i$ and $r \sum_{i \in I} r_i e_i = \sum_{i \in I} r r_i e_i$. $\square$

Corollary 10.5.2. *Given any set X there exists a free left R-module with basis X, and it is unique up to isomorphism.*

Proof. Let $(e_x)_{x \in X}$ be the standard basis of $\bigoplus_{x \in X} {}_R R$. There is a bijection θ of $(e_x)_{x \in X}$ onto X. Extend θ to a bijection $\bigoplus_{x \in X} {}_R R \longrightarrow F$, where F is a set which contains X and has $|\bigoplus_{x \in X} {}_R R|$ elements. Left R-module operations can then be defined on F so that θ is an isomorphism. The resulting module F is free on X. $\square$

For example, there exists a free abelian group F on any set X; it is isomorphic to a direct sum of $|X|$ copies of $\mathbb{Z}$; every element a of F can be written uniquely as a linear combination $a = \sum_{x \in X} n_x x$ of elements of X with integer coefficients.

Another consequence of Proposition 10.5.1 is that free modules are bimodules. Since R is an R-R-bimodule, every direct sum $\bigoplus_{i \in I} {}_R R$ is an R-R-bimodule; the right action of R on $\bigoplus_{i \in I} {}_R R$ is given by $(r_i)_{i \in I} r = (r_i r)_{i \in I}$. By Proposition 10.5.1:

Corollary 10.5.3. *Let M be a free left R-module. For each basis* $(e_i)_{i \in I}$ *of M there is an R-R-bimodule structure on M; the right action of R on M is given by* $(\sum_{i \in I} x_i e_i) r = \sum_{i \in I} x_i r e_i$.

The right action of R in Corollary 10.5.3 is defined by means of coordinates and usually depends on the basis. However, if R is commutative, then every left R-module M is an R-R-bimodule; this provides the same right action of R as Corollary 10.5.3, so in this case the bimodule structure does not depend on the basis.

10.5.c. Universal Property

Proposition 10.5.4. *Let F be a free left R-module and X be a basis of F. Every mapping of X into a left R-module M extends uniquely to a module homomorphism of F into M:*

Proof. Let $\varphi : X \longrightarrow M$ be a mapping. A homomorphism $\psi : F \longrightarrow M$ which extends φ must preserve linear combinations and send $\sum_{x \in X} r_x \, x$ to $\sum_{x \in X} r_x \, \psi(x) = \sum_{x \in X} r_x \, \varphi(x)$. Since X is a basis of F, there is a unique mapping ψ of F into M such that $\psi(\sum_{x \in X} r_x \, x) = \sum_{x \in X} r_x \, \varphi(x)$ whenever $r_x = 0$ for almost all x; ψ is a homomorphism, since $\sum_{x \in X} r_x \, x \longmapsto (r_x)_{x \in X}$ is an isomorphism. $\qquad\square$

This universal property characterizes free modules up to isomorphism (see the exercises). It is a particular case of the universal property of direct sums.

Corollary 10.5.5. *A left R-module which is generated by a subset X is a homomorphic image of the free left R-module on X.*

Proof. By Corollary 10.5.2 there is a module F which is free on X. By Proposition 10.5.4, the inclusion mapping $X \longrightarrow M$ extends to a homomorphism φ of F into M; φ is surjective, since M is generated by $X \subseteq \operatorname{Im} \varphi$. $\qquad\square$

10.5.d. Homomorphisms. Some properties of vector spaces extend to all free modules. For instance, homomorphisms of one free R-module into another can be represented by matrices with coefficients in R.

An $I \times J$ **matrix** with coefficients in R is a mapping of $I \times J$ into R, where I and J are sets, and is usually written as a family $A = (a_{ij})_{i \in I, j \in J}$ and thought of as a rectangular array in which I indexes rows and J indexes columns. Thus the (i, j) **entry** a_{ij} lies in the i-th row and the j-th column; the i-th **row** is the family $(a_{ij})_{j \in J}$; and the j-th **column** is the family $(a_{ij})_{i \in I}$. The **size** of the matrix is $|I| \times |J|$ (the number of rows times the number of columns).

For technical reasons we begin with right R-modules. Let M and N be free right R-modules and $(e_j)_{j \in J}$, $(f_i)_{i \in I}$ be bases of M and N respectively, written as families (with $e_j \neq e_h$, $f_i \neq f_k$ whenever $j \neq h$, $i \neq k$). Let $\varphi : M \longrightarrow N$ be a homomorphism. Each $\varphi(e_j) \in N$ can be written uniquely as a linear combination

$$\varphi(e_j) = \sum_{i \in I} f_i \, r_{ij}$$

(with $r_{ij} \in R$ and $r_{ij} = 0$ for almost all $i \in I$). The **matrix** of φ in the given bases $(e_j)_{j \in J}$ and $(f_i)_{i \in I}$ is the $I \times J$ matrix $M(\varphi) = (r_{ij})_{i \in I, j \in J}$; its j-th column gives the coordinates of $\varphi(e_j)$ in the basis $(f_i)_{i \in I}$. The matrix of φ is **column finitary**, meaning that each column $(r_{ij})_{i \in I}$ has only finitely many nonzero entries.

Proposition 10.5.6. *Let M and N be free right R-modules with bases $(e_j)_{j \in J}$ and $(f_i)_{i \in I}$ respectively. There is a one-to-one correspondence between module homomorphisms of M into N and column finitary matrices $(r_{ij})_{i \in I, j \in J}$ with coefficients in R.*

Proof. We just saw that the matrix $M(\varphi)$ of a module homomorphism $\varphi : M \longrightarrow N$ (in the given bases) is a column finitary matrix. Conversely, each column finitary matrix $(r_{ij})_{i \in I, j \in J}$ (with coefficients in R) determines a mapping $e_j \longmapsto \sum_{i \in I} f_i r_{ij}$ of $X = (e_j)_{j \in J}$ into N which, by Proposition 10.5.4, extends uniquely to a homomorphism $\varphi : M \longrightarrow N$, namely

$$\varphi \left(\sum_{j \in J} e_j x_j \right) = \sum_{j \in J} \left(\sum_{i \in I} f_i r_{ij} \right) x_j = \sum_{i \in I, j \in J} f_i r_{ij} x_j . \qquad (10.1)$$

It is immediate that $(r_{ij})_{i \in I, j \in J}$ is the matrix of φ and that we have constructed a one-to-one correspondence between homomorphisms and matrices. $\qquad \square$

Operations on matrices are defined to match the operations on homomorphisms (addition and composition). The **sum** of two matrices $A = (a_{ij})_{i \in I, j \in J}$ and $B = (b_{ij})_{i \in I, j \in J}$ is the matrix $C = A + B = (c_{ij})_{i \in I, j \in J}$ defined by

$$c_{ij} = a_{ij} + b_{ij}$$

for all i and j. If $\varphi : M \longrightarrow N$ and $\psi : M \longrightarrow N$ are homomorphisms of free right R-modules, then, relative to any given bases, $M(\varphi + \psi) = M(\varphi) + M(\psi)$.

The **product** of two column finitary matrices $A = (a_{ij})_{i \in I, j \in J}$ and $B = (b_{jk})_{j \in J, k \in K}$ is the matrix $C = AB = (c_{ik})_{i \in I, k \in K}$ defined by

$$c_{ik} = \sum_{j \in J} a_{ij} b_{jk}$$

for all i and k; the sum $\sum_{j \in J} a_{ij} b_{jk}$ exists in R, since B is column finitary. The sizes of A and B ensure that the columns of A correspond to the rows of B; the rows of AB correspond to the rows of A, and the columns of AB to the columns of B. (The product of any two matrices A and B of sizes $I \times J$ and $J \times K$ is defined whenever, for each i and k, $a_{ij} b_{jk} = 0$ for almost all j.)

If $\varphi : M \longrightarrow N$ is a homomorphism, where M and N are free right R-modules, and the coordinates $(x_j)_{j \in J}$ of $x \in M$ are arranged in a one-column matrix X, then the product $M(\varphi) X$ gives the coordinates of $\varphi(x)$, by equation (10.1). If

$\varphi : M \longrightarrow N$ and $\psi : N \longrightarrow P$ are homomorphisms of free right R-modules, then, relative to any given bases of M, N, P, $M(\psi \circ \varphi) = M(\psi)M(\varphi)$ (see the exercises).

(We defined $M(\varphi)$ so that the the coordinates of $\varphi(e_j)$ constitute the j-th column of $M(\varphi)$. Exchanging rows and columns in $M(\varphi)$ yields a $J \times I$ matrix $M'(\varphi)$ in which the coordinates of $\varphi(e_j)$ constitute the j-th row rather than the j-th column; with this arrangement the matrix of a composition is $M'(\psi \circ \varphi) = M'(\varphi)M'(\psi)$ rather than $M(\psi \circ \varphi) = M(\psi)M(\varphi)$.)

Corollary 10.5.7. *Under matrix addition and multiplication, the set of all column finitary $I \times I$ matrices $(a_{ij})_{i,j \in I}$ with coefficients in R is a ring $M_I(R)$, which is isomorphic to $\mathrm{End}_R(\bigoplus_{i \in I} R_R)$.*

If $I = \{1, 2, \ldots, n\}$, then $M_I(R)$ is denoted by $M_n(R)$ and consists of all $n \times n$ matrices with coefficients in R. Note that R is an arbitrary ring in the construction of $M_I(R)$ (even the identity element could be dispensed with).

Now let M and N be free left R-modules, with bases $(e_j)_{j \in J}$ and $(f_i)_{i \in I}$ respectively. Then M and N are free right R^{op}-modules (with the same bases), and an R-module homomorphism $\varphi : M \longrightarrow N$ is an R^{op}-module homomorphism and has a column finitary $I \times J$ matrix $M(\varphi)$ in the given bases, with entries in R^{op}. Now $M(\varphi)$ looks just like a matrix with entries in R, but the matrix product in $M(\psi \circ \varphi) = M(\psi)M(\varphi)$ must now be calculated with R^{op} rather than R. In particular, $\mathrm{End}_R(\bigoplus_{i \in I} {}_R R) \cong M_I(R^{\mathrm{op}})$ (not $M_I(R)$).

10.5.e. The Rank. Some free modules have bases with different numbers of elements (see the exercises). We note two cases where this does not happen.

Proposition 10.5.8. *When F is a free left R-module with an infinite basis, then all bases of F have the same number of elements.*

Proof. We note that a basis X of F is minimal among the subsets of F which generate F: if F were generated by $Y \subsetneq X$, then any $x \in X \backslash Y$ would be a linear combination of elements of Y and X would not be linearly independent.

Assume that F has an infinite basis X, and let Y be another basis of F. Each $x \in X$ is a linear combination of elements of Y; hence there is for each $x \in X$ a finite subset S_x of Y such that x is a linear combination of elements of S_x. Then F is generated by $\bigcup_{x \in X} S_x$; since Y is minimal with this property, it follows that $Y = \bigcup_{x \in X} S_x$. Hence $|Y| \leqslant \aleph_0 |X|$; by Corollary B.5.12, $|Y| \leqslant \aleph_0 |X| = |X|$.

Similarly each $y \in Y$ is a linear combination of elements of X, and there is for each $y \in Y$ a finite subset T_y of X such that y is a linear combination of elements of T_y. As above, this implies $X = \bigcup_{y \in Y} T_y$. Therefore Y is infinite (otherwise, X would be finite), and $|X| \leqslant \aleph_0 |Y| = |Y|$. $\qquad \square$

Proposition 10.5.9. *If R is commutative, then all bases of a free R-module have the same number of elements.*

Proof. Let F be a left R-module with a basis X. By Proposition 10.5.8, we may assume that X is finite. Then the result holds if R is a field (then F is a finite-dimensional vector space). In general, let $\mathfrak{m}$ be a maximal ideal of R so that $R/\mathfrak{m}$ is a field. If $r_x \in \mathfrak{m}$ for all $x \in X$ (and $r_x = 0$ for almost all x), then $\sum_{x \in X} r_x\, x \in \mathfrak{m}F$. Conversely, $\mathfrak{m}F$ is generated by elements rt of F with $r \in \mathfrak{m}$, whose coordinates in the basis X are all in $\mathfrak{m}$; hence $\sum_{x \in X} r_x\, x \in \mathfrak{m}F$ implies $r_x \in \mathfrak{m}$ for all x. In particular, $x - y \notin \mathfrak{m}F$ when $x, y \in X$ are different.

Consider the left R-module $F/\mathfrak{m}F$. We have $\mathfrak{m} \subseteq \operatorname{Ann}(F/\mathfrak{m}F)$. Hence the left R-module structure $R \longrightarrow \operatorname{End}(F/\mathfrak{m}F)$, whose kernel contains $\mathfrak{m}$, factors through the projection $R \longrightarrow R/\mathfrak{m}$. Thus there is a left $R/\mathfrak{m}$-module structure on $F/\mathfrak{m}F$; the action of $R/\mathfrak{m}$ on $F/\mathfrak{m}F$ satisfies $(r + \mathfrak{m})(t + \mathfrak{m}F) = rt + \mathfrak{m}F$.

We show that $\{ x + \mathfrak{m}F ; x \in X \}$ is a basis of the left $R/\mathfrak{m}$-module $F/\mathfrak{m}F$; hence $|X|$ is the dimension of the vector space $F/\mathfrak{m}F$ and is uniquely determined by F. First, $\{ x + \mathfrak{m}F ; x \in X \}$ is linearly independent: if $\sum_{x \in X} (r_x + \mathfrak{m})(x + \mathfrak{m}F) = 0$ in $F/\mathfrak{m}F$, then $\sum_{x \in X} r_x\, x \in \mathfrak{m}F$, which we saw implies that $r_x \in \mathfrak{m}$ for all x. Also each $t \in F$ can be written in the form $t = \sum_{x \in X} r_x\, x$; hence each $t + \mathfrak{m}F \in F/\mathfrak{m}F$ can be written in the form $t + \mathfrak{m}F = \sum_{x \in X} (r_x + \mathfrak{m})(x + \mathfrak{m}F)$. $\square$

When all the bases of a free left R-module F have the same number of elements, that number is the **rank** of F (also called the **dimension** of F). For example, every free abelian group has a rank.

10.5.f. Presentations. Let M be a left R-module and X be a subset of M which generates M. By Corollary 10.5.5, M is a homomorphic image of the free module F_X on X and can be constructed as $M \cong F_X/K$, where K is a submodule of F_X. This describes M as a set of linear combinations of elements of X between which K determines a number of equalities called relations.

As in Section 2.6 we formally define a module **relation** between elements of X as an ordered pair (u, v) of elements of the free left R-module F_X on X. The relation (u, v) is normally written like an equality $u = v$, or in the equivalent form $u - v = 0$. When φ is a mapping of X into a left R-module M, the relation $u = v$ **holds in** M (**via** φ) in case $\psi(u) = \psi(v)$, where $\psi : F_X \longrightarrow M$ is the module homomorphism which extends φ; equivalently, in case $u - v \in \operatorname{Ker} \psi$.

A **presentation** of a module M consists of a set X, a set $S \subseteq F_X \times F_X$ of module relations between the elements of X, and an isomorphism $M \cong \langle X; S \rangle$ of M onto the quotient module $\langle X; S \rangle = F_X/K$, where F_X is the free left R-module on X and K is the submodule of F_X generated by $\{ u - v ; (u, v) \in S \}$. The inclusion mapping $X \longrightarrow F_X$ induces a mapping $\iota : X \longrightarrow \langle X; S \rangle$ which sends $x \in X$ to $x + K \in \langle X; S \rangle$. The homomorphism $F_X \longrightarrow \langle X; S \rangle$ which extends ι is precisely the projection $F_X \longrightarrow F_X/K$. Therefore $\langle X; S \rangle$ has the following properties:

Proposition 10.5.10. $\langle X; S \rangle$ *is generated by* $\iota(X)$, *and every relation* $u = v$ *in* S *holds in* $\langle X; S \rangle$ *(via* ι*).*

Due to Proposition 10.5.10, $\langle X;S \rangle$ is called the left R-module **generated by X subject to S**; the elements of S are the **defining relations** of $\langle X;S \rangle$. As in the case of groups, one should keep in mind that ι need not be injective and that there many modules generated by $\iota(X)$ in which every relation $u = v$ in S holds. However, $\langle X;R \rangle$ is the "largest" such module, as shown by its universal property, which skeptical readers will eagerly verify:

Proposition 10.5.11. *Let S be a set of module relations between the elements of a set X. Let M be a left R-module and $\varphi : X \longrightarrow M$ be a mapping such that every relation $u = v$ in S holds in M (via φ). There is a homomorphism $\psi : \langle X;S \rangle \longrightarrow M$ unique such that the diagram*

commutes. If M is generated by $\varphi(X)$, then ψ is surjective.

Exercises

1. Show that $\mathbb{Q}$ is not a free $\mathbb{Z}$-module.

2. Show that $(e_i)_{i \in I}$ is a basis of a left R-module M if and only if $(r_i)_{i \in I} \longmapsto \sum_{i \in I} r_i e_i$ is an isomorphism of $\bigoplus_{i \in I} {}_R R$ onto M.

3. Show that the universal property in Proposition 10.5.4 characterizes free left R-modules up to isomorphism. (Let F be a free left R-module, with a basis B. Show that a left R-module M is isomorphic to F if and only if there exists a mapping $\iota : B \longrightarrow M$ such that every mapping $\varphi : B \longrightarrow N$ of B into a left R-module factors uniquely through ι; that is, $\varphi = \psi \circ \iota$ for some unique module homomorphism $\psi : M \longrightarrow N$.)

4. Prove that a direct sum of free modules is free.

5. Let M and N be free left R-modules with bases $(e_i)_{i \in I}$ and $(f_j)_{j \in J}$ respectively. Show that M^t defines a one-to-one correspondence between module homomorphisms of M into N and row finitary matrices $(r_{ji})_{j \in J, i \in I}$ with coefficients in R.

6. Verify that $M(\psi \circ \varphi) = M(\psi) M(\varphi)$.

7. Prove directly that the product of column finitary matrices is associative ($(AB)C = A(BC)$ whenever AB and BC are defined).

8. Prove directly that column finitary $I \times I$ matrices $(a_{ij})_{i,j \in I}$ (with coefficients in R) form a ring.

9. Let K be a field and V be a vector space over K with an infinite basis $e_0, \ldots, e_n, \ldots$ and $R = \mathrm{End}_K(V)$. Show that $\{1\}$ and $\{\alpha, \beta\}$ are bases of ${}_R R$, where $\alpha(e_{2n}) = e_n$, $\alpha(e_{2n+1}) = 0$ and $\beta(e_{2n+1}) = e_n$, $\beta(e_{2n}) = 0$ for all n.

10. Prove that the module ${}_R R$ in the previous exercise has a basis with m elements for every integer $m > 0$.

11. Prove Proposition 10.5.11.

10.6. VECTOR SPACES

A **division ring** is a ring in which every nonzero element is a unit; division rings are also called **skew fields**. For example, a commutative division ring is a field; the quaternion algebra in Section A.7 is a division ring but not a field.

In this section we show that some basic properties of vector spaces over a field extend to modules over division rings, which are also called **vector spaces**.

10.6.a. Bases and Dimension

Theorem 10.6.1. *Every vector space over a division ring has a basis. In fact every linearly independent subset of a vector space is contained in a basis.*

Proof. Let V be a vector space over a division ring R. The union X of a chain $(X_i)_{i \in I}$ of linearly independent subsets of V is a linearly independent subset of V: if $\sum_{x \in X} r_x\, x = 0$ (with $r_x \in R$ and $r_x = 0$ for almost all x), then $\{x \in X \,;\, r_x \neq 0\}$, which is finite, is contained in some X_i, and it follows that $r_x = 0$ for all x. By Zorn's Lemma, every linearly independent subset of V is contained in a maximal linearly independent subset of V. The following lemma shows that a maximal linearly independent subset also generates V. $\qquad\square$

Lemma 10.6.2. *Let X be a linearly independent subset of a vector space V and $y \in V \backslash X$. Then $X \cup \{y\}$ is linearly independent if and only if y is not a linear combination of elements of X.*

Proof. If $y \notin X$ is a linear combination of elements of X, then $X \cup \{y\}$ is not linearly independent. Conversely, assume that $X \cup \{y\}$ is not linearly independent. Then $r_y y + \sum_{x \in X} r_x\, x = 0$ for some $r_y, r_x \in R$, at least one of which is not zero. Since X is linearly independent, we have $r_y \neq 0$; multiplying on the left by r_y^{-1} shows that y is a linear combination of elements of X. $\qquad\square$

Theorem 10.6.3. *All bases of a vector space have the same number of elements.*

Proof. Let V be a vector space over a division ring R. By Proposition 10.5.8, we may assume that V has a finite basis. Then Proposition 10.5.8 implies that every basis of V is finite, and we need only prove that any two finite bases X and Y of V have the same number of elements. This follows from the **exchange property** in Lemma 10.6.4 below. Successive applications of Lemma 10.6.4 construct a basis of V in which every element of X has been replaced by an element of Y. This implies $|Y| \geqslant |X|$. Exchanging X and Y yields $|X| = |Y|$. $\qquad\square$

Lemma 10.6.4. *Let X and Y be finite bases of a vector space V over a division ring R. For each $x \in X$ there exists $y \in Y$, with either $y = x$ or $y \notin X$, such that $(X \backslash \{x\}) \cup \{y\}$ is a basis of V.*

Proof. The result is trivial if $x \in Y$, and we may assume that $x \notin Y$. Let $X' = X \setminus \{x\}$. If every element of Y is a linear combination of elements of X', then every element of V is a linear combination of elements of X' and x is a linear combination of elements of X', a contradiction since X is linearly independent. Therefore some $y \in Y$ is not a linear combination of elements of X'. Then $X' \cup \{y\}$ is linearly independent by Lemma 10.6.2.

If x is not a linear combination of elements of $X' \cup \{y\}$, then $X \cup \{y\} = X' \cup \{y\} \cup \{x\}$ is linearly independent by Lemma 10.6.2, a contradiction since $y \notin X$ is a linear combination of elements of X. Therefore every element of X is a linear combination of elements of $X' \cup \{y\}$; hence so is every element of V. □

The rank of a vector space V over a division ring R is its **dimension**, which we denote by dim V or $\dim_R V$.

10.6.b. Subspaces. The submodules of a vector space V are called **subspaces**.

Proposition 10.6.5. *Every subspace S of a vector space V is a direct summand of V; furthermore* dim S + dim V/S = dim V.

Proof. By Theorem 10.6.1, S has a basis X, which is a linearly independent subset of V and is therefore contained in a basis Z of V. Let T be the subspace of V generated by $Y = Z \setminus X$. Since $Y \subseteq Z$ is linearly independent, it is a basis of T. Hence dim S + dim T = dim V. We prove that $V = S \oplus T$; then $V/S \cong T$ (by the Second Isomorphism Theorem) and dim V/S = dim T.

Since Z is a basis of V, every element of V can be written in the form $\sum_{z \in Z} r_z z = \sum_{x \in X} r_x x + \sum_{y \in Y} r_y y$. Hence $V = S + T$. Assume that $\sum_{x \in X} r_x x = \sum_{y \in Y} r_y y \in S \cap T$. Then $\sum_{x \in X} r_x x - \sum_{y \in Y} r_y y = 0$; since Z is linearly independent this implies $\sum_{x \in X} r_x x = \sum_{y \in Y} r_y y = 0$. Thus $S \cap T = 0$. □

Proposition 10.6.6. *Let S be a subspace of a vector space V. If* dim V *is finite and* dim S = dim V, *then $S = V$.*

The proof is an exercise.

10.6.c. Division rings. Our last two results are properties of division rings.

Theorem 10.6.7 (Wedderburn). *Every finite division ring is a field.*

Proof. Let D be a finite division ring. Let

$$K = \{ x \in D \, ; \, xy = yx \text{ for all } y \in D \}$$

be the center of D. It is immediate that K is a subfield of D, and that D is a vector space over K. If $|K| = q$, then $|D| = q^n$, where $n = \dim_K D$.

We use the class equation of the multiplicative group $D^* = D \setminus \{0\}$ to show that $n = 1$. The center of D^* is $K^* = K \setminus \{0\}$. For each $a \in D^*$,

$$L = \{x \in D; \, xa = ax\}$$

is a subdivision ring of D which contains K. Hence D is a vector space over L, and as with field extensions, $\dim_K D = (\dim_K L)(\dim_L D)$ (see the exercises). It follows that L has q^d elements, where d divides n, and $d < n$ if $a \notin K$. Then the centralizer $L \setminus \{0\}$ of a in D^* has order $q^d - 1$. Hence the class equation of D^* reads

$$q^n - 1 = (q - 1) + \sum \frac{q^n - 1}{q^d - 1},$$

where the summation has one term for each nontrivial conjugacy class in which d divides n and $d < n$.

Let $\Phi_n(X)$ be the n-th cyclotomic polynomial. By Lemma 7.7.1, $\Phi_n(X)$ divides $X^n - 1$ and $(X^n - 1)/(X^d - 1)$ for each divisor $d < n$ of n, since

$$X^n - 1 = \prod_{c|n} \Phi_c(X) = \Phi_n(X) \left(\prod_{c|n, \, c<n, \, c \nmid d} \Phi_c(X) \right) (X^d - 1).$$

Hence $\Phi_n(q)$ divides $q^n - 1$ and $(q^n - 1)/(q^d - 1)$ for each divisor $d < n$ of n. The class equation then shows that $\Phi_n(q)$ divides $q - 1$. But $\Phi_n(q) > q - 1$ if $n \geqslant 2$, by Lemma 7.7.5. Therefore $n = 1$, and $D = K$ is a field. $\square$

Lemma 10.6.8. *Let D be a division ring which has finite dimension over a subfield K. If K is algebraically closed, then $D = K$.*

Proof. This is clear if D is a field. In general, let $\alpha \in D$. Since D has finite dimension over K, the powers $1, \alpha, \alpha^2, \ldots, \alpha^n, \ldots$ of α cannot be linearly independent. Therefore there exists a relation $f(\alpha) = 0$, where $f \in K[X]$, $f \neq 0$. Since K is algebraically closed, f has a factorization $f(X) = a_n(X - r_1) \ldots (X - r_n)$, where $n > 0$ and $a_n, r_1, \ldots, r_n \in K$, $a_n \neq 0$. Then $a_n(\alpha - r_1) \ldots (\alpha - r_n) = 0$, which in the division ring D implies $\alpha = r_i \in K$ for some i. $\square$

Exercises

1. Let S be a subspace of a vector space V over a division ring. Assume that $\dim V$ is finite and that $\dim S = \dim V$. Prove that $S = V$.

2. Let V and W be vector spaces over a division ring and $\varphi : V \longrightarrow W$ be a homomorphism. Prove that $\dim \operatorname{Ker} \varphi + \dim \operatorname{Im} \varphi = \dim V$.

3. Let S and T be subspaces of a vector space over a division ring. Prove that $\dim (S \cap T) + \dim (S + T) = \dim S + \dim T$.

4. Let $K \subseteq L \subseteq D$ be subdivision rings of a division ring D. Prove that $(\dim_K L)(\dim_L D) = \dim_K D$.

5. Let R be a ring which has an ideal I such that R/I is a division ring. Prove that all bases of a free R-module have the same number of elements.

6. Let R be a ring such that every left R-module is free. Prove that R is a division ring. (Show that R has no nonzero maximal left ideal.)

10.7. MODULES OVER PRINCIPAL IDEAL DOMAINS

A number of properties of abelian groups extend to modules over a principal ideal domain (PID). In this section we construct all finitely generated modules over a PID. The next section gives a nice application of this result.

10.7.a. Submodules of Free Modules

Theorem 10.7.1. *Let R be a principal ideal domain and F be a free R-module. Every submodule M of F is free, with $\operatorname{rank} M \leqslant \operatorname{rank} F$.*

Proof. Let X be a basis of F. For each subset Y of X, let F_Y be the submodule of F generated by Y, and let $M_Y = M \cap F_Y$. Let $\mathscr{S}$ be the set of all ordered pairs (Y, B) such that $Y \subseteq X$, M_Y is free, $|B| \leqslant |Y|$, and B is a basis of M_Y; $\mathscr{S}$ is not empty since $(\varnothing, \varnothing) \in \mathscr{S}$. Order $\mathscr{S}$ by $(Y, B) \leqslant (Z, C)$ if and only if $Y \subseteq Z$ and $B \subseteq C$. We show that $\mathscr{S}$ is inductive. Let $(Y_i, B_i)_{i \in I}$ be a chain of elements of $\mathscr{S}$. Then $Y = \bigcup_{i \in I} Y_i \subseteq X$. An element of F_Y is a linear combination of finitely many elements of Y, which are all contained in some Y_i; hence $F_Y = \bigcup_{i \in I} F_{Y_i}$ and $M_Y = \bigcup_{i \in I} M_{Y_i}$. Moreover $B = \bigcup_{i \in I} B_i$ is linearly independent, $|B| \leqslant |Y|$, and each element of M_Y belongs to some M_{Y_i} and is a linear combination of elements of $B_i \subseteq B$; hence B is a basis of M_Y. Thus $(Y, B) \in \mathscr{S}$ is an upper bound of $(Y_i, B_i)_{i \in I}$.

By Zorn's Lemma, $\mathscr{S}$ has a maximal element. We show that $Y = X$ when (Y, B) is maximal; then $M = M_Y$ is free and $\operatorname{rank} M = |B| \leqslant |X| = \operatorname{rank} F$.

Let $(Y, B) \in \mathscr{S}$, $Y \neq X$; we want to show that (Y, B) is not maximal. Let $x \in X \backslash Y$ and $Z = Y \cup \{x\}$. Then

$$M_Z = F_Z \cap M = \{ rx + t \in M \, ; \, r \in R, \, t \in F_Y \}.$$

Let

$$\mathfrak{a} = \{ r \in R \, ; \, rx + t \in M \text{ for some } t \in F_Y \}.$$

Then $\mathfrak{a}$ is an ideal of R. If $\mathfrak{a} = 0$, then $M_Z = M_Y$, $(Z, B) \in \mathscr{S}$, and (Y, B) is not maximal. Now assume that $\mathfrak{a} = Ra \neq 0$. Let $c = ax + p \in M$, where $p \in F_Y$. We show that $C = B \cup \{c\}$ is a basis of M_Z. Assume that $r_c c + \sum_{b \in B} r_b b = 0$ (with $r_b, r_c \in R$ and $r_b = 0$ for almost all b). Then $r_c ax \in F_Y$; since X is linearly independent, this implies $r_c a = 0$ and $r_c = 0$. Then $\sum_{b \in B} r_b b = 0$ and $r_b = 0$ for all $b \in B$. Thus C is linearly independent. If $q = rx + t \in M_Z$, then $r = sa$ for some $s \in R$ and $q - sc \in M_Y$; hence M_Z is generated by C. Thus C is a basis of M_Z. Then $(Z, C) \in \mathscr{S}$ and (Y, B) is not maximal. $\quad\square$

Theorem 10.7.1 can be improved if F is finitely generated.

Lemma 10.7.2. *Let R be a principal ideal domain, F be a finitely generated free R-module, and M be a submodule M of F. There exists a basis $e_1,\ldots,e_n$ of F, $r \leqslant n$, and nonzero elements $a_1,\ldots,a_r$ of R such that a_i divides a_{i+1} for all $i < r$ and $a_1e_1,\ldots,a_re_r$ is a basis of M.*

Proof. The proof is by induction on the rank n of F. We may assume that $n > 1$ and that M and F are nonzero.

When $\varphi : F \longrightarrow {}_RR$ is a module homomorphism, then $\varphi(M)$ is an ideal of R. Since R is noetherian, there exists a homomorphism $\mu : F \longrightarrow {}_RR$ such that $\mu(M)$ is maximal among all ideals of the form $\varphi(M)$, where $\varphi : F \longrightarrow {}_RR$ is a homomorphism. Let $a \in R$ be a generator of $\mu(M)$ and $m \in M$ be such that $\mu(m) = a$. Then a divides $\mu(x)$ for every $x \in M$. Also:

(1) a divides $\varphi(m)$ for every module homomorphism $\varphi : F \longrightarrow {}_RR$. Indeed let $\varphi(m) = b$ and $d = \gcd(a,b)$. Since R is a PID we have $d = ua + vb$ for some $u,v \in R$. Then $\psi = u\mu + v\varphi : F \longrightarrow {}_RR$ is a module homomorphism, $\psi(m) = d$, $Ra \subseteq Rd \subseteq \psi(M)$, $Ra = Rd = \psi(M)$ by the choice of μ, and a divides $b = \varphi(m)$.

(2) $a \neq 0$ and a divides the coordinates of m in any basis of F. Indeed, in any basis of F, the i-th coordinate provides a module homomorphism $x \longmapsto x_i$ of F into ${}_RR$; by (1), a divides every coordinate of m. Since $M \neq 0$, coordinates also provide a homomorphism $\varphi : F \longrightarrow {}_RR$ such that $\varphi(M) \neq 0$; therefore $\mu(M) \neq 0$ and $a \neq 0$.

By (2), $m = ae$ for some $e \in F$. Then $\mu(e) = 1$. Hence $Re \cap \operatorname{Ker}\mu = 0$. Also $x - \mu(x)e \in \operatorname{Ker}\mu$ for all $x \in F$, whence $x = \mu(x)e + (x - \mu(x)e) \in Re + \operatorname{Ker}\mu$. Thus $F = Re \oplus \operatorname{Ker}\mu$. If $x \in M$, then a divides $\mu(x)$, so that $\mu(x)e \in Rm \subseteq M$ and $x - \mu(x)e \in M \cap \operatorname{Ker}\mu$; hence $M = Rm \oplus (M \cap \operatorname{Ker}\mu)$.

By Theorem 10.7.1, $\operatorname{Ker}\mu$ is free and finitely generated, and has rank $n - 1$ (since $F \cong {}_RR \oplus \operatorname{Ker}\mu$). Hence the induction hypothesis yields a nonempty basis $e_2,\ldots,e_n$ of $\operatorname{Ker}\mu$ and nonzero elements $a_2,\ldots,a_r$ of R such that $r \leqslant n$, a_i divides a_{i+1} for all $i < r$, and $a_2e_2,\ldots,a_re_r$ is a basis of $M \cap \operatorname{Ker}\mu$. (If $M \cap \operatorname{Ker}\mu = 0$, then $r = 1$.) Then $e,e_2,\ldots,e_n$ is a basis of F (since $F = Re \oplus \operatorname{Ker}\mu$) and $ae,a_2e_2,\ldots,a_re_r$ is a basis of M (since $M = Rae \oplus (M \cap \operatorname{Ker}\mu)$). It remains to show that a divides a_2. As in the proof of (2), let $d = \gcd(a,a_2)$. Since R is a PID, we have $d = ua + va_2$ for some $u,v \in R$. By Proposition 10.5.4 there is a homomorphism $\varphi : F \longrightarrow {}_RR$ such that $\varphi(e) = \varphi(e_2) = 1$ and $\varphi(e_i) = 0$ for all $i > 2$. Then $uae + va_2e_2 \in M$ and $\varphi(uae + va_2e_2) = ua + va_2 = d$; hence $Ra \subseteq Rd \subseteq \varphi(M)$, $Ra = Rd = \varphi(M)$ by the choice of μ, and a divides a_2. $\square$

In Lemma 10.7.2 the ideals $Ra_1,\ldots,Ra_r$ are unique, by Theorem 10.7.3 below.

10.7.b. Finitely Generated Modules. By Lemma 10.7.2, every finitely generated R-module is a direct sum of cyclic modules. There are two forms of this theorem:

Theorem 10.7.3. *Let R be a principal ideal domain. Every finitely generated R-module M is a direct sum of finitely many cyclic R-modules; namely M is isomorphic to the direct sum*

$$M \cong F \oplus R/Ra_1 \oplus \cdots \oplus R/Ra_t$$

of a finitely generated free R-module F and cyclic R-modules with annihilators $R \supsetneqq Ra_1 \supseteq \cdots \supseteq Ra_t \neq 0$. Moreover the rank of F, the number t, and the ideals $Ra_1, \ldots, Ra_t$ are unique.

Theorem 10.7.4. *Let R be a principal ideal domain. Every finitely generated R-module M is a direct sum*

$$M \cong F \oplus R/Rp_1^{k_1} \oplus \cdots \oplus R/Rp_s^{k_s}$$

of a finitely generated free R-module F and cyclic R-modules whose annihilators $Rp_1^{k_1}, \ldots, Rp_s^{k_s}$ are generated by positive powers of prime elements of R. Moreover the rank of F, the number s, and the ideals $Rp_1^{k_1}, \ldots, Rp_s^{k_s}$ are unique.

Theorem 10.7.4 implies the Fundamental Theorem of Finitely Generated Abelian Groups (Theorem 2.2.7). We prove 10.7.4 and 10.7.5 together.

Proof. By Corollary 10.5.5, there is an epimorphism $\varphi : F \longrightarrow M$ where F is a finitely generated free R-module. Then $M \cong F/\text{Ker}\,\varphi$. Lemma 10.7.2 provides a basis $e_1, \ldots, e_n$ of F, $r \leqslant n$, and nonzero elements $a_1, \ldots, a_r$ of R such that

$$F = Re_1 \oplus \cdots \oplus Re_r \oplus \cdots \oplus Re_n,$$

$$\text{Ker}\,\varphi = Ra_1 e_1 \oplus \cdots \oplus Ra_r e_r \oplus 0 \oplus \cdots \oplus 0,$$

and $Ra_i \supseteq Ra_{i+1}$ for all $i < r$. Hence

$$F/\text{Ker}\,\varphi \cong Re_1/Ra_1 e_1 \oplus \cdots \oplus Re_r/Ra_r e_r \oplus Re_{r+1} \oplus \cdots \oplus Re_n.$$

Zero terms may be deleted from this decomposition. Now $r \longmapsto re_i$ is an isomorphism ${}_R R \longrightarrow Re_i$ and sends Ra_i onto $Ra_i e_i$; therefore $Re_i/Ra_i e_i \cong R/Ra_i$. By Proposition 10.3.6, R/Ra_i is cyclic with annihilator Ra_i. This proves the existence part of Theorem 10.7.3.

Proving the uniqueness part requires additional concepts.

10.7.c. Torsion. An element x of a left R-module M is **torsion** in case $\text{Ann}\,(x) \neq 0$ ($rx = 0$ for some $r \in R$, $r \neq 0$). If R is a PID, then the **order** of x

is any generator of $\operatorname{Ann}(x)$ (e.g., the representative element which generates $\operatorname{Ann}(x)$). (If $R = \mathbb{Z}$, this defines the usual order in the abelian group M.) The element x is **torsion free** in case $\operatorname{Ann}(x) = 0$ ($rx = 0$ implies $r = 0$).

An R-module M is **torsion** in case $\operatorname{Ann}(x) \neq 0$ for every $x \in M$ (in case all its elements are torsion), and **torsion free** in case $\operatorname{Ann}(x) = 0$ for every $x \in M$, $x \neq 0$ (in case $rx = 0$ implies $r = 0$ or $x = 0$). If R has no zero divisors, then every free R-module $M \cong \bigoplus {}_R R$ is torsion free.

Proposition 10.7.5. *Let R be a principal ideal domain and M be an R-module. The torsion elements of M form a submodule $T(M)$ of M, the **torsion part** of M; moreover, $M/T(M)$ is torsion free. If M is finitely generated, then $M = F \oplus T(M)$, where F is free.*

Proof. The first part of the statement holds for any domain R. First, 0 is always torsion. If $r \neq 0$ and $rx = 0$, then $rsx = srx = 0$ and sx is torsion. If $r, s \neq 0$ and $rx = sy = 0$, then $rs \neq 0$, $rs(x - y) = srx - rsy = 0$, and $x - y$ is torsion. Hence the torsion elements of M form a submodule $T(M)$ of M. If $x \in M$ and $r \neq 0$, $rx \in T(M)$, then $srx = 0$ for some $s \neq 0$, $sr \neq 0$, and $x \in T(M)$; hence 0 is the only torsion element of $M/T(M)$.

Now let R be a PID and M be finitely generated. By Theorem 10.7.3,

$$M \cong F \oplus R/Ra_1 \oplus \cdots \oplus R/Ra_t,$$

where F is free and $a_1, \ldots, a_t \neq 0$. Now $T = R/Ra_1 \oplus \cdots \oplus R/Ra_t$ is torsion, since $a = a_1 \ldots a_t \neq 0$ and $ay = 0$ for all $y \in T$. Hence M is an internal direct sum $M = F' \oplus T'$ where F' is free and T' is torsion. Since F' is torsion free, $f' + t' \in M$ is torsion if and only if $f' = 0$; hence $T(M) = T'$ and $M = F' \oplus T(M)$. $\qquad\square$

Let $M \cong F \oplus R/Ra_1 \oplus \cdots \oplus R/Ra_t$ be as in Theorem 10.7.3. As above, $T(M) = R/Ra_1 \oplus \cdots \oplus R/Ra_t$. Then $F \cong M/T(M)$, by the Second Isomorphism Theorem. Hence the rank of F and the torsion module $R/Ra_1 \oplus \cdots \oplus R/Ra_t$ are, up to isomorphism, uniquely determined by M. To prove Theorem 10.7.3, it remains to show that the ideals $Ra_1 \supseteq \cdots \supseteq Ra_t$ are uniquely determined by M; it suffices to prove this when $M \cong R/Ra_1 \oplus \cdots \oplus R/Ra_t$ is torsion.

10.7.d. Torsion Modules. When $p \in R$ is prime, an R-module M is a *p*-**module** in case every element of M is torsion and its order is a power of p; equivalently, in case there is for each element x of M a power p^k of p such that $p^k x = 0$.

Every torsion R-module M contains a largest *p*-submodule, namely

$$M(p) = \{ x \in M ; \ p^k x = 0 \text{ for some } k \};$$

it is readily verified that $M(p)$ is a submodule of M.

In the following results we assume that representative elements have been selected in R (so that every principal ideal of R is generated by exactly one representative element).

Proposition 10.7.6. *Let R be a principal ideal domain and M be a torsion R-module. Then $M = \bigoplus_{p \in \mathscr{P}} M(p)$, where $\mathscr{P}$ is the set of all representative prime elements of R.*

If $M \cong R/Ra$ is cyclic, and $a = u p_1^{k_1} \ldots p_r^{k_r}$, where u is a unit and $p_1, \ldots, p_r$ are distinct representative prime elements, then $M(p) = 0$ if $p \nmid a$, $M(p_i) \cong R/Rp_i^{k_i}$, and $M \cong R/Rp_1^{k_1} \oplus \cdots \oplus R/Rp_r^{k_r}$.

Proof. Let $x \in M$ and $ax = 0$, where $0 \neq a \in R$. If $a = bc$, where $(b,c) = 1$, then $1 = rb + sc$ and $x = rbx + scx = y + z$, where $y = scx$ and $z = rbx$ satisfy $by = cz = 0$. If therefore $a = u p_1^{k_1} \ldots p_r^{k_r}$, where u is a unit and $p_1, \ldots, p_r$ are distinct representative prime elements, then $x = x_1 + \cdots + x_r$, where $p_i^{k_i} x_i = 0$, and $x \in M(p_1) + \cdots + M(p_r)$.

We now have $M = \sum_{p \in \mathscr{P}} M(p)$. To prove $M = \bigoplus_{p \in \mathscr{P}} M(p)$ we show that $M(p) \cap (\sum_{q \in \mathscr{P}, q \neq p} M(q)) = 0$ (Proposition 10.4.7). Let

$$x \in M(p) \cap \left(\sum_{q \in \mathscr{P}, q \neq p} M(q) \right).$$

Then $p^k x = 0$ for some k, and $x = \sum_{q \in \mathscr{P}, q \neq p} x_q$, where $q^{k_q} x_q = 0$ and $x_q = 0$ for almost all q. Let $b = \prod_{q \in \mathscr{P}, q \neq p, x_q \neq 0} q^{k_q}$. Then $bx = 0$ and $(b, p^k) = 1$. Hence $1 = rb + sp^k$ and $x = rbx + sp^k x = 0$.

Now let $M = R/Ra$ be cyclic. If $M(p) \neq 0$, then some $0 \neq x \in M$ has order p^k with $k > 0$; since $ax = 0$, this implies $p^k \mid a$ and $p \mid a$. Hence $M(p) = 0$ if $p \nmid a$. Let $a = u p_1^{k_1} \ldots p_r^{k_r}$, where u is a unit and $p_1, \ldots, p_r$ are distinct representative prime elements. By the first part of the proof, $M = M(p_1) + \cdots + M(p_r)$. Hence $M = M(p_1) \oplus \cdots \oplus M(p_r)$. It remains to show that $M(p_i) \cong R/Rp_i^{k_i}$. Let $b = u \prod_{j \neq i} p_j^{k_j}$, so that $bp_i^{k_i} = a$. If $x = y + Ra \in R/Ra$ has order $p_i^{k_i}$, then $p_i^{k_i} y \in Ra$, $a \mid p_i^{k_i} y$, and $b \mid y$. If, conversely, $b \mid y$, then $p_i^{k_i} y \in Ra$ and $y + Ra \in M(p_i)$. Thus $M(p_i) = Rb/Ra$. By the next lemma, $M(p_i) = Rb/Ra \cong R/Rp_i^{k_i}$. □

Lemma 10.7.7. $Rb/Rbc \cong R/Rc$ whenever R is a domain and $0 \neq b, c \in R$.

Proof. $\varphi : x \longmapsto bx + Rbc$ is a module homomorphism of R onto Rb/Rbc. We see that $\varphi(x) = 0$ if and only if $bc \mid bx$, if and only if $c \mid x$. Thus $\text{Ker } \varphi = Rc$ and $Rb/Ra \cong R/Rc$. □

10.7.e. p-Modules. Theorem 10.7.3 and Proposition 10.7.6 imply the existence part of Theorem 10.7.4. We now turn to uniqueness.

Let M be a finitely generated torsion R-module. By Theorem 10.7.3,

$$M \cong R/Ra_1 \oplus \cdots \oplus R/Ra_t,$$

where $R \supsetneq Ra_1 \supseteq \cdots \supseteq Ra_t \neq 0$. We have $a_t = u_t p_1^{k_{1t}} \ldots p_r^{k_{rt}}$ for some unit u_t and distinct representative prime elements $p_1, \ldots, p_r$. Since each a_j divides a_t, we have $a_j = u_j p_1^{k_{1j}} \ldots p_r^{k_{rj}}$ for some unit u_j, where the exponents $k_{ij} \leqslant k_{it}$ may be zero. Since $Ra_1 \supseteq \cdots \supseteq Ra_t$, we have $k_{i1} \leqslant \cdots \leqslant k_{it}$ for each i. By Proposition 10.7.6,

$$M(p_i) \cong R/Rp_i^{k_{i1}} \oplus \cdots \oplus R/Rp_i^{k_{it}},$$

and $M(p) = 0$ if p does not divide a_t. Hence $p_1, \ldots, p_r$ and the direct sums $R/Rp_i^{k_{i1}} \oplus \cdots \oplus R/Rp_i^{k_{it}}$ are uniquely determined by M. Uniqueness in Theorems 10.7.3 and 10.7.4 then follows from:

Lemma 10.7.8. *Let R be a principal ideal domain, $p \in R$ be prime, and*

$$M \cong R/Rp^{k_1} \oplus \cdots \oplus R/Rp^{k_t}$$

be a finitely generated p-module, where $0 < k_1 \leqslant \cdots \leqslant k_t$. Then t and $k_1, \ldots, k_t$ are uniquely determined by M.

Proof. Let $C = R/Ra$ be a cyclic R-module, where $p \mid a$. By the First Isomorphism Theorem, $C/pC = (R/Ra)/(Rp/Ra) \cong R/Rp$. Therefore M/pM is the direct sum of t copies of R/Rp. Since R/Rp is a field, t is the dimension of M/pM as an R/Rp-module and is uniquely determined by M.

We have $\mathrm{Ann}(M) = Rp^{k_t}$. Hence k_t is uniquely determined by M.

Let $C = R/Rp^k$ be a cyclic p-module, where $k > 0$. By Lemma 10.7.7, $pC = Rp/Rp^k \cong R/Rp^{k-1}$. If $k = 1$, then $pC = 0$. Then we can show by induction on k_t that $k_1, \ldots, k_t$ are uniquely determined by M. If $k_t = 1$, then $k_1 = \cdots = k_t = 1$. If $k_t > 1$, then

$$pM \cong R/Rp^{k_1-1} \oplus \cdots \oplus R/Rp^{k_t-1}.$$

All zero terms may be removed from this direct sum. The number of nonzero terms is uniquely determined by pM and by M. Hence the number of k_i's which equal 1 is uniquely determined by M. The remaining k_i's are uniquely determined by M, by the induction hypothesis. $\square$

Exercises

1. Let R be a PID and M be an R-module which is generated by r elements. Prove that every submodule of M can be generated by at most r elements.
2. Let R be a PID and M be a finitely generated torsion R-module. Show that $M(p) = 0$ for almost all representative prime elements p.
3. Let R be a PID and $M \cong R/Ra$ be a cyclic R-module with annihilator Ra. Show that every submodule of M is cyclic, with annihilator $Rb \supseteq Ra$.
4. Let R be a commutative ring such that every submodule of a free R-module is free. Prove that R is a PID.

10.8. JORDAN FORM OF MATRICES

In this section V is a finite-dimensional vector space over a field K and $T : V \longrightarrow V$ is a linear transformation. The main result of this section puts the matrix of T in Jordan form if K is algebraically closed. This is an application of the structure theorem for finitely generated modules over PIDs (Theorem 10.7.4).

10.8.a. $K[X]$-**Modules.** We begin by making V a $K[X]$-module. The ring $\mathrm{End}_K(V)$ of all linear transformations of V [into itself] is also a vector space over K. There is a homomorphism of K into $\mathrm{End}_K(V)$ which sends $a \in K$ to the linear transformation $a = aI : v \longmapsto av$ of V, where $I = 1_V$ is the identity on V. By the universal property of $K[X]$, this homomorphism $K \longrightarrow \mathrm{End}_K(V)$ extends to a homomorphism $\varphi_T : K[X] \longrightarrow \mathrm{End}_K(V)$ which sends X to T; when $f(X) = a_n X^n + \cdots + a_1 X + a_0 \in K[X]$, then $\varphi_T(f)$ is denoted by $f(T)$ and given by

$$f(T) = a_n T^n + \cdots + a_1 T + a_0 I,$$

where $T^2 v = T(Tv)$, $T^3 v = T(T(Tv))$, and so on. Thus we can evaluate a polynomial at a linear transformation of V; the result is a linear transformation of V.

The homomorphism φ_T is a $K[X]$-module structure on V, which we say is **induced** by T; the action of $f \in K[X]$ on $v \in V$ is given by

$$f \cdot v = f(T)v = a_n T^n v + \cdots + a_1 Tv + a_0 v$$

when $f(X) = a_n X^n + \cdots + a_1 X + a_0$. In particular, $X \cdot v = Tv$ for all $v \in V$.

The $K[X]$-module V is finitely generated, since V is already finitely generated as a K-module. We see that $S \subseteq V$ is a submodule of the $K[X]$-module V if and only if S is a subspace of the vector space V and $TS \subseteq S$ ($Ts \in S$ for all $s \in S$). In particular, T has a restriction $T_{|S}$ to S. Obviously $f(T)s = f(T_{|S})s$ for all $s \in S$; hence the $K[X]$-module structure on the submodule S coincides with the $K[X]$-module structure induced by $T_{|S}$.

10.8.b. Minimal Polynomials. The annihilator

$$\mathrm{Ann}(V) = \mathrm{Ker}\, \varphi_T = \{ f(X) \in K[X] \, ; \, f(T) = 0 \}$$

of the $K[X]$-module V is a ideal of the principal ideal domain $K[X]$. Since $\dim_K K[X]$ is infinite and $\dim_K V$ is finite, $\varphi_T : K[X] \longrightarrow \mathrm{End}_K(V)$ cannot be injective, and $\mathrm{Ann}(V) \neq 0$. Hence $\mathrm{Ann}(V)$ is generated by a unique monic polynomial $m_T(X)$; $m_T(X)$ is the **minimal polynomial** of T. By definition, $f(T) = 0$ if and only if m_T divides f.

Let S be a submodule of V; by the above, $\mathrm{Ann}(S)$ is generated (as an ideal of $K[X]$) by the minimal polynomial of $T_{|S}$.

By Theorem 10.7.4, the $K[X]$-module V is the direct sum of a free $K[X]$-module F and of cyclic $K[X]$-modules $S_1,\ldots,S_t$ whose annihilators $(q_1^{k_1}),\ldots,(q_t^{k_t})$ are uniquely determined and are generated by positive powers of prime elements of $K[X]$. In other words, $q_1,\ldots,q_t \in K[X]$ are irreducible polynomials, which may be chosen monic. Then $q_i^{k_i}$ is the minimal polynomial of the restriction of T to S_i. Also $F = 0$, since $F \subseteq V$ is a direct sum of copies of $K[X]$ but has finite dimension over K. We have proved:

Theorem 10.8.1. *Let V be a finite-dimensional vector space over a field K and $T : V \longrightarrow V$ be a linear transformation. Then V is the direct sum*

$$V = S_1 \oplus \cdots \oplus S_t$$

of cyclic $K[X]$-submodules $S_1,\ldots,S_t$ such that the minimal polynomial of $T_{|S_i}$ is a positive power $q_i^{k_i}$ of a monic irreducible polynomial $q_i \in K[X]$. Moreover the number t, the polynomials $q_1,\ldots,q_t$ of K, and the numbers $k_1,\ldots,k_t$ are unique.

Corollary 10.8.2. *In Theorem 10.8.1, the minimal polynomial of T is the least common multiple of $q_1^{k_1},\ldots,q_t^{k_t}$.*

Proof. $f(T) = 0$ if and only if $f(T_{|S_i}) = 0$ for all i. $\qquad\square$

10.8.c. Cyclic Submodules. By Theorem 10.8.1, our study of T can be restricted to the case where V is a cyclic $K[X]$-module.

Lemma 10.8.3. *Let V be a vector space of finite dimension n over a field K and $T : V \longrightarrow V$ be a linear transformation. If V is a cyclic $K[X]$-module generated by $e \in V$, then $n = \dim_K V = \deg m_T$ and $e, Te, T^2e, \ldots, T^{n-1}e$ is a basis of V.*

Proof. Assume that $V = K[X]e$. Then every element of V has the form $f(T)e$ for some $f(X) \in K[X]$, and is a linear combination (with coefficients in K) of $e, Te, T^2e, \ldots$. Now $f = m_T q + r$, where $\deg r < \deg m_T$, so that $f(T)e = r(T)e$ is a linear combination of $e, Te, \ldots, T^{m-1}e$, where $m = \deg m_T$. If furthermore $a_0 e + a_1 Te + \cdots + a_{m-1}T^{m-1}e = 0$, then $g(X) = a_0 + a_1 X + \cdots + a_{m-1}X^{m-1} \in K[X]$ satisfies $g(T)e = 0$, $g(T)v = 0$ for all $v = f(T)e \in V$, and $g(T) = 0$; therefore m_T divides g and $g = 0$ (since $\deg g < \deg m_T$). Thus $e, Te, \ldots, T^{m-1}e$ is a basis of V. In particular, $n = \dim_K V = m$. $\qquad\square$

When Lemma 10.8.3 is applied to one of the cyclic modules $S_1,\ldots,S_t$ in Theorem 8.1, then m_T is a power of an irreducible polynomial. If K is algebraically closed, then $m_T(X) = (X - \lambda)^n$ for some $\lambda \in K$ and Lemma 10.8.3 can be sharpened.

Lemma 10.8.4. *Let V be a finite dimensional vector space over a field K and $T : V \longrightarrow V$ be a linear transformation. If V is a cyclic $K[X]$-module and $m_T(X) = (X - \lambda)^n$ for some $\lambda \in K$ and $n > 0$, then V has a basis $e_1, \ldots, e_n$ such that $Te_i = \lambda e_i + e_{i+1}$ for all $i < n$ and $Te_n = \lambda e_n$.*

Proof. Assume that $m_T(X) = (X - \lambda)^n$, and let e generate V as a $K[X]$-module. By Lemma 10.8.3, $e, Te, \ldots, T^{n-1}e$ is a basis of V. Then

$$e_1 = e, \ e_2 = (T - \lambda I)e, \ldots, \ e_i = (T - \lambda I)^{i-1}e, \ldots, \ e_n = (T - \lambda I)^{n-1}e$$

is also a basis of V: indeed the matrix of $e, (T - \lambda I)e, \ldots, (T - \lambda I)^{n-1}e$ in the basis $e, Te, \ldots, T^{n-1}e$ is triangular, with 1's on the diagonal, and is invertible. We see that $(T - \lambda I)e_i = e_{i+1}$ for all $i < n$; also $(T - \lambda I)e_n = (T - \lambda I)^n e = 0$. $\qquad\square$

In Lemma 10.8.4, the matrix of T in the basis $e_1, \ldots, e_n$ (with the coordinates of $Te_1, \ldots, Te_n$ arranged in columns) is

$$
\begin{pmatrix}
\lambda & 0 & 0 & 0 & \cdots & 0 \\
1 & \lambda & 0 & 0 & \cdots & 0 \\
0 & 1 & \lambda & 0 & \cdots & 0 \\
\vdots & \ddots & \ddots & \ddots & \ddots & \vdots \\
0 & \cdots & 0 & 1 & \lambda & 0 \\
0 & \cdots & 0 & 0 & 1 & \lambda
\end{pmatrix}.
$$

A square matrix of this form is a **Jordan block**. (Jordan blocks are also defined with 1's above the diagonal rather than below.)

10.8.d. Jordan Form. If K is algebraically closed in Theorem 10.8.1, then the irreducible polynomial q_i is of the form $q_i(X) = X - \lambda$ for some $\lambda \in K$. Then Lemma 10.8.4 can be applied to each of the $K[X]$-submodules S_i and to the restriction T_i of T to S_i. Combining the bases of $S_1, \ldots, S_t$ yields a basis $e_1, \ldots, e_n$ of V in which the matrix of T is in **Jordan form**,

$$
\begin{pmatrix}
J_1 & 0 & \cdots & 0 \\
0 & J_2 & & 0 \\
\vdots & & \ddots & \vdots \\
0 & 0 & \cdots & J_t
\end{pmatrix}
$$

that is, consists of Jordan blocks arranged along the diagonal and zeros elsewhere. We have proved:

Theorem 10.8.5. *Let V be a finite-dimensional vector space over an algebraically closed field K, and let $T : V \longrightarrow V$ be a linear transformation. There is a basis of V in which the matrix of T is in Jordan form. Moreover all such matrices of T contain the same Jordan blocks.*

Corollary 10.8.6. *When the matrix of T is in Jordan form:*

(1) *the diagonal entries are the eigenvalues of T;*
(2) *the minimal polynomial of T is $(X - \lambda_1)^{\ell_1} \ldots (X - \lambda_r)^{\ell_r}$, where $\lambda_1, \ldots, \lambda_r$ are the distinct eigenvalues of T and ℓ_i is the size of the largest Jordan block with λ_i on the diagonal;*
(3) *for each eigenvalue λ of T, the dimension of the corresponding eigenspace is the number of Jordan blocks with λ on the diagonal.*

We leave the proofs as exercises.

Corollary 10.8.7. *When K is algebraically closed, T is diagonalizable (there is a basis of V in which the matrix of T is diagonal) if and only if the minimal polynomial of T is separable.*

Proof. This follows from part (3) of Corollary 10.8.6.   □

10.8.e. The Characteristic Polynomial. Recall that the **characteristic polynomial** of T is $c_T(X) = \det (T - XI) \in K[X]$; the characteristic polynomial of T is also the characteristic polynomial of the matrix of T in any basis.

Theorem 10.8.8 (Cayley-Hamilton). *Let V be a finite-dimensional vector space over a field K. Then $c_T(T) = 0$ for every linear transformation T of V.*

Proof. First let K be algebraically closed. By Theorem 10.8.5, there is a basis of V in which the matrix of T is in Jordan form. Then $c_T(X) = (\lambda_1 - X)^{n_1} \ldots (\lambda_r - X)^{n_r}$, where $\lambda_1, \ldots, \lambda_r$ are the distinct eigenvalues of T and n_i is the number of appearances of λ_i on the diagonal. By part (2) of Corollary 10.8.6, m_T divides c_T; hence $c_T(T) = 0$.

For an arbitrary field K we note that Theorem 10.8.8 is equivalent to its matrix form, $c_M(M) = 0$ for every $M \in M_n(K)$. Now a matrix with coefficients in K can be viewed as a matrix with coefficients in the algebraic closure $\overline{K}$ of K. This does not alter the characteristic polynomial $c_M(X) = \det (M - XI)$. Therefore $c_M(M) = 0$ in $M_n(\overline{K})$ yields $c_M(M) = 0$ in $M_n(K)$.   □

Exercises

1. State the theorem obtained by using Theorem 10.7.3 rather than Theorem 10.7.4 in the proof of Theorem 10.8.1.

2. Define the minimal polynomial of an $n \times n$ matrix with coefficients in a field K. Prove the following: if E is a field extension of K, then the minimal polynomial of A remains the same when A is viewed as an $n \times n$ matrix with coefficients in E. (You may use Exercise 1).

3. In Lemma 10.8.3, find the matrix of T in the basis $e_1, \dots, e_n$, using the cofficients of m_T.

4. Let V be a finite-dimensional vector space over a field K and $T : V \longrightarrow V$ be a linear transformation. Assume that V has a basis $e_1, \dots, e_n$ such that $Te_i = \lambda e_i + e_{i+1}$ for all $i < n$ and $Te_n = \lambda e_n$, where $\lambda \in K$ and $n > 0$. Prove that V is a cyclic $K[X]$-module and that $m_T(X) = (X - \lambda)^n$.

5. Let the matrix of T be in Jordan form. Prove that the minimal polynomial of T is $(X - \lambda_1)^{\ell_1} \dots (X - \lambda_r)^{\ell_r}$, where $\lambda_1, \dots, \lambda_r$ are the distinct eigenvalues of T and ℓ_i is the size of the largest Jordan block with λ_i on the diagonal.

6. Let the matrix of T be in Jordan form. Prove that for each eigenvalue λ of T, the dimension of the corresponding eigenspace is the number of Jordan blocks with λ on the diagonal.

10.9. CHAIN CONDITIONS

This section contains basic properties of noetherian and artinian modules which will be used in later chapters.

10.9.a. Noetherian Modules. A module is **noetherian** in case its submodules satisfy the ascending chain condition. For instance, a commutative ring R is noetherian if and only if the R-module ${}_R R$ is noetherian.

By Proposition B.2.1, noetherian modules are characterized by any of the following properties:

(1) Every infinite ascending sequence of submodules

$$S_1 \subseteq \dots \subseteq S_n \subseteq S_{n+1} \subseteq \dots$$

terminates (there exists $m > 0$ such that $S_n = S_m$ for all $n \geqslant m$).

(2) There is no strictly ascending infinite sequence of submodules

$$S_1 \subsetneq \dots \subsetneq S_n \subsetneq S_{n+1} \subsetneq \dots.$$

(3) Every nonempty set of submodules (partially ordered by inclusion) has a maximal element.

We note one more characterization.

Proposition 10.9.1. *A left R-module M is noetherian if and only if every submodule of M is finitely generated.*

Proof. Let M be noetherian and S be a submodule of M. Since M is noetherian, the set of all finitely generated submodules of S has a maximal element T. For any $s \in S \backslash T$, the submodule $T + Rs$ of T is finitely generated (by s and the generators of T); since $T \subseteq T + Rs$, it follows from the choice of T that $T = T + Rs$ and $s \in T$. Thus $S = T$ and S is finitely generated.

Conversely, assume that every submodule of M is finitely generated. Let

$$S_1 \subseteq \cdots \subseteq S_n \subseteq S_{n+1} \subseteq \cdots$$

be an ascending sequence of submodules of M. Then $S = \bigcup_{n>0} S_n$ is a submodule of M and is generated by a finite subset X. Since X is finite, X is contained in some S_m. Then $S_m \subseteq S$, and the inclusions $S_m \subseteq S_n \subseteq S \subseteq S_m$ show that $S_n = S_m$ for all $n \geqslant m$. □

Proposition 10.9.2. *Let M be a left R-module and N be a submodule of M; M is noetherian if and only if N and M/N are noetherian.*

Proof. Assume that N and M/N are noetherian. Let

$$S_1 \subseteq \cdots \subseteq S_n \subseteq S_{n+1} \subseteq \cdots$$

be an ascending sequence of submodules of M. Then

$$S_1 \cap N \subseteq \cdots \subseteq S_n \cap N \subseteq S_{n+1} \cap N \subseteq \cdots$$

is an ascending sequence of submodules of N and

$$(S_1 + N)/N \subseteq \cdots \subseteq (S_n + N)/N \subseteq (S_{n+1} + N)/N \subseteq \cdots$$

is an ascending sequence of submodules of M/N. Since both sequences terminate, there exists $m > 0$ such that $S_n \cap N = S_m \cap N$ and $(S_n + N)/N = (S_m + N)/N$ for all $n \geqslant m$. Then $S_n + N = S_m + N$ for all $n \geqslant m$. This implies $S_n = S_m$ for all $n \geqslant m$: if $x \in S_n$, then $x \in S_m + N$, $x = y + z$ for some $y \in S_m$ and $z \in N$, $z = x - y \in S_n \cap N$, $z \in S_m$, and $x = y + z \in S_m$. Thus M is noetherian. We leave the converse as an exercise. □

A ring R is **left noetherian** in case the left R-module $_R R$ is noetherian; equivalently, in case the left ideals of R satisfy the ascending chain condition; equivalently (by Proposition 10.9.1), in case every left ideal of R is finitely generated.

Proposition 10.9.3. *If R is left noetherian, then every finitely generated left R-module is noetherian.*

Note that a noetherian module must be finitely generated, by Proposition 10.9.2.

Proof. We show that every finitely generated free left R-module F is noetherian; then the result follows from Corollary 10.5.5. Let $e_1, \ldots, e_n$ be a basis of F, so that $F = Re_1 \oplus \cdots \oplus Re_n$. The proof is by induction on n. If $n = 0$, then $F = 0$ is noetherian. If $n > 0$, then $S = Re_1 \oplus \cdots \oplus Re_{n-1}$ is noetherian by the induction hypothesis, $F/S \cong Re_n \cong {}_R R$ is noetherian, and F is noetherian by Proposition 10.9.2. □

10.9.b. Artinian Modules. A module is **artinian** in case its submodules satisfy the descending chain condition. We saw (Proposition 10.3.11) that a module has finite length (has a composition series) if and only if it is noetherian and artinian.

By Proposition B.2.2, artinian modules are characterized by any of the following properties:

(1) Every infinite descending sequence of submodules

$$S_1 \supseteq \cdots \supseteq S_n \supseteq S_{n+1} \supseteq \cdots$$

terminates (there exists $m > 0$ such that $S_n = S_m$ for all $n \geqslant m$).

(2) There is no strictly descending infinite sequence of submodules

$$S_1 \supsetneq \cdots \supsetneq S_n \supsetneq S_{n+1} \supsetneq \cdots.$$

(3) Every nonempty set of submodules (partially ordered by inclusion) has a minimal element.

A ring R is **left artinian** in case the left R-module ${}_R R$ is artinian; equivalently, in case the left ideals of R satisfy the descending chain condition.

The following properties are exercises:

Proposition 10.9.4. *Let M be a left R-module and N be a submodule of N; M is artinian if and only if N and M/N are artinian.*

Proposition 10.9.5. *If R is left artinian, then every finitely generated R-module is artinian.*

Exercises

1. Let M be a noetherian left R-module and N be a submodule of N. Show that N and M/N are noetherian.

2. Let M be an artinian left R-module and N be a submodule of N. Show that N and M/N are artinian.

3. Let M be a left R-module and N be a submodule of N. If N and M/N are artinian, then prove that M is artinian.

4. Let R be a left artinian ring. Show that every finitely generated R-module is artinian.

5. Let E be an infinite extension of a field K (e.g., $E = K(X)$). Let R be the set of all upper triangular matrices

$$\begin{pmatrix} \alpha & \beta \\ 0 & c \end{pmatrix}$$

with $\alpha, \beta \in E$ and $c \in K$. Show that R is a ring and is left noetherian and left artinian but is neither right noetherian nor right artinian.

11

COMMUTATIVE RINGS

This chapter is a introduction to Commutative Algebra. It brings additional properties of commutative rings, especially of noetherian rings, and is directed toward classic results of Dedekind and Hilbert on two subjects that had much to do with the development of abstract algebra: algebraic sets, and rings of algebraic integers. In many ways this is a continuation of Chapter 5, except that we can now use Galois theory, valuation rings, and modules.

Required results: Chapter 5, Sections 5.1, 5.5, 5.6, 5.7, and Proposition 5.2.9; Chapter 7, Sections 7.1, 7.2, 7.3, 7.4; Chapter 10, Sections 10.1, 10.2, 10.3, 10.9; and properties of valuation rings in Section 9.4.

All rings in this chapter are commutative rings with identity.

11.1. LOCALIZATION

Localization generalizes the construction of the field of fractions of a domain but applies to any commutative ring. In this section we mostly shuttle ideals back and forth between a ring and its localizations, as a tool for later sections.

11.1.a. Rings of Fractions. Let R be a ring (a commutative ring with identity element). A **multiplicative subset** of R is a subset S of R which contains the identity element and is closed under multiplication (if $s, t \in S$, then $st \in S$); equivalently, a submonoid of $(R, \cdot)$. We call S **proper** if $0 \notin S$.

It is straightforward that an equivalence relation $\equiv$ on $R \times S$ is defined by

$$(a, s) \equiv (b, t) \iff atu = bsu \qquad \text{for some} \quad u \in S.$$

For instance, $(a, s) \equiv (at, st)$ for all $t \in S$; $(a, 1) \equiv (0, 1)$ if and only if $as = 0$ for some $s \in S$. The equivalence class of $(a, s) \in R \times S$ is a **fraction**; we denote it by a/s. The **ring of fractions** of R with denominators in S is the set $S^{-1}R = (R \times S)/\equiv$ of all fractions, with operations given by

$$(a/s) + (b/t) = (at + bs)/st, \qquad (a/s)(b/t) = ab/st.$$

It is straightforward that the operations on $S^{-1}R$ are well defined and that $S^{-1}R$ is a ring, with zero element $0/1$ and identity element $1/1$. For all $s,t \in S$, s/t is a unit in $S^{-1}R$, with $(s/t)^{-1} = t/s$.

Proposition 11.1.1. *Let R be a ring and S be a proper multiplicative subset of R. Then $S^{-1}R$ is a ring; $\iota : a \longmapsto a/1$ is a ring homomorphism, which is injective if R is a domain, and $\iota(s)$ is a unit in $S^{-1}R$ for every $s \in S$.*

If, for instance, R is a domain, then $S = R\backslash\{0\}$ is a proper multiplicative subset and $S^{-1}R = Q(R)$ is the field of fractions of R. In general, $S^{-1}R$ has a similar universal property:

Proposition 11.1.2. *Let R be a ring and S be a proper multiplicative subset of R. When $\varphi : R \longrightarrow R'$ is a ring homomorphism such that $\varphi(s)$ is a unit of R' for every $s \in S$, then φ factors uniquely through $\iota : R \longrightarrow S^{-1}R$: there exists a unique homomorphism $\psi : S^{-1}R \longrightarrow R'$ such that $\psi \circ \iota = \varphi$,*

given by $\psi(a/s) = \varphi(a)\varphi(s)^{-1}$ for all $a \in R$, $s \in S$.

Proof. First, $a/s=b/t$ implies $atu = bsu$ for some $u \in S$, $\varphi(a)\varphi(t)\varphi(u) = \varphi(b)\varphi(t)\varphi(u)$, $\varphi(a)\varphi(t) = \varphi(b)\varphi(t)$, since $\varphi(u)$ is a unit, and $\varphi(a)\varphi(s)^{-1} = \varphi(b)\varphi(t)^{-1}$. Hence there exists a mapping $\psi : S^{-1}R \longrightarrow R'$ such that $\psi(a/s) = \varphi(a)\varphi(s)^{-1}$ for all $a/s \in S^{-1}R$. It is immediate that ψ is a homomorphism and that $\psi \circ \iota = \varphi$.

Conversely, let $\chi : S^{-1}R \longrightarrow R'$ be a homomorphism (of rings with identity elements). For each $s \in S$, $\chi(\iota(s))$ is a unit of R', since $\iota(s)$ is a unit of $S^{-1}R$, and $\chi(\iota(s))^{-1} = \chi(\iota(s)^{-1})$. Since $a/s = (a/1)(1/s) = \iota(a)\iota(s)^{-1}$ in $S^{-1}R$, we have $\chi(a/s) = \chi(\iota(a))\,\chi(\iota(s))^{-1}$. If $\chi \circ \iota = \varphi$, then $\chi(a/s) = \varphi(a)\varphi(s)^{-1} = \psi(a/s)$. Hence ψ is unique. □

We note that ι is in general not injective; in fact $\iota(a) = 0$ if and only if $sa = 0$ for some $s \in S$. If ι is injective, then R may be identified with the subring $\iota(R)$ of $S^{-1}R$ so that $R \subseteq S^{-1}R$.

If R is a domain, then ι is injective. In this case R has a field of fractions Q, and the homomorphism $\psi : S^{-1}R \longrightarrow Q$ provided by Proposition 11.1.2 is injective, since $a/s = b/t$ holds in $S^{-1}R$ if and only if $a/s = b/t$ holds in Q. Hence $S^{-1}R$ may be identified with the subring $\{a/s \in Q ; a \in R, s \in S\}$ of Q so that $R \subseteq S^{-1}R \subseteq Q$.

11.1.b. Modules of Fractions. Propositions 11.1.1 and 11.1.2 extend to modules. Let S be a proper multiplicative subset of R and M be an R-module.

Define an equivalence relation $\equiv$ on $M \times S$ by

$$(x,s) \equiv (y,t) \iff utx = usy \qquad \text{for some} \quad u \in S.$$

We denote the equivalence class of $(x,s) \in M \times S$ by x/s. The **module of fractions** of M with denominators in S is the set $S^{-1}M = (M \times S)/\equiv$ of all fractions; the addition and left action of $S^{-1}R$ on M are given by

$$(x/s) + (y/t) = (tx + sy)/st, \qquad (a/s)(x/t) = ax/st.$$

It is straightforward that $\equiv$ is an equivalence relation, that the operations above are well defined, and that $S^{-1}M$ is an $S^{-1}R$-module. Since $\iota : R \longrightarrow S^{-1}R$ is a ring homomorphism, $S^{-1}M$ is also an R-module in which $r(x/s) = \iota(r)(x/s) = rx/s$.

Proposition 11.1.3. *Let R be a ring, S be a proper multiplicative subset of R, and M be an R-module. Then $S^{-1}M$ is an $S^{-1}R$-module and $\iota : x \longmapsto x/1$ is an R-module homomorphism. Moreover every R-homomorphism of M into an $S^{-1}R$-module factors uniquely through ι.*

We leave the proof to the reader. Again ι is not in general injective; in fact $\iota(x) = 0$ if and only if $sx = 0$ for some $s \in S$.

Proposition 11.1.4. *Let R be a ring and S be a proper multiplicative subset of R. Let M be an R-module and N be a submodule of M. Then $S^{-1}N$ is isomorphic to the submodule $(S^{-1}R)N = \{x/s \in S^{-1}M ; x \in N, s \in S\}$ of $S^{-1}M$.*

Proof. It is immediate that $(S^{-1}R)N = \{x/s \in S^{-1}M ; x \in N, s \in S\}$. By the universal property of $S^{-1}N$, there is an $S^{-1}R$-module homomorphism $\kappa : S^{-1}N \longrightarrow S^{-1}M$ such that the square

$$\begin{array}{ccc}
N & \xrightarrow{\ \iota\ } & S^{-1}N \\
\scriptstyle\subseteq \downarrow & & \downarrow \scriptstyle\kappa \\
M & \xrightarrow{\ \iota\ } & S^{-1}M
\end{array}$$

commutes; by definition, κ sends $x/1 \in S^{-1}N$ to $x/1 \in S^{-1}M$, and $x/s = (1/s)(x/1) \in S^{-1}N$ to $(1/s)(x/1) = x/s$ in $S^{-1}M$. If $x \in N$ and $x/s = 0$ in $S^{-1}M$, then $ux = 0$ for some $u \in S$ and $x/s = 0$ in $S^{-1}N$; hence κ is injective. $\qquad\square$

When N is a submodule of M, Proposition 11.1.4 allows us to identify $x/s \in S^{-1}N$ and $x/s \in S^{-1}M$ when $x \in N$, and to identify $S^{-1}N$ and $(S^{-1}R)N = \{x/s \in S^{-1}M ; x \in N, s \in S\}$.

11.1.c. Localization. In what follows $\mathfrak{p}$ is a prime ideal of R ($\mathfrak{p} \neq R$, and $ab \in \mathfrak{p}$ implies $a \in \mathfrak{p}$ or $b \in \mathfrak{p}$). Then $S = R \backslash \mathfrak{p}$ is a proper multiplicative subset of R. The ring of fractions $R_\mathfrak{p} = (R \backslash \mathfrak{p})^{-1} R$ is the **localization** of R at $\mathfrak{p}$. If $\iota : R \longrightarrow R_\mathfrak{p}$ is injective (e.g., if R is a domain), then R may be identified with the subring $\iota(R)$ of $R_\mathfrak{p}$.

Every ring is isomorphic to a ring of fractions: if S is contained in the group of units of R, then $S^{-1} R \cong R$ (see the exercises). On the other hand, not every ring is isomorphic to a localization.

Proposition 11.1.5. *Let R be a ring and $\mathfrak{p}$ be a prime ideal of R. Then $\mathfrak{M} = \{ a/s \in R_\mathfrak{p} \,;\, a \in \mathfrak{p} \}$ is the only maximal ideal of $R_\mathfrak{p}$; moreover $x \in R_\mathfrak{p}$ is a unit if and only if $x \notin \mathfrak{M}$.*

Proof. Let $x = a/s \in R_\mathfrak{p}$. If $a \notin \mathfrak{p}$, then x is a unit, with $x^{-1} = s/a$. If, conversely, x is a unit of $R_\mathfrak{p}$, with $x^{-1} = b/t$, then $ab/st = 1$, $uab = ust \notin \mathfrak{p}$ for some $u \notin \mathfrak{p}$, and $a \notin \mathfrak{p}$. Thus x is a unit of $R_\mathfrak{p}$ if and only if $a \notin \mathfrak{p}$.

If $a \notin \mathfrak{p}$, then $x = a/s \notin \mathfrak{M}$, since $a/s = p/t \in \mathfrak{M}$ (with $p \in \mathfrak{p}$) implies $uta = usp \in \mathfrak{p}$ for some $u \notin \mathfrak{p}$ and $a \in \mathfrak{p}$ (since $ut \notin \mathfrak{p}$). If, conversely, $x \notin \mathfrak{M}$, then $a \notin \mathfrak{p}$. Therefore $x = a/s$ is a unit of $R_\mathfrak{p}$ if and only if $x \notin \mathfrak{M}$. Hence the ideal $\mathfrak{M}$ of $R_\mathfrak{p}$ is a maximal ideal. $\qquad\square$

A commutative ring R is **local** in case it has only one maximal ideal $\mathfrak{m}$. Then the group of units of R is $R \backslash \mathfrak{m}$ (see the exercises). For example, valuation rings are local (Proposition 9.4.1). Proposition 11.1.5 states that $R_\mathfrak{p}$ is a local ring, whence the name "localization." Local rings derive their name from algebraic geometry, where such rings arise at points of algebraic varieties (see Proposition 11.10.3).

11.1.d. Ideals. Let S be a proper multiplicative subset of R. If $\mathfrak{a}$ is an ideal of R, then Proposition 11.1.4 shows that

$$\mathfrak{a}^E = S^{-1} \mathfrak{a} = \{ a/s \in S^{-1} R \,;\, a \in \mathfrak{a},\, s \in S \}$$

is an ideal of $S^{-1} R$; $\mathfrak{a}^E$ is the **expansion** of $\mathfrak{a}$. For example, the maximal ideal of $R_\mathfrak{p}$ is $\mathfrak{p}^E$.

Proposition 11.1.6. *Every ideal of $S^{-1} R$ is the expansion of some ideal $\mathfrak{a}$ of R.*

Proof. Let $\mathfrak{A}$ be an ideal of $S^{-1} R$. Then $\mathfrak{a} = \iota^{-1}(\mathfrak{A})$ is an ideal of R. Also $a \in \mathfrak{a}$ implies $a/s = (1/s)\iota(a) \in \mathfrak{A}$. Conversely, $a/s \in \mathfrak{A}$ implies $\iota(a) = s(a/s) \in \mathfrak{A}$ and $a \in \mathfrak{a}$. Thus $\mathfrak{A} = \mathfrak{a}^E$. $\qquad\square$

When $\mathfrak{A}$ is an ideal of $S^{-1} R$, the ideal $\mathfrak{A}^C = \iota^{-1}(\mathfrak{A})$ of R is the **contraction** of $\mathfrak{A}$. If $\iota : R \longrightarrow S^{-1} R$ is injective, and R is identified with $\iota(R) \subseteq S^{-1} R$, then $\mathfrak{A}^C = \mathfrak{A} \cap R$. In general, the proof of Proposition 11.1.6 shows that $\mathfrak{A} = (\mathfrak{A}^C)^E$.

The following properties are exercises:

Proposition 11.1.7. *For all ideals* $\mathfrak{a}, \mathfrak{b}$ *of* R:

(1) $\mathfrak{a}^E = S^{-1}R$ *if and only if* $\mathfrak{a} \cap S \neq \emptyset$; *thus* $\mathfrak{a}^E = R_{\mathfrak{p}}$ *if and only if* $\mathfrak{a} \not\subseteq \mathfrak{p}$;
(2) *if* $\mathfrak{A}$ *is an ideal of* $S^{-1}R$ *and* $\mathfrak{a} = \mathfrak{A}^C$, *then* $\mathfrak{A} = \mathfrak{a}^E$;
(3) $(\mathfrak{a} + \mathfrak{b})^E = \mathfrak{a}^E + \mathfrak{b}^E$, $(\mathfrak{a} \cap \mathfrak{b})^E = \mathfrak{a}^E \cap \mathfrak{b}^E$, *and* $(\mathfrak{a}\mathfrak{b})^E = \mathfrak{a}^E \mathfrak{b}^E$.

Proposition 11.1.8. *If* R *is noetherian, then* $S^{-1}R$ *is noetherian.*

Proof. Let

$$\mathfrak{A}_1 \subseteq \cdots \subseteq \mathfrak{A}_n \subseteq \mathfrak{A}_{n+1} \subseteq \cdots$$

be an ascending sequence of ideals of $S^{-1}R$. Let $\mathfrak{a}_n = \mathfrak{A}_n^C$. Then

$$\mathfrak{a}_1 \subseteq \cdots \subseteq \mathfrak{a}_n \subseteq \mathfrak{a}_{n+1} \subseteq \cdots$$

is an ascending sequence of ideals of R. If R is noetherian, this sequence terminates: there exists $m > 0$ such that $\mathfrak{a}_n = \mathfrak{a}_m$ for all $n \geqslant m$. Then $\mathfrak{A}_n = \mathfrak{a}_n^E = \mathfrak{a}_m^E = \mathfrak{A}_m$ for all $n \geqslant m$, by Proposition 11.1.7. $\qquad\square$

11.1.e. Prime and Primary Ideals

Proposition 11.1.9. *Let* S *be a proper multiplicative subset of* R. *Then* $\mathfrak{a} \longmapsto \mathfrak{a}^E$, $\mathfrak{A} \longmapsto \mathfrak{A}^C$ *is a one-to-one correspondence between the prime ideals of* $S^{-1}R$ *and the prime ideals of* R *contained in* $R\backslash S$.

Proof. Let $\mathfrak{a} \subseteq R\backslash S$ be a prime ideal of R. If $x/s \in \mathfrak{a}^E$, then $x/s = a/t$ for some $a \in \mathfrak{a}$, $utx = usa \in \mathfrak{a}$ for some $u \in S$, and $x \in \mathfrak{a}$, since $\mathfrak{a}$ is prime and $ut \notin \mathfrak{a} \subseteq R\backslash S$. Thus $x/s \in \mathfrak{a}^E$ if and only if $x \in \mathfrak{a}$. In particular, $\mathfrak{a} = (\mathfrak{a}^E)^C$. Moreover $(a/s)(b/t) = ab/st \in \mathfrak{a}^E$ implies $ab \in \mathfrak{a}$, so that at least one of a/s, b/t is in $\mathfrak{a}^E$. Thus $\mathfrak{a}^E$ is a prime ideal of $S^{-1}R$. If, conversely, $\mathfrak{A}$ is a prime ideal of $S^{-1}R$, then $\mathfrak{a} = \mathfrak{A}^C$ is a prime ideal of R, and $\mathfrak{A} = \mathfrak{a}^E$ by Proposition 11.1.7. $\qquad\square$

Proposition 11.1.10. *Let* S *be a proper multiplicative subset of* R. *Then* $\mathfrak{a} \longmapsto \mathfrak{a}^E$, $\mathfrak{A} \longmapsto \mathfrak{A}^C$ *is a one-to-one correspondence between the primary ideals of* $S^{-1}R$ *and the primary ideals of* R *contained in* $R\backslash S$, *and preserves radicals.*

Proof. Let $\mathfrak{a} \subseteq R\backslash S$ be a primary ideal of R and $\mathfrak{r} = \mathrm{Rad}\,\mathfrak{a}$. Then $\mathfrak{r} \subseteq R\backslash S$ (since $s \in S$ implies $s^n \in S$ and $s^n \notin \mathfrak{a}$). If $x/s \in \mathfrak{a}^E$, then $x/s = a/t$ for some $a \in \mathfrak{a}$, $utx = usa \in \mathfrak{a}$ for some $u \in S$, and $x \in \mathfrak{a}$, since $\mathfrak{a}$ is primary and $ut \notin \mathfrak{r} \subseteq R\backslash S$. Thus $x/s \in \mathfrak{a}^E$ if and only if $x \in \mathfrak{a}$. In particular, $\mathfrak{a} = (\mathfrak{a}^E)^C$. Moreover $(x/s)^n \in \mathfrak{a}^E$ if and only if $x^n \in \mathfrak{a}$, so that $\mathrm{Rad}\,(\mathfrak{a}^E) = (\mathrm{Rad}\,\mathfrak{a})^E$. If finally $(a/s)(b/t) = ab/st \in \mathfrak{a}^E$ and $a/s \notin \mathfrak{r}^E$, then $ab \in \mathfrak{a}$ and $a \notin \mathfrak{r}$, so that $b \in \mathfrak{a}$ and $b/t \in \mathfrak{a}^E$. Thus $\mathfrak{a}^E$ is a primary ideal of $S^{-1}R$.

Conversely, let $\mathfrak{A}$ be a primary ideal of $S^{-1}R$ and $\mathfrak{R} = \mathrm{Rad}\,\mathfrak{A}$. Then $\mathfrak{r} = \mathfrak{R}^C$ is the radical of $\mathfrak{a} = \mathfrak{A}^C$, and it is then immediate that $\mathfrak{a}$ is $\mathfrak{r}$-primary. Also $\mathfrak{A} = \mathfrak{a}^E$ by Proposition 11.1.7. $\square$

If $\mathfrak{m}$ is a maximal ideal of R, then $\mathfrak{m}^n$ is $\mathfrak{m}$-primary: indeed $\mathfrak{m}$ is the radical of $\mathfrak{m}^n$, and $ab \in \mathfrak{m}^n$, $a \notin \mathfrak{m}$ implies $\mathfrak{m} + Ra = R$, $1 = ra + m$ for some $r \in R$ and $m \in \mathfrak{m}$, $1 = (ra + m)^n = sa + m^n$ for some $s \in R$, and $b = sab + m^n b \in \mathfrak{m}^n$. With localization we can extend this construction to all prime ideals.

Proposition 11.1.11. *Let $\mathfrak{p}$ be a prime ideal of R. Then*

$$\mathfrak{p}^{(n)} = ((\mathfrak{p}^E)^n)^C = \{\, a \in R \,;\, sa \in \mathfrak{p}^n \text{ for some } s \in R\backslash\mathfrak{p} \,\}$$

is a $\mathfrak{p}$-primary ideal of R, and $\mathfrak{p}^{(n)} \supseteq \mathfrak{p}^{(n+1)}$.

Proof. Since $\mathfrak{p}$ is prime, $a \in ((\mathfrak{p}^E)^n)^C$ is equivalent to $a/1 \in (\mathfrak{p}^E)^n = (\mathfrak{p}^n)^E$, $a/1 = b/s$ for some $b \in \mathfrak{p}^n$ and $s \in R\backslash\mathfrak{p}$, and $tsa \in \mathfrak{p}^n$ for some $ts \in R\backslash\mathfrak{p}$. Since $\mathfrak{p}^E$ is a maximal ideal of $R_\mathfrak{p}$, $(\mathfrak{p}^n)^E = (\mathfrak{p}^E)^n$ is $\mathfrak{p}^E$-primary and $\mathfrak{p}^{(n)}$ is $\mathfrak{p}$-primary by Proposition 11.1.10. $\square$

That $\mathfrak{p}^{(n)}$ is $\mathfrak{p}$-primary can be verified directly (see the exercises). The ideal $\mathfrak{p}^{(n)}$ of R is the n-th **symbolic power** of $\mathfrak{p}$. It will be used in Section 11.7.

Exercises

1. Let R be a ring and S be a proper multiplicative subset of R. Verify that $S^{-1}R$ is a ring.

2. Let R be a ring and S be a proper multiplicative subset of R which is contained in the group of units of R. Show that $\iota : R \longrightarrow S^{-1}R$ is an isomorphism.

3. Let R be a ring, S be a proper multiplicative subset of R, and M be an R-module. Verify that $S^{-1}M$ is an $S^{-1}R$-module.

4. Show that an R-module M is an $S^{-1}R$-module if and only if each $s \in S$ acts on M by automorphisms.

5. Let R be a (local) ring with only one maximal ideal $\mathfrak{m}$. Prove that the group of units of R is $R\backslash\mathfrak{m}$.

6. Let R be a local ring. Prove that there is a commutative ring R' and a prime ideal $\mathfrak{p}$ of R' such that $R \cong R'_\mathfrak{p}$.

7. Let R be a ring and $\mathfrak{p}$ be a prime ideal of R. Show that $R_\mathfrak{p}/\mathfrak{p}^E$ is isomorphic to the field of quotients of $R/\mathfrak{p}$.

8. Describe the local ring $\mathbb{Z}_{(p)}$.

9. Prove that $\mathfrak{a}^E = S^{-1}R$ if and only if $\mathfrak{a} \cap S \neq \emptyset$.

10. Show that $(\mathfrak{a} + \mathfrak{b})^E = \mathfrak{a}^E + \mathfrak{b}^E$.

11. Show that $(\mathfrak{a} \cap \mathfrak{b})^E = \mathfrak{a}^E \cap \mathfrak{b}^E$.

12. Show that $(\mathfrak{a}\mathfrak{b})^E = \mathfrak{a}^E \mathfrak{b}^E$.

13. Let $\mathfrak{p}$ be a prime ideal of R, $n > 0$, and $\mathfrak{q}\ (= \mathfrak{p}^{(n)}) = \{a \in R\,;\, sa \in \mathfrak{p}^n$ for some $s \in R\backslash\mathfrak{p}\,\}$. Prove directly that $\mathfrak{q}$ is a $\mathfrak{p}$-primary ideal of R.

14. Let $\mathfrak{p}$ be a finitely generated prime ideal of R. Show that every $\mathfrak{p}$-primary ideal of R contains a symbolic power of $\mathfrak{p}$.

11.2. RING EXTENSIONS

Ring extensions share a number of properties with field extensions, even though they are much more general.

11.2.a. Definition. Recall that all rings in this chapter are commutative rings with identity elements. A **ring extension** of a ring R is a ring A of which R is a subring (in particular, R contains the identity element of A). A ring extension A of R is (in particular) an R-module, on which R acts by multiplication in A; A is a faithful R-module, since $ab = 0$ for all $b \in A$ implies $a = a1 = 0$.

Proposition 11.2.1. *Let A be a ring extension of R and S be a subset of A. The subring $R[S]$ of A generated by $R \cup S$ is the set of all linear combinations with coefficients in R of products of powers of elements of S.*

Proof. This is proved like Proposition 6.1.9. Let $(X_s)_{s \in S}$ be a family of indeterminates indexed by S (with $X_s \neq X_t$ if $s \neq t$). By the universal property of $R[(X_s)_{s \in S}]$ (Proposition 4.3.12), there is a unique homomorphism $\varphi : R[(X_s)_{s \in S}] \longrightarrow A$ such that $\varphi(a) = a$ for all $a \in R$ and $\varphi(X_s) = s$ for all $s \in S$; φ sends $\sum_m a_m X_{s_1}^{m_{s_1}} \ldots X_{s_t}^{m_{s_t}} \in K[(X_s)_{s \in S}]$ to $\sum_m a_m s_1^{m_{s_1}} \ldots s_t^{m_{s_t}} \in A$, and $\operatorname{Im}\varphi$ is the smallest subring of A which contains R and all $s \in S$. Thus $R[S] = \operatorname{Im}\varphi$ and consists of all linear combinations with coefficients in R of products of powers of elements of S. $\qquad\square$

Corollary 11.2.2. *In a ring extension, $\alpha \in R[s_1, \ldots, s_n]$ if and only if $\alpha = f(s_1, \ldots, s_n)$ for some $f(X_1, \ldots, X_n) \in R[X_1, \ldots, X_n]$; $\alpha \in R[S]$ if and only if $\alpha \in R[s_1, \ldots, s_n]$ for some $s_1, \ldots, s_n \in S$.*

A ring extension A of R is **finitely generated** in case $A = R[s_1, \ldots, s_n]$ for some finitely many elements $s_1, \ldots, s_n$ of A.

11.2.b. Integral Elements

Proposition 11.2.3. *Let A be a ring extension of R and $\alpha \in A$. The following conditions are equivalent:*

(1) $f(\alpha) = 0$ *for some monic polynomial $f(X) \in R[X]$.*
(2) $R[\alpha]$ *is a finitely generated R-module.*

(3) *There exists a faithful $R[\alpha]$-module which is a finitely generated R-module.*

Proof. (1) implies (2). Assume that $f(\alpha) = 0$, where $f(X) \in R[X]$ is monic. Let $\beta \in R[\alpha]$. By Corollary 11.2.2, $\beta = g(\alpha)$ for some $g \in R[X]$. Since f is monic, $g \in R[X]$ can be divided by $f : g = fq + r$, where $\deg r < n = \deg f$. Then $\beta = g(\alpha) = r(\alpha)$. Thus $R[\alpha]$ is generated by $\alpha^{n-1}, \ldots, \alpha, 1$ as an R-module.

(2) implies (3) since $R[\alpha]$ is a faithful $R[\alpha]$-module.

(3) implies (1). Let M be a faithful $R[\alpha]$-module which is generated as an R-module by finitely many elements $x_1, \ldots, x_n$. For each i we have $\alpha x_i \in M$, and there exists $a_{i1}, \ldots, a_{in} \in R$ such that $\alpha x_i = a_{i1} x_1 + \cdots + a_{in} x_n$. Then

$$-a_{i1} x_1 - \cdots - a_{i,i-1} x_{i-1} + (\alpha - a_{ii}) x_i - a_{i,i+1} x_{i+1} - \cdots - a_{in} x_n = 0$$

for all i. By Lemma 11.2.4 below, the determinant

$$D = \begin{vmatrix} \alpha - a_{11} & -a_{12} & \cdots & -a_{1n} \\ -a_{21} & \alpha - a_{22} & \cdots & -a_{2n} \\ \vdots & \vdots & \ddots & \vdots \\ -a_{n1} & -a_{n2} & \cdots & \alpha - a_{nn} \end{vmatrix}$$

satisfies $Dx_i = 0$ for all i. Since $x_1, \ldots, x_n$ generate M, this implies $Dx = 0$ for all $x \in M$. Since M is faithful, $D = 0$. Expanding D shows that $D = f(\alpha)$ for some monic polynomial $f(X) \in R[X]$. $\square$

Lemma 11.2.4. *Let M be an R-module. Assume that $x_1, \ldots, x_n \in M$ and $a_{ij} \in R$ $(1 \leqslant i, j \leqslant n)$ satisfy $\sum_{1 \leqslant j \leqslant n} a_{ij} x_j = 0$ for all i. Then $D = \det(a_{ij})$ satisfies $Dx_i = 0$ for all i.*

Proof. Expanding D by columns yields cofactors c_{ik} $(1 \leqslant i, k \leqslant n)$ such that $\sum_{1 \leqslant k \leqslant n} c_{ik} a_{kj} = D$ if $i = j$, 0 if $i \neq j$. Hence $Dx_i = \sum_{1 \leqslant k \leqslant n} c_{ik} a_{ki} x_i = \sum_{1 \leqslant j, k \leqslant n} c_{ik} a_{kj} x_j = 0$ for all i. $\square$

An element α of a ring extension of R is **integral** over R in case it satisfies the equivalent conditions in Proposition 11.2.3. The following result, whose proof is an exercise, relates integral elements to algebraic elements.

Proposition 11.2.5. *Let R be a domain and F be its field of fractions. Then α is algebraic over F if and only if $r\alpha$ is integral over R for some $r \in R$, $r \neq 0$.*

11.2.c. Integral Extensions. A ring extension $R \subseteq A$ is **integral** in case every element of A is integral over R.

Proposition 11.2.6. *Let A be a ring extension of R.*

(1) *If A is a finitely generated R-module, then A is integral over R.*

(2) *If $A = R[\alpha_1,\ldots,\alpha_n]$ and $\alpha_1,\ldots,\alpha_n$ are integral over R, then A is a finitely generated R-module (hence is integral over R).*

(3) *If $A = R[S]$ and every $s \in S$ is integral over R, then A is integral over R.*

Proof. (1) Let $\alpha \in A$. Then A is a faithful $R[\alpha]$-module (since A has an identity element); if A is finitely generated as an R-module, then α is integral over R by Proposition 11.2.3.

(2) is shown by induction on n. The result is trivial if $n = 0$. In general, let A be a ring extension of R which is finitely generated as an R-module and α be integral over R. Then α is integral over A and $A[\alpha]$ is a finitely generated A-module. If every element of A is a linear combination of $\beta_1,\ldots,\beta_s$ with coefficients in R, and every element of $A[\alpha]$ is a linear combination of $\gamma_1,\ldots,\gamma_t$ with coefficients in A, then every element of $A[\alpha]$ is a linear combination of $\beta_1\gamma_1,\ldots,\beta_s\gamma_t$ with coefficients in R. Hence $A[\alpha]$ is a finitely generated R-module.

(3) follows from (1), (2), and Corollary 11.2.2. $\square$

The proofs of the following properties make neat exercises:

Proposition 11.2.7. *Let R, A, B be rings.*

(1) *If $R \subseteq A \subseteq B$ and B is integral over R, then A is integral over R and B is integral over A. If, conversely, A is integral over R and B is integral over A, then B is integral over R.*

(2) *If B is integral over A and $R[B]$ exists, then $R[B]$ is integral over $R[A]$.*

(3) *If A is integral over R and $\varphi : A \longrightarrow B$ is a ring homomorphism, then $\varphi(A)$ is integral over $\varphi(B)$.*

(4) *If A is integral over R, then $S^{-1}A$ is integral over $S^{-1}R$ for every proper multiplicative subset S of R.*

Corollary 11.2.8. *Let $R \subseteq A$ be rings. The elements of A which are integral over R constitute a subring of A.*

Exercises

1. Let R be a domain and F be its field of fractions. Prove that α is algebraic over F if and only if $r\alpha$ is integral over R for some $r \in R$, $r \neq 0$.

2. Let $R \subseteq A \subseteq B$ be rings. Show that B is integral over R if and only if A is integral over R and B is integral over A.

3. If B is integral over A and $R[B]$ exists, show that $R[B]$ is integral over $R[A]$.

4. If A is integral over R and $\varphi : A \longrightarrow B$ is a ring homomorphism, show that $\varphi(A)$ is integral over $\varphi(R)$.

5. If A is integral over R, and S is a proper multiplicative subset of R, show that $S^{-1}A$ is integral over $S^{-1}R$.

6. Let $R \subseteq A$ be rings. Prove that the elements of A which are integral over R constitute a subring of R.

7. Let R be a domain and F be its field of fractions. Show that $a \in R$ is a unit of R if and only if $1/a \in F$ is integral over R.

8. If A is integral over R, then show that $A[X_1, \ldots, X_n]$ is integral over $R[X_1, \ldots, X_n]$.

9. Let A be a domain which is integral over R. Prove that A is a field if and only if R is a field.

11.3. INTEGRALLY CLOSED RINGS

The results in Section 11.2 suggest a new class of domains.

11.3.a. Definition. The **integral closure** of R in $A \supseteq R$ is the ring $\overline{R}$ of all elements of A that are integral over R. The ring R is **integrally closed in** $A \supseteq R$ in case $\overline{R} = R$, equivalently, in case no $\alpha \in A \backslash R$ is integral over R.

A domain R is **integrally closed** in case it is integrally closed in its field of fractions. Integrally closed domains are not rare; in fact every domain R is a subring of an integrally closed domain: if Q is the field of fractions of R and $\overline{R}$ is the integral closure of R in Q, then Q is the field of fractions of $\overline{R}$ and $\overline{R}$ is integrally closed in Q by Proposition 11.2.7; also $\overline{R}$ is integral over R.

Proposition 11.3.1. *Every UFD is integrally closed.*

Proof. Let R be a UFD and Q be its field of fractions. Let $a/b \in Q$ be integral over R. We may assume that no irreducible element of R divides both a and b. We have $f(a/b) = 0$ for some monic polynomial $f(X) = X^n + r_{n-1}X^{n-1} + \cdots + r_0 \in R[X]$ and

$$a^n + r_{n-1}a^{n-1}b + \cdots + r_0 b^n = 0.$$

If $p \in R$ is irreducible and divides b, then p divides a^n and p divides a, a contradiction. Therefore b is a unit of R and $a/b \in R$. □

An integrally closed domain which is not a UFD will be seen shortly.

Proposition 11.3.2. *If R is integrally closed, then $S^{-1}R$ is integrally closed for every proper multiplicative subset S of R.*

Proof. Since R is a domain we may identify $S^{-1}R$ with the subring $\{a/s \in F ; a \in R, s \in S\}$ of the field of fractions F of R, so that $R \subseteq S^{-1}R \subseteq F$. If $\alpha \in F$ is integral over $S^{-1}R$, then

$$\alpha^n + (r_{n-1}/s_{n-1})\alpha^{n-1} + \cdots + (r_0/s_0) = 0$$

for some $r_{n-1}, \ldots, r_0 \in R$ and $s_{n-1}, \ldots, s_0 \in S$. Let $s = s_{n-1} \ldots s_0 \in S$. Multiplying by s^n shows that $s\alpha$ is integral over R. Therefore $s\alpha \in R$ and $\alpha \in S^{-1}R$. ☐

11.3.b. Algebraic Integers. Let R be a domain with field of fractions K and E be an algebraic field extension of K. An **algebraic integer** (or just **integer**) in E is an element of E which is integral over R. By Corollary 11.2.8, the algebraic integers of E form a ring extension of R (the integral closure of R in E).

Proposition 11.3.3. *Let R be a domain, K be its field of fractions, E be an algebraic field extension of K, and A be the integral closure of R in E. Then A is integrally closed, and its field of fractions is E.*

Proof. Every $\alpha \in E$ is algebraic over K; hence $r\alpha$ is integral over R for some $r \in R$, $r \neq 0$, by Proposition 11.2.5. Thus $\alpha = r\alpha/r$ with $r, r\alpha \in A$. ☐

The algebraic integers of an extension E of $\mathbb{Q}$ are readily determined if E is a quadratic extension of $\mathbb{Q}$ (if $[E : \mathbb{Q}] = 2$). Then $E = \mathbb{Q}(\sqrt{d})$ where we may assume that $d \in \mathbb{Z}$ and d is **square free** (if n^2 divides d, then $n = \pm 1$).

Proposition 11.3.4. *Let $d \in \mathbb{Z}$ be square free and $x, y \in \mathbb{Q}$. If d is even, or if $d \equiv 3$ (mod 4), then $x + y\sqrt{d}$ is an algebraic integer in $\mathbb{Q}[\sqrt{d}]$ if and only if $x, y \in \mathbb{Z}$, and $\mathbb{Z}[\sqrt{d}]$ is integrally closed.*

The case where $d \equiv 1$ (mod 4) is an exercise.

Proof. First $\sqrt{d}$ is integral over $\mathbb{Z}$; hence $x + y\sqrt{d}$ is integral over $\mathbb{Z}$ for all $x, y \in \mathbb{Z}$, by Corollary 11.2.8.

For the converse we note that $z \in \mathbb{Q}$ and $dz^2 \in \mathbb{Z}$ implies $z \in \mathbb{Z}$. Indeed $z = p/q$, where $p, q \in \mathbb{Z}$, and we may assume that $(p, q) = 1$; since $dp^2/q^2 = dz^2 \in \mathbb{Z}$, we see that q^2 divides dp^2, so that q^2 divides d and $q = 1$.

Let $E = Q(\sqrt{d})$ and $\alpha = x + y\sqrt{d} \in E$ be integral over $\mathbb{Z}$. There is a $\mathbb{Q}$-automorphism of E which sends $\sqrt{d}$ onto $-\sqrt{d}$. Therefore $\beta = x - y\sqrt{d}$ is also integral over $\mathbb{Z}$. By Corollary 11.2.8, $2x = \alpha + \beta$ and $x^2 - dy^2 = \alpha\beta$ are integral over $\mathbb{Z}$. Since $\mathbb{Z}$ is integrally closed by Proposition 11.3.1, we have $u = 2x \in \mathbb{Z}$, $x^2 - dy^2 \in \mathbb{Z}$, $u^2 - 4dy^2 = 4x^2 - 4dy^2 \in \mathbb{Z}$, and $4y^2d \in \mathbb{Z}$; since d is square free, $v = 2y \in \mathbb{Z}$. Also $(u^2 - dv^2)/4 = x^2 - dy^2 \in \mathbb{Z}$, so that $u^2 - dv^2$ is divisible by 4.

Assume that $x \notin \mathbb{Z}$. Then u is odd, dv^2 is odd, and d, v are odd. Modulo 4 we then have $dv^2 \equiv u^2$, $u^2 \equiv v^2 \equiv 1$, and $d \equiv 1$. Since we assumed that d is even or that $d \equiv 3$, it follows that $x \in \mathbb{Z}$. Then $dy^2 \in \mathbb{Z}$ and $y \in \mathbb{Z}$. ☐

For instance, $\mathbb{Z}[\sqrt{-5}]$ is integrally closed, but is not a UFD (see the exercises). The following result will be used to show that $\mathbb{Z}[\sqrt{-5}]$ is a Dedekind domain.

Proposition 11.3.5. *Let R be an integrally closed domain, K be its field of fractions, E be a finite separable extension of K, and A be the integral closure of R in E. Then A is contained in a finitely generated R-submodule of E.*

Proof. By the Primitive Element Theorem (Theorem 7.2.10), $E = K(\alpha)$ for some $\alpha \in E$. By Proposition 11.2.5, $r\alpha$ is integral over R for some $r \in R$, $r \neq 0$; then $E = K(r\alpha)$, and we may assume that α is integral over R. Since E is separable over K, α has $n = [E : K]$ distinct conjugates $\alpha_1 = \alpha, \alpha_2, \dots, \alpha_n$ in $\overline{K} \supseteq E$ which are, like α, integral over R; $F = K(\alpha_1, \dots, \alpha_n)$ is Galois over K.

For each $\beta \in A \subseteq E$ there is a polynomial $f(X) = b_0 + b_1 X + \cdots + b_{n-1} X^{n-1} \in K[X]$ such that $\beta = f(\alpha)$. The conjugates of β are $\beta_j = f(\alpha_j) = \sum_i b_i \alpha_j^i$ and are, like β, integral over R. Let δ be the Vandermonde determinant

$$\delta = \begin{vmatrix} 1 & 1 & \cdots & 1 \\ \alpha_1 & \alpha_2 & \cdots & \alpha_n \\ \vdots & \vdots & \ddots & \vdots \\ \alpha_1^{n-1} & \alpha_2^{n-1} & \cdots & \alpha_n^{n-1} \end{vmatrix}$$

By Proposition 5.2.9, $\delta = \prod_{i>j} (\alpha_i - \alpha_j) \neq 0$. Expanding δ by rows yields cofactors γ_{jk} $(1 \leqslant j, k \leqslant n)$ such that $\sum_j \alpha_j^i \gamma_{jk} = \delta$ if $i = k$, 0 if $i \neq k$. Hence $\sum_j \beta_j \gamma_{jk} = \sum_{ij} b_i \alpha_j^i \gamma_{jk} = b_k \delta$ and

$$\sum_j \beta_j \gamma_{jk} \delta = b_k \delta^2.$$

We use δ^2 because $\delta^2 = \prod_{i>j} (\alpha_i - \alpha_j)^2$ is the discriminant of $\mathrm{Irr}_K(\alpha) \in K[X]$ (whose roots are $\alpha_1, \dots, \alpha_n$) and belongs to K. (Alternately, δ^2 is fixed under every K-automorphism σ of $F = K(\alpha_1, \dots, \alpha_n)$, for σ permutes $\alpha_1, \dots, \alpha_n$ and sends $\delta = \prod_{i>j} (\alpha_i - \alpha_j)$ to $\pm\delta$.) Then $b_k \delta^2 \in K$. Moreover $b_k \delta^2 = \sum_j \beta_j \gamma_{jk} \delta$ is integral over R, since δ and γ_{jk} are, like $\alpha_1, \dots, \alpha_n$ and $\beta_1, \dots, \beta_n$, integral over R. Therefore $b_k \delta^2 \in R$ for every k. Hence every $\beta = \sum_i b_i \alpha^i \in A$ belongs to the submodule of E generated by $1/\delta^2, \alpha/\delta^2, \dots, \alpha^{n-1}/\delta^2$. $\qquad\square$

Exercises

1. Show that $\mathbb{Z}[\sqrt{-5}]$ is not a UFD.

2. Show that $\mathbb{Q}[\sqrt{5}]$ contains an algebraic integer $x + y\sqrt{5}$ with $x, y \notin \mathbb{Z}$ so that $\mathbb{Z}[\sqrt{5}]$ is not integrally closed.

3. Let $d \in \mathbb{Z}$ be squarefree and $x, y \in \mathbb{Q}$. If $d \equiv 1 \pmod 4$, then when is $x + y\sqrt{d}$ an algebraic integer?

4. Let F be the field of fractions of an integrally closed domain A. Let f and g be monic polynomials in $F[X]$ such that $fg \in A[X]$. Show that $f, g \in A[X]$.

5. Let F be the field of fractions of an integrally closed domain A. Let α be integral over A. Show that $\text{Irr}_F(\alpha) \in A[X]$.

6. Let R be an integrally closed domain, K be its field of fractions, E be a finite separable extension of K, and A be the integral closure of R in E. If R is noetherian, then show that A is noetherian and is a finitely generated R-submodule of E.

7. Let R be an integrally closed domain, K be its field of fractions, E be a finite separable extension of K, and A be the integral closure of R in E. If R is a PID, then show that there is a basis of E over K which generates A as an R-module.

11.4. INTEGRAL EXTENSIONS

This section studies prime ideals in integral extensions of R and how they relate to prime ideals of R.

11.4.a. Lying Over. Let $R \subseteq A$ be a ring extension. If $\mathfrak{P}$ is a prime ideal of A, then $\mathfrak{P} \cap R$ is a prime ideal of R. A prime ideal $\mathfrak{P}$ of A **lies over** a prime ideal $\mathfrak{p}$ of R in case $\mathfrak{P} \cap R = \mathfrak{p}$.

Proposition 11.4.1 (Lying Over). *If A is integral over R, then for each prime ideal $\mathfrak{p}$ of R there exists a prime ideal $\mathfrak{P}$ of A which lies over $\mathfrak{p}$. More generally, for every ideal $\mathfrak{A}$ of A such that $\mathfrak{A} \cap R \subseteq \mathfrak{p}$, there exists a prime ideal $\mathfrak{P}$ of A which contains $\mathfrak{A}$ and lies over $\mathfrak{p}$.*

Proof. $S = R \backslash \mathfrak{p}$ is a proper multiplicative subset of R, hence also of A. Let $\mathfrak{A}$ be an ideal of A such that $\mathfrak{A} \cap R \subseteq \mathfrak{p}$ (e.g., $\mathfrak{A} = 0$). Then $\mathfrak{A} \cap S = \varnothing$. Let $\mathscr{S}$ be the set of all ideals $\mathfrak{B}$ of A such that $\mathfrak{A} \subseteq \mathfrak{B}$ and $\mathfrak{B} \cap S = \varnothing$. Then $\mathscr{S}$ is a nonempty inductive set. By Zorn's Lemma, $\mathscr{S}$ has a maximal element $\mathfrak{P}$.

We have $1 \notin \mathfrak{P}$, since $1 \in S$. If $\alpha, \beta \notin \mathfrak{P}$, then $(\mathfrak{P} + A\alpha) \cap S \neq \varnothing$, by maximality of $\mathfrak{P}$, and $p + a\alpha \in S$ for some $p \in \mathfrak{P}$, $a \in A$; similarly $q + b\beta \in S$ for some $q \in \mathfrak{P}$, $b \in A$, so that $(p + a\alpha)(q + b\beta) \in S$, $(p + a\alpha)(q + b\beta) \notin \mathfrak{P}$, and $\alpha\beta \notin \mathfrak{P}$. Thus $\mathfrak{P}$ is a prime ideal.

Since $\mathfrak{P} \cap S = \varnothing$, we have $\mathfrak{P} \cap R \subseteq \mathfrak{p}$. Assume that $\mathfrak{P} \cap R \subsetneqq \mathfrak{p}$. Let $c \in \mathfrak{p} \backslash \mathfrak{P}$. As above we have $p + \alpha c = s \in S$ for some $p \in \mathfrak{P}$ and some $\alpha \in A$. Since A is integral over R, we have

$$\alpha^n + r_{n-1}\alpha^{n-1} + \cdots + r_0 = 0$$

for some $r_{n-1}, \ldots, r_0 \in R$. Multiplying by c^n yields

$$(s - p)^n + cr_{n-1}(s - p)^{n-1} + \cdots + c^n r_0 = c^n\alpha^n + cr_{n-1}c^{n-1}\alpha^{n-1} + \cdots + c^n r_0 = 0.$$

Since $c \in \mathfrak{p}$ and $p \in \mathfrak{P}$ this implies $s^n \in \mathfrak{P}$, so that $s \in \mathfrak{P}$ and $s \in \mathfrak{P} \cap R \subseteq \mathfrak{p}$, contradicting $s \in S = R \backslash \mathfrak{p}$. Therefore $\mathfrak{P} \cap R = \mathfrak{p}$. $\qquad\square$

11.4.b. Properties

Proposition 11.4.2 (Maximality). *Let A be integral over R and $\mathfrak{P}$ be a prime ideal of A which lies over $\mathfrak{p}$. Then $\mathfrak{P}$ is a maximal ideal of A if and only if $\mathfrak{p}$ is a maximal ideal of R.*

Proof. Since $\mathfrak{P} \cap R = \mathfrak{p}$, the inclusion $R \longrightarrow A$ induces an injective homomorphism $R/\mathfrak{p} \longrightarrow A/\mathfrak{P}$, by which $R/\mathfrak{p}$ may be identified with a subring of $A/\mathfrak{P}$. Then $A/\mathfrak{P}$ is integral over $R/\mathfrak{p}$, since A is integral over R.

Assume that $\mathfrak{p}$ is maximal, so that $R/\mathfrak{p}$ is a field. Let $\alpha \in A/\mathfrak{P}$, $\alpha \neq 0$. Then α is integral over $R/\mathfrak{p}$ and is algebraic over $R/\mathfrak{p}$; hence $(R/\mathfrak{p})[\alpha] \subseteq A/\mathfrak{P}$ is a field and α is invertible in $A/\mathfrak{P}$. Thus $A/\mathfrak{P}$ is a field, and $\mathfrak{P}$ is maximal.

Now assume that $\mathfrak{p}$ is not maximal. Then $R/\mathfrak{p}$ has a maximal ideal $\mathfrak{q}$. By Proposition 11.4.1, there is a prime ideal $\mathfrak{Q}$ of $A/\mathfrak{P}$ which lies over $\mathfrak{q}$, in particular $\mathfrak{Q} \neq 0, A/\mathfrak{P}$. Therefore $\mathfrak{P}$ is not maximal. $\qquad\square$

Proposition 11.4.3 (Incomparability). *Let A be integral over R and $\mathfrak{P}, \mathfrak{Q}$ be prime ideals of A which lie over $\mathfrak{p}$. If $\mathfrak{P} \subseteq \mathfrak{Q}$, then $\mathfrak{P} = \mathfrak{Q}$.*

Proof. Let $\alpha \in \mathfrak{Q}$. Since A is integral over R, we have $f(\alpha) = 0$, in particular, $f(\alpha) \in \mathfrak{Q}$, for some monic polynomial $f(X) \in R[X]$. Let $f(X) = X^n + r_{n-1}X^{n-1} + \cdots + r_0 \in R[X]$ be a monic polynomial such that

$$f(\alpha) = \alpha^n + r_{n-1}\alpha^{n-1} + \cdots + r_1\alpha + r_0 \in \mathfrak{Q},$$

with the least possible degree n. Since $\alpha \in \mathfrak{Q}$, we have $r_0 \in \mathfrak{Q} \cap R = \mathfrak{p}$ and

$$\alpha\,(\alpha^{n-1} + r_{n-1}\alpha^{n-2} + \cdots + r_1) \in \mathfrak{P}.$$

But $\alpha^{n-1} + r_{n-1}\alpha^{n-2} + \cdots + r_1 \notin \mathfrak{P}$, by the choice of n. Therefore $\alpha \in \mathfrak{P}$. $\qquad\square$

The proof of Proposition 11.4.1 shows that an ideal $\mathfrak{A}$ of A, which is maximal such that $\mathfrak{A} \cap R \subseteq \mathfrak{p}$, is a prime ideal which lies over $\mathfrak{p}$. The proof of Proposition 11.4.3 shows the converse: if $\mathfrak{P}$ lies over $\mathfrak{p}$, then $\mathfrak{P}$ is maximal among the ideals $\mathfrak{A}$ of A such that $\mathfrak{A} \cap R \subseteq \mathfrak{p}$.

11.4.c. Extending Homomorphisms.
The previous results imply homomorphism extension properties which will be used in Section 11.9.

Proposition 11.4.4. *Let A be integral over R. Every homomorphism of R into an algebraically closed field can be extended to A.*

Proof. If R is a field, then A is algebraic over R, and the result follows from Theorem 6.3.4.

Now let R be local. Let K be an algebraically closed field and $\varphi : R \longrightarrow K$ be a homomorphism whose kernel is the maximal ideal $\mathfrak{m}$ of R. By Propositions 11.4.1 and 11.4.2, there is a maximal ideal $\mathfrak{M}$ of A lying over $\mathfrak{m}$. Then $A/\mathfrak{M}$ is a field; the inclusion $R \longrightarrow A$ induces a homomorphism $R/\mathfrak{m} \longrightarrow A/\mathfrak{M}$,

so that the field $R/\mathfrak{m}$ can be identified with a subfield of $A/\mathfrak{M}$; then $A/\mathfrak{M}$ is integral over $R/\mathfrak{m}$ and is an algebraic extension of $R/\mathfrak{m}$. Since $\operatorname{Ker}\varphi = \mathfrak{m}$, φ factors through the projection $R \longrightarrow R/\mathfrak{m}$. The resulting homomorphism $R/\mathfrak{m} \longrightarrow K$ can be extended to $A/\mathfrak{M}$, and the commutative diagram above shows that φ has been extended to A.

We can now consider the general case. Given $\varphi : R \longrightarrow K$, then $\mathfrak{p} = \operatorname{Ker}\varphi$ is a prime ideal of R, $S = R\backslash\mathfrak{p}$ is a multiplicative subset of R, and $S^{-1}A$ is integral over $S^{-1}R$ by Proposition 11.2.7. Now $S^{-1}R = R_{\mathfrak{p}}$ is local; by Proposition 11.1.2, φ factors through $\iota : R \longrightarrow S^{-1}R$.

$$
\begin{array}{ccc}
A & \xrightarrow{\ \iota\ } & S^{-1}A \\
\scriptstyle\subseteq\ \big\uparrow & & \scriptstyle\subseteq\ \big\uparrow \\
R & \xrightarrow[\ \iota\ ]{} & S^{-1}R \longrightarrow K
\end{array}
$$

By the local case the new homomorphism $S^{-1}R \longrightarrow K$ extends to $S^{-1}A$, and the commutative diagram above shows that φ has been extended to A. $\square$

As a consequence of Proposition 11.4.4, we obtain a property of fields, which will be used in Section 11.9.

Proposition 11.4.5. *Every homomorphism of a field K into an algebraically closed field can be extended to every finitely generated ring extension of K.*

Proof. Let L be algebraically closed, $\varphi : K \longrightarrow L$ be a homomorphism, and let R be a ring extension of K which is finitely generated as a ring extension, so that $R = K[\alpha_1,\ldots,\alpha_n]$ for some $\alpha_1,\ldots,\alpha_n \in R$.

First assume that R is a field. We may assume that R is not algebraic over K. Let $\beta_1,\ldots,\beta_t$ be a transcendence base of R over K. Each $\alpha \in R$ is algebraic over $K(\beta_1,\ldots,\beta_t)$ and satisfies a polynomial equation $a_k\alpha^k + \cdots + a_1\alpha + a_0 = 0$ with coefficients $a_k,\ldots,a_0 \in K(\beta_1,\ldots,\beta_t)$, $a_k \neq 0$. Multiplying by a common denominator yields a polynomial equation

$$
b_k\alpha^k + \cdots + b_1\alpha + b_0 = 0
$$

with coefficients $b_k,\ldots,b_0 \in K[\beta_1,\ldots,\beta_t]$, $b_k \neq 0$. Hence α is integral over $K[\beta_1,\ldots,\beta_t, 1/b_k]$. Applying this to $\alpha_1,\ldots,\alpha_n$ yields $c_1,\ldots,c_n \in K[\beta_1,\ldots,\beta_t]$,

$c_1,\ldots,c_n \neq 0$ such that $\alpha_1,\ldots,\alpha_n$ are integral over $K[\beta_1,\ldots,\beta_t,\ 1/c_1,\ldots,1/c_n]$. Then $c = c_1 c_2 \ldots c_n \neq 0$ in $K[\beta_1,\ldots,\beta_t]$ and $1/c_1,\ldots,1/c_n \in K[\beta_1,\ldots,\beta_t,\ 1/c]$, so that $\alpha_1,\ldots,\alpha_n$ are integral over $K[\beta_1,\ldots,\beta_t,\ 1/c]$. Hence R is integral over $K[\beta_1,\ldots,\beta_t,\ 1/c]$.

Let c^φ be the image of c under the homomorphism

$$K[\beta_1,\ldots,\beta_t] \cong K[X_1,\ldots,X_t] \longrightarrow L[X_1,\ldots,X_t].$$

Since L is infinite and $c^\varphi \neq 0$, there exist $\gamma_1,\ldots,\gamma_t \in L$ such that $c^\varphi(\gamma_1,\ldots,\gamma_t) \neq 0$ (see the exercises). There is a homomorphism $\psi: K[\beta_1,\ldots,\beta_t] \longrightarrow L$ which extends φ and sends $\beta_1,\ldots,\beta_t$ to $\gamma_1,\ldots,\gamma_t$. Then $\mathfrak{p} = \operatorname{Ker}\psi$ is a prime ideal of $K[\beta_1,\ldots,\beta_t]$ and Proposition 11.1.2 extends ψ to the local ring $K[\beta_1,\ldots,\beta_t]_\mathfrak{p}$. Now $c \notin \mathfrak{p}$; hence $K[\beta_1,\ldots,\beta_t,1/c] \subseteq K[\beta_1,\ldots,\beta_t]_\mathfrak{p}$ and φ extends to $K[\beta_1,\ldots, \beta_t,\ 1/c]$. Then Proposition 11.4.4 extends φ to R, which proves the result when R is a field.

Now let $R = K[\alpha_1,\ldots,\alpha_n]$ be any finitely generated ring extension of K. Let $\mathfrak{m}$ be a maximal ideal of R and $\pi: R \longrightarrow R/\mathfrak{m}$ be the projection. Then $R/\mathfrak{m}$ is a field, $\pi K \cong K$, and $R/\mathfrak{m} = \pi K[\pi\alpha_1,\ldots,\pi\alpha_n]$ is finitely generated over πK:

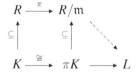

By the first part of the proof every homomorphism of πK into L extends to $R/\mathfrak{m}$. Therefore every homomorphism of $K \cong \pi K$ into L extends to R. $\qquad\square$

Exercises

1. Prove the *Going Up Theorem*: let A be integral over R, and let $\mathfrak{p} \subsetneq \mathfrak{q}$ be prime ideals of R; if $\mathfrak{P}$ is a prime ideal of A which lies over $\mathfrak{p}$, then $\mathfrak{P} \subsetneq \mathfrak{Q}$ for some prime ideal $\mathfrak{Q}$ of A which lies over $\mathfrak{q}$.

2. Find all prime ideals of $\mathbb{Z}[\sqrt{-5}]$ which lie over the prime ideal (5) of $\mathbb{Z}$.

3. Find all prime ideals of $\mathbb{Z}[\sqrt{-5}]$ which lie over the prime ideal (2) of $\mathbb{Z}$.

4. Find all prime ideals of $\mathbb{Z}[\sqrt{-5}]$ which lie over the prime ideal (3) of $\mathbb{Z}$.

5. Prove the following: when K is an infinite field and $f \in K[X_1,\ldots,X_n]$, $f \neq 0$, then $f(x_1,\ldots,x_n) \neq 0$ for some $x_1,\ldots,x_n \in K$. (Start with $n = 1$ and proceed by induction.)

6. Let $\mathfrak{p}_1,\ldots,\mathfrak{p}_r$ be prime ideals of a commutative ring. Show that an ideal $\mathfrak{a}$ which is contained in $\mathfrak{p}_1 \cup \cdots \cup \mathfrak{p}_r$ is contained in some $\mathfrak{p}_i$.

7. Let R be an integrally closed domain, K be its field of fractions, E be a finite Galois extension of K, and A be the integral closure of R in E. Let $\mathfrak{P}$ and $\mathfrak{Q}$ be prime ideals of A which lie over the same prime ideal $\mathfrak{p}$ of R. Prove that $\mathfrak{P}$ and $\mathfrak{Q}$ are conjugate ($\mathfrak{Q} = \sigma\mathfrak{P}$ for some $\sigma \in \operatorname{Gal}_K(E)$). (Hence there are only finitely many prime ideals of A which lie over $\mathfrak{p}$.) (Use the norm in E and the previous exercise.)

8. Prove the *Going Down Theorem*: let R be an integrally closed domain, K be its field of fractions, E be a finite separable extension of K, and A be the integral closure of R in E; let $\mathfrak{p} \subsetneq \mathfrak{q}$ be prime ideals of R. If $\mathfrak{Q}$ is a prime ideal of A which lies over $\mathfrak{q}$, then $\mathfrak{P} \subsetneq \mathfrak{Q}$ for some prime ideal $\mathfrak{P}$ of A which lies over $\mathfrak{p}$. (Use the previous exercise.)

11.5. DEDEKIND DOMAINS

This section is a continuation of Section 5.7 and brings additional characterizations and examples of Dedekind domains.

11.5.a. Fractional Ideals. Let R be a domain and Q be its field of fractions. Recall that a fractional ideal of R is a subset of Q of the form

$$\mathfrak{a}/c = \{ a/c \in Q \, ; \, a \in \mathfrak{a} \},$$

where $\mathfrak{a}$ is an ideal of R and $c \in R$, $c \neq 0$.

Proposition 11.5.1. *Let R be a domain and Q be its field of fractions. A subset $\mathfrak{A}$ of Q is a fractional ideal of R if and only if it is an R-submodule of Q and $c\mathfrak{A} \subseteq R$ for some $c \in R$, $c \neq 0$. Then $\mathfrak{A}$ is isomorphic as an R-module to an ideal of R.*

Proof. Let $\mathfrak{A}$ be a submodule of Q. If $c\mathfrak{A} \subseteq R$ for some $c \in R$, $c \neq 0$, then $\mathfrak{a} = c\mathfrak{A}$ is an ideal (= a submodule) of R, $x \longmapsto cx$ is a module isomorphism of $\mathfrak{A}$ onto $\mathfrak{a}$, and $\mathfrak{A} = \mathfrak{a}/c$ is a fractional ideal of R. The converse is clear. $\square$

By Lemma 5.7.1, every finitely generated submodule of Q is a fractional ideal. If R is noetherian, then conversely every fractional ideal of R is a finitely generated R-module (since it is isomorphic to an ideal of R).

Recall that a fractional ideal $\mathfrak{A}$ of a domain R is invertible in case there exists a fractional ideal $\mathfrak{B}$ of R such that $\mathfrak{A}\mathfrak{B} = R$. Then $\mathfrak{B}$ is unique. By Lemma 5.7.3, every invertible fractional ideal is a finitely generated R-module.

11.5.b. Valuation Rings. Recall (Proposition 9.4.6) that a ring R is a discrete valuation ring if and only if R is a PID with a unique prime ideal $\mathfrak{p} \neq 0, R$. In particular, a discrete valuation ring is a Dedekind domain.

Lemma 11.5.2. *Let R be a noetherian, integrally closed domain R with only one nonzero prime ideal. Then R is a discrete valuation ring.*

Proof. First R has only one maximal ideal $\mathfrak{p}$ and is local. We need to prove that every ideal of R is principal. Let Q be the field of fractions of R. First we show that $\mathfrak{A} : \mathfrak{A} = \{ x \in Q \, ; \, x\mathfrak{A} \subseteq \mathfrak{A} \} = R$ for every nonzero fractional ideal $\mathfrak{A}$ of R. We already have $R \subseteq \mathfrak{A} : \mathfrak{A}$. Now $\mathfrak{A} : \mathfrak{A}$ is a fractional ideal of R and is a finitely generated R-module since R is noetherian. Therefore every element of $\mathfrak{A} : \mathfrak{A}$ is integral over R; since R is integrally closed, it follows that $\mathfrak{A} : \mathfrak{A} = R$.

Next we show that

$$\mathfrak{p}' = \{x \in Q \, ; \, x\mathfrak{p} \subseteq R\} \supsetneq R.$$

First $R \subseteq \mathfrak{a}'$ for every ideal $\mathfrak{a}$ of R. If $a \in \mathfrak{p}$, $a \neq 0$, then a is not a unit, $1/a \notin R$, $1/a \in (Ra)'$, and $R \subsetneq (Ra)'$. Since R is noetherian, there is an ideal $\mathfrak{a} \neq 0$ of R which is maximal such that $R \subsetneq \mathfrak{a}'$. Assume that $ab \in \mathfrak{a}$, $b \notin \mathfrak{a}$. By the choice of $\mathfrak{a}$, there exists $c \in \mathfrak{a}' \backslash R$, and $(Rb + \mathfrak{a})' = R$. Then $ac(Rb + \mathfrak{a}) \subseteq c\mathfrak{a} \subseteq R$, $ac \in (Rb + \mathfrak{a})' = R$, $c(Ra + \mathfrak{a}) \subseteq R$, $c \in (Ra + \mathfrak{a})'$, $R \subsetneq (Ra + \mathfrak{a})'$, and $a \in \mathfrak{a}$ by the choice of $\mathfrak{a}$. Thus $\mathfrak{a}$ is a prime ideal. Therefore $\mathfrak{a} = \mathfrak{p}$, and $R \subsetneq \mathfrak{p}'$.

It follows that $\mathfrak{p}$ is invertible. Indeed $\mathfrak{p} \subseteq \mathfrak{p}\mathfrak{p}' \subseteq R$; since $\mathfrak{p}$ is maximal, either $\mathfrak{p}\mathfrak{p}' = \mathfrak{p}$ or $\mathfrak{p}\mathfrak{p}' = R$, and $\mathfrak{p}\mathfrak{p}' = \mathfrak{p}$ would imply $\mathfrak{p}' \subseteq \mathfrak{p} : \mathfrak{p} = R$.

Next we show that $\mathfrak{i} = \bigcap_{n>0} \mathfrak{p}^n = 0$ (this also follows from Theorem 11.7.1 below): since $\mathfrak{p}\mathfrak{p}' = R$, we have $\mathfrak{p}^n \mathfrak{p}' = \mathfrak{p}^{n-1}$, $\mathfrak{i}\mathfrak{p}' \subseteq \mathfrak{i}$, $R \subsetneq \mathfrak{p}' \subseteq \mathfrak{i} : \mathfrak{i}$, and $\mathfrak{i} = 0$ by the beginning of the proof.

We show that $\mathfrak{p}$ is principal. First $\mathfrak{p}^2 \subsetneq \mathfrak{p}$; otherwise, $\mathfrak{p}^n = \mathfrak{p}$ for all n and $\mathfrak{i} = \mathfrak{p} \neq 0$. Let $p \in \mathfrak{p}\backslash\mathfrak{p}^2$. Then $p\mathfrak{p}' \subseteq \mathfrak{p}\mathfrak{p}' = R$ and $p\mathfrak{p}' \not\subseteq \mathfrak{p}$ (otherwise, $p \in p\mathfrak{p}\mathfrak{p}' \subseteq \mathfrak{p}^2$). Since R is local, every proper ideal of R is contained in $\mathfrak{p}$; therefore $p\mathfrak{p}'$ is not proper, $p\mathfrak{p}' = R$, and $\mathfrak{p} = p\mathfrak{p}'\mathfrak{p} = Rp$.

Finally every ideal $\mathfrak{a} \neq R, 0$ of R is principal. Indeed $\mathfrak{a} \subseteq \mathfrak{p}$ since R is local, but $\mathfrak{a}$ is not contained in all $\mathfrak{p}^n$ since $\mathfrak{i} = \bigcap_{n>0} \mathfrak{p}^n = 0$. Hence $\mathfrak{a} \subseteq \mathfrak{p}^n = Rp^n$ and $\mathfrak{a} \not\subseteq \mathfrak{p}^{n+1}$ for some n. Let $a \in \mathfrak{a}\backslash\mathfrak{p}^{n+1}$. Then $a = rp^n$ for some $r \in R$; $r \notin \mathfrak{p}$; r is a unit; $Rp^n = Ra \subseteq \mathfrak{a}$; and $\mathfrak{a} = Rp^n$. $\qquad \square$

11.5.c. Dedekind Domains

Theorem 11.5.3. *For a domain R the following conditions are equivalent:*

(1) *R is a Dedekind domain.*

(2) *R is noetherian, every nonzero prime ideal of R is maximal, and R is integrally closed.*

(3) *R is noetherian and $R_\mathfrak{p}$ is a PID for every prime ideal $\mathfrak{p} \neq 0$ of R.*

Proof. (1) implies (2). Let R be Dedekind with field of fractions Q and $x \in Q$ be integral over R. Then $R[x]$ is a finitely generated R-module. By Lemma 5.7.1, $R[x]$ is a fractional ideal. Also $(R[x])(R[x]) = R[x]$; since $R[x]$ is invertible, this implies $R[x] = R$ and $x \in R$. Thus R is integrally closed. The rest of (2) follows from Theorem 5.7.4.

(2) implies (3). Assume (2) and let $\mathfrak{p} \neq 0$ be a prime ideal of R. Then $R_\mathfrak{p}$ is noetherian (Proposition 11.1.8) and integrally closed (Proposition 11.3.2), and it follows from Proposition 11.1.9 that $\mathfrak{p}^E$ is the only prime ideal of $R_\mathfrak{p}$. Hence $R_\mathfrak{p}$ is a PID (actually a discrete valuation ring) by Lemma 11.5.2.

(3) implies (1). Assuming (3), we show that every nonzero ideal $\mathfrak{a}$ of R is invertible. We have $\mathfrak{a}\mathfrak{a}' \subseteq R$. Assume that $\mathfrak{a}\mathfrak{a}' \subsetneq R$. Then $\mathfrak{a}\mathfrak{a}' \subseteq \mathfrak{m}$ for some

maximal ideal $\mathfrak{m}$ of R. In $R_\mathfrak{m}$ the ideal $\mathfrak{a}^E$ is principal by (3): $\mathfrak{a}^E = R_\mathfrak{m}(a/s)$ for some $a \in \mathfrak{a}$, $s \in R\backslash\mathfrak{m}$. Also $\mathfrak{a} = Ra_1 + \cdots + Ra_n$ is finitely generated, since R is noetherian. For each i, $a_i/1 \in \mathfrak{a}^E$ and $a_i/1 = (x_i/s_i)(a/s)$ for some $x_i \in R$, $s_i \in R\backslash\mathfrak{m}$. Then $s_i s a_i = x_i a$ holds in R, and $a_i = x_i a/s_i s$ holds in Q. Hence $t = s_1 \ldots s_n s \in R\backslash\mathfrak{m}$, and $(t/a)a_i = tx_i/s_i s \in R$ holds in Q. Therefore $t/a \in \mathfrak{a}'$ and $t = (t/a)a \in \mathfrak{a}'\mathfrak{a} \subseteq \mathfrak{m}$; this is the required contradiction. $\square$

Corollary 11.5.4. *A domain R is a discrete valuation ring if and only if it is a Dedekind domain with only one nonzero prime ideal.*

It follows from Theorem 11.5.3 that the integral closure of $\mathbb{Z}$ in any finite field extension of $\mathbb{Q}$ is a Dedekind domain. More generally:

Theorem 11.5.5. *Let R be a Dedekind domain with fraction field K. The integral closure of R in any finite field extension of K is a Dedekind domain.*

Proof. Let E be a finite extension of K. By Proposition 11.3.5, the integral closure A of R in E is contained in a finitely generated R-submodule M of E. Since R is noetherian, M is noetherian (Proposition 10.9.3), A is a finitely generated submodule of M, A is a noetherian R-module, and A is a noetherian ring (since every ideal of A is an R-submodule). By Proposition 11.3.3, A is integrally closed. Finally every prime ideal of A lies over a prime ideal of R and is zero or maximal by Proposition 11.4.2. Hence it follows from Theorem 11.5.3 that A is a Dedekind domain. $\square$

Exercises

1. Give an example of a Dedekind domain which is not a UFD.
2. Let R be a noetherian integrally closed domain and $\mathfrak{p}$ be a minimal (nonzero) prime ideal of R. Show that the only $\mathfrak{p}$-primary ideals are the symbolic powers of $\mathfrak{p}$.
3. Show that a Dedekind domain with finitely many prime ideals is a PID.

11.6. GALOIS GROUPS

In this section we use properties of rings of algebraic integers to obtain information about Galois groups. The main result concerns irreducible polynomials $f \in \mathbb{Z}[X]$, for which we obtain elements of the Galois group with known cycle structures. This helps find Galois groups over $\mathbb{Q}$.

11.6.a. Lying Over. In Section 11.4 we studied prime ideals of an integral extension A of a ring R. Additional properties obtain when R is integrally closed and A is the integral closure of R in a finite Galois extension of its field of fractions.

Proposition 11.6.1. *Let R be an integrally closed domain, K be its field of fractions, E be a finite Galois extension of K, and A be the integral closure of R in E. Any two prime ideals $\mathfrak{P}$ and $\mathfrak{Q}$ of A which lie over the same prime ideal $\mathfrak{p}$ of R are conjugate in E ($\mathfrak{Q} = \sigma\mathfrak{P}$ for some $\sigma \in \mathrm{Gal}_K(E)$). Therefore there are only finitely many prime ideals of A which lie over $\mathfrak{p}$.*

Proof. Let $G = \mathrm{Gal}\,(E/K)$ be the Galois group. When $\alpha \in E$ is integral over R, every conjugate $\sigma\alpha$ of α is integral over R and the norm $N(\alpha) = \prod_{\sigma \in G} \sigma\alpha$ is integral over R. Also $N(\alpha) \in K$. Since R is integrally closed, we have $N(\alpha) \in R$.

Now let $\mathfrak{P}$ and $\mathfrak{Q} \subseteq A$ lie over $\mathfrak{p}$. If $\alpha \in \mathfrak{Q}$, then $N(\alpha) \in \mathfrak{Q} \cap R \subseteq \mathfrak{P}$ and $\sigma\alpha \in \mathfrak{P}$ for some $\sigma \in G$. Therefore $\mathfrak{Q} \subseteq \bigcup_{\sigma \in G} \sigma\mathfrak{P}$. By Lemma 11.6.2 below, $\mathfrak{Q} \subseteq \sigma\mathfrak{P}$ for some $\sigma \in \mathfrak{P}$, and $\mathfrak{Q} = \sigma\mathfrak{P}$ by Proposition 11.4.3. $\qquad\square$

Lemma 11.6.2. *Let $\mathfrak{p}_1,\dots,\mathfrak{p}_r$ be prime ideals of a commutative ring. An ideal $\mathfrak{a}$ which is contained in $\mathfrak{p}_1 \cup \cdots \cup \mathfrak{p}_r$ is contained in some $\mathfrak{p}_i$.*

Proof. We may assume that $\mathfrak{a}$ is not contained in the union of any proper subset of $\{\mathfrak{p}_1,\dots,\mathfrak{p}_r\}$. Then $\mathfrak{p}_i \not\subseteq \mathfrak{p}_j$ when $i \neq j$. If $\mathfrak{a} \cap (\bigcap_{i \neq j} \mathfrak{p}_i) \not\subseteq \mathfrak{p}_j$ for every j, and $a_j \in (\mathfrak{a} \cap (\bigcap_{i \neq j} \mathfrak{p}_i)) \setminus \mathfrak{p}_j$, then $a = a_1 + \cdots + a_r \in \mathfrak{a} \setminus (\mathfrak{p}_1 \cup \cdots \cup \mathfrak{p}_r)$, a contradiction. Therefore $\mathfrak{a} \cap (\bigcap_{i \neq j} \mathfrak{p}_i) \subseteq \mathfrak{p}_j$ for some j; then $\mathfrak{a}\mathfrak{p}_1 \dots \mathfrak{p}_{j-1} \mathfrak{p}_{j+1} \dots \mathfrak{p}_r \subseteq \mathfrak{p}_j$, one of $\mathfrak{a}, \mathfrak{p}_1,\dots,\mathfrak{p}_{j-1},\mathfrak{p}_{j+1},\dots,\mathfrak{p}_r$ is contained in $\mathfrak{p}_j$, and $\mathfrak{a} \subseteq \mathfrak{p}_j$. $\qquad\square$

11.6.b. Galois Extensions of $\mathbb{Q}$. We now let $R = \mathbb{Z}$ and obtain still more properties.

Proposition 11.6.3. *Let E be a finite Galois extension of $\mathbb{Q}$ and A be its ring of algebraic integers. There is a basis of E over $\mathbb{Q}$ which is also a basis of A over $\mathbb{Z}$.*

Proof. By Proposition 11.3.5, A is contained in a finitely generated $\mathbb{Z}$-submodule S of E. Since $E \supseteq \mathbb{Q}$ has characteristic 0, S is torsion free and is a finitely generated free $\mathbb{Z}$-module. Therefore $A \subseteq S$ is a finitely generated free $\mathbb{Z}$-module.

Every basis $\beta_1,\dots,\beta_n$ of A over $\mathbb{Z}$ is a basis of E over $\mathbb{Q}$. Indeed $\beta_1,\dots,\beta_n$ are linearly independent over $\mathbb{Q}$: if $\sum_i q_i \beta_i = 0$ for some rational numbers $q_i \in \mathbb{Q}$, then $q_1,\dots,q_n$ have a common denominator m, so that $mq_i \in \mathbb{Z}$ for all i, and $\sum_i mq_i \beta_i = 0$ implies $mq_i = 0$ for all i and $q_i = 0$ for all i. Moreover, $\beta_1,\dots,\beta_n$ span E: if $\alpha \in E$, then $m\alpha \in A$ for some $m \in \mathbb{Z}$, $m \neq 0$, $m\alpha = \sum_i n_i \beta_i$ for some $n_i \in \mathbb{Z}$, and $\alpha = \sum_i (n_i/m)\beta_i$ for some $(n_i/m) \in \mathbb{Q}$. $\qquad\square$

Lemma 11.6.4. *Let E be a finite Galois extension of $\mathbb{Q}$ of degree n, A be its ring of algebraic integers, and $\mathfrak{P}_1,\dots,\mathfrak{P}_r$ be the prime ideals of A lying over $p\mathbb{Z}$.*

(1) *$A/\mathfrak{P}_i \cong A/\mathfrak{P}_j$ for all i,j; $\overline{E} = A/\mathfrak{P}_i$ is a finite Galois extension of $\mathbb{Z}_p$, and $\mathrm{Gal}\,(\overline{E}/\mathbb{Z}_p)$ is cyclic.*

(2) *$kr \leqslant n$, where $|\overline{E}| = p^k$.*

Part (2) could be proved by extending valuations (as in Section 9.5).

Proof. First $\mathfrak{P} = \mathfrak{P}_1$ is a maximal ideal of A by Proposition 11.4.2, so that $\overline{E} = A/\mathfrak{P}$ is a field. Since $\mathfrak{P} \cap \mathbb{Z} = p\mathbb{Z}$, $\mathbb{Z}_p = \mathbb{Z}/p\mathbb{Z}$ is a subfield of $\overline{E}$; in particular, $\overline{E}$ has characteristic p. Also $A/\mathfrak{P}_i \cong A/\mathfrak{P}$ for all i, since Proposition 11.6.1 provides an automorphism $\sigma \in \mathrm{Gal}(E/\mathbb{Q})$ such that $\sigma\mathfrak{P}_i = \mathfrak{P}$ (and $\sigma A = A$). By the Chinese Remainder Theorem (Proposition 5.6.2), the projections $A \longrightarrow A/\mathfrak{P}_i$ induce a surjective homomorphism $A \longrightarrow A/\mathfrak{P}_1 \times \cdots \times A/\mathfrak{P}_r$, whose kernel is $\mathfrak{A} = \mathfrak{P}_1 \cap \cdots \cap \mathfrak{P}_r$. Hence $A/\mathfrak{A} \cong \overline{E}^r$. But $p\mathbb{Z} \subseteq \mathfrak{A}$, so that $pA \subseteq \mathfrak{A}$, and Proposition 11.6.3 yields $A \cong \mathbb{Z}^n$, whence $A/pA \cong (\mathbb{Z}/p\mathbb{Z})^n$ is finite, with p^n elements. Therefore $A/\mathfrak{A}$ is finite and $\overline{E}$ is finite. Hence $|\overline{E}| = p^k$ for some k. Moreover $|\overline{E}^r| = |A/\mathfrak{A}| \leqslant |A/pA| = p^n$ and $kr \leqslant n$.

Since $\overline{E}$ is finite, it is the splitting field of the separable polynomial $X^{p^k} - X \in \mathbb{Z}_p[X]$ and is a Galois extension of $\mathbb{Z}_p$. Also $\tau : \alpha \longmapsto \alpha^p$ is an automorphism of $\overline{E}$. We see that $\tau\alpha = \alpha$ for all $\alpha \in \mathbb{Z}_p$. Since $X^p - X \in \mathbb{Z}_p[X]$ has p roots in $\overline{E}$, $\mathbb{Z}_p$ is the fixed field of $\langle \tau \rangle \subseteq \mathrm{Gal}(\overline{E}/\mathbb{Z}_p)$. Therefore $\mathrm{Gal}(\overline{E}/\mathbb{Z}_p) = \langle \tau \rangle$. $\qquad\square$

We can now show:

Proposition 11.6.5. *Let $q \in \mathbb{Z}[X]$ be a monic irreducible polynomial, p be a prime number, and $\overline{q}$ be the image of q in $\mathbb{Z}_p[X]$, which is the product of monic irreducible polynomials $q_1, \ldots, q_s \in \mathbb{Z}_p[X]$. If $q_1, \ldots, q_s$ are distinct of degrees $d_1, \ldots, d_s$, then the Galois group of q contains a product of pairwise disjoint cycles of orders $d_1, \ldots, d_s$.*

Proof. Let $\alpha_1, \ldots, \alpha_m$ be the roots of q, which are integral over $\mathbb{Z}$, since q is monic. Let $E = \mathbb{Q}(\alpha_1, \ldots, \alpha_m)$ be the splitting field of q. Let $\mathfrak{P}_1, \ldots, \mathfrak{P}_r$, $\overline{E}$, and k, r, n be as in Lemma 11.6.4. Let $\mathfrak{P} = \mathfrak{P}_1$ and $\alpha \longmapsto \overline{\alpha}$ denote the projection $A \longrightarrow \overline{E} = A/\mathfrak{P}$. Since $q = (X - \alpha_1)\ldots(X - \alpha_m)$ in $E[X]$, we have

$$q_1 \ldots q_s = \overline{q} = (X - \overline{\alpha}_1)\ldots(X - \overline{\alpha}_m)$$

in $\overline{E}[X]$ (since $\mathfrak{P} \cap \mathbb{Z} = p\mathbb{Z}$, $q \in \mathbb{Z}[X]$ has the same projection in $\overline{E}[X]$ as in $\mathbb{Z}_p[X]$). Since $\overline{E}$ is Galois over $\mathbb{Z}_p$, $q_1, \ldots, q_s$ are separable and have no multiple roots in $\overline{E}$. Moreover $q_1, \ldots, q_s$ are distinct and have no common root in $\overline{E}$. Therefore $\overline{\alpha}_1, \ldots, \overline{\alpha}_m$ are all distinct.

Let $G = \mathrm{Gal}(E/\mathbb{Q})$, $\overline{G} = \mathrm{Gal}(\overline{E}/\mathbb{Z}_p)$, and $H = \{\sigma \in G ; \sigma\mathfrak{P} = \mathfrak{P}\}$ be the stabilizer of $\mathfrak{P}$ under the action of G. By Proposition 11.6.1, the orbit of $\mathfrak{P}$ is $\mathfrak{P}_1, \ldots, \mathfrak{P}_r$; hence $[G : H] = r$ and $|H| = n/r \geqslant k$. Each $\sigma \in H$ satisfies $\sigma\mathfrak{P} = \mathfrak{P}$ and $\sigma A = A$, and induces an automorphism $\overline{\sigma}$ of $A/\mathfrak{P} = \overline{E}$ such that $\overline{\sigma}\,\overline{\alpha} = \overline{\sigma\alpha}$ for all $\alpha \in E$. This defines a homomorphism $\varphi : \sigma \longmapsto \overline{\sigma}$ of H into $\overline{G}$. If $\overline{\sigma} = 1$, then $\overline{\alpha}_i = \overline{\sigma}\,\overline{\alpha}_i = \overline{\sigma\alpha}_i$ and $\sigma\alpha_i = \alpha_i$ for all i (since $\overline{\alpha}_1, \ldots, \overline{\alpha}_m$ are distinct); hence φ is injective. Therefore $|H| \leqslant |\overline{G}| = k$. Hence $|H| = k$ and φ is an isomorphism. (Also, $n = kr$.) In particular, every $\overline{\tau} \in \overline{G}$ is induced by some $\sigma \in H$.

We may identify $\sigma \in H$ with a permutation of $\alpha_1, \ldots, \alpha_n$, and $\overline{\sigma} \in \overline{G}$ with a permutation of $\overline{\alpha}_1, \ldots, \overline{\alpha}_n$. Since $\overline{\alpha}_1, \ldots, \overline{\alpha}_m$ are distinct, σ and $\overline{\sigma}$ have the same cycle structure. Let $\overline{\tau}$ be a generator of $\overline{G}$ (which is cyclic by Lemma 11.6.4). If q_j has degree $d = d_j$ and roots $\beta_1, \ldots, \beta_d$, then $\overline{\tau}$ must permute $\beta_1, \ldots, \beta_d$. Moreover $\overline{\tau}$ cannot permute any proper subset, say $\{\beta_1, \ldots, \beta_\ell\}$, of $\{\beta_1, \ldots, \beta_d\}$; otherwise, $f = (X - \beta_1) \ldots (X - \beta_\ell)$ and $g = (X - \beta_{\ell+1}) \ldots (X - \beta_d)$ are fixed by $\overline{\tau}$, f and g are fixed by $\overline{G}$, $f, g \in \mathbb{Z}_p[X]$, and f divides the irreducible polynomial q_j, a contradiction. Therefore the restriction of $\overline{\tau}$ to $\{\beta_1, \ldots, \beta_d\}$ is a d-cycle. Thus $\overline{\tau}$ is a product of pairwise disjoint cycles, whose orders are the degrees $d_1, \ldots, d_s$ of $q_1, \ldots, q_s$. Then $\overline{\tau}$ is induced by $\tau \in H$, which has the same cycle structure. $\qquad\square$

For example, let $q = X^5 - X + 1 \in \mathbb{Z}[X]$. The reader will show that $\overline{q}$ is irreducible in $\mathbb{Z}_3[X]$, so that q is irreducible in $\mathbb{Z}[X]$, and that

$$X^5 - X + 1 = (X^2 + X + 1)(X^3 + X^2 + 1) \quad \text{in} \quad \mathbb{Z}_2[X].$$

By Proposition 11.6.5, the Galois group G of q over $\mathbb{Q}$ (viewed as a subgroup of S_5) contains a 5-cycle and contains the product of a 2-cycle and a disjoint 3-cycle. Hence G contains a 5-cycle and a transposition, and it is all of S_5.

Exercises

1. Verify that $q = X^5 - X + 1 \in \mathbb{Z}[X]$ and $\overline{q} = X^5 - X + 1 \in \mathbb{Z}_3[X]$ are irreducible, and that the Galois group of q over $\mathbb{Q}$ is [isomorphic to] S_5.

2. Find the Galois group of $X^4 + 2X^2 + X + 1$ over $\mathbb{Q}$.

3. Find the Galois group of $X^4 + X + 1$ over $\mathbb{Q}$.

11.7. NOETHERIAN RINGS

This section continues Sections 5.5 and 5.6 with remarkable properties of prime ideals. The main result implies that the prime ideals of a noetherian ring satisfy the descending chain condition.

11.7.a. The Krull Intersection Theorem

Theorem 11.7.1 (Krull Intersection Theorem). *Let R be noetherian and $\mathfrak{a}$ be an ideal of R. Let $\mathfrak{i} = \bigcap_{n>0} \mathfrak{a}^n$. Then $\mathfrak{a}\mathfrak{i} = \mathfrak{i}$ and $(1 - a)\mathfrak{i} = 0$ for some $a \in \mathfrak{a}$. If R is a domain or R is local, and $\mathfrak{a} \neq R$, then $\mathfrak{i} = 0$.*

Proof. We have $\mathfrak{a}\mathfrak{i} \subseteq \mathfrak{i}$. Let $\mathfrak{q}$ be a primary ideal of R which contains $\mathfrak{a}\mathfrak{i}$ and $\mathfrak{p}$ be the radical of $\mathfrak{q}$. By Lemma 5.6.6, $\mathfrak{p}^n \subseteq \mathfrak{q}$ for some n. If $\mathfrak{i} \not\subseteq \mathfrak{q}$, then $\mathfrak{a}\mathfrak{i} \subseteq \mathfrak{q}$ implies $\mathfrak{a} \subseteq \mathfrak{p}$, since $\mathfrak{q}$ is primary, and $\mathfrak{i} \subseteq \mathfrak{a}^n \subseteq \mathfrak{q}$. This contradiction shows that

$i \subseteq q$. Thus every primary ideal of R which contains $\mathfrak{a}i$ must contain i. Since $\mathfrak{a}i$ is an intersection of primary ideals, this implies $i \subseteq \mathfrak{a}i$. Thus $\mathfrak{a}i = i$.

Applying the following lemma to the finitely generated R-module i yields $(1 - a)i = 0$ for some $a \in \mathfrak{a}$. If R is a domain and $\mathfrak{a} \neq R$, then $1 - a \neq 0$ and $i = 0$. If R is local and $\mathfrak{a} \neq R$, then $1 - a$ is a unit and again $i = 0$. $\square$

Lemma 11.7.2. *Let M be a finitely generated R-module. If $\mathfrak{a}M = M$ for some ideal $\mathfrak{a}$ of R, then $(1 - a)M = 0$ for some $a \in \mathfrak{a}$.*

Proof. We have $M = Rx_1 + \cdots + Rx_n$ for some $x_1, \ldots, x_n \in M$. Then $\mathfrak{a}M = \mathfrak{a}x_1 + \cdots + \mathfrak{a}x_n$: if indeed $a \in \mathfrak{a}$ and $m \in M$, then $m = r_1 x_1 + \cdots + r_n x_n$ for some $r_1, \ldots, r_n \in R$ and $am = ar_1 x_1 + \cdots + ar_n x_n \in \mathfrak{a}x_1 + \cdots + \mathfrak{a}x_n$; therefore $\mathfrak{a}M \subseteq \mathfrak{a}x_1 + \cdots + \mathfrak{a}x_n$, and the converse inclusion is clear.

If $M = \mathfrak{a}M$, then there are equalities

$$x_i = a_{i1}x_1 + \cdots + a_{in}x_n$$

with $a_{ij} \in \mathfrak{a}$. This yields an $n \times n$ matrix $A = (a_{ij})_{1 \leq i,j \leq n}$ such that the matrix $B = I - A = (b_{ij})_{1 \leq i,j \leq n}$ satisfies $b_{i1}x_1 + \cdots + b_{in}x_n = 0$ for all i. By Lemma 11.2.4, the determinant d of B satisfies $dx_i = 0$ for all i. Then $dM = 0$. Since $B = I - A$ and $a_{ij} \in \mathfrak{a}$ for all i, j, we have $d = 1 - a$ for some $a \in \mathfrak{a}$. $\square$

11.7.b. Isolated Prime Ideals

Proposition 11.7.3. *Let $\mathfrak{a}$ be an ideal of any commutative ring R. Every prime ideal which contains $\mathfrak{a}$ contains a prime ideal which contains $\mathfrak{a}$ and is minimal with this property.*

Proof. We show that the intersection $i = \bigcap_{i \in I} \mathfrak{p}_i$ of a chain $(\mathfrak{p}_i)_{i \in I}$ of prime ideals of R is a prime ideal of R. Let $a, b \notin i$. Then $a \notin \mathfrak{p}_i$, $b \notin \mathfrak{p}_j$ for some $i, j \in I$. If $\mathfrak{p}_i \subseteq \mathfrak{p}_j$, then $a, b \notin \mathfrak{p}_i$ and $ab \notin \mathfrak{p}_i$; if $\mathfrak{p}_j \subseteq \mathfrak{p}_i$, then $a, b \notin \mathfrak{p}_j$ and $ab \notin \mathfrak{p}_j$; either way, $ab \notin i$.

Let $\mathfrak{p} \supseteq \mathfrak{a}$ be a prime ideal of R and $\mathscr{P}$ be the set of all prime ideals of R which contain $\mathfrak{a}$ and are contained in $\mathfrak{p}$. By the first half of the proof, $\mathscr{P}$, ordered by reverse inclusion ($\mathfrak{p}_i \leq \mathfrak{p}_j$ if and only if $\mathfrak{p}_i \supseteq \mathfrak{p}_j$), is inductive. By Zorn's Lemma, $\mathscr{P}$ has a maximal element (under reverse inclusion). $\square$

A prime ideal $\mathfrak{p}$ of R is **minimal over** $\mathfrak{a}$, or is an **isolated** prime ideal of $\mathfrak{a}$, in case $\mathfrak{p}$ is minimal among all prime ideals of R which contain $\mathfrak{a}$.

Proposition 11.7.4. *Let $\mathfrak{a}$ be an ideal of a noetherian ring R. There are only finitely many prime ideals of R that are minimal over $\mathfrak{a}$.*

Proof. This follows from the primary decomposition of $\mathfrak{a}$ (Theorem 5.6.11; see the exercises). Here is a direct proof.

Assume that the result is false. By the ascending chain condition, there exists an ideal i of R which is maximal with the property that there are infinitely many prime ideals that are minimal over i. Then i is not prime, and there exist ideals $i \subsetneq \mathfrak{a}, \mathfrak{b}$ such that $\mathfrak{a}\mathfrak{b} \subseteq i$. By the choice of i, there are only finitely many prime ideals of R that are minimal over $\mathfrak{a}$ and finitely many prime ideals of R that are minimal over $\mathfrak{b}$. If now a prime ideal $\mathfrak{p}$ is minimal over i, then $\mathfrak{a}\mathfrak{b} \subseteq i \subseteq \mathfrak{p}$, $\mathfrak{a} \subseteq \mathfrak{p}$ or $\mathfrak{b} \subseteq \mathfrak{p}$, and $\mathfrak{p}$ is minimal over $\mathfrak{a}$ or minimal over $\mathfrak{b}$; hence there are only finitely many prime ideals that are minimal over i, a contradiction. □

11.7.c. The Hauptidealsatz. A prime ideal $\mathfrak{p}$ **has finite height** in case there is an integer $n > 0$ such that every finite strictly descending sequence

$$\mathfrak{p} = \mathfrak{p}_0 \supsetneq \mathfrak{p}_1 \supsetneq \cdots \supsetneq \mathfrak{p}_m$$

of prime ideals has length at most n ($m \leqslant n$). Then the **height** of $\mathfrak{p}$ is the smallest such integer n. Thus $\mathfrak{p}$ has height at most n if and only if every finite strictly descending sequence $\mathfrak{p} = \mathfrak{p}_0 \supsetneq \mathfrak{p}_1 \supsetneq \cdots \supsetneq \mathfrak{p}_m$ has length $m \leqslant n$.

Krull's Hauptidealsatz (Principal Ideal Theorem) states:

Theorem 11.7.5 (Krull's Hauptidealsatz). *In a noetherian ring, a prime ideal which is minimal over a principal ideal has height at most* 1.

Proof. We begin the proof with two lemmas.

Lemma 11.7.6. *If* $\mathfrak{m}$ *is a maximal ideal of a noetherian ring R, then $R/\mathfrak{m}^n$ is an artinian R-module for every $n > 0$.*

Proof. We show that $M = \mathfrak{m}^{n-1}/\mathfrak{m}^n$ is an artinian R-module for all $n > 0$. Since $(R/\mathfrak{m}^n)/(\mathfrak{m}^{n-1}/\mathfrak{m}^n) \cong R/\mathfrak{m}^{n-1}$, it follows from Proposition 10.9.4, by induction on n, that $R/\mathfrak{m}^n$ is artinian.

We have $\mathfrak{m}M = 0$. Hence M is an $R/\mathfrak{m}$-module, in which $(r + \mathfrak{m})m = rm$. We see that the sub-$R/\mathfrak{m}$-modules of M are its sub-R-modules. Now $\mathfrak{m}^{n-1}$ is a finitely generated R-module; hence M is a finitely generated R-module, and a finitely generated $R/\mathfrak{m}$-module. Since $R/\mathfrak{m}$ is a field, M is an artinian $R/\mathfrak{m}$-module, and M is an artinian R-module. □

Lemma 11.7.7. *Let R be a noetherian local ring, $\mathfrak{m}$ be its maximal ideal, and $a \in \mathfrak{m}$. Assume that R/Ra is an artinian R-module. Then there is at most one prime ideal $\mathfrak{p} \subseteq \mathfrak{m}$ which does not contain a.*

Proof. Assume $a \notin \mathfrak{p} \subseteq \mathfrak{m}$. Proposition 11.1.11 yields a descending sequence

$$\mathfrak{p}^{(1)} \supseteq \cdots \supseteq \mathfrak{p}^{(n)} \supseteq \mathfrak{p}^{(n+1)} \supseteq \cdots$$

of $\mathfrak{p}$-primary ideals $\mathfrak{p}^{(n)} = ((\mathfrak{p}^E)^n)^C$. Let $\mathfrak{b} = \mathfrak{p} \cap Ra$. Then $\mathfrak{p}/\mathfrak{b} \cong (\mathfrak{p} + Ra)/Ra \subseteq R/Ra$ is an artinian R-module. Hence the descending sequence

$$\mathfrak{p}^{(1)}/\mathfrak{b} \supseteq \cdots \supseteq (\mathfrak{p}^{(n)} + \mathfrak{b})/\mathfrak{b} \supseteq (\mathfrak{p}^{(n+1)} + \mathfrak{b})/\mathfrak{b} \supseteq \cdots$$

terminates, and there is some $m > 0$ such that $\mathfrak{p}^{(n)} + \mathfrak{b} = \mathfrak{p}^{(m)} + \mathfrak{b}$ for all $n \geqslant m$. We show that $\mathfrak{p}^{(n)} = \mathfrak{p}^{(m)}$ for all $n \geqslant m$.

Let $n \geqslant m$. Every $x \in \mathfrak{p}^{(m)} \subseteq \mathfrak{p}^{(n)} + Ra$ can be written in the form $x = y + ra$ with $y \in \mathfrak{p}^{(n)}$ and $r \in R$. Now $a \notin \mathfrak{p}$ and $ra = x - y \in \mathfrak{p}^{(m)}$; hence $r \in \mathfrak{p}^{(m)}$ and $x \in \mathfrak{p}^{(n)} + \mathfrak{p}^{(m)}a$. Thus $\mathfrak{p}^{(m)} = \mathfrak{p}^{(m)} + \mathfrak{p}^{(m)}a$. Hence the finitely generated R-module $M = \mathfrak{p}^{(m)}/\mathfrak{p}^{(n)}$ satisfies $M = aM = (Ra)M$. By Lemma 11.7.2, $(1 - b)M = 0$ for some $b \in Ra$. Then $b \in \mathfrak{m}$, $1 - b \notin \mathfrak{m}$, $1 - b$ is a unit of R, since R is local, and $M = 0$. Thus $\mathfrak{p}^{(m)} = \mathfrak{p}^{(n)}$ for all $n \geqslant m$.

By Theorem 11.7.1, applied to the local ring $R_\mathfrak{p}$, $\bigcap_{n \geqslant m} (\mathfrak{p}^E)^n = 0$. Hence $\mathfrak{p}^{(m)} = \bigcap_{n \geqslant m} \mathfrak{p}^{(n)} = 0$ and $\mathfrak{p} = \mathrm{Rad}\,\mathfrak{p}^{(m)} = \mathrm{Rad}\,0$. $\square$

We now turn to Theorem 11.7.5. First we consider the case where R is local and $\mathfrak{p}$ is the maximal ideal of R. Assume that $\mathfrak{p}$ is minimal over a principal ideal Ra. The radical of Ra is a prime ideal contained in $\mathfrak{p}$; therefore $\mathfrak{p}$ is the radical of Ra. By Lemma 5.6.6, $\mathfrak{p}^n \subseteq Ra$ for some $n > 0$; by Lemma 11.7.6, $R/\mathfrak{p}^n$ is an artinian R-module; by Proposition 10.9.4, R/Ra is an artinian R-module. If $\mathfrak{q} \subsetneqq \mathfrak{p}$ is a prime ideal, then $a \notin \mathfrak{q}$, since $\mathfrak{p}$ is minimal over Ra. By Lemma 11.7.7, there is at most one such prime ideal $\mathfrak{q}$. Thus $\mathfrak{p}$ has height at most 1.

Now let R be any noetherian ring and $\mathfrak{p}$ be a prime ideal of R which is minimal over a principal ideal Ra. Then $R_\mathfrak{p}$ is a local noetherian ring, $\mathfrak{p}^E$ is the maximal ideal of $R_\mathfrak{p}$, and $\mathfrak{p}^E$ is minimal over the principal ideal $(Ra)^E = R_\mathfrak{p}\iota(a)$ by Proposition 11.1.9. Hence $\mathfrak{p}^E$ has height at most 1 in $R_\mathfrak{p}$, and it follows from Proposition 11.1.9 that $\mathfrak{p}$ has height at most 1 in R. $\square$

11.7.d. Main Result

Theorem 11.7.8 (Krull). *In a noetherian ring, every prime ideal has finite height. In fact, if $\mathfrak{p}$ is minimal over an ideal with r generators, then $\mathfrak{p}$ has height at most r.*

Proof. First we show:

Lemma 11.7.9. *Let R be a noetherian ring, $\mathfrak{q}_1, \ldots, \mathfrak{q}_r$ be prime ideals of R, and $\mathfrak{p}_0 \supsetneqq \mathfrak{p}_1 \supsetneqq \cdots \supsetneqq \mathfrak{p}_m$ be a chain of prime ideals of R. If $\mathfrak{p}_0$ is contained in no $\mathfrak{q}_j$, then there exists a chain $\mathfrak{p}_0 \supsetneqq \mathfrak{p}'_1 \supsetneqq \cdots \supsetneqq \mathfrak{p}'_{m-1} \supsetneqq \mathfrak{p}_m$ of prime ideals of R such that $\mathfrak{p}'_1, \ldots, \mathfrak{p}'_{m-1}$ are contained in no $\mathfrak{q}_j$.*

Proof. The proof is by induction on m. The result is trivial if $m \leqslant 1$. For $m > 1$, the induction hypothesis yields a chain $\mathfrak{p}_0 \supsetneqq \mathfrak{p}'_1 \supsetneqq \cdots \supsetneqq \mathfrak{p}'_{m-2} \supsetneqq \mathfrak{p}_{m-1}$ of prime ideals of R such that $\mathfrak{p}'_1, \ldots, \mathfrak{p}'_{m-2}$ are contained in no $\mathfrak{q}_j$. By Lemma

11.6.2, $\mathfrak{p}'_{m-2} \not\subseteq \mathfrak{p}_m \cup \mathfrak{q}_1 \cup \cdots \cup \mathfrak{q}_r$, and there exists $a \in \mathfrak{p}'_{m-2} \setminus (\mathfrak{p}_m \cup \mathfrak{q}_1 \cup \cdots \cup \mathfrak{q}_r)$. By Proposition 11.7.3, $\mathfrak{p}'_{m-2}$ contains a prime ideal $\mathfrak{p}'_{m-1}$ which is minimal over $Ra + \mathfrak{p}_m$. Then $\mathfrak{p}'_{m-1} \not\subseteq \mathfrak{q}_1, \ldots, \mathfrak{q}_r$. Also $\mathfrak{p}'_{m-1} \supsetneq \mathfrak{p}_m$, since $a \notin \mathfrak{p}_m$, and the Hauptidealsatz (Theorem 11.7.5) implies $\mathfrak{p}'_{m-2} \supsetneq \mathfrak{p}'_{m-1}$: in the noetherian ring $R/\mathfrak{p}_m$, $\mathfrak{p}'_{m-1}/\mathfrak{p}_m$ is minimal over the principal ideal $(Ra + \mathfrak{p}_m)/\mathfrak{p}_m$, but $\mathfrak{p}'_{m-2}/\mathfrak{p}_m$ has height at least 2 and cannot be minimal over $(Ra + \mathfrak{p}_m)/\mathfrak{p}_m$. $\qquad\square$

We now prove Theorem 11.7.8. Let R be noetherian, $\mathfrak{a} = Rx_1 + \cdots + Rx_r$ be an ideal of R with r generators, and $\mathfrak{p}$ be minimal over $\mathfrak{a}$. We show by induction on r that $\mathfrak{p}$ has height at most r. If $r = 0$, then $\mathfrak{a} = 0$ and $\mathfrak{p}$ is a minimal prime ideal. Now assume that $r > 0$. Let $\mathfrak{b} = Rx_1 + \cdots + Rx_{r-1} \subseteq \mathfrak{a}$. If $\mathfrak{p}$ is minimal over $\mathfrak{b}$, then $\mathfrak{p}$ has height at most $r - 1$ by the induction hypothesis. Otherwise, there are finitely many prime ideals $\mathfrak{q}_1, \ldots, \mathfrak{q}_s$ that are minimal over $\mathfrak{b}$, by Proposition 11.7.4, and $\mathfrak{p} \supseteq \mathfrak{b}$ is contained in no $\mathfrak{q}_j$. Let

$$\mathfrak{p} = \mathfrak{p}_0 \supsetneq \mathfrak{p}_1 \supsetneq \cdots \supsetneq \mathfrak{p}_m$$

be a chain of prime ideals. By Lemma 11.7.9, we may assume that $\mathfrak{p}_1, \ldots, \mathfrak{p}_{m-1}$ are contained in no $\mathfrak{q}_j$. Now $\mathfrak{p}$ is minimal over $\mathfrak{a} = Rx_r + \mathfrak{b}$. In the noetherian ring $R/\mathfrak{b}$, $\mathfrak{q}_1/\mathfrak{b}, \ldots, \mathfrak{q}_s/\mathfrak{b}$ are minimal over 0; $\mathfrak{p}/\mathfrak{b}$ is minimal over the principal ideal $(Rx_r + \mathfrak{b})/\mathfrak{b}$, but not minimal over 0, and has height 1 by Theorem 11.7.5. Now $(\mathfrak{p}_{m-1} + \mathfrak{b})/\mathfrak{b}$ is contained in $\mathfrak{p}/\mathfrak{b}$ but is contained in no $\mathfrak{q}_j/\mathfrak{b}$; therefore $\mathfrak{p}/\mathfrak{b}$, which has height 1, is minimal over $(\mathfrak{p}_{m-1} + \mathfrak{b})/\mathfrak{b}$. Then $\mathfrak{p}$ is minimal over $\mathfrak{p}_{m-1} + \mathfrak{b}$, and $\mathfrak{p}/\mathfrak{p}_{m-1}$ is minimal over $(\mathfrak{p}_{m-1} + \mathfrak{b})/\mathfrak{p}_{m-1}$. Now $(\mathfrak{p}_{m-1} + \mathfrak{b})/\mathfrak{p}_{m-1}$, like $\mathfrak{b}$, has $r - 1$ generators. By the induction hypothesis, $\mathfrak{p}/\mathfrak{p}_{m-1}$ has height at most $r - 1$. Hence $m - 1 \leqslant r - 1$, and $m \leqslant r$. $\qquad\square$

Exercises

1. Let $\mathfrak{a}$ be an ideal of a noetherian ring R with a reduced primary decomposition $\mathfrak{a} = \mathfrak{q}_1 \cap \cdots \cap \mathfrak{q}_r$. Show that every prime ideal of R which is minimal over $\mathfrak{a}$ is the radical of some $\mathfrak{q}_i$. (Hence there are only finitely many prime ideals of R which are minimal over $\mathfrak{a}$.)

2. Let $\mathfrak{p}$ be a prime ideal of height r in a noetherian ring R. Show that there exists an ideal $\mathfrak{a}$ of R with r generators over which $\mathfrak{p}$ is minimal. (Construct $a_1, \ldots, a_r \in \mathfrak{p}$ by induction so that the prime ideals that are minimal over $Ra_1 + \cdots + Ra_i$ have height i, for every $i \leqslant r$.)

3. Let $\mathfrak{p}$ be a prime ideal of R. Show that the height of $\mathfrak{p}$ is the dimension of $R_{\mathfrak{p}}$.

4. Let $\mathfrak{p} \neq 0$ be a prime ideal of R. Show that $\dim R \geqslant 1 + \dim R/\mathfrak{p}$.

11.8. POLYNOMIAL RINGS

In this section we consider chains of prime ideals of $R[X_1, \ldots, X_n]$ and show that such chains have length at most n when R is a field.

11.8.a. Dimension. The **spectrum** of a noetherian ring R is the set of all the prime ideals of R, partially ordered by inclusion. The **Krull dimension** or **dimension** of R is finite in case there is an integer $n > 0$ such that every chain of its spectrum has length at most n (equivalently, in case every prime ideal of R has height at most n); then the dimension $\dim R$ of R is the smallest such integer n. If no such n exists, then R has infinite dimension.

For example, a field has dimension 0; if, conversely, R has dimension 0, then 0 is the only maximal ideal of R and R is a field. Dedekind domains and PIDs have dimension 1, since all their nonzero prime ideals are maximal. The main result in this section implies that $K[X_1, \ldots, X_n]$ has dimension n when K is a field. Thus there are noetherian rings of arbitrary finite dimension.

The following properties are exercises: when $\mathfrak{p}$ is a prime ideal of R,

(1) the height of $\mathfrak{p}$ is the dimension of $R_\mathfrak{p}$;
(2) if $\mathfrak{p} \neq 0$, then $\dim R \geqslant 1 + \dim R/\mathfrak{p}$.

11.8.b. Main Result

Theorem 11.8.1. *Let R be a noetherian domain. If R has dimension n, then $R[X]$ has dimension $n + 1$.*

Proof. We begin with a lemma. As usual, we regard the elements of R as constant polynomials, so that $R \subseteq R[X]$.

Lemma 11.8.2. *Let R be a domain and $\mathfrak{P} \neq 0$ be a prime ideal of $R[X]$. If $\mathfrak{P} \cap R = 0$, then $\mathfrak{P}$ is a minimal prime ideal.*

Proof. Let Q be the field of fractions of R. Then $Q = S^{-1}R$, where $S = R \backslash \{0\}$, and we see that $Q[X] \cong S^{-1}R[X]$. Let $\mathfrak{Q} \subseteq \mathfrak{P}$ be a prime ideal of $R[X]$. If $\mathfrak{Q} \cap R = \mathfrak{P} \cap R = 0$, then $\mathfrak{P}^E, \mathfrak{Q}^E \neq Q[X]$ and $\mathfrak{P}^E$ and $\mathfrak{Q}^E$ are prime ideals of $Q[X]$ by Proposition 11.1.7. If $\mathfrak{Q} \neq 0$, then $\mathfrak{Q}^E \neq 0$, $\mathfrak{Q}^E$ is a maximal ideal (since $Q[X]$ is a PID), $\mathfrak{Q}^E = \mathfrak{P}^E$, and $\mathfrak{Q} = \mathfrak{P}$ by Proposition 11.1.9. $\square$

Now let R be a noetherian domain of dimension n. Then $R[X]/(X) \cong R$ and $\dim R[X] \geqslant n + 1$ by (2). We prove by induction on n that $\dim R[X] \leqslant n + 1$. If $n = 0$, then R is a field, $R[X]$ is a PID, and $\dim R[X] = 1$. Let $n > 0$ and

$$\mathfrak{P}_0 \supsetneqq \mathfrak{P}_1 \supsetneqq \cdots \supsetneqq \mathfrak{P}_m$$

be a chain of prime ideals of $R[X]$. We want to show that $m \leqslant n + 1$. Since $n \geqslant 1$ we may assume that $m \geqslant 2$.

Assume that $\mathfrak{P}_{m-1} \cap R = 0$. Then $\mathfrak{P}_{m-2} \cap R \neq 0$ by Lemma 11.8.2 and there exists $0 \neq a \in \mathfrak{P}_{m-2} \cap R$. Now $\mathfrak{P}_{m-2}$ has height at least 2 and is not minimal over (a) by the Hauptidealsatz. However, $\mathfrak{P}_{m-2}$ contains a prime ideal $\mathfrak{P}'_{m-1}$

which is minimal over (a). Then

$$\mathfrak{P}_0 \supsetneqq \mathfrak{P}_1 \supsetneqq \cdots \supsetneqq \mathfrak{P}_{m-2} \supsetneqq \mathfrak{P}'_{m-1} \supsetneqq 0$$

is a chain of prime ideals of $R[X]$ in which $\mathfrak{P}'_{m-1} \cap R \supseteq (a) \neq 0$. Therefore we may assume that $\mathfrak{P}_{m-1} \cap R \neq 0$.

Then $\mathfrak{p} = \mathfrak{P}_{m-1} \cap R$ is a nonzero prime ideal of R. By (2), $\dim R/\mathfrak{p} \leqslant \dim R - 1 = n - 1$; by the induction hypothesis, $\dim(R/\mathfrak{p})[X] \leqslant n$. Now the projection $R \longrightarrow R/\mathfrak{p}$ induces a homomorphism $R[X] \longrightarrow (R/\mathfrak{p})[X]$; its kernel $\mathfrak{P} = \mathfrak{p}[X] \subseteq \mathfrak{P}_{m-1}$ is a nonzero prime ideal of $R[X]$, and $(R[X])/\mathfrak{P} \cong (R/\mathfrak{p})[X]$ has dimension at most n. Then the chain

$$\mathfrak{P}_0/\mathfrak{P} \supsetneqq \mathfrak{P}_1/\mathfrak{P} \supsetneqq \cdots \supsetneqq \mathfrak{P}_{m-1}/\mathfrak{P}$$

of prime ideals of $(R[X])/\mathfrak{P}$ shows that $m - 1 \leqslant n$, and $m \leqslant n + 1$. $\square$

Corollary 11.8.3. *When K is a field, then $K[X_1,\ldots,X_n]$ has dimension n.*

11.9. ALGEBRAIC SETS

This section contains Hilbert's Nullstellensatz and the basic properties of algebraic sets which it implies.

11.9.a. Definition. Let K be a field and $\overline{K}$ be its algebraic closure. A **zero** of set S of polynomials $S \subseteq K[X_1,\ldots,X_n]$ is an element $(x_1,\ldots,x_n)$ of $\overline{K}^n$ such that $f(x_1,\ldots,x_n) = 0$ for all $f \in S$. The **zero set** of S is the set $Z(S)$ of all zeros of S. An **algebraic set** in $\overline{K}^n$ (also known as an **affine** algebraic set) with coefficients in K is the zero set of some set of polynomials $S \subseteq K[X_1,\ldots,X_n]$; then S is a set of **equations** of the algebraic set.

The straight line $x + y - 4 = 0$ and circle $x^2 + y^2 - 10 = 0$ are examples of algebraic sets in $\mathbb{C}^2$ with coefficients in $\mathbb{R}$. In $\mathbb{C}^2$, algebraic sets with a single equation (=zero sets of single polynomials) are known as **algebraic curves**. There are other types of algebraic sets in $\mathbb{C}^2$: the zero set of $(X + Y - 4)(X^2 + Y^2 - 10)$ consists (= is the union) of two algebraic curves; the zero set of $\{X + Y - 4, X^2 + Y^2 - 10\}$ consists of two points, $(1,3)$ and $(3,1)$; finally $\emptyset = Z(1)$ and $\mathbb{C}^2 = Z(0)$ are algebraic sets.

The study and classification of algebraic sets, known as **Algebraic Geometry**, requires large quantities of algebra. In fact the systematic study of fields, rings, and modules was initiated largely because extensive knowledge of these objects is necessary for a proper understanding of algebraic sets. The few simple properties obtained here require most of the results in this chapter.

First we note that the zero set of $S \subseteq K[X_1,\ldots,X_n]$ coincides with the zero set of the ideal of $K[X_1,\ldots,X_n]$ generated by S. Therefore every algebraic set is the zero set of some ideal. Since $K[X_1,\ldots,X_n]$ is noetherian, it follows that every algebraic set is the zero set of a finite set of polynomials (= can be defined by finitely many equations).

Proposition 11.9.1. *Every intersection of algebraic sets is an algebraic set. The union of finitely many algebraic sets is an algebraic set.*

Proof. If $A_i = Z(\mathfrak{a}_i)$ is the zero set of an ideal $\mathfrak{a}_i$, then $\bigcap_{i \in I} A_i = Z(\sum_{i \in I} \mathfrak{a}_i)$. Let A and B be the zero sets of $\mathfrak{a}$ and $\mathfrak{b}$. Every $x \in A \cup B$ is a zero of $\mathfrak{a} \cap \mathfrak{b}$. Conversely, let $x \in Z(\mathfrak{a} \cap \mathfrak{b})$. Assume that $x \notin \mathfrak{a}$. Then $f(x) \neq 0$ for some $f \in \mathfrak{a}$. Hence $g \in \mathfrak{b}$ implies $fg \in \mathfrak{a} \cap \mathfrak{b}$, $f(x)g(x) = 0$, and $g(x) = 0$, so that $x \in B$. $\square$

11.9.b. The Nullstellensatz. Nullstellensatz means "theorem of zero points." There are several versions. The original version, due to Hilbert, reads:

Theorem 11.9.2 (Hilbert's Nullstellensatz). *Let K be a field, $\mathfrak{a}$ be an ideal of $K[X_1,\ldots,X_n]$, and $f \in K[X_1,\ldots,X_n]$. If every zero of $\mathfrak{a}$ is a zero of f, then $\mathfrak{a}$ contains some power of f.*

Proof. This proof, due to Lang, is based on the homomorphism properties in Section 11.4. Assume that $\mathfrak{a}$ contains no power of f. Then Proposition 5.6.4 provides a prime ideal $\mathfrak{p}$ which contains $\mathfrak{a}$ but not f. Let $R = K[X_1,\ldots,X_n]/\mathfrak{p}$ and $\pi : K[X_1,\ldots,X_n] \longrightarrow K[X_1,\ldots,X_n]/\mathfrak{p}$ be the projection. Then $\pi K \cong K$, R is a domain, and R is finitely generated over πK, by $\alpha_1 = \pi X_1,\ldots,\alpha_n = \pi X_n$. Moreover $y = f^\pi(\alpha_1,\ldots,\alpha_n) = \pi(f) \neq 0$, since $f \notin \mathfrak{p}$ (f^π is the image of f under the homomorphism $K[X_1,\ldots,X_n] \longrightarrow R[X_1,\ldots,X_n]$ induced by $\pi_{|K} : K \longrightarrow R$).

By Proposition 11.4.5, the homomorphism $\pi K \cong K \subseteq \overline{K}$ extends to the subring $(\pi K)[\alpha_1,\ldots,\alpha_n, 1/y]$ of the fraction field of R. This provides a homomorphism $\psi : (\pi K)[\alpha_1,\ldots,\alpha_n, 1/y] \longrightarrow \overline{K}$ such that $\psi\pi$ is the identity on K (since ψ extends $\pi K \cong K \subseteq \overline{K}$) and $\psi y \neq 0$. Now $f(\psi\alpha_1,\ldots,\psi\alpha_n) = f^{\psi\pi}(\psi\alpha_1,\ldots,\psi\alpha_n) = \psi y \neq 0$, whereas

$$g(\psi\alpha_1,\ldots,\psi\alpha_n) = g^{\psi\pi}(\psi\pi X_1,\ldots,\psi\pi X_n) = \psi(\pi(g)) = 0$$

for all $g \in \mathfrak{a} \subseteq \mathfrak{p}$. Thus $(\psi\alpha_1,\ldots,\psi\alpha_n)$ is a zero of $\mathfrak{a}$ but not of f. $\square$

11.9.c. Consequences of the Nullstellensatz

Corollary 11.9.3. *Every proper ideal of $K[X_1,\ldots,X_n]$ has at least one zero in $\overline{K}^n$.*

Corollary 11.9.4. $\mathfrak{m}$ *is a maximal ideal of $K[X_1,\ldots,X_n]$ if and only if $\mathfrak{m} = \{ f \in K[X_1,\ldots,X_n] ; f(\alpha) = 0 \}$ for some $\alpha \in \overline{K}^n$.*

The proofs of these corollaries make fun exercises.

The next consequence of the Nullstellensatz is that every algebraic set is the zero set of a unique semiprime ideal. More precisely, to each algebraic set $A \subseteq \overline{K}^n$, there corresponds the ideal

$$I(A) = \{ f \in K[X_1,\ldots,X_n] \; ; \; A \subseteq Z(f) \}$$

of all polynomials that are zero on A; then:

Corollary 11.9.5. *I and Z induce an order-reversing one-to-one correspondence between algebraic sets in $\overline{K}^n$ and semiprime ideals of $K[X_1,\ldots,X_n]$.*

Proof. The radical $\mathfrak{r} = \operatorname{Rad}\mathfrak{a}$ of an ideal $\mathfrak{a}$ is a semiprime ideal, and $Z(\mathfrak{r}) = Z(\mathfrak{a})$, since $f^r(x) = 0$ in $\overline{K}$ is equivalent to $f(x) = 0$. Similarly $I(A)$ is a semiprime ideal. The definitions imply that $A \subseteq Z(I(A))$ for every $A \subseteq \overline{K}^n$ and $\mathfrak{a} \subseteq I(Z(\mathfrak{a}))$ for every ideal $\mathfrak{a} \subseteq K[X_1,\ldots,X_n]$. If $A = Z(\mathfrak{a})$ is algebraic, then $\mathfrak{a} \subseteq I(A)$ and $Z(I(A)) \subseteq Z(\mathfrak{a}) = A$; therefore $Z(I(A)) = A$. If $\mathfrak{a}$ is an ideal, then $I(Z(\mathfrak{a})) \subseteq \operatorname{Rad}\mathfrak{a}$ by the Nullstellensatz; if $\mathfrak{a}$ is semiprime, this yields $I(Z(\mathfrak{a})) = \mathfrak{a}$. $\qquad\square$

In the noetherian ring $K[X_1,\ldots,X_n]$ every ideal $\mathfrak{a}$ is a reduced intersection $\mathfrak{a} = \mathfrak{q}_1 \cap \cdots \cap \mathfrak{q}_r$ of primary ideals with unique radicals $\mathfrak{p}_1,\ldots,\mathfrak{p}_r$. If $\mathfrak{a}$ is semiprime, then applying Rad to $\mathfrak{a} = \mathfrak{q}_1 \cap \cdots \cap \mathfrak{q}_r$ yields $\mathfrak{a} = \mathfrak{p}_1 \cap \cdots \cap \mathfrak{p}_r$; hence every semiprime ideal of $K[X_1,\ldots,X_n]$ is uniquely an irredundant intersection of distinct prime ideals.

An **irreducible** algebraic set (also called an **affine algebraic variety**) is the zero set of a prime ideal; equivalently, an algebraic set is irreducible if and only if it is not the union of two smaller algebraic sets (see the exercises). By the above:

Corollary 11.9.6. *Every algebraic set is uniquely an irredundant finite union of irreducible algebraic sets.* $\qquad\square$

Corollary 11.9.6 reduces the study of algebraic sets to that of irreducible algebraic sets; equivalently, prime ideals of $K[X_1,\ldots,X_n]$.

By Corollary 11.8.3, $K[X_1,\ldots,X_n]$ has dimension n: the longest chain of prime ideals of $K[X_1,\ldots,X_n]$ has length n. Therefore the longest chain of irreducible algebraic sets in $\overline{K}^n$ has length n. For example, a prime ideal of $\mathbb{R}[X,Y]$ is either maximal or minimal or 0; hence an irreducible algebraic set in $\mathbb{C}^2$ is either a single point, or an algebraic curve with an irreducible equation, or $\mathbb{C}^2$ itself.

The **dimension** of an irreducible algebraic set A is the length of the longest chain of algebraic sets contained in A. If A is the zero set of a prime ideal $\mathfrak{p}$, then the dimension of A is the length of the longest chain $\mathfrak{p}_0 \supsetneq \mathfrak{p}_1 \supsetneq \cdots \supsetneq \mathfrak{p}_r = \mathfrak{p}$ of prime ideals of $K[X_1,\ldots,X_n]$. For instance, points have dimension 0;

irreducible algebraic curves in $\mathbb{C}^2$ have dimension 1; irreducible algebraic surfaces in $\mathbb{C}^3$ have dimension 2.

11.9.d. The Zariski Topology. Since $\emptyset = Z(1)$ and $\overline{K}^n = Z(\emptyset)$ are algebraic, it follows from Proposition 11.9.1 that the algebraic sets in $\overline{K}^n$ with coefficients in K are the closed sets of a topology; this is the **Zariski topology** on $\overline{K}^n$ (relative to K). This provides a different view of algebraic sets.

A topological space is **noetherian** in case its open sets satisfy the ascending chain condition. With the Zariski topology, $\overline{K}^n$ is noetherian: a descending sequence of closed subsets of $\overline{K}^n$ corresponds to an ascending sequence of semiprime ideals of $K[X_1,\dots,X_n]$ and terminates, since $K[X_1,\dots,X_n]$ is noetherian.

Corollary 11.9.6 is a purely topological result. A closed set C is irreducible in case it is not empty and not the union $C = A \cup B$ of two proper closed subsets $A, B \subsetneqq C$. In a noetherian topological space, every closed set is uniquely an irredundant finite union of distinct irreducible closed sets (see the exercises). The dimension of an irreducible algebraic set A is also purely topological: it is the length of the longest chain of irreducible closed subsets of A.

Exercises

1. Show that every proper ideal of $K[X_1,\dots,X_n]$ has at least one zero in $\overline{K}^n$.

2. Let K be a field. Show that $\mathfrak{m}$ is a maximal ideal of $K[X_1,\dots,X_n]$ if and only if there exist $\alpha_1,\dots,\alpha_n \in \overline{K}$ such that

$$\mathfrak{m} = \{ f \in K[X_1,\dots,X_n]; \ f(\alpha_1,\dots,\alpha_n) = 0 \}.$$

3. Let K be a field. Show that the minimal prime ideals of $K[X_1,\dots,X_n]$ are the principal ideals generated by irreducible polynomials.

4. Show that an algebraic set A is irreducible if and only if it is not the union $A = B \cup C$ of two smaller algebraic sets $B, C \subsetneqq A$.

5. Show that a closed set in a noetherian topological space is uniquely an irredundant finite union of distinct irreducible closed sets.

11.10. RATIONAL FUNCTIONS

Algebraic geometry studies algebraic sets only up to isomorphism. In this section we define isomorphisms of algebraic sets. We also construct for each irreducible algebraic set A a ring $C(A)$ which determines A up to isomorphism.

In what follows K is an algebraically closed field.

11.10.a. The Coordinate Ring. We begin with the ring $C(A)$ and define isomorphisms later. Let $A \subseteq K^n$ be an algebraic set. Every polynomial $f \in$

$K[X_1,\ldots,X_n]$ induces a mapping $f_{|A} : A \longrightarrow K$; the value of $f_{|A}$ at $x = (x_1,\ldots,x_n) \in A$, also denoted by $f(x)$, is obtained by evaluating $f(X_1,\ldots,X_n)$ at $(x_1,\ldots,x_n)$. For instance, every $a \in K$ induces a constant mapping $x \longmapsto a$ on A; each indeterminate X_i induces a "coordinate mapping" which sends $x = (x_1,\ldots,x_n) \in A$ to its i-th coordinate x_i.

Under pointwise addition and multiplication, the mappings $A \longrightarrow K$ induced by polynomials form a ring $C(A)$, the **coordinate ring** of A (so called because it is generated by the coordinate mappings). We identify $a \in K$ and the constant mapping $x \longmapsto a$ on A so that K is a subring of $C(A)$. Multiplication in $C(A)$ then makes $C(A)$ a vector space over K; then the multiplication on $C(A)$ is bilinear. Anticipating on Chapter 16, we say that $C(A)$ is a K-algebra.

Two polynomials $f, g \in K[X_1,\ldots,X_n]$ induce the same polynomial mapping $A \longrightarrow K$ if and only if $f - g \in I(A) = \{ h \in K[X_1,\ldots,X_n]; h(x) = 0$ for all $x \in A \}$. Hence there is an isomorphism $C(A) \cong K[X_1,\ldots,X_n]/I(A)$, which sends $f_{|A} \in C(A)$ to $f + I(A) \in K[X_1,\ldots,X_n]/I(A)$.

Proposition 11.10.1. $C(A)$ *is a commutative, finitely generated ring extension of K (= a finitely generated K-algebra) and has trivial nilradical; if A is irreducible, then $C(A)$ is a domain. Conversely, every commutative, finitely generated ring extension R of K with trivial nilradical is isomorphic to the coordinate ring of some algebraic set; if R is a domain, then R is isomorphic to the coordinate ring of some irreducible algebraic set.*

Proof. Since $K[X_1,\ldots,X_n]$ is generated, as a ring extension of K, by $X_1,\ldots,$ X_n, $C(A)$ is generated (as a ring extension of K) by the coordinate mappings induced by $X_1,\ldots,X_n$. Moreover $C(A) \cong K[X_1,\ldots,X_n]/I(A)$ has trivial nilradical, since $I(A)$ is a semiprime ideal; if A is irreducible, then $I(A)$ is a prime ideal and $C(A)$ is a domain.

Conversely, let R be a finitely generated ring extension of K with trivial nilradical (or which is a domain). Then R is the quotient of a polynomial algebra $K[X_1,\ldots,X_n]$ (for some n) by some semiprime (or prime) ideal $\mathfrak{a}$ and is then isomorphic to the coordinate ring of the corresponding algebraic set $Z(\mathfrak{a})$. $\square$

The points (= the elements) of an irreducible algebraic set A can be recovered from $C(A)$. For each $x \in A$ let

$$\mathfrak{m}_x = \{ f \in C(A); f(x) = 0 \}.$$

Proposition 11.10.2. *When A is an irreducible algebraic set, the mapping $x \longmapsto \mathfrak{m}_x$ is a one-to-one correspondence between the points of A and the maximal ideals of $C(A)$.*

Proof. By Corollary 11.9.4, there is a one-to-one correspondence $x \longmapsto I(x)$ between the elements of K^n and the maximal ideals of $K[X_1,\ldots,X_n]$. By Corollary 11.9.5, $x \in A = Z(I(A))$ if and only if $I(A) \subseteq I(x)$. Hence there is a one-to-one correspondence between the elements of A and the maximal ideals of $K[X_1,\ldots,X_n]$ containing $I(A)$, which in turn correspond to the maximal ideals of $K[X_1,\ldots,X_n]/I(A)$; to $x \in A$ corresponds $I(x)/I(A)$, which the isomorphism $K[X_1,\ldots,X_n]/I(A) \cong C(A)$ takes to $\mathfrak{m}_x$. ☐

The dimension of an irreducible algebraic set can also be recovered from its coordinate ring. When $A = Z(\mathfrak{p})$ is irreducible (with $\mathfrak{p}$ prime), the dimension of A is the length of the longest chain of prime ideals of $K[X_1,\ldots,X_n]$ that contain $\mathfrak{p}$; hence the dimension of A is the dimension of $C(A) \cong K[X_1,\ldots,X_n]/\mathfrak{p}$. The field of fractions of the domain $C(A)$ is the field of **rational mappings** of A or the **function field** of A; it is a finitely generated field extension of K and is purely transcendental, since K is algebraically closed. It can be shown that the dimension of A also equals the transcendence degree over K of its function field.

The main result of this section is that an irreducible algebraic set is determined up to isomorphism by its coordinate ring.

11.10.b. Regular Mappings. First we give an equivalent definition of $C(A)$. Let A be an irreducible algebraic set. We equip A with the Zariski topology (induced by the Zariski topology on K^n). Since A is irreducible, every nonempty open subset U of A is dense in A: if we had $U \cap V = \varnothing$ for some open subset $V \neq \varnothing$ of A, then $A = (A \backslash U) \cup (A \backslash V)$ would not be irreducible.

We saw that every polynomial $f \in K[X_1,\ldots,X_n]$ induces a polynomial mapping $A \longrightarrow K$. If now $f = g/h \in K(X_1,\ldots,X_n)$ is a rational fraction, then $A \backslash Z(h)$ is an open subset of A, and f induces a mapping of $A \backslash Z(h)$ into K. Let U be a nonempty open subset of A (perhaps A itself). A mapping $f : U \longrightarrow K$ is **regular** in case it is locally induced by rational fractions: that is, for each $x \in U$ the restriction of f to some open neighborhood V of x in U is induced by a rational fraction g/h (there exist $g, h \in K[X_1,\ldots,X_n]$ such that $h(v) \neq 0$ and $f(v) = g(v)/h(v)$ for all $v \in V$).

The reader will verify that under pointwise operations the regular mappings of U into K constitute a domain $R(U)$.

A polynomial mapping is regular on all of A. We now prove the converse.

Lemma 11.10.3. *Regular mappings are continuous.*

Proof. Let $f : A \longrightarrow K$ be regular. In the Zariski topology, closed sets on K (other that K itself) are finite. Hence it suffices to show that $f^{-1}(a)$ is closed in A for every $a \in K$. Let $x \in A \backslash f^{-1}(a)$. There is an open neighborhood U of x in A on which f is induced by a rational fraction $g/h \in K(X_1,\ldots,X_n)$ (in particular, $U \subseteq A \backslash Z(h)$). Now

$$f^{-1}(a) \cap U = \{\, x \in U \,;\, g(x)/h(x) = a \,\} = U \cap Z(g - ah).$$

Then $V = U \backslash Z(g - ah)$ is open, $x \in V$, and $V \subseteq A \backslash f^{-1}(a)$. ☐

Regular mappings are, by definition, assembled from mappings of open subsets induced by rational fractions. We now show that a regular mapping on A can be assembled from smaller pieces, one for each point of A. Thus regular mappings can be used to study an algebraic set at any one of its points.

Let A be an irreducible algebraic set and $x \in A$. Consider ordered pairs (f, U) where U is an open neighborhood of x in A and $f : U \longrightarrow K$ is regular. Say that (f, U) and (g, V) are **equivalent at** x in case f and g agree on $U \cap V$. A **germ of regular mappings** at x is an equivalence class $\mathrm{cls}(f, U)$ of pairs (f, U) (with U an open neighborhood of x in A and $f : U \longrightarrow K$ regular on U). Our stalwart reader will verify that the above is indeed an equivalence relation and that the germs of regular mappings at x form a ring $R_x(A)$ under pointwise operations. There is an injective homomorphism $f \longmapsto \mathrm{cls}(f, A)$ of $R(A)$ into $R_x(A)$.

$R_x(A)$ is a local ring. In fact:

Lemma 11.10.4. $R_x(A) \cong C(A)_{\mathfrak{m}_x}$.

Proof. The elements of $C(A)_{\mathfrak{m}_x}$ are fractions $f = g/h$ where g and h are polynomial mappings on A and $h \notin \mathfrak{m}_x$ $(h(x) \neq 0)$. Then $U = A \backslash Z(h)$ is an open neighborhood of x and f is regular on U. Moreover equal fractions yield pairs that are equivalent at x. Hence there is a homomorphism $\varphi : f \longmapsto \mathrm{cls}(f_{|U}, U)$ of $C(A)_{\mathfrak{m}_x}$ into $R_x(A)$. The definition of germs shows that φ is surjective. If $\varphi(f) = 0$, then (f, U) is equivalent to $(0, V)$ for some neighborhood V of x, f is 0 on some open neighborhood of x, and $f = 0$, since f is continuous by Lemma 11.10.4, and nonempty open subsets of A are dense in S; thus φ is injective. $\qquad \square$

Proposition 11.10.5. $R(A) = C(A)$.

Proof. $C(A) \subseteq R(A)$, since polynomial mappings are regular. Lemma 11.10.4 also yields for each $x \in A$ an injective homomorphism $\varphi : R(A) \longrightarrow R_x(A) \longrightarrow C(A)_{\mathfrak{m}_x}$ which can be described as follows. A regular mapping f is induced in some neighborhood V of x by a rational fraction g/h, with h nonzero on V. The isomorphism $R_x(A) \cong C(A)_{\mathfrak{m}_x}$ takes $\mathrm{cls}(f, U)$ to $g_{|A}/h_{|A}$, where $U = A \backslash Z(h) \supseteq V$. Hence φ takes f to $\mathrm{cls}(f, A) = \mathrm{cls}(f, U)$ and thence to $g_{|A}/h_{|A}$.

We may identify $R(A)$, $C(A)$, and their localizations with subrings of the fraction field of $R(A)$. By the above there are inclusions $C(A) \subseteq R(A) \subseteq C(A)_{\mathfrak{m}_x}$ for each $x \in A$, so that

$$C(A) \subseteq R(A) \subseteq \bigcap_{x \in A} C(A)_{\mathfrak{m}_x} = \bigcap_{\mathfrak{m}} C(A)_{\mathfrak{m}},$$

where the last intersection runs over all maximal ideals of $C(A)$ by Proposition 11.10.2. Then the following lemma yields $\bigcap_{\mathfrak{m}} C(A)_{\mathfrak{m}} = C(A)$ and $C(A) = R(A)$: $\qquad \square$

Lemma 11.10.6. *Let R be a domain and Q be its fraction field. If $x \in Q$ belongs to $R_\mathfrak{m}$ for every maximal ideal $\mathfrak{m}$ of R, then $x \in R$.*

Proof. Since $x \in R_\mathfrak{m}$, there is for each maximal ideal $\mathfrak{m}$ of R a fraction $a/b = x$ with $a, b \in R$, $b \notin \mathfrak{m}$. Hence $D = \{d \in R; x = c/d \text{ for some } c \in R\}$ is not contained in a maximal ideal of R, and neither is the ideal $\mathfrak{d}$ of R generated by D. Therefore $\mathfrak{d} = R$ and there exist $d_1, \dots, d_k \in D$, $r_1, \dots, r_k \in R$ such that $\sum_i r_i d_i = 1$. Then $x = \sum_i r_i d_i x \in R$, since $xd \in R$ for every $d \in D$. □

11.10.c. Morphisms. The definition of regular mappings can be generalized as follows. A mapping $f = (f_1, \dots, f_m)$ of A into K^m is **regular** in case each of its components f_j is a regular mapping of A into K; equivalently (by Proposition 11.10.5) in case $f_1, \dots, f_m$ are induced by polynomials. Still more generally, a mapping $f = (f_1, \dots, f_m)$ of A into another irreducible algebraic set $B \subseteq K^m$ is **regular** in case each of its components f_j is a regular mapping of A into K. Regular mappings of A into B are also called **morphisms** of A into B.

The identity mapping on A is regular. If $f : A \longrightarrow B$ and $g : B \longrightarrow C$ are regular, then $g \circ f : A \longrightarrow C$ is regular: indeed $f_1(x_1, \dots, x_n), \dots, f_m(x_1, \dots, x_n)$ are induced by polynomials, every $g_k(y_1, \dots, y_m)$ is induced by a polynomial, hence every component $g_k(f_1(x_1, \dots, x_n), \dots, f_m(x_1, \dots, x_n))$ of $g \circ f$ is induced by a polynomial.

Proposition 11.10.7. *When A and B are irreducible algebraic sets, there is a one-to-one correspondence between regular mappings of A into B and K-algebra homomorphisms of $C(B)$ into $C(A)$.*

This implies an equivalence of categories as defined in Chapter 17.

Proof. Let $f : A \longrightarrow B$ be regular. If $g : B \longrightarrow K$ is regular, then $g \circ f : A \longrightarrow K$ is regular. This provides a mapping $\overline{f} : g \longmapsto g \circ f$ of $C(B)$ into $C(A)$, which respects constants and pointwise operations and so is a K-algebra homomorphism (a ring homomorphism which preserves the action of K).

Conversely, let $\varphi : C(B) \longrightarrow C(A)$ be a K-algebra homomorphism. Let q_j be the j-th coordinate mapping $(y_1, \dots, y_m) \longmapsto y_j$ on B. Then $q_j \in C(B)$, $\varphi(q_j) \in C(A)$ is regular, and $\overline{\varphi} = (\varphi(q_1), \dots, \varphi(q_m))$ is a regular mapping of A into K^m. In fact $\overline{\varphi}$ is a regular mapping of A into B. Indeed let $x \in A$ and $f = \sum_k c_k Y_1^{k_1} \dots Y_m^{k_m} \in I(B)$. Then $f(q_1, \dots, q_m) = 0$ in $C(B)$ and

$$f(\overline{\varphi}(x)) = f(\varphi(q_1)(x), \dots, \varphi(q_m)(x)) = \sum_k c_k \, \varphi(q_1)^{k_1}(x) \dots \varphi(q_m)^{k_1}(x)$$

$$= \varphi \left(\sum_k c_k q_1^{k_1} \dots q_m^{k_m} \right)(x) = \varphi(f(q_1, \dots, q_m))(x) = 0,$$

since φ is a homomorphism. This shows $\overline{\varphi}(x) \in Z(I(B)) = B$.

Finally $\overline{\varphi} = \overline{\psi}$ implies $\varphi = \psi$, since $q_1,\ldots,q_m$ generate $C(B)$; hence $\varphi \longmapsto \overline{\varphi}$ is injective. If, conversely, $\varphi = \overline{f}$, then $\overline{\varphi} = (f \circ q_1,\ldots,f \circ q_m) = f$; hence $\varphi \longmapsto \overline{\varphi}$ is bijective, and $f \longmapsto \overline{f}$ is the inverse bijection. $\qquad\square$

The definition of $\overline{f}$ shows that $\overline{1_A} = 1_{C(A)}$ and $\overline{g \circ f} = \overline{g} \circ \overline{f}$.

11.10.d. Isomorphisms. Two irreducible algebraic sets A and B are **isomorphic** in case there exist mutually inverse regular bijections $A \longrightarrow B$ and $B \longrightarrow A$.

For example, in K^2 the straight line $y = 0$ is isomorphic to the parabola $y = x^2$, since the mutually inverse bijections $(x,y) \longmapsto (x, y + x^2)$ and $(x,y) \longmapsto (x, y - x^2)$ are regular. This shows that isomorphism has little respect for shape or degree. But the following consequence of Proposition 11.10.7 shows that isomorphisms respect other essential properties:

Theorem 11.10.8. *Two irreducible algebraic sets are isomorphic if and only if their coordinate rings are isomorphic.*

Theorem 11.10.8 redefines Algebraic Geometry as the study of finitely generated commutative K-algebras without zero divisors.

Exercises

1. Let U be an open subset of an irreducible algebraic set. Verify that the regular mappings on U form a domain under pointwise operations.
2. For an irreducible algebraic set, verify that the germs of regular mappings at x form a ring under pointwise operations.
3. Show that every straight line in K^2 is isomorphic to K.
4. Show that every conic (irreducible curve of degree 2) in K^2 is isomorphic to either K or $K \backslash 0$.
5. Show that K and $K \backslash 0$ are not isomorphic.

In the following exercises, a **quasi-affine algebraic variety** is a nonempty open subset of an irreducible algebraic set. A **regular** mapping of a quasi-affine variety $U \subseteq K^n$ into a quasi-affine variety $V \subseteq K^m$ is a mapping $f = (f_1,\ldots,f_m)$ of U into V such that every component f_j of f is a regular mapping of U into K. Two quasi-affine varieties U and V are **isomorphic** in case there exist mutually inverse regular bijections $U \longrightarrow V$ and $V \longrightarrow U$.

6. Let U be a quasi-affine algebraic variety and B be an irreducible algebraic set. Prove that there is a one-to-one correspondence between regular mappings of U into B and K-algebra homomorphisms of $C(B)$ into $R(U)$.
7. Let U and V be quasi-affine varieties. Prove that $U \cong V$ implies $R(U) \cong R(V)$; prove the converse in case U or V is an affine variety.
8. Find $R(U)$ when U is any nonempty open subset of K.

12

SEMISIMPLICITY

The main result of this chapter is the Artin-Wedderburn Theorem; it is a fine example of how the structure of a ring affects the structure of its modules and vice versa.

Required results: Chapter 4, Sections 4.1 and 4.2; Chapter 10, Sections 10.1, 10.2, 10.3, 10.4, and 10.6.

All rings in this chapter have an identity element, and all modules are unital.

12.1. SEMISIMPLE MODULES

Semisimple modules are a class of left R-modules that are easily constructed from the ring R.

12.1.a. Definition. Recall that a module M is **simple** in case $M \neq 0$ and M has no submodule $N \neq 0, M$. Simple left R-modules are readily constructed from R and its maximal left ideals:

Proposition 12.1.1. *A left R-module is simple if and only if it is isomorphic to R/L for some maximal left ideal L of R.*

Proof. A simple left R-module S is necessarily cyclic: if $a \in S$, $a \neq 0$, then $Ra \neq S$ is impossible. If $S = Ra$ is cyclic, then $r \longmapsto ra$ is a homomorphism of $_RR$ onto S; its kernel L is a left ideal of R (the annihilator of a), so that $S \cong R/L$. The submodules of R/L are of the form L'/L, where $L' \supseteq L$ is a left ideal of R; hence R/L is simple if and only if L is a maximal left ideal of R. $\qquad\square$

Proposition 12.1.2. *For a module M the following conditions are equivalent:*

(1) *M is a sum of simple submodules.*

(2) *M is a direct sum of simple submodules.*

(3) *Every submodule of M is a direct summand of M.*

Proof. Clearly (2) implies (1). We show that (1) implies (2) and (3). Let M be a sum $M = \sum_{i \in I} S_i$ of simple submodules $(S_i)_{i \in I}$ of M and N be a submodule of M. Let $\mathscr{S}$ be the set of all subsets J of I such that $N + \sum_{i \in J} S_i = N \oplus \bigoplus_{i \in J} S_i$; equivalently, such that $N \cap \sum_{i \in J} S_i = 0$ and $S_j \cap (N + \sum_{i \in J, i \neq j} S_i) = 0$ for all $j \in J$. The union of a chain of elements of $\mathscr{S}$ is an element of $\mathscr{S}$, since $x \in \sum_{i \in J} S_i$ implies $x \in \sum_{i \in K} S_i$ for some finite subset K of J. Hence $\mathscr{S}$, ordered by inclusion, is inductive. By Zorn's Lemma, $\mathscr{S}$ has a maximal element.

Let $J \in \mathscr{S}$ be maximal and $N_J = N \oplus \bigoplus_{j \in J} S_j$. For each $i \in I$ we have $S_i \cap N_J = 0$ or $S_i \cap N_J = S_i$, since S_i is simple. If $S_i \cap N_J = 0$, then $N_J + S_i = N_J \oplus S_i$ and $J \cup \{i\} \in \mathscr{S}$, contradicting the maximality of J. Therefore $S_i \subseteq N_J$. Since this holds for all $i \in I$, we have $N_J = N \oplus \bigoplus_{j \in J} S_j = M$. This proves (3) and, with $N = 0$, (2).

Now assume (3). We show that every nonzero cyclic submodule Ra of M contains a simple submodule. There is a module homomorphism $\varphi : r \longmapsto ra$ of $_R R$ onto Ra, whose kernel $K \neq R$ is a left ideal of R. By Zorn's Lemma, K is contained in a maximal left ideal $L \neq R$ of R. Then $La = \varphi(L)$ is a maximal submodule of Ra and Ra/La is simple. By (3), $M = La \oplus N$ for some submodule N. Then $Ra = La \oplus (Ra \cap N)$: if $x = \ell a + n \in Ra$, with $\ell \in L$ and $n \in N$, then $n = x - \ell a \in Ra \cap N$, which shows that $Ra = La + (Ra \cap N)$. Hence $Ra \cap N \cong Ra/La$ is simple and contained in Ra.

It follows that (3) implies (1). Let N be the sum of all simple submodules of M. By (3), $M = N \oplus P$ for some submodule P. If $P \neq 0$, then P contains a simple submodule S, contradicting $P \cap N = 0$. Therefore $P = 0$ and $M = N$. $\quad\square$

In Proposition 12.1.2, there is a uniqueness statement if $M = S_1 \oplus \cdots \oplus S_n$ is a direct sum of finitely many simple modules. Then

$$0 \subsetneq S_1 \subsetneq S_1 \oplus S_2 \subsetneq \cdots \subsetneq S_1 \oplus \cdots \oplus S_n = M$$

is a composition series of M. By the Jordan-Hölder Theorem, the number n and (up to isomorphism) the modules $S_1, \ldots, S_n$ are unique (this also follows from the Krull-Schmidt Theorem); also M has finite length (every chain of submodules has length at most n).

A module is **semisimple** in case it satisfies the equivalent conditions in Proposition 12.1.2. Semisimple modules are also called **completely reducible**. They are readily constructed from R and its maximal left ideals. For example, a vector space of dimension 1 is simple; hence every vector space is semisimple. On the other hand, most cyclic abelian groups are not semisimple $\mathbb{Z}$-modules.

12.1.b. Properties. The following general properties make great exercises.

Proposition 12.1.3. (1) *A direct sum of semisimple modules is semisimple.*
(2) *Every submodule of a semisimple module is semisimple.*
(3) *Every homomorphic image of a semisimple module is semisimple.*

12.1.c. Endomorphisms. Our next properties (for use in the next section) concern the ring of endomorphisms $\text{End}_R(M)$ of a left R-module M (with endomorphisms written on the left) and the similar ring $E^{\text{op}} = \text{End}_R^{\text{op}}(M)$ obtained by writing endomorphisms on the right. The operations on $\text{End}_R(M)$ are defined by $(\eta + \zeta)x = \eta x + \zeta x$, $(\eta\zeta)x = \eta(\zeta x)$ for all $x \in M$; the operations on $\text{End}_R^{\text{op}}(M)$, by $x(\eta + \zeta) = x\eta + x\zeta$, $x(\eta\zeta) = (x\eta)\zeta$ for all $x \in M$.

Lemma 12.1.4. *There is an isomorphism $R \cong \text{End}_R^{\text{op}}(_R R)$ which assigns to each $c \in R$ the endomorphism $x \longmapsto xc$ of $_R R$.*

Proof. Let $\eta \in \text{End}_R^{\text{op}}(_R R)$ and $1\eta = c$; then $x\eta = (x1)\eta = x(1\eta) = xc$ for all $x \in R$. If $\zeta \in \text{End}_R^{\text{op}}(_R R)$ and $1\zeta = d$, then $x\zeta = xd$ and $1\eta\zeta = c\zeta = cd$. Thus evaluation at $1 \in R$ is a ring homomorphism of $\text{End}_R^{\text{op}}(_R R)$ into R, which is injective, since we saw that η is determined by 1η, and surjective, since $x \longmapsto xc$ is an endomorphism of $_R R$ for each $c \in R$. $\qquad\square$

Lemma 12.1.5 (Schur). *Let S and T be simple left R-modules. A module homomorphism of S into T is either 0 or an isomorphism. In particular, $\text{End}_R(S)$ is a division ring.*

Proof. If $\varphi : S \longrightarrow T$ is not zero, then $\text{Ker}\,\varphi \neq S$ and $\text{Im}\,\varphi \neq 0$; hence $\text{Ker}\,\varphi = 0$, $\text{Im}\,\varphi = T$, and φ is an isomorphism. $\qquad\square$

Let M be a left R-module M and $E = \text{End}_R^{\text{op}}(M)$. Then E acts on M by evaluation, so that M is a right E-module. In fact M is a left R-, right E-bimodule (since $(rx)\eta = r(x\eta)$ for all $r \in R$, $x \in M$, $\eta \in E$). We write the E-endomorphisms of M on the left; by definition, $\xi \in \text{End}_E(M)$ if and only if ξ is an additive endomorphism of M ($\xi(x + y) = \xi x + \xi y$ for all $x, y \in M$) and $\xi(x\eta) = (\xi x)\eta$ for all $x \in M$, $\eta \in E$ (equivalently, ξ commute with every $\eta \in E$).

For each $r \in R$, $x \longmapsto rx$ is an E-endomorphism of M (since $r(x\eta) = (rx)\eta$ for all $\eta \in E$). This defines a canonical ring homomorphism $R \longrightarrow \text{End}_E(M)$. The reader will show:

Lemma 12.1.6. *Let $E = \text{End}_R^{\text{op}}(M)$. If $M = _R R$, then $R \longrightarrow \text{End}_E(M)$ is an isomorphism.*

12.1.d. The Density Theorem. Let M be semisimple and $E = \text{End}_R(M)$. The Jacobson Density Theorem states that the image of $R \longrightarrow \text{End}_E(M)$ is dense in the following sense:

Theorem 12.1.7 (Jacobson Density Theorem). *Let M be a semisimple left R-module and $E = \text{End}_R^{\text{op}}(M)$. For every $\xi \in \text{End}_E(M)$ and $x_1, \ldots, x_n \in M$, there exists $r \in R$ such that $\xi x_i = r x_i$ for all i.*

If, for instance, K is a field and V is a vector space over K with a finite basis $e_1, \ldots, e_n$, then V is semisimple and Theorem 12.1.7 implies that the center of $M_n(K)$ consists of scalar matrices: if T is in the center of $E = \text{End}_K^{\text{op}}(V) \cong M_n(K)^{\text{op}}$ and we write T on the left, then $T(x\eta) = (Tx)\eta$ for all $\eta \in E$, $T \in \text{End}_E(V)$, and there exists $\lambda \in K$ such that $T(e_i) = \lambda x_i$ for all i. Then $T(x) = \lambda x$ for all $x \in V$. This also follows from Proposition 3.6.5.

Proof. Let $\xi \in \text{End}_E(M)$ and $x \in M$. Since M is semisimple, we have $M = Rx \oplus N$ for some submodule N. The projection $\pi : M \longrightarrow Rx$ can be viewed as an R-endomorphism of M. Then $\pi \in E$ and $\xi x = \xi(x\pi) = (\xi x)\pi \in Rx$. This proves the theorem in case $n = 1$.

In general, let $M^n = \bigoplus_{i=1,\ldots,n} M$ and $F = \text{End}_R(M^n)$. Then M^n is a semisimple left R-module and ξ induces $\xi^n = \bigoplus_{i=1,\ldots,n} \xi : M^n \longrightarrow M^n$, $\xi^n(x_1, \ldots, x_n) = (\xi x_1, \ldots, \xi x_n)$. By Lemma 12.1.8 below, $\xi^n \in \text{End}_F(M^n)$. By the case $n = 1$, there exists for each $(x_1, \ldots, x_n) \in M^n$ some $r \in R$ such that $\xi^n(x_1, \ldots, x_n) = r(x_1, \ldots, x_n)$. $\square$

Lemma 12.1.8. *Let M be a left R-module and $M^n = \bigoplus_{i=1,2,\ldots,n} M$. Let $E = \text{End}_R^{\text{op}}(M)$ and $F = \text{End}_R^{\text{op}}(M^n)$. If $\xi : M \longrightarrow M$ is an E-endomorphism, then $\xi^n = \bigoplus_{i=1,2,\ldots,n} \xi : M^n \longrightarrow M^n$ is an F-endomorphism; moreover $\xi \longmapsto \xi^n$ is an isomorphism $\text{End}_E(M) \cong \text{End}_F(M^n)$.*

Lemma 12.1.8 by itself implies that the center of $M_n(D)$ consists of scalar matrices for every division ring D (see the exercises).

Proof. Let $\eta \in F$. Composing η with the i-th injection $\iota_i : M \longrightarrow M^n$ and j-th projection $\pi_j : M^n \longrightarrow M$ yields an R-endomorphism $\eta_{ij} : M \longrightarrow M$. By definition, the j-th component of $(x_1, \ldots, x_n)\eta = \sum_i \iota_i(x_i)\eta$ is $\sum_i x_i \eta_{ji}$:

$$(x_1, \ldots, x_n)\eta = \left(\sum_i x_i \eta_{1i}, \ldots, \sum_i x_i \eta_{ni} \right).$$

If $\xi \in \text{End}_E(M)$, then ξ commutes with every η_{ij}; hence $\xi^n = \bigoplus_{i=1,2,\ldots,n} \xi : M^n \longrightarrow M^n$, $\xi^n(x_1, \ldots, x_n) = (\xi x_1, \ldots, \xi x_n)$, commutes with every $\eta \in F$ and $\xi^n \in \text{End}_F(N)$. Then $\Theta : \xi \longmapsto \xi^n$ is an injective homomorphism of $\text{End}_E(M)$ into $\text{End}_F(M^n)$. Conversely, an F-endomorphism of M^n must commute with its many R-endomorphisms; we use this to show that Θ is surjective.

Let $\omega \in \text{End}_F(M^n)$. Composing ω with the injections $M \longrightarrow M^n$ and projections $M^n \longrightarrow M$ yields $\mathbb{Z}$-endomorphisms $\omega_{ij} : M \longrightarrow M$ such that

$$\omega(x_1, \ldots, x_n) = \left(\sum_i \omega_{1i} x_i, \ldots, \sum_i \omega_{ni} x_i \right).$$

Now $\omega \in \mathrm{End}_F(M^n)$ commutes with the R-endomorphism $\zeta_{jk} = \iota_k \circ \pi_j$ which sends $(x_1, \ldots, x_n)$ to $(0, \ldots, 0, x_j, 0, \ldots, 0)$ (with x_j as k-th component). We have

$$\omega((x_1, \ldots, x_n)\zeta_{jk}) = (\omega_{1k}x_j, \ldots, \omega_{nk}x_j),$$

$$(\omega(x_1, \ldots, x_n))\zeta_{jk} = \left(0, \ldots, 0, \sum_i \omega_{ji}x_i, 0, \ldots, 0\right)$$

(with $\sum_i \omega_{ji}x_i$ as k-th component). Therefore $\omega_{jk} = 0$ whenever $j \neq k$ and $\omega(x_1, \ldots, x_n) = (\omega_{11}x_1, \ldots, \omega_{nn}x_n)$. Similarly ω commutes with the R-endomorphism $\zeta_\sigma : (x_1, \ldots, x_n) \longmapsto (x_{\sigma 1}, \ldots, x_{\sigma n})$, where σ is any permutation of $1, 2, \ldots, n$; therefore $\omega_{jj} = \omega_{kk}$ for all j, k. Thus $\omega = \xi^n$, where $\xi = \omega_{ii}$; moreover $\xi \in \mathrm{End}_E(M)$ (ξ commutes with every $\eta \in E$), since $\omega = \xi^n$ commutes with $\eta^n \in F$. $\square$

Exercises

1. Prove that every direct sum of semisimple modules is semisimple.
2. Prove that every submodule of a semisimple module is semisimple.
3. Prove that every homomorphic image of a semisimple module is semisimple.
4. Let $E = \mathrm{End}_R^{\mathrm{op}}(M)$. Show that $R \longrightarrow \mathrm{End}_E(M)$ is an isomorphism when $M = {}_R R$.
5. Let S be a simple left R-module and $D = \mathrm{End}_R^{\mathrm{op}}(S)$. Assume that $\mathrm{Ann}(S) = 0$ and that S has finite dimension over D. Use the Jacobson Density Theorem to prove that $R \cong \mathrm{End}_D(S)$.
6. Let D be a division ring. Use Lemma 12.1.8 to show that the center of $M_n(D)$ consists of all scalar matrices λI with λ in the center of D.

12.2. THE ARTIN-WEDDERBURN THEOREM

The main result in this section constructs all rings R such that every left R-module is semisimple.

12.2.a. Semisimple Rings. An element e of a ring R is **idempotent** in case $e^2 = e$. (Then $e^n = e$ for all $n > 0$.)

Proposition 12.2.1. *For a ring R the following properties are equivalent:*

(1) *Every left R-module is semisimple.*

(2) *${}_R R$ is semisimple.*

(3) *Every left ideal of R is generated by an idempotent.*

Proof. Obviously (1) implies (2). (2) implies (1) by Proposition 12.1.3: if ${}_R R$ is semisimple, then every free left R-module is semisimple, and every homomorphic image of a free left R-module is semisimple.

(2) implies (3). Let L be a left ideal of R. By (2), $R = L \oplus M$ for some left ideal M. Then $1 = e + f$ for some $e \in L$ and $f \in F$. Then $Re \subseteq L$. If, conversely, $x \in L$, then $x = x1 = xe + xf$, which in $R = L \oplus M$ implies $x = xe$ (and $xf = 0$). It follows that e is idempotent and that $L \subseteq Re$, so that $L = Re$.

(3) implies (2). Let $L = Re$ be a left ideal generated by an idempotent e. Let $M = R(1 - e)$. For each $x \in R$ we have $x = xe + x(1 - e) \in L + M$. Also $x = re = s(1 - e) \in Re \cap R(1 - e)$ implies $xe = ree = re = x$ and $x = xe = s(1 - e)e = s(e - e^2) = 0$. Thus $R = L \oplus M$. Hence (3) implies that every submodule of $_R R$ is a direct summand, so that $_R R$ is semisimple. $\square$

We call a ring R **left semisimple** in case every left R-module is semisimple. These rings are normally called **semisimple artinian**, a terminology which will be understood in Section 12.4. Fields are left semisimple; less trivial examples will be seen shortly.

A left semisimple ring is a direct sum of minimal (nonzero) left ideals: indeed a left ideal is minimal if and only if it is simple as a left R-module. The remaining simple left R-modules are determined by the next result.

Proposition 12.2.2. *Let R be a left semisimple ring.*

(1) *Let S be a simple left R-module and L be a minimal left ideal of R. If $S \ncong L$, then $LS = 0$.*

(2) *Every simple left R-module is isomorphic to a minimal left ideal of R.*

Proof. (1) Assume that $LS \neq 0$. Then $La \neq 0$ for some $a \in S$; since S is simple, $S = La$. There is a nonzero module homomorphism $r \longmapsto ra$ of L into La; by Schur's Lemma 12.1.5, it is an isomorphism, and $S = La \cong L$.

(2) Let S be a simple left R-module. Since R is a sum of minimal left ideals, we cannot have $LS = 0$ for every minimal left ideal L, lest $S = RS = 0$. Hence $LS \neq 0$ for some L, and $S \cong L$ by (1). $\square$

12.2.b. Examples

Proposition 12.2.3. *Let D be a division ring. Then $R = M_n(D)$ is the direct sum of n minimal left ideals and is left semisimple; all simple left R-modules are faithful; all simple left R-modules are isomorphic to each other and have dimension n over D.*

Proof. Let L_i be the set of all $n \times n$ matrices in which all entries are zero outside of the i-th column; L_i has dimension n over D. Matrix multiplication shows that each L_i is a left ideal of R. The reader will readily show that L_i is a minimal left ideal. We see that $R = \sum_{i \in I} L_i$; hence R is left semisimple.

Let E_{ij} be the matrix in which the (i, j) entry is 1, and all other entries are 0. Then $E_{ij} \in L_j$ and the matrix product AE_{ij} contains the i-th column of A. Hence $AB = 0$ for all $B \in L_j$ implies $A = 0$. Thus L_j is a faithful R-module.

For each i, j, the mapping $\theta : L_i \longrightarrow L_j$ which moves column i to column j is readily seen to be an R-isomorphism. If a simple left R-module S is not isomorphic to any L_i, then $L_i S = 0$ for all i by Proposition 12.2.2, and $RS = 0$, contradicting $S \neq 0$; therefore S is isomorphic to all L_i. In particular, S is faithful. We identify $a \in D$ with the scalar matrix $aI \in M_n(D)$. As a module over $D \subseteq M_n(D)$, $S \cong L_i$ has dimension n. $\qquad \square$

The next results use direct products $R = R_1 \times R_2 \times \cdots \times R_n$ of finitely many rings (with componentwise operations). It is convenient to view R as an internal direct sum $R = R_1 \oplus \cdots \oplus R_n$ (of abelian groups); then each R_i is a subring of R.

Proposition 12.2.4. *A direct product of finitely many left semisimple rings is left semisimple.*

Proof. If each R_i is left semisimple, then each R_i is a direct sum of minimal left ideals of R_i, which by the next lemma are minimal left ideals of $R = R_1 \oplus \cdots \oplus R_n$; hence R is a direct sum of minimal left ideals. $\qquad \square$

Lemma 12.2.5. *Let $R = R_1 \oplus \cdots \oplus R_n$ be a direct product of finitely many rings. The left (right, two-sided) ideals of R_i coincide with the left (right, two-sided) ideals of R contained in R_i. The minimal left (right, two-sided) ideals of R_i coincide with the minimal left (right, two-sided) ideals of R contained in R_i.*

Proof. With componentwise operations we have $R_i R_j = 0$ whenever $i \neq j$.

A left ideal L of R which is contained in R_i is a left ideal of R_i. Conversely, a left ideal L of R_i is a left ideal of R: if $x = \sum_{j \in I} x_j \in R$ (with $x_j \in R_j$), then $xL = \sum_{j \in I} x_j L = x_i L \subseteq L$. Therefore a left ideal of R_i is minimal in R_i if and only if it is minimal in R. Similar arguments apply to right ideals and to two-sided ideals. $\qquad \square$

12.2.c. Structure. A ring R is **simple** in case $R \neq 0$ and R has no two-sided ideal $I \neq 0, R$. We now show that a left semisimple ring is a direct product of finitely many simple rings.

Call two minimal left ideals of a ring R **isomorphic** in case they are isomorphic as left R-modules. This is an equivalence relation, and it partitions the set of all minimal left ideals of R into isomorphy classes $(C_i)_{i \in I}$.

Theorem 12.2.6. *Let R be a left semisimple ring. Let $(C_i)_{i \in I}$ be the family of all isomorphy classes of minimal left ideals of R. For each class C_i let R_i be the sum of all minimal left ideals $L \in C_i$.*

(1) *I is finite (there are only finitely many nonisomorphic simple R-modules).*

(2) *R is isomorphic to the direct product of the finitely many rings R_i.*

(3) *Each R_i is a minimal two-sided ideal of R.*

(4) *Each R_i is a simple ring and a left semisimple ring; moreover all simple left R_i-modules are isomorphic.*

Proof. Since R is left semisimple, we have $R = \sum_{i \in I} R_i$. Each R_i is a left ideal of R. By Proposition 12.2.2, $R_i R_j = 0$ whenever $i \neq j$. Hence $R_i R = R_i \sum_{j \in I} R_j = R_i R_i \subseteq R_i$, and R_i is a two-sided ideal of R.

Let $A \neq 0$ be a two-sided ideal of R. By Proposition 12.1.3, A is a semi-simple R-module and contains a simple submodule, that is, a minimal left ideal L of R. Say $L \in C_i$, and let $L' \in C_i$. Since L is a direct summand of R, composing the projection $R \longrightarrow L$, the isomorphism $L \longrightarrow L'$, and the inclusion $L' \longrightarrow R$ yields a endomorphism η of $_R R$ such that $L\eta = L'$. By Lemma 12.1.4, there exists $c \in R$ such that $r\eta = rc$ for all $r \in R$. Hence $L' = L\eta = Lc \subseteq A$. It follows that $R_i \subseteq A$. This proves (3).

Since $R = \sum_{i \in I} R_i$, we have $1 = \sum_{i \in I} e_i$, where $e_i \in R_i$ and $e_i = 0$ for almost all i. If $x \in R_i$, then $x = x1 = \sum_{j \in I} x e_j = x e_i$, since $R_i R_j = 0$ whenever $i \neq j$; similarly $e_i x = x$. Thus e_i is an identity element of R_i, and R_i is a ring. Then $e_i = 0$ implies $R_i = 0$; however, $R_i \neq 0$, since $C_i \neq \emptyset$. Therefore $e_i \neq 0$ for all i. Therefore I is finite, which proves (1).

We have $R_i \cap (\sum_{j \neq i} R_j) = 0$, since $x \in R_i$ implies $x = e_i x$ and $x \in \sum_{j \neq i} R_j$ implies $e_i x = 0$. Hence $R = \bigoplus_{i \in I} R_i$ (as an R-module). Moreover

$$(\textstyle\sum_{i \in I} x_i)(\sum_{j \in I} y_j) = \sum_{i,j \in I} x_i y_j = \sum_{i \in I} x_i y_i$$

(when $x_i, y_i \in R_i$). This proves (2).

By Lemma 12.2.5 and (3), R_i is a simple ring. Also the minimal left ideals of R_i coincide with the minimal left ideals $L \subseteq R_i$ of R; hence R_i is a sum of minimal left ideals of R_i. Moreover the minimal left ideals of R_i are all in C_i, since $L \in C_j$ implies $L \subseteq R_j$ and $L \not\subseteq R_i$ when $j \neq i$. Thus any two minimal left ideals of R_i are isomorphic as left R-modules, hence also as left R_i-modules. This proves (4). $\qquad\square$

Corollary 12.2.7. *A left semisimple ring R is simple if and only if all simple left R-modules are isomorphic. $M_n(D)$ is simple for every division ring D.*

By Proposition 12.2.4 and Theorem 12.2.6, a ring is left semisimple if and only if it is a direct product of finitely many rings that are both left semisimple and simple.

12.2.d. Simple Rings. We now prove:

Theorem 12.2.8 (Wedderburn). *A ring is both simple and left semisimple if and only if it is isomorphic to a ring of $n \times n$ matrices over a division ring.*

Proof. When D is a division ring, $M_n(D)$ is both simple and left semisimple, by Proposition 12.2.3.

Conversely, let R be both simple and left semisimple. Then R is a direct sum $R = \bigoplus_{i \in I} L_i$ of minimal left ideals $(L_i)_{i \in I}$. Now $1 = \sum_{i \in I} c_i$, where $c_i \in L_i$ and $c_i = 0$ for almost all i. If $x \in L_j$, then $x = x1 = \sum_{i \in I} x c_i \in \bigoplus_{i \in I} L_i$ and $x = x c_j$. Therefore $c_j \neq 0$ (otherwise, $L_j = 0$). It follows that I is finite. Since $L_i \cong S$ for all i, we have ${}_R R \cong S^m$ for some $m > 0$.

By Schur's Lemma, $D = \mathrm{End}_R^{\mathrm{op}}(S)$ is a division ring. Let $E = \mathrm{End}_R^{\mathrm{op}}(S^m)$. By Lemma 12.1.6, $\mathrm{End}_E(S^m) \cong \mathrm{End}_E({}_R R) \cong R$. Then Lemma 12.1.8 yields $R \cong \mathrm{End}_E(S^m) \cong \mathrm{End}_D(S)$. Moreover S is a finite-dimensional right D-module; otherwise, $\mathrm{End}_D(S)$ would contain a two-sided ideal $A \neq 0, \mathrm{End}_D(S)$ consisting of all endomorphisms of finite rank. Hence $R \cong \mathrm{End}_D(S) \cong M_n(D)$ for some n. $\qquad\square$

There is a uniqueness property for Theorem 12.2.8:

Proposition 12.2.9. *Let C and D be division rings. If $M_m(C) \cong M_n(D)$, then $m = n$ and $C \cong D$.*

Proof. Let V be a vector space over D of finite dimension n (with D acting on the right) and $R = \mathrm{End}_D(V) \cong M_n(D)$. Every $x \in V$, $x \neq 0$ is part of some basis of V; hence for each $y \in X$ there is some linear transformation $\eta \in R$ of V such that $\eta x = y$; thus $Rx = V$. This shows that V is a simple R-module, and is R-isomorphic to the minimal left ideals of R.

Let $v \in V$, $v \neq 0$. By the Jacobson Density Theorem (Theorem 12.1.7), there is for each $\xi \in \mathrm{End}_R^{\mathrm{op}}(V)$ some $a \in D$ such that $v\xi = va$. Since $Rv = V$ this implies $x\xi = xa$ for every $x \in V$. Thus the homomorphism $D \longrightarrow \mathrm{End}_R^{\mathrm{op}}(V)$ is surjective; since D is a division ring, its kernel is 0, and we obtain $\mathrm{End}_R^{\mathrm{op}}(V) \cong D$. This property extends to every simple R-module.

If now $R \cong M_n(D) \cong M_m(C)$, then $D \cong \mathrm{End}_R^{\mathrm{op}}(V) \cong C$. Furthermore the isomorphism $D \cong C$ makes a commutative triangle

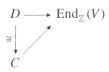

with the D and C-module structures on V. Therefore V has the same dimension as a D-module and as a C-module, and $m = n$. $\qquad\square$

12.2.e. Semisimple Rings. Theorems 12.2.6 and 12.2.8 now yield:

Theorem 12.2.10 (Artin-Wedderburn). *A ring is left semisimple if and only if it is isomorphic to a direct product $M_{n_1}(D_1) \times \cdots \times M_{n_r}(D_r)$ of finitely many finite-dimensional matrix rings over division rings.*

It follows from Proposition 12.2.9 that in Theorem 12.2.10 the integers $n_1, \ldots, n_r$ and division rings $D_1, \ldots, D_r$ are unique (see the exercises).

Wedderburn's Theorem has another consequence. We know that matrices with coefficients in a field can be transposed so that $(AB)^t = B^t A^t$. When D is a division ring and $A \in M_n(D)$, we regard A^t as a matrix over the opposite division ring D^{op}, and then $(AB)^t = B^t A^t$ holds in $M_n(D^{op})$ for every $A, B \in M_n(D)$. This shows that $M_n(D)^{op} \cong M_n(D^{op})$. If now R is a left semisimple ring, then by Theorem 12.2.10 so is R^{op}.

We call a ring R **right semisimple** in case every right R-module is semisimple. We have proved:

Corollary 12.2.11. *A ring is left semisimple if and only if it is right semisimple.*

In all the above we have considered only rings with an identity element. Semisimple modules and rings can be defined without this requirement; Theorem 12.2.10 still holds but requires a more delicate proof.

Exercises

1. Let D be a division ring and $R = M_n(D)$. Let $L_i \subseteq R$ be the set of all matrices in which all entries are zero outside of the i-th column. Prove that L_i is a minimal left ideal of R.

2. Show that, in Theorem 12.2.10, the integers $n_1, \ldots, n_r$ and division rings $D_1, \ldots, D_r$ are unique (up to isomorphism).

3. Prove that a commutative ring is left semisimple if and only if it is isomorphic to a direct product of finitely many fields.

4. Let $R = \bigoplus_{i \in I} R_i$ be a left semisimple ring, written as a direct product of finitely many left simple rings $(R_i)_{i \in I}$. Prove that every two-sided ideal of R has the form $\bigoplus_{i \in J} R_i$ for some $J \subseteq I$.

12.3. THE JACOBSON RADICAL

The Jacobson radical provides a completely different approach to semisimplicity, as we will see in the next section. This section contains general properties.

12.3.a. Definition. The (left) **Jacobson radical** $J(R)$ of a ring R is the intersection of all the maximal left ideals of R.

Proposition 12.3.1. *$J(R)$ is the intersection of all the annihilators of simple left R-modules. Hence $J(R)$ is a two-sided ideal of R.*

Proof. If L is maximal left ideal of R, then $S = {}_R R/L$ is a simple left R-module and $\text{Ann}(S) \subseteq L$. Hence the intersection of all $\text{Ann}(S)$ is contained in $J(R)$. Conversely, let $r \in J(R)$ and S be a simple left R-module. If $x \in S$, $x \neq 0$, then $Rx = S$, $_R R/\mathrm{Ann}\,(x) \cong Rx$ is simple, Ann (x) is a maximal left ideal of R, and $r \in \mathrm{Ann}\,(x)$. Since this holds for every nonzero $x \in S$, we have $r \in \mathrm{Ann}\,(S)$, for every simple left R-module S.   □

We now give more elementary definitions of $J(R)$.

Proposition 12.3.2. $x \in J(R)$ *if and only if* $1 + tx$ *has a left inverse for every* $t \in R$.

Proof. If $x \in J(R)$ and $t \in R$, then tx belongs to every maximal left ideal of R and $1 + tx$ belongs to no maximal left ideal of R. Hence $R(1 + tx) = R$; otherwise, Zorn's Lemma would provide a maximal left ideal $L \supseteq R(1 + tx)$, and $1 + tx$ has a left inverse. But if $x \notin J(R)$, then $x \notin L$ for some maximal left ideal L, $L + Rx = R$, since L is maximal, and $1 = \ell + tx$ for some $\ell \in L$ and $t \in R$; we see that $1 - tx \in L$ has no left inverse.   □

Proposition 12.3.3. $J(R)$ *is the largest two-sided ideal A of R such that $1 + x$ is a unit of R for every $x \in A$.*

Proof. Let $x \in J(R)$. By Proposition 12.3.2, $1 + x$ has a left inverse y, and then $y = 1 - yx$ has a left inverse z. But y already has a right inverse $1 + x$. Hence $z = z(y(1 + x)) = (zy)(1 + x) = 1 + x$, and $1 + x = z$ has an inverse y, for every $x \in J(R)$.

Conversely, let A be a two-sided ideal of R such that $1 + x$ is a unit of R for every $x \in A$. If $x \in A$, then $1 + tx$ is a unit for every $t \in R$, and $x \in J(R)$ by Proposition 12.3.2; thus $A \subseteq J(R)$.   □

Proposition 12.3.3 implies:

Corollary 12.3.4. $J(R) = J(R^{\mathrm{op}})$.

Thus $J(R)$ is also the intersection of all the maximal right ideals of R, and the intersection of all the annihilators of simple right R-modules. Since the left Jacobson radical of R coincides with the right Jacobson radical of R, it is called simply the Jacobson radical of R.

12.3.b. Properties

Proposition 12.3.5. $J(R/J(R)) = 0$.

Proof. If L is a maximal left ideal of R, then $J(R) \subseteq L$ and $L/J(R)$ is a maximal left ideal of $R/J(R)$; conversely, every maximal left ideal of $R/J(R)$ is of this form. (Thus every simple R-module is a simple $R/J(R)$-module, and conversely.) Hence the intersection of all maximal left ideals of $R/J(R)$ is $J(R)/J(R) = 0$.   □

A left, right, or two-sided ideal N of R is **nilpotent** in case $N^m = 0$ for some $m > 0$, equivalently, in case some $m > 0$ exists such that $x_1 x_2 \ldots x_m = 0$ for all $x_1, \ldots, x_m \in N$.

Proposition 12.3.6. *$J(R)$ contains all nilpotent ideals of R.*

Proof. Let N be a nilpotent left ideal and S be a simple left R-module. If $NS \neq 0$, then $NS = S$ and $S = NS = N^2 S = \cdots = N^m S = 0$, a contradiction. Therefore $NS = 0$, and $N \subseteq \mathrm{Ann}(S)$. Thus $N \subseteq J(R)$. Then $J(R) = J(R^{\mathrm{op}})$ contains all nilpotent right ideals of R. □

Sharper results will be found in the exercises.

An element of R is **nilpotent** in case $x^n = 0$ for some $n > 0$. If R is commutative, then the element x is nilpotent if and only if the ideal Rx is nilpotent; hence:

Corollary 12.3.7. *If R is commutative, then $J(R)$ contains every nilpotent element of R.*

Lemma 12.3.8 (Nakayama). *Let M be a finitely generated left R-module. If N is a submodule of M and $N + J(R)M = M$, then $N = M$.*

Proof. First consider the case $N = 0$. Assume that $J(R)M = M$. Since M is finitely generated, M has a minimal generating subset X which is finite. Then $M = \sum_{x \in X} Rx$ implies $M = J(R)M = \sum_{x \in X} J(R)x$. If $y \in X$, then $y = \sum_{x \in X} r_x x$ for some $r_x \in J(R)$, $(1 - r_y)y = \sum_{x \in X, x \neq y} r_x x$, and $y = \sum_{x \in X, x \neq y} (1 - r_y)^{-1} r_x x$ by Proposition 12.3.3. Therefore M is generated by $X \setminus \{y\}$. This contradiction shows that $X = \emptyset$. Hence $M = 0$.

In the general case, M/N is finitely generated (if X generates M, then $\{x + N ; x \in X\}$ generates M/N). If $N + J(R)M = M$, then $J(R)(M/N) = M/N$, $M/N = 0$, and $N = M$. □

The exercises give some neat applications of Nakayama's Lemma.

Exercises

An element r of a ring R is **quasiregular** in case $1 - r$ has an inverse (is a unit of R); r is **left quasiregular** in case $1 - r$ has a left inverse.

1. Prove that a nilpotent element is quasiregular.
2. Prove that $J(R)$ is the largest left ideal L of R such that every element of L is left quasiregular.
3. Prove that $J(R)$ contains no idempotent $e^2 = e \neq 0$.
4. Prove the following: if a ring R has only one maximal left ideal L, then L is a two-sided ideal and a maximal right ideal.

5. Let $e \in R$ be idempotent. Prove that eRe is a ring. Prove that $J(eRe) = J(R) \cap eRe$. (When S is a simple left R-module, show that eS is either 0 or a simple eRe-module.)

6. Let R be a ring. Show that a matrix $A \in M_n(R)$ is in $J(M_n(R))$ if and only if every entry of A is in $J(R)$. (When S is a simple left R-module, show that S^n is a simple $M_n(R)$-module under matrix multiplication, if the elements of S^n are treated as column matrices.)

7. Show that every R-module homomorphism $\varphi : M \longrightarrow M'$ induces a module homomorphism $\overline{\varphi} : M/J(R)M \longrightarrow M'/J(R)M'$. If $\overline{\varphi}$ is surjective and M' is finitely generated, prove that φ is surjective.

8. Let M be a finitely generated left R-module and $\pi : M \longrightarrow M/J(R)M$ be the projection. If $\pi(x_1), \ldots, \pi(x_n)$ generate $M/J(R)M$, then prove that $x_1, \ldots, x_n$ generate M.

12.4. ARTINIAN RINGS

This section connects our two approaches to semisimplicity, with some applications to ring theory.

12.4.a. Definition. A ring R is **left noetherian** in case its left ideals satisfy the ascending chain condition; equivalently, in case the module $_R R$ is noetherian. Likewise R is **left artinian** in case its left ideals satisfy the descending chain condition; equivalently, in case the module $_R R$ is artinian. In a left artinian ring, every descending sequence of left ideals terminates, and every nonempty set of left ideals has a minimal element.

12.4.b. Properties

Proposition 12.4.1. *If R is left artinian, then $J(R)$ is nilpotent; hence $J(R)$ is the largest nilpotent two-sided ideal of R.*

Proof. Let $J = J(R)$. The sequence $J \supseteq J^2 \supseteq J^3 \supseteq \cdots$ terminates at some J^n ($J^n = J^m$ for all $m \geq n$). Assume that $J^n \neq 0$. Since $J^n R = J^n \neq 0$, there exists a left ideal L minimal such that $J^n L \neq 0$. Then $J^n a \neq 0$ for some $0 \neq a \in L$, $J^n a \subseteq L$, $J^n(J^n a) \neq 0$, and $J^n a = L$ by the choice of L. Hence $ra = a$ for some $r \in J^n \subseteq J$, $(1 - r)a = 0$, and Proposition 12.3.3 yields $a = 0$, a contradiction. $\square$

Lemma 12.4.2. *If R is left artinian, then $J(R)$ is the intersection of finitely many maximal left ideals of R.*

Proof. Let $\mathscr{I}$ be the set of all intersections of finitely many maximal left ideals of R. By the descending chain condition, $\mathscr{I}$ has a minimal element J. Then J is contained in every maximal left ideal L, since $J \cap L \subsetneqq J$ is not possible. Therefore $J = J(R)$. $\square$

Theorem 12.4.3. *A ring R is left semisimple if and only if R is left artinian and $J(R) = 0$.*

Proof. Let R be a left artinian ring such that $J(R) = 0$. By Lemma 12.4.2, there exists finitely many maximal left ideals $L_1, \ldots, L_n$ such that $L_1 \cap \cdots \cap L_n = J(R) = 0$. Let $\varphi : R \longrightarrow R/L_1 \times \cdots \times R/L_n$ be the module homomorphism induced by the projections $R \longrightarrow R/L_i$. Then $\operatorname{Ker} \varphi \subseteq L_1 \cap \ldots \cap L_n$ and φ is injective. Hence $_R R$ is isomorphic to a submodule of the semisimple module $R/L_1 \times \cdots \times R/L_n$ and is semisimple.

We now prove the converse. If $R = M_n(D)$ for some division ring D, then by Proposition 12.2.3 R is the direct sum of n minimal left ideals and has a faithful simple left R-module. Hence $J(R) = 0$ by Proposition 12.3.1.

If now R is left semisimple, then R is isomorphic to a direct product $R_1 \times \cdots \times R_k$ of rings $R_i \cong M_{n_i}(D_i)$. Proposition 12.3.2 shows that

$$J(R_1 \times \cdots \times R_n) = J(R_1) \times \cdots \times J(R_n);$$

hence $J(R) = 0$. Also R is a direct sum of finitely many minimal left ideals. Hence the module $_R R$ has a composition series. By Proposition 10.3.11, R is left artinian (and left noetherian). $\qquad \square$

A ring R is **semisimple** in case $J(R) = 0$. (Semisimple rings are also called **semiprimitive**.) Thus Theorem 12.4.3 states that a ring is left semisimple if and only if it is left artinian and semisimple. Semisimple left artinian rings are right artinian by Corollary 12.2.11, and are called **semisimple artinian**.

Let R be any left artinian ring. Then $R/J(R)$ is semisimple by Proposition 12.3.5, and it inherits the descending chain condition on left ideals from R. Thus there is a nilpotent ideal $J(R)$ such that $R/J(R)$ has a known structure. This can be used to prove properties of left artinian rings in general. We give two examples.

Proposition 12.4.4. *When R is left artinian, a left R-module M is semisimple if and only if $J(R)M = 0$.*

Proof. Let $J = J(R)$. Since $J \subseteq \operatorname{Ann}(S)$ for every simple R-module S, we have $JS = 0$, and $JM = 0$ whenever M is semisimple.

Conversely, assume that $JM = 0$. Then $J \subseteq \operatorname{Ann}(M)$ and the ring homomorphism $R \longrightarrow \operatorname{End}(M)$ induces a ring homomorphism $R/J \longrightarrow \operatorname{End}(M)$. This makes M an R/J-module. We see that the submodules of $_R M$ coincide with the submodules of $_{R/J} M$. Since R/J is a left semisimple ring, $_{R/J} M$ is semisimple, every submodule of $_{R/J} M$ is a direct summand, every submodule of $_R M$ is a direct summand, and M is semisimple. $\qquad \square$

Theorem 12.4.5. *Every left artinian ring is left noetherian.*

Proof. The proof of Theorem 12.4.3 shows that $_R R$ has a composition series when R is left artinian and semisimple. If R is any left artinian ring, then $R/J(R)$ is semisimple and the left R-module $R/J(R)$ has a composition series.

We show that $J = J(R)$ also has this property; then $_RR$ has a composition series, and it follows from Proposition 10.3.11 that R is left noetherian.

By Proposition 12.4.1, J has a descending sequence

$$J \supseteq J^2 \supseteq J^3 \supseteq \cdots \supseteq J^n = 0.$$

Every $J^k \subseteq _RR$ is an artinian R-module; hence every J^k/J^{k+1} is an artinian R-module. Also J^k/J^{k+1} is semisimple by Proposition 12.4.4. Therefore J^k/J^{k+1} is the direct sum of finitely many simple R-modules and has a composition series (since a direct sum of infinitely many simple modules contains an infinite strictly descending sequence of submodules). Hence the R-module J has a composition series. $\square$

Exercises

1. Show that every simple left R-module is a simple left $R/J(R)$-module, and conversely.
2. Prove that a left artinian ring is isomorphic to a direct product of finitely many division rings if and only if it contains no nonzero nilpotent element.
3. Show that every nil left or right ideal of a left artinian ring is nilpotent. (An ideal L is **nil** when every element of L is nilpotent.)

13

REPRESENTATIONS OF FINITE GROUPS

Matrix representations and characters are one of the basic tools of finite group theory. This chapter contains a few basic results, which provide a nice application of semisimplicity, and a proof that groups of order $p^m q^n$ are solvable.

Required results: Chapter 10, Sections 10.1 through 10.4 and Lemma 10.6.8; Chapter 12, Sections 12.1 and 12.2. Section 13.3 uses general properties of solvable groups from Section 3.7, general properties of integral elements from Section 11.2, and Proposition 11.3.1.

13.1. THE GROUP ALGEBRA

If this section we show that matrix representations of a group can be viewed as modules over its group algebra and are direct sums of irreducible representations.

13.1.a. The Group Algebra. A K-**algebra** is a vector space A over K with a bilinear associative multiplication; we also require an identity element. In particular, A is a ring. We identify K with the subalgebra $\{\lambda 1; \lambda \in K\}$ of A; since the multiplication on A is bilinear, K is **central** in A (every $\lambda \in K$ commutes with every element of A). A **homomorphism** of K-algebras is a linear transformation which preserves products and identity elements.

Algebras in general are studied in Chapter 16. For now we note that the ring $M_n(K)$ of all $n \times n$ matrices with entries in K is a K-algebra; so is the ring $\mathrm{End}_K(V)$ of all linear transformations of a vector space V over K, which is isomorphic to $M_n(K)$ when V has dimension n.

The simplest way to construct a bilinear multiplication on a vector space is to define it on a basis and then extend it by bilinearity. In detail, a K-algebra $K[G]$ can be constructed from any monoid G (or from any group G) as follows. As a vector space, $K[G]$ is any vector space over K with basis G, so that every element of $K[G]$ can be written uniquely as a linear combination $x = \sum_{g \in G} x_g g$ with $x_g \in K$ and $x_g = 0$ for almost all $g \in G$. (One may construct $K[G]$ as the vector space $\bigoplus_{g \in G} K$ of all families $x = (x_g)_{g \in G}$ of elements of K such that $x_g = 0$ for almost all g; the standard basis can be identified with G.) The multiplication on G extends to a bilinear multiplication on $K[G]$:

$$(\textstyle\sum_{g \in G} x_g g)(\sum_{h \in G} y_h h) = \sum_{k \in G} z_k k \,, \qquad \text{where} \quad z_k = \sum_{g,h \in G,\, gh=k} x_g y_h \,.$$

It is immediate that this defines a K-algebra with identity element $1 \in G$; $K[G]$ is the **semigroup algebra** or **semigroup ring** of G over K (the **group algebra** or **group ring** of G if G is a group). We identify K with the subalgebra $\{\lambda 1 \,;\, \lambda \in K\}$ of $K[G]$; then K is central in $K[G]$. For example, the polynomial ring $K[(X_i)_{i \in I}]$ is the semigroup algebra of a free commutative monoid on the set I.

Since G is a basis of $K[G]$, every mapping φ of G into a vector space V over K extends uniquely to a linear transformation $\overline{\varphi} : K[G] \longrightarrow V$; namely

$$\overline{\varphi}(\textstyle\sum_{g \in G} x_g g) = \sum_{g \in G} x_g \varphi(g).$$

Proposition 13.1.1. *Every multiplicative homomorphism of G into a K-algebra A extends uniquely to an algebra homomorphism of $K[G]$ into A.*

Proof. Let $\varphi : G \longrightarrow A$ be a multiplicative homomorphism (a mapping φ such that $\varphi(gh) = \varphi(g)\varphi(h)$ for all $g, h \in G$ and $\varphi(1) = 1$). Then the linear transformation $\overline{\varphi} : K[G] \longrightarrow A$ which extends φ is an algebra homomorphism. Indeed $\overline{\varphi}(1) = \varphi(1) = 1$; if $z_k = \sum_{g,h \in G,\, gh=k} x_g y_h$, then

$$\overline{\varphi}((\textstyle\sum_{g \in G} x_g g)(\sum_{h \in G} y_h h)) = \overline{\varphi}(\sum_{k \in G} z_k k)$$

$$= \textstyle\sum_{k \in G} z_k \varphi(k)$$

$$= \textstyle\sum_{g,h \in G} x_g y_h \varphi(g)\varphi(h)$$

$$= \overline{\varphi}(\textstyle\sum_{g \in G} x_g g)\,\overline{\varphi}(\sum_{h \in G} y_h h). \qquad \square$$

In what follows we denote $\overline{\varphi}$ by φ. This notation automatically extends mappings $G \longrightarrow V$ to linear transformations $K[G] \longrightarrow V$ whenever V is a vector space over K, and multiplicative homomorphisms $G \longrightarrow A$ to algebra homomorphisms $K[G] \longrightarrow A$ whenever A is a K-algebra.

13.1.b. Representations. Traditionally, a **representation** of a group G over a field K is a representation of G by invertible linear transformations; that is, a homomorphism $\rho : G \longrightarrow GL(V)$, where V is a vector space over some field K. The dimension of V is the **degree** or **dimension** of the representation.

Equivalently, a **matrix representation** of G is a homomorphism of G into the group $GL(n,K) \subseteq M_n(K)$ of all invertible $n \times n$ matrices with entries in the field K, which is isomorphic to $GL(V)$ when V has dimension n. (Recall that $M_n(K)$ consists of column finitary matrices when n is infinite.)

A group G always has at least two representations: the **trivial** representation $\tau(g) = 1_K$, which has dimension 1, and the **regular** representation $\rho_r :$ $G \longrightarrow GL(K[G])$, $\rho_r(g)(x) = gx$ for all $x \in K[G]$, which has dimension $|G|$.

Two representations $\rho_1 : G \longrightarrow GL(V_1)$ and $\rho_2 : G \longrightarrow GL(V_2)$ of G are **equivalent** in case there exists an invertible linear transformation T such that $\rho_2(g) = T\,\rho_1(g)\,T^{-1}$ for all $g \in G$.

For a more modern view of the representations of G we use the group algebra $K[G]$ constructed above. A representation $\rho : G \longrightarrow GL(V)$ of G is a multiplicative homomorphism of G into the K-algebra $\mathrm{End}_K(V)$ and extends uniquely to an algebra homomorphism $\rho : K[G] \longrightarrow \mathrm{End}_K(V)$, by Proposition 13.1.1. This makes V a $K[G]$-module. In V, $g \bullet v = \rho(g)(v)$; more generally, $\rho(\sum_{g \in G} x_g\, g) = \sum_{g \in G} x_g\, \rho(g)$, so that

$$(\textstyle\sum_{g \in G} x_g\, g) \bullet v = \sum_{g \in G} x_g\, \rho(g)(v).$$

Proposition 13.1.2. *There is a one-to-one correspondence between representations of a group G over a field K and $K[G]$-modules. Moreover two representations are equivalent if and only if the corresponding $K[G]$-modules are isomorphic.*

Proof. If $\rho : G \longrightarrow GL(V)$ of G is a representation, then as above $\rho :$ $K[G] \longrightarrow \mathrm{End}_K(V)$ is a $K[G]$-module structure on V. Conversely, let V be a $K[G]$-module. The action of $K[G]$ on V induces an action of $K \subseteq K[G]$ on V, so that V is a vector space over K. Moreover every $x \in K[G]$ acts on V by linear transformations. Indeed $x \bullet (v + w) = x \bullet v + x \bullet w$ since V is a $K[G]$-module, and $x \bullet \lambda v = (x\lambda) \bullet v = (\lambda x) \bullet v = \lambda(x \bullet v)$ for all $\lambda \in K$, since K is central in $K[G]$. In particular, $\rho(g) : v \longmapsto g \bullet v$ is a linear transformation. We have $\rho(1) = 1$ and $\rho(gh) = \rho(g)\rho(h)$ for all $g, h \in G$, since $1 \bullet v = v$ and $g \bullet (h \bullet v) = gh \bullet v$, and the linear transformation $\rho(g) : v \longmapsto g \bullet v$ is invertible, with inverse $\rho(g^{-1}) : v \longmapsto g^{-1} \bullet v$. Hence $\rho : G \longrightarrow GL(V)$ is a representation of G. We see that this provides a one-to-one correspondence between $K[G]$-modules and representations of G.

Let $\rho_1 : G \longrightarrow GL(V_1)$ and $\rho_2 : G \longrightarrow GL(V_2)$ be two representations of G. If ρ_1 and ρ_2 are equivalent, then there exists an invertible linear transformation $T : V_1 \longrightarrow V_2$ such that $\rho_2(g) = T \circ \rho_1(g) \circ T^{-1}$ for all $g \in G$. Then $g \bullet T(v) = \rho_2(g)(T(v)) = T(\rho_1(g)(v)) = T(g \bullet v)$ for all $v \in V_1$, which implies that

$$T((\textstyle\sum_{g \in G} x_g\, g) \bullet v) = T(\textstyle\sum_{g \in G} x_g\,(g \bullet v)) = \sum_{g \in G} x_g\, T(g \bullet v)$$

$$= \textstyle\sum_{g \in G} x_g\, g \bullet T(v) = (\sum_{g \in G} x_g\, g) \bullet T(v).$$

Hence T is a $K[G]$-module isomorphism. If, conversely, $T : V_1 \longrightarrow V_2$ is a $K[G]$-module isomorphism, then T is a K-module isomorphism and $\rho_2(g)$ $(T(v)) = g \cdot T(v) = T(g \cdot v) = T(\rho_1(g)(v))$, so that ρ_1 and ρ_2 are equivalent. $\square$

The proof of Proposition 13.1.2 shows that the $K[G]$-module structure on V is determined by its vector space structure and by the action of G. Accordingly $K[G]$-modules are also called G-**modules**.

13.1.c. Maschke's Theorem. The first important result on representations of finite groups is Maschke's Theorem, which constructs all $K[G]$-modules from minimal ideals of $K[G]$.

Theorem 13.1.3 (Maschke). *Let G be a finite group and K be a field. If K has characteristic 0, or if K has characteristic $p \neq 0$ that does not divide the order of G, then $K[G]$ is semisimple artinian.*

Proof. We prove that every submodule V of a $K[G]$-module W is a direct summand of W. Since W is a vector space, there is a subspace V' of W such that $W = V \oplus V'$ as a vector space. The projection $\pi : W \longrightarrow V$ is a linear transformation. For each $w \in W$ let

$$\varphi(w) = \frac{1}{n} \sum_{g \in G} g^{-1} \cdot \pi(g \cdot w) \in V,$$

where $n = |G| \neq 0$ in K by the choice of K. Then φ is a $K[G]$-homomorphism: indeed φ is a linear transformation and

$$\varphi(h \cdot w) = \frac{1}{n} \sum_{g \in G} g^{-1} \cdot \pi(gh \cdot w) = \frac{1}{n} \sum_{g \in G} h(gh)^{-1} \cdot \pi(gh \cdot w)$$

$$= \frac{1}{n} \sum_{k \in G} h k^{-1} \cdot \pi(k \cdot w) = h \cdot \varphi(w)$$

for all $h \in G$. If $w \in V$, then $\pi(w) = w$ and

$$\varphi(w) = \frac{1}{n} \sum_{g \in G} g^{-1} \cdot \pi(g \cdot w) = \frac{1}{n} \sum_{g \in G} g^{-1} \cdot (g \cdot w) = w,$$

since $|G| = n$. Thus $\varphi \circ \iota = 1_V$, where $\iota : V \longrightarrow W$ is the inclusion, and Corollary 10.4.10 yields $W = V \oplus \operatorname{Ker} \varphi$ (as a $K[G]$-module). $\square$

Corollary 13.1.4. *Let G be a finite group and K be a field whose characteristic does not divide the order of G. Up to isomorphism, there are only finitely many simple $K[G]$-modules, and they all have finite dimensions over K.*

Proof. The first assertion is part of Theorem 12.2.6. Every simple $K[G]$-module S is isomorphic to a minimal left ideal L of $K[G]$; since G is finite, $K[G]$ has finite dimension over K, and so do L and S. $\square$

13.1.d. Irreducible Representations. Assume that G is finite and that the characteristic of K does not divide the order of G, so that $K[G]$ is semisimple artinian. Then every $K[G]$-module is a direct sum of simple $K[G]$-modules, and a $K[G]$-module S is simple ($S \neq 0$ and S has no submodule $\neq 0, S$) if and only if it indecomposable ($S \neq 0$, and $S = A \oplus B$ implies $A = 0$ or $B = 0$). These basic results can be expressed in terms of matrix representations, as follows.

The **direct sum** of a family $(\rho_i)_{i \in I}$ of representations $\rho_i : G \longrightarrow GL(V_i)$ of G is the representation $\rho = \bigoplus_{i \in I} \rho_i : G \longrightarrow GL(\bigoplus_{i \in I} V_i)$ defined by $\rho(g) = \bigoplus_{i \in I} \rho_i(g) : (v_i)_{i \in I} \longmapsto (\rho_i(g)(v_i))_{i \in I}$. This makes $\bigoplus_{i \in I} V_i$ a direct sum of $K[G]$-modules. If B_i is a basis of V_i over K, then the disjoint union $B = \bigcup_{i \in I} B_i$ is a basis of V over K; in this basis, the matrix $M(g)$ of $\rho(g)$ is made of diagonal blocks

$$M(g) = \begin{pmatrix} M_i(g) & 0 & 0 & \cdots \\ 0 & M_j(g) & 0 & \cdots \\ 0 & 0 & M_k(g) & \cdots \\ \vdots & \vdots & \vdots & \ddots \end{pmatrix}$$

where $M_i(g)$ is the matrix of $\rho_i(g)$ in the basis B_i.

A representation $\rho : G \longrightarrow GL(V)$ is **irreducible** if and only if it is indecomposable ($V \neq 0$, and $\rho = \rho_1 \oplus \rho_2$ implies $V_1 = 0$ or $V_2 = 0$); equivalently, if and only if V is an indecomposable $K[G]$-module.

Maschke's Theorem and Corollary 13.1.4 can then be restated as follows. Assume that G is finite and that the characteristic of K does not divide the order of G. Then every representation of G over K is a direct sum of irreducible representations, and a representation $\rho : G \longrightarrow GL(V)$ is irreducible if and only if V is simple as a $K[G]$-module. Up to equivalence, there are only finitely many irreducible representations of G, and all are finite dimensional.

13.1.e. Simple $K[G]$-Modules. Assume that G is finite and that the characteristic of K does not divide the order of G. For more precise handling of simple $K[G]$-modules, we write $K[G]$ as in Section 12.2 as a direct product of finitely many simple rings $K[G] = R_1 \oplus \cdots \oplus R_s$. Up to isomorphism there is one simple $K[G]$-module S_i for each R_i; we let S_i be any minimal left ideal of R_i, so that $K[G]$ acts on S_i by left multiplication in $K[G]$.

Each R_i is isomorphic to a ring $M_{n_i}(D_i)$ of $n_i \times n_i$ matrices over a division ring D_i. We can say more if K is algebraically closed.

Proposition 13.1.5. *Let G be a finite group and K be a field whose characteristic does not divide the order of G. If K is algebraically closed, then $R_i \cong M_{d_i}(K)$, where $d_i = \dim_K S_i$, and $\sum_{1 \leq i \leq s} d_i^2 = |G|$.*

Proof. $R_i \cong M_{n_i}(D_i)$ for some division ring D_i. Now D_i is finite dimensional over K. By Lemma 10.6.8, $D_i = K$, since K is algebraically closed. Hence $R_i \cong M_{n_i}(K)$ for some $n_i > 0$. Any minimal left ideal S_i of R_i is a simple $K[G]$-module; then $n_i = \dim_K S_i = d_i$ (Proposition 12.2.3). Finally R_i has dimension n_i^2 over K, whence $\sum_{1 \leqslant i \leqslant s} n_i^2 = \dim_K K[G] = |G|$. $\qquad \square$

Proposition 13.1.6. *Let G be a finite group and K be a field whose characteristic does not divide the order of G. If K is algebraically closed and G is abelian, then every simple $K[G]$-module has dimension 1 over K.*

The proof is an exercise.

Exercises

1. Verify that the multiplication on $K[G]$ is well defined and associative.
2. Let G be a totally ordered group (a group with a total order $\leqslant$ such that $g \leqslant h$ implies $kg \leqslant kh$ and $gk \leqslant hk$ for all k). Show that $K[G]$ has no zero divisors.
3. Let G be a finite abelian group and K be an algebraically closed field whose characteristic does not divide the order of G. Show that every simple $K[G]$-module has dimension 1 over K.

13.2. CHARACTERS

In many cases a representation of a group G over a field K is determined by its character, which is a mapping of G into K. This has made characters a major tool of finite group theory.

13.2.a. Definition. Recall that a linear transformation $T : V \longrightarrow V$ of a finite dimensional vector space V has a **trace** $\mathrm{Tr}\, T$, which is the trace of the matrix of T in any basis of V. The **character** of a finite-dimensional representation $\rho : G \longrightarrow GL(V)$ is the mapping $\chi = \chi_\rho : G \longrightarrow K$ defined by

$$\chi(g) = \mathrm{Tr}\, \rho(g).$$

We see that $\chi(1) = \mathrm{Tr}\, 1_V = \dim_K V$ is the dimension of the representation. Also χ is a **class function**, meaning that it is constant on conjugacy classes: indeed $\chi(hgh^{-1}) = \chi(g)$, since $\rho(hgh^{-1}) = \rho(h)\rho(g)\rho(h)^{-1}$ has the same trace as $\rho(g)$.

Equivalent representations have the same character, since $\rho(g)$ and $T\rho(g)T^{-1}$ have the same trace.

A **character** of G over K is the character of a finite-dimensional representation of V over K. For example, the **trivial** character χ_τ is the character of the trivial representation τ; $\chi_\tau(g) = 1$ for all $g \in G$, since the identity mapping on K has trace 1. More generally, the characters of one dimensional representations are the homomorphisms of G into $K^* = K \backslash \{0\}$. The **regular** character of a finite group G is the character χ_r of its regular representation; the reader will show:

Proposition 13.2.1. *The regular character* χ_r *satisfies* $\chi_r(1) = |G|$ *and* $\chi_r(g) = 0$ *for all* $g \neq 1$.

Every character $\chi : G \longrightarrow K$ extends uniquely to a linear transformation $\chi : K[G] \longrightarrow K$. For every $x = \sum_{g \in G} x_g\, g \in K[G]$,

$$\chi_\rho(x) = \sum_{g \in G} x_g\, \chi_\rho(g) = \sum_{g \in G} x_g \operatorname{Tr}\rho(g) = \operatorname{Tr}(\sum_{g \in G} x_g\, \rho(g)) = \operatorname{Tr}\rho(x).$$

13.2.b. Irreducible Characters. An **irreducible** character is the character of an irreducible representation. For example, the trivial character is irreducible.

When Maschke's Theorem applies, every finite-dimensional representation is a finite direct sum $\rho = \bigoplus_{i \in I} \rho_i$ of irreducible representations. Then χ_ρ is a pointwise sum $\sum_{i \in I} \chi_{\rho_i}$ of irreducible characters, since the trace of the finite direct sum $\bigoplus_{i \in I} \rho_i(g)$ is the sum of the traces of its components $\rho_i(g)$; and there are only finitely many irreducible characters, by Corollary 13.1.4. Thus:

Proposition 13.2.2. *Let* G *be a finite group and* K *be a field whose characteristic does not divide the order of* G. *There are only finitely many irreducible characters, and every character is a [pointwise] sum of irreducible characters.*

Proposition 13.2.3. *Let* G *be a finite group and* K *be an algebraically closed field whose characteristic does not divide the order of* G. *The values of any character of* G *over* K *are sums of roots of unity in* K.

Proof. Let $g \in G$ and $H = \langle g \rangle$ be the cyclic subgroup of G generated by g. Let χ be the character of a representation $\rho : G \longrightarrow GL(V)$ of G. Then $\rho_{|H} : H \longrightarrow GL(V)$ is a representation of H and $\chi_{|H}$ is a character of H. By Proposition 13.1.6, the irreducible representations of H over K are all one dimensional. Hence the irreducible characters of H are homomorphisms of H into K^*; their values are roots of unity since every element of H has finite order. By Proposition 13.2.2, $\chi(g) = \chi_{|H}(g)$ is a sum of roots of unity. $\square$

For more precise results we write $K[G]$ as in Section 13.1 as a direct product of finitely many simple rings $K[G] = R_1 \oplus \cdots \oplus R_s$. In such a direct product $R_i R_j = 0$ whenever $i \neq j$. The identity element 1 of R is the sum $1 = e_1 + \cdots + e_s$ of the identity elements of $R_1, \ldots, R_s$, and $e_i e_j = 0$ whenever $i \neq j$. Up to isomorphism there is one simple $K[G]$-module S_i for each R_i; we let S_i be any minimal left ideal of R_i so that $K[G]$ acts on S_i by left multiplication in $K[G]$ ($x \cdot s = xs$ for all $x \in K[G]$, $s \in S_i$). We denote by d_i the dimension of S_i over K.

By the above there are, up to equivalence, s irreducible representations $\rho_i : G \longrightarrow GL(S_i)$, which have s irreducible characters χ_i.

Lemma 13.2.4. *The irreducible characters have the following properties:*

(1) $\chi_i(x) = 0$ *whenever* $x \in R_j$ *and* $j \neq i$.

(2) $\chi_i(e_j) = 0$ *if* $j \neq i$, *and* $\chi_i(e_i) = \chi_i(1) = d_i$.

(3) $\chi_i(e_i x) = \chi_i(x)$ *for all* $x \in K[G]$.

Proof. (1) When $x \in R_j$, $j \neq i$, we have $\rho_i(x)(s) = x \bullet s = xs = 0$ for all $s \in S_i$, so that $\rho_i(x) = 0$ and $\chi_i(x) = \text{Tr}\,\rho_i(x) = 0$.

(2) By (1), $d_i = \chi_i(1) = \chi_i(\sum_{1 \leqslant j \leqslant s} e_j) = \sum_{1 \leqslant j \leqslant s} \chi_i(e_j) = \chi_i(e_i)$.

(3) Since e_i is the identity element of R_i, $\rho_i(e_i)$ is the identity on S_i and

$$\chi_i(e_i x) = \text{Tr}\,\rho_i(e_i x) = \text{Tr}(\rho_i(e_i)\,\rho_i(x)) = \text{Tr}\,\rho_i(x) = \chi_i(x). \qquad \square$$

Lemma 13.2.5. *Let* $\rho : G \longrightarrow GL(V)$ *be a finite-dimensional representation. If* $V \cong \bigoplus_{1 \leqslant i \leqslant s} S_i^{m_i}$, *then* $\chi = \sum_{1 \leqslant i \leqslant s} m_i \chi_i$.

Proof. Let $V \cong \bigoplus_{1 \leqslant i \leqslant s} S_i^{m_i}$ (where $m_i \geqslant 0$ is the number of times that S_i appears in the direct sum). Then ρ has, up to equivalence, a similar direct sum decomposition $\rho = \bigoplus_{1 \leqslant i \leqslant s} \rho_i^{m_i}$ with m_i copies of ρ_i. Taking traces yields $\chi = \sum_{1 \leqslant i \leqslant s} m_i \chi_i$. $\qquad \square$

Theorem 13.2.6. *Let* G *be a finite group and* K *be a field of characteristic* 0.

(1) *Every character can be written uniquely as a linear combination of irreducible characters with nonnegative integer coefficients.*

(2) *The irreducible characters are linearly independent over* K.

(3) *Two representations are equivalent if and only if they have the same character.*

Proof. Let $\rho : G \longrightarrow GL(V)$ be any finite-dimensional representation. Then the $K[G]$-module V is a direct sum of simple $K[G]$-modules, which has only finitely many terms since V has finite dimension over K. Thus $V \cong \bigoplus_{1 \leqslant i \leqslant s} S_i^{m_i}$, where $m_i \geqslant 0$ is the number of times that S_i appears in the direct sum. By Lemma 13.2.5, $\chi = \sum_{1 \leqslant i \leqslant s} m_i \chi_i$. (This proves Proposition 13.2.2 again.)

By Lemma 13.2.4, $\chi(e_i) = \sum_{1 \leqslant j \leqslant s} \chi_j(e_i) = m_i d_i$. If K has characteristic 0, then $d_i \neq 0$ in K and $m_i = 1/d_i\, \chi(e_i)$ is uniquely determined by χ. This proves (1).

More generally, assume that $\sum_{1 \leqslant i \leqslant s} \lambda_i \chi_i = 0$ (i.e., $\sum_{1 \leqslant i \leqslant s} \lambda_i \chi_i(g) = 0$ for all $g \in G$), where $\lambda_1, \ldots, \lambda_s \in K$. Then $\lambda_j d_j = \sum_{1 \leqslant i \leqslant s} \lambda_i \chi_i(e_j) = 0$ and $\lambda_j = 0$ for all j. This proves (2).

Finally two representations with the same character have the same multiplicities m_i, so that the corresponding $K[G]$-modules are both isomorphic to $\bigoplus_{1 \leqslant i \leqslant s} S_i^{m_i}$; hence two representations with the same character are equivalent. □

13.2.c. Conjugacy Classes. Each conjugacy class C of G has a **sum** $c = \sum_{g \in C} g$ in $K[G]$. Since the conjugacy classes of G form a partition of G, their sums are linearly independent in $K[G]$.

Lemma 13.2.7. *The sums of the conjugacy classes of G constitute a basis of the center $Z(K[G])$ of $K[G]$.*

Proof. The sum $c = \sum_{g \in C} g$ of a conjugacy class C commutes with every $\lambda \in K$, and with every $h \in G$: $hch^{-1} = \sum_{g \in C} hgh^{-1} = \sum_{g \in C} g$, since $g \longmapsto hgh^{-1}$ is a permutation of C. Therefore c commutes with every $x \in K[G]$.

Conversely, let $z = \sum_{g \in G} z_g g \in Z(K[G])$. Then $hzh^{-1} = z$ for every $h \in G$. Since G is a basis of $K[G]$, $\sum_{g \in G} z_g hgh^{-1} = \sum_{g \in G} z_g g$ implies $z_g = z_{hgh^{-1}}$ for all $h \in G$, so that the mapping $g \longmapsto z_g$ is constant on every conjugacy class. Hence z is a linear combination of sums of conjugacy classes. □

Theorem 13.2.8. *Let G be a finite group and K be a field whose characteristic does not divide the order of G. If K is algebraically closed, then the number of irreducible characters of G equals the number of conjugacy classes of G.*

Proof. The center $Z(K[G])$ of $K[G] = R_1 \oplus \cdots \oplus R_s$ is the direct sum $Z(K[G]) = Z(R_1) \oplus \cdots \oplus Z(R_s)$. By Proposition 13.1.5, $R_i \cong M_{d_i}(K)$, where $d_i = \dim_K S_i$. Now the center of $M_n(K)$ consists of all the scalar matrices (e.g., by Proposition 3.6.5) and is isomorphic to K. Hence $Z(K[G]) \cong K^s$. Then G has s conjugacy classes, by Lemma 13.2.7. □

The following properties will be used in the next section:

Lemma 13.2.9. *Let G be a finite group and K be an algebraically closed field whose characteristic does not divide the order of G.*

(1) $\rho_i : K[G] \longrightarrow \mathrm{End}_K(S_i)$ *is surjective.*
(2) *If C is a conjugacy class with sum $c \in K[G]$, then $\rho_i(c)$ is a scalar linear transformation $\rho_i(c) : v \longmapsto c_i v$, and $c_i d_i = |C| \chi_i(g)$ for all $g \in C$.*

Proof. (1) Let $E = \mathrm{End}_{K[G]}^{\mathrm{op}}(S_i)$. Since K is central in $K[G]$, the scalar linear transformations $\lambda 1 : x \longmapsto \lambda s$ of S_i are $K[G]$-endomorphisms of S_i and constitute a subfield $K' \cong K$ of E. Now E is a division ring by Schur's Lemma (Lemma 12.1.4). By Lemma 10.6.8, $E = K'$. Thus E contains only scalar linear transformations (and $E \cong K$). Hence $\mathrm{End}_E(S_i) = \mathrm{End}_K(S_i)$.

Let $T \in \text{End}_K(S_i)$ and $s_1, \ldots, s_{d_i}$ be a basis of S_i over K. By the Jacobson Density Theorem (Theorem 12.1.7), there exists $x \in K[G]$ such that $T(s_j) = x \cdot s_j$ for all j. Then $T(s) = x \cdot v$ for all $v \in S_i$, that is, $T = \rho(x)$.

(2) Since c commutes with every $x \in K[G]$, $\rho_i(c)$ commutes with every linear transformation in $\text{Im}\,\rho_i$, which by (1) is every linear transformation in $\text{End}_K(S_i)$. Therefore $\rho_i(c)$ is a scalar linear transformation: $\rho_i(c) : v \longmapsto \lambda v$ for some $\lambda \in K$. The trace of $\rho_i(c) = \sum_{g \in C} \rho_i(g)$ is $\sum_{g \in C} \chi_i(g) = |C|\,\chi_i(g)$ for any $g \in C$; hence $d_i \lambda = |C|\,\chi_i(g)$. (By Lemma 13.2.10 below, $d_i \neq 0$ in K.) $\qquad\square$

13.2.d. Orthogonality. Recall that d_i denotes $\dim_K S_i$. If K is algebraically closed, then $R_i \cong M_{d_i}(K)$ and $\sum_{1 \leqslant i \leqslant s} d_i^2 = |G|$ (Proposition 13.1.5).

Lemma 13.2.10. *Let G be a finite group and K be an algebraically closed field whose characteristic does not divide the order of G. The regular character is $\chi_r = \sum_{1 \leqslant i \leqslant s} d_i \chi_i$. Moreover $d_i \neq 0$ in K, and $e_i = \sum_{g \in G} e_{ig}\, g$, where*

$$e_{ig} = \frac{1}{|G|}\,\chi_r(e_i g^{-1}) = \frac{d_i}{|G|}\,\chi_i(g^{-1}).$$

Proof. By Proposition 13.1.5, $R_i \cong M_{d_i}(K) \cong S_i^{d_i}$ (as a $K[G]$-module) and $K[G] \cong \bigoplus_{1 \leqslant i \leqslant s} S_i^{d_i}$. By Lemma 13.2.5, $\chi_r = \sum_{1 \leqslant i \leqslant s} d_i \chi_i$. Then

$$\chi_r(e_i g^{-1}) = \chi_r(\textstyle\sum_{h \in G} e_{ih} h g^{-1}) = \sum_{h \in G} e_{ih}\,\chi_r(h g^{-1}) = e_{ig}|G|$$

by Proposition 13.2.1. Since $\chi_j(x) = 0$ for all $x \in R_i$ when $i \neq j$, we also have

$$e_{ig}|G| = \chi_r(e_i g^{-1}) = \textstyle\sum_{1 \leqslant j \leqslant s} d_j \chi_j(e_i g^{-1}) = d_i \chi_i(e_i g^{-1}) = d_i \chi_i(g^{-1})$$

by Lemma 13.2.4. This implies $d_i \neq 0$ in K, since $e_i \neq 0$. $\qquad\square$

Theorem 13.2.11. *Let G be a finite group and K be an algebraically closed field whose characteristic does not divide the order of G. Then*

$$\sum_{g \in G} \chi_i(g)\,\chi_j(g^{-1}) = \begin{cases} |G| & \text{if } i = j, \\ 0 & \text{if } i \neq j. \end{cases}$$

Proof. By Lemma 13.2.10, $e_i = \sum_{g \in G} (d_i/|G|)\chi_i(g^{-1})\,g$. Hence

$$\chi_j(e_i) = \sum_{g \in G} \frac{d_i}{|G|}\,\chi_i(g^{-1})\,\chi_j(g).$$

If $j \neq i$, then $\chi_j(e_i) = 0$, and we obtain $\sum_{g \in G} \chi_i(g^{-1})\,\chi_j(g) = 0$. If $j = i$, then $\chi_j(e_i) = d_i \neq 0$ and we obtain $\sum_{g \in G} \chi_i(g^{-1})\,\chi_j(g) = |G|$. $\qquad\square$

Exercises

1. Let G be a finite group and χ_r be its regular character. Show that $\chi_r(1) = |G|$ and that $\chi_r(g) = 0$ for all $g \neq 1$.
2. When K has characteristic $p \neq 0$, show that two representations that have the same character are not necessarily equivalent.
3. Let G be a finite group and K be an algebraically closed field whose characteristic $p \neq 0$ does not divide the order of G. Show that the irreducible characters are linearly independent over K.
4. Find all irreducible characters of V_4 over $\mathbb{C}$.
5. Find all irreducible characters of S_3 over $\mathbb{C}$.
6. Let K be any field. Show that every normal subgroup of a finite group G is the kernel of a representation of G over K.

13.3. BURNSIDE'S THEOREM

In this section we use characters to prove Burnside's Theorem that every group of order $p^m q^n$ (with p and q prime) is solvable.

13.3.a. Complex Characters. As in Section 13.2, G is any finite group, the irreducible characters of G over $\mathbb{C}$ are $\chi_1, \ldots, \chi_s$, and the dimensions of the corresponding irreducible representations $\rho_1, \ldots, \rho_s$ are $d_1, \ldots, d_s$. Since $\mathbb{C}$ is an algebraically closed field of characteristic 0, complex characters (characters over $\mathbb{C}$) have all the properties in Section 13.2. They also have properties of their own.

Lemma 13.3.1. *Let G be a finite group and χ be the character of a complex representation $\rho : G \longrightarrow GL(V)$ of G.*

(1) $\rho(g)$ *is diagonalizable for all $g \in G$.*
(2) $\chi(g)$ *is a sum of $\chi(1)$ roots of unity, so that $|\chi(g)| \leqslant \chi(1)$. Moreover $\chi(g) = \chi(1)$ if and only if $\rho(g) = 1$; $|\chi(g)| = \chi(1)$ if and only if $\rho(g) = \lambda 1$ for some $\lambda \in \mathbb{C}$, and then λ is a root of unity.*
(3) $\chi(g^{-1}) = \overline{\chi(g)}$.

Proof. (1) Let $H = \langle g \rangle \subseteq G$. The representation $\rho_{|H} : H \longrightarrow GL(V)$ is equivalent to a direct sum of irreducible representations, which by Proposition 13.1.6 are all one dimensional. Hence there is a basis of V in which the matrix of $\rho(g) = \rho_{|H}(g)$ consists of one-dimensional diagonal blocks and is a diagonal matrix

$$
M = \begin{pmatrix}
\zeta_1 & 0 & \cdots & 0 \\
0 & \zeta_2 & \cdots & 0 \\
\vdots & \vdots & \ddots & \vdots \\
0 & 0 & \cdots & \zeta_d
\end{pmatrix},
$$

where $d = \dim_{\mathbb{C}} V = \chi(1)$. Since g has finite order in G, M^k is the identity matrix for some k, and $\zeta_1, \ldots, \zeta_d$ are roots of unity in $\mathbb{C}$.

(2) We have $\chi(g) = \operatorname{Tr} M = \zeta_1 + \cdots + \zeta_d$, and $\chi(g)$ is a sum of d roots of unity (as in Proposition 13.2.3). Hence $|\chi(g)| \leqslant |\zeta_1| + \cdots + |\zeta_d| = d = \chi(1)$. Moreover $\chi(g) = d$ if and only if $\zeta_1 = \cdots = \zeta_d = 1$, which is equivalent to $\rho(g) = 1$. Similarly $|\chi(g)| = d$ if and only if $\zeta_1 = \cdots = \zeta_d$, which is equivalent to $\rho(g) = \zeta 1$, where $\zeta = \zeta_1 = \cdots = \zeta_d$ is a root of unity.

(3) In the same basis, the matrix of $\rho(g^{-1}) = (\rho(g))^{-1}$ is

$$
M^{-1} = \begin{pmatrix} \zeta_1^{-1} & 0 & \cdots & 0 \\ 0 & \zeta_2^{-1} & \cdots & 0 \\ \vdots & \vdots & \ddots & \vdots \\ 0 & 0 & \cdots & \zeta_d^{-1} \end{pmatrix}.
$$

Since each ζ_i is a root of unity, we have $\zeta_i^{-1} = \overline{\zeta_i}$. Hence $\chi(g^{-1}) = \zeta_1^{-1} + \cdots + \zeta_d^{-1} = \overline{\zeta_1} + \cdots + \overline{\zeta_d} = \overline{\chi(g)}$. $\qquad \square$

The **kernel** of a complex character χ is

$$
K(\chi) = \{ g \in G;\ \chi(g) = \chi(1) \}.
$$

By Lemma 13.3.1, the kernel of χ is the kernel of the corresponding representation and so is a normal subgroup of G. Conversely, every normal subgroup of G is the kernel of some character (see the exercises).

Similar to the kernel of χ is the larger subgroup

$$
Z(\chi) = \{ g \in G;\ |\chi(g)| = \chi(1) \}.
$$

Lemma 13.3.2. $Z(\chi) \lhd G$ and $Z(\chi)/K(\chi) \subseteq Z(G/K(\chi))$.

Proof. Let $\rho : G \longrightarrow GL(V)$ be the corresponding representation. By Lemma 13.3.1, $g \in Z(\chi)$ if and only if there exists $\lambda \in \mathbb{C}$ such that $\rho(g)(v) = \lambda v$ for all $v \in V$. In other words, $Z(\chi) = \{ g \in G;\ \rho(g) \in Z(GL(V)) \}$ (Proposition 3.6.5). Therefore $Z(\chi) \lhd G$; moreover $\rho(Z(\chi))$ is contained in the center of $\rho(G)$, so that $Z(\chi)/K(\chi) \cong \rho(Z(\chi))$ is contained in the center of $G/K(\chi) \cong \rho(G)$. $\qquad \square$

13.3.b. Integrality. By Proposition 13.2.3, any value $\chi(g)$ of a complex character is a sum of roots of unity and is integral over $\mathbb{Z}$.

Let $C_1, \ldots, C_s$ be the conjugacy classes of G and $c_1, \ldots, c_s \in K[G]$ be their sums. By Lemma 13.2.9, $\rho_i(c_j)$ is a scalar linear transformation $\rho_i(c) : v \longmapsto c_{ij} v$, where $c_{ij} = |C_j| \chi_i(g)/d_i$ for any $g \in C_j$.

Lemma 13.3.3. c_{ij} is integral over $\mathbb{Z}$.

Proof. For any j, k, $c_j c_k = (\sum_{h' \in C_j} h')(\sum_{h'' \in C_k} h'') = \sum_{g \in C_j C_k} n_g g$, where n_g is the number of pairs $(h', h'') \in C_j \times C_k$ such that $h'h'' = g$. Now $C_j C_k$ is a union of conjugacy classes. Moreover $n_g = n_{g'}$ whenever g and g' are conjugate (since $g = h'h''$ if and only if $tgt^{-1} = (th't^{-1})(th''t^{-1})$). Hence $c_j c_k$ is a linear combination of $c_1, \ldots, c_s$ with integer coefficients.

Take a basis of S_i, and let M_ℓ be the matrix of $\rho_i(c_\ell)$, which is a scalar matrix with $c_{i\ell}$ on the diagonal. By the above, $c_j c_k$ is a linear combination of $c_1, \ldots, c_s$ with integer coefficients; hence $\rho_i(c_j) \rho_i(c_k)$ is a linear combination of $\rho_i(c_1), \ldots, \rho_i(c_s)$ with integer coefficients; $M_j M_k$ is a linear combination of $M_1, \ldots, M_s$ with integer coefficients; and $c_{ij} c_{ik}$ is a linear combination of $c_{i1}, \ldots, c_{is}$ with integer coefficients. Hence the additive subgroup A of $\mathbb{C}$ generated by 1 and $c_{i1}, \ldots, c_{is}$ is closed under multiplication and is a ring. Therefore every element of A is integral over $\mathbb{Z}$, by Proposition 11.2.6. $\square$

Proposition 13.3.4. *If d_i and $|C_j|$ are relatively prime, then $\chi_i(g) = 0$ for all $g \in C_j \backslash Z(\chi_i)$.*

Proof. Let $\alpha = \chi_i(g)/d_i$. Since $g \notin Z(\chi_i)$ we have $|\chi_i(g)| < d_i$ and $|\alpha| < 1$. Also $ud_i + v|C_j| = 1$ for some integers u, v; hence $\alpha = (ud_i + v|C_j|)[\chi_i(g)/d_i] = u\chi_i(g) + vc_{ij}$ is integral over $\mathbb{Z}$, by Proposition 13.2.3 and Lemma 13.3.3.

Let E be a Galois extension of $\mathbb{Q}$ which contains α. For every $\sigma \in \mathrm{Gal}(E/\mathbb{Q})$, $\sigma\chi_i(g)$ is, like $\chi_i(g)$, a sum of d_i roots of unity, so that $|\sigma\chi_i(g)| \leqslant d_i$ and $|\sigma\alpha| \leqslant 1$. Then $N(\alpha) = \prod_{\sigma \in \mathrm{Gal}(E/\mathbb{Q})} \sigma\alpha$ has $|N(\alpha)| < 1$. Moreover, $N(\alpha) \in \mathbb{Z}$, since $N(\alpha) \in \mathbb{Q}$ is integral over $\mathbb{Z}$ and $\mathbb{Z}$ is integrally closed by Proposition 11.3.1. Therefore $N(\alpha) = 0$, $\alpha = 0$, and $\chi_i(g) = 0$. $\square$

13.3.c. Burnside's Theorem. We can now prove:

Theorem 13.3.5 (Burnside). *Let p and q be prime numbers. Every group of order $p^m q^n$ is solvable.*

Proof. We show that a group G of order $p^m q^n$ cannot be a simple nonabelian group. Since every simple factor of G has order $p^k q^\ell$ for some $k \leqslant m$, $\ell \leqslant n$, it follows that all simple factors of G are abelian and that G is solvable.

Let G be a simple nonabelian group of order $p^m q^n$. Since p-groups are solvable we may assume that $p \neq q$ and that $m, n > 0$.

We show that $Z(\chi_i) = 1$ unless χ_i is the trivial character. Assume that $Z(\chi_i) \neq 1$. Since G is simple, $Z(\chi_i) = G$ by Lemma 13.3.2. Then $|\chi_i(g)| = d_i$ for all $g \in G$. By Lemma 13.3.1 and Theorem 13.2.11,

$$|G| \, d_i^2 = \sum_{g \in G} \chi_i(g) \, \overline{\chi_i(g)} = \sum_{g \in G} \chi_i(g) \, \chi_i(g^{-1}) = |G|.$$

Hence $d_i = 1$ and $\rho_i : G \longrightarrow \mathbb{C}^*$. Now $\mathrm{Ker}\, \rho_i \neq 1$, since G is nonabelian, so $\mathrm{Ker}\, \rho_i = G$ and χ_i is trivial ($\chi_i(g) = 1$ for all $g \in G$).

Let S be a Sylow q-subgroup of G. Then $Z(S) \neq 1$, and there exists $h \in Z(S)$, $h \neq 1$. Then $h \notin Z(\chi_i)$ unless χ_i is the trivial character. Moreover the centralizer of h contains S, its index is a power p^k of p, and the conjugacy class of h has p^k elements. Let χ_r be the regular character of G.

We have $\chi_r = \sum_{1 \leqslant i \leqslant s} d_i \chi_i$ (Lemma 13.2.10) and $\chi_r(h) = 0$, since $h \neq 1$ (Proposition 13.2.1). Number $\chi_1, \ldots, \chi_s$ so that χ_1 is the trivial character. Then

$$0 = \sum_{1 \leqslant i \leqslant s} d_i \, \chi_i(h) = 1 + \sum_{i>1, \, p \nmid d_i} d_i \, \chi_i(h) + \sum_{i>1, \, p \mid d_i} d_i \, \chi_i(h) . \quad (13.1)$$

If $i > 1$ and $p \nmid d_i$, then $\chi_i(h) = 0$ by Proposition 13.3.4, since $h \notin Z(\chi_i)$; otherwise, $\chi_i(h)$ is integral over $\mathbb{Z}$ by Proposition 13.2.3. Hence (13.1) yields an equality $1 + p\alpha = 0$, where α is integral over $\mathbb{Z}$. Then $\alpha = -1/p \notin \mathbb{Z}$ is integral over $\mathbb{Z}$, contradicting Proposition 11.3.1. $\qquad\qquad\square$

14

MORE MODULES

This chapter continues Chapter 10 and studies a number of basic constructions and some important classes of modules. The last sections introduce, informally, a few concepts from category theory (functors, natural homomorphisms, exactness), and form an introduction to Chapters 15 and 17 (and to homological algebra proper).

Required results: Chapter 10, Sections 10.1 through 10.5. Sections 14.4 and 14.5 also use properties from Sections 10.7 and 12.4.

All rings in this chapter have an identity element; all modules and bimodules are unital.

14.1. PULLBACKS AND PUSHOUTS

Pullbacks and pushouts are commutative squares with universal properties. These basic constructions will be used in the next sections. We study them with left R-modules, but the results extend to right R-modules and to bimodules.

14.1.a. Pullbacks. A **pullback** of left R-modules is a commutative square

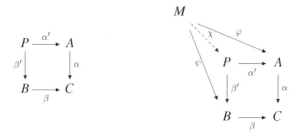

$\alpha \circ \alpha' = \beta \circ \beta'$ of modules and homomorphisms, with the following universal property: for every commutative square $\alpha \circ \varphi = \beta \circ \psi$ (with the same α and β), there exists a unique homomorphism χ such that $\varphi = \alpha' \circ \chi$ and $\psi = \beta' \circ \chi$.

411

For example,

is a pullback (where π and ρ are the projections). Other examples include intersections and inverse images (see the exercises).

Proposition 14.1.1. *Given homomorphisms* $\alpha : A \longrightarrow C$ *and* $\beta : B \longrightarrow C$ *of left R-modules, there exists a pullback* $\alpha \circ \alpha' = \beta \circ \beta'$, *and it is unique up to isomorphism.*

Proof. Uniqueness follows from the universal property. Let $\alpha \circ \alpha' = \beta \circ \beta'$ and $\alpha \circ \alpha'' = \beta \circ \beta''$ be pullbacks. Then there exists a homomorphism χ such that $\alpha'' = \alpha' \circ \chi$ and $\beta'' = \beta' \circ \chi$, and a homomorphism χ' such that $\alpha' = \alpha'' \circ \chi'$ and $\beta' = \beta'' \circ \chi'$. Then $\alpha' = \alpha' \circ \chi \circ \chi'$ and $\beta' = \beta' \circ \chi \circ \chi'$; by uniqueness in the universal property of $\alpha \circ \alpha' = \beta \circ \beta'$, $\chi \circ \chi'$ must be the identity. Similarly $\alpha'' = \alpha'' \circ \chi' \circ \chi$ and $\beta'' = \beta'' \circ \chi' \circ \chi$; by uniqueness in the universal property of $\alpha \circ \alpha'' = \beta \circ \beta''$, $\chi' \circ \chi$ must be the identity. Thus χ and χ' are mutually inverse isomorphisms.

Existence is proved by constructing a pullback. Pairs of homomorphisms $\varphi, \psi : M \longrightarrow A, B$ can be described by single homomorphisms $M \longrightarrow A \times B$; this suggests that P can be retrieved from the direct product $A \times B$. Indeed:

Proposition 14.1.2. *Given homomorphisms* $\alpha : A \longrightarrow C$ *and* $\beta : B \longrightarrow C$ *of left R-modules, let*

$$P = \{ (a,b) \in A \oplus B \,;\, \alpha(a) = \beta(b) \} \,;$$

let $\alpha' : P \longrightarrow A$, $\beta' : P \longrightarrow B$ *be induced by the projections. Then* $\alpha \circ \alpha' = \beta \circ \beta'$ *is a pullback.*

Proof. The construction insures that $\alpha \circ \alpha' = \beta \circ \beta'$. Now let $\varphi : M \longrightarrow A$ and $\psi : M \longrightarrow B$ be homomorphisms such that $\alpha \circ \varphi = \beta \circ \psi$. Then $\chi(x) = (\varphi(x), \psi(x)) \in P$ for every $x \in M$; this defines a homomorphism $\chi : M \longrightarrow P$ such that $\alpha' \circ \chi = \varphi$ and $\beta' \circ \chi = \psi$. If $\chi' : M \longrightarrow P$ is another homomorphism such that $\alpha' \circ \chi' = \varphi$ and $\beta' \circ \chi' = \psi$, then for each $x \in M$ the components of $\chi'(x) \in A \oplus B$ are $\varphi(x)$ and $\psi(x)$; hence $\chi' = \chi$. $\qquad\square$

By Proposition 14.1.1, this construction yields every pullback, up to isomorphism.

14.1.b. Properties. The following properties are readily proved, using either the definition of a pullback or the construction in Proposition 14.1.2:

Proposition 14.1.3 (Transfer). *Let $\alpha \circ \alpha' = \beta \circ \beta'$ be a pullback.*

(1) *If α is a monomorphism (= is injective), then so is β'.*

(2) *If α is an epimorphism (= is surjective), then so is β'.*

Proposition 14.1.4 (Juxtaposition). *In the commutative diagram*

$$
\begin{array}{ccccc}
Q & \xrightarrow{\gamma'} & P & \xrightarrow{\beta'} & A \\
{\scriptstyle\alpha''}\downarrow & & {\scriptstyle\alpha'}\downarrow & & \downarrow{\scriptstyle\alpha} \\
C & \xrightarrow{\gamma} & B & \xrightarrow{\beta} & D
\end{array}
$$

(1) *if $\alpha \circ \beta' = \beta \circ \alpha'$ and $\alpha' \circ \gamma' = \gamma \circ \alpha''$ are pullbacks, then $\alpha \circ (\beta' \circ \gamma') = (\beta \circ \gamma) \circ \alpha''$ is a pullback;*

(2) *if $\alpha \circ \beta' = \beta \circ \alpha'$ and $\alpha \circ (\beta' \circ \gamma') = (\beta \circ \gamma) \circ \alpha''$ are pullbacks, then $\alpha' \circ \gamma' = \gamma \circ \alpha''$ is a pullback.*

14.1.c. Pushouts. A **pushout** of left R-modules is a commutative square

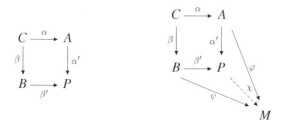

$\alpha' \circ \alpha = \beta' \circ \beta$ of modules and homomorphisms, with the following universal property: for every commutative square $\varphi \circ \alpha = \psi \circ \beta$ (with the same α and β), there exists a unique homomorphism χ such that $\varphi = \chi \circ \alpha'$ and $\psi = \chi \circ \beta'$.

For example,

$$
\begin{array}{ccc}
0 & \longrightarrow & A \\
\downarrow & & \downarrow{\scriptstyle\iota} \\
B & \xrightarrow{\kappa} & A \oplus B
\end{array}
$$

is a pushout (where ι and κ are the injections). The exercises give other examples.

Proposition 14.1.5. *Given homomorphisms $\alpha : C \longrightarrow A$ and $\beta : C \longrightarrow B$ of left R-modules, there exists a pushout $\alpha' \circ \alpha = \beta' \circ \beta$, and it is unique up to isomorphism.*

As with Proposition 14.1.1, uniqueness follows from the universal property, and existence is proved by constructing a pushout. Since every pair of homomorphisms $\varphi, \psi : A, B \longrightarrow M$ can be described by a single homomorphism $A \oplus B \longrightarrow M$, P can be retrieved from the direct sum $A \oplus B$ as follows:

Proposition 14.1.6. *Given homomorphisms* $\alpha : C \longrightarrow A$ *and* $\beta : C \longrightarrow B$ *of left R-modules, let* $\iota : A \longrightarrow A \oplus B$ *and* $\kappa : B \longrightarrow A \oplus B$ *be the injections,* $P = (A \oplus B)/\mathrm{Im}\,(\iota \circ \alpha - \kappa \circ \beta)$, *with projection* $\pi : A \oplus B \longrightarrow P$, *and* $\alpha' = \pi \circ \iota$, $\beta' = \pi \circ \kappa$. *Then* $\alpha \circ \alpha' = \beta \circ \beta'$ *is a pushout.*

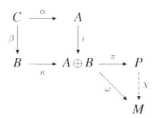

Proof. By definition, $\mathrm{Ker}\,\pi = \mathrm{Im}\,(\iota \circ \alpha - \kappa \circ \beta)$; hence $\alpha' \circ \alpha = \pi \circ \iota \circ \alpha = \pi \circ \kappa \circ \beta = \beta' \circ \beta$. Let $\varphi : A \longrightarrow M$ and $\psi : B \longrightarrow M$ be homomorphisms. By the universal property of the direct sum there is a unique homomorphism $\omega : A \oplus B \longrightarrow M$ such that $\omega \circ \iota = \varphi$ and $\omega \circ \kappa = \psi$. If $\varphi \circ \alpha = \psi \circ \beta$, then $\omega(\iota(\alpha(c))) = \omega(\kappa(\beta(c)))$ for all $c \in C$, $\mathrm{Ker}\,\pi = \mathrm{Im}\,(\iota \circ \alpha - \kappa \circ \beta) \subseteq \mathrm{Ker}\,\omega$, and ω factors uniquely through π: $\omega = \chi \circ \pi$ for some unique $\chi : P \longrightarrow M$. Then $\varphi = \omega \circ \iota = \chi \circ \alpha'$ and $\psi = \omega \circ \kappa = \chi \circ \beta'$. If, conversely, $\varphi = \chi' \circ \alpha'$ and $\psi = \chi' \circ \beta'$, then $\chi' \circ \pi \circ \iota = \varphi = \chi \circ \pi \circ \iota$ and $\chi' \circ \pi \circ \kappa = \psi = \chi \circ \pi \circ \kappa$; by the uniqueness in the universal property of $A \oplus B$, this implies that $\chi' \circ \pi = \chi \circ \pi$ and $\chi' = \chi$. $\square$

By Proposition 14.1.5, this construction yields every pushout, up to isomorphism.

14.1.d. Properties. Pushouts are similar to pullbacks; in fact pushouts can be defined by reversing all arrows in the definition of pullbacks. This process is explained in Chapter 17 and can be further observed in the following properties, whose proofs will entertain our reader:

Proposition 14.1.7 (Transfer). *Let* $\alpha' \circ \alpha = \beta' \circ \beta$ *be a pushout.*

(1) *If* α *is a monomorphism, then so is* β'.
(2) *If* α *is an epimorphism, then so is* β'.

Proposition 14.1.8 (Juxtaposition). *In the commutative diagram*

$$
\begin{array}{ccccc}
D & \xrightarrow{\beta} & B & \xrightarrow{\gamma} & C \\
\downarrow{\alpha} & & \downarrow{\alpha'} & & \downarrow{\alpha''} \\
A & \xrightarrow{\beta'} & P & \xrightarrow{\gamma'} & Q
\end{array}
$$

(1) *if* $\alpha' \circ \beta = \beta' \circ \alpha$ *and* $\alpha'' \circ \gamma = \gamma' \circ \alpha'$ *are pushouts, then* $\alpha'' \circ (\gamma \circ \beta) = (\gamma' \circ \beta') \circ \alpha$ *is a pushout;*

(2) *if* $\alpha' \circ \beta = \beta' \circ \alpha$ *and* $\alpha'' \circ (\gamma \circ \beta) = (\gamma' \circ \beta') \circ \alpha$ *are pushouts, then* $\alpha'' \circ \gamma = \gamma' \circ \alpha'$ *is a pushout.*

Exercises

1. Let K and L be submodules of A. Show that

$$
\begin{array}{ccc}
K \cap L & \longrightarrow & K \\
\downarrow & & \downarrow \\
L & \longrightarrow & A
\end{array}
$$

is a pullback (of inclusion mappings).

2. Let $\varphi : A \longrightarrow B$ be a module homomorphism. Show that

$$
\begin{array}{ccc}
\text{Ker}\,\varphi & \longrightarrow & 0 \\
\subseteq \downarrow & & \downarrow \\
A & \xrightarrow{\varphi} & B
\end{array}
$$

is a pullback.

3. Let $\varphi : A \longrightarrow B$ be a module homomorphism and C be a submodule of B. Show that

$$
\begin{array}{ccc}
\varphi^{-1}(C) & \longrightarrow & C \\
\subseteq \downarrow & & \downarrow \subseteq \\
A & \xrightarrow{\varphi} & B
\end{array}
$$

is a pullback.

4. Let $\alpha \circ \alpha' = \beta \circ \beta'$ be a pullback. If α is a monomorphism, then prove that β' is a monomorphism.

5. Let $\alpha \circ \alpha' = \beta \circ \beta'$ be a pullback. If α is an epimorphism, then prove that β' is an epimorphism.

6. In the commutative diagram

$$
\begin{array}{ccccc}
Q & \xrightarrow{\gamma'} & P & \xrightarrow{\beta'} & A \\
\alpha'' \downarrow & & \alpha' \downarrow & & \downarrow \alpha \\
C & \xrightarrow{\gamma} & B & \xrightarrow{\beta} & D
\end{array}
$$

(a) if $\alpha \circ \beta' = \beta \circ \alpha'$ and $\alpha' \circ \gamma' = \gamma \circ \alpha''$ are pullbacks, then prove that $\alpha \circ (\beta' \circ \gamma') = (\beta \circ \gamma) \circ \alpha''$ is a pullback;

(b) if $\alpha \circ \beta' = \beta \circ \alpha'$ and $\alpha \circ (\beta' \circ \gamma') = (\beta \circ \gamma) \circ \alpha''$ are pullbacks, then prove that $\alpha' \circ \gamma' = \gamma \circ \alpha''$ is a pullback.

7. Let K and L be submodules of A. Show that

$$
\begin{array}{ccc}
A & \longrightarrow & A/K \\
\downarrow & & \downarrow \\
A/L & \longrightarrow & A/(K+L)
\end{array}
$$

is a pushout (of projections).

8. Let $\varphi : A \longrightarrow B$ be a module homomorphism. Show that

$$
\begin{array}{ccc}
A & \stackrel{\varphi}{\longrightarrow} & B \\
\downarrow & & \downarrow \\
0 & \longrightarrow & B/\mathrm{Im}\,\varphi
\end{array}
$$

is a pushout.

9. Let $\alpha' \circ \alpha = \beta' \circ \beta$ be a pushout. If α is an epimorphism, then prove that β' is an epimorphism.

10. Let $\alpha' \circ \alpha = \beta' \circ \beta$ be a pushout. If α is a monomorphism, then prove that β' is a monomorphism.

11. In the commutative diagram

$$
\begin{array}{ccccc}
D & \stackrel{\beta}{\longrightarrow} & B & \stackrel{\gamma}{\longrightarrow} & C \\
\alpha \downarrow & & \alpha' \downarrow & & \alpha'' \downarrow \\
A & \underset{\beta'}{\longrightarrow} & P & \underset{\gamma'}{\longrightarrow} & Q
\end{array}
$$

(a) if $\alpha' \circ \beta = \beta' \circ \alpha$ and $\alpha'' \circ \gamma = \gamma' \circ \alpha'$ are pushouts, then prove that $\alpha'' \circ (\gamma \circ \beta) = (\gamma' \circ \beta') \circ \alpha$ is a pushout;

(b) if $\alpha' \circ \beta = \beta' \circ \alpha$ and $\alpha'' \circ (\gamma \circ \beta) = (\gamma' \circ \beta') \circ \alpha$ are pushouts, then prove that $\alpha'' \circ \gamma = \gamma' \circ \alpha'$ is a pushout.

12. Define a pullback of sets. Prove that pullbacks of sets exist and are unique up to isomorphism.

13. Define a pushout of (not necessarily abelian) groups. Show that a free product of two groups amalgamating a subgroup is a pushout.

14.2. PROJECTIVE MODULES

Projective modules are an important class of modules. This section contains their basic properties, with some applications to ring theory. The results are stated for left R-modules and extend to right R-modules.

14.2.a. Definition. A left R-module P is **projective** in case every homomorphism with domain P can be factored or **lifted** through every epimorphism: that is, if M and N are left R-modules, $\varphi : P \longrightarrow N$ is a homomorphism, and $\rho : M \longrightarrow N$ is an epimorphism (= a surjective homomorphism), then there exists a homomorphism $\psi : P \longrightarrow M$ such that $\varphi = \rho \circ \psi$:

$$
\begin{array}{ccc}
 & & P \\
 & \psi \swarrow & \downarrow \varphi \\
M & \xrightarrow{\ \rho\ } & N
\end{array}
$$

This factorization need not be unique (ψ need not be unique in the above).

If P is projective, then every epimorphism $\rho : M \longrightarrow P$ **splits** ($\rho \circ \mu = 1_P$ for some $\mu : P \longrightarrow M$), since 1_P can be lifted through ρ. Then $M = \operatorname{Ker} \rho \oplus \operatorname{Im} \mu$ by Corollary 10.4.10, and P is isomorphic to a direct summand of M. In fact:

Proposition 14.2.1. *A left R-module P is projective if and only if every epimorphism $\rho : M \longrightarrow P$ splits.*

Proof. Let $\varphi : P \longrightarrow N$ and $\rho : M \longrightarrow N$ be homomorphisms, with ρ surjective. Construct the pullback $\varphi \circ \rho' = \rho \circ \varphi'$:

$$
\begin{array}{ccc}
Q & \xrightarrow{\ \rho'\ } & P \\
\varphi' \downarrow & \xleftarrow{\ \nu\ } & \downarrow \varphi \\
M & \xrightarrow{\ \rho\ } & N
\end{array}
$$

By Proposition 14.1.3, ρ' is an epimorphism. Hence ρ' splits: $\rho' \circ \nu = 1_P$ for some $\nu : P \longrightarrow Q$. Then $\varphi = \varphi \circ \rho' \circ \nu = \rho \circ \varphi' \circ \nu$ can be lifted through ρ. □

14.2.b. Examples

Proposition 14.2.2. *Every free module is projective.*

Proof. Let F be a free left R-module, with a basis $(e_i)_{i \in I}$, and $\rho : M \longrightarrow N$ be an epimorphism. We show that every homomorphism $\varphi : F \longrightarrow N$ can be lifted through ρ. Since ρ is surjective, there is for each $i \in I$ some $m_i \in M$ such that $\rho(m_i) = \varphi(e_i)$. By the universal property of F, there is a homomorphism $\psi : F \longrightarrow M$ such that $\psi(e_i) = m_i$ for all i. Then $\rho(\psi(e_i)) = \varphi(e_i)$ for all i, and it follows from the uniqueness in the universal property of F that $\rho \circ \psi = \varphi$. □

Proposition 14.2.3. *A module is projective if and only if it is isomorphic to a direct summand of a free module.*

Proof. Let P be projective. By Corollary 10.5.5, there is an epimorphism $F \longrightarrow P$ where F is free. This splits by Proposition 14.2.1. Therefore P is isomorphic to a direct summand of F.

Conversely, let F be free and A be a direct summand of F. Then $F = A \oplus B$ for some submodule B. Let $\iota : A \longrightarrow F$ be the injection and $\pi : F \longrightarrow A$ be the projection. Let $\varphi : A \longrightarrow N$ be a homomorphism and $\rho : M \longrightarrow N$ be an epimorphism:

$$
\begin{array}{ccc}
F & \underset{\iota}{\overset{\pi}{\rightleftarrows}} & A \\
{\scriptstyle\psi}\downarrow & & \downarrow{\scriptstyle\varphi} \\
M & \underset{\rho}{\longrightarrow} & N
\end{array}
$$

By Proposition 14.2.2, $\varphi \circ \pi$ can be lifted through ρ : $\varphi \circ \pi = \rho \circ \psi$ for some homomorphism $\psi : F \longrightarrow M$. Since $\pi \circ \iota = 1_A$, this implies $\rho \circ \psi \circ \iota = \varphi$, and φ can be lifted through ρ. Thus A is projective; so is any module $P \cong A$. $\square$

From previous sections we know that all modules over a division ring are free, hence projective (Theorem 10.6.1), and that all projective modules over a PID are free (Theorem 10.7.1). The following result is readily proved using Theorem 12.4.3:

Proposition 14.2.4. *For a ring R the following conditions are equivalent: (1) R is semisimple artinian; (2) every left R-module is projective; (3) every right R-module is projective.*

Proposition 14.2.4 gives ready examples of projective modules which are not free. Another example is given in the exercises.

14.2.c. Properties. The following property is an exercise:

Proposition 14.2.5. $\bigoplus_{i \in I} M_i$ *is projective if and only if every M_i is projective.*

Proposition 14.2.5 implies that a direct product of finitely many projective modules is projective. This property does not extend to infinite products: by the next result, the direct product of countably many copies of $\mathbb{Z}$ is not a projective $\mathbb{Z}$-module.

Proposition 14.2.6 (Baer). *The direct product of countably many copies of $\mathbb{Z}$ is not a free abelian group.*

Proof. Let $A = \prod_{n>0} \mathbb{Z}$; the elements of A are all infinite sequences $(x_n)_{n>0}$ of integers. We see that $|A| = \aleph_0^{\aleph_0} \geqslant 2^{\aleph_0} > \aleph_0$, and A is not countable. (Actually $|A| = |\mathbb{R}|$.) Let p be any prime number and B be the subgroup of all $(x_n)_{n>0} \in A$ such that, for every $k > 0$, x_n is divisible by p^k for almost all n. For example, $(p^n x_n)_{n>0} \in B$ for every $(x_n)_{n>0} \in A$; therefore B is not countable.

We show that B/pB is countable. Let R be the set of all sequences $(x_n)_{n>0} \in A$ such that $0 \leqslant x_n < p$ for all n and $x_n = 0$ for almost all n. We see that $R \subseteq B$ and that R is countable. For each $x = (x_n)_{n>0} \in B$, division by p yields $x_n = py_n + r_n$, where $0 \leqslant r_n < p$ for all n, and $r_n = 0$ for almost all n, since p divides

x_n for almost all n. Hence $r = (r_n)_{n>0} \in R$. Moreover each p^k divides y_n for almost all n, since p^{k+1} divides x_n for almost all n, and $y = (y_n)_{n>0} \in B$. Thus every coset $x + pB \subseteq B$ contains an element of R; hence B/pB is countable.

Now assume that A is free. Then B is free by Theorem 10.7.1. The rank of B equals the dimension of the vector space B/pB over $\mathbb{Z}_p$ and is countable. Hence B is the direct sum of countably many copies of $\mathbb{Z}$ and is countable; this is the required contradiction.   □

14.2.d. Dedekind Domains. We complete this section with a characterization of Dedekind domains in terms of projective modules.

Proposition 14.2.7. *A nonzero fractional ideal of a domain R is invertible if and only if it is a projective R-module.*

Proof. Let $\mathfrak{A} \neq 0$ be a fractional ideal of R.

If $\mathfrak{A}\mathfrak{B} = R$, then $1 = a_1 b_1 + \cdots + a_n b_n$ for some $a_1,\ldots,a_n \in \mathfrak{A}$ and $b_1,\ldots,b_n \in \mathfrak{B}$. Hence $\mathfrak{A} = Ra_1 + \cdots + Ra_n$, since $a = ab_1 a_1 + \cdots + ab_n a_n$ for every $a \in \mathfrak{A}$. Let $M = Re_1 \oplus \cdots \oplus Re_n$ be the free R-module with basis $e_1,\ldots,e_n$. We have module homomorphisms $\rho: M \longrightarrow \mathfrak{A}$, $r_1 e_1 + \cdots + r_n e_n \longmapsto r_1 a_1 + \cdots + r_n a_n$, and $\kappa: \mathfrak{A} \longrightarrow M$, $a \longmapsto ab_1 e_1 + \cdots + ab_n e_n$. We see that $\rho \circ \kappa = 1_{\mathfrak{A}}$. Therefore $\mathfrak{A}$ is isomorphic to a direct summand $\kappa(\mathfrak{A})$ of M and is projective.

Conversely, assume that $\mathfrak{A} \neq 0$ is a projective R-module. Write $\mathfrak{A}$ as a direct summand of a free module, as follows. Let $(g_i)_{i \in I}$ generate $\mathfrak{A}$ as an R-module so that every element a of $\mathfrak{A}$ can be written in the form $a = \sum_{i \in I} r_i g_i$ with $r_i \in R$ and $r_i = 0$ for almost all i. Let $M = \bigoplus_{i \in I} R$ be the free R-module on the set I. Then $\rho: M \longrightarrow \mathfrak{A}$, $(r_i)_{i \in I} \longmapsto \sum_{i \in I} r_i g_i$ is an epimorphism. Since $\mathfrak{A}$ is projective, ρ splits, and there exists a homomorphism $\kappa: \mathfrak{A} \longrightarrow M$ such that $\rho \circ \kappa = 1_{\mathfrak{A}}$. Let $\pi_i: M = \bigoplus_{i \in I} R \longrightarrow R$ be the i-th projection. For each $a \in \mathfrak{A}$ denote $\pi_i(\kappa(a))$ by a_i; thus $a_i = 0$ for almost all i, and $\kappa(a) = (a_i)_{i \in I}$.

Let $a, c \in \mathfrak{A}$, $c \neq 0$. In the field of fractions Q of R, we have $ra, sc \in R$ for some $r, s \in R$, $r, s \neq 0$. Since $\varphi_i = \pi_i \circ \kappa: \mathfrak{A} \longrightarrow R$ is a module homomorphism, we have $r\varphi_i(sac) = \varphi_i(rsac) = s\varphi_i(rac)$ and

$$ac_i = a\varphi_i(c) = ra\varphi_i(c)/r = \varphi_i(rac)/r$$
$$= \varphi_i(sac)/s = sc\varphi_i(a)/s = c\varphi_i(a) = ca_i.$$

Hence $ac_i/c = a_i \in R$ for all $a \in A$ and

$$c_i/c \in \mathfrak{A}' = \{x \in Q \,;\, x\mathfrak{A} \subseteq R\}$$

for all i. Moreover

$$c = \rho(\kappa(c)) = \rho((c_i)_{i \in I}) = \sum_{i \in I} c_i g_i = c\left(\sum_{i \in I}(c_i/c)g_i\right)$$

and $1 = \sum_{i \in I}(c_i/c)g_i \in \mathfrak{A}'\mathfrak{A}$. Hence $\mathfrak{A}\mathfrak{A}' = R$.   □

Proposition 14.2.8. *For a domain R the following conditions are equivalent:*

(1) *R is a Dedekind domain.*

(2) *Every fractional ideal of R is projective.*

(3) *Every ideal of R is projective.*

Proof. By definition (Theorem 5.7.4), R is a Dedekind domain if and only if every nonzero fractional ideal of R is invertible. Hence (1) and (2) are equivalent by Proposition 14.2.7. (2) implies (3) since every ideal of R is a fractional ideal of R; (3) implies (2) since every fractional ideal of R is isomorphic to an ideal of R by Proposition 10.5.1. ☐

Exercises

1. Prove that a direct summand of a projective module is projective.

2. Prove that a direct sum of projective modules is projective.

3. Prove that a ring R is semisimple artinian if and only if every left R-module is projective, if and only if every right R-module is projective.

4. Let $m > 1$ and $n > 1$ be relatively prime. Show that $\mathbb{Z}_m$ is a projective, but not free, $\mathbb{Z}_{mn}$-module.

5. Give another example of a projective module which is not free.

14.3. INJECTIVE MODULES

Injective modules are another important class of modules. The basic results in this section are stated for left R-modules but extend to right R-modules.

14.3.a. Definition. A left R-module J is **injective** in case every homomorphism into J can be factored or **extended** through every monomorphism: that is, if M and N are left R-modules, $\varphi : M \longrightarrow J$ is a homomorphism, and $\mu : M \longrightarrow N$ is a monomorphism, then there exists a homomorphism $\psi : N \longrightarrow J$ such that $\varphi = \psi \circ \mu$:

This factorization need not be unique (ψ need not be unique in the above). If $M \subseteq N$ and $\mu : M \longrightarrow N$ is the inclusion homomorphism, then $\psi_{|M} = \varphi$ and ψ extends φ in the usual sense.

If J is injective, then every monomorphism $\mu : J \longrightarrow M$ **splits** ($\rho \circ \mu = 1_J$ for some $\rho : M \longrightarrow J$), since 1_J can be extended through μ. Then $M = \operatorname{Ker} \rho \oplus \operatorname{Im} \mu$ by Corollary 10.4.10, and J is isomorphic to a direct summand of M. In fact:

Proposition 14.3.1. *For a left R-module J the following are equivalent:*

(1) *J is injective.*

(2) *Every monomorphism $J \longrightarrow M$ splits.*

(3) *J is a direct summand of every left R-module $M \supseteq J$.*

Proof. We show that (2) implies (1). Let $\varphi : M \longrightarrow J$ and $\mu : M \longrightarrow N$ be homomorphisms, with μ injective. Construct the pushout $\varphi' \circ \mu = \mu' \circ \varphi$:

$$
\begin{array}{ccc}
J & \underset{\rho}{\overset{\mu'}{\rightleftarrows}} & P \\
\varphi \uparrow & & \uparrow \varphi' \\
M & \underset{\mu}{\longrightarrow} & N
\end{array}
$$

By Proposition 14.1.7, μ' is a monomorphism. By (2), μ' splits: $\rho \circ \mu' = 1_J$ for some $\rho : P \longrightarrow J$. Then $\varphi = \rho \circ \mu' \circ \varphi = \rho \circ \varphi' \circ \mu$ can be extended through μ. $\qquad\square$

14.3.b. Examples. Vector spaces are injective. More generally, Theorem 12.4.3 readily implies the following result.

Proposition 14.3.2. *For a ring R the following conditions are equivalent:* (1) *R is semisimple artinian;* (2) *every left R-module is injective;* (3) *every right R-module is injective.*

More interesting examples are provided by the next result.

Proposition 14.3.3 (Baer's Criterion). *For a left R-module J the following conditions are equivalent:*

(1) *J is injective.*

(2) *Every R-homomorphism of a left ideal of R into J can be extended to $_R R$.*

(3) *For each R-homomorphism φ of a left ideal L of R into J, there exists $m \in J$ such that $\varphi(r) = rm$ for all $r \in L$.*

Proof. (1) implies (2); the equivalence of (2) and (3) is an exercise. We show that (2) implies (1). Let J have property (2). We want to prove that every module homomorphism $\varphi : M \longrightarrow J$ can be extended through every monomorphism $\mu : M \longrightarrow N$. It suffices to prove this in case M is a submodule of N and μ is the inclusion homomorphism.

Let $\mathscr{S}$ be the set of all ordered pairs (A, α) in which A is a submodule of N containing M and $\alpha : A \longrightarrow J$ is a homomorphism which extends φ. Order $\mathscr{S}$ by: $(A, \alpha) \leqslant (B, \beta)$ if and only if $A \subseteq B$ and β extends α. In $\mathscr{S}$ every chain $(A_i, \alpha_i)_{i \in I}$ has an upper bound (A, α), where $A = \bigcup_{i \in I} A_i$ and $\alpha(x) = \alpha_i(x)$ whenever $x \in A_i$. Thus $\mathscr{S}$ is inductive, and Zorn's Lemma provides a maximal element (C, γ) of $\mathscr{S}$. We show that $C = N$.

Assume that $(A, \alpha) \in \mathscr{S}$ and $A \neq N$. Let $b \in N \backslash A$ and $B = Rb + A$. Then

$$L = \{ r \in R ; rb \in A \}$$

is a left ideal of R and $r \longmapsto \alpha(rb)$ is an R-homomorphism of L into J. By (2) there is a homomorphism $\chi : {}_R R \longrightarrow J$ such that $\chi(r) = \alpha(rb)$ for all $r \in L$. If $rb + a = r'b + a'$, then $(r - r')b = a' - a \in A$, $r - r' \in L$, $\chi(r - r') = \alpha(rb - r'b)$, and $\chi(r) + \alpha(rb) = \chi(r') + \alpha(r'b)$. Hence a mapping $\beta : B \longrightarrow J$ is defined by

$$\beta(rb + a) = \chi(r) + \alpha(a)$$

for all $r \in R$ and $a \in A$. We see that β is an R-homomorphism and extends α. Since $A \subsetneq B$ this shows that (A, α) is not maximal in $\mathscr{S}$. Therefore the maximal element (C, γ) of $\mathscr{S}$ satisfies $C = N$, and φ can be extended to all of N. $\square$

A left R-module M is **divisible** in case $rM = M$ for all $r \neq 0$, $r \in R$ (in case the equation $rx = m$ has a solution in M for every $0 \neq r \in R$ and $m \in M$).

Corollary 14.3.4. *When R is a principal ideal domain, an R-module is injective if and only if it is divisible.*

Proof. Let J be injective, $a \in J$, and $0 \neq r \in R$. Since R is a domain, every element of Rr can be written in the form tr for some unique $t \in R$, and there is a module homomorphism $\varphi : tr \longmapsto ta$ of Rr into J. Since J is injective, there exists $m \in J$ such that $\varphi(s) = sm$ for all $s \in R$. Then $a = \varphi(r) = rm$. Thus J is divisible.

Conversely, let M be a divisible R-module. Let Rr be any (left) ideal of R and $\varphi : Rr \longrightarrow M$ be a module homomorphism. If $r = 0$, then φ extends to the zero homomorphism ${}_R R \longmapsto M$. Otherwise, let $a = \varphi(r) \in M$. Since M is divisible, we have $a = rb$ for some $b \in M$. Hence $\varphi(tr) = ta = trb$ for all $tr \in Rr$, and M is injective by Baer's criterion. $\square$

14.3.c. Injective Abelian Groups. By Corollary 14.3.4, an abelian group is injective if and only if it is divisible. For example, $\mathbb{Q}$ (under addition) is injective.

We construct another divisible abelian group. For each prime p let

$$\mathbb{Z}_{p^\infty} = \langle a_1, \ldots, a_n, \ldots ; \ pa_1 = 0, \ pa_n = a_{n-1} \text{ for all } n > 1 \rangle.$$

Proposition 14.3.5. $\mathbb{Z}_{p^\infty}$ *is the union of cyclic subgroups $C_1 \subseteq C_2 \subseteq \cdots \subseteq C_n \subseteq \cdots$ of orders $p, p^2, \ldots, p^n, \ldots$ and is a divisible abelian group.*

Proof. First we find a model of $\mathbb{Z}_{p^\infty}$. Let U be the multiplicative group of all complex roots of unity and $\alpha_n = e^{2i\pi/p^n} \in U$; α_n is a primitive p^n-th root of unity. Also $\alpha_1^p = 1$ and $\alpha_n^p = \alpha_{n-1}$ for all $n > 1$. Hence there is a homomorphism $\varphi : \mathbb{Z}_{p^\infty} \longrightarrow U$ such that $\varphi(a_n) = \alpha_n$ for all n. (Actually φ is an isomorphism of $\mathbb{Z}_{p^\infty}$ onto $U(p) = \{\omega \in \mathbb{C}; \omega^{p^k} = 1 \text{ for some } k > 0\}$; see the exercises.)

Now a_n has order at least p^n, since $\alpha_n = \varphi(a_n)$ has order p^n in U. But an easy induction shows that $p^n a_n = 0$ for all n. Hence a_n has order p^n, and $C_n = \langle a_n \rangle \subseteq \mathbb{Z}_{p^\infty}$ has order p^n. Also $C_n \subseteq C_{n+1}$, since $pa_{n+1} = a_n$. Each $x \in \mathbb{Z}_{p^\infty}$ is a linear combination $x = \sum_{n>0} x_n a_n$, with $x_n \in \mathbb{Z}$, $x_n = 0$ for almost all n; since $a_n \in C_m$ for all $n \leqslant m$, we have $x \in C_m$ if $x_n = 0$ for all $n > m$. Therefore $\mathbb{Z}_{p^\infty} = \bigcup_{n>0} C_n$.

Let $m \neq 0$. Then $m = p^k \ell$, where p does not divide ℓ. We show separately that every $x \in \mathbb{Z}_{p^\infty}$ is divisible by p^k and divisible by ℓ; it follows that every $x \in \mathbb{Z}_{p^\infty}$ is divisible by m. First $x \in C_n$ for some n, so that $x = ta_n = tpa_{n+1} = \cdots = tp^k a_{n+k}$ is divisible by p^k. Next $up^n + v\ell = 1$ for some $u, v \in \mathbb{Z}$; then $p^n x = 0$ and $x = up^n x + v\ell x = v\ell x$ is divisible by ℓ. $\qquad\square$

Theorem 14.3.6. *An abelian group is divisible if and only if it is a direct sum of copies of $\mathbb{Q}$ and $\mathbb{Z}_{p^\infty}$ (for various primes p).*

Proof. A direct sum of copies of $\mathbb{Q}$ and $\mathbb{Z}_{p^\infty}$ is divisible. Conversely, let A be a divisible abelian group. The torsion part

$$T = \{x \in A;\ nx = 0 \text{ for some } n \neq 0\}$$

of A is divisible: if $n \neq 0$ and $nx = a \in T$, then $x \in T$. Hence T is injective, and $A = T \oplus B$, where $B \cong A/T$ is torsion free and divisible like A. In B each equation $nx = b$ (with $n \neq 0$) has a unique solution. Hence B is a $\mathbb{Q}$-module, where $(m/n)b$ is the solution of $nx = mb$. Then B is a direct sum of copies of $\mathbb{Q}$.

By Proposition 10.7.6, T is a direct sum of p-groups $T = \bigoplus_{p \text{ prime}} T(p)$,

$$T(p) = \{x \in T;\ p^k x = 0 \text{ for some } k > 0\}.$$

Each $T(p)$ is divisible since T is divisible.

We show that a divisible p-group A is a direct sum of copies of $\mathbb{Z}_{p^\infty}$. We show that each $b \in A$, $b \neq 0$ belongs to a subgroup $B \cong \mathbb{Z}_{p^\infty}$ of A. Let $p^k > 1$ be the order of b. Construct a sequence $b_1, \ldots, b_n, \ldots$ of elements of A as follows. If $j < k$, then $b_j = p^{k-j}b$; also $b_k = b$. Since A is divisible there exist $b_{k+1}, \ldots$ such that $b_j = pb_{j+1}$ for all $j \geqslant k$. Then $b_j = pb_{j+1}$ for all $j \geqslant 1$; by induction,

b_j has order p^j. Let B be the subgroup of A generated by $b_1, \ldots, b_j, \ldots$. Since $pb_1 = 0$ and $pb_j = b_{j-1}$ for all $j > 0$, there is a homomorphism φ of $\mathbb{Z}_{p^\infty}$ onto B such that $\varphi(a_j) = b_j$ for all j. Since b_j has order p^j, φ is injective on each $\langle a_j \rangle$ and is injective, since $\mathbb{Z}_{p^\infty} = \bigcup_{j>0} \langle a_j \rangle$. Thus $B \cong \mathbb{Z}_{p^\infty}$.

Let $\mathscr{I}$ be the set of all families $(B_i)_{i \in I}$ of subgroups of A such that $B_i \cong \mathbb{Z}_{p^\infty}$ for all i and $\sum_{i \in I} B_i = \bigoplus_{i \in I} B_i$ (equivalently, $B_j \cap (\sum_{i \in I, i \neq j} B_i) = 0$ for all $j \in I$). By Zorn's Lemma, $\mathscr{I}$ has a maximal element. Now every $\bigoplus_{i \in I} B_i$ is divisible; therefore $A = (\bigoplus_{i \in I} B_i) \oplus C$ for some subgroup C. If $\bigoplus_{i \in I} B_i \neq A$, then $C \neq 0$ contains a subgroup $B \cong \mathbb{Z}_{p^\infty}$; then $(B_i)_{i \in I} \cup \{B\}$ is in $\mathscr{I}$ and $(B_i)_{i \in I}$ is not maximal. Therefore $A = \bigoplus_{i \in I} B_i$ when $(B_i)_{i \in I}$ is maximal. $\quad\square$

14.3.d. Properties

Proposition 14.3.7. *Let $(M_i)_{i \in I}$ be a family of left R-modules. The direct product $\prod_{i \in I} M_i$ is injective if and only if every M_i is injective.*

The proof is an exercise. On the other hand, a direct sum of injective modules need not be injective; see Theorem 14.3.10 below.

Theorem 14.3.8. *Every left R-module can be embedded into an injective left R-module.*

We prove this when $R = \mathbb{Z}$; the general case will be proved in Section 14.6, using modules of homomorphisms. For abelian groups, Theorem 14.3.8 reads:

Proposition 14.3.9. *Every abelian group can be embedded into a divisible abelian group.*

Proof. Let A be an abelian group. There is an epimorphism $F \longrightarrow A$ where F is a free abelian group. Now F is a direct sum of copies of $\mathbb{Z}$ and can be embedded into a direct sum D of copies of $\mathbb{Q}$. Construct the pushout

By Proposition 14.1.7, $A \longrightarrow B$ is a monomorphism, and $D \longrightarrow B$ is an epimorphism; since D is divisible, this implies that B is divisible. $\quad\square$

14.3.e. Left Noetherian Rings

Theorem 14.3.10. *A ring R is left noetherian if and only if every direct sum of injective left R-modules is injective.*

Proof. Let R be left noetherian and $J = \bigoplus_{i \in I} J_i$ be the direct sum of a family of injective left R-modules $(J_i)_{i \in I}$. Let L be a left ideal of R and $\varphi : L \longrightarrow J$ be an R-homomorphism. Since R is left noetherian, L is finitely generated: $L = Rt_1 + \cdots + Rt_k$ for some $t_1, \ldots, t_k \in L$. Each $\varphi(t_h) \in \bigoplus_{i \in I} J_i$ has finitely many nonzero components in the direct sum and belongs to a finite direct sum $\bigoplus_{i \in S_h} J_i$ (=with $S_h \subseteq I$ finite). Hence $\operatorname{Im} \varphi$ is contained in a finite direct sum $\bigoplus_{i \in S} J_i$ (= with $S \subseteq I$ finite). Since S is finite, $\bigoplus_{i \in S} J_i$ is injective by Proposition 14.3.7; hence $\varphi : L \longrightarrow \bigoplus_{i \in S} J_i$ extends to a module homomorphism $R \longrightarrow \bigoplus_{i \in S} J_i \subseteq \bigoplus_{i \in I} J_i$. Therefore J is injective by Baer's criterion.

Conversely, assume that every direct sum of injective left R-modules is injective. Let $L_1 \subseteq L_2 \subseteq \cdots \subseteq L_k \subseteq \cdots$ be an ascending sequence of left ideals of R. Then $L = \bigcup_{k > 0} L_k$ is a left ideal of R. For each $k > 0$ the left R-module R/L_k is by Theorem 14.3.8 a submodule of an injective R-module J_k. By the hypothesis, $J = \bigoplus_{k > 0} J_k$ is injective. Let $\varphi : L \longrightarrow J$ be the homomorphism

$$\varphi(x) = (\pi_k(x))_{k > 0}$$

induced by the projections $\pi_k : R \longrightarrow R/L_k \subseteq J_k$ (note that $x \in L$ implies $x \in L_m$ and $\pi_k(x) = 0$ for all $k \geqslant m$). Since J is injective, φ extends to a homomorphism $\psi : R \longrightarrow J$. Then $a = \psi(1) \in \bigoplus_{k > 0} J_k$ has finitely many nonzero components and $a \in \bigoplus_{0 < k < n} J_k$ for some $n > 0$. For every $x \in L$ we now have $\varphi(x) = \psi(x1) = xa \in \bigoplus_{0 < k < n} J_k$; in particular, $\pi_n(x) = 0$ and $x \in L_n$. Thus $L = L_n$, and $L_k = L_n$ for all $k \geqslant n$. $\qquad\square$

Exercises

1. Prove that a ring R is semisimple artinian if and only if every left R-module is injective, if and only if every right R-module is injective.

2. Show that the following two conditions on a left R-module J are equivalent: every R-homomorphism of a left ideal of R into J can be extended to $_RR$; for each R-homomorphism φ of a left ideal L of R into J, there exists $m \in J$ such that $\varphi(r) = rm$ for all $r \in L$. (This completes the proof of Proposition 14.3.3.)

3. Prove that the subgroups $C_1, C_2, \ldots, C_n, \ldots$ in Proposition 14.3.5 are the only nontrivial proper subgroups of $\mathbb{Z}_{p^\infty}$.

4. Show that $\mathbb{Z}_{p^\infty}$ is indecomposable.

5. Let U be the multiplicative group of all complex roots of unity. Show that $U(p) \cong \mathbb{Z}_{p^\infty}$.

6. Prove that $\mathbb{Q}/\mathbb{Z} \cong \bigoplus_{p \text{ prime}} \mathbb{Z}_{p^\infty}$.

7. Let $(M_i)_{i \in I}$ be a family of left R-modules. Prove that $\prod_{i \in I} M_i$ is injective if and only if every M_i is injective.

8. Can you extend Theorem 14.3.6 to modules over any principal ideal domain?

9. Let J is an injective left R-module and $a \in J$, $r \in R$ be such that $\operatorname{Ann}(r) \subseteq \operatorname{Ann}(a)$ (if $t \in R$ and $tr = 0$, then $ta = 0$). Prove that $a = rx$ for some $x \in J$.

10. Let R be a domain and F be its field of fractions. Prove that F is an injective R-module.

14.4. THE INJECTIVE HULL

In Section 14.5 we learned that every module M has an injective extension. In this section we show that every module M has, up to isomorphism, a smallest injective extension; this is the injective hull of M. The results are stated for left R-modules but extend to right R-modules.

14.4.a. Essential Submodules. A submodule S of a left R-module M is **essential** in case $S \cap T \neq 0$ for every submodule $T \neq 0$ of M; equivalently, in case for each $a \in M$, $a \neq 0$, there exists $r \in R$ such that $0 \neq ra \in S$. Then M is an **essential extension** of S. Essential submodules are also called **large**.

A monomorphism $\mu : S \longrightarrow M$ is **essential** in case $\operatorname{Im} \mu$ is an essential submodule of M.

When μ is an essential monomorphism and $\varphi \circ \mu$ is injective, then $\operatorname{Ker} \varphi \cap \operatorname{Im} \mu = 0$ (since $\varphi(\mu(x)) = 0$ implies $x = 0$), $\operatorname{Ker} \varphi = 0$, and φ is injective.

Proposition 14.4.1. *Let $M \xrightarrow{\mu} N \xrightarrow{\nu} P$ be monomorphisms. Then $\nu \circ \mu$ is essential if and only if μ and ν are essential.*

The proof is an exercise.

14.4.b. Essential Extensions

Proposition 14.4.2. *Let $\mu : M \longrightarrow N$ and $\nu : M \longrightarrow J$ be monomorphisms. If μ is essential and J is injective, then there exists a monomorphism $\xi : N \longrightarrow J$ such that $\nu = \xi \circ \mu$.*

Proof. Since J is injective, $\nu = \xi \circ \mu$ for some $\xi : N \longrightarrow J$; ξ is a monomorphism, since μ is essential, and $\xi \circ \mu$ is a monomorphism. $\qquad\square$

Proposition 14.4.2 states that every essential extension can be embedded into every injective extension.

Proposition 14.4.3. *A left R-module J is injective if and only if every essential monomorphism $J \longrightarrow M$ is an isomorphism.*

Proof. Let J be injective and $\mu : J \longrightarrow M$ be an essential monomorphism. Since J is injective, μ splits: $\sigma \circ \mu = 1_J$ for some homomorphism $\sigma : M \longrightarrow J$. Then σ is surjective, and injective since μ is essential; hence σ is an isomorphism, and $\mu = \sigma^{-1}$ is also an isomorphism.

Conversely, assume that every essential monomorphism $\mu : J \longrightarrow M$ is an isomorphism. We show that J is a direct summand of every extension $M \supseteq J$. By Zorn's Lemma, there is a submodule K of M maximal such that $J \cap K = 0$. Let $\mu : J \longrightarrow M/K$ be induced by the projection $\pi : M \longrightarrow M/K$. Then μ is a monomorphism: if $\mu(x) = 0$, then $x \in J \cap K = 0$. Moreover μ is essential: a nonzero submodule of M/K has the form L/K where $K \subsetneq L \subseteq M$; since K is maximal, $J \cap L \neq 0$ and $\mu(J) \cap (L/K) = \mu(J \cap L) \neq 0$. By the hypothesis, μ is an isomorphism; hence $\mu(J) = M/K$, $J + K = M$, and $M = J \oplus K$. $\square$

14.4.c. The Injective Hull

Theorem 14.4.4. *For every left R-module M there exists an essential mono-morphism $\mu : M \longrightarrow J$ into an injective R-module J, and μ is unique up to isomorphism.*

Proof. By Theorem 14.3.8 there is an extension $M \subseteq K$ with K injective. Let $\mathscr{S}$ be the set of all submodules $M \subseteq S \subseteq K$ in which M is essential. If $(S_i)_{i \in I}$ is a chain of submodules of K and M is essential in each S_i, then M is essential in $S = \bigcup_{i \in I} S_i$: if $N \neq 0$ is a submodule of S, then $N \cap S_i \neq 0$ for some i and $M \cap N = M \cap N \cap S_i \neq 0$, since M is essential in S_i. By Zorn's Lemma, $\mathscr{S}$ has a maximal element J. If J had a proper essential extension, then by Proposition 14.4.2, J would have a proper essential extension $J \subsetneq J' \subseteq K$ and would not be maximal. Therefore J is injective by Proposition 14.4.3.

Now let $\mu : M \longrightarrow J$ and $\nu : M \longrightarrow K$ be essential monomorphisms into injective modules J and K. By Proposition 14.4.2, there is a monomorphism $\theta : J \longrightarrow K$ such that $\nu = \theta \circ \mu$. Then θ is essential by Proposition 14.4.1, since ν is essential, and θ is an isomorphism by Proposition 14.4.3, since J is injective. $\square$

The injective module J in Theorem 14.4.4, which is unique up to isomorphism, is the **injective hull** or **injective envelope** of M. Up to isomorphism, J is an injective, essential extension of M; J is a maximal essential extension of M (by Proposition 14.4.3) as well as a minimal injective extension of M (by Proposition 14.4.2). Proposition 14.4.2 also shows that up to isomorphism, J is the largest essential extension of M and the smallest injective extension of M.

Exercises

1. Let $M \xrightarrow{\mu} N \xrightarrow{\nu} P$ be monomorphisms. If μ and ν are essential, then prove that $\nu \circ \mu$ is essential.

2. Let $M \xrightarrow{\mu} N \xrightarrow{\nu} P$ be monomorphisms. If $\nu \circ \mu$ is essential, then prove that μ and ν are essential.

3. Prove that every nonzero ideal of a [commutative] domain R is an essential sub-module of R.

4. Let N be a submodule of M. Prove that there exists a submodule C of M which is maximal such that $N \cap C = 0$. Prove that $N + C$ is essential in M.

5. Show that $\bigoplus_{i \in I} M_i$ need not be essential in $\prod_{i \in I} M_i$.

6. Show that $\mathbb{Q}$ is the injective hull of $\mathbb{Z}$.

7. Show that $\mathbb{Z}_{p^\infty}$ is the injective hull of $\mathbb{Z}_{p^n}$ for every $n > 0$.

8. Let R be a PID. Show that the field of fractions of R (viewed as an R-module) is the injective hull of $_R R$.

14.5. EXACT SEQUENCES

This section introduces two more basic tools of module theory: exact sequences, and proofs by "diagram chasing."

In what follows, R is a ring and "module" means "left R-module," but (unless explicitly stated otherwise) the results apply equally to right R-modules and to R-S-bimodules.

14.5.a. Definition. A finite or infinite sequence

$$\cdots M_i \xrightarrow{\varphi_i} M_{i+1} \xrightarrow{\varphi_{i+1}} M_{i+2} \cdots$$

of module homomorphisms is **null** in case $\varphi_{i+1} \circ \varphi_i = 0$ for all i (equivalently, in case $\operatorname{Im} \varphi_i \subseteq \operatorname{Ker} \varphi_{i+1}$ for all i); it is **exact** in case $\operatorname{Im} \varphi_i = \operatorname{Ker} \varphi_{i+1}$ for all i.

For example, a module homomorphism $\varphi : M \longrightarrow N$ is a monomorphism if and only if the sequence $0 \longrightarrow M \xrightarrow{\varphi} N$ is exact; $\varphi : M \longrightarrow N$ is an epimorphism if and only if the sequence $M \xrightarrow{\varphi} N \longrightarrow 0$ is exact; $\varphi : M \longrightarrow N$ is an isomorphism if and only if the sequence $0 \longrightarrow M \xrightarrow{\varphi} N \longrightarrow 0$ is exact. For every module homomorphism $\varphi : M \longrightarrow N$ there are exact sequences $0 \longrightarrow \operatorname{Ker} \varphi \xrightarrow{\subseteq} M \xrightarrow{\varphi} N$ and $M \xrightarrow{\varphi} N \longrightarrow \operatorname{Coker} \varphi \longrightarrow 0$, where $\operatorname{Coker} \varphi = N/\operatorname{Im} \varphi$ is the cokernel of φ.

A **short exact sequence** is an exact sequence of the form

$$0 \longrightarrow A \xrightarrow{\mu} B \xrightarrow{\rho} C \longrightarrow 0;$$

equivalently, a sequence (μ, ρ) in which μ is a monomorphism, ρ is an epimorphism, and $\operatorname{Im} \mu = \operatorname{Ker} \rho$. Exact sequences of the form

$$0 \longrightarrow A \xrightarrow{\mu} B \xrightarrow{\rho} C$$

are sometimes called **left exact**; exact sequences of the form

$$A \xrightarrow{\mu} B \xrightarrow{\rho} C \longrightarrow 0$$

are sometimes called **right exact**.

14.5.b. Split Exact Sequences. For any two left R-modules A and C, there is a short exact sequence

$$0 \longrightarrow A \xrightarrow{\iota} A \oplus C \xrightarrow{\pi} C \longrightarrow 0,$$

where ι is the injection and π is the projection. Short exact sequences that are of this form (up to isomorphism) can be characterized as follows:

Proposition 14.5.1. *For a short exact sequence*

$$0 \longrightarrow A \xrightarrow{\ \mu\ } B \xrightarrow{\ \rho\ } C \longrightarrow 0$$

the following conditions are equivalent:

(1) μ *splits (there exists a homomorphism* $B \xrightarrow{\ \sigma\ } A$ *such that* $\sigma \circ \mu = 1_A$).

(2) ρ *splits (there exists a homomorphism* $C \xrightarrow{\ \nu\ } B$ *such that* $\rho \circ \nu = 1_C$).

(3) *There exists an isomorphism* $B \cong A \oplus C$ *such that the following diagram*

$$
\begin{array}{ccccccccc}
0 & \longrightarrow & A & \xrightarrow{\ \mu\ } & B & \xrightarrow{\ \rho\ } & C & \longrightarrow & 0 \\
& & \Big\| & & \Big\updownarrow{\scriptstyle\cong} & & \Big\| & & \\
0 & \longrightarrow & A & \xrightarrow[\iota]{\ \ } & A \oplus C & \xrightarrow[\pi]{\ \ } & C & \longrightarrow & 0
\end{array}
$$

commutes, where ι *is the injection and* π *is the projection.*

Proof. (1) implies (3). Assume that $\sigma \circ \mu = 1_A$. Let $\pi' : A \oplus C \longrightarrow A$ be the projection. By the universal property of direct products, there is a homomorphism $\zeta : B \longrightarrow A \oplus C$ such that $\pi' \circ \zeta = \sigma$ and $\pi \circ \zeta = \rho$; namely $\zeta(b) = (\sigma(b), \rho(b))$:

$$
\begin{array}{ccccccccc}
0 & \longrightarrow & A & \underset{\sigma}{\overset{\mu}{\rightleftarrows}} & B & \xrightarrow{\ \rho\ } & C & \longrightarrow & 0 \\
& & \Big\| & & \Big\downarrow{\scriptstyle\zeta} & & \Big\| & & \\
0 & \longrightarrow & A & \underset{\pi'}{\overset{\iota}{\rightleftarrows}} & A \oplus C & \xrightarrow{\ \pi\ } & C & \longrightarrow & 0
\end{array}
$$

Then $\zeta(\mu(a)) = (a, 0)$ and $\zeta \circ \mu = \iota$. It remains to show that ζ is an isomorphism. (This also follows from Lemma 14.5.4 below.) If $\zeta(b) = 0$, then $\rho(b) = 0$, $b = \mu(a)$ for some $a \in A$, $a = \sigma(\mu(a)) = \sigma(b) = 0$, and $b = 0$; thus ζ is injective. If $a \in A$ and $c \in C$, then $c = \rho(b)$ for some $b \in B$, and $b' = b - \mu(\sigma(b)) + \mu(a)$ satisfies $\sigma(b') = a$ and $\rho(b') = \rho(b) = c$, since $\sigma \circ \mu = 1_A$ and $\rho \circ \mu = 0$; thus ζ is surjective.

(2) implies (3). Assume that $\rho \circ \nu = 1_C$. Let $\iota' : C \longrightarrow A \oplus C$ be the injection. By the universal property of direct sums, there is a homomorphism $\theta : A \oplus C \longrightarrow B$ such that $\theta \circ \iota = \mu$ and $\theta \circ \iota' = \nu$; namely $\theta(a, c) = \mu(a) + \nu(c)$:

$$
\begin{array}{ccccccccc}
0 & \longrightarrow & A & \xrightarrow{\ \mu\ } & B & \underset{\nu}{\overset{\rho}{\rightleftarrows}} & C & \longrightarrow & 0 \\
& & \Big\| & & \Big\uparrow{\scriptstyle\theta} & & \Big\| & & \\
0 & \longrightarrow & A & \xrightarrow{\ \iota\ } & A \oplus C & \underset{\iota'}{\overset{\pi}{\rightleftarrows}} & C & \longrightarrow & 0
\end{array}
$$

Then $\rho(\theta(a,c)) = \rho(\nu(c)) = c$ and $\rho \circ \theta = \pi'$. It remains to show that θ is an isomorphism (or to quote Lemma 14.5.4 again). If $\theta(a,c) = \mu(a) + \nu(c) = 0$, then $c = \rho(\theta(a,c)) = 0$, and then $a = 0$ (since μ is injective); thus θ is injective. If $b \in B$, then $b - \nu(\rho(b)) = \mu(a)$ for some $a \in A$, since $\rho(b - \nu(\rho(b))) = 0$, and $\theta(a, \rho(b)) = \mu(a) + \nu(\rho(b)) = b$; thus θ is surjective.

(3) implies (1) and (2). If θ us an isomorphism and $\theta \circ \iota = \mu$, $\rho \circ \theta = \pi$, then

$$
\begin{array}{ccccccccc}
0 & \longrightarrow & A & \underset{\sigma}{\overset{\mu}{\rightleftarrows}} & B & \underset{\nu}{\overset{\rho}{\rightleftarrows}} & C & \longrightarrow & 0 \\
 & & \| & & \theta\uparrow & & \| & & \\
0 & \longrightarrow & A & \underset{\pi'}{\overset{\iota}{\rightleftarrows}} & A \oplus C & \underset{\iota'}{\overset{\pi}{\rightleftarrows}} & C & \longrightarrow & 0
\end{array}
$$

$\sigma = \pi' \circ \theta^{-1}$ and $\nu = \theta \circ \iota'$ satisfy $\sigma \circ \mu = \pi' \circ \iota = 1_A$ and $\rho \circ \nu = \pi \circ \iota' = 1_C$ (by Corollary 10.4.5). $\qquad\square$

A short exact sequence **splits** in case it satisfies the equivalent conditions in Proposition 14.5.1.

Not all exact sequences split. Propositions 14.2.1 and 14.3.1 imply:

Proposition 14.5.2. *A left R-module P is projective if and only if every short exact sequence* $0 \longrightarrow A \longrightarrow B \longrightarrow P \longrightarrow 0$ *splits.*

Proposition 14.5.3. *A left R-module J is projective if and only if every short exact sequence* $0 \longrightarrow J \longrightarrow B \longrightarrow C \longrightarrow 0$ *splits.*

14.5.c. Diagram Chasing. Short exact sequences lead to a method of proof known as **diagram chasing**, in which module elements are "chased" all over a commutative diagram. We give a typical example.

Lemma 14.5.4 (The Short Five Lemma). *Let*

$$
\begin{array}{ccccccccc}
0 & \longrightarrow & A & \overset{\mu}{\longrightarrow} & B & \overset{\rho}{\longrightarrow} & C & \longrightarrow & 0 \\
 & & \alpha\downarrow & & \beta\downarrow & & \gamma\downarrow & & \\
0 & \longrightarrow & A' & \underset{\mu'}{\longrightarrow} & B' & \underset{\rho'}{\longrightarrow} & C' & \longrightarrow & 0
\end{array}
$$

be a commutative diagram with exact rows. If α and γ are isomorphisms, then so is β.

Proof. Assume that $\beta(b) = 0$. Then $\gamma(\rho(b)) = \rho'(\beta(b)) = 0$ and $\rho(b) = 0$. Since the top row is exact, $b = \mu(a)$ for some $a \in A$. Now $\mu'(\alpha(a)) = \beta(\mu(a)) = \beta(b) = 0$. Hence $\alpha(a) = 0$, $a = 0$, and $b = \mu(a) = 0$. Thus β is injective.

Let $b' \in B'$. Then $\rho'(b') = \gamma(c)$ for some $c \in C$ and $c = \rho(b)$ for some $b \in B$. Hence $\rho'(\beta(b)) = \gamma(\rho(b)) = \rho'(b')$. Since the bottom row is exact, $b' - \beta(b) = \mu'(a')$ for some $a' \in A'$. Now $a' = \alpha(a)$ for some $a \in A$. Hence $b' = \beta(b) + \mu'(\alpha(a)) = \beta(b + \mu(a))$. Thus β is surjective.   □

Sharper versions of Lemma 14.5.2 will be found in the exercises.

Lemma 14.5.5 (The Nine Lemma). *Let*

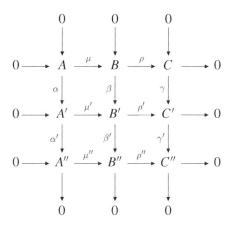

be a commutative diagram with exact columns. If the first two rows are exact, then the last row is exact. If the last two rows are exact, then the first row is exact.

The proof makes a dandy exercise.

Exercises

1. Show that $A \xrightarrow{\varphi} B \xrightarrow{\psi} C$ is null if and only if φ factors though $\operatorname{Ker} \psi$, if and only if ψ factors through $\operatorname{Coker} \varphi$.

2. Let

$$A \xrightarrow{\varphi} B \xrightarrow{\chi} C \xrightarrow{\psi} D \xrightarrow{\omega} E$$

with vertical maps $\alpha, \beta, \gamma, \delta, \varepsilon$ to

$$A' \xrightarrow{\varphi'} B' \xrightarrow{\chi'} C' \xrightarrow{\psi'} D' \xrightarrow{\omega'} E'$$

be a commutative diagram with exact rows.

(a) If α is an epimorphism and β and δ are monomorphisms, prove that γ is a monomorphism.

(b) If ε is a monomorphism and β and δ are epimorphisms, prove that γ is an epimorphism.

3. Let

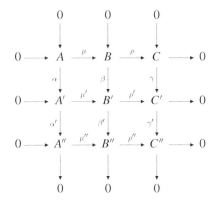

be a commutative diagram with exact columns.

(a) If the first two rows are exact, then prove that the last row is exact.

(b) If the last two rows are exact, then prove that the first row is exact.

4. Let

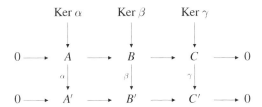

be a commutative diagram with exact rows. Show that there is an exact sequence
$0 \longrightarrow \operatorname{Ker} \alpha \longrightarrow \operatorname{Ker} \beta \longrightarrow \operatorname{Ker} \gamma$ which keeps the diagram commutative.

5. Let

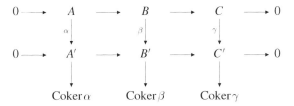

be a commutative diagram with exact rows. Show that there is an exact sequence
$\operatorname{Coker} \alpha \longrightarrow \operatorname{Coker} \beta \longrightarrow \operatorname{Coker} \gamma \longrightarrow 0$ which keeps the diagram commutative.

6. Given an exact sequence $0 \longrightarrow A \longrightarrow B \longrightarrow C \longrightarrow 0$ and a homomorphism
$C' \longrightarrow C$, show that there is a commutative diagram

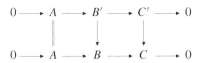

in which the right square is a pullback and the top row is exact.

7. Given an exact sequence $0 \longrightarrow A \longrightarrow B \longrightarrow C \longrightarrow 0$ and a homomorphism $A \longrightarrow A'$, show that there is a commutative diagram

$$
\begin{array}{ccccccccc}
0 & \longrightarrow & A' & \longrightarrow & B' & \longrightarrow & C & \longrightarrow & 0 \\
 & & \uparrow & & \uparrow & & \| & & \\
0 & \longrightarrow & A & \longrightarrow & B & \longrightarrow & C & \longrightarrow & 0
\end{array}
$$

in which the left square is a pushout and the top row is exact.

14.6. GROUPS OF HOMOMORPHISMS

This section contains basic properties of the abelian group $\mathrm{Hom}_R(A,B)$ of all module homomorphisms $A \longrightarrow B$, when A and B are left R-modules. We leave to the reader the similar properties of the group $\mathrm{Hom}_{R,S}(A,B)$ of bimodule homomorphisms when A and B are R-S-bimodules.

14.6.a. Module Structures. Let A and B be left R-modules. Recall that the pointwise sum $(\varphi + \psi)(a) = \varphi(a) + \psi(a)$ of two module homomorphisms $\varphi, \psi : A \longrightarrow B$ is a module homomorphism. Under this operation the set of all module homomorphisms $A \longrightarrow B$ is an abelian group $\mathrm{Hom}_R(A,B)$.

Additional module structures on A and B induce module structures on $\mathrm{Hom}_R(A,B)$, as follows:

Proposition 14.6.1. *If A is an R-S-bimodule and B is an R-T-bimodule, then $\mathrm{Hom}_R(A,B)$ is an S-T-bimodule, in which $(s\varphi)(a) = \varphi(as)$ and $(\varphi t)(a) = \varphi(a)t$ for all $a \in A$, $s \in S$, $t \in T$, and $\varphi \in \mathrm{Hom}_R(A,B)$.*

It is tempting to write homomorphisms on the right just to make this read $a(s\varphi) = (as)\varphi$ and $a(\varphi t) = (a\varphi)t$.

Proof. We see that $s\varphi$ and φt are homomorphisms, since A is an R-S-bimodule and B is an R-T-bimodule. Also $s(\varphi + \psi) = s\varphi + s\psi$ and $s(s'\varphi) = (ss')\varphi$, since

$$(s(s'\varphi))(a) = (s'\varphi)(as) = \varphi((as)s') = \varphi(a(ss')) = ((ss')\varphi)(a).$$

Thus $\mathrm{Hom}_R(A,B)$ is a left S-module. Similarly $\mathrm{Hom}_R(A,B)$ is a right T-module, and $(s\varphi)t = s(\varphi t)$. $\qquad\square$

If, for instance, M is a left R-module, then $\mathrm{Hom}_R({}_RR,M)$ is a left R-module (since R is an R-R-bimodule). In fact the reader will easily show:

Proposition 14.6.2. *Let M be a R-module. For each $m \in M$, $r \longmapsto rm$ is a homomorphism of ${}_RR$ into M. This constructs every homomorphism of ${}_RR$ into M. Hence there is a module isomorphism $\mathrm{Hom}_R(R,M) \cong M$.*

14.6.b. Homomorphisms. Let M, A, and B be left R-modules. Each homomorphism $\varphi : A \longrightarrow B$ induces a mapping

$$\varphi_* = \operatorname{Hom}_R(M, \varphi) : \operatorname{Hom}_R(M, A) \longrightarrow \operatorname{Hom}_R(M, B)$$

defined for all $\alpha : M \longrightarrow A$ by

$$\varphi_*(\alpha) = \varphi \circ \alpha .$$

Proposition 14.6.3. *Let $\varphi : A \longrightarrow B$ be a homomorphism of left R-modules. Then $\varphi_* = \operatorname{Hom}_R(M, \varphi) : \operatorname{Hom}_R(M, A) \longrightarrow \operatorname{Hom}_R(M, B)$ is a homomorphism of abelian groups.*

If M is a R-S-bimodule and A, B are R-T-bimodules, and φ is a bimodule homomorphism, then φ_* is a bimodule homomorphism.

If φ is the identity on A, then φ_* is the identity on $\operatorname{Hom}_R(M, A)$. If $\psi : B \longrightarrow C$ is another homomorphism, then $(\psi \circ \varphi)_* = \psi_* \circ \varphi_*$. If $\varphi' : A \longrightarrow B$ is another homomorphism, then $(\varphi + \varphi')_* = \varphi_* + \varphi'_*$.

The proof is straightforward. Proposition 14.6.3 can be restated as follows. A **covariant functor** (or just a **functor**) from left R-modules to abelian groups is a construct F which assigns to each left R-module A an abelian group $F(A)$, and to each module homomorphism $\varphi : A \longrightarrow B$ an abelian group homomorphism $F(\varphi) : F(A) \longrightarrow F(B)$, so that $F(1_A)$ is the identity on $F(A)$ and $F(\psi \circ \varphi) = F(\psi) \circ F(\varphi)$ whenever $A \xrightarrow{\varphi} B \xrightarrow{\psi} C$. Functors in general are defined in Section 17.2.

For each left R-module M, there is by Proposition 14.6.3 a covariant functor $\operatorname{Hom}_R(M, -)$ from left R-modules to abelian groups which assigns to each left R-module A the abelian group $\operatorname{Hom}_R(M, A)$, and to each module homomorphism $\varphi : A \longrightarrow B$ the additive homomorphism $\operatorname{Hom}_R(M, \varphi)$. If M is a R-S-bimodule, there is a similar functor, also denoted by $\operatorname{Hom}_R(M, -)$, from R-T-bimodules to S-T-bimodules, which assigns to each R-T-bimodule A the S-T-bimodule $\operatorname{Hom}_R(M, A)$, and to each bimodule homomorphism $\varphi : A \longrightarrow B$ the bimodule homomorphism $\operatorname{Hom}_R(M, \varphi)$.

Similarly each homomorphism $\varphi : A \longrightarrow B$ induces a mapping

$$\varphi^* = \operatorname{Hom}_R(\varphi, M) : \operatorname{Hom}_R(B, M) \longrightarrow \operatorname{Hom}_R(A, M)$$

defined for all $\beta : B \longrightarrow M$ by

$$\varphi^*(\beta) = \beta \circ \varphi .$$

Proposition 14.6.4. *Let $\varphi : A \longrightarrow B$ be a homomorphism of left R-modules. Then $\varphi^* = \operatorname{Hom}_R(\varphi, M) : \operatorname{Hom}_R(B, M) \longrightarrow \operatorname{Hom}_R(A, M)$ is a homomorphism of abelian groups.*

If A, B are R-S-bimodules and M is a R-T-bimodule, and φ is a bimodule homomorphism, then φ^* is a bimodule homomorphism.

If φ is the identity on A, then φ^* is the identity on $\mathrm{Hom}_R(M,A)$. If $\psi : B \longrightarrow C$ is another homomorphism, then $(\psi \circ \varphi)^* = \varphi^* \circ \psi^*$. If $\varphi' : A \longrightarrow B$ is another homomorphism, then $(\varphi + \varphi')^* = \varphi^* + \varphi'^*$.

The reversal of arrows and compositions in Proposition 14.6.4 can be expressed as follows. A **contravariant functor** from left R-modules to abelian groups is a construct F which assigns to each left R-module A an abelian group $F(A)$, and to each module homomorphism $\varphi : A \longrightarrow B$ an abelian group homomorphism $F(\varphi) : F(B) \longrightarrow F(A)$, so that $F(1_A)$ is the identity on $F(A)$ and $F(\psi \circ \varphi) = F(\varphi) \circ F(\psi)$ whenever $A \xrightarrow{\varphi} B \xrightarrow{\psi} C$. By Proposition 14.6.4, there is for each left R-module M a contravariant functor $\mathrm{Hom}_R(-,M)$ from left R-modules to abelian groups, which assigns to each left R-module A the abelian group $\mathrm{Hom}_R(A,M)$, and to each homomorphism $\varphi : A \longrightarrow B$ the homomorphism $\mathrm{Hom}_R(\varphi,M) : \mathrm{Hom}_R(B,M) \longrightarrow \mathrm{Hom}_R(A,M)$. For each R-T-bimodule M, there is a similar contravariant functor $\mathrm{Hom}_R(-,M)$ from R-S-bimodules to S-T-bimodules.

Proposition 14.6.5. *When $\alpha : A \longrightarrow B$ and $\varphi : M \longrightarrow N$ are module homomorphisms, the following diagram commutes:*

$$
\begin{array}{ccc}
\mathrm{Hom}_R(M,A) & \xrightarrow{\mathrm{Hom}_R(M,\alpha)} & \mathrm{Hom}_R(M,B) \\[2mm]
{\scriptstyle \mathrm{Hom}_R(\varphi,A)}\uparrow & & \uparrow{\scriptstyle \mathrm{Hom}_R(\varphi,B)} \\[2mm]
\mathrm{Hom}_R(N,A) & \xrightarrow[\mathrm{Hom}_R(N,\alpha)]{} & \mathrm{Hom}_R(N,B)
\end{array}
$$

Proof. For all $\xi : N \longrightarrow A$, $\varphi^*(\alpha_*(\xi)) = \alpha \circ \xi \circ \varphi = \alpha_*(\varphi^*(\xi))$. $\square$

When F and G are covariant functors from, say, left R-modules to abelian groups, a homomorphism $\tau_M : F(M) \longrightarrow G(M)$ (which depends on M) is **natural** in M in case the square

$$
\begin{array}{ccc}
F(M) & \xrightarrow{\tau_M} & G(M) \\[2mm]
{\scriptstyle F(\varphi)}\downarrow & & \downarrow{\scriptstyle G(\varphi)} \\[2mm]
F(N) & \xrightarrow[\tau_N]{} & G(N)
\end{array}
$$

commutes for every homomorphism $\varphi : M \longrightarrow N$. There is a similar definition for contravariant functors. Naturality is one way to express the idea that τ_M is constructed from M in some sort of canonical fashion.

Proposition 14.6.5 shows that the homomorphism $\mathrm{Hom}_R(M,\alpha)$ is natural in M, for each α. The same square shows that $\mathrm{Hom}_R(\varphi,A)$ is natural in A, for each φ.

14.6.c. Left Exactness

Proposition 14.6.6. *If* $0 \longrightarrow A \xrightarrow{\mu} B \xrightarrow{\rho} C$ *is exact, then*

$$0 \longrightarrow \mathrm{Hom}_R(M,A) \xrightarrow{\mathrm{Hom}_R(M,\mu)} \mathrm{Hom}_R(M,B) \xrightarrow{\mathrm{Hom}_R(M,\rho)} \mathrm{Hom}_R(M,C)$$

is exact.

Proof. Let $\alpha \in \mathrm{Hom}_R(M,A)$. If $\mu_*(\alpha) = 0$, then $\mu \circ \alpha = 0$, $\mathrm{Im}\,\alpha \subseteq \mathrm{Ker}\,\mu = 0$, and $\alpha = 0$. This proves exactness at $\mathrm{Hom}_R(M,A)$.

$$
\begin{array}{ccccccc}
0 \longrightarrow & A & \xrightarrow{\ \mu\ } & B & \xrightarrow{\ \rho\ } & C \\
 & & {\scriptstyle \alpha}\nwarrow & \uparrow{\scriptstyle \beta} & & \\
 & & & M & & \\
\end{array}
$$

Similarly $\rho_*(\mu_*(\alpha)) = \rho \circ \mu \circ \alpha = 0$. Conversely, assume that $\rho_*(\beta) = 0$, where $\beta \in \mathrm{Hom}_R(M,B)$. Then $\rho \circ \beta = 0$ and $\mathrm{Im}\,\beta \subseteq \mathrm{Ker}\,\rho = \mathrm{Im}\,\mu$. Since μ is injective, β factors through μ: $\beta = \mu(\alpha)$ for some $\alpha : M \longrightarrow A$, and $\beta \in \mathrm{Im}\,\mu_*$. This proves exactness at $\mathrm{Hom}_R(M,B)$. $\qquad\square$

Proposition 14.6.6 does not extend to all exact sequences: in fact the sequence $0 \longrightarrow \mathrm{Hom}_R(M,A) \longrightarrow \mathrm{Hom}_R(M,B) \longrightarrow \mathrm{Hom}_R(M,C) \longrightarrow 0$ need not be exact even if $0 \longrightarrow A \longrightarrow B \longrightarrow C \longrightarrow 0$ is exact (see the exercises and Proposition 14.6.7 below).

A covariant functor is **left exact** in case it transforms left exact sequences into left exact sequences, and **exact** in case it transforms short exact sequences into short exact sequences. (We will show in Chapter 17 that an exact functor transforms all exact sequences into exact sequences.) Thus $\mathrm{Hom}_R(M,-)$ is left exact but in general not exact. The reader will show that:

Proposition 14.6.7. *For a left R-module P the following are equivalent:*

(1) *P is projective.*

(2) $\mathrm{Hom}_R(P,\rho)$ *is an epimorphism for every epimorphism* ρ.

(3) $\mathrm{Hom}_R(P,-)$ *is exact.*

Proposition 14.6.8. *If* $A \xrightarrow{\mu} B \xrightarrow{\rho} C \longrightarrow 0$ *is exact, then*

$$0 \longrightarrow \mathrm{Hom}_R(C,M) \xrightarrow{\mathrm{Hom}_R(\rho,M)} \mathrm{Hom}_R(B,M) \xrightarrow{\mathrm{Hom}_R(\mu,M)} \mathrm{Hom}_R(A,M)$$

is exact.

The proof is an exercise; again the result does not extend to short exact sequences. A contravariant functor is **left exact** in case it transforms right exact sequences into left exact sequences. $\mathrm{Hom}_R(-,M)$ is left exact, but not exact:

Proposition 14.6.9. *For a left R-module J the following are equivalent:*

(1) *J is injective.*
(2) $\mathrm{Hom}_R(\mu,J)$ *is an epimorphism for every monomorphism μ.*
(3) $\mathrm{Hom}_R(-,J)$ *is exact.*

14.6.d. Direct Sums and Products. The universal property of direct products yields the following result:

Proposition 14.6.10. *There is an isomorphism*

$$\mathrm{Hom}_R(A, \textstyle\prod_{i\in I} B_i) \cong \textstyle\prod_{i\in I} \mathrm{Hom}_R(A,B_i)$$

which is natural in A and $(B_i)_{i\in I}$ and makes the diagram

$$\mathrm{Hom}_R(A,\textstyle\prod_{i\in I} B_i) \xrightarrow{\;\cong\;} \textstyle\prod_{i\in I} \mathrm{Hom}_R(A,B_i)$$

$$\mathrm{Hom}_R(A,\pi_i) \searrow \qquad \downarrow \rho_i$$

$$\mathrm{Hom}_R(A,B_i)$$

commute for every i, where ρ_i and $\pi_i : \prod_{i\in I} B_i \longrightarrow B_i$ are the projections.

Proof. Let $B = \prod_{i\in I} B_i$. The projections $\pi_i : B \longrightarrow B_i$ induce homomorphisms $\pi_{i*} = \mathrm{Hom}_R(A,\pi_i) : \mathrm{Hom}_R(A,B) \longrightarrow \mathrm{Hom}_R(A,B_i)$ and $\pi_* : \mathrm{Hom}_R(A,B) \longrightarrow \prod_{i\in I} \mathrm{Hom}_R(A,B_i)$, namely $\pi_*(\varphi) = (\pi_{i*}(\varphi))_{i\in I} = (\pi_i \circ \varphi)_{i\in I}$. The universal property of $B = \prod_{i\in I} B_i$ precisely states that π_* is bijective.

That π_* is natural in A means that every homomorphism $\alpha : A \longrightarrow A'$ induces homomorphisms $\alpha^* = \mathrm{Hom}_R(\alpha,B)$ and $\alpha_i^* = \mathrm{Hom}_R(\alpha,B_i)$ such that the square

$$\mathrm{Hom}_R(A,B) \xrightarrow{\;\pi_*\;} \textstyle\prod_{i\in I} \mathrm{Hom}_R(A,B_i)$$

$$\alpha^* \uparrow \qquad\qquad \uparrow \textstyle\prod_{i\in I} \alpha_i^*$$

$$\mathrm{Hom}_R(A',B) \xrightarrow[\;\pi_*\;]{} \textstyle\prod_{i\in I} \mathrm{Hom}_R(A',B_i)$$

commutes. Indeed, for each $\varphi \in \mathrm{Hom}_R(A,B)$,

$$\pi_*(\alpha_*(\varphi)) = \pi_*(\varphi \circ \alpha) = (\pi_i \circ \varphi \circ \alpha)_{i\in I} = (\alpha_i^*(\pi_i \circ \varphi))_{i\in I}$$

$$= (\textstyle\prod_{i\in I} \alpha_i^*)((\pi_i \circ \varphi)_{i\in I}) = (\textstyle\prod_{i\in I} \alpha_i^*)(\pi_*(\varphi)).$$

Similarly, that π_* is natural in $(B_i)_{i\in I}$ means that every family of homomorphisms $\beta_i : B_i \longrightarrow B_i'$ $(i \in I)$ induces homomorphisms $\beta = \prod_{i\in I} \beta_i : B \longrightarrow B'$

$= \prod_{i\in I} B'_i$, $\beta_* = \mathrm{Hom}_R(A,\beta)$ and $\beta_{i*} = \mathrm{Hom}_R(A,\beta_i)$ such that the square

$$\begin{array}{ccc}
\mathrm{Hom}_R(A,B) & \xrightarrow{\pi_*} & \prod_{i\in I}\, \mathrm{Hom}_R(A,B_i) \\
\beta_* \downarrow & & \downarrow \prod_{i\in I}\,\beta_{i*} \\
\mathrm{Hom}_R(A,B') & \xrightarrow{\pi_*} & \prod_{i\in I}\, \mathrm{Hom}_R(A,B'_i)
\end{array}$$

commutes. Let $\pi'_i : B' \longrightarrow B'_i$ be the projection. Recall that $\pi'_i \circ (\prod_{i\in I}\beta_i) = \beta_i \circ \pi_i$ for all i. Hence, for each $\varphi \in \mathrm{Hom}_R(A,B)$,

$$\pi'_*(\beta_*(\varphi)) = (\pi'_i \circ \beta \circ \varphi)_{i\in I} = (\beta_i \circ \pi_i \circ \varphi)_{i\in I} = (\beta_{i*}(\pi_i \circ \varphi))_{i\in I}$$

$$= (\prod_{i\in I}\beta_{i*})((\pi_i \circ \varphi)_{i\in I}) = (\prod_{i\in I}\beta_{i*})(\pi_*(\varphi)). \qquad \square$$

There is a similar property for direct sums, whose proof is left to the reader:

Proposition 14.6.11. *There is an isomorphism*

$$\mathrm{Hom}_R(\bigoplus_{i\in I} A_i, B) \cong \prod_{i\in I}\, \mathrm{Hom}_R(A_i, B)$$

which is natural in $(A_i)_{i\in I}$ *and* B *and makes the diagram*

$$\begin{array}{ccc}
\mathrm{Hom}_R(\bigoplus_{i\in I} A_i, B) & \xrightarrow{\cong} & \prod_{i\in I}\, \mathrm{Hom}_R(A_i, B) \\
& {\scriptstyle \mathrm{Hom}_R(\iota_i,B)} \searrow & \downarrow \rho_i \\
& & \mathrm{Hom}_R(A_i, B)
\end{array}$$

commute for every i, *where* ρ_i *is the projection and* $\iota_i : A_i \longrightarrow \bigoplus_{i\in I} A_i$ *is the injection.*

In particular, both $\mathrm{Hom}_R(M,-)$ and $\mathrm{Hom}_R(-,M)$ preserve finite direct sums. Pullbacks and pushouts are considered in the exercises.

14.6.e. Injective Modules. As an application of Hom and its properties, we prove Theorem 14.3.8, that every left R-module can be embedded into an injective left R-module. First we show that:

Lemma 14.6.12. *For each abelian group A and left R-module M, there is an isomorphism* $\mathrm{Hom}_R(M,\mathrm{Hom}_{\mathbb{Z}}(R,A)) \cong \mathrm{Hom}_{\mathbb{Z}}(M,A)$ *which is natural in M.*

Proof. Let $\xi \in \mathrm{Hom}_R(M,\mathrm{Hom}_{\mathbb{Z}}(R,A))$. Then $\xi(x) \in \mathrm{Hom}_{\mathbb{Z}}(R,A)$ for each $x \in M$ and

$$(\xi(x))(r) = (\xi(x))(1r) = (r\xi(x))(1) = \xi(rx)(1)$$

for all $x \in M$ and $r \in R$. Thus ξ is determined by the additive homomorphism $\overline{\xi} : x \longmapsto (\xi(x))(1)$ of M into A. In other words, $\Theta : \xi \longmapsto \overline{\xi}$ is an injective mapping of $\operatorname{Hom}_R(M, \operatorname{Hom}_{\mathbb{Z}}(R, A))$ into $\operatorname{Hom}_{\mathbb{Z}}(M, A)$; Θ is an additive homomorphism, since $(\xi + \zeta)(x) = \xi(x) + \zeta(x)$.

Let $\psi \in \operatorname{Hom}_{\mathbb{Z}}(M, A)$. Define $\xi(x) \in \operatorname{Hom}_{\mathbb{Z}}(R, A)$ by $(\xi(x))(r) = \psi(rx)$. Then $\xi(x + y) = \xi(x) + \xi(y)$ and $(r\xi(x))(s) = (\xi(x))(sr) = \psi(srx) = (\xi(rx))(s)$ for all $x, y \in M$ and $r, s \in R$. Thus $\xi \in \operatorname{Hom}_R(M, \operatorname{Hom}_{\mathbb{Z}}(R, A))$. Clearly $\overline{\xi} = \psi$. Thus Θ is surjective.

For every module homomorphism $\varphi : M \longrightarrow N$ the square

$$\begin{array}{ccc} \operatorname{Hom}_R(M, \operatorname{Hom}_{\mathbb{Z}}(R, A)) & \xrightarrow{\Theta_M} & \operatorname{Hom}_{\mathbb{Z}}(M, A) \\ \varphi^* \uparrow & & \uparrow \varphi^* \\ \operatorname{Hom}_R(N, \operatorname{Hom}_{\mathbb{Z}}(R, A)) & \xrightarrow{\Theta_N} & \operatorname{Hom}_{\mathbb{Z}}(N, A) \end{array}$$

commutes. Indeed, for every $\xi \in \operatorname{Hom}_R(N, \operatorname{Hom}_{\mathbb{Z}}(R, A))$, $\varphi^*(\Theta_N(\xi)) = \overline{\xi} \circ \varphi \in \operatorname{Hom}_{\mathbb{Z}}(M, A)$ sends $x \in M$ to $\overline{\xi}(\varphi(x)) = (\xi(\varphi(x)))(1)$, and $\Theta_M(\varphi^*(\xi)) = \overline{\xi \circ \varphi}$ sends $x \in M$ to $(\overline{\xi \circ \varphi})(x) = (\xi(\varphi(x)))(1)$; thus $\varphi^*(\Theta_N(\xi)) = \Theta_M(\varphi^*(\xi))$. Thus Θ is natural in M. (Eager readers will prove naturality in A.) $\square$

Lemma 14.6.13. *If D is a divisible abelian group, then the left R-module $\operatorname{Hom}_{\mathbb{Z}}(R, D)$ is injective.*

Proof. Let $A = D$ be divisible and $\varphi : M \longrightarrow N$ be any monomorphism of left R-modules. Since A is an injective $\mathbb{Z}$-module, $\varphi^* : \operatorname{Hom}_{\mathbb{Z}}(N, A) \longrightarrow \operatorname{Hom}_{\mathbb{Z}}(M, A)$ is surjective, by Proposition 6.9. Since the above diagram commutes,

$$\varphi^* : \operatorname{Hom}_R(N, \operatorname{Hom}_{\mathbb{Z}}(R, A)) \longrightarrow \operatorname{Hom}_R(M, \operatorname{Hom}_{\mathbb{Z}}(R, A))$$

is surjective. Therefore $\operatorname{Hom}_{\mathbb{Z}}(R, A)$ is an injective R-module, by Proposition 14.6.9. $\square$

We can now prove Theorem 14.3.8. Let M be any left R-module. By Proposition 14.3.9, there is an additive monomorphism $\mu : M \longrightarrow D$ into a divisible abelian group D. Let $\nu : M \longrightarrow \operatorname{Hom}_{\mathbb{Z}}(R, D)$ send $a \in M$ to $\nu(m) : r \longmapsto \mu(rm)$. Then $\nu(m)$ and ν are additive homomorphisms. Also $\nu(sm)(r) = \mu(rsm) = \nu(m)(rs) = (s\nu(m))(r)$, so that ν is a module homomorphism. Finally $\nu(m) = 0$ implies $\mu(m) = \nu(m)(1) = 0$ and $m = 0$, so that ν is injective and embeds M into $\operatorname{Hom}_{\mathbb{Z}}(R, D)$, which is an injective R-module by Lemma 14.6.13. $\square$

Exercises

1. Prove that there is for each left R-module M a module isomorphism $\operatorname{Hom}_R(R, M) \cong M$ which is natural in M.

2. Let M be a R-S-bimodule, A and B be S-T-bimodules, and $\varphi : A \longrightarrow B$ be a bimodule homomorphism. Show that $\mathrm{Hom}_R(M, \varphi)$ is a bimodule homomorphism.

3. If M is a left R-module and $A \xrightarrow{\ \mu\ } B \xrightarrow{\ \rho\ } C \longrightarrow 0$ is exact, then prove that

$$0 \longrightarrow \mathrm{Hom}_R(C, M) \xrightarrow{\ \mathrm{Hom}_R(\rho, M)\ } \mathrm{Hom}_R(B, M) \xrightarrow{\ \mathrm{Hom}_R(\mu, M)\ } \mathrm{Hom}_R(A, M)$$

is exact.

4. Prove that a left R-module P is projective if and only if $\mathrm{Hom}_R(P, -)$ is exact.

5. When P is projective, show that $\mathrm{Hom}_R(P, A) \longrightarrow \mathrm{Hom}_R(P, B) \longrightarrow \mathrm{Hom}_R(P, C)$ is exact whenever $A \longrightarrow B \longrightarrow C$ is exact.

6. Give an example of a left R-module M and a short exact sequence $0 \longrightarrow A \longrightarrow B \longrightarrow C \longrightarrow 0$ such that

$$0 \longrightarrow \mathrm{Hom}_R(M, A) \longrightarrow \mathrm{Hom}_R(M, B) \longrightarrow \mathrm{Hom}_R(M, C) \longrightarrow 0$$

is not exact.

7. Prove that a left R-module J is injective if and only if $\mathrm{Hom}_R(-, J)$ is exact.

8. Give an example of a left R-module M and a short exact sequence $0 \longrightarrow A \longrightarrow B \longrightarrow C \longrightarrow 0$ such that

$$0 \longrightarrow \mathrm{Hom}_R(C, M) \longrightarrow \mathrm{Hom}_R(B, M) \longrightarrow \mathrm{Hom}_R(A, M) \longrightarrow 0$$

is not exact.

9. Prove that there is an isomorphism

$$\mathrm{Hom}_R\left(\bigoplus_{i \in I} A_i, B\right) \cong \prod_{i \in I} \mathrm{Hom}_R(A_i, B)$$

which is natural in $(A_i)_{i \in I}$ and B.

10. Prove that a sequence of left R-modules $0 \longrightarrow A \xrightarrow{\ \mu\ } B \xrightarrow{\ \rho\ } C$ is exact if and only if

$$0 \longrightarrow \mathrm{Hom}_R(M, A) \xrightarrow{\ \mathrm{Hom}_R(M, \mu)\ } \mathrm{Hom}_R(M, B) \xrightarrow{\ \mathrm{Hom}_R(M, \rho)\ } \mathrm{Hom}_R(M, C)$$

is exact for every left R-module M.

11. Prove that a sequence of left R-modules $A \xrightarrow{\ \mu\ } B \xrightarrow{\ \rho\ } C \longrightarrow 0$ is exact if and only if

$$0 \longrightarrow \mathrm{Hom}_R(C, M) \xrightarrow{\ \mathrm{Hom}_R(\rho, M)\ } \mathrm{Hom}_R(B, M) \xrightarrow{\ \mathrm{Hom}_R(\mu, M)\ } \mathrm{Hom}_R(A, M)$$

is exact for every left R-module M.

12. Show that $\mathrm{Hom}_R(M, -)$ preserves pullbacks (if $\alpha \circ \alpha' = \beta \circ \beta'$ is a pullback of left R-modules, then $\mathrm{Hom}_R(M, \alpha) \circ \mathrm{Hom}_R(M, \alpha') = \mathrm{Hom}_R(M, \beta) \circ \mathrm{Hom}_R(M, \beta')$ is a pullback of abelian groups).

13. Show that $\mathrm{Hom}_R(-, M)$ sends pushouts to pullbacks (if $\alpha' \circ \alpha = \beta' \circ \beta$ is a pushout of left R-modules, then $\mathrm{Hom}_R(\alpha, M) \circ \mathrm{Hom}_R(\alpha', M) = \mathrm{Hom}_R(\beta, M) \circ \mathrm{Hom}_R(\beta', M)$ is a pullback of abelian groups).

14.7. DIRECT LIMITS

Direct limits are basic constructions which, unlike previous constructions in this chapter, are not specific to modules but apply to many algebraic systems.

14.7.a. Direct Systems. A **preordered set** (also called a **quasiordered set**) consists of a set I and a binary relation $\leqslant$ on I which is reflexive ($i \leqslant i$ for all $i \in I$) and transitive (if $i \leqslant j$ and $j \leqslant k$, then $i \leqslant k$) but not necessarily antisymmetric. For example, every partially ordered set is a preordered set.

A preordered set I is **directed upward** (or just **directed**) in case for each $i, j \in I$ there exists $k \in I$ such that $i \leqslant k$ and $j \leqslant k$. Then every finite subset of I has an upper bound: for every $i_1, \ldots, i_n \in I$, there exists $k \in I$ such that $i_t \leqslant k$ for all t. For instance, every chain (= totally ordered set) is directed upward.

Let I be a preordered set which is directed upward. A **direct system** of left R-modules **over** I consists of left R-modules $(A_i)_{i \in I}$ and homomorphisms $\alpha_{ij} : A_i \longrightarrow A_j$ (one for each pair (i, j) such that $i \leqslant j$ in I) such that, for all $i, j, k \in I$:

(1) α_{ii} is the identity on A_i;

(2) if $i \leqslant j \leqslant k$, then $\alpha_{jk} \circ \alpha_{ij} = \alpha_{ik}$.

One may think of a direct system over I as a large commutative diagram of modules, with one module for each $i \in I$ and arrows running upward.

For example, a chain of submodules can be viewed as a direct system. More generally, a family $(A_i)_{i \in I}$ of submodules of a module M is **directed** [upward] in case for each $i, j \in I$ there exists $k \in I$ such that $A_i \subseteq A_k$ and $A_j \subseteq A_k$. Every directed family of submodules $(A_i)_{i \in I}$ can be viewed as a direct system, as follows. Preorder I so that $i \leqslant j$ if and only if $A_i \subseteq A_j$. Then I is directed upward. When $i \leqslant j$, $\alpha_{ij} : A_i \longrightarrow A_j$ is the inclusion homomorphism. The union $\bigcup_{i \in I} A_i$ of a nonempty directed family of submodules of M is a **directed union**; it is a submodule of M.

14.7.b. Direct Limits. Direct limits are also called **inductive limits** and **directed colimits**.

Let $\mathscr{A} : (A_i)_{i \in I}, (\alpha_{ij})_{i, j \in I, i \leqslant j}$ be a direct system of left R-modules over a directed [upward] preordered set I. A **cone** $(\varphi_i)_{i \in I} : \mathscr{A} \longrightarrow M$ from $\mathscr{A}$ to a left R-module M is a family $(\varphi_i)_{i \in I}$ of homomorphisms $\varphi_i : A_i \longrightarrow M$ such that $\varphi_j \circ \alpha_{ij} = \alpha_i$ whenever $i \leqslant j$ in I:

$$A_j \xrightarrow{\varphi_j} M$$
$$\alpha_{ij} \Big\uparrow \quad \nearrow \varphi_i$$
$$A_i$$

Thus, adding a cone to a direct system yields an even larger commutative diagram. For instance, if B contains the union $\bigcup_{i \in I} A_i$ of a directed family $\mathscr{A} = (A_i)_{i \in I}$, then the inclusion homomorphisms $A_i \longrightarrow B$ constitute a cone from $\mathscr{A}$ to B.

In general, if $(\varphi_i)_{i \in I} : \mathscr{A} \longrightarrow M$ is a cone and $\psi : M \longrightarrow N$ is a homomorphism, then $(\psi \circ \varphi_i)_{i \in I} : \mathscr{A} \longrightarrow N$ is a cone.

A **direct limit** of the direct system $\mathscr{A}$ consists of a left R-module $A = \varinjlim_{i \in I} A_i$ and a **limit cone**, which is a cone $(\alpha_i)_{i \in I} : \mathscr{A} \longrightarrow A$ with the following universal property: for every module M and every cone $(\varphi_i)_{i \in I} : \mathscr{A} \longrightarrow M$ there exists a unique homomorphism $\varphi : A \longrightarrow M$ such that $\varphi_i = \varphi \circ \alpha_i$ for all $i \in I$.

$$A \dashrightarrow^{\varphi} M$$
$$\alpha_i \Big\uparrow \quad \nearrow \varphi_i$$
$$A_i$$

If $(\alpha_i)_{i \in I} : \mathscr{A} \longrightarrow A$ is a limit cone of $\mathscr{A}$, and $\theta : A \longrightarrow M$ is an isomorphism, then $(\theta \circ \alpha_i)_{i \in I} : \mathscr{A} \longrightarrow M$ is another limit cone of $\mathscr{A}$: if $(\varphi_i)_{i \in I} : \mathscr{A} \longrightarrow N$ is

$$A \xrightarrow{\theta} M$$
$$\alpha_i \Big\uparrow \quad \searrow^{\varphi} \quad \Big\downarrow$$
$$A_i \xrightarrow{\varphi_i} N$$

any cone, there is a homomorphism $\varphi : A \longrightarrow N$ unique such that $\varphi \circ \alpha_i = \varphi_i$ for all i, and $\varphi \circ \theta^{-1} : M \longrightarrow N$ is unique such that $(\varphi \circ \theta^{-1}) \circ (\theta \circ \alpha_i) = \varphi_i$ for all i. The next result, which follows from the universal property in the usual way, states that all limit cones of $\mathscr{A}$ are obtained in this fashion.

Proposition 14.7.1. *The direct limit of a direct system of left R-modules is unique up to isomorphism.*

14.7.c. Examples. At this point the main examples of direct limits are directed unions. Indeed the reader will verify that the union of a directed family of submodules is its direct limit. For example, it follows from Proposition 14.3.5 that $\mathbb{Z}_{p^\infty} \cong \varinjlim_{n \to \infty} \mathbb{Z}_{p^n}$. It is tempting to describe $\mathbb{Z}_{p^\infty}$ as a directed union of the groups $\mathbb{Z}_{p^n}$. Unfortunately, the latter are not actually subgroups of each other. Instead of trying to fit them inside each other, one can use the existing monomorphisms $\mathbb{Z}_{p^m} \longrightarrow \mathbb{Z}_{p^n}$ (where $m \leqslant n$) to construct $\mathbb{Z}_{p^\infty}$ cleanly as a direct limit.

The finitely generated submodules of a module M are a directed family: if $A, B \subseteq M$ are generated by finite subsets X, Y, then $A + B$ contains A and B and is generated by the finite subset $X \cup Y$. Since M is the union of its finitely generated submodules, we obtain:

Proposition 14.7.2. *Every module is the direct limit of its finitely generated submodules.*

Other examples will be found in the exercises.

14.7.d. Construction

Proposition 14.7.3. *Every direct system of left R-modules has a direct limit.*

Proof. Let $\mathscr{A} : (A_i)_{i \in I}$, $(\alpha_{ij})_{i,j \in I, i \leqslant j}$ be a direct system of left R-modules. We construct a direct limit of $\mathscr{A}$ in two steps. First we ignore the module structures and construct a limit cone $(\alpha_i)_{i \in I} : \mathscr{A} \longrightarrow A$ of $\mathscr{A}$, viewing $\mathscr{A}$ as a direct system of sets and mappings. Then we show that there is exactly one left R-module structure on A such that every α_i is a homomorphism, and that $(\alpha_i)_{i \in I} : \mathscr{A} \longrightarrow A$ is a limit cone of $\mathscr{A}$. It will be clear that similar arguments yield direct limits of groups, rings, fields, and so on.

Any cone $\varphi : \mathscr{A} \longrightarrow M$ sends $x \in A_i$ and $\alpha_{ij}(x) \in A_j$ to the same element $\varphi_i(x) = \varphi_j(\alpha_{ij}(x))$ of M. More generally, if $x \in A_i$ and $y \in A_j$ have a common higher image in $\mathscr{A}$ (when $\alpha_{ik}(x) = \alpha_{jk}(y)$ for some $k \geqslant i, j$), then φ sends x and y to the same element $\varphi_i(x) = \varphi_k(\alpha_{ik}(x) = \varphi_k(\alpha_{jk}(y) = \varphi_j(y)$ of M. In particular, a limit cone must identify elements with a common higher image in $\mathscr{A}$.

It turns out that we can construct a limit cone which identifies only such elements. In detail, let S be the set of all pairs (x, i) where $i \in I$ and $x \in A_i$. (Up to bijections, S is the disjoint union of all A_i.) Define a relation $\sim$ on S by

$$(x, i) \sim (y, j) \qquad \text{if and only if} \quad \alpha_{ik}(x) = \alpha_{jk}(y) \quad \text{for some} \quad k \geqslant i, j$$

(i.e., $(x, i) \sim (y, j)$ if and only if x and y have a higher image in $\mathscr{A}$). Then $\alpha_{i\ell}(x) = \alpha_{k\ell}(\alpha_{ik}(x)) = \alpha_{k\ell}(\alpha_{jk}(y)) = \alpha_{j\ell}(y)$ for all $\ell \geqslant k$. We see that $(x, i) \sim (\alpha_{ij}(x), j)$ whenever $i \leqslant j$ and $x \in A_i$.

The relation $\sim$ is an equivalence relation. It is reflexive, since $\alpha_{ii}(x) = x$, symmetric, and transitive: if $\alpha_{i\ell}(x) = \alpha_{j\ell}(y)$ and $\alpha_{jm}(y) = \alpha_{km}(z)$, then I contains some $n \geqslant \ell, m$, since I is directed, and $\alpha_{in}(x) = \alpha_{jn}(y) = \alpha_{kn}(z)$. Let $\mathrm{cls}(x, i)$ denote the $\sim$-class of (x, i).

Let $A = S/\sim$ be the set of all $\sim$-classes and $\alpha_i : A_i \longrightarrow A$, $\alpha_i(x) = \mathrm{cls}(x, i)$. We have $\alpha_j \circ \alpha_{ij} = \alpha_i$ whenever $i \leqslant j$, since $(x, i) \sim (\alpha_{ij}(x), j)$.

Let $(\varphi_i)_{i \in I} : \mathscr{A} \longrightarrow M$ be a family of mappings $\varphi_i : A_i \longrightarrow M$ such that $\varphi_j \circ \alpha_{ij} = \varphi_i$ whenever $i \leqslant j$ in I. If $\mathrm{cls}\,(x,i) = \mathrm{cls}\,(y,j)$, then $\alpha_{ik}(x) = \alpha_{jk}(y)$ for some $k \geqslant i,j$ and $\varphi_i(x) = \varphi_k(\alpha_{ik}(x)) = \varphi_k(\alpha_{jk}(y)) = \varphi_j(y)$. Therefore there is a mapping $\varphi : A \longrightarrow M$ unique such that $\varphi(\mathrm{cls}\,(x,i)) = \varphi_i(x)$ whenever $x \in A_i$; equivalently, such that $\varphi \circ \alpha_i = \varphi_i$ for all i. (Thus $(\alpha_i)_{i \in I} : \mathscr{A} \longrightarrow A$ is a limit cone of $\mathscr{A}$ if $\mathscr{A}$ is viewed as a direct system of sets and mappings.)

We see that $A = \bigcup_{i \in I} \mathrm{Im}\,\alpha_i$. Moreover this is a directed union, since $i \leqslant j$ implies $\alpha_j \circ \alpha_{ij} = \alpha_i$ and $\mathrm{Im}\,\alpha_i \subseteq \mathrm{Im}\,\alpha_j$. Hence every finite sequence $a_1, \ldots, a_n$ of elements of A is contained in some $\mathrm{Im}\,\alpha_i$ and can be written in the form $a_1 = \mathrm{cls}\,(x_1,i), \ldots, a_n = \mathrm{cls}\,(x_n,i)$ for some $i \in I$ and $x_1, \ldots, x_n \in A_i$. We use this to place a left R-module structure on A.

Let $a,b \in A$. Then $a = \mathrm{cls}\,(x,i)$, $b = \mathrm{cls}\,(y,i)$ for some $i \in I$ and $x,y \in A_i$. We show that $a + b \in A$ can be defined by $a + b = \mathrm{cls}\,(x + y,i)$. Assume that $a = \mathrm{cls}\,(x,i) = \mathrm{cls}\,(u,j)$ and $b = \mathrm{cls}\,(y,i) = \mathrm{cls}\,(v,j)$. Then $\alpha_{i\ell}(x) = \alpha_{j\ell}(u)$ and $\alpha_{im}(y) = \alpha_{jm}(v)$ for some $\ell,m \geqslant i,j$; since I is directed there exists $k \geqslant \ell,m$ and we have $\alpha_{ik}(x) = \alpha_{jk}(u)$, $\alpha_{ik}(y) = \alpha_{jk}(v)$,

$$(x + y,i) \sim (\alpha_{ik}(x + y),k) = (\alpha_{ik}(x) + \alpha_{ik}(y),k)$$
$$= (\alpha_{jk}(u) + \alpha_{jk}(v),k) = (\alpha_{jk}(u + v),k) \sim (u + v,j),$$

and $\mathrm{cls}\,(x + y,i) = \mathrm{cls}\,(u + v,j)$. Therefore an addition $A \times A \longrightarrow A$ on A is well defined by $a + b = \mathrm{cls}\,(x + y,i)$ whenever $a = \mathrm{cls}\,(x,i)$, $b = \mathrm{cls}\,(y,i)$. Clearly this is the only addition on M which is preserved by all α_i. Since every A_i is an abelian group, it is immediate that A is an abelian group; in A, $\mathrm{cls}\,(0,i) = \mathrm{cls}\,(0,j)$ is the identity element, and $\mathrm{cls}\,(-x,i)$ is the opposite of $\mathrm{cls}\,(x,i)$.

If $a = \mathrm{cls}\,(x,i) = \mathrm{cls}\,(y,j)$ and $r \in R$, then $\alpha_{ik}(x) = \alpha_{jk}(y)$ for some $k \geqslant i,j$ and

$$(rx,i) \sim (\alpha_{ik}(rx),k) = (r\,\alpha_{ik}(x),k) = (r\,\alpha_{jk}(y),k) = (\alpha_{jk}(ry),k) \sim (ry,j).$$

Therefore a left action of R on A is well defined by $ra = \mathrm{cls}\,(rx,i)$ whenever $a = \mathrm{cls}\,(x,i)$. It is immediate that this is the only action preserved by all α_i and that it makes A a left R-module. Thus there is exactly one left R-module structure on A such that every α_i is a homomorphism.

Let $(\varphi_i)_{i \in I} : \mathscr{A} \longrightarrow M$ be a cone. We saw that there exists a unique mapping $\varphi : A \longrightarrow M$ such that $\varphi_i = \varphi \circ \alpha_i$ for all $i \in I$. It is immediate that φ is a homomorphism: for every $a = \alpha_i(x), b = \alpha_i(y) \in A$ and $r \in R$, we have

$$\varphi(a + b) = \varphi(\alpha_i(x + y)) = \varphi_i(x + y) = \varphi_i(x) + \varphi_i(y) = \varphi(a) + \varphi(b)$$

and

$$\varphi(ra) = \varphi(\alpha_i(rx)) = \varphi_i(rx) = r\,\varphi_i(x) = r\,\varphi(a).$$

Hence φ is the only module homomorphism such that $\varphi_i = \varphi \circ \alpha_i$ for all $i \in I$. Thus $(\alpha_i)_{i \in I}$ is a limit cone of the given direct system $\mathscr{A}$ of left R-modules. $\square$

The direct limit constructed in this proof has a number of interesting properties, which can be used for a simpler characterization.

Proposition 14.7.4. *Let* $\mathscr{A} : (A_i)_{i \in I}$, $(\alpha_{ij})_{i,j \in I, i \leqslant j}$ *be a direct system of left R-modules. A cone* $\varphi_i : \mathscr{A} \longrightarrow M$ *is a limit cone of* $\mathscr{A}$ *if and only if:*

(i) $M = \bigcup_{i \in I} \operatorname{Im} \varphi_i$; *and*

(ii) *for every i,* $\operatorname{Ker} \varphi_i = \bigcup_{j \geqslant i} \operatorname{Ker} \alpha_{ij}$.

Then

(iii) $\varphi_i(x) = \varphi_j(y)$ *if and only if* $\alpha_{ik}(x) = \alpha_{jk}(y)$ *for some* $k \geqslant i, j$.

Note that the unions in (i) and (ii) are directed unions.

Proof. Let $(\alpha_i)_{i \in I} : \mathscr{A} \longrightarrow A$ be the limit cone of $\mathscr{A}$ constructed in the proof of Proposition 14.7.3. We saw that $A = \bigcup_{i \in I} \operatorname{Im} \alpha_i$ and that $\alpha_i(x) = \alpha_j(y)$ if and only if $\alpha_{ik}(x) = \alpha_{jk}(y)$ for some $k \geqslant i, j$. Thus $(\alpha_i)_{i \in I}$ has properties (i) and (iii). If $(\varphi_i)_{i \in I} : \mathscr{A} \longrightarrow M$ is another limit cone, then there is an isomorphism $\theta : A \longrightarrow M$ such that $\varphi_i = \theta \circ \alpha_i$ for all i; therefore $(\varphi_i)_{i \in I}$ has properties (i) and (iii); (iii) implies (ii).

Conversely, let $(\varphi_i)_{i \in I} : \mathscr{A} \longrightarrow M$ be a cone such that (i) and (ii) hold. There is a homomorphism $\theta : A \longrightarrow M$ such that $\varphi_i = \theta \circ \alpha_i$ for all i. We have $\theta(A) = \theta(\bigcup_{i \in I} \operatorname{Im} \alpha_i) = \bigcup_{i \in I} \operatorname{Im} \varphi_i = M$; hence θ is surjective. If $\theta(a) = 0$, then $a = \alpha_i(x)$ for some $i \in I$ and $x \in A_i$, $\varphi_i(x) = \theta(\alpha_i(x)) = 0$, $\alpha_{ij}(x) = 0$ for some $j \geqslant i$ by (ii), and $a = \alpha_j(\alpha_{ij}(x)) = 0$. Thus θ is an isomorphism. Therefore $(\varphi_i)_{i \in I} = (\theta \circ \alpha_i)_{i \in I}$ is a limit cone. $\square$

To our reader's undoubted delight, there is no further need for the construction of direct limits: indeed (i) and (ii) provide a simpler characterization.

If every α_{ij} is injective, then every α_i is injective by (ii), and $A = \varinjlim_{i \in I} A_i$ is the directed union of all $\operatorname{Im} \alpha_i \cong A_i$. Thus direct limits allow us to assemble disparate objects into a directed union. Then we may think of A_i as an approximation of $\varinjlim_{i \in I} A_i$, which becomes more and more accurate as i increases in I.

14.7.e. Homomorphisms. Let

$$\mathscr{A} : (A_i)_{i \in I}, \; (\alpha_{ij})_{i,j \in I, i \leqslant j} \qquad \text{and} \qquad \mathscr{B} : (B_i)_{i \in I}, \; (\beta_{ij})_{i,j \in I, i \leqslant j}$$

be direct systems of left R-modules over the same directed preordered set I. A **homomorphism** of $\mathscr{A}$ into $\mathscr{B}$ is a family $(\varphi_i)_{i \in I}$ of homomorphisms

$\varphi_i : A_i \longrightarrow B_i$ such that the square

$$
\begin{array}{ccc}
A_j & \xrightarrow{\varphi_j} & B_j \\
{\scriptstyle \alpha_{ij}} \uparrow & & \uparrow {\scriptstyle \beta_{ij}} \\
A_i & \xrightarrow[\varphi_i]{} & B_i
\end{array}
$$

commutes for every $i \leqslant j$ in I. For example, a cone $(\varphi_i)_{i \in I} : \mathscr{A} \longrightarrow M$ is a homomorphism of $\mathscr{A}$ into the **constant** direct system $\mathscr{M}$ (also denoted by M) in which $M_i = M$ for all i and $\mu_{ij} = 1_M$ whenever $i \leqslant j$.

Every homomorphism of direct systems induces a homomorphism of direct limits. In the above let $(\alpha_i)_{i \in I} : \mathscr{A} \longrightarrow A$ and $(\beta_i)_{i \in I} : \mathscr{B} \longrightarrow B$ be limit cones. Then $(\beta_i \circ \varphi_i)_{i \in I}$ is a cone for $\mathscr{A}$: if $i \leqslant j$, then $\beta_j \circ \varphi_j \circ \alpha_{ij} = \beta_j \circ \beta_{ij} \circ \varphi_i = \beta_i \circ \varphi_i$. Hence there is a homomorphism $\varphi = \varinjlim_{i \in I} \varphi_i : A \longrightarrow B$ unique such that

$$
\begin{array}{ccc}
A & \xrightarrow{\varphi} & B \\
{\scriptstyle \alpha_i} \uparrow & & \uparrow {\scriptstyle \beta_i} \\
A_i & \xrightarrow[\varphi_i]{} & B_i
\end{array}
$$

$\beta_i \circ \varphi_i = \varphi \circ \alpha_i$ for all i, and we have proved:

Proposition 14.7.5. *Let $\mathscr{A}$ and $\mathscr{B}$ be direct systems of left R-modules over the same directed preordered set I. Every homomorphism $(\varphi_i)_{i \in I} : \mathscr{A} \longrightarrow \mathscr{B}$ of direct systems induces a homomorphism $\varphi = \varinjlim_{i \in I} \varphi_i : \varinjlim_{i \in I} A_i \longrightarrow \varinjlim_{i \in I} B_i$ unique such that $\varphi \circ \alpha_i = \beta_i \circ \varphi_i$ for all i.*

The following properties are exercises:

Proposition 14.7.6. *If φ_i is the identity on A_i for every i, then $\varinjlim_{i \in I} \varphi_i$ is the identity on $\varinjlim_{i \in I} A_i$. If $(\varphi_i)_{i \in I} : \mathscr{A} \longrightarrow \mathscr{B}$ and $(\psi_i)_{i \in I} : \mathscr{B} \longrightarrow \mathscr{C}$ are homomorphisms of direct systems, then $\varinjlim_{i \in I} \psi_i \circ \varphi_i = (\varinjlim_{i \in I} \psi_i) \circ (\varinjlim_{i \in I} \varphi_i)$.*

Thus $\varinjlim_{i \in I}$ is in effect a functor from direct systems over I to left R-modules.

14.7.g. Properties. A subset J of a preordered set I is **cofinal** in case for each $i \in I$ we have $i \leqslant j$ for some $j \in J$.

Proposition 14.7.7. *Let $\mathscr{A} : (A_i)_{i \in I}, (\alpha_{ij})_{i,j \in I, i \leqslant j}$ be a direct system of left R-modules. If J is a cofinal subset of I, then $\varinjlim_{i \in I} A_i = \varinjlim_{i \in J} A_i$. More pre-*

cisely, if $(\alpha_i)_{i \in I} : \mathcal{A} \longrightarrow A$ *is a limit cone of* $\mathcal{A}$*, then* $(\alpha_i)_{i \in J} : \mathcal{A} \longrightarrow A$ *is a limit cone of* $\mathcal{A}_{|J} : (A_i)_{i \in J}, (\alpha_{ij})_{i,j \in J, i \leqslant j}$.

Proof. We use Proposition 14.7.4. First J is directed upward, since I is directed upward and J is cofinal in I. Hence $\mathcal{A}_{|J}$ is a direct system. Since J is cofinal, we have $\bigcup_{i \in I} \operatorname{Im} \alpha_i = \bigcup_{i \in J} \operatorname{Im} \alpha_i$. Also let $j \in J$. If $\alpha_j(x) = 0$, then $\alpha_{jk}(x) = 0$ for some $k \geqslant j$ in I, $k \leqslant \ell$ for some $\ell \in J$, and $\alpha_{j\ell}(x) = \alpha_{k\ell}(\alpha_{jk}(x)) = 0$; conversely, $\alpha_{j\ell}(x) = 0$ implies $\alpha_j(x) = 0$. Hence $(\alpha_i)_{i \in J}$ is a limit cone of $\mathcal{A}_{|J}$. $\square$

The next property is specific to modules.

Proposition 14.7.8 (Exactness). *Let* $\mathcal{A}$*,* $\mathcal{B}$*, and* $\mathcal{C}$ *be direct systems of left* R*-modules and* $(\varphi_i)_{i \in I} : \mathcal{A} \longrightarrow \mathcal{B}$ *and* $(\psi_i)_{i \in I} : \mathcal{B} \longrightarrow \mathcal{C}$ *be homomorphisms. Let* $\varphi = \varinjlim_{i \in I} \varphi_i$ *and* $\psi = \varinjlim_{i \in I} \psi_i$*. If* $A_i \xrightarrow{\varphi_i} B_i \xrightarrow{\psi_i} C_i$ *is exact for every* $i \in I$*, then*

$$\varinjlim_{i \in I} A_i \xrightarrow{\varphi} \varinjlim_{i \in I} B_i \xrightarrow{\psi} \varinjlim_{i \in I} C_i$$

is exact.

Proof. Let $(\alpha_i)_{i \in I} : \mathcal{A} \longrightarrow A$, $(\beta_i)_{i \in I} : \mathcal{B} \longrightarrow B$, and $(\gamma_i)_{i \in I} : \mathcal{C} \longrightarrow C$ be limit cones. By definition, the following diagram commutes for every $i \in I$:

$$
\begin{array}{ccccc}
A & \xrightarrow{\varphi} & B & \xrightarrow{\psi} & C \\
\uparrow{\scriptstyle \alpha_i} & & \uparrow{\scriptstyle \beta_i} & & \uparrow{\scriptstyle \gamma_i} \\
A_i & \xrightarrow{\varphi_i} & B_i & \xrightarrow{\psi_i} & C_i
\end{array}
$$

Exactness of the top row is proved by diagram chasing, using Proposition 14.7.4. We have $\psi \circ \varphi = 0$, since $\psi_i \circ \varphi_i = 0$ for all i. Hence $\operatorname{Im} \varphi \subseteq \operatorname{Ker} \psi$. Conversely, let $b \in \operatorname{Ker} \psi$. By Proposition 14.7.4, $b = \beta_i(b_i)$ for some $i \in I$ and $b_i \in B_i$. Then $\gamma_i(\psi_i(b_i)) = \psi(\beta_i(b_i)) = 0$; by Proposition 14.7.4, $\gamma_{ij}(\psi_i(b_i)) = 0$ for some $j \geqslant i$. Then $\psi_j(\beta_{ij}(b_i)) = \gamma_{ij}(\psi_i(b_i)) = 0$; by exactness, $\beta_{ij}(b_i) = \varphi_j(a_j)$ for some $a_j \in A_j$. Hence $b = \beta_j(\beta_{ij}(b_i)) = \beta_j(\varphi_j(a_j)) = \varphi(\alpha_j(a_j)) \in \operatorname{Im} \varphi$. $\square$

In particular, a direct limit of short exacts sequences is a short exact sequence.

Corollary 14.7.9. *A direct limit of monomorphisms (of epimorphisms) is a monomorphism (an epimorphism).*

Other properties will be found in the exercises.

Exercises

1. Prove that the direct limit of a direct system is unique up to isomorphism.
2. Prove that the direct limit of a nonempty directed family of submodules is its union.
3. Prove that a free left R-module is a direct limit of finitely generated free submodules.
4. Prove that every direct sum $\bigoplus_{i \in I} M_i$ is the direct limit of finite direct sums: $\bigoplus_{i \in I} M_i \cong \varinjlim_{J \subseteq I, J \text{ finite}} \bigoplus_{i \in J} M_i$.
5. Show that $\mathbb{Z}_{p^\infty} \cong \varinjlim_{n>0} \mathbb{Z}_{p^n}$.
6. Define and construct direct limits of (not necessarily abelian) groups.
7. Define and construct direct limits of fields; show that a direct limit of fields is a field.
8. Let $(\varphi_i)_{i \in I} : \mathscr{A} \longrightarrow \mathscr{B}$ and $(\psi_i)_{i \in I} : \mathscr{B} \longrightarrow \mathscr{C}$ be homomorphisms of direct systems. Prove that $\varinjlim (\psi_i \circ \varphi_i) = (\varinjlim \psi_i) \circ (\varinjlim \varphi_i)$.
9. Prove, without using exactness, that a direct limit of monomorphisms is a monomorphism.
10. Prove, without using exactness, that a direct limit of epimorphisms is an epimorphism.
11. Prove that direct limits of left R-modules preserve finite direct sums: show that $\varinjlim(A_i \oplus B_i) \cong (\varinjlim A_i) \oplus (\varinjlim B_i)$ whenever $\mathscr{A} : (A_i)_{i \in I}$, $(\alpha_{ij})_{i,j \in I, i \leqslant j}$ and $\mathscr{B} : (B_i)_{i \in I}$, $(\beta_{ij})_{i,j \in I, i \leqslant j}$ are direct systems of left R-modules over the same directed preordered set I.
12. Prove that direct limits of left R-modules preserve pullbacks.
13. Prove that direct limits of left R-modules preserve pushouts.
14. Prove that a ring R is left noetherian if and only if every left R-module which is the union of a directed family of injective submodules is injective.

14.8. INVERSE LIMITS

Inverse limits are the last of our basic constructions. They are very similar to the general limits in Chapter 17.

14.8.a. Construction. A preordered set I is **directed downward** in case for each $i, j \in I$ there exists $k \in I$ such that $k \leqslant i$ and $k \leqslant j$. For instance, every chain is directed downward.

Let I be a preordered set which is directed downward. An **inverse system** $\mathscr{A}$ of left R-modules over I consists of modules $(A_i)_{i \in I}$ and homomorphisms $\alpha_{ij} : A_i \longrightarrow A_j$ (one for each pair (i, j) such that $i \leqslant j$ in I) such that for all $i, j, k \in I$:

(1) α_{ii} is the identity on A_i;
(2) if $i \leqslant j \leqslant k$, then $\alpha_{jk} \circ \alpha_{ij} = \alpha_{ik}$.

Like direct systems, inverse systems can be viewed as large commutative diagrams, with arrows running upward. A chain of submodules $(A_i)_{i \in I}$ can be viewed as an inverse system: I can be preordered so that $i \leqslant j$ if and only if $A_i \subseteq A_j$, and then $\alpha_{ij} : A_i \longrightarrow A_j$ is the inclusion homomorphism.

Let $\mathscr{A} : (A_i)_{i \in I}, (\alpha_{ij})_{i,j \in I, i \leqslant j}$ be an inverse system of left R-modules over a preordered set I which is directed downward. A **cone** [of homomorphisms] of a left R-module M into $\mathscr{A}$ is a family $(\varphi_i)_{i \in I}$ of homomorphisms $\varphi_i : M \longrightarrow A_i$ such that $\alpha_{ij} \circ \varphi_i = \varphi_j$ whenever $i \leqslant j$ in I.

The **inverse limit** (also called **projective limit**) of an inverse system $\mathscr{A}$: $(A_i)_{i \in I}, (\alpha_{ij})_{i,j \in I, i \leqslant j}$ consists of a left R-module $A = \varprojlim_{i \in I} A_i$ and a **limit cone**, which is a cone $(\alpha_i)_{i \in I} : A \longrightarrow \mathscr{A}$ with the following universal property: for each cone $(\varphi_i)_{i \in I} : M \longrightarrow \mathscr{A}$, there exists a homomorphism $\varphi : M \longrightarrow A$ unique such that $\varphi_i = \alpha_i \circ \varphi$ for all $i \in I$.

For example, the inverse limit of a chain of submodules is its intersection (see the exercises).

Proposition 14.8.1. *Every inverse system of left R-modules has an inverse limit, which is unique up to isomorphism.*

Proof. Uniqueness follows from the universal property in the usual fashion. Existence is established by constructing an inverse limit for any inverse system $\mathscr{A} : (A_i)_{i \in I}, (\alpha_{ij})_{i,j \in I, i \leqslant j}$ of left R-modules, as follows. Let $P = \prod_{i \in I} A_i$, with projections $\pi_i : P \longrightarrow A_i$. Let

$$A = \{ (x_i)_{i \in I} ; x_j = \alpha_{ij}(x_i) \text{ whenever } i \leqslant j \}.$$

Equivalently, $A = \bigcap_{i,j \in I, i \leqslant j} \mathrm{Ker}\,(\pi_j - \alpha_{ij} \circ \pi_i)$; hence A is a submodule of P. Let $\alpha_i = \pi_{i|A} : A \longrightarrow A_i$: if $(x_i)_{i \in I} \in A$, then $\alpha_i((x_i)_{i \in I}) = x_i$. We show that $(\alpha_i)_{i \in I} : A \longrightarrow \mathscr{A}$ is a limit cone for $\mathscr{A}$.

By definition of A, $\alpha_j = \alpha_{ij} \circ \alpha_i$ whenever $i \leqslant j$. Let $(\varphi_i)_{i \in I} : M \longrightarrow \mathscr{A}$ be any cone (so that $\varphi_j = \alpha_{ij} \circ \varphi_i$ whenever $i \leqslant j$ in I). By the universal property of the direct product, there is a homomorphism $\varphi : M \longrightarrow P$ unique such that $\pi_i \circ \varphi = \varphi_i$ for all i; namely $\varphi(x) = (\varphi_i(x))_{i \in I}$ for all $x \in M$. We have $\varphi(x) \in A$ for all $x \in M$, since $\varphi_j(x) = \alpha_{ij}(\varphi_i(x))$ whenever $i \leqslant j$. Hence φ can be viewed as a homomorphism of M into A. Then $\alpha_i(\varphi(x)) = \varphi_i(x)$ for all $x \in M$. If, conversely, $\psi : M \longrightarrow A$ is a homomorphism such that $\alpha_i \circ \psi = \varphi_i$ for all i, then the i-th component of $\psi(x)$ is $\varphi_i(x)$ and $\psi(x) = \varphi(x)$ for all $x \in M$. $\qquad \square$

Again we note that the construction of inverse limits is essentially a set construction, and can be used for inverse limits of sets, groups, and so on (see the exercises).

14.8.b. Homomorphisms. Let

$$\mathscr{A} : (A_i)_{i \in I}, \; (\alpha_{ij})_{i,j \in I, i \leqslant j} \qquad \text{and} \qquad \mathscr{B} : (B_i)_{i \in I}, \; (\beta_{ij})_{i,j \in I, i \leqslant j}$$

be inverse systems of left R-modules over the same directed downward preordered set I. A **homomorphism** of $\mathscr{A}$ into $\mathscr{B}$ is a family $(\varphi_i)_{i \in I}$ of homomorphisms $\varphi_i : A_i \longrightarrow B_i$ such that the square

$$
\begin{array}{ccc}
A_j & \xrightarrow{\;\varphi_j\;} & B_j \\
\alpha_{ij} \uparrow & & \uparrow \beta_{ij} \\
A_i & \xrightarrow[\;\varphi_i\;]{} & B_i
\end{array}
$$

commutes for every $i \leqslant j$ in I. For example, a cone $(\varphi_i)_{i \in I} : M \longrightarrow \mathscr{A}$ is a homomorphism of the **constant** inverse system $\mathscr{M}$ (also denoted by M) in which $M_i = M$ for all i and $\mu_{ij} = 1_M$ whenever $i \leqslant j$.

Every homomorphism of inverse systems induces a homomorphism of inverse limits. In the above let $(\alpha_i)_{i \in I} : A \longrightarrow \mathscr{A}$ and $(\beta_i)_{i \in I} : B \longrightarrow \mathscr{B}$ be limit cones. Then $(\varphi_i \circ \alpha_i)_{i \in I}$ is a cone from A to $\mathscr{B}$: $\beta_{ij} \circ \varphi_i \circ \alpha_i = \varphi_j \circ \alpha_{ij} \circ \alpha_i = \varphi_j \circ \alpha_j$ whenever $i \leqslant j$. Hence there is a unique homomorphism $\varphi = \varprojlim_{i \in I} \varphi_i : A \longrightarrow B$

$$
\begin{array}{ccc}
A_i & \xrightarrow{\;\varphi_i\;} & B_i \\
\alpha_i \uparrow & & \uparrow \beta_i \\
A & \xrightarrow[\;\varphi\;]{} & B
\end{array}
$$

such that $\beta_i \circ \varphi = \varphi_i \circ \alpha_i$ for all i, and we have proved:

Proposition 14.8.2. *Let $\mathscr{A}$ and $\mathscr{B}$ be inverse systems of left R-modules over the same directed downward preordered set I. Every homomorphism $(\varphi_i)_{i \in I} :$ $\mathscr{A} \longrightarrow \mathscr{B}$ of inverse systems induces a homomorphism $\varphi = \varprojlim_{i \in I} \varphi_i :$ $\varprojlim_{i \in I} A_i \longrightarrow \varprojlim_{i \in I} B_i$ unique such that $\beta_i \circ \varphi = \varphi_i \circ \alpha_i$ for all i.*

The following properties are exercises:

Proposition 14.8.3. *If φ_i is the identity on A_i for every i, then $\varprojlim_{i \in I} \varphi_i$ is the identity on $\varprojlim_{i \in I} A_i$. If $(\varphi_i)_{i \in I} : \mathscr{A} \longrightarrow \mathscr{B}$ and $(\psi_i)_{i \in I} : \mathscr{B} \longrightarrow \mathscr{C}$ are homomorphisms of inverse systems, then $\varprojlim_{i \in I} \psi_i \circ \varphi_i = (\varprojlim_{i \in I} \psi_i) \circ (\varprojlim_{i \in I} \varphi_i)$.*

Thus $\varprojlim_{i \in I}$ is a functor from inverse systems over I to left R-modules.

14.8.c. Properties. The following properties are exercises:

Proposition 14.8.4. *Let* $\mathscr{A} : (A_i)_{i \in I}$, $(\alpha_{ij})_{i,j \in I, i \leqslant j}$ *be an inverse system of left R-modules. There is an isomorphism* $\mathrm{Hom}_R(M, \varprojlim A_i) \cong \varprojlim \mathrm{Hom}_R(M, A_i)$ *which is natural in* $\mathscr{A}$ *and M.*

Proposition 14.8.5. *Let* $\mathscr{A} : (A_i)_{i \in I}$, $(\alpha_{ij})_{i,j \in I, i \leqslant j}$ *be a direct system of left R-modules. There is an isomorphism* $\mathrm{Hom}_R(\varinjlim A_i, M) \cong \varprojlim \mathrm{Hom}_R(A_i, M)$ *which is natural in* $\mathscr{A}$ *and M.*

A left exactness property will also be found in the exercises.

Exercises

1. Prove that the inverse limit of an inverse system is unique up to isomorphism.
2. Prove that the inverse limit of a chain of submodules is its intersection.
3. Define and construct inverse limits of groups.
4. Let $\mathscr{A} : (A_i)_{i \in I}, (\alpha_{ij})_{i,j \in I, i \leqslant j}$ be an inverse system of left R-modules. Let $\mathscr{A}_S$ be the inverse system of sets with the same sets A_i and mappings α_{ij}. Let $(\alpha_i)_{i \in I} : A \longrightarrow \mathscr{A}_S$ be a limit cone of $\mathscr{A}_S$. Show that there is exactly one left R-module structure on A such that every α_i is a homomorphism, and then $(\alpha_i)_{i \in I}$ is a limit cone of $\mathscr{A}$.
5. Let $(\varphi_i)_{i \in I} : \mathscr{A} \longrightarrow \mathscr{B}$ and $(\psi_i)_{i \in I} : \mathscr{B} \longrightarrow \mathscr{C}$ be homomorphisms of inverse systems. Prove that $\varprojlim_{i \in I} (\psi_i \circ \varphi_i) = (\varprojlim_{i \in I} \psi_i) \circ (\varprojlim_{i \in I} \varphi_i)$.
6. Let $\mathscr{A}$, $\mathscr{B}$, and $\mathscr{C}$ be inverse systems and $(\mu_i)_{i \in I} : \mathscr{A} \longrightarrow \mathscr{B}$ and $(\rho_i)_{i \in I} : \mathscr{B} \longrightarrow \mathscr{C}$ be homomorphisms, with $\mu = \varprojlim_{i \in I} \mu_i$, $\rho = \varprojlim_{i \in I} \rho_i$. If

$$0 \longrightarrow A_i \xrightarrow{\mu_i} B_i \xrightarrow{\rho_i} C_i$$

is exact for every $i \in I$, then prove that

$$0 \longrightarrow \varprojlim_{i \in I} A_i \xrightarrow{\mu} \varprojlim_{i \in I} B_i \xrightarrow{\rho} \varprojlim_{i \in I} C_i$$

is exact.
7. Let $\mathscr{A} : (A_i)_{i \in I}$, $(\alpha_{ij})_{i,j \in I, i \leqslant j}$ be an inverse system of left R-modules. Prove that $\mathrm{Hom}_R(M, \varprojlim A_i) \cong \varprojlim \mathrm{Hom}_R(M, A_i)$.
8. Let $\mathscr{A} : (A_i)_{i \in I}$, $(\alpha_{ij})_{i,j \in I, i \leqslant j}$ be a direct system of left R-modules. Prove that $\mathrm{Hom}_R(\varinjlim A_i, M) \cong \varprojlim \mathrm{Hom}_R(A_i, M)$.
9. Let $\mathscr{A} : (A_i)_{i \in I}$, $(\alpha_{ij})_{i,j \in I, i \leqslant j}$ and $\mathscr{B} : (B_i)_{i \in I}$, $(\beta_{ij})_{i,j \in I, i \leqslant j}$ be inverse systems of left R-modules. Prove that $\varprojlim(A_i \oplus B_i) \cong (\varprojlim A_i) \oplus (\varprojlim B_i)$.
10. Show that every Galois group is an inverse limit of finite groups. (Such groups are called **profinite**.)

15

TENSOR PRODUCTS

Tensor products are another basic construction for modules. This chapter contains basic properties, and applications to dual modules and to flat modules; applications to algebras are in Chapter 16.

Required results: Chapter 10; Chapter 14, Sections 14.2, 14.5, and 14.6. Direct limits (Section 14.7) are required for Section 15.4.

All rings in this chapter have an identity element, and all modules and bimodules are unital.

15.1. CONSTRUCTION

15.1.a. Bilinear Mappings. Let R be a commutative ring. When A,B,C are R-modules, a **bilinear mapping** $\beta : A \times B \longrightarrow C$ is a mapping such that

$$\beta(a + a',b) = \beta(a,b) + \beta(a',b),$$

$$\beta(a,b + b') = \beta(a,b) + \beta(a,b'), \qquad \text{and}$$

$$\beta(ra,b) = r\beta(a,b) = \beta(a,rb),$$

for all $a,a' \in A$, $b,b' \in B$, $r \in R$.

Bilinear mappings $\beta : A \times B \longrightarrow C$ can be characterized by properties of the mappings $\beta(a,-) : b \longmapsto \beta(a,b)$, $B \longrightarrow C$ and $\beta(-,b) : a \longmapsto \beta(a,b)$, $A \longrightarrow C$. By definition, β is bilinear if and only if $\beta(a,-)$ and $\beta(-,b)$ are module homomorphisms for all $a \in A$ and $b \in B$. The following result is straightforward:

Lemma 15.1.1. *Let R be a commutative ring and A,B,C be R-modules. For a mapping $\beta : A \times B \longrightarrow C$ the following conditions are equivalent:*

(1) *β is bilinear.*

(2) *$a \longmapsto \beta(a,-)$ is a module homomorphism of A into $\mathrm{Hom}_R(B,C)$.*

(3) *$b \longmapsto \beta(-,b)$ is a module homomorphism of B into $\mathrm{Hom}_R(A,C)$.*

15.1.b. Bihomomorphisms. The definition of bilinear mappings could be extended directly to left modules over an arbitrary ring R. However, it is more fruitful to extend Lemma 15.1.1. Then properties (2) and (3) require A and B to be modules on opposite sides: if B is a left R-module, then $\mathrm{Hom}(B,C)$ is a right R-module and A should be a right R-module; if A is a right R-module, then $\mathrm{Hom}(A,C)$ is a left R-module, and B should be a left R-module.

Lemma 15.1.2. *Let A be a right R-module, B be a left R-module, and C be an abelian group. For a mapping $\beta : A \times B \longrightarrow C$ the following conditions are equivalent:*

(1) *For all $a,a' \in A$, $b,b' \in B$, $r \in R$,*

$$\beta(a + a',b) = \beta(a,b) + \beta(a',b),$$
$$\beta(a,b + b') = \beta(a,b) + \beta(a,b') \qquad (\beta \text{ is } \textbf{biadditive}), \text{ and}$$
$$\beta(ar,b) = \beta(a,rb) \qquad (\beta \text{ is } \textbf{balanced}).$$

(2) $a \longmapsto \beta(a,-)$ *is a module homomorphism of A into $\mathrm{Hom}_{\mathbb{Z}}(B,C)$.*
(3) $b \longmapsto \beta(-,b)$ *is a module homomorphism of B into $\mathrm{Hom}_{\mathbb{Z}}(A,C)$.*

Proof. Recall that addition on $\mathrm{Hom}_{\mathbb{Z}}(B,C)$ is pointwise and that the right action of R on $\mathrm{Hom}_{\mathbb{Z}}(B,C)$ is given by $(\varphi r)(b) = \varphi(rb)$ (for all $r \in R$, $\varphi \in \mathrm{Hom}_{\mathbb{Z}}(B,C)$, and $b \in B$). The three conditions in part (1) mean, respectively, that $\beta(a + a',-) = \beta(a,-) + \beta(a',-)$, $\beta(a,-) \in \mathrm{Hom}_{\mathbb{Z}}(B,C)$, and $\beta(ar,-) = \beta(a,-)r$. Thus (1) and (2) are equivalent. The equivalence of (2) and (3) is similar. $\qquad \square$

When A is a right R-module, B is a left R-module, and C is an abelian group, we call a mapping $\beta : A \times B \longrightarrow C$ a **bihomomorphism** of modules in case it satisfies the equivalent conditions in Lemma 15.1.2. Bihomomorphisms are also called **middle linear** mappings, **R-biadditive** mappings, **balanced biadditive** mappings, and **balanced products**. The terminology "bihomomorphism" works well when Lemma 15.1.1 is extended to bimodules.

Lemma 15.1.3. *Let A be a Q-R-bimodule, B be an R-S-bimodule, and C be a Q-S-bimodule. For a mapping $\beta : A \times B \longrightarrow C$ the following are equivalent:*

(1) *For all $a,a' \in A$, $b,b' \in B$, $q \in Q$, $r \in R$, $s \in S$,*

$$\beta(a + a',b) = \beta(a,b) + \beta(a',b),$$
$$\beta(a,b + b') = \beta(a,b) + \beta(a,b'),$$
$$\beta(qa,b) = q\,\beta(a,b),$$
$$\beta(ar,b) = \beta(a,rb), \qquad and$$
$$\beta(a,bs) = \beta(a,b)\,s.$$

(2) $a \longmapsto \beta(a, -)$ is a bimodule homomorphism of A into $\text{Hom}_S(B,C)$.

(3) $b \longmapsto \beta(-,b)$ is a bimodule homomorphism of B into $\text{Hom}_Q(A,C)$.

Proof. This is proved like Lemma 15.1.2. Since B is an R-S-bimodule and C is a Q-S-bimodule, $\text{Hom}_S(B,C)$ is an Q-S-bimodule. The conditions in part (1) mean that $\beta(a + a', -) = \beta(a, -) + \beta(a', -)$, $\beta(a, -) \in \text{Hom}_S(B,C)$, $\beta(ar, -) = \beta(a, -)r$, and $\beta(sa, -) = s\beta(a, -)$, so that (1) and (2) are equivalent. Similarly (1) and (3) are equivalent. $\square$

When A is a Q-R-bimodule, B is an R-S-bimodule, and C is a Q-S-bimodule, we say that a mapping $\beta : A \times B \longrightarrow C$ is a **bihomomorphism** of bimodules in case it satisfies the equivalent conditions in Lemma 15.1.3. If R is commutative and $Q = S = R$, then Lemma 15.1.3 reduces to Lemma 15.1.1, and bimodule bihomomorphisms coincide with bilinear mappings. Lemma 15.1.3 reduces to Lemma 15.1.2 when $Q = S = \mathbb{Z}$.

15.1.c. The Tensor Product. Let A be a right R-module and B be a left R-module. If $\tau : A \times B \longrightarrow T$ is a bihomomorphism and $\gamma : T \longrightarrow C$ is a homomorphism (of abelian groups), then $\gamma \circ \tau : A \times B \longrightarrow C$ is a bihomomorphism. The **tensor product** of A and B is an abelian group $A \otimes_R B$ together with a bihomomorphism $\tau : A \times B \longrightarrow A \otimes_R B$ which has the following universal property: for every abelian group C and bihomomorphism $\beta : A \times B \longrightarrow C$, there is a unique additive homomorphism $A \otimes_R B \longrightarrow C$ such that the following diagram commutes:

Then the bihomomorphism τ is the **tensor map**; τ is sometimes denoted by $\otimes$, and $\tau(a,b)$ is denoted by $a \otimes b$.

Proposition 15.1.4. *Let A be a right R-module and B be a left R-module. There exists a tensor product of A and B, and it is unique up to isomorphism.*

Proof. The uniqueness of tensor products follows from their universal property. Existence is proved as follows. Let T be the abelian group generated by all ordered pairs $(a,b) \in A \times B$, subject to all defining relations $(a + a', b) = (a,b) + (a',b)$, $(a, b + b') = (a,b) + (a,b')$, $(ar, b) = (a, rb)$, where $a, a' \in A$, $b, b' \in B$, $r \in R$. We see that the injection $\tau : A \times B \longrightarrow T$ is a bihomomorphism. More generally, a mapping $\beta : A \times B \longrightarrow C$ is a bihomomorphism if and only if the defining relations of T hold in C via β; then β factors uniquely through τ (Proposition 10.5.11). $\square$

This construction is not particularly enlightening. In fact there are only a few cases where we will know what $A \otimes_R B$ actually looks like. Generally, tensor products are manipulated through their properties, not their construction.

Still, the proof of Proposition 15.1.4 shows that $A \otimes_R B \cong T$ is generated by elements of the form $a \otimes b$ with $a \in A$, $b \in B$. Moreover

$$(a + a') \otimes b = a \otimes b + a' \otimes b,$$

$$a \otimes (b + b') = a \otimes b + a \otimes b', \qquad \text{and}$$

$$ar \otimes b = a \otimes rb,$$

for all $a, a' \in A$, $b, b' \in B$, $r \in R$. Every element of $A \otimes_R B$ is a sum of these generators. Examples show that $A \otimes_R B$ usually contains elements which are proper sums of generators and not of the form $a \otimes b$.

15.1.d. Bimodules. The tensor product of modules also serves for bimodules.

Proposition 15.1.5. *Let A be a Q-R-module and B be an R-S-bimodule.*

(1) *There is a unique Q-S-bimodule structure on the abelian group $A \otimes_R B$ such that $q(a \otimes b) = qa \otimes b$ and $(a \otimes b)s = a \otimes bs$ for all $a \in A$, $b \in B$, $q \in Q$, $s \in S$.*

(2) *The tensor map $A \times B \longrightarrow A \otimes_R B$ is a bihomomorphism of bimodules.*

(3) *For every Q-S-bimodule C and bihomomorphism $\beta : A \times B \longrightarrow C$ of bimodules, there is a unique bimodule homomorphism $A \otimes_R B \longrightarrow C$ such that the following diagram commutes:*

$$
\begin{array}{ccc}
 & A \times B & \\
{\scriptstyle \otimes}\downarrow & & \searrow{\scriptstyle \beta} \\
A \otimes_R B & \dashrightarrow & C
\end{array}
$$

Proof. Let $q \in Q$. Then $\beta_q : (a, b) \longmapsto qa \otimes b$ is a bihomomorphism of modules of $A \times B$ into the abelian group $A \otimes_R B$. Therefore there is a unique additive homomorphism $\overline{\beta}_q : A \otimes_R B \longrightarrow A \otimes_R B$ such that $\overline{\beta}_q(a \otimes b) = qa \otimes b$ for all $a \in A$, $b \in B$. We already have $\overline{\beta}_q(t + u) = \overline{\beta}_q(t) + \overline{\beta}_q(u)$ for all $t, u \in A \otimes_R B$; the identities $\overline{\beta}_{q+q'}(t) = \overline{\beta}_q(t) + \overline{\beta}_{q'}(t)$ and $\overline{\beta}_{qq'}(t) = \overline{\beta}_q(\overline{\beta}_{q'}(t))$ hold for all $t \in A \otimes_R B$ since they hold whenever $t = a \otimes b$. Hence $qt = \overline{\beta}_q(t)$ defines a left S-module structure on $A \otimes_R B$, unique such that $q(a \otimes b) = qa \otimes b$.

Similarly there is a unique right S-module structure on $A \otimes_R B$ such that $(a \otimes b)s = a \otimes bs$. The identity $q(ts) = (qt)s$ holds for all $t \in A \otimes_R B$, since it holds whenever $t = a \otimes b$. This proves (1); (2) follows from (1).

Let C be a Q-S-bimodule and $\beta : A \times B \longrightarrow C$ be a bihomomorphism of bimodules. There is a unique additive homomorphism $\overline{\beta} : A \otimes_R B \longrightarrow C$ such that $\beta(a,b) = \overline{\beta}(a \otimes b)$ for all $a \in A$, $b \in B$. The identities $\overline{\beta}(qt) = q\overline{\beta}(t)$ and $\overline{\beta}(ts) = \overline{\beta}(t)s$ hold whenever $t = a \otimes b$ by the choice of β, and therefore hold for all $t \in A \otimes_R B$. Hence $\overline{\beta}$ is a bimodule homomorphism, and (3) holds. $\square$

Corollary 15.1.6. *Let R be a commutative ring and A, B be R-modules.*

(1) *There is a unique R-module structure on $A \otimes_R B$ such that $r(a \otimes b) = ra \otimes b = a \otimes rb$ for all $a \in A$, $b \in B$, $r \in R$.*

(2) *The tensor map $A \times B \longrightarrow A \otimes_R B$ is bilinear.*

(3) *For every bilinear mapping $\beta : A \times B \longrightarrow C$ there is a unique module homomorphism $A \otimes_R B \longrightarrow C$ such that the following diagram commutes:*

By Proposition 15.1.5, subsequent properties of the tensor product $A \otimes_R B$ may be stated for bimodules, with A an Q-R-bimodule and B an R-S-bimodule. Then just let $Q = \mathbb{Z}$ if A is only a right R-module, and $S = \mathbb{Z}$ if B is only a left R-module.

15.1.e. Homomorphisms. Module homomorphisms induce homomorphisms of tensor products, as follows.

Proposition 15.1.7. *Let $\varphi : A \longrightarrow A'$ be a homomorphism of Q-R-bimodules and $\psi : B \longrightarrow B'$ be a homomorphism of R-S-bimodules. There exists a unique bimodule homomorphism $\varphi \otimes \psi$ such that the following diagram commutes:*

$$
\begin{array}{ccc}
A \otimes_R B & \xrightarrow{\varphi \otimes \psi} & A' \otimes_R B' \\
\otimes \uparrow & & \uparrow \otimes \\
A \times B & \xrightarrow{\varphi \times \psi} & A' \times B'
\end{array}
$$

equivalently, $(\varphi \otimes \psi)(a \otimes b) = \varphi(a) \otimes \psi(b)$ for all $a \in A$, $b \in B$. The following holds:

(1) $1_A \otimes 1_B = 1_{A \otimes_R B}$.

(2) *If $\varphi \circ \varphi'$ and $\psi \circ \psi'$ are defined, then $(\varphi \circ \varphi') \otimes (\psi \circ \psi') = (\varphi \otimes \psi) \circ (\varphi' \otimes \psi')$.*

(3) *If $\varphi' : A \longrightarrow A'$ and $\psi' : B \longrightarrow B'$ are module homomorphisms, then $(\varphi + \varphi') \otimes \psi = (\varphi \otimes \psi) + (\varphi' \otimes \psi)$ and $\varphi \otimes (\psi + \psi') = (\varphi \otimes \psi) + (\varphi \otimes \psi')$.*

Proof. $(a,b) \longmapsto \varphi(a) \otimes \psi(b)$ is a bihomomorphism of bimodules. Properties (1), (2), and (3) are left as exercises. $\qquad\square$

Thus tensor products yield for each Q-R-bimodule A a (covariant) functor $A \otimes_R -$ which assigns to each R-S-bimodule B the Q-S-bimodule $A \otimes_R B$, and to each bimodule homomorphism $\psi : B \longrightarrow B'$ the bimodule homomorphism $A \otimes_R \psi = 1_A \otimes \psi$. Similarly there is for every R-S-bimodule B a functor $- \otimes_R B$ from Q-R-bimodules to Q-S-bimodules.

15.1.f. Free Modules. We now give two examples of tensor products; more will be found in the exercises. In our first example, $A = F$ is a free right R-module.

Proposition 15.1.8. (1) *Let F be a free right R-module with basis $(e_i)_{i \in I}$ and B be a left R-module. There is an isomorphism $F \otimes_R B \cong \bigoplus_{i \in I} B$ which is natural in B and sends $\sum_{i \in I} (e_i \otimes b_i)$ to $(b_i)_{i \in I}$ whenever $b_i = 0$ for almost all i.*

(2) *Let A be a right R-module and F be a free left R-module with basis $(e_i)_{i \in I}$. There is an isomorphism $A \otimes_R F \cong \bigoplus_{i \in I} A$ which is natural in A and sends $\sum_{i \in I} (a_i \otimes e_i)$ to $(a_i)_{i \in I}$ whenever $a_i = 0$ for almost all i.*

Proof. We prove (1), leaving (2) to the reader. Every $x \in F$ can be written uniquely as a linear combination $x = \sum_{i \in I} e_i x_i$, with $x_i \in R$ for all i and $x_i = 0$ for almost all i. Define $\tau : F \times B \longrightarrow \bigoplus_{i \in I} B$ by: $\tau(\sum_{i \in I} e_i x_i, b) = (x_i b)_{i \in I}$. Then τ is a bihomomorphism. We show that every bihomomorphism $\beta : F \times B \longrightarrow C$ factors uniquely through τ. Indeed $\overline{\beta} : (b_i)_{i \in I} \longmapsto \sum_{i \in I} \beta(e_i, b_i)$ is an additive homomorphism of $\bigoplus_{i \in I} B$ into C, and

$$\beta(\textstyle\sum_{i \in I} e_i x_i, b) = \sum_{i \in I} \beta(e_i, x_i b) = \overline{\beta}((x_i b)_{i \in I}) = \overline{\beta}(\tau(\sum_{i \in I} e_i x_i, b))$$

for all $\sum_{i \in I} e_i x_i \in F$ and $b \in B$. Thus $\overline{\beta} \circ \tau = \beta$. Now $\operatorname{Im} \tau$ contains $\operatorname{Im} \iota_i$ for every injection $\iota_i : B \longrightarrow \bigoplus_{i \in I} B$, since $\tau(e_i, b) = \iota_i(b)$. Hence $\operatorname{Im} \tau$ generates $\bigoplus_{i \in I} B$. Therefore $\overline{\beta}$ is the only homomorphism such that $\overline{\beta} \circ \tau = \beta$.

Thus τ is a tensor product of F and B, and there is an isomorphism of abelian groups $\theta : F \otimes_R B \longrightarrow \bigoplus_{i \in I} B$ such that $\theta((\sum_{i \in I} e_i x_i) \otimes b) = (x_i b)_{i \in I}$. This implies $\theta(e_i \otimes b) = \iota_i(b)$ and $\theta(\sum_{i \in I} (e_i \otimes b_i)) = \sum_{i \in I} \iota_i(b_i) = (b_i)_{i \in I}$.

Recall that the basis $(e_i)_{i \in I}$ of F gives rise to an R-R-bimodule structure on F, in which $r \sum_{i \in I} e_i x_i = \sum_{i \in I} e_i r x_i$. Hence $F \otimes_R B$ is a left R-module, in which $r((\sum_{i \in I} e_i x_i) \otimes b) = (\sum_{i \in I} e_i r x_i) \otimes b$. We see that $\theta(r((\sum_{i \in I} e_i x_i) \otimes b)) = (r x_i b)_{i \in I} = r \theta((\sum_{i \in I} e_i x_i) \otimes b)$; therefore θ is an isomorphism of left R-modules. The reader will readily show that θ is natural in B. $\qquad\square$

Corollary 15.1.9. *There is an isomorphism $R \otimes_R B \cong B$ which sends $1 \otimes b$ to b and is natural in B, and an isomorphism $A \otimes_R R \cong A$ which sends $a \otimes 1$ to a and is natural in A.* $\qquad\square$

Proposition 15.1.7 can also be proved from Corollary 15.1.9 and Proposition 15.2.6 below.

Corollary 15.1.10. *If E is a free right R-module with basis $(e_i)_{i \in I}$, and F is a free left R-module with basis $(f_j)_{j \in J}$, then $E \otimes_R F$ is a free left R-module with basis $(e_i \otimes f_j)_{i \in I, j \in J}$.*

Proof. By Proposition 15.1.8, every element of $E \otimes_R F$ can be written uniquely in the form $\sum_{i \in I} (e_i \otimes b_i)$; with $b_i = \sum_{j \in J} y_{ij} f_j \in F$, $\sum_{i \in I} (e_i \otimes b_i) = \sum_{i \in I} (e_i \otimes (\sum_{j \in J} y_{ij} f_j)) = \sum_{i \in I, j \in J} (e_i y_{ij} \otimes f_j) = \sum_{i \in I, j \in J} y_{ij} (e_i \otimes f_j)$. □

When R is a division ring, Corollary 15.1.10 constructs all tensor products; then $\dim E \otimes F = (\dim E)(\dim F)$. Tensor products were first considered in this case (with $R = \mathbb{R}$ or $\mathbb{C}$).

15.1.g. Change of Rings. The reader knows that a vector space over $\mathbb{R}$ can be enlarged to a vector space over $\mathbb{C}$: if $(e_i)_{i \in I}$ is a basis of V over $\mathbb{R}$, then V consists of all linear combinations $\sum_{i \in I} x_i e_i$ with real coefficients and can be enlarged to the vector space of all linear combinations $\sum_{i \in I} x_i e_i$ with complex coefficients. Tensor products provide a general construction, which does not depend on the choice of a basis or even on the existence of a basis.

Let $\rho : R \longrightarrow S$ be a ring homomorphism. A left S-module M is also a left R-module, in which $rx = \rho(r)x$ for all $r \in R$, $x \in M$; the R-module structure $R \longrightarrow \mathrm{End}(M)$ on M is the composition $R \xrightarrow{\rho} S \longrightarrow \mathrm{End}(M)$. (For example, using the inclusion homomorphism $\mathbb{R} \longrightarrow \mathbb{C}$, a vector space over $\mathbb{C}$ is also a vector space over $\mathbb{R}$.) Tensor products provide a converse construction, which transforms left R-modules into left S-modules (e.g., enlarges vector spaces over $\mathbb{R}$ into vector spaces over $\mathbb{C}$).

Proposition 15.1.11. *Let $\rho : R \longrightarrow S$ be a ring homomorphism.*

(1) *For each left R-module M there is a left S-module N and an R-module homomorphism $\iota : M \longrightarrow N$ such that every R-module homomorphism of M into a left S-module factors uniquely through ι.*

(2) *Up to isomorphism, $N = S \otimes_R M$ and $\iota(x) = 1 \otimes x$.*

(3) *If M is a free left R-module with basis $(e_i)_{i \in I}$, then N is a free left S-module with basis $(\iota(e_i))_{i \in I}$.*

Proof. First, S is an S-R-bimodule, with $sr = s\rho(r)$ for all r, s. Hence $N = S \otimes_R M$ is defined and is a left S-module, in which $s(t \otimes x) = st \otimes x$. Let $\iota(x) = 1 \otimes x$ for all $x \in M$. Then $\iota : M \longrightarrow N$ is a homomorphism of left R-modules. If φ is an R-module homomorphism of M into a left S-module P, then $\beta : S \times M \longrightarrow P$, $(s, x) \longmapsto s\varphi(x)$ is a bihomomorphism, and there is an additive homomorphism $\psi : N \longrightarrow P$ such that $\psi(s \otimes x) = s\varphi(x)$ for all $s \in S$, $x \in M$;

in particular, $\psi \circ \iota = \varphi$. We see that ψ is a homomorphism of left S-modules. If, conversely, $\chi : N \longrightarrow P$ is an S-homomorphism such that $\chi \circ \iota = \varphi$, then $\chi(s \otimes x) = s\chi(1 \otimes x) = s\varphi(x) = \psi(s \otimes x)$ and $\chi = \psi$. Thus ι has the required universal property. This proves (1); (2) follows from (1) since the universal property determines N and ι uniquely up to isomorphism.

Now assume that M is free, with basis $(e_i)_{i \in I}$. By Proposition 15.1.8, there is an isomorphism $\theta : S \otimes_R M \longrightarrow \bigoplus_{i \in I} S$ of left S-modules which sends $\sum_{i \in I} (s_i \otimes e_i)$ to $(s_i)_{i \in I}$. Hence every element of $S \otimes_R M$ can be written uniquely in the form $\sum_{i \in I} s_i \otimes e_i = \sum_{i \in I} s_i \iota(e_i)$. This proves (3). $\qquad \square$

If M is free with a known basis, then N can be constructed from its basis in part (3) as a module with the same basis as M and coordinates in S.

15.1.h. Longer Tensor Products. Proposition 15.1.4 constructs tensor products of two modules. A similar construction yields tensor products of more than two modules.

Let R be a commutative ring. When $A_1, \ldots, A_n, C$ are R-modules, a **multilinear** or **n-linear mapping** $\mu : A_1 \times \cdots \times A_n \longrightarrow C$ is a mapping such that

$$\mu(a_1, \ldots, a_{i-1}, a_i + a'_i, a_{i+1}, \ldots, a_n) = \mu(a_1, \ldots, a_{i-1}, a_i, a_{i+1}, \ldots, a_n)$$
$$+ \mu(a_1, \ldots, a_{i-1}, a'_i, a_{i+1}, \ldots, a_n),$$

$$\mu(a_1, \ldots, a_{i-1}, ra_i, a_{i+1}, \ldots, a_n) = r\mu(a_1, \ldots, a_{i-1}, a_i, a_{i+1}, \ldots, a_n)$$

for all $a_1, \ldots, a_n \in A$ and $r \in R$. Multilinear mappings $\mu : A_1 \times \cdots \times A_n \longrightarrow C$ can be characterized by properties of the mappings

$$\mu(a_1, \ldots, a_{i-1}, -, a_{i+1}, \ldots, a_n) : a_i \longmapsto \mu(a_1, \ldots, a_{i-1}, a_i, a_{i+1}, \ldots, a_n) :$$

a mapping $\mu : A_1 \times \cdots \times A_n \longrightarrow C$ is multilinear if and only if, for every i,

$$(a_1, \ldots, a_{i-1}, a_{i+1}, \ldots, a_n) \longmapsto \mu(a_1, \ldots, a_{i-1}, -, a_{i+1}, \ldots, a_n)$$

is a multilinear mapping of $A_1 \times \cdots \times A_{i-1} \times A_{i+1} \times \cdots \times A_n$ into $\mathrm{Hom}_R(A_i, C)$.

Multihomomorphisms generalize multilinear mappings to bimodules over arbitrary rings. We state this when $n = 3$ and leave the general case to the reader. The following result is straightforward:

Lemma 15.1.12. *Let A be a Q-R-bimodule, B be an R-S-bimodule, C be an S-T-bimodule, and D be a Q-T-bimodule. For a mapping $\tau : A \times B \times C \longrightarrow D$ the following conditions are equivalent:*

(1) for all $a, a' \in A$, $b, b' \in B$, $c, c' \in C$, $q \in Q$, $r \in R$, $s \in S$, $t \in T$,

$$\tau(a + a', b, c) = \tau(a, b, c) + \tau(a', b, c),$$

$$\tau(a, b + b', c) = \tau(a, b, c) + \tau(a, b', c),$$

$$\tau(a, b, c + c') = \tau(a, b, c) + \tau(a, b, c'),$$

$$\tau(qa, b, c) = q\tau(a, b, c),$$

$$\tau(ar, b, c) = \tau(a, rb, c),$$

$$\tau(a, bs, c) = \tau(a, b, sc), \qquad and$$

$$\tau(a, b, ct) = \tau(a, b, c)t.$$

(2) $(a, b) \longmapsto \tau(a, b, -)$ *is a bihomomorphism (of bimodules) of $A \times B$ into* $\mathrm{Hom}_T(C, D)$.

(3) $(b, c) \longmapsto \tau(-, b, c)$ *is a bihomomorphism of $B \times C$ into* $\mathrm{Hom}_Q(A, D)$.

When A is a Q-R-bimodule, B is an R-S-bimodule, C is an S-T-bimodule, and D is a Q-T-bimodule, we call a mapping $\tau : A \times B \times C \longrightarrow D$ is a **trihomomorphism** in case it satisfies the equivalent conditions in Lemma 15.1.12. If R is commutative and $Q = S = T = R$, then trihomomorphisms coincide with trilinear mappings.

If $\tau : A \times B \times C \longrightarrow M$ is a trihomomorphism and $\gamma : M \longrightarrow D$ is a bimodule homomorphism, then $\gamma \circ \tau : A \times B \times C \longrightarrow D$ is a trihomomorphism. The **tensor product** of A B and C is a Q-T-bimodule $A \otimes_R B \otimes_S C$ together with a trihomomorphism $\tau : A \times B \times C \longrightarrow A \otimes_R B \otimes_S C$ with the following universal property: for every Q-T-bimodule D and trihomomorphism $\beta : A \times B \times C \longrightarrow D$, there is a unique bimodule homomorphism $A \otimes_R B \otimes_S C \longrightarrow D$ such that the following diagram commutes:

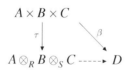

The trihomomorphism τ is the **tensor map**; $\tau(a, b, c)$ is denoted by $a \otimes b \otimes c$.

Proposition 15.1.13. *Let A be a Q-R-bimodule, B be an R-S-bimodule, and C be an S-T-bimodule. There exists a tensor product of A, B, and C, and it is unique up to isomorphism.*

Uniqueness follows from the universal property. Existence is proved by constructing $A \otimes_R B \otimes_S C$ as the abelian group generated by all ordered triples $(a, b, c) \in A \times B \times C$, subject to all defining relations $(a + a', b, c) = (a, b, c) + (a', b, c)$, $(a, b + b', c) = (a, b, c) + (a, b', c)$, $(a, b, c + c') = (a, b, c) + (a, b, c')$, $(ar, b, c) = (a, rb, c)$, $(a, bs, c) = (a, b, sc)$. A Q-T-bimodule structure on $A \otimes_R$

$B \otimes_S C$ is defined as in Proposition 15.1.5. The reader will verify that the canonical mapping $\tau : A \times B \times C \longrightarrow A \otimes_R B \otimes_S C$ is a trihomomorphism with the required universal property.

In Proposition 15.1.13, we also have tensor products $(A \otimes_R B) \otimes_S C$ and $A \otimes_R (B \otimes_S C)$; in the next section we will show that $(A \otimes_R B) \otimes_S C \cong A \otimes_R (B \otimes_S C) \cong A \otimes_R B \otimes_S C$ (Proposition 15.2.2).

If R is commutative and A, B, C are R-modules, Proposition 15.1.13 yields an R-module $A \otimes_R B \otimes_R C$ and a trilinear mapping $\tau : A \times B \times C \longrightarrow A \otimes_R B \otimes_R C$ such that every trilinear mapping $A \times B \times C \longrightarrow D$ factors uniquely through τ. More generally:

Proposition 15.1.14. *Let R be a commutative ring and $A_1, \ldots, A_n, C$ be R-modules. There exists an R-module $A_1 \otimes_R \cdots \otimes_R A_n$ and an n-linear mapping $\tau : (a_1, \ldots, a_n) \longmapsto a_1 \otimes \cdots \otimes a_n$ such that every n-linear mapping $A_1 \times \cdots \times A_n \longrightarrow C$ factors uniquely through τ. Moreover $A_1 \otimes_R \ldots \otimes_R A_n$ and τ are unique up to isomorphism.*

Exercises

1. Let A be a right R-module, B be a left R-module, and C be an abelian group. Show that there is a one-to-one correspondence between bihomomorphisms $A \times B \longrightarrow C$ and module homomorphisms $B \longrightarrow \mathrm{Hom}_{\mathbb{Z}}(A, C)$.

2. Let A be a right R-module, B be a left R-module. Prove that the tensor product of A and B is unique up to isomorphism.

3. If $\varphi \circ \varphi'$ and $\psi \circ \psi'$ are defined, then prove that $(\varphi \circ \varphi') \otimes (\psi \circ \psi') = (\varphi \otimes \psi) \circ (\varphi' \otimes \psi')$.

4. If $\varphi, \varphi' : A \longrightarrow A'$ and $\psi, \psi' : B \longrightarrow B'$ are bimodule homomorphisms, then prove that $(\varphi + \varphi') \otimes \psi = (\varphi \otimes \psi) + (\varphi' \otimes \psi)$ and $\varphi \otimes (\psi + \psi') = (\varphi \otimes \psi) + (\varphi \otimes \psi')$.

5. Show that $A \otimes_R \varphi$ is natural in A.

6. Let B be a left R-module. Prove directly that there is an isomorphism $R \otimes_R B \cong B$, $1 \otimes b \longmapsto b$ which is natural in B.

7. Let A be a right R-module and F be a free left R-module with basis $(f_i)_{i \in I}$. Show that there is an isomorphism $A \otimes_R F \cong \bigoplus_{i \in I} A$ which is natural in A and sends $\sum_{i \in I} (a_i \otimes f_i)$ to $(a_i)_{i \in I}$ whenever $a_i = 0$ for almost all i.

8. Let M be a left R-module and I be an ideal of R. Show that there is an isomorphism $(R/I) \otimes_R M \cong M/IM$ which is natural in M.

9. Show that $\mathbb{Z}_m \otimes_{\mathbb{Z}} \mathbb{Z}_n \cong \mathbb{Z}_d$, where d is the g.c.d. $d = (m, n)$.

10. Let $\rho : R \longrightarrow S$ be a ring homomorphism and A, B be left S-modules. Show that there is a monomorphism $\mathrm{Hom}_S({}_S A, {}_S B) \longrightarrow \mathrm{Hom}_R({}_R A, {}_R B)$ which is natural in A and B.

11. Let $\rho : R \longrightarrow S$ be a ring homomorphism, A be a right S-module, and B be a left S-module. Show that there is an epimorphism $A_R \otimes_R {}_R B \longrightarrow A_S \otimes_S {}_S B$ which is natural in A and B.

12. Let R be a commutative ring, S be a proper multiplicative subset of R, and M be an R-module. Show that there is an isomorphism $S^{-1} R \otimes_R M \cong S^{-1} M$ which is natural in M. In detail, show that $S^{-1} M$ is an $S^{-1} R$-module; $\iota : x \longmapsto x/1$ is an R-module

homomorphism; and there is an $S^{-1}R$-module isomorphism $\theta : (a/s) \otimes x \longmapsto ax/s$ of $S^{-1}R \otimes_R M$ onto $S^{-1}M$ which is natural in M and makes the following triangle commute:

$$
\begin{array}{ccc}
M & \xrightarrow{\ \iota\ } & S^{-1}M \\
 & \searrow & \ \ \downarrow{\scriptstyle \theta} \\
 & & S^{-1}R \otimes_R M
\end{array}
$$

13. Give a general definition of multihomomorphisms of bimodules. Define and construct tensor products of finite sequences of bimodules.

15.2. PROPERTIES

The main properties of tensor products are commutativity, associativity, adjoint associativity, right exactness, and preservation of direct sums and direct limits. We state the results for bimodules; they apply to modules since a left R-module is an R-$\mathbb{Z}$-bimodule and a right R-module is a $\mathbb{Z}$-R-bimodule.

15.2.a. Commutativity and Associativity. We begin with properties of the tensor product as an operation on modules: commutativity and associativity. Corollary 15.1.9 shows that R acts as an identity element.

Commutativity follows from a left-right symmetry in the definition of bi-homomorphisms. Recall that a right R-module A is a left R^{op}-module, in which $r * a = ar$ (using $*$ for the actions of opposite rings). Let A be an Q-R-bimodule and B be an R-S-bimodule. Then B is a S^{op}-R^{op}-bimodule (in which $t * b = bt$, $b * r = rb$), A is an R^{op}-Q^{op}-bimodule, and $\beta : A \times B \longrightarrow C$ is a bihomomorphism if and only if $\beta^{\mathrm{op}}(b,a) = \beta(a,b)$ is a bihomomorphism $\beta^{\mathrm{op}} : B \times A \longrightarrow C$.

Proposition 15.2.1. *Let A be an Q-R-bimodule and B be an R-S-bimodule. Then B is a S^{op}-R^{op}-bimodule, A is an R^{op}-Q^{op}-bimodule, and there is an isomorphism $B \otimes_{R^{\mathrm{op}}} A \cong A \otimes_R B$, $b \otimes a \longmapsto a \otimes b$, which is natural in A and B.*

Proof. First, $A \otimes_R B$ is an Q-S-bimodule; $B \otimes_{R^{\mathrm{op}}} A$ is a S^{op}-Q^{op}-bimodule and is also an Q-S-bimodule, in which $q(b \otimes a) = (b \otimes a) * q = b \otimes (a * q) = b \otimes qa$ and $(b \otimes a)s = bs \otimes a$. We saw that $\beta : A \times B \longrightarrow C$ is a bihomomorphism (of bimodules) if and only if $\beta^{\mathrm{op}}(b,a) = \beta(a,b)$ is a bihomomorphism $\beta^{\mathrm{op}} : B \times A \longrightarrow C$. Now $\tau(a,b) = a \otimes b$ is a bihomomorphism; hence τ^{op} is a bihomomorphism, and there is a bimodule homomorphism $\theta : B \otimes_{R^{\mathrm{op}}} A \longrightarrow A \otimes_R B$ such that $\theta(b \otimes a) = a \otimes b$ for all $a \in A$ and $b \in B$. Similarly there is a bimodule homomorphism $\zeta : A \otimes_R B \longrightarrow B \otimes_{R^{\mathrm{op}}} A$ such that $\zeta(a \otimes b) = b \otimes a$ for all a, b. The uniqueness in the universal property of tensor products implies that θ and ζ are mutually inverse isomorphisms. Naturality is left to the reader. $\qquad\square$

When R is commutative, Proposition 15.2.1 implies that $B \otimes_R A \cong A \otimes_R B$ for all R-modules A and B.

Proposition 15.2.2 (Associativity). *Let A be a Q-R-bimodule, B be an R-S-bimodule, and C be an S-T-bimodule. There are **associativity** isomorphisms of bimodules*

$$(A \otimes_R B) \otimes_S C \cong A \otimes_R B \otimes_S C \cong A \otimes_R (B \otimes_S C),$$

$(a \otimes b) \otimes c \longmapsto a \otimes b \otimes c \longmapsto a \otimes (b \otimes c)$, *which are natural in A, B, and C.*

Proof. By Lemma 15.1.12, $\beta : (a,b) \longmapsto a \otimes b \otimes -$ is a bihomomorphism of $A \times B$ into $\mathrm{Hom}_T(C, A \otimes_R B \otimes_S C)$. Hence there is a homomorphism $\overline{\beta}$ of $A \otimes_R B$ into $\mathrm{Hom}_T(C, A \otimes_R B \otimes_S C)$ such that $\overline{\beta}(a \otimes b) = \beta(a,b)$. By Lemma 15.1.3, $(u,c) \longmapsto \overline{\beta}(u)(c)$ is a bihomomorphism of $(A \otimes_R B) \times C$ into $A \otimes_R B \otimes_S C$, and there is a homomorphism θ of $(A \otimes_R B) \otimes_S C$ into $A \otimes_R B \otimes_S C$ such that $\theta(u \otimes c) = (\overline{\beta}(u))(c)$ for all $u \in A \otimes_R B$ and $c \in C$. In particular, $\theta((a \otimes b) \otimes c) = a \otimes b \otimes c$ for all a,b,c. Conversely, the mapping $(a,b,c) \longmapsto (a \otimes b) \otimes c$ of $A \times B \times C$ into $(A \otimes_R B) \otimes_S C$ is a trihomomorphism; hence there is a bimodule homomorphism $\zeta : A \otimes_R B \otimes_S C \longrightarrow (A \otimes_R B) \otimes_S C$ such that $\zeta(a \otimes b \otimes c) = (a \otimes b) \otimes c$ for all a,b,c. Uniqueness in the universal properties then implies that θ and ζ are mutually inverse isomorphisms. Thus $(A \otimes_R B) \otimes_S C \cong A \otimes_R B \otimes_S C$. The other isomorphism, and naturality in A, B, C, are left to the reader. $\qquad\square$

If R is commutative and $A_1, \ldots, A_n$ are R-modules, it follows from Proposition 15.2.2 that all tensor products $(\ldots((A_1 \otimes_R A_2) \otimes_R A_3)\ldots) \otimes_R A_n$, $(\ldots(A_1 \otimes_R (A_2 \otimes_R A_3))\ldots) \otimes_R A_n$, and so on, of $A_1, \ldots, A_n$ (in that order) are naturally isomorphic to $A_1 \otimes \cdots \otimes A_n$; they are customarily identified to $A_1 \otimes \cdots \otimes A_n$.

15.2.b. Adjoint Associativity. When A and B are R-S-bimodules, we denote by $\mathrm{Hom}_{RS}(A,B)$ the set of all bimodule homomorphisms of A into B. Under pointwise addition $\mathrm{Hom}_{RS}(A,B)$ is an abelian group.

Proposition 15.2.3 (Adjoint Associativity). *Let A be a Q-R-bimodule, B be an R-S-bimodule, and C be a Q-S-bimodule. There are **adjoint associativity** isomorphisms of abelian groups*

$$\Theta : \mathrm{Hom}_{QS}(A \otimes_R B, C) \cong \mathrm{Hom}_{QR}(A, \mathrm{Hom}_S(B,C)),$$

$$\Psi : \mathrm{Hom}_{QS}(A \otimes_R B, C) \cong \mathrm{Hom}_{RS}(B, \mathrm{Hom}_Q(A,C)),$$

which are natural in A, B, and C. If $\varphi : A \otimes_R B \longrightarrow C$, then $(\Theta(\varphi)(a))(b) = (\Psi(\varphi)(b))(a) = \varphi(a \otimes b)$.

Proof. Proposition 15.1.5 and Lemma 15.1.3 provide bijections between bimodule homomorphisms $A \otimes_R B \longrightarrow C$, bihomomorphisms $A \times B \longrightarrow C$, and bimodule homomorphisms $A \longrightarrow \operatorname{Hom}_S(B,C)$, which send a bimodule homomorphism $\varphi : A \otimes_R B \longrightarrow C$ to the bihomomorphism $\beta_\varphi : (a,b) \longmapsto \varphi(a \otimes b)$ and to the bimodule homomorphism $a \longmapsto \beta_\varphi(a,-)$ of A into $\operatorname{Hom}_S(B,C)$. This yields a bijection $\Theta : \operatorname{Hom}_{QS}(A \otimes_R B, C) \longrightarrow \operatorname{Hom}_{QR}(A, \operatorname{Hom}_S(B,C))$; by definition, $(\Theta(\varphi)(a))(b) = \beta_\varphi(a,b) = \varphi(a \otimes b)$ for all $a \in A$, $b \in B$. It is immediate that Θ preserves pointwise addition. The reader will verify that Θ is natural in A, B, and C. The isomorphism Ψ is similar. $\qquad\qquad\square$

If R is commutative and $Q = S = R$, then A, B, C are arbitrary R-modules and the adjoint associativity isomorphisms are R-module isomorphisms

$$\operatorname{Hom}_R(A \otimes_R B, C) \cong \operatorname{Hom}_R(A, \operatorname{Hom}_R(B,C)) \cong \operatorname{Hom}_R(B, \operatorname{Hom}_R(A,C)).$$

15.2.c. Right Exactness. Adjoint associativity is used to prove other properties.

Proposition 15.2.4 (Right exactness). *If*

$$A \longrightarrow B \longrightarrow C \longrightarrow 0$$

is an exact sequence of R-S-bimodules, then

$$M \otimes_R A \longrightarrow M \otimes_R B \longrightarrow M \otimes_R C \longrightarrow 0 \qquad and$$

$$A \otimes_S N \longrightarrow B \otimes_S N \longrightarrow C \otimes_S N \longrightarrow 0$$

are exact for all M, N.

One says that the functors $M \otimes_R -$ and $- \otimes_S N$ are **right exact**.

Proof. Let $A \xrightarrow{\;\mu\;} B \xrightarrow{\;\sigma\;} C \longrightarrow 0$ be exact and M be a right R-module. (If right exactness holds whenever M is a right R-module, it holds whenever M is a Q-R-bimodule.) By Proposition 15.1.7, the sequence

$$M \otimes_R A \xrightarrow{\;\bar\mu\;} M \otimes_R B \xrightarrow{\;\bar\sigma\;} M \otimes_R C \longrightarrow 0,$$

where $\bar\mu = M \otimes_R \mu$ and $\bar\sigma = M \otimes_R \sigma$, is null ($\bar\sigma \circ \bar\mu = 0$).

To prove exactness we construct for any right S-module G the diagram

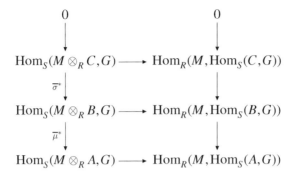

in which $\overline{\mu}^* = \operatorname{Hom}_{\mathbb{Z}}(\overline{\mu}, G)$, $\overline{\sigma}^* = \operatorname{Hom}_{\mathbb{Z}}(\overline{\sigma}, G)$, and the horizontal arrows are adjoint associativity isomorphisms (from Proposition 15.2.3, with $Q = \mathbb{Z}$). The diagram commutes by naturality of the adjoint associativity isomorphism, and the right column is exact by Propositions 14.6.8 and 14.6.6. Hence the left column is exact: $\overline{\sigma}^*$ is injective and $\operatorname{Ker} \overline{\mu}^* = \operatorname{Im} \overline{\sigma}^*$, for every right S-module G.

We can now show that the null sequence $\bullet \xrightarrow{\overline{\mu}} \bullet \xrightarrow{\overline{\sigma}} \bullet \longrightarrow 0$ is exact. We already have $\operatorname{Im} \overline{\mu} \subseteq \operatorname{Ker} \overline{\sigma}$. Let $G = (M \otimes_R C)/\operatorname{Im} \overline{\sigma}$ and $\pi : M \otimes_R C \longrightarrow G$ be the projection. Then $\overline{\sigma}^*(\pi) = \pi \circ \overline{\sigma} = 0$; hence $\pi = 0$, $\operatorname{Im} \overline{\sigma} = M \otimes_R C$, and $\overline{\sigma}$ is surjective. Similarly let $G = (M \otimes_R B)/\operatorname{Im} \overline{\mu}$ and $\rho : M \otimes_R B \longrightarrow G$ be the projection. Then $\overline{\mu}^*(\rho) = \rho \circ \overline{\mu} = 0$; hence $\rho = \overline{\sigma}^*(\chi) = \chi \circ \overline{\sigma}$ and $\overline{\sigma}(t) = 0$ implies $\rho(t) = 0$ and $t \in \operatorname{Im} \overline{\mu}$. Thus $\operatorname{Ker} \overline{\sigma} \subseteq \operatorname{Im} \overline{\mu}$.

The exactness of

$$A \otimes_S N \longrightarrow B \otimes_S N \longrightarrow C \otimes_S N \longrightarrow 0$$

can be proved similarly but follows from Proposition 15.2.1.   □

The functor $M \otimes_R -$ is in general not exact; that is, if $0 \longrightarrow A \xrightarrow{\mu} B \xrightarrow{\sigma} C \longrightarrow 0$ is a short exact sequence, then the sequence

$$0 \longrightarrow M \otimes_R A \longrightarrow M \otimes_R B \longrightarrow M \otimes_R C \longrightarrow 0$$

is generally not exact ($M \otimes_R A \longrightarrow M \otimes_R B$ need not be injective; see the exercises). Modules M such that $- \otimes_R M$ is exact are studied in Section 15.4.

15.2.d. Direct Sums. The functors $A \otimes_R -$ and $- \otimes_R B$ preserve direct sums.

Proposition 15.2.5. *There are natural isomorphisms*

$$A \otimes_R \left(\bigoplus_{i \in I} B_i \right) \cong \bigoplus_{i \in I} (A \otimes_R B_i), \qquad \left(\bigoplus_{i \in I} A_i \right) \otimes_R B \cong \bigoplus_{i \in I} (A_i \otimes_R B),$$

which send $a \otimes (b_i)_{i \in I}$ to $(a \otimes b_i)_{i \in I}$ and $(a_i)_{i \in I} \otimes b$ to $(a_i \otimes b)_{i \in I}$.

Proof. First let A be a right R-module and $(B_i)_{i \in I}$ be a family of left R-modules. Let $\iota_i : B_i \longrightarrow \bigoplus_{i \in I} B_i$ be the injection and $\overline{\iota}_i = A \otimes_R \iota_i : A \otimes_R B_i \longrightarrow A \otimes_R (\bigoplus_{i \in I} B_i)$. For every abelian group G there is a commutative diagram

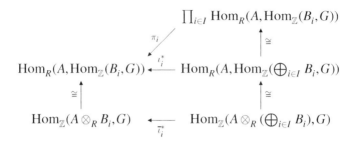

where $\iota_i^*, \overline{\iota}_i^*$ are induced by ι_i and $\overline{\iota}_i$, the top triangle is provided by Propositions 14.6.10 and 14.6.11, and the remaining vertical arrows are adjoint associativity isomorphisms.

If $(\varphi_i)_{i \in I}$ is a family of homomorphisms $\varphi_i : A \otimes_R B_i \longrightarrow G$, then the diagram shows that there is a unique homomorphism $\varphi : A \otimes_R (\bigoplus_{i \in I} B_i) \longrightarrow G$ such that $\varphi_i = \overline{\iota}_i^*(\varphi) = \varphi \circ \overline{\iota}_i$ for all i. This universal property characterizes the direct sum; therefore there is an additive isomorphism $\theta : A \otimes_R (\bigoplus_{i \in I} B_i) \cong \bigoplus_{i \in I} (A \otimes_R B_i)$ which makes the following triangle commute for every i:

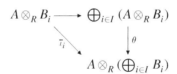

in which the horizontal arrow is the injection. Then $\theta((a \otimes b_i)_{i \in I}) = a \otimes (b_i)_{i \in I}$.

If A is a Q-R-bimodule and $(B_i)_{i \in I}$ is a family of R-S-bimodules, then $\theta((a \otimes b_i)_{i \in I}) = a \otimes (b_i)_{i \in I}$ shows that θ is a bimodule homomorphism. Naturality is straightforward. Proposition 15.2.1 provides the other isomorphism. $\square$

Exercises give another proof of Proposition 15.2.5, based on Proposition 10.4.4.

15.2.e. Direct Limits. The functors $A \otimes_R -$ and $- \otimes_R B$ preserve direct limits. The following result is proved like Proposition 15.2.5.

Proposition 15.2.6. *There are natural isomorphisms*

$$A \otimes_R (\varinjlim_{i \in I} B_i) \cong \varinjlim_{i \in I} (A \otimes_R B_i), \qquad (\varinjlim_{i \in I} A_i) \otimes_R B \cong \varinjlim_{i \in I} (A_i \otimes_R B).$$

Exercises

1. Show that the isomorphism $B \otimes_{R^{\mathrm{op}}} A \cong A \otimes_R B$ (which sends $b \otimes a$ to $a \otimes b$) is natural in A and B.

2. Let A be a Q-R-bimodule, B be an R-S-bimodule, and C be an S-T-bimodule. Prove directly that there is a bimodule isomorphism

$$(A \otimes_R B) \otimes_S C \cong A \otimes_R (B \otimes_S C)$$

which sends $(a \otimes b) \otimes c$ to $a \otimes (b \otimes c)$ and is natural in A, B, and C.

3. Show that the adjoint associativity isomorphism

$$\mathrm{Hom}_{QS}(A \otimes_R B, C) \cong \mathrm{Hom}_{QR}(A, \mathrm{Hom}_S(B, C))$$

is natural in A, B, and C.

4. Let A be a right R-module, B be a R-S-bimodule, and C be a T-S-bimodule. Show that the adjoint associativity isomorphism

$$\mathrm{Hom}_S(A \otimes_R B, C) \cong \mathrm{Hom}_R(A, \mathrm{Hom}_S(B, C))$$

is an isomorphism of left T-modules.

5. Give a direct proof that the functor $- \otimes_R M$ is right exact for every right R-module M.

6. Find a monomorphism $A \longrightarrow B$ of abelian groups such that $\mathbb{Z}_2 \otimes_{\mathbb{Z}} A \longrightarrow \mathbb{Z}_2 \otimes_{\mathbb{Z}} B$ is not a monomorphism.

7. Let M be an R-module and I be an ideal of R. Use right exactness to prove that there is an isomorphism $(R/I) \otimes_R M \cong M/IM$ which is natural in M.

8. If φ and ψ are epimorphisms, then prove that $\varphi \otimes \psi$ is an epimorphism.

9. If A and B are finitely generated, then prove that $A \otimes_R B$ is finitely generated.

10. Prove Proposition 15.2.5, using Proposition 10.4.4.

11. Let A be a right R-module and $\mathscr{B} : (B_i)_{i \in I}, (\beta_{ij})_{i,j \in I, i \leqslant j}$ be a direct system of left R-modules. Prove that there is an isomorphism $A \otimes_R (\varinjlim_{i \in I} B_i) \cong \varinjlim_{i \in I} (A \otimes_R B_i)$ which is natural in A and $\mathscr{B}$.

12. Prove that $M \otimes_R -$ preserves pushouts (if $\gamma \circ \alpha = \delta \circ \beta$ is a pushout, then $(M \otimes \gamma) \circ (M \otimes \alpha) = (M \otimes \delta) \circ (M \otimes \beta)$ is a pushout).

15.3. DUAL MODULES

The dual of an R-module M is $M^* = \mathrm{Hom}_R(M, R)$. The results in this section extend to modules some familiar properties of vector spaces; they provide additional examples of tensor products, which will be used in the next section.

15.3.a. Definition. The **dual** of a left R-module M is the right R-module $M^* = \mathrm{Hom}_R(M, {}_R R)$ in which $(\alpha r)(x) = \alpha(rx)$ for all $\alpha \in M^*$, $r \in R$, $x \in M$.

Similarly, the **dual** of a right R-module M is the left R-module $M^* = \text{Hom}_R(M, R_R)$ in which $(r\alpha)(x) = \alpha(xr)$ for all $\alpha \in M^*$, $r \in R$, $x \in M$.

For example, $({}_RR)^* \cong R_R$ and $(R_R)^* \cong {}_RR$ (see the exercises).

15.3.b. Properties. Dual modules are groups of homomorphisms and inherit properties from Section 14.6. Every homomorphism $\varphi : M \longrightarrow N$ of left (or right) R-modules induces a **dual** homomorphism of right (or left) R-modules $\varphi^* = \text{Hom}_R(\varphi, R) : N^* \longrightarrow M^*$; namely $\varphi^*(\alpha) = \alpha \circ \varphi$ for all $\alpha \in N^*$. Thus we have contravariant functors $-^* = \text{Hom}_R(-, R)$.

If $A \xrightarrow{\mu} B \xrightarrow{\rho} C \longrightarrow 0$ is exact, then

$$0 \longrightarrow C^* \xrightarrow{\rho^*} B^* \xrightarrow{\mu^*} A^*$$

is exact by Proposition 14.6.8. By Proposition 14.6.11, there is an isomorphism

$$(\bigoplus_{i \in I} A_i)^* \cong \prod_{i \in I} A_i^*$$

which is natural in $(A_i)_{i \in I}$. In particular, $(\bigoplus_{i \in I} A_i)^* \cong \bigoplus_{i \in I} A_i^*$ when I is finite.

This last property implies that the dual of a finitely generated free module is a finitely generated free module. More precisely:

Proposition 15.3.1. *If the left (or right) R-module F is free with a finite basis $(e_i)_{i \in I}$, then F^* is free with a finite basis $(e_i^*)_{i \in I}$ such that $e_i^*(e_i) = 1$, $e_i^*(e_j) = 0$ for all $j \neq i$; $(e_i^*)_{i \in I}$ is the **dual basis** of the given basis $(e_i)_{i \in I}$.*

The proof is an exercise. The result does not extend to all free modules; for instance, $(\bigoplus_{i \in I} \mathbb{Z})^* \cong \prod_{i \in I} \mathbb{Z}$ is not free if I is countable (Proposition 14.2.6).

Corollary 15.3.2. *If P is a finitely generated projective left R-module, then P^* is a finitely generated projective right R-module.*

Proof. There is a finitely generated free left R-module F and an epimorphism $F \longrightarrow P$ which splits since P is projective, so that $F \cong P \oplus Q$ for some submodule Q of F. By Proposition 15.3.1, F^* is finitely generated and free, and $F^* \cong P^* \oplus Q^*$ shows that P^* is finitely generated and projective. $\quad\square$

15.3.c. Double Duals. The **double dual** of M is $M^{**} = (M^*)^*$. The following result is straightforward:

Proposition 15.3.3. *For every left (or right) R-module M there is an evaluation homomorphism $\varepsilon_M : M \longrightarrow M^{**}$ which is natural in M: namely, $(\varepsilon_M(x))(\alpha) = \alpha(x)$ for all $x \in M$ and $\alpha \in M^*$.*

Proposition 15.3.4. *If the left (or right) R-module M is finitely generated and projective, then the evaluation homomorphism $M \longrightarrow M^{**}$ is an isomorphism.*

Proof. If F is free with a finite basis $(e_i)_{i \in I}$, then applying Proposition 15.3.1 twice yields a basis $(e_i^{**})_{i \in I}$ of M^{**} such that $e_i^{**}(e_i^*) = 1$ and $e_i^{**}(e_j^*) = 0$ for all $j \neq i$. We see that $e_i^{**} = \varepsilon_F(e_i)$. Therefore ε_F is an isomorphism.

Now let P be finitely generated and projective. Then P is a direct summand of a finitely generated free module F, and there are homomorphisms $\iota : P \longrightarrow F$ and $\pi : F \longrightarrow P$ such that $\pi \circ \iota = 1_P$. Hence there is a commutative diagram

$$
\begin{array}{ccc}
P & \xrightarrow{\ \varepsilon_P\ } & P^{**} \\
\pi \big\uparrow\big\downarrow \iota & & \pi^{**} \big\uparrow\big\downarrow \iota^{**} \\
F & \xrightarrow[\ \varepsilon_F\ ]{} & F^{**}
\end{array}
$$

in which $\pi^{**} \circ \iota^{**} = 1_{P^{**}}$ and ε_F is an isomorphism. Then $\iota^{**} \circ \varepsilon_P = \varepsilon_F \circ \iota$ is injective, ε_P is injective, $\varepsilon_P \circ \pi = \pi^{**} \circ \varepsilon_F$ is surjective, and ε_P is surjective. $\square$

15.3.d. Tensor Products

Proposition 15.3.5. *Let A and B be left R-modules. There is a homomorphism*

$$
\zeta : A^* \otimes_R B \longrightarrow \mathrm{Hom}_R(A, B)
$$

which is natural in A and B such that $(\zeta(\alpha \otimes b))(a) = \alpha(a)b$ for all $\alpha \in A^*$, $a \in A$, $b \in B$. If A is finitely generated and projective, then ζ is an isomorphism.

Proof. For each $\alpha \in A^*$ and $b \in B$, the mapping $\beta_{\alpha,b}(a) = \alpha(a)b$ of A into B is a module homomorphism. We see that $\beta : A^* \times B \longrightarrow \mathrm{Hom}_R(A, B)$, $(\alpha, b) \longmapsto \beta_{\alpha,b}$ is a bihomomorphism. Hence there is a unique homomorphism $\zeta : A^* \otimes_R B \longrightarrow \mathrm{Hom}_R(A, B)$ such that $(\zeta(\alpha \otimes b))(a) = \alpha(a)b$ for all α, a, b. Our tireless reader will show that ζ is natural in A and B.

If A is free with a finite basis $(e_i)_{i \in I}$, then A^* is free with the dual basis $(e_i^*)_{i \in I}$; by Proposition 15.1.8, every element of $A^* \otimes_R B$ can be written in the form $\sum_{i \in I} e_i^* \otimes b_i$ for some unique $b_i \in B$. Then $\zeta(\sum_{i \in I} e_i^* \otimes b_i)(e_j) = \sum_{i \in I} e_i^*(e_j)b_i = b_j$ and $\zeta(\sum_{i \in I} e_i^* \otimes b_i)$ is the homomorphism of A into B which sends e_j to b_j for every j. The universal property of free modules states precisely that ζ is bijective.

Finally let A be finitely generated and projective. There exists a finitely generated free left R-module F and homomorphisms $\iota : A \longrightarrow F$ and $\pi : F \longrightarrow A$ such that $\pi \circ \iota = 1_A$. Since ζ is natural, there is a commutative diagram

$$
\begin{array}{ccc}
A^* \otimes_R B & \xrightarrow{\ \zeta_A\ } & \mathrm{Hom}_R(A, B) \\
\iota' \big\uparrow\big\downarrow \pi' & & \iota'' \big\uparrow\big\downarrow \pi'' \\
F^* \otimes_R B & \xrightarrow[\ \zeta_F\ ]{} & \mathrm{Hom}_R(F, B)
\end{array}
$$

where ζ_F is an isomorphism and $\iota' = \iota^* \otimes B$, $\pi' = \pi^* \otimes B$, $\iota'' = \mathrm{Hom}_R(\iota, B)$, $\pi'' = \mathrm{Hom}_R(\pi, B)$. Then $\iota^* \circ \pi^* = 1$, $\iota' \circ \pi' = 1$, and $\iota'' \circ \pi'' = 1$. Hence ζ_A is injective, since $\pi'' \circ \zeta_A = \zeta_F \circ \pi'$ is injective, and ζ_A is surjective, since $\zeta_A \circ \iota' = \iota'' \circ \zeta_F$ is surjective. $\qquad\square$

Corollary 15.3.6. *When A is a finitely generated projective right R-module, there is an isomorphism*

$$A \otimes_R B \cong \mathrm{Hom}_R(A^*, B)$$

which is natural in A and B.

Proof. A^* is a finitely generated projective left R-module, and $A^{**} \cong A$; apply Proposition 15.3.5 to A^*. $\qquad\square$

Corollary 15.3.7. *If R is commutative and A, B are R-modules, with A finitely generated and projective, then there is an isomorphism*

$$A^* \otimes_R B^* \cong (A \otimes_R B)^*$$

which is natural in A and B.

Proof. Propositions 15.3.5 and 15.2.3 provide natural isomorphisms

$$A^* \otimes_R B^* \cong \mathrm{Hom}_R(A, B^*) = \mathrm{Hom}_R(A, \mathrm{Hom}_R(B, R)) \cong \mathrm{Hom}_R(A \otimes_R B, R).$$

$\qquad\square$

Exercises

1. Prove that $({}_R R)^* \cong R_R$ and $(R_R)^* \cong {}_R R$.

2. Let F be a free left R-module with a finite basis $(e_i)_{i \in I}$. Prove that F^* is free with a finite basis $(e_i^*)_{i \in I}$ such that $e_i^*(e_i) = 1$, $e_i^*(e_j) = 0$ for all $j \neq i$.

3. Let E and F be free left R-modules with finite bases $(e_i)_{i \in I}$ and $(f_j)_{j \in J}$. Let $\varphi : E \longrightarrow F$ be a homomorphism and M be the matrix of φ in the given bases. Show that the matrix of φ^* in the dual bases is the transpose of M.

4. Let N be a submodule of M and $N^\perp = \{ \alpha \in M^* ; \alpha(N) = 0 \}$. Show that $(M/N)^* \cong N^\perp$ and there is a monomorphism $M^*/N^\perp \longrightarrow N^*$.

5. Verify that $(\varepsilon(x))(\alpha) = \alpha(x)$ defines a homomorphism $\varepsilon : M \longrightarrow M^{**}$ which is natural in M.

6. Show that M^* is isomorphic to a direct summand of M^{***}.

7. Prove that the homomorphism $\zeta : A^* \otimes_R B \longrightarrow \mathrm{Hom}_R(A, B)$ in Proposition 15.3.5 is natural in A and B.

8. Let η be the isomorphism in Corollary 15.3.7. Show that $(\eta(\alpha \otimes \beta))(a \otimes b) = \alpha(a)\beta(b)$ for all α, β, a, b.

15.4. FLAT MODULES

A left R-module M is **flat** in case the functor $- \otimes_R M$ is exact (preserves short exact sequences): that is, if, for every short exact sequence

$$0 \longrightarrow A \longrightarrow B \longrightarrow C \longrightarrow 0$$

of right R-modules, the sequence

$$0 \longrightarrow A \otimes_R M \longrightarrow B \otimes_R M \longrightarrow C \otimes_R M \longrightarrow 0$$

is exact. By Proposition 15.2.4, M is flat if and only if $\mu \otimes M$ is a monomorphism whenever $\mu : A \longrightarrow B$ is a monomorphism of right R-modules.

15.4.a. Examples

Proposition 15.4.1. *Every projective module is flat.*

Proof. Every free right R-module is flat by Proposition 15.1.8. A projective right R-module P is a direct summand of a free module F and there are homomorphisms $\iota : P \longrightarrow F$ and $\pi : F \longrightarrow P$ such that $\pi \circ \iota = 1_P$. For each monomorphism $\mu : A \longrightarrow B$ there is a commutative diagram

$$
\begin{array}{ccc}
A \otimes P & \xrightarrow{\mu \otimes P} & B \otimes P \\
\pi' \big\uparrow\!\big\downarrow \iota' & & \pi'' \big\uparrow\!\big\downarrow \iota'' \\
A \otimes F & \xrightarrow[\mu \otimes F]{} & B \otimes F
\end{array}
$$

where $\pi' = A \otimes \pi$, $\iota' = A \otimes \iota$, $\pi'' = B \otimes \pi$, $\iota'' = B \otimes \iota$. Then ι' is injective, since $\pi' \circ \iota' = 1$, and $\mu \otimes F$ is injective since F is free. Therefore $\mu \otimes P$ is injective. $\square$

Proposition 15.4.2. *A direct sum $\bigoplus_{i \in I} M_i$ of modules is flat if and only if every summand M_i is flat.*

Proof. If every M_i is flat, then $\bigoplus_{i \in I} M_i$ is flat by Proposition 15.2.6. The converse holds since a direct summand of a flat module is flat (see the exercises). $\square$

Proposition 15.4.3. *A direct limit of flat modules is flat.*

Proof. This follows from Proposition 15.2.6, since by Corollary 14.7.9 a direct limit of monomorphisms is a monomorphism. $\square$

Proposition 15.4.4. *An abelian group is flat if and only if it is torsion free.*

Proof. Infinite cyclic groups are flat (as $\mathbb{Z}$-modules) by Proposition 15.4.1. Let C be a finite cyclic group of order m. Then $\mu(x) = mx$ is a monomorphism

$\mu : \mathbb{Z} \longrightarrow \mathbb{Z}$, but $\mu \otimes_{\mathbb{Z}} C = 0$, since $\mu(x \otimes c) = mx \otimes c = x \otimes mc = 0$ for all $x \in \mathbb{Z}$, $c \in C$. Hence C is not flat.

Now let A be any abelian group. If A is torsion free, then every finitely generated subgroup of A is flat and A, which is the direct limit of its finitely generated subgroups by Proposition 14.7.2, is flat by Proposition 15.4.3. Now assume that A is not torsion free. Then A contains a finite cyclic subgroup C. We saw that there is a monomorphism $\mu : \mathbb{Z} \longrightarrow \mathbb{Z}$ such that $\mu \otimes C$ is not a monomorphism. Let $\iota : C \longrightarrow A$ be the inclusion. In the commutative diagram

$$
\begin{array}{ccc}
\mathbb{Z} \otimes C & \xrightarrow{\mu \times C} & \mathbb{Z} \otimes C \\
{\scriptstyle \mathbb{Z} \otimes \iota} \downarrow & & \downarrow {\scriptstyle \mathbb{Z} \otimes \iota} \\
\mathbb{Z} \otimes A & \xrightarrow[\mu \times A]{} & \mathbb{Z} \otimes A
\end{array}
$$

$\mathbb{Z} \otimes \iota$ is injective (since $\mathbb{Z}$ is flat) and $\mu \otimes C$ is not injective; therefore $\mu \otimes A$ is not injective. $\square$

Proposition 15.4.4 shows that a flat module need not be projective.

Proposition 15.4.5. *A left R-module M is flat if and only if the right R-module* $\mathrm{Hom}_{\mathbb{Z}}(M, \mathbb{Q}/\mathbb{Z})$ *is injective.*

Proof. First $\mathbb{Q}/\mathbb{Z}$ is a divisible abelian group and is an injective $\mathbb{Z}$-module. Hence $\mathrm{Hom}_{\mathbb{Z}}(-, \mathbb{Q}/\mathbb{Z})$ is exact. If M is flat, then $- \otimes_R M$ is exact, $\mathrm{Hom}_{\mathbb{Z}}(- \otimes_R M, \mathbb{Q}/\mathbb{Z})$ is exact, $\mathrm{Hom}_R(-, \mathrm{Hom}_{\mathbb{Z}}(M, \mathbb{Q}/\mathbb{Z}))$ is exact by Proposition 15.2.3, and $\mathrm{Hom}_{\mathbb{Z}}(M, \mathbb{Q}/\mathbb{Z})$ is injective.

We derive the converse from the following property of $\mathbb{Q}/\mathbb{Z}$: if g is some element of an abelian group G, and $\varphi(g) = 0$ for every homomorphism $\varphi : G \longrightarrow \mathbb{Q}/\mathbb{Z}$, then $g = 0$. Indeed assume that $g \neq 0$. There is a homomorphism $\psi : \langle g \rangle \longrightarrow \mathbb{Q}/\mathbb{Z}$ such that $\psi(g) \neq 0$: if g has finite order $n > 0$, let $\psi(g) = 1/n + \mathbb{Z}$; otherwise, let $\psi(g) = 1/2 + \mathbb{Z}$. Since $\mathbb{Q}/\mathbb{Z}$ is injective, ψ can be extended to a homomorphism $\varphi : G \longrightarrow \mathbb{Q}/\mathbb{Z}$, and $\varphi(g) = \psi(g) \neq 0$.

We use this property to prove that the exactness of

$$
\mathrm{Hom}_{\mathbb{Z}}(C, \mathbb{Q}/\mathbb{Z}) \xrightarrow{\beta^*} \mathrm{Hom}_{\mathbb{Z}}(B, \mathbb{Q}/\mathbb{Z}) \xrightarrow{\alpha^*} \mathrm{Hom}_{\mathbb{Z}}(A, \mathbb{Q}/\mathbb{Z})
$$

implies the exactness of $A \xrightarrow{\alpha} B \xrightarrow{\beta} C$. If $c = \beta(\alpha(a)) \in \mathrm{Im}\,(\beta \circ \alpha)$, then $\varphi(c) = \varphi(\beta(\alpha(a))) = (\alpha^*(\beta^*(\varphi)))(a) = 0$ for every homomorphism $\varphi : C \longrightarrow \mathbb{Q}/\mathbb{Z}$; therefore $c = 0$, and $\beta \circ \alpha = 0$. Conversely, let $b \in \mathrm{Ker}\,\beta$. Let $\pi : B \longrightarrow B/\mathrm{Im}\,\alpha$ be the projection. For every homomorphism $\varphi : B/\mathrm{Im}\,\alpha \longrightarrow \mathbb{Q}/\mathbb{Z}$, we have $\alpha^*(\varphi \circ \pi) = \varphi \circ \pi \circ \alpha = 0$,

$$
\begin{array}{ccccc}
A & \xrightarrow{\alpha} & B & \xrightarrow{\beta} & C \\
& & {\scriptstyle \pi} \downarrow & & \downarrow {\scriptstyle \psi} \\
& & B/\mathrm{Im}\,\alpha & \xrightarrow[\varphi]{} & \mathbb{Q}/\mathbb{Z}
\end{array}
$$

hence $\varphi \circ \pi \in \operatorname{Im} \beta^*$, $\varphi \circ \pi = \psi \circ \beta$ for some homomorphism $\psi : C \longrightarrow \mathbb{Q}/\mathbb{Z}$, and $\varphi(\pi(b)) = \psi(\beta(b)) = 0$. Therefore $\pi(b) = 0$ and $b \in \operatorname{Im} \alpha$.

If now $\operatorname{Hom}_{\mathbb{Z}}(M, \mathbb{Q}/\mathbb{Z})$ is injective, then $\operatorname{Hom}_R(-, \operatorname{Hom}_{\mathbb{Z}}(M, \mathbb{Q}/\mathbb{Z}))$ is exact, $\operatorname{Hom}_{\mathbb{Z}}(- \otimes_R M, \mathbb{Q}/\mathbb{Z})$ is exact by Proposition 15.2.3, and $- \otimes_R M$ is exact by the above. $\square$

15.4.b. Lazard's Theorem. Lazard's Theorem characterizes flat modules as direct limits of free modules.

The proof requires two lemmas. Call a left R-module A **finitely presented** in case there exists an exact sequence

$$F_1 \longrightarrow F_2 \longrightarrow A \longrightarrow 0$$

in which F_1 and F_2 are finitely generated free left R-modules; equivalently, in case A has a presentation $A \cong F_1/K$ with finitely many generators (F_1 is finitely generated) and finitely many defining relations (K is finitely generated).

Lemma 15.4.6. *For every $t \in A \otimes_R B$ there exists a finitely generated free module F and a homomorphism $\varphi : F \longrightarrow A$ such that $t \in \operatorname{Im}(\varphi \otimes B)$.*

Proof. By Proposition 14.7.2, A is the direct limit of its finitely generated submodules S. Hence $A \otimes_R B$ is the direct limit of all $S \otimes_R B$, and we have $t \in \operatorname{Im}(\iota \otimes B)$ for some inclusion homomorphism $\iota : S \longrightarrow A$, where S is finitely generated. Then there is an epimorphism $\pi : F \longrightarrow S$, where F is free and finitely generated, and $t \in \operatorname{Im}((\iota \circ \pi) \otimes B)$, since $\pi \otimes B$ is an epimorphism by Proposition 15.2.4. $\square$

Lemma 15.4.7. *Every homomorphism of a finitely presented module into a flat module factors through a finitely generated free module.*

Proof. Let $\varphi : A \longrightarrow M$ be a homomorphism, where M is a flat left R-module and A is finitely presented, with an exact sequence

$$F_1 \xrightarrow{\tau} F_2 \xrightarrow{\sigma} A \longrightarrow 0$$

in which F_1 and F_2 are finitely generated free left R-modules. For the dual modules there is an exact sequence $0 \longrightarrow K \xrightarrow{\kappa} F_2^* \xrightarrow{\tau^*} F_1^*$. Then Proposition 15.3.5 yields a commutative diagram

$$
\begin{array}{ccccc}
K \otimes_R M & \xrightarrow{\bar{\kappa}} & F_2^* \otimes_R M & \xrightarrow{\bar{\tau}^*} & F_1^* \otimes_R M \\
& & \Big\downarrow{\zeta} & & \Big\downarrow{\zeta} \\
& & \operatorname{Hom}_R(F_2, M) & \xrightarrow{\tau'} & \operatorname{Hom}_R(F_1, M)
\end{array}
$$

in which $\overline{\kappa} = \kappa \otimes M$, $\overline{\tau}^* = \tau^* \otimes M$, $\tau' = \mathrm{Hom}_R(\tau, M)$, the vertical arrows are isomorphisms, and the top row is exact since M is flat. Since $\tau'(\varphi \circ \sigma) = \varphi \circ \sigma \circ \tau = 0$, we have $\zeta^{-1}(\varphi \circ \sigma) = \overline{\kappa}(t)$ for some $t \in K \otimes_R M$.

By Lemma 15.4.6, there exists a monomorphism $\mu : F \longrightarrow K$ such that F is finitely generated free and $t \in \mathrm{Im}\,\mu \otimes M$. Since $F \cong F^{**}$, there exists a finitely generated free module $F' \cong F^*$ and a monomorphism $\nu : F'^* \longrightarrow K$ such that $t = \overline{\nu}(t')$ for some $t' \in F'^* \otimes_R M$, where $\overline{\nu} = \nu \otimes M$. Since $F'^{**} \cong F'$ and $F_2^{**} \cong F_2$, there is a homomorphism $\xi : F_2 \longrightarrow F'$ such that $\xi^* = \kappa \circ \nu$. Then

$$\overline{\xi}^*(t') = \overline{\kappa}(\overline{\nu}(t')) = \overline{\kappa}(t) = \zeta^{-1}(\varphi \circ \sigma)$$

(where $\overline{\xi}^* = \xi^* \otimes M$). Also $\tau^* \circ \xi^* = \tau^* \circ \kappa \circ \nu = 0$, so that $\xi \circ \tau = 0$ and ξ factors through $\sigma : \xi = \rho \circ \sigma$ for some $\rho : A \longrightarrow F'$:

$$
\begin{array}{ccccccc}
F_1 & \xrightarrow{\ \tau\ } & F_2 & \xrightarrow{\ \sigma\ } & A & \longrightarrow & 0 \\
& & \xi \downarrow & \swarrow_{\rho} & \downarrow_{\varphi} & & \\
& & F' & \dashrightarrow{}_{\psi} & M & &
\end{array}
$$

We now have a commutative diagram

$$
\begin{array}{ccccc}
F'^* \otimes_R M & \xrightarrow{\ \overline{\xi}^*\ } & F_2^* \otimes_R M & \xrightarrow{\ \overline{\tau}^*\ } & F_1^* \otimes_R M \\
\zeta \downarrow & & \zeta \downarrow & & \downarrow \zeta \\
\mathrm{Hom}_R(F', M) & \xrightarrow{\ \xi'\ } & \mathrm{Hom}_R(F_2, M) & \xrightarrow{\ \tau'\ } & \mathrm{Hom}_R(F_1, M)
\end{array}
$$

in which $\xi' = \mathrm{Hom}_R(\xi, M)$ and the vertical arrows are isomorphisms. Then $\psi = \zeta(t') \in \mathrm{Hom}_R(F', M)$ satisfies $\psi \circ \rho \circ \sigma = \psi \circ \xi = \xi'(\psi) = \zeta(\overline{\xi}^*(t')) = \varphi \circ \sigma$ and $\psi \circ \rho = \varphi$. $\qquad\square$

Corollary 15.4.8. *Every finitely presented flat module is projective.*

Proof. The identity on such a module factors through a free module. $\qquad\square$

We can now prove:

Theorem 15.4.9 (Lazard). *For a left R-module M the following conditions are equivalent:*

(1) *M is flat.*

(2) *Every homomorphism of a finitely presented module into M factors through a finitely generated free module.*

(3) *M is a direct limit of finitely generated free left R-modules.*

(4) *M is a direct limit of free left R-modules.*

Proof. (3) implies (4), and (4) implies (1) by Proposition 15.3.3; (1) implies (2) by Lemma 15.4.7. Now assume (2).

Let $\pi : F \longrightarrow M$ be an epimorphism where F is free, with a basis $(e_i)_{i \in I}$. For each subset J of I, let F_J be the submodule of F generated by $(e_i)_{i \in J}$.

Let P be the set of all pairs $p = (J, S)$ where J is a finite subset of I and S is a finitely generated submodule of $F_J \cap \operatorname{Ker} \pi$. Order P by $(J, S) \leqslant (K, T)$ if and only if $J \subseteq K$ and $S \subseteq T$. Then P is directed upward. For each $p = (J, S) \in P$, there is a module $M_p = F_J/S$. If $p = (J, S) \leqslant (K, T) = q$ in P, then $S \subseteq T$ and there is a unique homomorphism $\varphi_{pq} : M_p \longrightarrow M_q$ such that the diagram

$$
\begin{array}{ccc}
F_J & \overset{\subseteq}{\longrightarrow} & F_K \\
\downarrow & & \downarrow \\
M_p = F_J/S & \underset{\varphi_{pq}}{\longrightarrow} & F_K/T = M_q
\end{array}
$$

where the vertical arrows are projections, commutes. This constructs a direct system $\mathscr{M}$ over P.

We show that $M = \varinjlim_{p \in P} M_p$. For each $p = (J, S) \in P$ we have $J \subseteq \operatorname{Ker} \pi$ and there is a unique homomorphism $\varphi_p : M_p \longrightarrow M$ such that the diagram

$$
\begin{array}{ccc}
F_J & \overset{\subseteq}{\longrightarrow} & F \\
\downarrow & & \downarrow{\scriptstyle \pi} \\
M_p = F_J/S & \underset{\varphi_p}{\longrightarrow} & M
\end{array}
$$

commutes. This constructs a cone $(\varphi_p)_{p \in P}$ which we show is a limit cone. Every element of M is the image under π of some element of F which belongs to some finitely generated submodule F_J; hence $M = \bigcup_{p \in P} \operatorname{Im} \varphi_p$. Let $p = (J, S) \in P$; assume that $\varphi_p(x + S) = 0$ for some $x \in F_J$. Then $x \in \operatorname{Ker} \pi$; there is a finite subset $J \subseteq K \subseteq I$ such that $x \in F_K$ and a finitely generated submodule $S \subseteq T \subseteq F_K \cap \operatorname{Ker} \pi$ which contains x. Then $\varphi_{pq}(x + S) = 0$, where $q = (K, T) \geqslant p$. Thus $\operatorname{Ker} \varphi_p = \bigcup_{q \geqslant p} \operatorname{Ker} \varphi_{pq}$. By Proposition 14.7.4, $(\varphi_p)_{p \in P}$ is a limit cone for $\mathscr{M}$.

To prove (3), we choose $\pi : F \longrightarrow M$ so that each $x \in M$ can be written in the form $\pi(e_i)$ for infinitely many $i \in I$. (For instance let $I = M \times \mathbb{N}$, F be free on I, and π be the homomorphism such that $\pi(e_{x,n}) = x$ for every $(x, n) \in I$.) Then we show that $Q = \{ p \in P; \ M_p \text{ is free} \}$ is cofinal in P. Let $p = (J, S) \in P$. By (2) the homomorphism $\varphi_p : M_p \longrightarrow M$ factors through a finitely generated free module F': $\varphi_p = \psi \circ \chi$, where $M_p \overset{\chi}{\longrightarrow} F' \overset{\psi}{\longrightarrow} M$ and F' has a finite basis B. By the choice of I there are for each $b \in B$ infinitely many $i \notin J$ such that $\pi(e_i) = \psi(b)$. Therefore there exists a finite subset L of I and an isomorphism $\theta : F_L \longrightarrow F'$ such that $L \cap J = \varnothing$ and $\psi(\theta(e_\ell)) = \pi(e_\ell)$ for all $\ell \in L$; namely the diagram

$$
\begin{array}{ccccc}
F_J & & F_L & \stackrel{\subseteq}{\longrightarrow} & F \\
\downarrow & & \theta \downarrow & & \downarrow \pi \\
M_p & \stackrel{}{\underset{\chi}{\longrightarrow}} & F' & \stackrel{}{\underset{\psi}{\longrightarrow}} & M
\end{array}
$$

commutes. The homomorphisms $F_J \longrightarrow M_p \longrightarrow F'$ and $F_L \cong F'$ induce an epimorphism $\rho : F_{J \cup L} \longrightarrow F'$ such that the diagram

$$
\begin{array}{ccc}
F_{J \cup L} & \stackrel{\subseteq}{\longrightarrow} & F \\
\rho \downarrow & & \downarrow \pi \\
F' & \stackrel{}{\underset{\psi}{\longrightarrow}} & M
\end{array}
$$

commutes. Then $T = \operatorname{Ker} \rho \subseteq \operatorname{Ker} \pi$, and T is finitely generated since ρ splits; hence $(J,S) \leqslant (J \cup L, T) \in P$. Also $(J \cup L, T) \in Q$ since $F_{J \cup L}/T \cong F'$ is free. Thus Q is cofinal in P. Hence $M = \varinjlim_{p \in P} M_p = \varinjlim_{p \in Q} M_p$ is a direct limit of finitely generated free modules. $\square$

Exercises

1. Prove that a direct summand of a flat module is flat.
2. Prove that a left R-module M is flat if and only if $\operatorname{Hom}_R(P, \rho)$ is surjective for every finitely presented module P and epimorphism $\rho : N \longrightarrow M$.
3. Prove that a left R-module M is flat if and only if $I \otimes_R M \longrightarrow R_R \otimes_R M$ is injective for every right ideal I of R. (Use Proposition 15.4.5.)
4. Let R be a commutative ring and S be a proper multiplicative subset of R. Show that $S^{-1}R$ is a flat R-module.
5. Let I and J be ideals of R. Show that $I \otimes_R J \cong IJ$ whenever I is a flat right R-module.

16

ALGEBRAS

Algebras are rings with a compatible vector space or module structure, and they are the last of the main structures in this book. Interest in algebras began with Hamilton's construction of quaternions (published in 1843) and Benjamin Pierce's paper *Linear Associative Algebras* (1864). Connections with algebraic sets and group representations were seen in Chapters 11 and 13. This chapter constructs various algebras with universal properties, studies simple algebras, and proves Frobenius's Theorem (published in 1877) which finds all division algebras over $\mathbb{R}$. The results add to our understanding of rings and provide some applications of tensor products.

Required results: tensor products (Chapter 15, Sections 15.1 and 15.2). Wedderburn's Theorem and related results from Section 12.2 are used heavily in Section 16.6, along with Lemma 10.6.8.

16.1. ALGEBRAS

16.1.a. Definition. In what follows, R is a commutative ring (with identity element). An **algebra** over R, also called an R-algebra, is an R-module A with a multiplication which is bilinear ($a(rb + sc) = rab + sac$, $(ra + sb)c = rac + sbc$ for all $r, s \in R$, $a, b, c \in A$), associative ($a(bc) = (ab)c$ for all $a, b, c \in A$), and has an identity element 1 ($1a = a = a1$ for all $a \in A$).

Equivalently, an R-algebra A is a ring (with identity element) with an R-module structure such that $(ra)b = a(rb) = r(ab)$ for all $r \in R$, $a, b \in A$.

The algebras defined above are also called **associative** algebras. Except in Section 16.7, all the algebras in this chapter are associative algebras (with identity element). A **nonassociative** algebra is defined as above but without the requirement that the multiplication be associative (or have an identity element); good examples will be seen in Section 16.7.

There is a third equivalent definition. In any R-algebra A, $r1 + s1 = (r + s)1$ and $(r1)(s1) = (rs)1$ for all $r, s \in R$ so that $\iota : r \longmapsto r1$ is a ring homomorphism. Moreover ι is a **central** homomorphism, that is, every $r1$ is **central** in A: $a(r1) = ra = (r1)a$ for all $a \in A$. In particular, $r \in R$ acts on A is multiplication by $\iota(r)$. If ι is injective (e.g., if R is a field), then $r1$ can be identified with r, which makes R a subring of A.

Conversely, every central ring homomorphism of R into a ring A determines an R-module structure on A and makes A is an R-algebra (see the exercises). Thus an R-algebra may be defined as a ring A with a central homomorphism $R \longrightarrow A$.

For example, the polynomial ring $R[(X_i)_{i \in I}]$ and the ring of formal power series $R[[(X_i)_{i \in I}]]$ with coefficients in R are R-algebras. $\mathbb{C}$ and the quaternion algebra $\mathbf{Q}$ are $\mathbb{R}$-algebras. Every ring is a $\mathbb{Z}$-algebra. The commutative ring R is an R-algebra (then $R \longrightarrow R$ is the identity on R); more generally, $R/\mathfrak{a}$ is an R-algebra for every ideal $\mathfrak{a}$ of R (then $R \longrightarrow R/\mathfrak{a}$ is the projection) and every (commutative) ring extension A of R is a R-algebra (then $R \longrightarrow A$ is the inclusion homomorphism).

The ring $M_n(R)$ of $n \times n$ matrices with entries in R is an R-algebra. More generally, $\mathrm{End}_R(M)$ is an R-algebra for every R-module M; then $R \longrightarrow \mathrm{End}_R(M)$ sends $r \in R$ to the scalar endomorphism $x \longmapsto rx$, which is central in $\mathrm{End}_R(M)$.

In Chapter 13, the group algebra $K[G]$ of a monoid or group G over a field K is a K-algebra; K can be replaced by any commutative ring in this construction.

16.1.b. Homomorphisms. A **homomorphism** of R-algebras is a ring homomorphism which is also an R-module homomorphism; equivalently, a mapping φ such that $\varphi(a + b) = \varphi(a) + \varphi(b)$, $\varphi(ra) = r\varphi(a)$, $\varphi(ab) = \varphi(a)\varphi(b)$, and $\varphi(1) = 1$, for all a, b and all $r \in R$; equivalently, a ring homomorphism such that the following triangle commutes:

For example, $\iota : R \longrightarrow A$ is the only algebra homomorphism of R into A. The Homomorphism Theorem extends to algebras, as follows.

A **subalgebra** of an R-algebra A is an R-algebra S such that $S \subseteq A$ and the inclusion mapping $S \longrightarrow A$ is an algebra homomorphism. Equivalently, $S \subseteq A$ is a subalgebra of A if and only if S is both a subring of A and a submodule of A; if and only if S is a subring of A and contains the image of $R \longrightarrow A$.

An **ideal** of an R-algebra A is an ideal I of the ring A; since $ra = (r1)a$, I is also a submodule of A. The quotient ring A/I is also a quotient module and thus is an R-algebra, the **quotient algebra** A/I; $R \longrightarrow A \longrightarrow A/I$ is central since $R \longrightarrow A$ is central. The projection $A \longrightarrow A/I$ is an algebra homomorphism.

Proposition 16.1.1 (Homomorphism Theorem for Algebras). *When $\varphi : A \longrightarrow B$ is a homomorphism of R-algebras, then $\mathrm{Im}\, \varphi$ is a subalgebra of B,*

Ker φ *is an ideal of A, and there exists an algebra isomorphism* $\theta : A/\text{Ker } \varphi \longrightarrow$ Im φ *such that the following diagram commutes:*

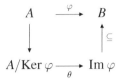

This follows from the Homomorphism Theorems for rings and modules: when φ is an algebra homomorphism, both theorems yield the same diagram, with the same mappings; hence θ is an algebra isomorphism.

16.1.c. Graded Algebras. A **graded** R-algebra is an R-algebra A with submodules A_n ($n \geqslant 0$) such that $A = \bigoplus_{n \geqslant 0} A_n$, $1 \in A_0$, and $A_m A_n \subseteq A_{m+n}$ for all $m,n \geqslant 0$. The elements of A_n are **homogeneous** of **degree** n.

For example, the polynomial algebra $R[(X_i)_{i \in I}]$ is a graded R-algebra; the homogeneous elements of $R[(X_i)_{i \in I}]$ are the homogeneous polynomials in the usual sense, with the usual degree. But we are interested in graded algebras largely because most of the algebras constructed in this chapter are graded algebras.

In a graded algebra $A = \bigoplus_{n \geqslant 0} A_n$, every element can be written uniquely as a sum $a = \sum_{n \geqslant 0} a_n$ of homogeneous elements $a_n \in A_n$ (with $a_n = 0$ for almost all n); a_n is the n-th **homogeneous component** of a. If $a \neq 0$, then the **degree** of a is the largest n such that $a_n \neq 0$; the degree of 0 is, say, $-\infty$. For example, let $f(X,Y) = X^2 + Y^2 + 2X - 4Y + 3 \in R[X,Y]$; the homogeneous components of f are $X^2 + Y^2$, $2X - 4Y$, and 3; f has degree 2 but is not homogeneous.

In general, if $a = \sum_{n \geqslant 0} a_n$ and $b = \sum_{n \geqslant 0} b_n$, with $a_n, b_n \in A_n$, then $a + b = \sum_{n \geqslant 0} a_n + b_n$ and $ab = \sum_{n \geqslant 0} c_n$, where $c_n = \sum_{i+j=n} a_i b_j$.

A **homomorphism** of graded algebras of $A = \bigoplus_{n \geqslant 0} A_n$ into $B = \bigoplus_{n \geqslant 0} B_n$ is an algebra homomorphism which sends A_n into B_n for every $n \geqslant 0$.

There is a Homomorphism Theorem for graded algebras.

When $A = \bigoplus_{n \geqslant 0} A_n$ is a graded algebra, a **graded subalgebra** of A is a graded algebra S such that $S \subseteq A$, and the inclusion mapping $S \longrightarrow A$ is a homomorphism of graded algebras. Equivalently, $S \subseteq A$ is a subalgebra of A if and only if S is a subalgebra and $S = \bigoplus_{n \geqslant 0} S_n$, where S_n is a submodule of A_n. Then $S_n = S \cap A_n$, and $S_m S_n \subseteq S_{m+n}$ for all m,n.

A **graded ideal** of a graded R-algebra $A = \bigoplus_{n \geqslant 0} A_n$ is an ideal I of A such that $I = \bigoplus_{n \geqslant 0} I_n$, where I_n is a submodule of A_n. Then $I_n = I \cap A_n$, and $A_m I_n \subseteq I_{m+n}$, $I_n A_m \subseteq I_{n+m}$ for all m,n; the quotient algebra A/I is a graded R-algebra, with $A/I = \bigoplus_{n \geqslant 0} A_n/I_n$; the projection $A \longrightarrow A/I$ is a homomorphism of graded algebras.

For example, the ideal I of $R[X,Y]$ generated by $X - Y$ is a graded ideal: since $X - Y$ is homogeneous, a polynomial $f \in R[X,Y]$ is divisible by $X - Y$ if and only if every homogeneous component of f is divisible by $X - Y$; hence

$I = \bigoplus_{n \geqslant 0} I_n$, where I_n is the set of all homogeneous elements of I of degree n. On the other hand, the ideal J of $R[X,Y]$ generated by $X^2 - Y$ is not a graded ideal: there are no nonzero homogeneous multiples of $X^2 - Y$, so the set J_n of all homogeneous elements of J of degree n contains only 0 and $J \neq \bigoplus_{n \geqslant 0} J_n$.

Proposition 16.1.2 (Homomorphism Theorem for Graded Algebras).
When $\varphi : A \longrightarrow B$ is a homomorphism of graded R-algebras, $\mathrm{Im}\, \varphi$ is a graded subalgebra of B, $\mathrm{Ker}\, \varphi$ is a graded ideal of A, and there exists a graded algebra isomorphism $\theta : A/\mathrm{Ker}\, \varphi \longrightarrow \mathrm{Im}\, \varphi$ such that the following diagram commutes:

$$
\begin{array}{ccc}
A & \overset{\varphi}{\longrightarrow} & B \\
\downarrow & & \uparrow \subseteq \\
A/\mathrm{Ker}\, \varphi & \underset{\theta}{\longrightarrow} & \mathrm{Im}\, \varphi
\end{array}
$$

The proof is an exercise.

Exercises

R is a commutative ring (with identity element).

1. Show that $\mathrm{End}_R(M)$ is an R-algebra for every R-module M.
2. Let I be an ideal of R. Verify that R/I is an R-algebra.
3. Let A be a ring and $\varphi : R \longrightarrow A$ be a ring homomorphism which is central ($\varphi(r)a = a\varphi(r)$ for all $r \in R$ and $a \in A$). Show that φ determines an R-module structure on A such that A is an R-algebra.
4. Let A be an R-module. Show that A is an R-algebra if and only if there exist module homomorphisms $\iota : R \longrightarrow A$ and $\mu : A \otimes A \longrightarrow A$ such that the following diagrams commute:

$$
\begin{array}{ccccc}
A \otimes A & \overset{\mu}{\longrightarrow} & A & \overset{\mu}{\longleftarrow} & A \otimes A \\
{\scriptstyle \iota \otimes 1}\downarrow & & \| & & \downarrow{\scriptstyle 1 \otimes \iota} \\
R \otimes A & \underset{\cong}{\longrightarrow} & A & \underset{\cong}{\longleftarrow} & A \otimes R
\end{array}
\qquad
\begin{array}{ccc}
A \otimes A \otimes A & \overset{\mu \otimes 1}{\longrightarrow} & A \otimes A \\
{\scriptstyle 1 \otimes \mu}\downarrow & & \downarrow{\scriptstyle \mu} \\
A \otimes A & \underset{\mu}{\longrightarrow} & A
\end{array}
$$

5. Prove the Homomorphism Theorem for graded algebras.
6. Show that $R[X]$ serves as the free R-algebra with one generator.
7. Show that $R[(X_i)_{i \in I}]$ serves as the free commutative R-algebra on the set I.

16.2. THE TENSOR ALGEBRA

The tensor algebra of a module is an algebra which is "freely" generated by that module.

16.2.a. Generation. In what follows, R is a commutative ring.

When A is an R-algebra, the subalgebra of A generated by a subset S is also generated by the submodule $M = \langle S \rangle$ of A generated by A. The latter is described by the next result, whose proof is an easy exercise:

Proposition 16.2.1. *Let A be an R-algebra and M be a submodule of A. The subalgebra of A generated by M consists of all sums of a multiple of 1 and finitely many products of elements of M.*

If A is generated by M, then, by Proposition 16.2.1, every element of A can be written as a sum $r1 + a_{11} \ldots a_{1n_1} + \cdots + a_{k1} \ldots a_{kn_k}$, where $r \in R$, $n_1, \ldots, n_k > 0$, and $a_{11}, \ldots, a_{kn_k} \in M$. This expression is generally not unique; thus the elements of M satisfy some relations in A. In this section we construct the tensor algebra $T(M)$ of a given R-module M, which is generated by M with the fewest possible relations between the elements of M, so that $T(M)$ is the "simplest" and "largest" R-algebra generated by M. Like similar constructions in previous chapters, there is a universal property which characterizes $T(M)$.

When M is a submodule of A, the product $a_1 \ldots a_n$ of n elements $a_1, \ldots, a_n$ of M is a multilinear function of $a_1, \ldots, a_n$ and induces an R-module homomorphism of $\bigotimes^n M = M \otimes \cdots \otimes M$ into A for every n. This suggests that $T(M)$ can be retrieved from the direct sum $\bigoplus_{n \geqslant 0} (\bigotimes^n M)$. In fact the direct sum serves.

16.2.b. Construction. Let M be an R-module. Let $T^0(M) = R$ and $T^1(M) = M$; for $n \geqslant 2$, let

$$T^n(M) = M \otimes \cdots \otimes M$$

be the tensor product $M_1 \otimes \cdots \otimes M_n$ where $M_1 = \cdots = M_n = M$. $T^n(M)$ is the n-th **tensor power** of M and is also denoted by $\bigotimes^n M$. Since $T^m(M) \otimes T^n(M) \cong T^{m+n}(M)$, there is a bilinear mapping

$$\otimes : T^m(M) \times T^n(M) \longrightarrow T^{m+n}(M)$$

which sends $(a_1 \otimes \cdots \otimes a_m, b_1 \otimes \cdots \otimes b_n)$ to $(a_1 \otimes \cdots \otimes a_m) \otimes (b_1 \otimes \cdots \otimes b_n) = a_1 \otimes \cdots \otimes a_m \otimes b_1 \otimes \cdots \otimes b_n$. If, say, $m = 0$, then the isomorphism $T^0(M) \otimes T^n(M) = R \otimes T^n(M) \cong T^n(M)$ sends $r \otimes t$ to rt; hence $\otimes : T^0(M) \times T^n(M) \longrightarrow T^n(M)$ sends (r, t) to $r \otimes t = rt$, for all $r \in R$ and $t \in T^n(M)$.

As an R-module, the **tensor algebra** of M is

$$T(M) = \bigoplus_{n \geqslant 0} T^n(M).$$

We identify $t \in T^n(M)$ and its image $\iota(t) \in T(M)$ under the direct sum injection $T^n(M) \longrightarrow \bigoplus_{n \geqslant 0} T^n(M)$, so that an element of $T(M)$ can be written uniquely as a sum $t = \sum_{n \geqslant 0} t_n$, with $t_n \in T^n(M)$ for all n and $t_n = 0$ for almost all n. The multiplication on $T(M)$ is the bilinear multiplication $\otimes$ induced by all tensor maps $\otimes : T^m(M) \times T^n(M) \longrightarrow T^{m+n}(M)$: if $t = \sum_{n \geqslant 0} t_n$ and $u = \sum_{n \geqslant 0} u_n \in$

$T(M)$, then $t \otimes u = \sum_{m,n \geqslant 0} t_m \otimes u_n = \sum_{n \geqslant 0} (\sum_{k+\ell=n} t_k \otimes u_\ell)$. This multiplication is bilinear. Associativity $t \otimes (u \otimes v) = (t \otimes u) \otimes v$ holds whenever $t \in T^\ell$, $u \in T^m$, $v \in T^n$ and therefore holds for all $t, u, v \in T(M)$; $1 \otimes t = t = t \otimes 1$ holds for all $t \in T^n(M)$ and therefore holds for all $t \in T(M)$. Thus $T(M)$ is a graded R-algebra. By Proposition 16.2.1, $T(M)$ is generated by its submodule M. Thus:

Proposition 16.2.2. *The tensor algebra $T(M)$ of a module M is a graded R-algebra and is generated by M.*

16.2.c. Properties. The tensor algebra has the desired universal property.

Proposition 16.2.3. *Let M be an R-module. Every module homomorphism of M into an R-algebra A extends uniquely to an algebra homomorphism of $T(M)$ into A:*

Proof. Let $\varphi : M \longrightarrow A$ be a module homomorphism. For each $n \geqslant 2$, the multilinear mapping $(a_1, \ldots, a_n) \longmapsto \varphi(a_1) \ldots \varphi(a_n)$ of $M \times \cdots \times M$ into A induces a module homomorphism $\psi_n : T^n(M) \longrightarrow A$ such that $\psi_n(a_1 \otimes \cdots \otimes a_n)$ $= \varphi(a_1) \ldots \varphi(a_n)$ for all $a_1, \ldots, a_n \in M$. Let $\psi_1 = \varphi : M \longrightarrow A$ and $\psi_0 : R \longrightarrow A$, $\psi_0(r) = r1$. The homomorphisms ψ_n $(n \geqslant 0)$ extend to a module homomorphism $\psi : T(M) = \bigoplus_{n \geqslant 0} T^n(M) \longrightarrow A$, $\psi(\sum_{n \geqslant 0} t_n) = \sum_{n \geqslant 0} \psi_n(t_n)$. We see that $\psi(1) = 1$; moreover $\psi(tu) = \psi(t)\psi(u)$ holds whenever t and u are homogeneous and therefore holds for all $t, u \in T(M)$. Thus ψ is an algebra homomorphism.

The homomorphism ψ is unique, since M generates $T(M)$: if ψ_1 and $\psi_2 :$ $T(M) \longrightarrow A$ extend φ, then $S = \{ t \in T(M); \psi_1(t) = \psi_2(t) \}$ is a subalgebra of $T(M)$ which contains M; therefore $S = T(M)$ and $\psi_1 = \psi_2$. $\square$

In the above, $\psi(T(M))$ is the subalgebra of A generated by $\varphi(M)$, by Proposition 16.2.1. Hence $T(M)$ is the "largest" R-algebra generated by M, in the sense that every R-algebra which is generated by M is a homomorphic image of $T(M)$.

Corollary 16.2.4. *If M is the free R-module on a set X, then $T(M)$ is the free R-algebra on X: every mapping of X into an R-algebra A extends uniquely to an algebra homomorphism of $T(M)$ into A.*

Proof. Every mapping of X into an R-algebra A extends uniquely to a module homomorphism of M into A and thence to an algebra homomorphism of $T(M)$ into A. $\square$

Proposition 16.2.5. *If M is the free R-module on a set X, then $T(M)$ is a free R-module, with a basis which consists of all $x_1 \otimes \cdots \otimes x_n$ with $n \geqslant 0$ and $x_1, \ldots, x_n \in X$.*

Proof. By convention, $x_1 \otimes \cdots \otimes x_n = 1 \in R$ if $n = 0$. For each $n \geqslant 2$, $T^n(M) = M \otimes \cdots \otimes M$ is a free R-module, whose basis consists of all $x_1 \otimes \cdots \otimes x_n$ with $x_1, \ldots, x_n \in M$. $\qquad \square$

Exercises

1. Show that $T(_R R) \cong R[X]$.
2. Let M be a free R-module with basis $(e_i)_{i \in I}$. Show that $T(M)$ is isomorphic to the polynomial algebra $R[(X_i)_{i \in I}]$ with noncommuting indeterminates constructed in Exercise 4.3.23.
3. Let M be an R-module. Show that there is a graded R-algebra $E = \bigoplus_{n \geqslant 0} \mathrm{End}_R(T^n(M))$ in which the product of $\eta \in \mathrm{End}_R(T^m(M))$ and $\zeta \in \mathrm{End}_R(T^n(M))$ is the endomorphism $\eta \otimes \zeta$ of $T^{m+n}(M) = T^m(M) \otimes T^n(M)$ such that $(\eta \otimes \zeta)(t \otimes u) = \eta(t) \otimes \zeta(u)$ for all $t \in T^m(M)$ and $u \in T^n(M)$.
4. Let M be an R-module and E be the graded R-algebra in the previous exercise. Construct a homomorphism $T(\mathrm{End}_R(M)) \longrightarrow E$; show that $T(\mathrm{End}_R(M)) \cong E$ if M is finitely generated and free.

16.3. THE SYMMETRIC ALGEBRA

The symmetric algebra is the commutative analogue of the tensor algebra.

16.3.a. Construction. As in Section 16.2 we want to construct the "simplest" and "largest" commutative R-algebra generated by a given R-module M (according to a suitable universal property). By Proposition 16.2.3, this algebra must be, up to isomorphism, a quotient algebra of $T(M)$; it is constructed so that the elements of M commute with each other.

By definition, the **symmetric algebra** of an R-module M is the quotient $S(M) = T(M)/I$, where I is the ideal of $T(M)$ generated by all $b \otimes a - a \otimes b$ with $a, b \in M$. We will show (Lemma 16.3.4) that I is a graded ideal. For now we note that $I \subseteq \bigoplus_{n \geqslant 2} T^n(M)$ (since the latter is an ideal of $T(M)$). Hence $I \cap M = 0$. We identify M with its image in $S(M) = T(M)/I$; the product $a_1 \ldots a_n$ of $a_1, \ldots, a_n \in M$ in $S(M)$ is the image of the product $a_1 \otimes \cdots \otimes a_n$ in $T(M)$ under the projection $T(M) \longrightarrow S(M)$. We see that M generates $S(M)$, and that $S(M)$ is commutative, since all its generators commute with each other.

Proposition 16.3.1. *The symmetric algebra $S(M)$ of a module M is a commutative R-algebra and is generated by M.*

16.3.b. Properties. The universal property of $S(M)$ is as follows.

Proposition 16.3.2. *Let M be an R-module. Every module homomorphism of M into a commutative R-algebra A extends uniquely to an algebra homomorphism of $S(M)$ into A.*

$$M \overset{\subseteq}{\longrightarrow} T(M)$$

(commutative diagram: $M \xrightarrow{\subseteq} T(M)$, with $\varphi : M \to A$, $\psi : T(M) \to A$ dashed, $\pi : T(M) \to S(M)$, and $\chi : S(M) \to A$)

Proof. Let $\varphi : M \longrightarrow A$ be a module homomorphism. By Proposition 16.2.3, φ extends uniquely to an algebra homomorphism $\psi : T(M) \longrightarrow A$. Since A is commutative, we have $\psi(b \otimes a) = \varphi(b)\varphi(a) = \varphi(a)\varphi(b) = \psi(a \otimes b)$ for all $a, b \in M$. Therefore $I \subseteq \mathrm{Ker}\,\psi$ and ψ factors uniquely through the projection $\pi : T(M) \longrightarrow S(M)$. $\square$

Corollary 16.3.3. *If M is the free R-module on a set X, then $S(M)$ is the free commutative R-algebra on X: every mapping of X into a commutative R-algebra A extends uniquely to an algebra homomorphism of $S(M)$ into A.*

Proof. Every mapping of X into a commutative R-algebra A extends uniquely to a module homomorphism of M into A and thence to an algebra homomorphism of $S(M)$ into A. $\square$

The following result shows that $S(M)$ is a graded algebra, and is sometimes used as the definition of $S(M)$:

Lemma 16.3.4. *I is a graded ideal $I = \bigoplus_{n \geqslant 2} I_n$, where I_n is the submodule of $T^n(M)$ generated by all $a_{\sigma 1} \otimes \cdots \otimes a_{\sigma n} - a_1 \otimes \cdots \otimes a_n$ where $a_1, \ldots, a_n \in M$ and σ is a permutation of $\{1, 2, \ldots, n\}$. Hence $S(M)$ is a graded R-algebra $S(M) = \bigoplus_{n \geqslant 0} S^n(M)$, where $S^n(M) = T^n(M)/I_n$.*

Proof. Let $n \geqslant 2$ and $a_1, \ldots, a_n \in M$. Since $S(M)$ is commutative, $a_{\sigma 1} \ldots a_{\sigma n} = a_1 \ldots a_n$ holds in $S(M)$ for every $\sigma \in S^n$, and $a_{\sigma 1} \otimes \cdots \otimes a_{\sigma n} - a_1 \otimes \cdots \otimes a_n \in I$ for every $\sigma \in S^n$. Hence $I_n \subseteq I$, and $\bigoplus_{n \geqslant 2} I_n \subseteq I$. To prove the converse inclusion, we note that

$$a_{\sigma 1} \otimes \cdots \otimes a_{\sigma n} \otimes b_{n+1} \otimes \cdots \otimes b_{n+m} - a_1 \otimes \cdots \otimes a_n \otimes b_{n+1} \otimes \cdots \otimes b_{n+m} \in I_{m+n}$$

for all $b_1, \ldots, b_m \in M$ (since $\sigma 1, \ldots, \sigma n, n+1, \ldots, n+m$ is a permutation of $1, 2, \ldots, n+m$); hence

$$a_{\sigma 1} \otimes \cdots \otimes a_{\sigma n} \otimes t - a_1 \otimes \cdots \otimes a_n \otimes t \in I_{m+n}$$

whenever $t \in T^m(M)$, and $i \otimes t \in I_{n+m}$ for all $i \in I_n$ and $t \in T^m(M)$. Similarly $t \otimes i \in I_{n+m}$ for all $i \in I_n$ and $t \in T^m(M)$. Hence $J = \bigoplus_{n \geqslant 2} I_n$ is an ideal of $T(M)$.

Since $b \otimes a - a \otimes b \in I_2 \subseteq J$ for all $a, b \in M$, it follows that $I \subseteq J$. Thus $I = J$; I is a graded ideal; and $S(M)$ is a graded algebra, with $S(M) \cong \bigoplus_{n \geqslant 0} T^n(M)/I_n$. $\square$

For the next result we note that the basis of a free module can be totally ordered (in fact, well ordered), by the Axiom of Choice.

Proposition 16.3.5. *If M is the free R-module on a totally ordered set X, then $S(M)$ is a free R-module, whose basis consists of all $x_1 \ldots x_n$ with $n \geqslant 0$ and $x_1 \leqslant \cdots \leqslant x_n \in X$. If $X = (e_i)_{i \in I}$, then $S(M) \cong R[(X_i)_{i \in I}]$.*

Proof. We let $X = (e_i)_{i \in I}$, where I is totally ordered, and take advantage of the existing commutative R-algebra $R[(X_i)_{i \in I}]$. By Proposition 16.3.2, there is a homomorphism $\rho : S(M) \longrightarrow R[(X_i)_{i \in I}]$ which sends $e_i \in S(M)$ to $X_i \in R[(X_i)_{i \in I}]$. Now every monomial $X_{j_1}^{m_1} \ldots X_{j_2}^{m_2} \ldots X_{j_k}^{m_k} \in R[(X_i)_{i \in I}]$ can be written uniquely in the form $X_{i_1} X_{i_2} \ldots X_{i_n}$ with $i_1 \leqslant i_2 \leqslant \cdots \leqslant i_n$. Hence these monoids form a basis of $R[(X_i)_{i \in I}]$ over R. Therefore the elements of $S(M)$ of the form $e_{i_1} e_{i_2} \ldots e_{i_n}$, with $i_1 \leqslant i_2 \leqslant \cdots \leqslant i_n$, are linearly independent in $S(M)$, since their images under ρ are linearly independent in $R[(X_i)_{i \in I}]$.

Since $S(M)$ is generated as an algebra by all e_i, $S(M)$ is generated as a module by all products $e_{i_1} e_{i_2} \ldots e_{i_n}$, with $i_1, \ldots, i_n \in I$ arbitrary. But $e_{i_1} e_{i_2} \ldots e_{i_n} = e_{i_{\sigma 1}} e_{i_{\sigma 2}} \ldots e_{i_{\sigma n}}$ for every permutation σ of $1, 2, \ldots, n$, since $S(M)$ is commutative. Therefore every product $e_{i_1} e_{i_2} \ldots e_{i_n}$ is equal to a product $e_{i_1} e_{i_2} \ldots e_{i_n}$ in which $i_1 \leqslant i_2 \leqslant \cdots \leqslant i_n$, and the latter generate $S(M)$ as a module. Therefore they constitute a basis of $S(M)$.

The homomorphism $\rho : S(M) \longrightarrow R[(X_i)_{i \in I}]$ now sends a basis of $S(M)$ onto a basis of $R[(X_i)_{i \in I}]$ and is therefore an isomorphism. $\square$

16.3.c. Symmetric Multilinear Mappings. Every n-linear mapping $M^n \longrightarrow N$ factors uniquely through the tensor map $M^n \longrightarrow T^n(M)$. There is a similar universal property for $S^n(M)$.

Let M and N be R-modules. An n-linear mapping $\mu : M^n \longrightarrow N$ is **symmetric** in case $\mu(a_{\sigma 1}, \ldots, a_{\sigma n}) = \mu(a_1, \ldots, a_n)$ for all $a_1, \ldots, a_n \in M$ and every permutation σ of $\{1, 2, \ldots, n\}$. For example, multiplication in $S(M)$ yields a symmetric n-linear mapping $\mu_n : M^n \longrightarrow S^n(M)$. The reader will show:

Proposition 16.3.6. *Every n-linear symmetric mapping $\mu : M^n \longrightarrow N$ factors uniquely through the multiplication map $\mu_n : M^n \longrightarrow S^n(M)$ (there is a unique module homomorphism $\varphi : S^n(M) \longrightarrow N$ such that $\varphi \circ \mu_n = \mu$).*

Exercises

1. Give a direct proof that $S({}_R R) \cong R[X]$.

2. Let M be a free R-module with basis $(e_i)_{i \in I}$. Give a direct proof that $S(M)$ is isomorphic to the usual polynomial algebra $R[(X_i)_{i \in I}]$.

3. Show that every symmetric n-linear mapping $\mu : M^n \longrightarrow N$ factors uniquely through the multiplication map $\mu_n : M^n \longrightarrow S^n(M)$ (there is a unique module homomorphism $\varphi : S^n(M) \longrightarrow N$ such that $\varphi \circ \mu_n = \mu$).

16.4. THE EXTERIOR ALGEBRA

The exterior algebra of a module is another algebra which can be constructed from the tensor algebra.

Exterior algebras arise from the calculus of differential forms in analysis. If $\omega = P\,dx + Q\,dy + R\,dz$ is a first-order differential form in three variables, then $d\omega = (R_y - Q_z)\,dy\,dz + (P_z - R_x)\,dz\,dx + (Q_x - P_y)\,dx\,dy$ can be calculated by substituting $dP = P_x\,dx + P_y\,dy + P_z\,dz$, $dQ = Q_x\,dx + Q_y\,dy + Q_z\,dz$, and $dR = R_x\,dx + R_y\,dy + R_z\,dz$ into ω and using the rules $dx\,dx = dy\,dy = dz\,dz = 0$ and $dz\,dy = -dy\,dz$, $dx\,dz = -dz\,dx$, $dy\,dx = -dx\,dy$. The same rules can be used to multiply differential forms together. Terms in these products are usually separated by wedges $\wedge$ to distinguish them from ordinary products. Then $\omega \wedge \omega = 0$ for every first-order differential form ω.

A submodule M of an R-algebra A is **anticommutative** in A in case $a^2 = 0$ for all $a \in M$. Then $ba = -ab$ for all $a, b \in M$, since $ab + ba = (a + b)^2 = 0$. (The converse holds if 2 is a unit in R.) The exterior algebra of M is the "largest" algebra generated by M in which M is anticommutative (according to a suitable universal property); examples will include the algebra of differential forms in n variables.

16.4.a. Construction. By definition, the **exterior algebra** of an R-module M (also called the **alternating algebra** of M) is the quotient $\bigwedge(M) = T(M)/J$, where J is the ideal of $T(M)$ generated by all $a \otimes a$ with $a \in M$; multiplication in $\bigwedge(M)$ is denoted by $\wedge$. We will show (Lemma 16.4.3) that J is a graded ideal. For now we note that $J \subseteq \bigoplus_{n \geqslant 2} T^n(M)$ (since the latter is an ideal of $T(M)$). Hence $J \cap M = 0$. We identify M with its image in $\bigwedge(M)$; the product $a_1 \wedge \cdots \wedge a_n$ of $a_1, \ldots, a_n \in M$ in $\bigwedge(M)$ is the image of the product $a_1 \otimes \cdots \otimes a_n$ in $T(M)$ under the projection $T(M) \longrightarrow \bigwedge(M)$. We see that M generates $\bigwedge(M)$, and that M is anticommutative in $\bigwedge(M)$.

Proposition 16.4.1. *The exterior algebra $\bigwedge(M)$ of a module M is an R-algebra which is generated by M and in which M is anticommutative.*

16.4.b. Properties. $\bigwedge(M)$ has the expected universal property:

Proposition 16.4.2. *Let M be an R-module. Let φ be a module homomorphism of M into an R-algebra A such that $\mathrm{Im}\ \varphi$ is anticommutative in A. Then φ extends uniquely to an algebra homomorphism of $\bigwedge(M)$ into A:*

Proof. Let $\varphi : M \longrightarrow A$ be a module homomorphism. By Proposition 16.2.3, φ extends uniquely to an algebra homomorphism $\psi : T(M) \longrightarrow A$. Since Im φ is anticommutative in A we have $\psi(a \otimes a) = \varphi(a)\varphi(a) = 0$ for all $a \in M$. Hence $J \subseteq \text{Ker } \psi$ and ψ factors uniquely through the projection $\pi : T(M) \longrightarrow \bigwedge(M)$. $\qquad\square$

To prove further properties we show that $\bigwedge(M)$ is a graded algebra. The following result is sometimes used as the definition of $\bigwedge(M)$.

Lemma 16.4.3. $J = \bigoplus_{n \geqslant 2} J_n$, where J_n is the submodule of $T^n(M)$ generated by all $a_1 \otimes \cdots \otimes a_n$ where $a_1, \ldots, a_n \in M$ and $a_i = a_j$ for some $i \neq j$. Hence $\bigwedge(M)$ is a graded R-algebra $\bigwedge(M) = \bigoplus_{n \geqslant 0} \bigwedge^n(M)$, where $\bigwedge^n(M) = T^n(M)/J_n$.

Proof. Let $n \geqslant 2$ and $a_1, \ldots, a_n \in M$. Since M is anticommutative in $\bigwedge(M)$, $a_i = a_j$ for some $i \neq j$ implies $a_1 \wedge \ldots \wedge a_n = \pm a_i \wedge a_j \wedge a_1 \wedge \cdots \wedge a_n = 0$ in $\bigwedge(M)$ and $a_1 \otimes \cdots \otimes a_n \in J$. Hence $J_n \subseteq J$, and $\bigoplus_{n \geqslant 2} J_n \subseteq J$. To prove the converse inclusion, we note that $a_i = a_j$ for some $i \neq j$ implies $a_1 \otimes \cdots \otimes a_n \otimes b_{n+1} \otimes \cdots \otimes b_{n+m} \in J_{m+n}$ for all $b_1, \ldots, b_m \in M$ and $a_1 \otimes \cdots \otimes a_n \otimes t \in J_{m+n}$ whenever $t \in T^m(M)$, and $j \otimes t \in J_{n+m}$ for all $j \in J_n$ and $t \in T^m(M)$. Similarly $t \otimes j \in J_{n+m}$ for all $j \in J_n$ and $t \in T^m(M)$. Hence $\bigoplus_{n \geqslant 2} J_n$ is an ideal of $T(M)$. Since $a \otimes a \in J_2$ for all $a, b \in M$, it follows that $J \subseteq \bigoplus_{n \geqslant 2} J_n$. Thus $J = \bigoplus_{n \geqslant 2} J_n$. This shows that J is a graded ideal. Hence $\bigwedge(M)$ is a graded algebra, with $\bigwedge(M) = \bigoplus_{n \geqslant 0} T^n(M)/J_n$. $\qquad\square$

We saw that $b \wedge a = -a \wedge b$ whenever $a, b \in M$ are homogeneous of degree 1. The reader will show that $v \wedge u = (-1)^{mn} u \wedge v$ whenever u and $v \in \bigwedge(M)$ are homogeneous of degrees m and n respectively. (Hence it is not true that $v \wedge u = -u \wedge v$ for all $u, v \in \bigwedge(M)$.)

For the next result we note, as before, that the basis of a free module can always be totally ordered.

Proposition 16.4.4. *If M is the free R-module on a totally ordered set X, then $\bigwedge^n(M)$ is a free R-module, whose basis consists of all $x_1 \wedge \cdots \wedge x_n$ with $x_1 < \cdots < x_n \in X$, and $\bigwedge(M)$ is a free R-module, whose basis consists of all $x_1 \wedge \cdots \wedge x_n$ with $n \geqslant 0$ and $x_1 < \cdots < x_n \in X$.*

Proof. We have $\bigwedge^1(M) = M$ and $x_1 \wedge \cdots \wedge x_n = 1$ if $n = 0$. Let $n \geqslant 2$. We know that $T^n(M)$ has a basis B which consists of all $x_1 \otimes \cdots \otimes x_n$ with $x_1, \ldots, x_n \in X$ (and n as given). Let Y_n be the set which consists of all $x_{\sigma 1} \otimes \cdots \otimes x_{\sigma n} - (\text{sgn}\,\sigma)\, x_1 \otimes \cdots \otimes x_n$, where $x_1, \ldots, x_n \in X$ all distinct and $\sigma \in S_n$ is a permutation of $1,\ldots,n$, and all $x_1 \otimes \cdots \otimes x_n$ with $x_1,\ldots,x_n \in X$ and a *duplication* ($x_i = x_j$ for some $i \neq j$). We show that J_n is the submodule of $T^n(M)$ generated by Y_n. Since M is alternating in $\bigwedge(M)$ we have $x_j \wedge x_i = -x_i \wedge x_j$ and

$$x_{\sigma 1} \wedge \cdots \wedge x_{\sigma n} = (\mathrm{sgn}\,\sigma)\, x_1 \wedge \cdots \wedge x_n$$

for all $x_1,\ldots,x_n \in X$; hence $x_{\sigma 1} \otimes \cdots \otimes x_{\sigma n} - (\mathrm{sgn}\,\sigma)\, x_1 \otimes \cdots \otimes x_n \in J_n$ for all $x_1,\ldots,x_n \in X$ and $\sigma \in S_n$. Thus $Y_n \subseteq J_n$. To show that J_n is generated by Y_n we show that every generator $a_1 \otimes \cdots \otimes a_n$ of J_n (with $a_j = a_k$ for some $j \neq k$) is a linear combination of elements of Y_n. We have $a_j = \sum_{i \in I} r_{ji} x_i$ and

$$a_1 \otimes \cdots \otimes a_n = \sum r_{1i_1} \ldots r_{ni_n}\, x_{i_1} \otimes \cdots \otimes x_{i_n},$$

where the sum runs over all $i_1,\ldots,i_n \in I$. In this sum, $x_{i_1} \otimes \cdots \otimes x_{i_n} \in Y_n$ whenever $x_{i_1},\ldots,x_{i_n}$ are not all distinct. If $a_1 = a_2$, then the remaining terms, where $x_{i_1},\ldots,x_{i_n}$ are all distinct, can be grouped in pairs

$$r_{1i_1} r_{2i_2} r_{3i_3} \ldots r_{ni_n}\, (x_{i_1} \otimes x_{i_2} \otimes x_{i_3} \otimes \cdots \otimes x_{i_n} + x_{i_2} \otimes x_{i_1} \otimes x_{i_3} \otimes \cdots \otimes x_{i_n}),$$

with $x_{i_1} \otimes x_{i_2} \otimes x_{i_3} \otimes \cdots \otimes x_{i_n} + x_{i_2} \otimes x_{i_1} \otimes x_{i_3} \otimes \cdots \otimes x_{i_n} \in Y_n$, since $r_{1i} = r_{2i}$ for all i. Thus $a_1 \wedge \cdots \wedge a_n$ is a linear combination of elements of Y_n. This also holds if $a_j = a_k$ for some $j \neq k$: the proof is the same, except for the notation which is much worse. Thus J_n is generated by Y_n.

The symmetric group S_n acts on the basis B of $T^n(M)$: if $b = x_1 \otimes \cdots \otimes x_n \in B$, then $\sigma b = x_{\sigma 1} \otimes \cdots \otimes x_{\sigma n}$. If $b = x_1 \otimes \cdots \otimes x_n$ has no duplication (if $x_1,\ldots,x_n$ are all distinct), then the orbit of b contains exactly one *strictly ascending* element $x_{\sigma 1} \otimes \cdots \otimes x_{\sigma n}$, that is, such that $x_{\sigma 1} < \cdots < x_{\sigma n}$. Let

$$C = \{c \in B\,;\ c \text{ is strictly ascending}\},$$

$$D = \{b \in B\,;\ b \text{ has a duplication}\},$$

$$B' = \{b' = b - \varepsilon c\,;\ b \in B,\ c = \sigma b \in C,\ \varepsilon = (\mathrm{sgn}\,\sigma),\ \sigma \in S_n\}.$$

Then $D \cup B' \cup C$ is a basis of $T^n(M)$ and J_n is generated by $D \cup B'$, by the first half of the proof; therefore $T^n(M) = J_n \oplus K_n$, where K_n is the submodule of $T^n(M)$ generated by C. Then C is a basis of K_n, and the projection $T^n(M) \longrightarrow T^n(M)/J_n = \bigwedge^n(M)$ induces an isomorphism $K_n \longrightarrow \bigwedge^n(M)$ and sends the basis C of K_n to a basis of $\bigwedge^n(M)$, which consists of all products $x_1 \wedge x_2 \wedge \cdots \wedge x_n$ with $x_1 < x_2 < \cdots < x_n$. $\square$

With Proposition 16.4.4 we can recover the algebra of differential forms in n variables as an exterior algebra. Let R be the ring of all suitable real valued functions of n variables (typically, with continuous partial derivatives in a given open set), with pointwise operations. First-order differential forms are the elements of a free R-module M of rank n, whose basis is denoted by $\{dx_1,\ldots,dx_n\}$. Then differential forms of order $k \leqslant n$ coincide with the elements of $\bigwedge^k(M)$; the wedge product of differential forms coincides with the operation in $\bigwedge(M)$.

16.4.c. Alternating Multilinear Mappings. Exterior algebras provide an alternative approach to determinants.

Proposition 16.4.5. *If M is free of rank n, with basis $e_1,\ldots,e_n$, then $\bigwedge^n(M)$ is free of rank 1, with basis $\{e_1 \wedge \cdots \wedge e_n\}$. If $a_1 = \sum_j r_{1j}e_j,\ldots,a_n = \sum_j r_{nj}e_j \in M$, then*

$$a_1 \wedge \cdots \wedge a_n = \det(r_{ij})\, e_1 \wedge \cdots \wedge e_n.$$

Proof. $X = \{e_1,\ldots,e_n\}$ can be totally ordered by $e_1 < e_2 < \cdots < e_n$ and has only one strictly ascending sequence of length n. Hence $\bigwedge^n(M)$ is free on $\{e_1 \wedge \cdots \wedge e_n\}$, by Proposition 16.4.4.

Recall that $\det(r_{ij}) = \sum_{\sigma \in S_n}(\mathrm{sgn}\,\sigma)\, r_{1,\sigma 1} \ldots r_{n,\sigma n}$. If $a_i = \sum_j r_{ij}e_j$, then

$$a_1 \wedge \cdots \wedge a_n = \sum r_{1j_1} \ldots r_{nj_n}\, e_{j_1} \wedge \cdots \wedge e_{j_n},$$

where the sum runs over all $j_1,\ldots,j_n \in \{1,2,\ldots,n\}$. If $j_1,\ldots,j_n$ are not all distinct, then $e_{j_1} \wedge \cdots \wedge e_{j_n} = 0$. If $j_1,\ldots,j_n$ are all distinct, then $j_1,\ldots,j_n$ is a permutation σ of $\{1,2,\ldots,n\}$ and $e_{\sigma 1} \wedge \cdots \wedge e_{\sigma n} = (\mathrm{sgn}\,\sigma)\, e_1 \wedge \cdots \wedge e_n$. Hence

$$a_1 \wedge \cdots \wedge a_n = \sum_{\sigma \in S_n} r_{1,\sigma 1} \ldots r_{n,\sigma n}\,(\mathrm{sgn}\,\sigma)\, e_1 \wedge \cdots \wedge e_n = \det(r_{ij})\, e_1 \wedge \cdots \wedge e_n.$$

$\square$

A similar formula calculates $a_1 \wedge \cdots \wedge a_k$ when $k \leqslant n$ (see the exercises).

More generally, let M and N be R-modules. An n-linear mapping $\mu : M^n \longrightarrow N$ is **alternating** in case $\mu(a_1,\ldots,a_n) = 0$ whenever $a_i = a_j$ for some $i \neq j$. (Then $\mu(a_{\sigma 1},\ldots,a_{\sigma n}) = (\mathrm{sgn}\,\sigma)\,\mu(a_1,\ldots,a_n)$ for every $a_1,\ldots,a_n \in M$ and every permutation σ of $\{1,2,\ldots,n\}$.) For example, multiplication in $\bigwedge(M)$ yields an alternating n-linear mapping $\mu_n : M^n \longrightarrow \bigwedge^n(M)$. The reader will show:

Proposition 16.4.6. *Every n-linear alternating mapping $\mu : M^n \longrightarrow N$ factors uniquely through the multiplication map $\mu_n : M^n \longrightarrow \bigwedge^n(M)$ (there is a unique module homomorphism $\varphi : \bigwedge^n(M) \longrightarrow N$ such that $\varphi \circ \mu_n = \mu$).*

With Propositions 16.4.5 and 16.4.6 we can construct determinants as follows. Let $n > 0$ and $M = (_R R)^n$. Since $\bigwedge^n(M) \cong {}_R R$, μ_n provides an n-linear alternating form δ, defined by: $a_1 \wedge \cdots \wedge a_n = \delta(a_1,\ldots,a_n)\, e_1 \wedge \cdots \wedge e_n$, where $e_1,\ldots,e_n$ is the standard basis of $(_R R)^n$; δ is not trivial, since $\delta(e_1,\ldots,e_n) = 1$. By Proposition 16.4.6, every n-linear alternating form is proportional to δ. Proposition 16.4.5 shows that δ is the usual determinant.

Exercises

1. Show that $v \wedge u = (-1)^{mn} u \wedge v$ whenever $u \in \bigwedge^m(M)$ and $v \in \bigwedge^n(M)$.
2. Let M be a free R-module of rank n. Show that $\bigwedge^k(M)$ has rank $\binom{n}{k}$, and $\bigwedge(M)$ has rank 2^n.
3. Let M be a free R-module of rank n and $e_1, \ldots, e_n$ be a basis of M. Let $a_1 = \sum_t r_{1t} e_t, \ldots, a_k = \sum_t r_{nt} e_t \in M$. Show that

$$a_1 \wedge \cdots \wedge a_k = \sum \det(r_{it_j}) \, e_{t_1} \wedge \cdots \wedge e_{t_k},$$

where the sum runs over all strictly increasing sequences $1 \leqslant t_1 < t_2 < \cdots < t_k \leqslant n$.

4. Let M and N be R-modules. Prove that every n-linear alternating mapping $\mu : M^n \longrightarrow N$ factors uniquely through the multiplication map $\mu_n : M^n \longrightarrow \bigwedge^n(M)$ (there is a unique module homomorphism $\varphi : \bigwedge^n(M) \longrightarrow N$ such that $\varphi \circ \mu_n = \mu$).

16.5. THE TENSOR PRODUCT OF ALGEBRAS

In this section we resume the study of algebras, with a look at their tensor products. In what follows R is commutative; all tensor products are over R.

16.5.a. Definition

Proposition 16.5.1. *When A and B are R-algebras, then $A \otimes B$ is an R-algebra in which $(a \otimes b)(a' \otimes b') = aa' \otimes bb'$.*

Proof. The mapping $(a, b, a', b') \longmapsto aa' \otimes bb'$ of $A \times B \times A \times B$ into $A \otimes B$ is multilinear, since $\otimes$ and the multiplications on A and B are bilinear. Hence there is a homomorphism $\mu : A \otimes B \otimes A \otimes B \longrightarrow A \otimes B$ such that $\mu(a \otimes b \otimes a' \otimes b') = aa' \otimes bb'$ for all $a, a' \in A$ and $b, b' \in B$. A bilinear multiplication on $A \otimes B$ is then defined by $tt' = \mu(t \otimes t')$ for all $t, t' \in A \otimes B$. In particular, $(a \otimes b)(a' \otimes b') = aa' \otimes bb'$ for all $a, a' \in A$ and $b, b' \in B$.

The bilinear multiplication on $A \otimes B$ is associative. Indeed

$$(a \otimes b)((a' \otimes b')(a'' \otimes b'')) = a(a'a'') \otimes b(b'b'')$$

$$= (aa')a'' \otimes (bb')b'' = ((a \otimes b)(a' \otimes b'))(a'' \otimes b'');$$

since every element t of $A \otimes B$ is a finite sum $t = \sum_i a_i \otimes b_i$, associativity holds in general. Also $1 \otimes 1$ is the identity element of $A \otimes B$: indeed $(1 \otimes 1)(a \otimes b) = a \otimes b = (a \otimes b)(1 \otimes 1)$ for all a, b; hence $(1 \otimes 1)t = t = t(1 \otimes 1)$ for all $t \in A \otimes B$. Thus $A \otimes B$ is an R-algebra. $\square$

The algebra $A \otimes B$ comes with a bimodule structure. Since R is central in A, A is a left A-, right R-bimodule. Similarly B is a left R-, right B-bimodule. Hence there is a left A-, right B-bimodule structure on $A \otimes B$ in which $a(a' \otimes b) = aa' \otimes b$, $(a \otimes b)b' = a \otimes bb'$.

16.5.b. Properties. The algebra $A \otimes B$ also comes with canonical homomorphisms $\iota : A \longrightarrow A \otimes B$ and $\kappa : B \longrightarrow A \otimes B$, defined by $\iota(a) = a \otimes 1$, $\kappa(b) = 1 \otimes b$, which are algebra homomorphisms but not necessarily injective and have the following universal property:

Proposition 16.5.2. *Let A and B be R-algebras and $\iota : A \longrightarrow A \otimes B$, $\kappa : B \longrightarrow A \otimes B$ be the canonical homomorphisms. Then $\iota(a)$ and $\kappa(b)$ commute for all $a \in A$ and $b \in B$. For every R-algebra C and homomorphisms $\varphi : A \longrightarrow C$, $\psi : B \longrightarrow C$ such that $\varphi(a)$ and $\psi(b)$ commute for all $a \in A$ and $b \in B$, there is a unique algebra homomorphism $\chi : A \otimes B \longrightarrow C$ such that $\chi \circ \iota = \varphi$ and $\chi \circ \kappa = \psi$:*

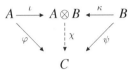

This is similar to the universal property of the direct sum of two groups.

Proof. If the diagram above commutes, then $\chi(a \otimes b) = \chi(\iota(a)\kappa(b)) = \varphi(a)\psi(b)$; therefore χ is unique.

Conversely, the mapping $(a,b) \longmapsto \varphi(a)\psi(b)$ of $A \times B$ into C is bilinear, since φ and ψ are algebra homomorphisms. Hence there is a module homomorphism $\chi : A \otimes B \longrightarrow C$ such that $\chi(a \otimes b) = \varphi(a)\psi(b)$ for all a,b. In particular, $\chi \circ \iota = \varphi$ and $\chi \circ \kappa = \psi$. Moreover χ is an algebra homomorphism, since

$$\chi((a \otimes b)(a' \otimes b')) = \varphi(aa')\psi(bb') = \varphi(a)\psi(b)\varphi(a')\psi(b') = \chi(a \otimes b)\chi(a' \otimes b')$$

and $\chi(1 \otimes 1) = \varphi(1)\psi(1) = 1$. $\square$

The reader will show:

Proposition 16.5.3. *If $\varphi : A \longrightarrow A'$ and $\psi : B \longrightarrow B'$ are algebra homomorphisms, then $\varphi \otimes \psi : A \otimes B \longrightarrow A' \otimes B'$ is an algebra homomorphism.*

Similarly the canonical module isomorphisms $R \otimes A \cong A$, $B \otimes A \cong A \otimes B$, $A \otimes (B \otimes C) \cong (A \otimes B) \otimes C$ are all algebras isomorphisms (exercises).

Proposition 16.5.4. *If A is a free R-module, with basis $(e_i)_{i \in I}$, then ι is injective and $A \otimes B$ is a free right B-module, with basis $(\iota(e_i))_{i \in I}$.*

Proof. Let A is a free R-module, with basis $(e_i)_{i \in I}$. Then $A = \bigoplus_{i \in I} Re_i$ and $A \otimes B = \bigoplus_{i \in I} Re_i \otimes B \cong \bigoplus_{i \in I} B$ (as R-modules); if $a = \sum_{i \in I} r_i e_i \in A$, then

the isomorphism $\theta : A \otimes B \longrightarrow \bigoplus_{i \in I} B$ sends $a \otimes b = (\sum_{i \in I} r_i e_i) \otimes b \in A \otimes B$
to $(r_i b)_{i \in I} \in \bigoplus_{i \in I} B$. We see that θ preserves the right action of B and is an
isomorphism of right B-modules; hence $A \otimes B$ is a free right B-module. Now
the free right B-module $\bigoplus_{i \in I} B$ has a standard basis $(f_j)_{j \in I}$, $f_j = (\delta_{ji})_{i \in I}$, where
$\delta_{ji} = 1 \in B$ if $i = j$ and $\delta_{ji} = 0$ if $i \neq j$. The B-isomorphism θ sends $\iota(e_j) = \sum_{i \in I} \delta_{ji} e_i \otimes 1$ to $(\delta_{ji})_{i \in I} = f_j$. Hence $(\iota(e_i))_{i \in I}$ is a basis of $A \otimes B$. In particular,
$\iota(\sum_{i \in I} r_i e_i) = \sum_{i \in I} r_i \iota(e_i) = 0$ implies $r_i = 0$ for all i and $\sum_{i \in I} r_i e_i = 0$, so that
ι is injective.   □

Corollary 16.5.5. *If K is a field, then for any K-algebras A and B the injections $A \longrightarrow A \otimes B$ and $B \longrightarrow A \otimes B$ are injective.*

Proposition 16.5.6. *If A and B are commutative, then $A \otimes B$ is a commutative
R-algebra and is also an A-algebra and a B-algebra.*

Proof. Commutativity is clear. Hence $\iota : A \longrightarrow A \otimes B$ and $\kappa : B \longrightarrow A \otimes B$ are central homomorphisms; they induce the given bimodule structure on
$A \otimes B$ since $(a \otimes 1)(a' \otimes b') = a(a' \otimes b')$, $(a \otimes b)(1 \otimes b') = (a \otimes b)b'$.   □

16.5.c. Bimodules. The tensor product of algebras has applications to modules and bimodules over algebras. When A is an R-algebra, every left A-module M is a left A-, right R-bimodule (since R is central in A) and the A-module
structure $A \longrightarrow \mathrm{End}_{\mathbb{Z}}(M)$ on M can be viewed as a homomorphism $A \longrightarrow \mathrm{End}_R(M)$ and is then an algebra homomorphism.

Now let A and B be R-algebras. We generally require of a left A-, right
B-bimodule M that the left action of R on M coincides with the right action
of R on M (so that M is an R-module in only one way). Then we have alge-
bra homomorphisms $\varphi : A \longrightarrow \mathrm{End}_R(M)$ and $\psi : B \longrightarrow \mathrm{End}_R^{\mathrm{op}}(M)$, equivalently $\psi : B^{\mathrm{op}} \longrightarrow \mathrm{End}_R(M)$. Moreover $\varphi(a)$ and $\psi(b)$ commute for all $a \in A$
and $b \in B$, since M is a bimodule. By Proposition 16.5.2, φ and ψ induce
an algebra homomorphism $\chi : A \otimes B^{\mathrm{op}} \longrightarrow \mathrm{End}_R(M)$ such that $\chi(a \otimes b) = \varphi(a)\psi(b)$; then M is a left $A \otimes B^{\mathrm{op}}$-module, in which $(a \otimes b)m = amb$. If, conversely, M is a left $A \otimes B^{\mathrm{op}}$-module, then the $A \otimes B^{\mathrm{op}}$-module structure $A \otimes B^{\mathrm{op}} \longrightarrow \mathrm{End}_R(M)$ induces module structures $A \longrightarrow A \otimes B^{\mathrm{op}} \longrightarrow \mathrm{End}_R(M)$
and $B^{\mathrm{op}} \longrightarrow A \otimes B^{\mathrm{op}} \longrightarrow \mathrm{End}_R(M)$, so that M is a left A-, right B-bimodule,
in which $am = (a \otimes 1)m$, $mb = (1 \otimes b)m$, and we have shown:

Proposition 16.5.7. *Let A and B be R-algebras. Every left A-, right B-bimod-
ule (with the same actions of R) is a left $A \otimes B^{\mathrm{op}}$-module, in which $(a \otimes b)x
= axb$ for all $a \in A$, $b \in B$, and $x \in M$. Conversely, every $A \otimes B^{\mathrm{op}}$-module is a
left A-, right B-bimodule.*

If for instance R and S are rings, then R and S are $\mathbb{Z}$-algebras and induce
the same action of $\mathbb{Z}$ on any module; hence a left R-, right S-bimodule (in the
usual sense) is a left $R \otimes_{\mathbb{Z}} S^{\mathrm{op}}$-module, and conversely.

Exercises

1. Let $\varphi : A \longrightarrow A'$ and $\psi : B \longrightarrow B'$ be algebra homomorphisms. Show that $\varphi \otimes \psi :$ $A \otimes B \longrightarrow A' \otimes B'$ is an algebra homomorphism.

2. Let A, B, C be R-algebras. Show that the canonical module isomorphisms $R \otimes A \cong A$, $B \otimes A \cong A \otimes B$, $A \otimes (B \otimes C) \cong (A \otimes B) \otimes C$ are algebras isomorphisms.

3. Let M and N be R-modules. Show that $S(M) \otimes S(N) \cong S(M \oplus N)$.

4. Let E and F be field extensions of a field K with a composite EF. Show that the following conditions are equivalent:

   (a) Every subset of E which is linearly independent over K is linearly independent over F.

   (b) Every subset of F which is linearly independent over K is linearly independent over E.

   (c) The homomorphism $E \otimes F \longrightarrow EF$, $x \otimes y \longmapsto xy$, is an isomorphism.

   (Such extensions are called **linearly disjoint** over K.)

16.6. SIMPLE ALGEBRAS

In this section we prove some remarkable properties of simple artinian algebras (= algebras whose underlying ring is simple artinian), and division algebras (= algebras whose underlying ring is a division ring) over a field K.

By Wedderburn's Theorem, a simple artinian K-algebra A is isomorphic, as a ring, to $M_n(D) \cong \mathrm{End}_D(S)$, where $n > 0$, S is any simple left A-module, and $D \cong \mathrm{End}_A^{\mathrm{op}}(S)$ is a division ring. In particular, the right D-module S is a K-module; hence D is a K-algebra, D is a division algebra, and the isomorphism $A \cong M_n(D)$ is an isomorphism of K-algebras.

Thus matrix rings provide examples of simple algebras. Field extensions also are simple algebras, in fact are division algebras. The quaternion algebra $\mathbf{Q}$ constructed in Section A.7 is a division algebra over $\mathbb{R}$.

16.6.a. The Center. Let A be a K-algebra. Since K is a field, the central homomorphism $K \longrightarrow A$ is injective. We identify each $x \in K$ with $x1 \in A$ so that K becomes a subalgebra of A. Then K is central in A (every $x \in K$ commutes with every $a \in A$). We denote by $[A : K]$ the dimension of A over K.

The **center** of a K-algebra A is

$$Z(A) = \{ z \in A; \ az = za \text{ for all } a \in A \}.$$

We see that $Z(A)$ is a subalgebra of A and that $K \subseteq Z(A)$.

Proposition 16.6.1. *The center of $M_n(D)$ is isomorphic to the center of D. Hence the center of a simple artinian K-algebra is a field.*

Proof. The proof is essentially an exercise. The reader will readily show (as in Proposition 3.6.5) that the center of $M_n(D)$ consists of scalar matrices (= matrices xI, where $x \in D$ and I is the identity matrix). Moreover two scalar matrices xI and yI commute if and only if the scalars x and y commute in D. Thus $x \longmapsto xI$ provides the required isomorphism $Z(D) \cong Z(M_n(D))$. The center of the division ring D is a sub division ring of D and is a field.   □

16.6.b. Central Simple Algebras. Let A be a simple artinian K-algebra. By Proposition 16.6.1, the center C of A is a field (which contains K). Then A is a C-algebra. We see that A is a simple artinian C-algebra with center C.

We saw that in general $K \subseteq Z(A)$. A K-algebra A is **central** in case $Z(A) = K$. By the above, a simple artinian K-algebra is a simple artinian central C-algebra for some field extension C of K.

We now concentrate on the central case. First we show that the tensor product of two central simple K-algebras is central simple.

Theorem 16.6.2. *Let A and B be K-algebras. If B is simple and central, then:*

(1) *every ideal of $A \otimes B$ has the form $I \otimes B$, where I is an ideal of A; if A is simple, then $A \otimes B$ is simple;*

(2) *$Z(A \otimes B) = Z(A)$; if A is central, then $A \otimes B$ is central.*

First we prove a lemma.

Lemma 16.6.3. *Let A and B be K-algebras. If B is simple and central, and $J \neq 0$ is an ideal of $A \otimes B$, then $J \cap A \neq 0$.*

Proof. By Proposition 16.5.5 we may view A and B as subalgebras of $A \otimes B$. Since $J \neq 0$ there exists $t = \sum_{1 \leq i \leq m} a_i \otimes b_i \in J \setminus \{0\}$. Choose $t \in J \setminus \{0\}$ so that m is as small as possible. Then $a_1, \ldots, a_m$ are linearly independent over K, and so are $b_1, \ldots, b_m$. In particular, $b_m \neq 0$.

We show that we can arrange $b_m = 1$. Since B is simple, we have $\sum_j x_j b_m y_j = 1$ for some $x_j, y_j \in B$. Let $c_i = \sum_j x_j b_i y_j \in B$ and

$$u = \sum_j (1 \otimes x_j) t (1 \otimes y_j) = \sum_{i,j} a_i \otimes x_j b_i y_j = \sum_{1 \leq i \leq m} a_i \otimes c_i.$$

Then $u \in J$. Moreover $u \neq 0$. Indeed $a_1, \ldots, a_m$ are linearly independent over K and are part of a basis of A over K. By Proposition 16.5.6 $a_1 \otimes 1, \ldots, a_m \otimes 1$ are part of a basis of $A \otimes B$ over B and are linearly independent over B. Hence $u = \sum_{1 \leq i \leq n} a_i \otimes c_i \neq 0$, since $c_m = 1 \neq 0$.

For any $b \in B$ we now have

$$v = (1 \otimes b)u - u(1 \otimes b)$$

$$= \sum_{1 \leq i \leq m} a_i \otimes (b c_i - c_i b) = \sum_{1 \leq i \leq m-1} a_i \otimes (b c_i - c_i b),$$

since $c_m = 1$. Since $v \in J$ this implies $v = 0$, by the choice of m. Since $a_1 \otimes 1, \ldots, a_m \otimes 1$ are linearly independent over B, it follows that $b c_i - c_i b = 0$ for all $i \leqslant m$ and all $b \in B$. Thus $c_1, \ldots, c_m \in Z(B) = K$ and

$$u = \sum_{1 \leqslant i \leqslant m} a_i \otimes c_i = \sum_{1 \leqslant i \leqslant m} a_i c_i \otimes 1 \in A. \qquad \square$$

We can now prove Theorem 16.6.2. Let J be an ideal of $A \otimes B$. Then $I = J \cap A$ is an ideal of A and $I \otimes B \subseteq J$. We show that $J = I \otimes B$. As a vector space over K, $A = I \oplus I'$ for some subspace I', so that $A \otimes B = (I \otimes B) \oplus (I' \otimes B)$ and there are vector space isomorphisms $(A \otimes B)/(I \otimes B) \cong I' \otimes B \cong (A/I) \otimes B$. Now the vector space homomorphism

$$\pi : A \otimes B \longrightarrow (A \otimes B)/(I \otimes B) \cong (A/I) \otimes B,$$

whose kernel is $I \otimes B$, is induced by the projection $A \longrightarrow A/I$ and is an algebra homomorphism. Hence $\pi(J)$ is an ideal of $(A/I) \otimes B$. Since $J \cap A = I$, we have $\pi(J) \cap (A/I) = 0$: if $t \in J$ and $\pi(t) \in A/I$, then $\pi(t) = \pi(a)$ for some $a \in A$, $t - a \in \operatorname{Ker} \pi \subseteq J$, $a = t - (t - a) \in J \cap A$, and $\pi(t) = \pi(a) = 0$. By Lemma 16.6.3, $\pi(J) = 0$, and $J = I \otimes B$. This proves (1).

If $t \in Z(A)$, then $t \in Z(A \otimes B)$, since every $a \in A$ and $b \in B$ commute in $A \otimes B$. Conversely, let $t \in Z(A \otimes B)$. Then $t = \sum_{1 \leqslant i \leqslant m} a_i \otimes b_i$ for some $a_i \in A$, $b_i \in B$. As in the proof of Lemma 16.6.2 we can choose a_i and b_i so that m is as small as possible; then $a_1, \ldots, a_m$ are linearly independent over K, and $a_1 \otimes 1, \ldots, a_m \otimes 1$ are linearly independent over B. For all $b \in B$,

$$\sum_{1 \leqslant i \leqslant m} a_i \otimes (b b_i - b_i b) = \sum_{1 \leqslant i \leqslant m} (1 \otimes b) t - t (1 \otimes b) = 0;$$

hence $b b_i - b_i b = 0$ for all $b \in B$ and $b_i \in Z(B) = K$ for all i. Then

$$t = \sum_{1 \leqslant i \leqslant m} a_i \otimes b_i = \sum_{1 \leqslant i \leqslant m} a_i b_i \otimes 1 \in A$$

and $t \in Z(A)$, which proves (2). $\qquad \square$

Corollary 16.6.4. *Let A be a simple central K-algebra. If $[A : K]$ is finite, then $[A : K]$ is a square.*

Proof. Assume that $[A : K]$ is finite. Then A is left artinian, since left ideals of A are subspaces. Hence $A \cong M_n(D)$ for some division algebra D, whose center is K by Proposition 16.6.1, and $[A : K] = n^2 [D : K]$.

Let $\overline{K}$ be the algebraic closure of K. By Theorem 16.6.2, $\overline{D} = \overline{K} \otimes D$ is a simple K-algebra whose center is $\overline{K}$. Moreover $[\overline{D} : \overline{K}] = [D : K]$ by Proposition 16.5.4. Hence $\overline{D}$ is a left artinian, simple $\overline{K}$-algebra and $\overline{D} \cong M_m(E)$ for some division $\overline{K}$-algebra E. Since $\overline{K}$ is algebraically closed and $[E : \overline{K}]$ is finite, Lemma 10.6.8 yields $E = \overline{K}$. Hence $[D : K] = [\overline{D} : \overline{K}] = m^2$ is a square. $\qquad \square$

16.6.c. The Skolem-Noether Theorem

Theorem 16.6.5 (Skolem-Noether). *Let A be a simple central K-algebra of finite dimension over K, and B be a simple K-algebra. Any two homomorphisms $\varphi, \psi : B \longrightarrow A$ are conjugate (there exists a unit u of A such that $\psi(b) = u\,\varphi(b)\,u^{-1}$ for all $b \in B$).*

Proof. First, φ and ψ are injective since B is simple. Also $A \cong \mathrm{End}_D(V)$, where V is any simple left A-module and $D = \mathrm{End}_A^{\mathrm{op}}(V)$ is a division K-algebra; then V is a faithful A-module (Proposition 12.2.3), a finite-dimensional right D-module, and a left A-, right D-bimodule. Then $B \xrightarrow{\varphi} A$ makes V a left B-, right D-bimodule and a left $B \otimes_K D^{\mathrm{op}}$-module, in which $(b \otimes d)v = bvd = \varphi(b)vd$ (Proposition 16.5.7). Similarly $B \xrightarrow{\psi} A$ yields another B-, D-bimodule structure on V, and another left $B \otimes D^{\mathrm{op}}$-module structure on V.

We now have two left $B \otimes D^{\mathrm{op}}$ modules V_1 and V_2 which have the same dimension over K. Now $R = B \otimes D^{\mathrm{op}}$ is simple by Theorem 16.6.2, and left artinian since it has finite dimension over K. Hence there is, up to isomorphism, only one simple left R-module S, and every R-module is a direct sum of copies of S. Thus $V_1 \cong S^k$ and $V_2 \cong S^\ell$; since V_1 and V_2 have the same dimension over K, this implies $k = \ell$ and $V_1 \cong V_2$ as R-modules, hence also as B-,D-bimodules.

The isomorphism $\theta : V_1 \longrightarrow V_2$ is a D-automorphism of V such that $\theta(\varphi(b)v) = \theta(bv) = b\theta(v) = \psi(b)\theta(v)$ for all $b \in B$, $v \in V$. The isomorphism $\mathrm{End}_D(V) \cong A$ sends θ to a unit u of A such that $u\,\varphi(b)v = \psi(b)uv$ for all $b \in B$, $v \in V$; since V is a faithful A-module, this implies $u\,\varphi(b) = \psi(b)u$ for all $b \in B$. $\qquad\qquad\square$

16.6.d. Centralizers.

The **centralizer** of a subset S of a K-algebra A is

$$\{\, a \in A; \ as = sa \text{ for all } s \in S \,\};$$

it is a subalgebra of A.

Lemma 16.6.6. *Let A be a simple central K-algebra of finite dimension over K, B be a simple subalgebra of A, and C be the centralizer of B. Then C is a simple subalgebra, B is the centralizer of C, and $[A : K] = [B : K]\,[C : K]$.*

Proof. As in the proof of Theorem 16.6.5, $A \cong \mathrm{End}_D(V)$, where V is any simple left A-module and $D = \mathrm{End}_A^{\mathrm{op}}(V)$ is a division K-algebra; then V is an A-, D-bimodule and a left $A \otimes D^{\mathrm{op}}$-module (Proposition 16.5.7). We may also view V as a B-, D-bimodule and as a left $B \otimes D^{\mathrm{op}}$-module. Also $R = B \otimes D^{\mathrm{op}}$ is simple and $R \cong \mathrm{End}_E(W)$, where W is any simple left R-module and $E = \mathrm{End}_R^{\mathrm{op}}(W)$ is a division K-algebra.

Let $\theta : A \cong \mathrm{End}_D(V)$ be the isomorphism. We see that $a \in A$ is in C if and only if $\theta(a)$ commutes with $\theta(b)$ for all $b \in B$, if and only if $\theta(a)$ is a B-, D-bimodule endomorphism of V. Hence θ induces an isomorphism

$C \cong \text{End}_R(V)$. Since R is simple, we have $V \cong W^r$ for some r (as R-modules) and $C \cong \text{End}_R(W^r) \cong M_r(\text{End}_R(W)) = M_r(E)$. Hence C is simple.

Since $A \cong \text{End}_D(V)$ and $C \cong M_r(E)$, we have

$$[A : K] = (\dim_D V)^2 [D : K] \qquad \text{and} \qquad [C : K] = r^2 [E : K] .$$

Similarly $R \cong \text{End}_E(W)$ yields

$$[B : K] [D : K] = [R : K] = (\dim_E W)^2 [E : K]$$

by Proposition 16.5.4. Now $\dim_K(W) = [E : K] (\dim_E W)$, and

$$(\dim_D V) [D : K] = \dim_K V = r [E : K] (\dim_E W)$$

since $V \cong W^r$. Hence

$$[B : K] [C : K] = \frac{(\dim_E W)^2 [E : K]}{[D : K]} r^2 [E : K] = (\dim_D V)^2 [D : K] = [A : K] .$$

If now C' is the centralizer of C, then $B \subseteq C'$ and $[C' : K] = [A : K] / [C : K] = [B : K]$; hence $C' = B$. $\qquad\square$

Proposition 16.6.7. *Let D be a division ring, K be the center of D, and $F \supseteq K$ be a maximal subfield of D. If D has finite dimension over K, then K is its own centralizer in D and $[D : K] = [F : K]^2$.*

Proof. D is a simple artinian central K-algebra and F is a simple subalgebra of D. Now the centralizer C of F in D contains F. In fact $C = F$: since each $c \in C$ commutes with every element of F, the subdivision ring $F(c)$ of D generated by F and c is a field, and the maximality of F yields $F(c) = F$ and $c \in F$. By Lemma 16.6.6, $[D : K] = [F : K]^2$. $\qquad\square$

16.6.e. Frobenius's Theorem. With the preceding results we can prove:

Theorem 16.6.8 (Frobenius). *A division $\mathbb{R}$-algebra which has finite dimension over $\mathbb{R}$ is isomorphic to $\mathbb{R}$, $\mathbb{C}$, or $\mathbf{Q}$.*

Proof. Let D be a division $\mathbb{R}$-algebra with center K and $F \supseteq K$ be a maximal subfield of D. Then $\mathbb{R} \subseteq K \subseteq F \subseteq D$ and $[D : K] = [F : K]^2$ by Proposition 16.6.7. Since $[F : \mathbb{R}]$ is finite, either $F = \mathbb{R}$ or (up to isomorphism) $F = \mathbb{C}$. If $F = \mathbb{R}$, then $K = \mathbb{R}$ and $D = \mathbb{R}$, since $[D : K] = [F : K]^2 = 1$. If $F = \mathbb{C}$, then either $K = \mathbb{R}$ or $K = \mathbb{C}$; if $K = \mathbb{C}$, then again $D = K$ and $D = \mathbb{C}$.

If $F = \mathbb{C}$ and $K = \mathbb{R}$, then $[D : K] = 4$. We have two algebra homomorphisms of $\mathbb{C}$ into D, the inclusion map and complex conjugation. By the Skolem-Noether Theorem, these are conjugate in D: there exists a unit $u \in D$ such that $\bar{z} = uzu^{-1}$ for all $z \in \mathbb{C}$. Then $u^2 z u^{-2} = z$ for all $z \in \mathbb{C}$ and u^2 is in the centralizer of $\mathbb{C}$; by Proposition 16.6.7, $u^2 \in \mathbb{C}$. In fact $u^2 \in \mathbb{R}$, since $\bar{u}^2 = u\,u^2 u^{-1} = u^2$, but $u \notin \mathbb{C}$, since $uiu^{-1} = \bar{i} \neq i$. Also $u^2 \geq 0$ would imply $u^2 = r^2$ for some $r \in \mathbb{R}$ and $u = \pm r \in \mathbb{C}$; hence $u^2 < 0$ and $u^2 = -r^2$ for some $r \in \mathbb{R}$, $r \neq 0$. Let $j = u/r$ and $k = ij$. Then $j \notin \mathbb{C}$, so that $\{1,j\}$ is a basis of D over $\mathbb{C}$ and $\{1,i,j,k\}$ is a basis of D over $\mathbb{R}$. Also $j^2 = -1$, and $jzj^{-1} = jrzr^{-1}j^{-1} = uzu^{-1} = \bar{z}$ for all $z \in \mathbb{C}$. Hence $ji = -ij = -k$, $k^2 = -i^2 j^2 = -1$, $ik = -j$, $ki = iji = -i^2 j = j$, $kj = -i$, and $jk = jij = -ij^2 = i$. Thus $D \cong \mathbf{Q}$. $\qquad\square$

The results in this section also yield another proof of Wedderburn's result (Theorem 10.6.7) that all finite division rings are fields (see the exercises).

Exercises

1. Let D be a division ring. Show that the center of $M_n(D)$ consists of scalar matrices.
2. Let A be a central simple K-algebra of finite dimension n over K. Show that $A \otimes A^{\mathrm{op}} \cong M_n(K)$.
3. Let A be a simple central K-algebra of finite dimension over K, B be a simple subalgebra of A, and C be the centralizer of B. Show that $B \otimes C \cong A$.
4. Show that a finite group is not the union of conjugates of any proper subgroups. (Hint: a proper subgroup H of a finite group G has at most $[G:H]$ conjugates.)
5. Use the Skolem-Noether Theorem and the previous exercise to prove that every finite division ring is a field.

16.7. LIE ALGEBRAS

Lie algebras were introduced by Sophus Lie, who showed that local problems in Lie groups (groups with a compatible analytic manifold structure) could be reduced to purely algebraic problems on this type of algebra. Our interest in Lie algebras lies in their relationship to ordinary algebras. This section defines Lie algebras and constructs their enveloping algebras. This provides an application of tensor and symmetric algebras.

16.7.a. Definition. Let R be a commutative ring [with identity element]. A **Lie algebra** over R is a nonassociative R-algebra L whose multiplication $(x,y) \longmapsto [x,y]$ satisfies

(1) $[x,x] = 0$,
(2) $[x,[y,z]] + [y,[z,x]] + [z,[x,y]] = 0$,

for all $x,y,z \in L$; (2) is the **Jacobi identity**. An identity element is not required. Since the multiplication is bilinear, it follows from (1) that

(3) $[y,x] = -[x,y]$

for all $x, y \in L$. Hence the Jacobi identity can be written

(4) $[x, [y, z]] = [y, [x, z]] - [z, [x, y]]$.

Homomorphisms, subalgebras, ideals, and quotient algebras of Lie algebras are defined as in Section 16.1.

The main algebraic examples of Lie algebras arise from the following construction. Let A be an associative R-algebra. The operation

$$[a, b] = ab - ba$$

has properties (1) and (2), as skeptical readers will verify; with this operation the R-module A becomes a Lie algebra $L(A)$.

A homomorphism $\varphi : A \longrightarrow B$ of associative algebras is also a homomorphism $\varphi : L(A) \longrightarrow L(B)$ of Lie algebras. If B is a subalgebra of A, then $L(B)$ is a subalgebra of $L(A)$. If I is an ideal of A, then I is an ideal of $L(A)$ and $L(A)/I = L(A/I)$ (see the exercises).

16.7.b. The Enveloping Algebra. The extent to which Lie algebras can be constructed as above from associative algebras is determined by the following construction. The **enveloping algebra** or **universal algebra** of a Lie algebra L over R is an associative R-algebra $U(L)$ [with an identity element], together with a homomorphism $\iota : L \longrightarrow L(U(L))$, with the following universal property: for every associative algebra A and homomorphism $\varphi : L \longrightarrow L(A)$, there is a unique homomorphism $\psi : U(L) \longrightarrow A$ such that $\varphi = \psi \circ \iota$.

The universal property implies that $U(L)$ and ι are unique up to isomorphism (if they are kind enough to exist).

Proposition 16.7.1. *Every Lie algebra has an enveloping algebra.*

Proof. Let L be a Lie algebra over R and $T = T(L)$ be the tensor algebra of the R-module L; we regard L as a submodule of T. Every Lie algebra homomorphism $\varphi : L \longrightarrow L(A)$ is a module homomorphism and extends to an algebra homomorphism $\overline{\varphi} : T \longrightarrow A$. Since $\varphi([x, y]) = [\varphi(x), \varphi(y)] = \varphi(x) \varphi(y) - \varphi(y) \varphi(x)$ for all $x, y \in L$, we have $\overline{\varphi}([x, y]) = \overline{\varphi}(x \otimes y) - \overline{\varphi}(y \otimes x)$ for all $x, y \in L$. This suggests that we retrieve $U(L)$ from T as follows.

Let J be the ideal of T generated by all

$$x \otimes y - y \otimes x - [x, y]$$

with $x, y \in L$. We show that $U = T/J$ serves as $U(L)$, with $\iota : L \longrightarrow U$ obtained by composing the inclusion homomorphism $L \longrightarrow T$ and the projection

$\pi : T \longrightarrow T/J$. First, T/J is an associative R-algebra, and ι is a homomorphism of L into $L(U)$, since

$$\iota(x)\iota(y) - \iota(y)\iota(x) - \iota([x,y]) = \pi(x \otimes y - y \otimes x - [x,y]) = 0$$

for all $x, y \in L$ by definition of J. If A is an associative algebra and $\varphi : L \longrightarrow L(A)$ is a homomorphism, then, as above, φ extends to an algebra homomorphism $\overline{\varphi} : T \longrightarrow A$ and $\overline{\varphi}([x,y]) = \overline{\varphi}(x \otimes y) - \overline{\varphi}(y \otimes x)$ for all $x, y \in L$:

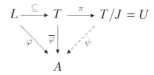

Therefore J is contained in the ideal $\operatorname{Ker}\overline{\varphi}$, and $\overline{\varphi}$ factors through the projection $\pi : T \longrightarrow U$. This provides a unique algebra homomorphism $\psi : U \longrightarrow A$ such that $\psi \circ \pi = \overline{\varphi}$. It is immediate that ψ is unique such that $\psi \circ \iota = \varphi$. □

16.7.c. The Graded Algebra of $U(L)$. The enveloping algebra can be described more precisely when L is free as an R-module. In this case we produce a basis of $U(L)$. This requires a couple of constructions.

$U = U(L) = T/J$ is not a graded algebra, but a graded algebra $G = G(L)$ can be constructed for it as follows. Let $T_n = \bigoplus_{i \leqslant n} T^i$ be the submodule of all elements of T of degree at most n. We see that $T_{-1} = 0$, $T_0 = R$, $T_{n-1} \subseteq T_n$, and $T_m T_n \subseteq T_{m+n}$ for all $m, n \geqslant 0$. Let $\pi : T \longrightarrow U$ be the projection. The submodules $U_n = \pi(T_n)$ of U satisfy $U_{-1} = 0$, $U_0 = R$, $U_{n-1} \subseteq U_n$, and $U_m U_n \subseteq U_{m+n}$ for all $m, n \geqslant 0$. Let

$$G^n = U_n/U_{n-1} \qquad \text{and} \qquad G = G(L) = \bigoplus_{n \geqslant 0} G^n .$$

The bilinear multiplication $U \times U \longrightarrow U$ on U induces bilinear mappings $U_m \times U_n \longrightarrow U_{m+n}$ and $G_m \times G_n \longrightarrow G_{m+n}$. Thus G becomes an [associative] graded algebra. The projection $\rho : U \longrightarrow G$ is an algebra homomorphism; the projection $\rho \circ \pi : T \longrightarrow G$ is a homomorphism of graded algebras ($\rho(\pi(T^n)) \subseteq G^n$).

Proposition 16.7.2. *There is a graded algebra homomorphism* $S(L) \longrightarrow G(L)$ *such that the following diagram commutes:*

$$\begin{array}{ccc} T(L) & \xrightarrow{\ \pi\ } & U(L) \\ {\scriptstyle \sigma}\downarrow & & \downarrow{\scriptstyle \rho} \\ S(L) & \longrightarrow & G(L) \end{array}$$

Proof. Note that σ and $\rho \circ \pi$ are homomorphisms of graded algebras. Now G is generated by all $\rho(\pi(x))$ with $x \in L$. When $x, y \in L$, $\pi(x \otimes y) = \pi(y \otimes x + [x,y])$ by definition of J, and $\rho(\pi(x \otimes y)) = \rho(\pi(y \otimes x))$ in G^2, since $\pi([x,y]) \in U_1$. Thus the generators of G commute, and G is commutative. Therefore $\rho(\pi(t)) = 0$ for all $t \in \operatorname{Ker} \sigma$, and $\rho \circ \pi$ factors through σ. $\qquad\square$

Theorem 16.7.3 (Poincaré-Birkhoff-Witt). *Let L be a Lie algebra over R. If L is a free R-module, then the canonical homomorphism $S(L) \longrightarrow G(L)$ is an isomorphism.*

16.7.d. Bases. The proof of Theorem 16.7.3 uses our second construction. In what follows, X is a basis of L and $S = S(L)$ is the symmetric algebra of L. We assume that X is totally ordered; by Proposition 16.3.5, S has a basis which consists of all $x_1 x_2 \ldots x_k$ with $k \geqslant 0$ and $x_1 \leqslant x_2 \leqslant \cdots \leqslant x_k$ in X.

Let P be the set of all ascending sequences $p : p_1 \leqslant p_2 \leqslant \cdots p_k$ of elements of X. For each $p \in P$ let $|p| = k$ denote the length of $p : p_1 \leqslant \cdots \leqslant p_k$, and let

$$p^* = p_1 p_2 \ldots p_k \in S.$$

Then $p \longmapsto p^*$ is a bijection of P onto the basis of S. If p is empty ($|p| = 0$), then $p^* = 1$ is the identity element of S.

Lemma 16.7.4. *There is a unique bilinear mapping $m : L \times S \longrightarrow S$, such that:*

(A) $m(x, p^*) = x p^*$ *whenever $x \leqslant p_1$ in X;*
(B) $m(x, p^*) - x p^*$ *has degree less than $|p|$ for all $x \in X$, $p \in P$;*
(C) $m(x, m(y, p^*)) - m(y, m(x, p^*)) = m([x,y], p^*)$ *for all $x, y \in X$, $p \in P$.*

Proof. Let S_n be the submodule of all $s \in S$ of degree at most n. We prove by induction on n that there exists a unique bilinear mapping $m_n : L \times S_n \longrightarrow S$ which satisfies (A) and (B) when $|p| \leqslant n$ and (C) when $|p| < n$ (then $m_n(x, p^*) \in S_n$, $m_n(x, p^*) \in S_n$ by (B), $m_n(y, p^*) \in S_n$, so that $m_n(x, m_n(y, p^*))$ and $m_n(y, m_n(x, p^*))$ are defined in (C)).

If $n = 0$, then $m_0(x, 1) = x$ by (A); this determines m_0, and then (B) holds and (C) is vacuous. Now let $n > 0$, and assume that the existence and uniqueness of m_{n-1} have been established. The restriction of m_n to $L \times S_{n-1}$ must coincide with m_{n-1}, since both satisfy (A) and (B) when $|p| \leqslant n - 1$ and (C) when $|p| < n - 1$. Consequently the bilinear mappings $m_n : L \times S_n \longrightarrow S$ will gracefully assemble into one bilinear mapping $m : L \times S \longrightarrow S$, and then m satisfies (A), (B), and (C) for all $p \in P$.

By bilinearity, m_n is determined by its values at all (x, p^*) with $x \in X$, $p \in P$, $|p| \leqslant n$. If $|p| < n$, then the above requires $m_n(x, p^*) = m_{n-1}(x, p^*)$. If $|p| = n$ and $x \leqslant p_1$, then (A) requires $m_n(x, p^*) = x p^*$.

Now assume that $|p| = n$, $x > p_1$. Let $y = p_1$ and $q \in P$ be the sequence $p_2 \leqslant \cdots \leqslant p_n$. Then $y < x$, $y \leqslant q_1$, $p^* = yq^* = m_{n-1}(y, q^*)$ by (A), $m_n(x, q^*) = m_{n-1}(x, q^*) = xq^* + s$, where $s \in S_{n-1}$ by (B), and (C) requires

$$m_n(x, p^*) = m_n(x, m_n(y, q^*))$$

$$= m_n(y, m_n(x, q^*)) + m_n([x,y], q^*)$$

$$= m_n(y, xq^*) + m_n(y, s) + m_n([x,y], q^*)$$

$$= yxq^* + m_{n-1}(y, s) + m_{n-1}([x,y], q^*),$$

since (A) requires $m_n(y, xq^*) = yxq^*$. This proves the uniqueness of m_n.

Conversely the above constructs m_n in such a way that (A) and (B) hold when $|p| \leqslant n$ and (C) holds when $|p| < n$ and $y < x$, $y \leqslant p_1$. When $|p| < n$, (C) also holds when $x < y$ and $x \leqslant p_1$, by (3), and when $x = y$, by (1); this leaves the case when $x, y > p_1$. In this case (C) is proved by computation. For clarity's sake we denote $m_n(x, s)$ by $x \bullet s$. We want to show that

$$x \bullet (y \bullet p^*) - y \bullet (x \bullet p^*) = [x, y] \bullet p^*$$

when $|p| < n$ and $x, y > p_1$. Since $\bullet$ is bilinear, we have $x \bullet (y \bullet s) - y \bullet (x \bullet s) = [x, y] \bullet s$ for all $s \in S_{n-2}$, by the induction hypothesis.

Let $z = p_1$ and $q \in P$ be the sequence $p_2 \leqslant \cdots \leqslant p_n$, so that $x, y > z$ and $z \leqslant q_1$. Then

$$y \bullet p^* = y \bullet (z \bullet q^*) = z \bullet (y \bullet q^*) + [y, z] \bullet q^*$$

by (C_{n-1}). As above $y \bullet q^* = yq^* + s$ for some $s \in S_{n-2}$, by (B). Now (C) holds for x, z, yq^*, since $z \leqslant q_1$, and for x, z, s, since $s \in S_{n-2}$; hence (C) holds for x, z, $y \bullet q^*$, and also for x, $[y, z]$, q^*, since $|q| < n - 1$, and

$$x \bullet (y \bullet p^*) = x \bullet (z \bullet (y \bullet q^*)) + x \bullet ([y, z] \bullet q^*)$$

$$= z \bullet (x \bullet (y \bullet q^*)) + [x, z] \bullet (y \bullet q^*) + [y, z] \bullet (x \bullet q^*) + [x, [y, z]] \bullet q^*.$$

Exchanging x and y yields

$$y \bullet (x \bullet p^*) = z \bullet (y \bullet (x \bullet q^*)) + [y, z] \bullet (x \bullet q^*) + [x, z] \bullet (y \bullet q^*) + [y, [x, z]] \bullet q^*.$$

Hence

$$x \bullet (y \bullet p^*) - y \bullet (x \bullet p^*)$$

$$= z \bullet (x \bullet (y \bullet q^*)) - z \bullet (y \bullet (x \bullet q^*)) + [x, [y, z]] \bullet q^* - [y, [x, z]] \bullet q^*$$

$$= z \bullet ([x, y] \bullet q^*) + [[x, y], z] \bullet q^*$$

$$= [x, y] \bullet (z \bullet q^*) + [z, [x, y]] \bullet q^* + [[x, y], z] \bullet q^*$$

$$= [x, y] \bullet (p^*),$$

by the induction hypothesis and (2). □

Corollary 16.7.5. *There is a homomorphism* $\varphi : T \longrightarrow \mathrm{End}_R(S)$ *such that*

(A') $\varphi(x)(p^*) = xp^*$ *whenever* $x \leqslant p_1$ *in* X;

(B') $\varphi(x)(p^*) - xp^*$ *has degree less than* $|p|$ *for all* $x \in X$, $p \in P$;

(B'') $\varphi(x_1 \otimes x_2 \otimes \cdots \otimes x_n)(1) - (x_1 x_2 \ldots x_n)$ *has degree less than n for all* $x_1, \ldots, x_n \in X$;

(C') $\varphi(u \otimes v) = \varphi(v \otimes u + [u,v])$ *for all* $u, v \in L$.

Hence $J \subseteq \mathrm{Ker}\,\varphi$.

Proof. The bilinear mapping $m : L \times S \longrightarrow S$ provides a module homomorphism $u \longmapsto (s \longmapsto m(u,s))$ of L into $\mathrm{End}_R(S)$, which extends to an algebra homomorphism $\varphi : T \longrightarrow \mathrm{End}_R(S)$. Then $\varphi(u)$ is the endomorphism $s \longmapsto m(u,s)$ of S and $\varphi(u)(s) = m(u,s)$ for all $u \in L$, $s \in S$. Hence (A') and (B') follow from (A) and (B); (B'') and (C') are exercises for the eager reader. (C') shows that the ideal $\mathrm{Ker}\,\varphi$ of T contains every generator of J. □

We can now prove Theorem 16.7.3, that $\tau : S \longrightarrow G$ is injective when L is free. Assume that $t \in T^n$, $\sigma(t) \in \mathrm{Ker}\,\tau$. Then $\pi(t) \in U_n$, but $\rho(\pi(t)) = \tau(\sigma(t)) = 0$. Hence $\pi(t) \in U_{n-1}$ and $\pi(t) = \pi(t')$ for some $t' \in T_{n-1}$. Then $t - t' \in J$. In T, $t - t'$ is a finite linear combination $\sum_p r_p x_1 \otimes x_2 \otimes \cdots \otimes x_k$, where each p is a sequence $x_1, \ldots, x_k$ of k elements of X with $k \leqslant n$. By (B''), $\varphi(t - t')(1) - \sum_p r_p x_1 x_2 \ldots x_k$ has degree less than n in S. Since $\varphi(t - t') = 0$, $\sigma(t - t') = \sum_p r_p x_1 x_2 \ldots x_k$ has degree less than n. But $\sigma(t)$ is homogeneous of degree n. Hence $\sigma(t) = 0$. □

Corollary 16.7.6. *Let L be a Lie algebra over R. If L is a free R-module, then $U(L)$ is a free R-module; if X is a totally ordered basis of L, then $U(L)$ has a basis which consists of all products $x_1 x_2 \ldots x_n \in U(L)$ in which $n \geqslant 0$ and $x_1 \leqslant x_2 \leqslant \cdots \leqslant x_n$ in X.*

Proof. For each $n \geqslant 0$ let B_n be the set of all products $x_1 x_2 \ldots x_n \in U_n$ in which $x_1 \leqslant x_2 \leqslant \cdots \leqslant x_n$ in X. Since $G \cong S$, $G_n \cong S_n$ has a basis which consists of all similar products $x_1 x_2 \ldots x_n \in G_n$ in which $x_1 \leqslant x_2 \leqslant \cdots \leqslant x_n$ in X. Now $\rho : U_n \longrightarrow G_n = U_n/U_{n-1}$ takes $x_1 x_2 \ldots x_n \in G_n$ to $x_1 x_2 \ldots x_n \in U_n$. Therefore B_n is linearly independent in U_n and generates a submodule V_n of U_n such that $U_n = U_{n-1} \oplus V_n$. It follows that $U_n = \bigoplus_{i \leqslant n} V_i$ and $U = \bigoplus_{n \geqslant 0} V_n$. Hence $\bigcup_{n \geqslant 0} B_n$ is a basis of U over R. □

Exercises

1. Let A be an associative algebra. Verify that $L(A)$ is a Lie algebra.

2. Let $A = M_n(R)$ and $E_{ij} \in M_n(R)$ be the matrix whose (i,j) entry is 1 and all other entries are 0. Verify that

$$[E_{ij}, E_{k\ell}] = \begin{cases} 0 & \text{if } k \neq j \text{ and } \ell \neq i, \\ E_{i\ell} & \text{if } k = j \text{ and } \ell \neq i, \\ -E_{kj} & \text{if } k \neq j \text{ and } \ell = i, \\ E_{ii} - E_{jj} & \text{if } k = j \text{ and } \ell = i. \end{cases}$$

3. Let A be an associative algebra and I be an ideal of A. Verify that I is an ideal of $L(A)$ and that $L(A)/I = L(A/I)$.

4. Verify that the enveloping algebra of a Lie algebra is unique up to isomorphism.

5. Let L be a Lie algebra and $\varphi : T(L) \longrightarrow \text{End}_R(S(L))$ be the algebra homomorphism in Corollary 16.7.5. Show that $\varphi(x_1 \otimes x_2 \otimes \cdots \otimes x_n)(1) - (x_1 x_2 \ldots x_n)$ has degree less than n for all $x_1, \ldots, x_n \in X$.

6. Let $\varphi : T(L) \longrightarrow \text{End}_R(S(L))$ be the algebra homomorphism in Corollary 16.7.5. Show that $\varphi(u \otimes v) = \varphi(v \otimes u + [u,v])$ for all $u, v \in L$.

17

CATEGORIES

Category theory was introduced by Eilenberg and MacLane in 1945 during the development of homological algebra. It unifies concepts from many parts of mathematics and is essential to conceptual understanding of algebra. It also gives quick access to a number of useful properties. This chapter is a short introduction to the subject, including functors, limits, abelian categories, adjoint functors, and tripleability.

Required results: this chapter draws examples from many of the previous chapters, particularly Chapters 2, 15, and 16. Free products with amalgamation (Section 2.8) are used in the proof of Proposition 17.1.3.

17.1. CATEGORIES

A characteristic feature of abstract algebra it that it ignores what groups, rings, modules, etc., are made of and studies only how their elements relate to each other (via operations, subgroups, etc.).

Category theory represents one more step in abstraction: it ignores elements and studies only how groups, rings, modules, etc., relate to each other. The fruitfulness of this approach is suggested by previous chapters.

17.1.a. Definition. Categories are defined so that all sets and mappings constitute a category, all groups and homomorphisms of groups constitute a category, all modules and homomorphisms of modules constitute a category, etc. This requires collections called **classes** that may be too large to be sets, and collections called **proper** classes which are too large to be sets (e.g., the class of all sets). Proper classes readily lead to paradoxes (to the extent that category theory is affectionately known as "abstract nonsense") and require careful handling. One must avoid unsafe practices, such as applying the Axiom of Choice to a proper class, or allowing proper classes to be elements.

Formally a **category** $\mathscr{C}$ consists of:

a class whose elements are the **objects** of $\mathscr{C}$;

a class whose elements are the **morphisms** or **arrows** of $\mathscr{C}$;

two functions which assign to each morphism of $\mathscr{C}$ a **domain** and a **codomain**, which are objects of $\mathscr{C}$; and

a partial operation which assigns to certain pairs (α, β) of morphisms of $\mathscr{C}$ their **composition** or **product** $\alpha\beta$ which is a morphism of $\mathscr{C}$,

such that the following axioms hold:

(1) $\alpha\beta$ is defined if and only if the domain of α is the codomain of β; then the domain of $\alpha\beta$ is the domain of β and the codomain of $\alpha\beta$ is the codomain of α. (We write morphisms "on the left.")

(2) For each object A of $\mathscr{C}$, there exists an **identity morphism** 1_A whose domain and codomain are A such that $\alpha 1_A = \alpha$ whenever α has domain A and $1_A\beta = \beta$ whenever β has codomain A.

(3) If $\alpha\beta$ and $\beta\gamma$ are defined, then $\alpha(\beta\gamma) = (\alpha\beta)\gamma$ (note that $\alpha(\beta\gamma)$ and $(\alpha\beta)\gamma$ are defined, by (1)).

Morphisms are denoted by arrows; $\alpha : A \longrightarrow B$ or $A \xrightarrow{\alpha} B$ means that the domain of α is A and its codomain is B; then we say that α is **from** A **to** B. Thus $\alpha\beta$ is defined if and only if $\bullet \xrightarrow{\beta} \bullet \xrightarrow{\alpha} \bullet$

In axiom (2), the identity morphism 1_A is unique for each object A. Consequently it is possible to define categories using morphisms only and no objects. The reader can discover precise axioms as an exercise.

17.1.b. Examples. Categories are defined so that sets and mappings are the objects and morphisms of a category. This requires a small technical proviso. In a category a morphism has only one domain and one codomain. Hence a mapping $\varphi : A \longrightarrow B$ must be distinguished from its compositions $A \xrightarrow{\varphi} B \xrightarrow{\subset} C$ with strict inclusions, which consist of the same ordered pairs but have different codomains. For instance, one may define a mapping $\varphi : A \longrightarrow B$ as an ordered triple consisting of the usual set of ordered pairs, the set A, and the set B. Then sets and mappings are the objects and morphisms of a category *Sets*; the identity morphism on a set is the usual identity mapping; composition in *Sets* is the usual composition of mappings.

With the same proviso, groups and homomorphisms are the objects and morphisms of a category *Grps*; abelian groups and homomorphisms are the objects and morphisms of a category *Abs*; rings [with an identity element] and ring homomorphisms are the objects and morphisms of a category *Rings*; left R-modules and module homomorphisms are the objects and morphisms of a category *R-Mods* (also denoted by $_R Mods$); R-algebras and algebra homomorphisms are the objects and morphisms of a category *R-Algs*; and so on.

A category is **small** in case its class of objects is a set and its class of morphisms is a set (the latter is sufficient, since there is a one-to-one correspondence between objects and identity morphisms). The next examples are of small categories.

Recall that a preordered set is a set with a reflexive and transitive relation $\leqslant$; partially ordered sets are examples. Every preordered set I can be regarded as a category. The objects of I are its elements. The morphisms of I are all ordered pairs (i,j) of elements of I such that $i \leqslant j$ (thus the set of morphisms of I is its preorder relation). Domain and codomain are given by $(i,j) : i \longrightarrow j$ so that "arrows point upward," composition is given by $(j,k)(i,j) = (i,k)$.

17.1.c. Graphs. A graph is like a small category without composition. Formally a [small] **directed graph** $\mathscr{G}$ consists of:

a set whose elements are the **vertices** or **nodes** of $\mathscr{G}$;

a set whose elements are the **edges** or **arrows** of $\mathscr{G}$; and

two mappings which assign to each edge of $\mathscr{G}$ an **origin** and a **destination**, which are vertices of $\mathscr{G}$.

Edges are denoted by arrows; $a : i \longrightarrow j$ or $i \stackrel{a}{\longrightarrow} j$ mean that the origin of a is i and its destination is j; then we say that a is **from** i **to** j.

Graphs can be viewed as abstract diagrams, as with the **square** and **triangle** graphs:

Every small category is a graph. Conversely, every graph generates a small category, as follows. In a directed graph $\mathscr{G}$, a nonempty **path** from a vertex i to a vertex j is a sequence $(i,a_1,\dots,a_n,j)$

$$i \stackrel{a_1}{\longrightarrow} \bullet \stackrel{a_2}{\longrightarrow} \bullet \cdots \bullet \stackrel{a_n}{\longrightarrow} j$$

in which $n > 0$, i is the origin of a_1, the destination of each a_i is the origin of a_{i+1}, and the destination of a_n is j. For each vertex i of $\mathscr{G}$ the **empty path** from i to i is the sequence (i,i). Paths are composed by concatenation

$$(j,b_1,\dots,b_n,k)(i,a_1,\dots,a_m,j) = (i,a_1,\dots,a_m,b_1,\dots,b_n,k).$$

Proposition 17.1.1. *When $\mathscr{G}$ is a directed graph, the vertices and paths of $\mathscr{G}$ are the objects and morphisms of a small category $\widehat{\mathscr{G}}$.*

The proof is an exercise. $\widehat{\mathscr{G}}$ is the **free category** on $\mathscr{G}$, or the **category of paths** of $\mathscr{G}$; its universal property will be seen in Section 17.3.

17.1.d. Monomorphisms and Epimorphisms. In any category, composition properties of injective mappings can be used to define abstract analogues of injections, and similarly for surjections and bijections.

In a category, a **monomorphism** is a morphism μ such that $\mu\alpha = \mu\beta$ implies $\alpha = \beta$ (injective mappings have this property); an **epimorphism** is a morphism σ such that $\alpha\sigma = \beta\sigma$ implies $\alpha = \beta$ (surjective mappings have this property). An **isomorphism** is a morphism $\alpha : A \longrightarrow B$ which has an **inverse** $\beta : B \longrightarrow A$, such that $\alpha\beta = 1_B$ and $\beta\alpha = 1_A$.

In many categories, monomorphisms coincide with injective morphisms, and the two terms are used interchangeably.

Proposition 17.1.2. *In Grps a morphism is a monomorphism if and only if it is injective.*

Proof. Let $\mu : A \longrightarrow B$ be a monomorphism of groups. Assume that $\mu(x) = \mu(y)$. Since $\mathbb{Z}$ is free on $\{1\}$, there exist homomorphisms $\alpha : \mathbb{Z} \longrightarrow A$ and $\beta : \mathbb{Z} \longrightarrow A$ such that $\alpha(1) = x$ and $\beta(1) = y$. Hence $\mu(\alpha(1)) = \mu(\beta(1))$, $\mu \circ \alpha = \mu \circ \beta$, $\alpha = \beta$, and $x = y$. Thus μ is injective. The converse is clear. $\square$

The reader will establish similar results for *Sets*, *Rings*, *R-Mods*, and so on. Identification of epimorphisms with surjective morphisms is less successful.

Proposition 17.1.3. *In Grps a morphism is an epimorphism if and only if it is surjective.*

Proof. Surjective homomorphisms are epimorphisms. Conversely, let $\varphi : G \longrightarrow H$ be a homomorphism of groups which is not surjective. To prove that φ is not an epimorphism, we construct isomorphic copies H_1 and H_2 of H which contain $\operatorname{Im} \varphi$ as a subgroup and satisfy $H_1 \cap H_2 = \operatorname{Im} \varphi$. Then we embed the group amalgam $H_1 \cup H_2$ into its free product with amalgamation P (from Section 2.8). Composing each isomorphism $H \cong H_i$ and inclusion $H_i \subseteq P$ yields homomorphisms $\alpha_1, \alpha_2 : H \longrightarrow P$. We see that $\alpha_1 \circ \varphi = \alpha_2 \circ \varphi$ but $\alpha_1 \neq \alpha_2$. $\square$

A similar result is true for *Sets* and *R-Mods* but not for *Rings* or *R-Algs* (see the exercises). In such categories, using "epimorphism" to mean "surjective homomorphism" can be confusing.

17.1.e. Duality. The **dual** or **opposite** of a category $\mathscr{C}$ is the category $\mathscr{C}^{\mathrm{op}}$ constructed as follows. The objects and morphisms of $\mathscr{C}^{\mathrm{op}}$ are those of $\mathscr{C}$. However, the domain and codomain functions of $\mathscr{C}^{\mathrm{op}}$ are the codomain and domain functions of $\mathscr{C}$: if $A \xrightarrow{\alpha} B$ in $\mathscr{C}$, then $A \xleftarrow{\alpha} B$ in $\mathscr{C}^{\mathrm{op}}$. The composition $\alpha * \beta$ of α and β in $\mathscr{C}^{\mathrm{op}}$ is the composition $\beta\alpha$ of β and α in $\mathscr{C}$. Thus $\mathscr{C}^{\mathrm{op}}$ is constructed from $\mathscr{C}$ by reversing all arrows and all compositions. We see that $\mathscr{C}^{\mathrm{op}}$ is a category.

For example, *Sets*^{op} is the category of sets and mappings, but with mappings written on the right. If I is a preordered set, viewed as a category, then I^{op} is [the category arising from] the opposite preordered set, in which $i \leqslant j$ if and only if $i \geqslant j$ in I. If $\mathscr{G}^{\mathrm{op}}$ is the opposite graph of $\mathscr{G}$, then $(\widehat{\mathscr{G}})^{\mathrm{op}} = \widehat{\mathscr{G}^{\mathrm{op}}}$.

Since $\mathscr{C}^{\text{op}}$ is a category we have the following result:

Metatheorem 17.1.4. *A theorem which applies to all categories remains true if all arrows and compositions are reversed.*

A **metatheorem** states that certain unspecified results must be true. We illustrate Metatheorem 17.1.4 with the following result, whose proof is an exercise.

Proposition 17.1.5. *Let α and β be morphisms such that $\alpha\beta$ is defined. If α and β are monomorphisms, then $\alpha\beta$ is a monomorphism. If $\alpha\beta$ is a monomorphism, then β is a monomorphism.*

Reversing arrows and compositions transforms monomorphisms into epimorphisms. By Metatheorem 17.1.4, the following result holds, and requires no proof:

Proposition 17.1.6. *Let α and β be morphisms such that $\alpha\beta$ is defined. If α and β are epimorphisms, then $\alpha\beta$ is an epimorphism. If $\alpha\beta$ is an epimorphism, then α is an epimorphism.*

Definitions and results that obtain from each other by reversing arrows and compositions are **dual** of each other. For example, monomorphism and epimorphism are dual concepts; Propositions 17.1.5 and 17.1.6 are dual results. A definition or result is **self-dual** in case it remains unchanged when arrows and compositions are reversed. For example, isomorphism is a self-dual concept.

Careful readers will avoid applying Metatheorem 17.1.4 to *specific* categories. If Theorem T is true in one category $\mathscr{C}$, then the dual theorem T^{op} is true in $\mathscr{C}^{\text{op}}$, but it does not follow that T^{op} is true in $\mathscr{C}$: indeed T^{op} holds in $\mathscr{C}$ only if the hypothesis of T is broad enough to fit $\mathscr{C}^{\text{op}}$ as well as $\mathscr{C}$, so that T holds in $\mathscr{C}^{\text{op}}$.

Exercises

1. Find a set of axioms for categories using only morphisms. (Characterize identity morphisms as morphisms ε such that $\alpha\varepsilon = \alpha$ and $\varepsilon\beta = \beta$ whenever defined; these replace objects.)
2. Verify that every preordered set can be regarded as a category.
3. Show that a small category $\mathscr{C}$ is [the category which arises from] a preordered set if and only if for any objects i, j of $\mathscr{C}$ there is at most one morphism from i to j.
4. Show that a mapping is a monomorphism in *Sets* if and only if it is injective.
5. Show that a homomorphism of rings is a monomorphism in *Rings* if and only if it is injective.
6. Show that a homomorphism of left R-modules mapping is a monomorphism in *R-Mods* if and only if it is injective.
7. Show that a mapping is an epimorphism in *Sets* if and only if it is surjective.

8. Show that a homomorphism of left R-modules mapping is an epimorphism in *R-Mods* if and only if it is surjective.

9. Show that the inclusion homomorphism $\mathbb{Z} \longrightarrow \mathbb{Q}$ is an epimorphism in *Rings*.

10. Let I be a preordered set, regarded as a category. What are the monomorphisms of I? its epimorphisms? its isomorphisms?

11. Assume that $\alpha\beta$ is defined. Prove the following. If α and β are monomorphisms, then $\alpha\beta$ is a monomorphism. If $\alpha\beta$ is a monomorphism, then β is a monomorphism.

12. Assume that $\alpha\beta$ is defined. Give a direct proof of the following. If α and β are epimorphisms, then $\alpha\beta$ is an epimorphism. If $\alpha\beta$ is an epimorphism, then β is an epimorphism.

13. Let $\mathscr{G}$ be a directed graph. Show that $\widehat{\mathscr{G}^{\mathrm{op}}} = (\widehat{\mathscr{G}})^{\mathrm{op}}$ (where $\mathscr{G}^{\mathrm{op}}$ is the opposite graph of $\mathscr{G}$).

17.2. FUNCTORS

Functors are to categories what homomorphisms are to groups, rings, and the like.

17.2.a. Definition. A **functor** F, also called a **covariant functor**, from a category $\mathscr{A}$ to a category $\mathscr{B}$ assigns to each object A of $\mathscr{A}$ an object $F(A)$ of $\mathscr{B}$, and to each morphism α of $\mathscr{A}$ a morphism $F(\alpha)$ of $\mathscr{B}$, so that domains, codomains, identity morphisms, and composition are preserved; in detail:

(1) if $\alpha : A \longrightarrow B$, then $F(\alpha) : F(A) \longrightarrow F(B)$;

(2) $F(1_A) = 1_{F(A)}$;

(3) $F(\alpha\beta) = F(\alpha)F(\beta)$ whenever $\alpha\beta$ is defined.

This definition is self-dual: a functor from $\mathscr{A}$ to $\mathscr{B}$ is also a functor from $\mathscr{A}^{\mathrm{op}}$ to $\mathscr{B}^{\mathrm{op}}$.

17.2.b. Examples. Our first examples of functors were seen in Chapters 14 and 15. For every left R-module A, $\mathrm{Hom}_R(A, -)$ is a functor from *R-Mods* to the category *Abs* of abelian groups and homomorphisms. If A is an R-S-bimodule, there is also a functor $\mathrm{Hom}_R(A, -)$ from the category *R-T-Bimods* (also denoted by $_R Mods_T$) to the similar category *S-T-Bimods*.

For every right R-module A, $A \otimes_R -$ is a functor from *R-Mods* to *Abs*. If A is a Q-R-bimodule there is also a functor $A \otimes_R -$ from *R-T-Bimods* to *Q-T-Bimods*. There are similar functors $- \otimes_R B$.

The **forgetful** functor from *Grps* to *Sets* assigns to each group G its underlying set, which is also denoted by G, and to each homomorphism of groups φ the mapping φ. There are similar forgetful functors from *Rings*, *R-Mods*, *R-Algs* to *Sets*; from *Rings* and *R-Mods* to *Abs*; and so on.

The **free group** functor from *Sets* to *Grps* assigns to each set X the free group $F(X) = F_X$ on X constructed in Section 2.5, which comes with an injection $\iota_X : X \longrightarrow F_X$. When $\varphi : X \longrightarrow Y$ is a mapping, there is a unique homomorphism $F(\varphi) : F_X \longrightarrow F_Y$ such that $F(\varphi) \circ \iota_X = \iota_Y \circ \varphi$; this defines F on morphisms. Curious readers will verify that F is indeed a functor.

The definitions also show that a direct system is a functor from a preordered set which is directed upward; an inverse system is a functor from a preordered set which is directed downward.

For every objects A and B of a category $\mathscr{C}$, $\mathrm{Hom}_\mathscr{C}(A,B)$ denotes the class of all morphisms from A to B. (There are other notations, such as $\mathscr{C}(A,B)$.) A category $\mathscr{C}$ has **small Hom sets** in case $\mathrm{Hom}_\mathscr{C}(A,B)$ is a set for all A,B. For example, *Sets*, *Grps*, *Rings*, and so on, have small Hom sets; small categories have small Hom sets.

When $\mathscr{C}$ has small Hom sets, there is for each object A of $\mathscr{C}$ a functor $\mathrm{Hom}_\mathscr{C}(A,-)$ from $\mathscr{C}$ to *Sets*. ($\mathscr{C}$ is assumed to have small Hom sets because we may not arrange all classes and functions into a category.) The functor $\mathrm{Hom}_\mathscr{C}(A,-)$ assigns to each object B of $\mathscr{C}$ the set $\mathrm{Hom}_\mathscr{C}(A,B)$, and to each morphism $\alpha : B \longrightarrow C$ the mapping $\alpha_* = \mathrm{Hom}_\mathscr{C}(A,\alpha) : \mathrm{Hom}_\mathscr{C}(A,B) \longrightarrow \mathrm{Hom}_\mathscr{C}(A,C)$ defined by $\alpha_*(\beta) = \alpha\beta$:

17.2.c. Composition. Let $\mathscr{A}, \mathscr{B}, \mathscr{C}$ be categories. If F is a functor from $\mathscr{A}$ to $\mathscr{B}$, and G is a functor from $\mathscr{B}$ to $\mathscr{C}$, then $G \circ F$ is a functor from $\mathscr{A}$ to $\mathscr{C}$. This composition is associative. Moreover there is for each category $\mathscr{C}$ a "do nothing" **identity functor** $1_\mathscr{C}$ which is the identity on objects and morphisms of $\mathscr{C}$.

It now looks like categories and functors are the objects and morphisms of a category. Unfortunately, we may not collect proper classes into a class of objects, let alone collect all categories and functors into a category. However, this restriction does not apply to *small* categories and we obtain:

Proposition 17.2.1. *Small categories and their functors are the objects and morphisms of a category Cats.*

Two categories $\mathscr{A}$ and $\mathscr{B}$ are **isomorphic** in case there exist functors $F : \mathscr{A} \longrightarrow \mathscr{B}$ and $G : \mathscr{B} \longrightarrow \mathscr{A}$ such that $F \circ G = 1_\mathscr{B}$ and $G \circ F = 1_\mathscr{A}$. Such categories are essentially identical. A less restrictive concept of equivalence is defined below.

17.2.d. Contravariant Functors. For functors like $\mathrm{Hom}_R(-,A)$ which reverse arrows and compositions, we use dual categories. In general, a **contravariant functor** from a category $\mathscr{A}$ to a category $\mathscr{B}$ is a [covariant] functor from $\mathscr{A}^{\mathrm{op}}$ to $\mathscr{B}$. When restated in terms of $\mathscr{A}$, this definition states that a contravariant functor F from $\mathscr{A}$ to $\mathscr{B}$ assigns to each object A of $\mathscr{A}$ an object $F(A)$ of $\mathscr{B}$, and to each morphism α of $\mathscr{A}$ a morphism $F(\alpha)$ of $\mathscr{B}$ so that:

(1) if $\alpha : A \longrightarrow B$, then $F(\alpha) : F(B) \longrightarrow F(A)$;
(2) $F(1_A) = 1_{F(A)}$;
(3) $F(\alpha\beta) = F(\beta)F(\alpha)$ whenever $\alpha\beta$ is defined.

Equivalently, F is a functor from $\mathscr{A}$ to $\mathscr{B}^{\mathrm{op}}$.

If $\mathscr{C}$ has small Hom sets, there is for each object A of $\mathscr{C}$ a contravariant functor $\mathrm{Hom}_{\mathscr{C}}(-,A)$ from $\mathscr{C}$ to *Sets*. The functor $\mathrm{Hom}_{\mathscr{C}}(-,A)$ assigns to each object B of $\mathscr{C}$ the set $\mathrm{Hom}_{\mathscr{C}}(B,A)$, and to each morphism $\alpha : B \longrightarrow C$ of $\mathscr{C}$ the mapping $\alpha^* = \mathrm{Hom}_{\mathscr{C}}(\alpha,A) : \mathrm{Hom}_{\mathscr{C}}(C,A) \longrightarrow \mathrm{Hom}_{\mathscr{C}}(B,A)$ defined by $\alpha^*(\beta) = \beta\alpha$:

17.2.e. Natural Transformations. Let $\mathscr{A},\mathscr{B}$ be categories and F,G be functors from $\mathscr{A}$ to $\mathscr{B}$. A **natural transformation** $\tau : F \longrightarrow G$ from F to G assigns to each object A of $\mathscr{A}$ a morphism $\tau_A : F(A) \longrightarrow G(A)$ so that for each morphism $\alpha : A \longrightarrow B$ the following square commutes:

$$
\begin{array}{ccc}
F(A) & \xrightarrow{\ \tau_A\ } & G(A) \\
{\scriptstyle F(\alpha)}\downarrow & & \downarrow{\scriptstyle G(\alpha)} \\
F(B) & \xrightarrow[\ \tau_B\ ]{} & G(B)
\end{array}
$$

We also say that the morphism τ_A (which depends on A) is **natural in** A. If each morphism τ_A is an isomorphism, then the inverse isomorphism τ_A^{-1} is also natural in A; then τ is a **natural isomorphism**.

Examples of natural transformations were seen in Chapters 14 and 15. We note a similar example:

Lemma 17.2.2. *Let $\mathscr{C}$ have small Hom sets. For every morphisms $\alpha : A \longrightarrow B$ and $\gamma : C \longrightarrow D$ of $\mathscr{C}$, $\mathrm{Hom}_{\mathscr{C}}(\alpha,-) : \mathrm{Hom}_{\mathscr{C}}(B,-) \longrightarrow \mathrm{Hom}_{\mathscr{C}}(C,-)$ and $\mathrm{Hom}_{\mathscr{C}}(-,\gamma) : \mathrm{Hom}_{\mathscr{C}}(-,C) \longrightarrow \mathrm{Hom}_{\mathscr{C}}(-,D)$ are natural transformations.*

$$
\begin{array}{ccc}
\mathrm{Hom}_{\mathscr{C}}(A,C) & \xrightarrow{\ \mathrm{Hom}_{\mathscr{C}}(A,\gamma)\ } & \mathrm{Hom}_{\mathscr{C}}(A,D) \\
{\scriptstyle \mathrm{Hom}_{\mathscr{C}}(\alpha,C)}\uparrow & & \uparrow{\scriptstyle \mathrm{Hom}_{\mathscr{C}}(\alpha,D)} \\
\mathrm{Hom}_{\mathscr{C}}(B,C) & \xrightarrow[\ \mathrm{Hom}_{\mathscr{C}}(B,\gamma)\ ]{} & \mathrm{Hom}_{\mathscr{C}}(B,D)
\end{array}
$$

Proof. For every $\varphi \in \mathrm{Hom}_{\mathscr{C}}(B,C)$, $\gamma_*(\alpha^*(\varphi)) = \gamma\varphi\alpha = \alpha^*(\gamma_*(\varphi))$. $\qquad\square$

Direct systems in a category $\mathscr{C}$ over a given directed preordered set I provide another example. We saw that a direct system over I is a functor from I to $\mathscr{C}$; then homomorphisms of direct systems are natural transformations.

In general, two categories $\mathscr{A}$ and $\mathscr{B}$ are **equivalent** in case there exist functors $F : \mathscr{A} \longrightarrow \mathscr{B}$ and $G : \mathscr{B} \longrightarrow \mathscr{A}$ and natural isomorphisms $F \circ G \cong 1_{\mathscr{B}}$ and $G \circ F \cong 1_{\mathscr{A}}$; the pair (F, G) is an **equivalence** of categories. This concept is explored in the exercises.

Natural transformations compose: if $\sigma : F \longrightarrow G$ and $\tau : G \longrightarrow H$ are natural transformations, then so is $\tau\sigma : F \longrightarrow H$, which is defined by $(\tau\sigma)_A = \tau_A \sigma_A$. This composition is associative. Moreover there is for each functor F an identity natural transformation $1_F : F \longrightarrow F$, which is defined by $(1_F)_A = 1_{F(A)}$. For example, $\tau : F \longrightarrow G$ is a natural isomorphism if and only if there exists a natural transformation $\sigma : G \longrightarrow F$ such that $\sigma\tau = 1_F$ and $\tau\sigma = 1_G$ (then $\sigma_A = \tau_A^{-1}$ for all A).

These results suggest that functors from $\mathscr{A}$ to $\mathscr{B}$ are the objects of a category, but, as before, this is a forbidden construction unless $\mathscr{A}$ is small (so that functors from $\mathscr{A}$ to $\mathscr{B}$ and their natural transformations all are sets).

Proposition 17.2.3. *Let $\mathscr{B}$ be a category and $\mathscr{A}$ be a small category. The functors from $\mathscr{A}$ to $\mathscr{B}$ and their natural transformations are the objects and morphisms of a category* Func$(\mathscr{A}, \mathscr{B})$.

Func$(\mathscr{A}, \mathscr{B})$ is often denoted by $\mathscr{B}^{\mathscr{A}}$ or other notations and is a **functor category**. If, for example, I is a directed preordered set, then the objects of Func$(I, \mathscr{C})$ are direct systems in $\mathscr{C}$ over I, and its morphisms are homomorphisms of direct systems over I. Similar examples will be seen in the next section.

One can also compose natural transformations with functors. Let $\tau : F \longrightarrow G$ be a natural transformation, where $F, G : \mathscr{A} \longrightarrow \mathscr{B}$, and let $H : \mathscr{B} \longrightarrow \mathscr{C}$ be a functor. Then $H \circ \tau : H \circ F \longrightarrow H \circ G$ is a natural transformation: $H \circ \tau$ assigns $H(\tau_A)$ to each $A \in \mathscr{A}$ and the square

$$
\begin{array}{ccc}
H(F(A)) & \xrightarrow{H(\tau_A)} & H(G(A)) \\
{\scriptstyle H(F(\alpha))}\downarrow & & \downarrow{\scriptstyle H(G(\alpha))} \\
H(F(B)) & \xrightarrow[H(\tau_B)]{} & H(G(B))
\end{array}
$$

commutes for each $\alpha : A \longrightarrow B$, since τ is natural and H is a functor.

Similarly let $\tau : F \longrightarrow G$ be natural, where $F, G : \mathscr{A} \longrightarrow \mathscr{B}$, and $K : \mathscr{C} \longrightarrow \mathscr{A}$ be a functor. Then $\tau \circ K : F \circ K \longrightarrow G \circ K$ is a natural transformation: $\tau \circ K$ assigns $\tau_{K(C)}$ to each $C \in \mathscr{C}$ and is natural, since τ is natural.

17.2.f. Bifunctors. When $\mathscr{A}$ and $\mathscr{B}$ are categories, the cartesian **product** of $\mathscr{A}$ and $\mathscr{B}$ is the category $\mathscr{A} \times \mathscr{B}$ constructed as follows: an object of $\mathscr{A} \times \mathscr{B}$ is an ordered pair (A, B) of an object A of $\mathscr{A}$ and an object B of $\mathscr{B}$. A morphism of $\mathscr{A} \times \mathscr{B}$ is an ordered pair (α, β) of a morphism α of $\mathscr{A}$ and a morphism β of $\mathscr{B}$. Domains, codomains, and composition in $\mathscr{A} \times \mathscr{B}$ are defined componentwise: if $\alpha : A \longrightarrow A'$ in $\mathscr{A}$ and $\beta : B \longrightarrow B'$ in $\mathscr{B}$, then $(\alpha, \beta) : (A, B) \longrightarrow (A', B')$ in $\mathscr{A} \times \mathscr{B}$; $(\alpha, \beta)(\alpha', \beta')$ is defined in $\mathscr{A} \times \mathscr{B}$ if and only if $\alpha\alpha'$ is defined in $\mathscr{A}$ and $\beta\beta'$ is defined in $\mathscr{B}$, and then $(\alpha, \beta)(\alpha', \beta') = (\alpha\alpha', \beta\beta')$.

A **bifunctor** from categories $\mathscr{A}$ and $\mathscr{B}$ to a category $\mathscr{C}$ is a functor from $\mathscr{A} \times \mathscr{B}$ to $\mathscr{C}$.

Proposition 17.2.4. *Let F be a bifunctor from $\mathscr{A}$ and $\mathscr{B}$ to $\mathscr{C}$. For each object A of $\mathscr{A}$, $F(A, -)$ is a functor from $\mathscr{B}$ to $\mathscr{C}$. For each morphism $\alpha : A \longrightarrow A'$ of $\mathscr{A}$, $F(\alpha, -)$ is a natural transformation from $F(A, -)$ to $F(A', -)$, and $F(1_A, -) = 1_{F(A, -)}$, $F(\alpha\alpha', -) = F(\alpha, -)F(\alpha', -)$ for all α, α'.*

Conversely, for each object A of $\mathscr{A}$, let F_A be a functor from $\mathscr{B}$ to $\mathscr{C}$; for each morphism $\alpha : A \longrightarrow A'$ of $\mathscr{A}$, let F_α be a natural transformation from F_A to $F_{A'}$; assume that $F_{1_A} = 1_{F_A}$ and $F_\alpha F_{\alpha'} = F_{\alpha\alpha'}$ for all α, α'. There is a unique bifunctor F from $\mathscr{A}$ and $\mathscr{B}$ to $\mathscr{C}$ such that $F_A = F(A, -)$ and $F_\alpha = F(\alpha, -)$ for all A and α.

If $\mathscr{B}$ is small, this describes bifunctors from $\mathscr{A}$ and $\mathscr{B}$ to $\mathscr{C}$ as functors from $\mathscr{A}$ into $\text{Func}(\mathscr{B}, \mathscr{C})$.

Proof. Let F be a bifunctor. Then $F(A, -)$ assigns $F(A, B)$ to B and $F(A, \beta) = F(1_A, \beta)$ to β; $F(A, -)$ is a functor, since F is a functor. If $\alpha : A \longrightarrow A'$, then $F(\alpha, -) : F(A, -) \longrightarrow F(A', -)$ assigns $F(\alpha, B) = F(\alpha, 1_B) : F(A, B) \longrightarrow F(A', B)$ to each B. For each $\beta : B \longrightarrow B'$, the square

$$
\begin{array}{ccc}
F(A, B) & \xrightarrow{\ F(\alpha, B)\ } & F(A', B) \\
{\scriptstyle F(A, \beta)}\Big\downarrow & & \Big\downarrow{\scriptstyle F(A', \beta)} \\
F(A, B') & \xrightarrow{\ F(\alpha, B')\ } & F(A', B')
\end{array}
$$

commutes, since $F(\alpha, 1_{B'})F(1_A, \beta) = F(\alpha, \beta) = F(1_{A'}, \beta)F(\alpha, 1_B)$. Therefore $F(\alpha, -)$ is a natural transformation. It is immediate that $F(1_A, -) = 1_{F(A, -)}$ and $F(\alpha\beta, -) = F(\alpha, -)F(\beta, -)$.

Conversely, for each object A of $\mathscr{A}$, let F_A be a functor from $\mathscr{B}$ to $\mathscr{C}$; for each morphism $\alpha : A \longrightarrow A'$ of $\mathscr{A}$, let F_α be a natural transformation from F_A to $F_{A'}$; assume that $F_{1_A} = 1_{F_A}$ and $F_\alpha F_{\alpha'} = F_{\alpha\alpha'}$ for all α, α'. If F is a bifunctor from $\mathscr{A}$ and $\mathscr{B}$ to $\mathscr{C}$ such that $F_A = F(A, -)$ and $F_\alpha = F(\alpha, -)$ for all A and α, then $F(A, B) = F_A(B)$, and the above shows that $F(\alpha, \beta) = (F_\alpha)_{B'} F_A(\beta)$ (where A is the domain of α and B' is the codomain of β). Hence F is unique.

Conversely, define $F(A, B) = F_A(B)$ and $F(\alpha, \beta) = (F_\alpha)_{B'} F_A(\beta)$. Then $F(1_A, 1_B) = 1_{F(A, B)}$. If $\alpha : A \longrightarrow A'$, $\alpha' : A' \longrightarrow A''$, $\beta : B \longrightarrow B'$, $\beta' : B'$ $\longrightarrow$

B'' are morphisms, then the diagram

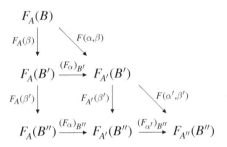

commutes; therefore $F(\alpha'\alpha, \beta'\beta) = F(\alpha',\beta')F(\alpha,\beta)$. Thus F is a bifunctor. $\square$

For example, if $\mathscr{C}$ has small Hom sets then $\mathrm{Hom}_{\mathscr{C}}(-,-)$ is a bifunctor from $\mathscr{C}^{\mathrm{op}}$ and $\mathscr{C}$ to *Sets*, by Lemma 17.2.2. Similarly, $\mathrm{Hom}_R(-,-)$ is a bifunctor from $(R\text{-}Mods)^{\mathrm{op}}$ and $R\text{-}Mods$ to *Abs*. Also $-\otimes_R -$ is a bifunctor from $R^{\mathrm{op}}\text{-}Mods$ (the category of right R-modules) and $R\text{-}Mods$ to *Abs* (or from $Q\text{-}R\text{-}Bimods$ and $R\text{-}S\text{-}Bimods$ to $Q\text{-}S\text{-}Bimods$). This follows from Proposition 17.2.4 (and previous results) and can also be verified directly.

Exercises

1. Verify that $\mathrm{Hom}_{\mathscr{C}}(A,-)$ and $\mathrm{Hom}_{\mathscr{C}}(-,A)$ are functors.
2. Verify that $\mathrm{Func}(\mathscr{A},\mathscr{B})$ is a category when $\mathscr{A}$ is small.
3. Show that equivalence of categories is reflexive symmetric and transitive.
4. A **skeleton** of a category $\mathscr{C}$ is a category $\mathscr{S}$ such that (a) every object of $\mathscr{S}$ is an object of $\mathscr{C}$; (b) every object A of $\mathscr{C}$ is isomorphic to a unique object S of $\mathscr{S}$, and there is a mapping which assigns to A an isomorphism $A \cong S$; (c) $\mathrm{Hom}_{\mathscr{S}}(S,T) = \mathrm{Hom}_{\mathscr{C}}(S,T)$ for all objects S,T of $\mathscr{S}$. Show that $\mathscr{C}$ and $\mathscr{S}$ are equivalent.
5. Let $\mathscr{A}$ be a category which has a skeleton $\mathscr{S}$ and $\mathscr{B}$ be a category which has a skeleton $\mathscr{T}$. Show that $\mathscr{A}$ and $\mathscr{B}$ are equivalent if and only if $\mathscr{S}$ and $\mathscr{T}$ are isomorphic.
6. Show that every preordered set is equivalent (as a category) to a partially ordered set.
7. Prove directly that $\mathrm{Hom}_{\mathscr{C}}(-,-)$ is a bifunctor whenever $\mathscr{C}$ has small Hom sets.
8. Prove **Yoneda's Lemma**: if $\mathscr{C}$ has small Hom sets, and $F : \mathscr{C} \longrightarrow Sets$ is a functor, then for each object C of $\mathscr{C}$ there is a bijection of $F(C)$ onto the class of all natural transformations $\mathrm{Hom}_{\mathscr{C}}(C,-) \longrightarrow F$ (in particular, the latter is a set); this bijection is natural in C and F.
9. Let $\mathscr{B}$ be a category and $\mathscr{A}$ be a small category. Show that evaluation $(A,F) \longmapsto F(A)$ is a bifunctor from $\mathscr{A}$ and $\mathrm{Func}(\mathscr{A},\mathscr{B})$ to $\mathscr{B}$.
10. State and prove a result similar to Proposition 17.2.4, which describes bifunctors from $\mathscr{A}$ and $\mathscr{B}$ to $\mathscr{C}$ as functors from $\mathscr{B}$ to $\mathrm{Func}(\mathscr{A},\mathscr{C})$ is $\mathscr{A}$ is small.

17.3. LIMITS AND COLIMITS

Limits provide one general model for many of the constructions in previous chapters: direct products, direct sums, pullbacks, kernels, direct limits, and others.

17.3.a. Diagrams. Let $\mathscr{C}$ be a category and $\mathscr{G}$ be a [small] directed graph. A **diagram** $D : \mathscr{G} \longrightarrow \mathscr{C}$ in $\mathscr{C}$ over $\mathscr{G}$ is a pair of mappings, both denoted by D, which assign to each vertex i of $\mathscr{G}$ an object D_i of $\mathscr{C}$, and to each edge $a : i \longrightarrow j$ of $\mathscr{G}$ a morphism $D_a : D_i \longrightarrow D_j$ of $\mathscr{C}$.

For example, a **square** (or a **triangle**) in a category $\mathscr{C}$ is a diagram in $\mathscr{C}$ over the square (or triangle) graph

When D and E are diagrams in a category $\mathscr{C}$ over the same graph $\mathscr{G}$, a **morphism** from D to E is a mapping, which we write as a family $\varphi = (\varphi_i)_{i \in \mathscr{G}}$, which assigns to each vertex i of $\mathscr{G}$ a morphism $\varphi_i : D_i \longrightarrow E_i$ so that the following square commutes for every edge $a : i \longrightarrow j$ of $\mathscr{G}$:

$$
\begin{array}{ccc}
D_i & \xrightarrow{\ \varphi_i\ } & E_i \\
D_a \downarrow & & \downarrow E_a \\
D_j & \xrightarrow{\ \varphi_j\ } & E_j
\end{array}
$$

(In this notation $(\varphi_i)_{i \in \mathscr{G}}$, "$i \in \mathscr{G}$" is short for "$i$ is a vertex of $\mathscr{G}$".)

Homomorphisms of diagrams compose: if $\varphi = (\varphi_i)_{i \in \mathscr{G}} : D \longrightarrow E$ and $\psi = (\psi_i)_{i \in \mathscr{G}} : E \longrightarrow F$ are morphisms of diagrams over $\mathscr{G}$, then so is $\psi\varphi = (\psi_i\varphi_i)_{i \in \mathscr{G}} : D \longrightarrow F$. Moreover there is for each diagram D over $\mathscr{G}$ an identity morphism 1_D, which assigns to $i \in \mathscr{G}$ the identity morphism 1_{D_i}. Thus:

Proposition 17.3.1. *Let $\mathscr{G}$ be a directed graph and $\mathscr{C}$ be a category. Diagrams in $\mathscr{C}$ over $\mathscr{G}$ and their morphisms are the objects and morphisms of a category* $\mathrm{Diag}(\mathscr{G}, \mathscr{C})$.

$\mathrm{Diag}(\mathscr{G}, \mathscr{C})$ is also denoted by $\mathscr{C}^{\mathscr{G}}$. This new category resembles a functor category and is in fact isomorphic to one:

Proposition 17.3.2. *Let $\mathscr{G}$ be a directed graph and $\mathscr{C}$ be a category.*

(1) *Every diagram $D : \mathscr{G} \longrightarrow \mathscr{C}$ extends uniquely to a functor $\widehat{D} : \widehat{\mathscr{G}} \longrightarrow \mathscr{C}$.*

(2) *Every morphism $\varphi : D \longrightarrow E$ of diagrams over $\mathscr{G}$ is a natural transformation $\varphi : \widehat{D} \longrightarrow \widehat{E}$.*

(3) *The categories* $\mathrm{Diag}(\mathscr{G},\mathscr{C})$ *and* $\mathrm{Func}(\widehat{\mathscr{G}},\mathscr{C})$ *are isomorphic.*

Proof. Recall that the objects of $\widehat{\mathscr{G}}$ are the vertices of $\mathscr{G}$ and that the morphisms of $\widehat{\mathscr{G}}$ are the empty paths (i,i) on each vertex and all nonempty paths $(j,a_1,\ldots,a_n,i)$. The extension $\widehat{D}$ of D to $\widehat{\mathscr{G}}$ is defined by

$$\widehat{D}(i) = D_i,$$

$$\widehat{D}(i,i) = 1_{D_i},$$

$$\widehat{D}(j,a_1,a_2,\ldots,a_n,i) = D_{a_1} D_{a_2} \ldots D_{a_n}.$$

It is immediate that $\widehat{D}$ is a functor. $\widehat{D}$ is the only functor which extends D, since every nonempty path is a composition in $\widehat{\mathscr{G}}$:

$$(j,a_1,\ldots,a_n,i) = (j,a_1,\bullet)(\bullet,a_2,\bullet)\ldots(\bullet,a_n,i). \qquad \square$$

With the construction of $\widehat{D}$ we can give a precise definition of commutative diagrams: a diagram D is **commutative** in case $\widehat{D}(\pi) = \widehat{D}(\pi')$ whenever π,π' : $i \longrightarrow j$ are nonempty paths with the same origin and destination.

17.3.b. Limits. Limits are a common generalization of direct products, pullbacks, and inverse limits.

A **cone** $\varphi : A \longrightarrow D$ from an object A to a diagram D over $\mathscr{G}$ is a family $\varphi = (\varphi_i)_{i \in \mathscr{G}}$ of morphisms $\varphi_i : A \longrightarrow D_i$ such that the triangle

commutes for every edge $a : i \longrightarrow j$ of $\mathscr{G}$. Equivalently, a cone from A to D is a morphism φ from $C(A)$ to D, where $C(A)$ is the **constant** diagram which assigns A to each vertex and 1_A to every edge.

Let D be a diagram in a category $\mathscr{C}$ over a directed graph $\mathscr{G}$. If $\lambda = (\lambda_i)_{i \in \mathscr{G}}$: $L \longrightarrow D$ is a cone from L to D and $\alpha : A \longrightarrow L$ is a morphism, then $\lambda\alpha = (\lambda_i\alpha)_{i \in \mathscr{G}} : A \longrightarrow D$ is a cone from A to D. A **limit cone** of D is a cone $\lambda : L \longrightarrow D$ with the following universal property: for every cone $\varphi : A \longrightarrow D$, there is a unique morphism $\overline{\varphi} : A \longrightarrow L$ such that $\varphi = \lambda\overline{\varphi}$; equivalently, such that commutes for every vertex i of $\mathscr{G}$. A **limit** of D is a pair of an object $L = \lim D$ and a limit cone $L \longrightarrow D$. The object L is also called a limit of D.

Proposition 17.3.3. *The limit cone and limit of a diagram are unique up to isomorphism (when they exist).*

Proof. This follows from the universal property in the usual way. Let $\lambda : L \longrightarrow D$ and $\mu : M \longrightarrow D$ be limit cones of D. By the universal property of limit cones, there exists a morphism $\theta : L \longrightarrow M$ such that $\lambda = \mu\theta$ and a morphism $\zeta : M \longrightarrow L$ such that $\mu = \lambda\zeta$. Then $\lambda = \lambda\zeta\theta$ and $\mu = \mu\theta\zeta$. By the uniqueness in the universal property of limit cones, $\zeta\theta = 1_L$ and $\theta\zeta = 1_M$, so that θ and ζ are mutually inverse isomorphisms. $\square$

Conversely, if $\lambda : L \longrightarrow D$ is a limit cone of D and $\theta : L' \longrightarrow L$ is an isomorphism, then $\lambda\theta : L' \longrightarrow D$ is a limit cone of D, since every cone to D factors uniquely through λ and therefore factors uniquely through $\lambda\theta$.

17.3.c. Examples. Limits include many constructions seen in previous chapters. Inverse limits are the most obvious example.

In a category $\mathscr{C}$, the **product** of a family $(D_i)_{i \in I}$ of objects of $\mathscr{C}$ consists of an object $P = \prod_{i \in I} D_i$ and of **projections** $(\pi_i)_{i \in I}$, $\pi_i : P \longrightarrow D_i$, with the following universal property: for every family $(\varphi_i)_{i \in I}$ of morphisms $\varphi_i : A \longrightarrow D_i$ of $\mathscr{C}$, there exists a unique morphism $\overline{\varphi} : A \longrightarrow P$ such that $\varphi_i = \pi_i\overline{\varphi}$ for all i:

For example, every family of sets has a product in *Sets*, which is the cartesian product with the usual projections. Similar results hold in *Grps*, *R-Mods*, and so on.

Products are limits of diagrams over certain graphs. A **discrete** graph is a graph without edges. A discrete graph I can be identified with the set I of its vertices. A diagram D over I is simply a family of objects $(D_i)_{i \in I}$ indexed by I. A cone $\varphi : A \longrightarrow D$ is simply a family $(\varphi_i)_{i \in I}$ of morphisms $\varphi_i : A \longrightarrow D_i$. Thus the object P and projections $\pi_i : P \longrightarrow D_i$ constitute a product of $(D_i)_{i \in I}$ if and only if $\pi : P \longrightarrow D$ is a limit cone of D.

The product of $(D_i)_{i \in I}$ is usually denoted by $\prod_{i \in I} D_i$; if $I = \{1, 2, \ldots, n\}$ is finite, then $\prod_{i \in I} D_i$ is denoted by $D_1 \times D_2 \times \cdots \times D_n$.

The reader will verify that products are associative: if $I = \bigcup_{j \in J} I_j$ is a partition of I, then there is a natural isomorphism $\prod_{i \in I} D_i \cong \prod_{j \in J} (\prod_{i \in I_j} D_i)$. For instance, $A \times B \times C$ is isomorphic to $(A \times B) \times C$ and to $A \times (B \times C)$.

The **equalizer** of two morphisms $\alpha, \beta : A \longrightarrow B$ in a category $\mathscr{C}$ is a morphism $\varepsilon : E \longrightarrow A$ such that $\alpha\varepsilon = \beta\varepsilon$, and that every morphism φ such that $\alpha\varphi = \beta\varphi$ factors uniquely through ε. Every equalizer is a monomorphism:

$\varepsilon\gamma = \varepsilon\delta$ implies $\gamma = \delta$, since $\varepsilon\gamma = \varepsilon\delta$ factors uniquely through ε:

$$E \xrightarrow{\varepsilon} A \underset{\beta}{\overset{\alpha}{\rightrightarrows}} B$$
$$\nwarrow \nearrow \varphi$$
$$C$$

For instance, in the category *R-Mods*, the equalizer of $\varphi : M \longrightarrow N$ and 0 is the inclusion homomorphism $\mathrm{Ker}\,\varphi \longrightarrow M$. The exercises give other examples.

We see that an equalizer of α and β is the morphism $\varepsilon : E \longrightarrow A$ in a limit cone of a diagram over the graph $\bullet \rightrightarrows \bullet$. A cone from C to D consists of two morphisms $\varphi : C \longrightarrow A$ and $\psi : C \longrightarrow B$ such that $\alpha\varphi = \beta\varphi = \psi$ and is uniquely determined by φ. Hence a cone (ε, η) is a limit cone of D if and only if ε is an equalizer of α and β and $\eta = \alpha\varepsilon = \beta\varepsilon$.

More generally, let A and B be two objects of a category $\mathscr{C}$ and S be a set of morphisms $\sigma : A \longrightarrow B$ from A to B. An **equalizer** of S is a morphism $\varepsilon : E \longrightarrow A$ with the following universal property: $\sigma\varepsilon = \tau\varepsilon$ for every $\sigma, \tau \in S$, and every morphism φ such that $\sigma\varphi = \tau\varphi$ for every $\sigma, \tau \in S$ factors uniquely through ε. As before ε must be a monomorphism. Equivalently, an equalizer of S is the morphism $\varepsilon : E \longrightarrow A$ in a limit cone of the diagram D which consists of A, B, and every $\sigma \in S$:

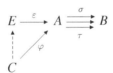

A **pullback** in a category $\mathscr{C}$ is a commutative square $\alpha\alpha' = \beta\beta'$ such that for every commutative square $\alpha\varphi = \beta\psi$ there is a unique morphism ξ such that $\varphi = \alpha'\xi$ and $\psi = \beta'\xi$. A pullback $\alpha\alpha' = \beta\beta'$ consists of a limit and some of the limit cone of the diagram D:

$$
\begin{array}{ccc}
 & & A \\
 & & \downarrow \alpha \\
B & \xrightarrow{\beta} & C
\end{array}
$$

Indeed a cone into D consists of morphisms $\varphi : X \longrightarrow A$, $\psi : X \longrightarrow B$, $\chi : X \longrightarrow C$ such that $\alpha\varphi = \beta\psi = \chi$, and is determined by the commutative square $\alpha\varphi = \beta\psi$. Hence (φ, ψ, χ) is a limit cone if and only if $\alpha\varphi = \beta\psi$ is a pullback.

17.3.d. Colimits. Let $\mathscr{G}$ be a [small] directed graph and $\mathscr{C}$ be a category. A diagram $D : \mathscr{G} \longrightarrow \mathscr{C}$ in $\mathscr{C}$ is also a diagram $D : \mathscr{G}^{\mathrm{op}} \longrightarrow \mathscr{C}^{\mathrm{op}}$ in $\mathscr{C}^{\mathrm{op}}$. The **colimit** and **colimit cone** of D in the given category $\mathscr{C}$ are its limit and limit cone in $\mathscr{C}^{\mathrm{op}}$ (if they exist). Thus limits and colimits are dual concepts.

In plainer terms, a **cone** $\varphi : D \longrightarrow A$ from D to an object A is a family $\varphi = (\varphi_i)_{i \in \mathscr{G}}$ of morphisms $\varphi_i : D_i \longrightarrow A$ such that the triangle

$$
\begin{array}{ccc}
D_j & \xrightarrow{\varphi_j} & A \\
D_a \uparrow & \nearrow \varphi_i & \\
D_i & &
\end{array}
$$

commutes for every edge $a : i \longrightarrow j$ of $\mathscr{G}$; equivalently, a morphism φ from D to the constant diagram $C(A)$. A colimit cone of D is a cone $\lambda : D \longrightarrow L$ with the following universal property: for every cone $\varphi : D \longrightarrow A$, there is a unique morphism $\overline{\varphi} : L \longrightarrow A$ such that $\varphi = \overline{\varphi} \lambda$; equivalently, such that the triangle

$$
\begin{array}{ccc}
D_i & \xrightarrow{\lambda_i} & L \\
\varphi_i \downarrow & \swarrow \overline{\varphi} & \\
A & &
\end{array}
$$

commutes for every vertex i of $\mathscr{G}$. The colimit of D is $L = \operatorname{colim} D$. By Proposition 17.3.3, the colimit and colimit cone of D are unique up to isomorphism (when they exist).

17.3.e. Examples. Colimits include many constructions seen in previous chapters. Direct limits are the most obvious example.

In a category $\mathscr{C}$, the **coproduct** of a family $(D_i)_{i \in I}$ of objects of $\mathscr{C}$ consists of an object $P = \coprod_{i \in I} D_i$ and of **injections** $(\iota_i)_{i \in I}$, $\iota_i : D_i \longrightarrow P$ with the following universal property: for every family $(\varphi_i)_{i \in I}$ of morphisms $\varphi_i : D_i \longrightarrow A$ of $\mathscr{C}$, there exists a unique morphism $\overline{\varphi} : P \longrightarrow A$ such that $\varphi_i = \overline{\varphi} \iota_i$ for all i:

$$
\begin{array}{ccc}
P & & \\
\iota_i \uparrow & \searrow \overline{\varphi} & \\
D_i & \xrightarrow[\varphi_i]{} & A
\end{array}
$$

For instance, the free product of a family of groups is its coproduct in *Grps*; the direct sum of a family of left R-modules is its coproduct in *R-Mods*; the coproduct of two commutative R-algebras is their tensor product. Other examples are given in the exercises.

Coproducts are colimits of diagrams over discrete graphs. A diagram D over a discrete graph I is a family $(D_i)_{i \in I}$ of objects; a cone $\varphi : D \longrightarrow A$ is a family $(\varphi_i)_{i \in I}$ of morphisms $\varphi_i : D_i \longrightarrow A$. Thus the object P and injections $\iota_i : D_i \longrightarrow P$ constitute a coproduct of $(D_i)_{i \in I}$ if and only if $\iota : D \longrightarrow P$ is a colimit cone of D.

We denote the coproduct of $(D_i)_{i \in I}$ by $\coprod_{i \in I} D_i$; if $I = \{1, 2, \ldots, n\}$ is finite, then $\coprod_{i \in I} D_i$ is denoted by $D_1 \coprod D_2 \coprod \cdots \coprod D_n$. A number of other symbols have been used for coproducts.

The reader will verify that coproducts are associative: if $I = \bigcup_{j \in J} I_j$ is a partition of I, then there is a natural isomorphism $\coprod_{i \in I} D_i \cong \coprod_{j \in J} (\coprod_{i \in I_j} D_i)$. For instance, $A \coprod B \coprod C$ is isomorphic to $(A \coprod B) \coprod C$ and to $A \coprod (B \coprod C)$.

The **coequalizer** of two morphisms $\alpha, \beta : A \longrightarrow B$ in a category $\mathscr{C}$ is a morphism $\gamma : B \longrightarrow C$ such that $\gamma \alpha = \gamma \beta$, and that every morphism φ such that $\varphi \alpha = \varphi \beta$ factors uniquely through γ. Every coequalizer is an epimorphism.

For instance, in the category R-*Mods*, the coequalizer of $\varphi : M \longrightarrow N$ and 0 is the projection homomorphism $M \longrightarrow \mathrm{Coker}\,\varphi = M/\mathrm{Im}\,\varphi$. Other examples and properties are given in the exercises.

The reader will verify that a coequalizer of α and β is the morphism $\gamma : A \longrightarrow C$ in a colimit cone of the diagram $A \overset{\alpha}{\underset{\beta}{\rightrightarrows}} B$.

A **pushout** in a category $\mathscr{C}$ is a commutative square $\alpha' \alpha = \beta' \beta$ such that for every commutative square $\varphi \alpha = \psi \beta$ there is a unique morphism ξ such that $\varphi = \xi \alpha'$ and $\psi = \xi \beta'$. A pushout $\alpha' \alpha = \beta' \beta$ consists of the colimit and morphisms $\alpha' : A \longrightarrow X$, $\beta' : B \longrightarrow X$ in a colimit cone of the diagram

$$\begin{array}{ccc} C & \overset{\alpha}{\longrightarrow} & A \\ {\scriptstyle \beta} \downarrow & & \\ B & & \end{array}$$

Exercises

1. For every category $\mathscr{C}$ and directed graph $\mathscr{G}$, show that there is a constant diagram functor from $\mathscr{C}$ to $\mathrm{Diag}(\mathscr{G}, \mathscr{C})$.

2. Let I be a preordered set. Show that every diagram in I is commutative. Show that preordered sets are the only small categories with this property.

3. Let I be a partially ordered set (regarded as a category). What is the product of a family of elements of I?

4. Let I be a preordered set. Show that the limit of any diagram in I is the product of the objects in the diagram.

5. Prove the following associativity property of products: if $I = \bigcup_{j \in J} I_j$ is a partition of I, then there is a natural isomorphism $\prod_{i \in I} D_i \cong \prod_{j \in J} (\prod_{i \in I_j} D_i)$.

6. Describe equalizers in *Sets*; *Grps*; *Rings*; *R-Mods*.

7. Show that every monomorphism in the category *Grps* is the equalizer of two homomorphisms.

8. Show that every monomorphism in the category *R-Mods* is the equalizer of two homomorphisms.

9. Show that not every monomorphism in the category *Rings* is an equalizer.

10. Describe coproducts in *Sets*.

11. Let *I* be a partially ordered set (regarded as a category). What is the coproduct of a family of elements of *I*?

12. Let *I* be a preordered set. Show that the colimit of any diagram in *I* is the coproduct of the objects in the diagram.

13. Prove the following associativity property of coproducts: if $I = \bigcup_{j \in J} I_j$ is a partition of *I*, then there is a natural isomorphism $\coprod_{i \in I} D_i \cong \coprod_{j \in J} (\coprod_{i \in I_j} D_i)$.

14. Construct coequalizers in *Sets*.

15. Construct coequalizers in *Grps*.

16. Prove that every epimorphism in the category *Grps* is a coequalizer.

17. Construct coequalizers in *R-Mods*. Is every epimorphism in *R-Mods* a coequalizer?

17.4. COMPLETENESS

This section contains constructions of limits and colimits, showing in particular that in *Sets*, *Grps*, *R-Mods*, and so on, every diagram has a limit and a colimit.

17.4.a. Completeness. A category $\mathscr{C}$ is **complete** in case every [small] diagram in $\mathscr{C}$ has a limit; we also say that $\mathscr{C}$ **has limits**.

Many applications require the stronger property that a limit can be assigned to every diagram in $\mathscr{C}$; that is, that there is for every graph $\mathscr{G}$ a mapping which assigns a limit object and a limit cone to every diagram in $\mathscr{C}$ over $\mathscr{G}$. By the Axiom of Choice, this condition is equivalent to completeness when $\mathscr{C}$ is small.

Proposition 17.4.1. *Sets is complete; in fact in Sets a limit can be assigned to every diagram.*

Proof. Let $D : \mathscr{G} \longrightarrow Sets$ be any diagram. Let $P = \prod_{i \in \mathscr{G}} D_i$ be the cartesian product and $\pi_i : P \longrightarrow D_i$ be the projection. We show that

$$L = \{ (x_i)_{i \in \mathscr{G}} \in P ; \ D_a(x_i) = x_j \text{ whenever } a : i \longrightarrow j \}$$

is a limit of *D*, with limit cone $\lambda_i = \pi_{i|L} : L \longrightarrow D_i$. If $\varphi : A \longrightarrow D$ is a cone, then $D_a(\varphi_i(a)) = \varphi_j(a)$ for all $a : i \longrightarrow j$ and $a \in A$. Therefore $\overline{\varphi}(a) = (\varphi_i(a))_{i \in \mathscr{G}} \in L$ for all $a \in A$. This defines a mapping $\overline{\varphi} : A \longrightarrow L$ such that $\lambda_i \circ \overline{\varphi} = \varphi_i$ for all *i*. Clearly $\overline{\varphi}$ is the only such mapping. □

We call the limit and limit cone constructed in this proof the **standard limit** of the given diagram of sets and mappings.

A functor $F : \mathscr{A} \longrightarrow \mathscr{B}$ **preserves** limits of diagrams over a graph $\mathscr{G}$ in case $F(\lambda) = (F(\lambda_i))_{i \in \mathscr{G}}$ is a limit cone of $F(D)$ in $\mathscr{B}$ whenever $D : \mathscr{G} \longrightarrow \mathscr{A}$ is a diagram over $\mathscr{G}$ and $\lambda = (\lambda_i)_{i \in \mathscr{G}}$ is a limit cone of D in $\mathscr{A}$.

Proposition 17.4.2. *Grps is complete; in fact, in Grps a limit can be assigned to every diagram, and the forgetful functor from Grps to Sets preserves all limits.*

Proof. Let $D : \mathscr{G} \longrightarrow Grps$ be any diagram. Let $P = \prod_{i \in \mathscr{G}} D_i$ be the direct product and $\pi_i : P \longrightarrow D_i$ be the projection. Let

$$L = \{ (x_i)_{i \in \mathscr{G}} \in P ; D_a(x_i) = x_j \text{ whenever } a : i \longrightarrow j \}$$

be the standard limit of D in *Sets*, with limit cone $\lambda_i = \pi_{i|L}$ $(i \in \mathscr{G})$. Since each D_a is a homomorphism, L is a subgroup of P; moreover each λ_i is a homomorphism. If $\varphi : A \longrightarrow D$ is a cone in *Grps*, then each φ_i is a homomorphism and $\overline{\varphi} : a \longmapsto (\varphi_i(a))_{i \in \mathscr{G}}$ is a homomorphism. We see that $\overline{\varphi}$ is the only homomorphism such that $\lambda_i \circ \overline{\varphi} = \varphi_i$ for all i.

This constructs a limit cone for D in *Grps* which is also a limit cone in *Sets*. By Proposition 17.3.3, every limit cone of D in *Grps* is isomorphic to the above and therefore is also a limit cone in *Sets*. $\quad\square$

We have in fact shown a stronger property. A functor $F : \mathscr{A} \longrightarrow \mathscr{B}$ **creates** limits of diagrams over a graph $\mathscr{G}$ in case the following holds for every diagram $D : \mathscr{G} \longrightarrow \mathscr{A}$: if $\mu : M \longrightarrow F(D)$ is a limit cone of $F(D)$ in $\mathscr{B}$, then there is a unique cone $\lambda : L \longrightarrow D$ such that $F(\lambda) = \mu$, and it is a limit cone of D in $\mathscr{A}$. A functor which creates limits must preserve limits.

Proposition 17.4.3. *The forgetful functor from Grps to Sets creates all limits.*

The proof of Proposition 17.4.2 extends readily to *Rings*, *R-Mods*, *R-Algs*, and so on: for each of these categories, the forgetful functor to *Sets* creates all limits, and in particular preserves all limits.

17.4.b. Hom Functors. We give an application of Proposition 17.4.1.

Proposition 17.4.4. *Let $\mathscr{C}$ be a category with small Hom sets. For each object A of $\mathscr{C}$, the functor $\mathrm{Hom}_{\mathscr{C}}(A, -)$ preserves all limits. In fact $\lambda : L \longrightarrow D$ is a limit cone of D if and only if $\mathrm{Hom}_{\mathscr{C}}(A, \lambda)$ is a limit cone of $\mathrm{Hom}_{\mathscr{C}}(A, D)$ for every object A of $\mathscr{C}$.*

Proof. Let $D : \mathscr{G} \longrightarrow \mathscr{C}$ be a diagram in $\mathscr{C}$ and A be an object of $\mathscr{C}$. Then $\mathrm{Hom}_{\mathscr{C}}(A, D)$ is a diagram of sets. The standard limit of $\mathrm{Hom}_{\mathscr{C}}(A, D)$ is

$$L_A = \{ (\varphi_i)_{i \in \mathscr{G}} \in P ; (D_a)_*(\varphi_i) = \varphi_j \text{ whenever } a : i \longrightarrow j \},$$

where $P = \prod_{i \in \mathscr{G}} \mathrm{Hom}_{\mathscr{C}}(A, D_i)$ and $(D_a)_* = \mathrm{Hom}_{\mathscr{C}}(A, D_a)$. We see that L_A is the set of all cones $\varphi : A \longrightarrow D$.

If $\lambda : L \longrightarrow D$ is a limit cone of D, then for each cone $\varphi : A \longrightarrow D$, there is a unique morphism $\overline{\varphi} : A \longrightarrow L$ such that $\varphi_i = \lambda_i \overline{\varphi}$ for all i. This states that the mapping $\lambda_A : \overline{\varphi} \longmapsto (\lambda_i \overline{\varphi})_{i \in \mathcal{G}} = (\mathrm{Hom}_{\mathcal{C}}(A, \lambda_i)(\overline{\varphi}))_{i \in \mathcal{G}}$ is a bijection of $\mathrm{Hom}_{\mathcal{C}}(A, L)$ onto L_A; hence the cone $\mathrm{Hom}_{\mathcal{C}}(A, \lambda) = (\mathrm{Hom}_{\mathcal{C}}(A, \lambda_i))_{i \in \mathcal{G}}$ is isomorphic to the standard limit cone of $\mathrm{Hom}_{\mathcal{C}}(A, D)$ and is a limit cone of $\mathrm{Hom}_{\mathcal{C}}(A, D)$.

Conversely, assume that $\mathrm{Hom}_{\mathcal{C}}(A, \lambda) = (\mathrm{Hom}_{\mathcal{C}}(A, \lambda_i))_{i \in \mathcal{G}}$ is a limit cone of $\mathrm{Hom}(A, D)$ for every A. Then $\lambda_A : \overline{\varphi} \longmapsto (\lambda_i \overline{\varphi})_{i \in \mathcal{G}} = (\mathrm{Hom}_{\mathcal{C}}(A, \lambda_i)(\overline{\varphi}))_{i \in \mathcal{G}}$ is a bijection of $\mathrm{Hom}_{\mathcal{C}}(A, L)$ onto the standard limit L_A. Hence for every cone $\varphi : A \longrightarrow D$ there is a unique morphism $\overline{\varphi} : A \longrightarrow L$ such that $\varphi_i = \lambda_i \overline{\varphi}$ for all i, and $\lambda : L \longrightarrow D$ is a limit cone of D. $\qquad \square$

Dually, $\mathrm{Hom}_{\mathcal{C}}(-, A)$ changes colimit cones to limit cones and colimits to limits. The properties $\mathrm{Hom}_R(A, \prod_{i \in I} B_i) \cong \prod_{i \in I} \mathrm{Hom}_R(A, B_i)$ and $\mathrm{Hom}_R(\bigoplus_{i \in I} A_i, B) \cong \prod_{i \in I} \mathrm{Hom}_R(A_i, B)$ of modules are particular cases of Proposition 17.4.4 and its dual.

17.4.c. Limits by Products and Equalizers. The next result is based on a general construction of limits, which is different from that in Proposition 17.4.1.

Proposition 17.4.5. *If $\mathcal{C}$ has products and equalizers, then $\mathcal{C}$ is complete. If a product can be assigned to every family of objects of $\mathcal{C}$, and an equalizer can be assigned to every pair $\alpha, \beta : A \longrightarrow B$, then a limit can be assigned to every diagram in $\mathcal{C}$.*

This provides alternate proofs of Propositions 17.4.1 and 17.4.2.

Proof. Let $\mathcal{G}$ be a graph; denote by E the set of edges of $\mathcal{G}$ and by o, d the origin and destination mappings of $\mathcal{G}$: if $a : i \longrightarrow j$ is an edge, then $o(a) = i$ and $d(a) = j$. Let D be a diagram over $\mathcal{G}$. If $\mathcal{C}$ has products, then there is a product $P = \prod_{i \in \mathcal{G}} D_i$ with projections $\pi_i : P \longrightarrow D_i$ and a product $Q = \prod_{a \in E} D_{d(a)}$ with projections $\rho_a : Q \longrightarrow D_{d(a)}$. The universal property of Q yields morphisms α and β such that $\rho_a \alpha = \pi_{d(a)}$ and $\rho_a \beta = D_a \pi_{o(a)}$ for all $a \in E$:

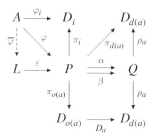

Let $\varepsilon : L \longrightarrow P$ be an equalizer of α and β and $\lambda_i = \pi_i \varepsilon$. We show that $\lambda = (\lambda_i)_{i \in \mathcal{G}} : L \longrightarrow D$ is a limit cone of D. First λ is a cone: if $a : i \longrightarrow j$ is an edge so that $o(a) = i$, $d(a) = j$, then $\alpha\varepsilon = \beta\varepsilon$ yields

$$\lambda_j = \pi_j\varepsilon = \rho_a\alpha\varepsilon = \rho_a\beta\varepsilon = D_a\pi_i\varepsilon = D_a\lambda_i.$$

Let $(\varphi_i)_{i \in \mathscr{G}} : A \longrightarrow D$ be any cone. There is a unique morphism $\varphi : A \longrightarrow P$ such that $\pi_i\varphi = \varphi_i$ for all i. Since $(\varphi_i)_{i \in \mathscr{G}}$ is a cone, we have, for every edge $a : i \longrightarrow j$,

$$\rho_a\alpha\varphi = \pi_j\varphi = \varphi_j = D_a\varphi_i = D_a\pi_i\varphi = \rho_a\beta\varphi.$$

Therefore $\alpha\varphi = \beta\varphi$ and $\varphi = \varepsilon\overline{\varphi}$ for some unique $\overline{\varphi} : A \longrightarrow L$. Then $\overline{\varphi}$ is unique such that $\varphi_i = \lambda_i\overline{\varphi}$. $\square$

A **finite** graph is a graph with finitely many objects and finitely many edges. A **finite** limit is a limit of a diagram over a finite graph.

Corollary 17.4.6. *If $\mathscr{C}$ has equalizers and finite products, then $\mathscr{C}$ has finite limits.*

17.4.d. Cocompleteness. A category $\mathscr{C}$ is **cocomplete** in case $\mathscr{C}^{\mathrm{op}}$ is complete; equivalently, in case every [small] diagram in $\mathscr{C}$ has a colimit. By Proposition 17.4.5, a category is cocomplete if and only if it has coproducts and coequalizers.

Proposition 17.4.7. *R-Mods is cocomplete; in fact a colimit can be assigned to every diagram in R-Mods.*

Proof. This follows from Proposition 17.4.5, but we give a direct proof. Let $D : \mathscr{G} \longrightarrow R\text{-Mods}$ be any diagram. Let $P = \bigoplus_{i \in \mathscr{G}} D_i$ be the direct sum and $\iota_i : D_i \longrightarrow P$ be the injection. Let K be the submodule of P generated by all $\iota_j(D_a(x)) - \iota_i(x)$, with $a : i \longrightarrow j$ and $x \in D_i$. Let $L = P/K$ and $\pi : P \longrightarrow L$ be the projection. We show that L is the colimit of D; the colimit cone is given by $\lambda_i = \pi \circ \iota_i : D_i \longrightarrow L$. Let $\varphi : D \longrightarrow A$ be a cone. Let $\varphi^* : P \longrightarrow A$ be the unique homomorphism such that $\varphi^* \circ \iota_i = \varphi_i$ for all i:

We have $\varphi_j(D_a(x)) = \varphi_i(x)$ for all $a : i \longrightarrow j$ and $x \in D_i$. Therefore $\mathrm{Ker}\,\varphi^*$ contains every generator of K, and φ^* factors uniquely through π. This yields a unique homomorphism $\overline{\varphi}$ such that $\overline{\varphi} \circ \pi \circ \iota_i = \varphi^* \circ \iota_i = \varphi_i$ for all i. $\square$

We leave to the reader similar arguments (using coproducts) showing that *Sets*, *Grps*, and the like, are cocomplete. Coproducts show that the forgetful functors to *Sets* do not preserve colimits, let alone create them.

17.4.e. Limit Functors. Let D and D' be diagrams over the same graph $\mathscr{G}$ and $\varphi = (\varphi_i)_{i \in \mathscr{G}} : D \longrightarrow D'$ be a morphism of diagrams. Let $\lambda : L \longrightarrow D$ and $\lambda' : L' \longrightarrow D'$ be limit cones of D and D'. Then $\varphi\lambda = (\varphi_i\lambda_i)_{i \in \mathscr{G}} : L \longrightarrow D'$ is a cone to D'; therefore there exists a unique morphism $\lim\varphi : L \longrightarrow L'$ such that $\varphi_i\lambda_i = \lambda'_i(\lim\varphi)$ for all i:

$$
\begin{array}{ccc}
L & \xrightarrow{\lambda_i} & D_i \\
{\scriptstyle \lim\varphi} \downarrow & & \downarrow {\scriptstyle \varphi_i} \\
L' & \xrightarrow[\lambda'_i]{} & D'_i
\end{array}
$$

For instance, let $(\alpha_i)_{i \in I}$ be a family of morphisms $\alpha_i : A_i \longrightarrow B_i$. If $\prod_{i \in I} A_i$ and $\prod_{i \in I} B_i$ exist, there is a unique morphism $\alpha = \prod_{i \in I} \alpha_i : \prod_{i \in I} A_i \longrightarrow \prod_{i \in I} B_i$ such that $\rho_i\alpha = \alpha_i\pi_i$ for all i (where $\pi_i : \prod_{i \in I} A_i \longrightarrow A_i$ and $\rho_i : \prod_{i \in I} B_i \longrightarrow B_i$ are the projections). Dually, if $\coprod_{i \in I} A_i$ and $\coprod_{i \in I} B_i$ exist, there is a unique morphism $\coprod_{i \in I} \alpha_i : \coprod_{i \in I} A_i \longrightarrow \coprod_{i \in I} B_i$ such that $(\coprod_{i \in I} \alpha_i)\iota_i = \kappa_i\alpha_i$ for all i (where $\iota_i : A_i \longrightarrow \coprod_{i \in I} A_i$ and $\kappa_i : B_i \longrightarrow \coprod_{i \in I} B_i$ are the injections).

In general, it is immediate that $\lim 1_D = 1_{\lim D}$ and that $\lim(\varphi\psi) = (\lim\varphi)(\lim\psi)$. We have proved:

Proposition 17.4.8. *Let $\mathscr{G}$ be a directed graph and $\mathscr{C}$ be a category. Assume that a limit can be assigned to every diagram in $\mathscr{C}$ over $\mathscr{G}$. Then there is a functor* $\mathrm{Diag}(\mathscr{G},\mathscr{C}) \longrightarrow \mathscr{C}$ *which assigns to each diagram $D : \mathscr{G} \longrightarrow \mathscr{C}$ its assigned limit $\lim D$ and to each morphism $\varphi : D \longrightarrow D'$ of diagrams the induced morphism* $\lim\varphi : \lim D \longrightarrow \lim D'$. $\square$

Dually (= by Metatheorem 17.1.4) there is a colimit functor $\mathrm{Diag}(\mathscr{G},\mathscr{C}) \longrightarrow \mathscr{C}$ if a colimit can be assigned to every diagram in $\mathscr{C}$ over $\mathscr{G}$.

For example, the categories *Sets*, *Grps*, *Rings*, *R-Mods*, and so on, have a limit functor for every graph. By Proposition 17.4.7, *R-Mods* has a colimit functor for every graph; so do *Sets*, *Grps*, *Rings*, *R-Algs*, and so on.

Exercises

1. Show that *R-Mods* is complete, and that the forgetful functor from *R-Mods* to *Sets* creates limits.

2. Show that the forgetful functor from *R-Mods* to *Abs* creates limits.

3. Show that a category is complete if and only if it has products and pullbacks.

4. Prove that *Sets* is cocomplete.

5. Prove that *Grps* is cocomplete.

6. Let $\mathscr{C}$ be a category with small Hom sets and D be a diagram in $\mathscr{C}$. Prove directly that $\lambda : D \longrightarrow L$ is a colimit cone of D if and only if $\mathrm{Hom}_{\mathscr{C}}(\lambda, A)$ is a colimit cone of $\mathrm{Hom}_{\mathscr{C}}(D, A)$ for every object A of $\mathscr{C}$.

7. Assume that a limit can be assigned to every diagram in $\mathscr{C}$. Prove that $\operatorname{Func}(\mathscr{A}, \mathscr{C})$ is complete for every small category $\mathscr{A}$; in fact a "pointwise" limit can be assigned to every diagram in $\operatorname{Func}(\mathscr{A}, \mathscr{C})$.

8. Prove that every limit functor preserves products.

9. Prove that every limit functor preserves equalizers.

17.5. ADDITIVE CATEGORIES

Additive categories and abelian categories are categories that share some remarkable features of *Abs*, *R-Mods*, and *R-S-Bimods*. The results in this section are not required in the succeeding sections.

17.5.a. Definition. An **additive category** is a category $\mathscr{A}$ with small Hom sets and with an abelian group operation on each $\operatorname{Hom}_{\mathscr{A}}(A, B)$ such that composition is biadditive ($(\alpha + \beta)\gamma = \alpha\gamma + \beta\gamma$ and $\alpha(\gamma + \delta) = \alpha\gamma + \alpha\delta$ whenever defined). (Additive categories are often required to also have a zero object and finite biproducts; we prefer the older definition.) This definition is self-dual: if $\mathscr{A}$ is an additive category, then $\mathscr{A}^{\mathrm{op}}$, with the same abelian group operations, is an additive category.

For example, *Abs*, *R-Mods*, and *R-S-Bimods* are additive categories.

The reader will verify that a ring [with an identity element] is precisely an additive category with just one object. Conversely, in an additive category $\mathscr{A}$, $\operatorname{End}_{\mathscr{A}}(A) = \operatorname{Hom}_{\mathscr{A}}(A, A)$ is a ring for each object A.

In an additive category $\mathscr{A}$, the **zero morphism** $0 : A \longrightarrow B$ is the zero element of $\operatorname{Hom}_{\mathscr{A}}(A, B)$ (such that $0 + \alpha = \alpha = \alpha + 0$ for all $\alpha : A \longrightarrow B$). If 0β is defined, then $0\beta = (0 + 0)\beta = 0\beta + 0\beta$ and $0\beta = 0$. Similarly $\gamma 0 = 0$ whenever $\gamma 0$ is defined.

In an additive category $\mathscr{A}$, a **zero object** of $\mathscr{A}$ is an object Z such that $\operatorname{Hom}_{\mathscr{A}}(A, Z) = \{0\}$ and $\operatorname{Hom}_{\mathscr{A}}(Z, A) = \{0\}$ for every object A; equivalently, such that there is only one morphism $A \longrightarrow Z$ and one morphism $Z \longrightarrow A$ for each object A. Necessarily $A \longrightarrow Z$ and $Z \longrightarrow B$ are zero morphisms; hence $\alpha : A \longrightarrow B$ is a zero morphism if and only if α factors through Z. A zero object (if it exists) is unique up to isomorphism, and is normally denoted by 0.

17.5.b. Biproducts. In an additive category, finite products (= products of finite families of objects) coincide with finite coproducts.

Proposition 17.5.1. *Let A, B, and P be objects of an additive category $\mathscr{A}$. The following conditions are equivalent:*

(1) *There exist morphisms $\pi : P \longrightarrow A$, $\rho : P \longrightarrow B$ such that P is a product of A and B with projections π and ρ.*

(2) *There exist morphisms $\iota : A \longrightarrow P$, $\kappa : B \longrightarrow P$ such that P is a coproduct of A and B with injections ι and κ.*

(3) *There exist morphisms*

$$A \underset{\pi}{\overset{\iota}{\rightleftarrows}} P \underset{\kappa}{\overset{\rho}{\rightleftarrows}} B$$

such that $\pi\iota = 1_A$, $\rho\kappa = 1_B$, $\pi\kappa = 0$, $\rho\iota = 0$, and $\iota\pi + \kappa\rho = 1_P$.

Proof. (1) implies (3). Let P be a product of A and B with projections π and ρ. By the universal property of products, there exists a morphism $\iota : A \longrightarrow P$ such that $\pi\iota = 1_A$ and $\rho\iota = 0$, and a morphism $\kappa : B \longrightarrow P$ such that $\pi\kappa = 0$ and $\rho\kappa = 1_B$. Then $\pi(\iota\pi + \kappa\rho) = \pi$ and $\rho(\iota\pi + \kappa\rho) = \rho$; therefore $\iota\pi + \kappa\rho = 1_P$.

(3) implies (1). Assume (3). Let $\alpha : C \longrightarrow A$ and $\beta : C \longrightarrow B$ be morphisms. Then $\gamma = \iota\alpha + \kappa\beta : C \longrightarrow P$ satisfies $\pi\gamma = \alpha$ and $\rho\gamma = \beta$. If $\delta : C \longrightarrow P$ and $\pi\delta = \alpha$ and $\rho\delta = \beta$, then $\delta = (\iota\pi + \kappa\rho)\delta = \iota\alpha + \kappa\beta = \gamma$. Thus P is a product of A and B.

Dually, (2) and (3) are equivalent. $\qquad\square$

Since products and coproducts are associative, Proposition 17.5.1 extends to all nonempty finite families of objects; the details are for our reader.

In Proposition 17.5.1, the object P, together with the morphisms π, ρ and/or ι, κ, is often called a **biproduct** of A and B, and is denoted by $A \oplus B$.

If $A \oplus B$ and $C \oplus D$ exist, then for any $\alpha : A \longrightarrow C$ and $\beta : B \longrightarrow D$ the induced morphisms $\alpha \times \beta$ and $\alpha \amalg \beta$ coincide:

$$
\begin{array}{ccc}
A \longleftarrow A \times C \longrightarrow C \\
{\scriptstyle\alpha}\downarrow \quad {\scriptstyle\alpha\times\beta}\downarrow \quad \downarrow{\scriptstyle\beta} \\
B \longleftarrow B \times D \longrightarrow D
\end{array}
\qquad
\begin{array}{ccc}
A \longrightarrow A \amalg C \longleftarrow C \\
{\scriptstyle\alpha}\downarrow \quad {\scriptstyle\alpha\amalg\beta}\downarrow \quad \downarrow{\scriptstyle\beta} \\
B \longrightarrow B \amalg D \longleftarrow D
\end{array}
$$

the proof is an exercise. The morphism $\alpha \times \beta = \alpha \amalg \beta$ is denoted by $\alpha \oplus \beta$.

We show that the addition on $\mathrm{Hom}_{\mathscr{A}}(A, B)$ is determined by biproducts, when the latter exist. Assume that $A \oplus A$ exists, with projections π, ρ and injections ι, κ. The **diagonal** morphism $\Delta_A : A \longrightarrow A \oplus A$ is the unique morphism such that $\pi\Delta_A = \rho\Delta_A = 1_A$. The **codiagonal** morphism $\nabla_A : A \oplus A \longrightarrow A$ is the unique morphism such that $\nabla_A\iota = \nabla_A\kappa = 1_A$.

Proposition 17.5.2. *When $\alpha, \beta : A \longrightarrow B$ are morphisms in an additive category, and $A \oplus A$, $B \oplus B$ exist, then $\alpha + \beta = \nabla_B(\alpha \oplus \beta)\Delta_A$.*

Proof. We have a commutative diagram

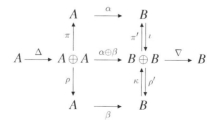

in which the vertical arrows are projections and injections. By definition, $\nabla_B \iota = \nabla_B \kappa = 1_B$ and $\pi \Delta_A = \rho \Delta_A = 1_A$. Hence

$$\nabla_B(\alpha \oplus \beta)\Delta_A = \nabla_B(\iota\pi' + \kappa\rho')(\alpha \oplus \beta)\Delta_A$$
$$= \pi'(\alpha \oplus \beta)\Delta_A + \rho'(\alpha \oplus \beta)\Delta_A = \alpha\pi\Delta_A + \beta\rho\Delta_A = \alpha + \beta. \quad \square$$

17.5.c. Additive Functors. When $\mathscr{A}$ and $\mathscr{B}$ are additive categories, a functor $F : \mathscr{A} \longrightarrow \mathscr{B}$ is **additive** in case $F(\alpha + \beta) = F(\alpha) + F(\beta)$ whenever $\alpha + \beta$ is defined.

For example, $\mathrm{Hom}_R(M, -)$ and $M \otimes_R -$ are additive functors. In any additive category, $\mathrm{Hom}_{\mathscr{A}}(A, -)$ and $\mathrm{Hom}_{\mathscr{A}}(-, B)$ are additive functors from $\mathscr{A}$ to *Abs*.

Proposition 17.5.3. *Let $\mathscr{A}$ and $\mathscr{B}$ be additive categories. If all biproducts exist in $\mathscr{A}$, then a functor $F : \mathscr{A} \longrightarrow \mathscr{B}$ is additive if and only if it preserves biproducts.*

Proof. If F preserves biproducts, then F preserves diagonal and codiagonal morphisms and it follows from Proposition 17.5.2 that F is additive. The converse follows from Proposition 17.5.1. $\quad \square$

17.5.d. Abelian Categories. In an additive category, a **kernel** of a morphism $\alpha : A \longrightarrow B$ is an equalizer $\kappa : K \longrightarrow A$ of α and 0. The domain of κ is also called a kernel of α. Thus $\alpha\kappa = 0$, and every morphism φ such that $\alpha\varphi = 0$ factors uniquely through κ:

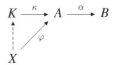

In particular, κ is a monomorphism and is unique up to isomorphism.

Dually, a **cokernel** of a morphism $\alpha : A \longrightarrow B$ is a coequalizer $\gamma : B \longrightarrow C$ of α and 0. The codomain of γ is also called a cokernel of α. Thus $\gamma\alpha = 0$, and every morphism φ such that $\varphi\alpha = 0$ factors uniquely through γ:

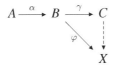

In particular, γ is an epimorphism and is unique up to isomorphism.

An **abelian category** is an additive category $\mathscr{A}$ such that:

(1) $\mathscr{A}$ has a zero object, biproducts, kernels, and cokernels;

(2) every monomorphism is a kernel and every epimorphism is a cokernel.

For example, *Abs*, *R-Mods*, and *R-S-Bimods* are abelian categories. We show that conversely many elementary properties of *Abs*, *R-Mods*, and *R-S-Bimods* extend to all abelian categories.

Proposition 17.5.4. *An abelian category has finite limits and colimits.*

Proof. This follows from Proposition 17.4.6 and (1). Since every two objects have a product, the associativity property of products provides a product for every nonempty finite family. The zero object is a product of the empty family. Hence $\mathscr{A}$ has finite products. It is immediate that a kernel of $\alpha - \beta$ is an equalizer of $\alpha, \beta : A \longrightarrow B$. Therefore $\mathscr{A}$ has finite limits. Dually $\mathscr{A}$ has finite colimits. $\qquad\square$

Some applications require a stronger condition than (1), namely that a limit and colimit can be asigned to every finite diagram.

Lemma 17.5.5. *In an abelian category: if σ is an epimorphism and μ is a kernel of σ, then σ is a cokernel of μ; if μ is a monomorphism and σ is a cokernel of μ, then μ is a kernel of σ.*

Proof. By (2), every epimorphism σ is a cokernel of some morphism α. If μ is a kernel of σ, then $\sigma\alpha = 0$ and α factors through μ: $\alpha = \mu\tau$. Now $\sigma\mu = 0$, and $\varphi\mu = 0$ implies $\varphi\alpha = 0$, so that φ factors uniquely through σ. Thus σ is a cokernel of μ. $\qquad\square$

Proposition 17.5.6. *In an abelian category: a morphism α is a monomorphism if and only if 0 is a kernel of α; α is an epimorphism if and only if 0 is a cokernel of α; α is an isomorphism if and only if it is both a monomorphism and an epimorphism.*

Proof. Let κ be a kernel of α. If α is a monomorphism, then $\alpha\kappa = 0 = \alpha 0$ implies $\kappa = 0$. If, conversely, $\kappa = 0$, and $\alpha\varphi = \alpha\psi$, then $\alpha(\varphi - \psi) = 0$, $\varphi - \psi$ factors through κ, and $\varphi - \psi = 0$.

Let $\alpha : A \longrightarrow B$ be a monomorphism and an epimorphism. Let γ be a cokernel of α. Then $\gamma = 0$, since α is an epimorphism. Also α is a kernel of γ by Lemma 17.5.5. Since $\gamma 1_B = 0$, 1_B factors through α : $1_B = \alpha\beta$ for some $\beta : B \longrightarrow A$. Then $\alpha\beta\alpha = \alpha = \alpha 1_A$ and $\beta\alpha = 1_A$. $\qquad\square$

Proposition 17.5.7. *In an abelian category, every morphism α has a decomposition $\alpha = \mu\sigma$ where μ is a monomorphism and σ is an epimorphism, and this decomposition is unique up to isomorphism.*

Proof. Let γ be a cokernel of α and μ be a kernel of γ. By Lemma 17.5.5, γ is a cokernel of μ. Since $\gamma\alpha = 0$, we have $\alpha = \mu\sigma$ for some unique σ:

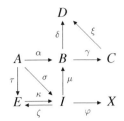

Assume that $\varphi\sigma = 0$. Let κ be a kernel of φ. Then σ factors through κ : $\sigma = \kappa\tau$. Let δ be a cokernel of $\mu\kappa$. Then $\delta\alpha = \delta\mu\kappa\tau = 0$ and δ factors through γ : $\delta = \xi\gamma$ for some ξ. Now $\mu\kappa$ is a kernel of δ by Lemma 17.5.5, since $\mu\kappa$ is a monomorphism. Since $\delta\mu = \xi\gamma\mu = 0$, μ factors through $\mu\kappa$: $\mu = \mu\kappa\zeta$ for some ζ. Since μ is a monomorphism this implies $\kappa\zeta = 1$. Hence $\varphi\kappa = 0$ implies $\varphi = \varphi\kappa\zeta = 0$. Therefore σ is an epimorphism.

Let $\alpha = \nu\tau$ be another decomposition where ν is a monomorphism and τ is an epimorphism. Then $\varphi\nu = 0$ implies $\varphi\alpha = 0$; if, conversely, $\varphi\alpha = \varphi\nu\tau = 0$, then $\varphi\nu = 0$, since τ is an epimorphism. Hence α and ν have the same cokernels. By Lemma 17.5.5, ν is a kernel of γ, and there is an isomorphism θ such that $\nu = \mu\theta$:

$$\begin{array}{ccccc}
A & \xrightarrow{\sigma} & I & \xrightarrow{\mu} & B & \xrightarrow{\gamma} & C \\
\| & & \uparrow{\scriptstyle\theta} & & \| & & \\
A & \xrightarrow{\tau} & J & \xrightarrow{\nu} & B & &
\end{array}$$

This implies $\mu\theta\tau = \nu\tau = \mu\sigma$ and $\theta\tau = \sigma$. $\qquad\square$

When $\alpha = \mu\sigma$, with μ a monomorphism and σ an epimorphism, μ is an **image** of α and σ is a **coimage** of α. The image and coimage of α are unique up to isomorphism. The proof of Proposition 17.5.7 shows that an image of α is a kernel of a cokernel of α; dually a coimage of α is a cokernel of a kernel of α.

Other properties will be found in the exercises.

Exercises

1. Verify that a ring [with an identity element] is precisely an additive category with just one object.

2. In an additive category $\mathscr{A}$, show that the following conditions are equivalent for an object Z: (a) Z is a zero object; (b) $1_Z = 0$; (c) $\mathrm{Hom}_{\mathscr{A}}(Z,Z) = 0$.

3. Extend Proposition 17.5.1 to all nonempty finite families $(A_i)_{i \in I}$ of objects.

4. Show that the diagonal morphism Δ_A is natural in A.

5. Assume that $A \oplus B$ and $C \oplus D$ exist. For any $\alpha : A \longrightarrow C$ and $\beta : B \longrightarrow D$, show that the induced morphisms $\alpha \times \beta$ and $\alpha \amalg \beta$ coincide.

6. In an abelian category, a morphism is a monomorphism if and only if its kernel is 0; a morphism is an epimorphism if and only if its cokernel is 0.

7. In an abelian category, show that a coimage of α is a cokernel of a kernel of α, and conversely.

8. In an abelian category, assume that $\alpha = \mu\sigma$, $\beta = \nu\tau$, where μ, ν are monomorphisms and σ, τ are epimorphisms. If the diagram

$$
\begin{array}{ccccc}
A & \xrightarrow{\sigma} & I & \xrightarrow{\mu} & B \\
\psi \downarrow & & \chi \downarrow & & \downarrow \varphi \\
C & \xrightarrow{\tau} & J & \xrightarrow{\nu} & D
\end{array}
$$

commutes, there is a unique morphism χ which keeps the diagram commutative.

9. Let $\alpha\beta' = \beta\alpha'$ be a pullback in an abelian category. Prove that if α is a monomorphism, then α' is a monomorphism.

10. Define exact sequences in an abelian category. Define split exact sequences by extending Proposition 15.5.1.

11. Show that the construction of pullbacks of modules extends to every abelian category.

12. Let $\mathscr{A}$ be a small category and $\mathscr{B}$ be an abelian category in which a limit and colimit can be assigned to every finite diagram in $\mathscr{B}$. Show that $\mathrm{Func}(\mathscr{A}, \mathscr{B})$ is abelian.

13. Let $\mathscr{A}$ be a small category and $\mathscr{B}$ be an abelian category in which a limit and colimit can be assigned to every finite diagram in $\mathscr{B}$. When is a sequence in $\mathrm{Func}(\mathscr{A}, \mathscr{B})$ exact?

17.6. ADJOINT FUNCTORS

Limits and colimits do not account for some universal properties encountered in previous chapters: free groups, free modules, tensor products, and others. These are particular cases of adjoint functors.

17.6.a. Definition. The definition of adjoint functors is modeled after the universal property of, say, the free group F_X on a set X and its injection $\iota_X : X \longrightarrow F_X$: for every mapping φ of X into a group G there is a unique homomorphism $\psi : F_X \longrightarrow G$ such that $\psi \circ \iota = \varphi$. We saw that F_X is the value at X of a free group functor from *Sets* to *Grps*, which assigns to a mapping $\varphi : X \longrightarrow Y$ the unique homomorphism $F(\varphi) : F_X \longrightarrow F_Y$ such that $F(\varphi) \circ \iota_X = \iota_Y \circ \varphi$.

A precise statement of this universal property in terms of categories requires the two categories *Sets* and *Grps* and the forgetful functor $S : Grps \longrightarrow Sets$. The inclusion mapping $\iota_X : X \longrightarrow F_X$ is a morphism from X to $S(F_X)$ and is natural in X by definition of the free group functor. The universal property of F_X now reads: for every group G and morphism $\varphi : X \longrightarrow S(G)$ (in *Sets*), there is a unique morphism $\psi : F_X \longrightarrow G$ (in *Grps*) such that $\varphi = S(\psi) \circ \iota_X$.

The general definition of adjoint functors is based on this formulation, its dual, and an equivalent property.

Proposition 17.6.1. *Let $\mathscr{A}$ and $\mathscr{C}$ have small Hom sets. For two functors $F : \mathscr{A} \longrightarrow \mathscr{C}$ and $G : \mathscr{C} \longrightarrow \mathscr{A}$ the following conditions are equivalent:*

(1) *There exists a natural transformation $\eta : 1_{\mathscr{A}} \longrightarrow G \circ F$ such that for each morphism $\alpha : A \longrightarrow G(C)$ in $\mathscr{A}$ there is a unique morphism $\gamma : F(A) \longrightarrow C$ in $\mathscr{C}$ such that $\alpha = G(\gamma)\eta_A$:*

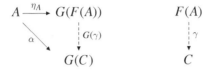

(2) *There exists a natural transformation $\varepsilon : F \circ G \longrightarrow 1_{\mathscr{C}}$ such that for each morphism $\gamma : F(A)C \longrightarrow$ in $\mathscr{C}$ there is a unique morphism $\alpha : A \longrightarrow G(C)$ in $\mathscr{A}$ such that $\gamma = \varepsilon_C F(\alpha)$:*

(3) *There exists a natural isomorphism $\operatorname{Hom}_{\mathscr{C}}(F(-),-) \cong \operatorname{Hom}_{\mathscr{A}}(-,G(-))$.*

Proof. (1) implies (3). The natural transformation $\eta : 1_{\mathscr{A}} \longrightarrow G \circ F$ provides, for every objects A of $\mathscr{A}$ and C of $\mathscr{C}$, a mapping $\theta_{A,C}$ of $\operatorname{Hom}_{\mathscr{A}}(F(A),C)$ into $\operatorname{Hom}_{\mathscr{C}}(A,G(C))$ defined by

$$\theta_{A,C}(\gamma) = G(\gamma)\eta_A.$$

By (1), $\theta_{A,C}$ is bijective for every A,C. Moreover $\theta_{A,C}$ is natural in A and C: if $\beta : A \longrightarrow B$ and $\delta : C \longrightarrow D$ are morphisms, then $\ell = \operatorname{Hom}_{\mathscr{C}}(F(\beta),\delta)$ sends $\gamma \in \operatorname{Hom}_{\mathscr{C}}(F(B),C)$ to $\delta\gamma F(\beta)$, $r = \operatorname{Hom}_{\mathscr{A}}(\beta,G(\delta))$ sends $\alpha \in \operatorname{Hom}_{\mathscr{A}}(B,G(C))$ to $G(\delta)\alpha\beta$; for every $\gamma \in \operatorname{Hom}_{\mathscr{C}}(F(B),C)$,

$$r(\theta_{B,C}(\gamma)) = G(\delta)G(\gamma)\eta_B\,\beta \qquad \text{and} \qquad \theta_{A,D}(\ell(\gamma)) = G(\delta)G(\gamma)G(F(\beta))\eta_A$$

are equal since η is a natural transformation, and the following square commutes:

$$
\begin{array}{ccc}
\operatorname{Hom}_{\mathscr{C}}(F(B),C) & \xrightarrow{\theta_{B,C}} & \operatorname{Hom}_{\mathscr{A}}(B,G(C)) \\
\ell \downarrow & & \downarrow r \\
\operatorname{Hom}_{\mathscr{C}}(F(A),D) & \xrightarrow[\theta_{A,D}]{} & \operatorname{Hom}_{\mathscr{A}}(A,G(D))
\end{array}
$$

(3) implies (1). Let $\theta : \mathrm{Hom}_{\mathscr{C}}(F(-), -) \longrightarrow \mathrm{Hom}_{\mathscr{A}}(-, G(-))$ be a natural isomorphism of bifunctors (so that the diagram above commutes). Appying $\theta_{A, F(A)}$ to $1_{F(A)}$ yields a morphism $\eta_A = \theta_{A, F(A)}(1_{F(A)})$ from A to $G(F(A))$. Since θ is natural we have, as above,

$$G(\delta)\theta_{B,C}(\gamma)\beta = r(\theta_{B,C}(\gamma)) = \theta_{A,D}(\ell(\gamma)) = \theta_{A,D}(\delta\gamma F(\beta)) \tag{$*$}$$

whenever $\beta : A \longrightarrow B$, $\gamma : F(B) \longrightarrow C$, and $\delta : C \longrightarrow D$. With $A = B$, $C = F(A) = F(B)$, $\beta = 1_A$, and $\gamma = 1_{F(A)}$, $(*)$ yields

$$G(\delta)\eta_A = \theta_{A,D}(\delta)$$

whenever $\delta : F(A) \longrightarrow D$. Then $(*)$ yields, with $C = D = F(B)$ and $\gamma = \delta = 1_{F(B)}$,

$$\eta_B\beta = \theta_{A, F(B)}(F(\beta)) = G(F(\beta))\eta_A$$

for all $\beta : A \longrightarrow B$. Thus $\eta : 1_{\mathscr{A}} \longrightarrow G \circ F$ is a natural transformation. For all A and C, $\theta_{A,C}$ sends $\gamma : F(A) \longrightarrow C$ to $G(\gamma)\eta_A : A \longrightarrow G(C)$ and is bijective; therefore (1) holds.

If we exchange F and G, and replace $\mathscr{A}$ and $\mathscr{C}$ by $\mathscr{C}^{\mathrm{op}}$ and $\mathscr{A}^{\mathrm{op}}$, then (1) becomes (2) but (3) remains unchanged. Therefore (2) and (3) are equivalent. $\square$

When the equivalent conditions in Proposition 17.6.1 hold, F and G are a pair of mutually **adjoint functors**; F is a **left adjoint** of G and G is a **right adjoint** of F. This terminology reflects the placement of F and G in the Hom functors of condition (3), and its analogy to the definition $\langle Fx, y \rangle = \langle x, Gy \rangle$ of adjoint linear transformations. An **adjunction** from $\mathscr{A}$ to $\mathscr{C}$ is more precisely defined as consisting of F, G, and at least one of η, ε, and θ.

17.6.b. Examples. As expected, the forgetful functor from *Grps* to *Sets* has a left adjoint, the free group functor. Similarly the forgetful functor from *R-Mods* to *Sets* has a left adjoint, the free left *R*-module functor.

We saw that tensor products have an adjoint associativity property: when A is a *Q-R*-bimodule, B is an *R-S*-bimodule, and C is a *Q-S*-bimodule, there is a natural isomorphism

$$\mathrm{Hom}_{QS}(A \otimes_R B, C) \cong \mathrm{Hom}_{QR}(A, \mathrm{Hom}_S(B, C)).$$

By part (3) of Proposition 17.6.1, $- \otimes_R B$ is a left adjoint of $\mathrm{Hom}_S(B, -)$, for every *R-S*-bimodule B.

Limit functors are right adjoints. Let $\mathscr{C}$ be a category in which a limit can be assigned to every diagram over some graph $\mathscr{G}$ so that there is a limit functor from $\mathrm{Diag}(\mathscr{G}, \mathscr{C})$ to $\mathscr{C}$. Let $C : \mathscr{C} \longrightarrow \mathrm{Diag}(\mathscr{G}, \mathscr{C})$ be the constant diagram functor. The limit cone $\lambda : C(\lim D) \longrightarrow D$ is readily seen to be natural in

D. For every morphism $\varphi : C(A) \longrightarrow D$ (= for every cone $A \longrightarrow D$) there is a unique morphism $\overline{\varphi} : A \longrightarrow \lim D$ such that $\varphi = \lambda C(\overline{\varphi})$; by part (2) of Proposition 17.6.1, $\lim$ is a right adjoint of C. Dually a colimit functor is a left adjoint of a constant diagram functor.

17.6.c. Properties. We note some easy consequences of the definition. The universal properties in conditions (1) and (2) readily imply:

Proposition 17.6.2. *Up to natural isomorphisms, a functor has at most one left adjoint, and at most one right adjoint.*

The existence of a left (or right) adjoint is considered in the next section.

Proposition 17.6.3. *Let $\mathscr{A}$ and $\mathscr{C}$ have small Hom sets. A functor $G : \mathscr{C} \longrightarrow \mathscr{A}$ has a left adjoint if and only if to each object A of $\mathscr{A}$ one can assign an object $F(A)$ of $\mathscr{C}$ and a morphism $\eta_A : A \longrightarrow G(F(A))$ such that for each morphism $\alpha : A \longrightarrow G(C)$ in $\mathscr{A}$ there is a unique morphism $\gamma : F(A) \longrightarrow C$ in $\mathscr{C}$ such that $\alpha = G(\gamma)\eta_A$.*

Proof. This condition is necessary by part (1) of Proposition 17.6.1. Now let F and η be as above. We extend F to a functor such that η is a natural transformation; then F is a left adjoint of G by Proposition 17.6.1. For each morphism $\alpha : A \longrightarrow B$ the morphism $\eta_B\alpha : A \longrightarrow G(F(B))$ yields a unique morphism $F(\alpha) : F(A) \longrightarrow F(B)$ such that $\eta_B\alpha = G(F(\alpha))\eta_A$:

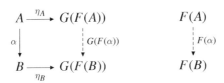

By uniqueness, $F(1_A) = 1_{F(A)}$ and $F(\alpha\beta) = F(\alpha)F(\beta)$ whenever $\alpha\beta$ is defined. Hence F is now a functor. The definition of $F(\alpha)$ shows that η is a natural transformation. $\qquad\square$

Lemma 17.6.4. *If $(F,G,\eta,\varepsilon,\theta)$ is an adjunction, then: $\theta_{A,C}(\gamma) = G(\gamma)\eta_A$; $\theta_{A,C}^{-1}(\alpha) = \varepsilon_C F(\alpha)$; and $G(\varepsilon_C)\eta_{G(C)} = 1_{G(C)}$, $\varepsilon_{F(A)}F(\eta_A) = 1_{F(A)}$ for all A,C,α,γ.*

Proof. The proof of Proposition 17.6.1 shows that if η has property (1), then $\theta_{A,C}(\gamma) = G(\gamma)\eta_A$ has property (3); conversely, if θ has property (3), then $\eta_A = \theta_{A,F(A)}(1_{F(A)})$ has property (1) and $\theta_{A,C}(\gamma) = G(\gamma)\eta_A$ holds.

Dually, if ε has property (2), then $\zeta_{A,C}(\alpha) = \varepsilon_C F(\alpha)$ is a natural isomorphism from $\mathrm{Hom}_{\mathscr{A}}(-,G(-))$ to $\mathrm{Hom}_{\mathscr{C}}(F(-),-)$, and $\theta = \zeta^{-1}$ has property (3). If, conversely, θ has property (3), then $\zeta = \theta^{-1}$ is a natural isomorphism, $\varepsilon_C = \theta_{G(C),C}^{-1}(1_{G(C)})$ has property (2), and $\theta_{A,C}^{-1}(\alpha) = \varepsilon_C F(\alpha)$ holds. Hence $1_{G(C)} = \theta_{G(C),C}(\varepsilon_C) = G(\varepsilon_C)\eta_{G(C)}$, and $1_{F(A)} = \theta_{A,F(A)}^{-1}(\eta_A) = \varepsilon_{F(A)}F(\eta_A)$. $\qquad\square$

Our next result follows from Proposition 17.4.4 and implies several results from previous chapters as well as Proposition 17.4.2.

Proposition 17.6.5. *Every right adjoint functor preserves limits. Dually every left adjoint functor preserves colimits.*

Proof. Let $F : \mathscr{A} \longrightarrow \mathscr{C}$ be a left adjoint of $G : \mathscr{C} \longrightarrow \mathscr{A}$ so that $\mathrm{Hom}_{\mathscr{C}}(F(-),-) \cong \mathrm{Hom}_{\mathscr{A}}(-,G(-))$. Let D be a diagram in $\mathscr{C}$ over some graph and $\lambda : C \longrightarrow D$ be a limit cone of D. For every object A of $\mathscr{A}$, $\mathrm{Hom}_{\mathscr{C}}(F(A),\lambda)$ is by Proposition 17.4.4 a limit cone of $\mathrm{Hom}_{\mathscr{C}}(F(A),D)$. Hence $\mathrm{Hom}_{\mathscr{A}}(A,G(\lambda))$ is a limit cone of $\mathrm{Hom}_{\mathscr{A}}(A,G(D))$. Since this holds for every A, $G(\lambda)$ is a limit cone of $G(D)$ by Proposition 17.4.4. $\square$

17.6.d. Morphisms. A **functorial transformation** (or **morphism**) from one adjunction to another is a pair of functors (P,Q) with the equivalent properties in the next result.

Proposition 17.6.6. *Let $(F,G,\eta,\varepsilon,\theta) : \mathscr{A} \longrightarrow \mathscr{C}$ and $(F',G',\eta',\varepsilon',\theta') : \mathscr{A}'$* $\longrightarrow \mathscr{C}'$ *be adjunctions. Let $P : \mathscr{A} \longrightarrow \mathscr{A}'$ and $Q : \mathscr{C} \longrightarrow \mathscr{C}'$ be functors such that $Q \circ F = F' \circ P$ and $P \circ G = G' \circ Q$:*

$$
\begin{array}{ccc}
\mathscr{A} \xrightarrow{\;F\;} \mathscr{C} & \qquad & \mathscr{C} \xrightarrow{\;G\;} \mathscr{A} \\
{\scriptstyle P}\downarrow \quad\quad \downarrow{\scriptstyle Q} & & {\scriptstyle Q}\downarrow \quad\quad \downarrow{\scriptstyle P} \\
\mathscr{A}' \xrightarrow[\;F'\;]{} \mathscr{C}' & & \mathscr{C}' \xrightarrow[\;G'\;]{} \mathscr{A}'
\end{array}
$$

The following conditions are equivalent:

(i) $P \circ \eta = \eta' \circ P$.

(ii) $Q \circ \varepsilon = \varepsilon' \circ Q$.

(iii) *The diagram*

$$
\begin{array}{ccc}
\mathrm{Hom}_{\mathscr{C}}(F(A),C) & \xrightarrow{\;\theta_{A,C}\;} & \mathrm{Hom}_{\mathscr{A}}(A,G(C)) \\
{\scriptstyle Q}\downarrow & & \downarrow{\scriptstyle P} \\
\mathrm{Hom}_{\mathscr{C}'}(F'(P(A)),Q(C)) & \xrightarrow[\;\theta'_{P(A),Q(C)}\;]{} & \mathrm{Hom}_{\mathscr{A}'}(P(A),G'(Q(C)))
\end{array}
$$

commutes for all objects A of $\mathscr{A}$ and C of $\mathscr{C}$.

Proof. Note that Q takes $\mathrm{Hom}_{\mathscr{C}}(F(A), C)$ to $\mathrm{Hom}_{\mathscr{C}'}(Q(F(A)), Q(C)) = \mathrm{Hom}_{\mathscr{C}'}(F'(P(A)),Q(C))$; P takes $\mathrm{Hom}_{\mathscr{A}}(A,G(C))$ to $\mathrm{Hom}_{\mathscr{A}'}(P(A),P(G(C))) = \mathrm{Hom}_{\mathscr{A}'}(P(A),G'(Q(C)))$. With $C = F(A)$, (iii) yields, by Lemma 17.6.4,

$$P(\eta_A) = P(\theta_{A,F(A)}(1_{F(A)}))$$

$$= \theta'_{P(A),Q(F(A))}(Q(1_{F(A)})) = \theta'_{P(A),F'(P(A))}(1_{F'(P(A))}) = \eta'_{P(A)},$$

which is (i). Conversely, (i) implies, again by Lemma 17.6.4, that

$$P(\theta_{A,C})(\gamma) = P(G(\gamma))P(\eta_A) = G'(Q(\gamma))\eta'_{P(A)} = \theta'_{P(A),Q(C)}(Q(\gamma))$$

for all $\gamma \in \mathrm{Hom}_{\mathscr{C}}(F(A),C)$. Thus (i) and (iii) are equivalent. Dually (ii) and (iii) are equivalent. □

Exercises

1. Prove that any two left adjoints of a functor are naturally isomorphic.
2. Prove that (F,G,η,ε) is an adjunction if and only if $G(\varepsilon_C)\eta_{G(C)} = 1_{G(C)}$ and $\varepsilon_{F(A)}F(\eta_A) = 1_{F(A)}$ for all A and C.
3. Show that every equivalence of categories is a pair of adjoint functors.
4. Show that a colimit functor is a left adjoint of the constant diagram functor.
5. For any set B show that $- \times B$ is a left adjoint of $\mathrm{Hom}(B,-)$.
6. Give examples of adjoint functors which have not been mentioned in this section.
7. Let (F,G) be a pair of adjoint functors between abelian categories. Show that F and G are additive functors.
8. Let (F,G,θ) be an adjunction between abelian categories. Show that θ is an isomorphism of abelian groups.
9. Show that adjunctions between small categories and their functorial transformations are the objects and morphisms of a category.

17.7. THE ADJOINT FUNCTOR THEOREM

In this section we take a very general look at universal properties and consider the existence of adjoint functors.

17.7.a. Initial and Terminal Objects. An **initial** object of a category $\mathscr{C}$ is an object C such that there is for every object A of $\mathscr{C}$ exactly one morphism from C to A. Dually a **terminal** object is an object C such that there is for every object A of $\mathscr{C}$ exactly one morphism from A to C.

Equivalently, a terminal object is a product of the empty family of objects; an initial object is a coproduct of the empty family of objects. Hence every complete category has a terminal object; every cocomplete category has an initial object.

For example, in the category *Sets*, $\emptyset$ is an initial object, and every one element set is a terminal object. A zero object in an additive category is both an initial and a terminal object. The exercises give other examples.

Proposition 17.7.1. *Any two initial objects of a category are isomorphic. Dually any two terminal objects of a category are isomorphic.*

Proof. Let A and B be initial objects. There exists a morphism $\alpha : A \longrightarrow B$ and a morphism $\beta : B \longrightarrow A$. Then 1_A and $\beta\alpha$ are morphisms from A to A;

since there is only one such morphism, $\beta\alpha = 1_A$. Similarly $\alpha\beta = 1_B$. Thus A and B are isomorphic. □

This proof should be familiar to the reader as the standard uniqueness argument for objects defined by universal properties. Indeed initial objects are defined by a particularly simple universal property. Conversely, examples (given below) indicate that *every* universal property can be expressed by the existence of an initial object (or a terminal object) in a suitable category. (The exercises outline another approach to universal properties, using representable functors.)

For example, let D be a diagram over a graph $\mathscr{G}$ in a category $\mathscr{A}$. A colimit cone of D is an initial object in the following cone category $\mathscr{C}$. The objects of $\mathscr{C}$ are the cones $\varphi : D \longrightarrow A$ from D. A morphism from $\varphi : D \longrightarrow A$ to $\psi : D \longrightarrow B$ is a morphism $\alpha : A \longrightarrow B$ of $\mathscr{A}$ such that $\alpha\varphi = \psi$ (such that $\alpha\varphi_i = \psi_i$ for all $i \in \mathscr{G}$). Composition is composition in $\mathscr{A}$. We see that $\mathscr{C}$ is a category and that a colimit cone of D is precisely an initial object of $\mathscr{C}$. A limit cone of D is a terminal object in the similar category of cones to D (see the exercises).

For another example, let $F : \mathscr{A} \longrightarrow \mathscr{C}$ be a left adjoint of $G : \mathscr{C} \longrightarrow \mathscr{A}$. Then $\eta_A : A \longrightarrow G(F(A))$ (natural in A) has the following universal property: for each morphism $\alpha : A \longrightarrow G(C)$ in $\mathscr{A}$, there is a unique morphism $\gamma : F(A) \longrightarrow C$ in $\mathscr{C}$ such that $\alpha = G(\gamma)\eta_A$:

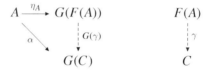

Let $\mathscr{C}_A$ be the following category. An object of $\mathscr{C}_A$ is an ordered pair (α, C) in which C is an object of $\mathscr{C}$ and α is a morphism in $\mathscr{A}$ from A to $G(C)$. A morphism from (α, C) to (β, D) is a morphism $\gamma : C \longrightarrow D$ in $\mathscr{C}$ such that $\beta = G(\gamma)\alpha$. Composition is composition in $\mathscr{C}$. The universal property of η_A states precisely that $(\eta_A, F(A))$ is an initial object of $\mathscr{C}_A$.

17.7.b. Existence. The following result gives a sufficient condition for the existence of an initial object:

Proposition 17.7.2. *Let $\mathscr{C}$ be a complete category with small Hom sets. Then $\mathscr{C}$ has an initial object if and only if it satisfies the* **solution set condition***:*

> *there exists a set $\mathscr{S}$ of objects of $\mathscr{C}$ such that for every object A of $\mathscr{C}$, there is a morphism from some $S \in \mathscr{S}$ to A.*

Proof. The solution set condition is necessary: if $\mathscr{C}$ has an initial object C, then $\{C\}$ is a solution set.

Conversely, assume that $\mathscr{C}$ has a solution set $\mathscr{S}$. By the hypothesis, $\mathscr{S}$ has a product P in $\mathscr{C}$. Then $\{P\}$ is a one element solution set: for every object A of $\mathscr{C}$, there is a morphism $P \longrightarrow S \longrightarrow A$. By the hypothesis, $\operatorname{Hom}_{\mathscr{C}}(P,P)$ is a set and has an equalizer $\varepsilon : E \longrightarrow P$. For every object A of $\mathscr{C}$, there is a morphism $\alpha : E \longrightarrow P \longrightarrow A$. We show that α is unique.

Assume that $\alpha, \beta : E \longrightarrow A$. Then α and β have an equalizer $\zeta : F \longrightarrow E$. There is a morphism $\varphi : P \longrightarrow F$. Then $\varepsilon \zeta \varphi \in \operatorname{Hom}_{\mathscr{C}}(P,P)$. By the choice of ε, $\varepsilon \zeta \varphi \varepsilon = 1_P \varepsilon = \varepsilon 1_E$; since ε is a monomorphism, this implies $\zeta \varphi \varepsilon = 1_E$.

$$F \xrightarrow{\ \zeta\ } E \underset{\beta}{\overset{\alpha}{\rightrightarrows}} A$$

with $\varphi : P \to F$ and $\varepsilon : E \to P$.

Then $\alpha \zeta = \beta \zeta$ implies $\alpha \zeta \varphi \varepsilon = \beta \zeta \varphi \varepsilon$ and $\alpha = \beta$. □

The proof of Proposition 17.7.2 gives an explicit construction of the initial object. If in Proposition 17.7.2 every diagram in $\mathscr{C}$ can be assigned a limit, then a specific initial object can be selected in $\mathscr{C}$.

17.7.c. The Adjoint Functor Theorem. Proposition 17.7.2 implies a sufficient condition for the existence of a left adjoint.

Theorem 17.7.3 (Adjoint Functor Theorem). *Let $\mathscr{A}$ and $\mathscr{C}$ have small Hom sets and $G : \mathscr{C} \longrightarrow \mathscr{A}$ be a functor. Assume that a limit can be assigned to every [small] diagram in $\mathscr{C}$. Then G has a left adjoint if and only if it preserves limits and satisfies the* **solution set condition***:*

> *to each object A of $\mathscr{A}$ can be assigned a set $\mathscr{S}$ of morphisms $\sigma : A \longrightarrow G(C_\sigma)$ such that every morphism $A \longrightarrow G(C)$ with $C \in \mathscr{C}$ can be written in the form $G(\gamma)\sigma$ for some $\sigma \in \mathscr{S}$ and $\gamma : C_\sigma \longrightarrow C$.*

Before proving the Adjoint Functor Theorem, we show its use by an example: we show how it implies the existence of free groups. We know that *Grps* is complete and has small Hom sets, and that the forgetful functor from *Grps* to *Sets* preserves limits (Proposition 17.4.2). The existence of free groups then follows from the existence of a solution set for each set X. We construct one as follows.

Every mapping φ of X into a group G factors through the subgroup H of G generated by $Y = \varphi(X)$. Now every element of H is a finite product of elements of $Z = Y \cup Y^{-1}$ (= a finite product of elements of Y and inverses of elements of Y). If Y is finite, then H is countable. If Y is infinite, then there are $|Y|$ elements of Z and $|Y|^n$ products of n elements of Z; by Corollary B.4.12, $|H| \leqslant |Y| + \cdots + |Y| + \cdots = \aleph_0 |Y| = |Y|$. In either case we have $|H| \leqslant \mathfrak{b}$, where $\mathfrak{b} = \max(|X|, \aleph_0)$ depends only on X.

For each cardinal number $\mathfrak{a} \leqslant \mathfrak{b}$, we can choose one set T of cardinality $\mathfrak{a}$ and place all possible group operations $T \times T \longrightarrow T$ on T. The result is a set $\mathscr{T}$ of groups such that every group H of cardinality $|H| \leqslant \mathfrak{b}$ is isomorphic to some $T \in \mathscr{T}$. By the above every mapping φ of X into any group factors through a group of cardinality $\leqslant \mathfrak{b}$, and factors through some $T \in \mathscr{T}$. Now there is only a set $\mathscr{S}$ of mappings $X \longrightarrow T \in \mathscr{T}$; we see that $\mathscr{S}$ is a solution set for X.

17.7.d. Comma Categories. To prove the Adjoint Functor Theorem, we apply Proposition 17.7.2 to a suitable category. This is the **comma category** $(A \downarrow G)$ which we previously denoted by $\mathscr{C}_A$.

Let $G : \mathscr{C} \longrightarrow \mathscr{A}$ be a functor and A be an object of $\mathscr{A}$. The objects of $(A \downarrow G)$, called **objects under A via G**, are the ordered pairs (α, C) in which C is an object of $\mathscr{C}$ and α is a morphism in $\mathscr{A}$ from A to $G(C)$. A morphism from (α, C) to (β, D) is a morphism $\gamma : C \longrightarrow D$ in $\mathscr{C}$ such that $\beta = G(\gamma)\alpha$. Composition is composition in $\mathscr{C}$. It is immediate that $(A \downarrow G)$ is a category. If $\mathscr{C}$ has small Hom sets, then so does $(A \downarrow G)$.

There is a **projection** functor from $(A \downarrow G)$ to $\mathscr{C}$ which assigns to each object (α, C) of $(A \downarrow G)$ the object C of $\mathscr{C}$.

Lemma 17.7.4. *Let $\mathscr{C}$ be a category with small Hom sets in which a limit can be assigned to every diagram, $G : \mathscr{C} \longrightarrow \mathscr{A}$ be a functor which preserves limits, and A be an object of $\mathscr{A}$. Then a limit can be assigned to every diagram in $(A \downarrow G)$, and the projection $(A \downarrow G) \longrightarrow \mathscr{C}$ preserves limits.*

Proof. Let D be a diagram in $(A \downarrow G)$ over a graph $\mathscr{G}$. For each $i \in \mathscr{G}$, $D_i = (\delta_i, C_i)$ is an object under A; $\delta_i : A \longrightarrow G(C_i)$ is a morphism in $\mathscr{A}$. For each $a : i \longrightarrow j$, $D_a : C_i \longrightarrow C_j$ is a morphism such that $\delta_j = G(D_a)\delta_i$. The objects $(C_i)_{i \in \mathscr{G}}$ and morphisms D_a constitute a diagram C over $\mathscr{G}$ (which is the projection of D in $\mathscr{C}$), and $\delta = (\delta_i)_{i \in I}$ is a cone from A into $G(C)$. Let $\lambda : L \longrightarrow C$ be the limit cone assigned to C:

$$
\begin{array}{ccc}
A \xrightarrow{\delta_i} G(C_i) & & C_i \\
{\scriptstyle \bar{\delta}}\downarrow \quad {}_{\delta_j}\searrow \quad \downarrow {\scriptstyle G(D_a)} & & {}^{\lambda_i}\nearrow \quad \downarrow {\scriptstyle D_a} \\
G(L) \xrightarrow[G(\lambda_j)]{} G(C_j) & & L \xrightarrow[\lambda_j]{} C_j
\end{array}
$$

Since G preserves limits, $G(\lambda) : G(L) \longrightarrow G(C)$ is a limit cone of $G(C)$. Therefore there is a unique morphism $\bar{\delta} : A \longrightarrow G(L)$ such that $\delta_i = G(\lambda_i)\bar{\delta}$ for all $i \in \mathscr{G}$.

We show that $(\bar{\delta}, L)$ is a limit of D; the limit cone is $\lambda = (\lambda_i)_{i \in \mathscr{G}} : (\bar{\delta}, L) \longrightarrow D$. We see that λ is a cone to D. Let (α, B) be an object of $(A \downarrow G)$ and φ be a cone from (α, B) to D. Then [the projection of] φ is a cone from B to C and

the diagram

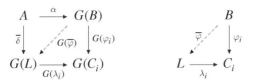

commutes. Since λ is a limit cone there is a unique morphism $\overline{\varphi} : B \longrightarrow L$ such that $\varphi_i = \lambda_i \overline{\varphi}$ for all i. Then $G(\lambda_i) G(\overline{\varphi}) \alpha = G(\lambda_i) \overline{\delta}$ for all i; since G preserves limits, this implies $G(\overline{\varphi}) \alpha = \overline{\delta}$. Thus $\overline{\varphi}$ is a morphism from (α, B) to $(\overline{\delta}, L)$, and is the only such morphism with $\varphi_i = \lambda_i \overline{\varphi}$ for all i. □

We can now prove Theorem 17.7.3. Let $\mathscr{C}$ and $G : \mathscr{C} \longrightarrow \mathscr{A}$ be as in Lemma 17.7.4. If G has a left adjoint, then $\{\eta_A\}$ is a solution set for each object A of $\mathscr{A}$. Conversely, assume that a solution set $\mathscr{S}$ of morphisms $\sigma : A \longrightarrow G(C_\sigma)$ can be assigned to each object A of $\mathscr{A}$ so that every morphism $\alpha : A \longrightarrow G(C)$ factors as $\alpha = G(\gamma) \sigma$ for some $\sigma \in \mathscr{S}$ and $\gamma : C_\sigma \longrightarrow C$. Then $\{(\sigma, C_\sigma); \sigma \in \mathscr{S}\}$ is a solution set in $(A \downarrow G)$ and $(A \downarrow G)$ has an initial object by Proposition 17.7.2; in fact the remark after Proposition 17.7.2 shows that an initial object $(\eta_A, F(A))$ can be assigned to $(A \downarrow G)$. Then Proposition 17.6.3 shows that G has a left adjoint. □

Exercises

1. Find all initial and terminal objects in the following categories: *Grps*; *Rings*; *R-Mods*; a partially ordered set I.

2. In an additive category, show that an initial object is a terminal object, and conversely.

3. Let D be a diagram over a graph G in a category $\mathscr{A}$. Define a category $\mathscr{C}$ of cones to D so that a limit cone of D is a terminal object in $\mathscr{C}$.

4. Use the Adjoint Functor Theorem to show that the forgetful functor from *R-Mods* to *Sets* has a left adjoint.

5. Use the Adjoint Functor Theorem to show that the forgetful functor from *Rings* to *Sets* has a left adjoint.

6. Let $\mathscr{C}$ be a category with small Hom sets. A functor $F : \mathscr{C} \longrightarrow Sets$ is **representable** in case it is naturally isomorphic to $\mathrm{Hom}_\mathscr{C}(C, -)$ for some object C of $\mathscr{C}$. Formulate a universal property for the object C, and show that C is unique up to isomorphism.

7. Let $\mathscr{C}$ be a category with small Hom sets. Show that $\mathscr{C}$ has an initial object if and only if the constant functor $\mathscr{C} \longrightarrow Sets$, $C \longmapsto \{1\}$, is representable.

8. Let D be a diagram in a category $\mathscr{C}$ with small Hom sets. Define a functor $F : \mathscr{C} \longrightarrow Sets$ such that $F(A)$ is the set of all cones from D to A. Show that D has a colimit if and only if F is representable.

9. Let $\mathscr{A}$ and $\mathscr{C}$ be categories with small Hom sets and $G : \mathscr{C} \longrightarrow \mathscr{A}$ be a functor. Show that G has a left adjoint if and only if $\mathrm{Hom}_\mathscr{A}(A, G(-))$ is representable for every object A of $\mathscr{A}$.

17.8. TRIPLES

Composing the free group functor from *Sets* to *Grps* with the forgetful functor from *Grps* to *Sets* yields a functor T from *Sets* to *Sets*. It turns out that groups can be defined using only the functor T and certain natural transformations. Moreover similar descriptions apply to most other algebraic systems. That abstract view of algebra is the subject of this section and the next.

17.8.a. Definition. In this section we write functors as left operators (we write FA instead of $F(A)$, FG instead of $F \circ G$, etc.).

Let $(F, G, \eta, \varepsilon)$ be an adjunction, where $F : \mathscr{A} \longrightarrow \mathscr{C}$, $G : \mathscr{C} \longrightarrow \mathscr{A}$ and the natural transformations $\eta : 1_{\mathscr{A}} \longrightarrow GF$, $\varepsilon : FG \longrightarrow 1_{\mathscr{C}}$ have the universal properties in Proposition 17.6.1.

Proposition 17.8.1. *Let* $(F, G, \eta, \varepsilon) : \mathscr{A} \longrightarrow \mathscr{C}$ *be an adjunction. Let* $T = GF$ *and* $\mu = G\varepsilon F : T \longrightarrow TT$. *The following diagrams commute:*

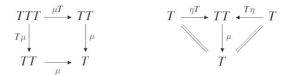

Proof. Since ε is a natural transformation, the square on the left

$$
\begin{array}{ccc}
FGFGC & \xrightarrow{\varepsilon FGC} & FGC \\
{\scriptstyle FG\varepsilon_C}\big\downarrow & & \big\downarrow{\scriptstyle \varepsilon_C} \\
FGC & \xrightarrow{\varepsilon_C} & C
\end{array}
\qquad
\begin{array}{ccc}
GFGFGFA & \xrightarrow{G\varepsilon FGFA} & GFGFA \\
{\scriptstyle GFG\varepsilon_{FA}}\big\downarrow & & \big\downarrow{\scriptstyle G\varepsilon_{FA}} \\
GFGFA & \xrightarrow{G\varepsilon_{FA}} & GFA
\end{array}
$$

commutes for every object C of $\mathscr{C}$. Letting $C = FA$ and applying G yields the square on the right and shows that $\mu_A \mu_{TA} = \mu_A (T\mu_A)$ for every object A of $\mathscr{A}$.

By Lemma 17.6.4, $(G\varepsilon_C)(\eta_{GC}) = 1_{GC}$ and $(\varepsilon_{FA})(F\eta_A) = 1_{FA}$ for all A, C. Hence $\mu_A(\eta_{TA}) = (G\varepsilon_{FA})(\eta_{GFA}) = 1_{GFA}$ and $\mu_A(T\eta_A) = (G\varepsilon_{FA})(GF\eta_A) = G1_{FA} = 1_{GFA}$ for all A. □

For example, let $F : Sets \longrightarrow Grps$ be the free group functor and $S : Grps \longrightarrow Sets$ be the forgetful (underlying set) functor. For each set X, $TX = SFX$ is the set of all reduced words in X; $\eta_X : X \longrightarrow TX$ is the inclusion mapping which sends each element x of X onto the one letter word x. For each group G,

$$
\theta_{X,G} : \mathrm{Hom}_{Grps}(FX, G) \longrightarrow \mathrm{Hom}_{Sets}(X, SG)
$$

sends a homomorphism $FX \longrightarrow G$ to its restriction to X. By Lemma 17.6.4, $\varepsilon_G = \theta_{SG,G}^{-1}(1_{SG})$ is the homomorphism $FSG \longrightarrow G$ which extends the identity on SG; ε_G takes a reduced word w whose letters are elements of G and sends it to the product of w as calculated in G. (We may think of ε_G as another way to describe the multiplication on G.) Then $\mu_X = S\varepsilon_{FX} : TTX \longrightarrow TX$ takes a reduced word $w \in TTX$ whose letters are reduced words in X and sends it to the product of w as calculated in FX. (This describes the multiplication on FX.)

In general, a **triple** (T, η, μ) in a category $\mathscr{A}$ consists of a functor $T : \mathscr{A} \longrightarrow \mathscr{A}$ and two natural transformations $\eta : 1_{\mathscr{A}} \longrightarrow T$ and $\mu : TT \longrightarrow T$ such that the following diagrams commute:

$$
\begin{array}{ccc}
TTT & \xrightarrow{\ \mu T\ } & TT \\
{\scriptstyle T\mu}\big\downarrow & & \big\downarrow{\scriptstyle \mu} \\
TT & \xrightarrow[\ \mu\ ]{} & T
\end{array}
\qquad\qquad
\begin{array}{ccc}
T & \xrightarrow{\ \eta T\ } TT \xleftarrow{\ T\eta\ } & T \\
 & \searrow\ \big\downarrow{\scriptstyle \mu}\ \swarrow & \\
 & T &
\end{array}
$$

Triples are also called **monads** and **standard constructions**. By Proposition 17.8.1, every adjunction between $\mathscr{A}$ and $\mathscr{C}$ induces a triple on $\mathscr{A}$.

17.8.b. T-**Algebras.** We show that conversely, every triple arises from a pair of adjoint functors. Let (T, η, μ) be a triple in a category $\mathscr{A}$. A T-**algebra** over $\mathscr{A}$ is a pair (A, φ) of an object A of $\mathscr{A}$ and a morphism $\varphi : TA \longrightarrow A$ (the T-algebra structure on A) such that the following diagrams commute:

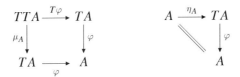

For example, let (T, η, μ) be the free group triple on *Sets*. When G is a group, the reader will verify that $(SG, S\varepsilon_G)$ is a T-algebra (this also follows from Proposition 17.8.3 below); SG is the underlying set of G and $S\varepsilon_G$ takes a reduced word $w \in FSG$ to its product as calculated in G. Thus the T-algebra structure $S\varepsilon_G : TSG \longrightarrow SG$ on SG is determined by its group structure. In the next section we will show that (with this particular triple) every T-algebra arises from a group by this construction, so that groups are (up to an isomorphism of categories) a particular case of algebras over a triple.

For another example, every ring R [with an identity element] induces a triple (T, η, μ) in *Abs*, in which $TA = R \otimes A$; $\eta_A : A \longrightarrow R \otimes A$ sends $a \in A$ onto $1 \otimes a$; and $\mu_A : R \otimes (R \otimes A) \longrightarrow R \otimes A$ sends $r \otimes (s \otimes a)$ onto $rs \otimes a$ (all tensor products are over $\mathbb{Z}$). A T-algebra is an abelian group A together with a homomorphism $\varphi : R \otimes A \longrightarrow A$ such that $\varphi \circ (R \otimes \varphi) = \varphi \circ \mu_A$ ($\varphi(r \otimes \varphi(s \otimes a)) = \varphi(rs \otimes a)$) and $\varphi \circ \eta_A = 1_A$ ($\varphi(1 \otimes a) = a$). There is a bijection between homomorphisms $\varphi : R \otimes A \longrightarrow A$ and biadditive mappings $\beta : R \times A \longrightarrow A$, which sends φ to $\beta(r, a) = \varphi(r \otimes a)$. Then φ is a T-algebra structure on A if and only if $\beta(r,\beta(s,a)) = \beta(rs,a)$ and $\beta(1,a) = a$ for all a,r,s, if and only if the biadditive mapping β is a left R-module structure on A. Thus left R-modules are (up to an isomorphism of categories) another particular case of T-algebras.

In general, a **morphism** (or **homomorphism**) of T-algebras from (A,φ) to (B,ψ) is a morphism $\alpha : A \longrightarrow B$ such that the following square commutes:

$$
\begin{array}{ccc}
TA & \xrightarrow{T\alpha} & TB \\
\varphi \downarrow & & \downarrow \psi \\
A & \xrightarrow{\ \ \alpha\ \ } & B
\end{array}
$$

Proposition 17.8.2. *Let (T,η,μ) be a triple in a category $\mathscr{A}$ with small Hom sets.*

(1) *T-algebras and their morphisms are the objects and morphisms of a category $\mathscr{A}^T$ with small Hom sets.*

(2) *There is an adjunction $(F^T,G^T,\eta^T,\varepsilon^T) : \mathscr{A} \longrightarrow \mathscr{A}^T$, where $G^T(A,\varphi) = A$, $F^T A = (TA,\mu_A)$, $\eta_A^T = \eta_A$, and $\varepsilon_{(A,\varphi)}^T = \varphi$.*

(3) *The triple induced by $(F^T,G^T,\eta^T,\varepsilon^T)$ is the given triple (T,η,μ).*

Proof. (1) is clear since T is a functor.

(2) The definition of a triple shows that $F^T A = (TA,\mu_A)$ is a T-algebra for every object A of $\mathscr{A}$. If $\alpha : A \longrightarrow B$ is a morphism in $\mathscr{A}$, then $F^T \alpha = T\alpha : (TA,\mu_A) \longrightarrow (TB,\mu_B)$ is a morphism of T-algebras, since μ is a natural transformation. This constructs a functor $F^T : \mathscr{A} \longrightarrow \mathscr{A}^T$. We also have a forgetful functor $G^T : \mathscr{A}^T \longrightarrow \mathscr{A}$.

We see that $G^T F^T = T$, so that $\eta : 1_{\mathscr{A}} \longrightarrow G^T F^T$. We show that for every morphism $\alpha : A \longrightarrow C = G^T(C,\psi)$ there is a morphism $\gamma : (TA,\mu_A) = F^T A \longrightarrow (C,\psi)$ (with $\gamma \mu_A = \psi(T\gamma)$) unique such that $\gamma \eta_A = (G^T \gamma)\eta_A = \alpha$:

$$
\begin{array}{ccc}
A \xrightarrow{\eta_A} TA & & TTA \xrightarrow{\mu_A} TA \\
{\scriptstyle\alpha}\searrow\ \ \downarrow{\scriptstyle\gamma} & & T\gamma\downarrow\ \ \ \ \downarrow\gamma \\
C & & TC \xrightarrow[\psi]{} C
\end{array}
$$

First, $\gamma \eta_A = \alpha$ implies $\gamma = \gamma \mu_A (T\eta_A) = \psi(T\gamma)(T\eta_A) = \psi(T\alpha)$. Conversely, $\gamma = \psi(T\alpha)$ satisfies $\gamma \mu_A = \psi(T\alpha)\mu_A = \psi\mu_C(TT\alpha) = \psi(T\psi)(TT\alpha) = \psi(T\gamma)$, since μ is natural and (C,ψ) is a T-algebra, so that γ is a morphism of T-algebras, and $\gamma \eta_A = \psi(T\alpha)\eta_A = \psi\eta_C \alpha = \alpha$, since η is natural and (C,ψ) is a T-algebra.

This universal property implies that F^T is a left adjoint of G^T, with $\eta^T = \eta$. The natural isomorphism

$$
\theta_{A,(C,\psi)}^T : \mathrm{Hom}_{\mathscr{A}^T}(F^T(A),(C,\psi)) \cong \mathrm{Hom}_{\mathscr{A}}(A,G^T(C,\psi))
$$

is given by $\theta^T_{A,(C,\psi)}(\gamma) = G^T(\gamma)\eta^T_A = \gamma\eta_A$; we saw that $\theta^T_{A,(C,\psi)}{}^{-1}(\alpha) = \psi(T\alpha)$. By Lemma 17.6.4, $\varepsilon^T_{(C,\psi)} = \theta^T_{C,(C,\psi)}{}^{-1}(1_C) = \psi$.

(3) We saw that $G^T F^T = T$ and $\eta^T = \eta$. Since $\varepsilon^T_{(C,\psi)} = \psi$, we have $\mu^T_A = G^T \varepsilon^T_{F^T A} = G^T \mu_A = \mu_A$. $\qquad\square$

17.8.c. Comparison. Which categories can be described as categories of T-algebras? This question will be considered in the next section, using the following comparison theorem:

Proposition 17.8.3. *Let* $(F,G,\eta,\varepsilon) : \mathscr{A} \longrightarrow \mathscr{C}$ *be an adjunction and* (T,η,μ) $= (GF,\eta,G\varepsilon F)$ *be the triple that it induces. There is a unique functor* $Q : \mathscr{C}$ $\longrightarrow \mathscr{A}^T$ *such that* $F^T = QF$ *and* $G^T Q = G$; *namely* $QC = (GC, G\varepsilon_C)$, $Q\gamma = G\gamma$ *for every object* C *and morphism* γ *of* $\mathscr{C}$.

Proof. We prove the uniqueness first. Let $Q : \mathscr{C} \longrightarrow \mathscr{A}^T$ be a functor such that $F^T = QF$ and $G^T Q = G$. Then $(1_{\mathscr{A}}, Q)$ is a functorial transformation of adjunctions as defined by Proposition 17.6.6, since $\eta^T = \eta$. Therefore $\varepsilon^T_{QC} = Q\varepsilon_C$ for every object C of $\mathscr{C}$.

Since $G^T Q = G$, we have $QC = (GC, \varphi_C)$ for some morphism $\varphi_C : GFGC$ $\longrightarrow GC$. By Proposition 17.8.2, $\varphi_C = \varepsilon^T_{QC} = Q\varepsilon_C = G^T Q\varepsilon_C = G\varepsilon_C$. Moreover $Q\gamma = G^T Q\gamma = G\gamma$ for every morphism γ of $\mathscr{C}$. Thus Q is unique.

Conversely, $(GC, G\varepsilon_C)$ is a T-algebra for every C: the diagrams

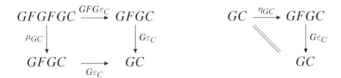

commute, the first since ε is a natural transformation, the second by Lemma 17.6.4. If $\gamma : C \longrightarrow D$ is a morphism in $\mathscr{C}$, then $G\gamma : GC \longrightarrow GD$ is a morphism of T-algebras: the square

$$\begin{array}{ccc} GFGC & \xrightarrow{GF\gamma} & GFGD \\ {\scriptstyle G\varepsilon_C}\downarrow & & \downarrow{\scriptstyle G\varepsilon_D} \\ C & \xrightarrow[\gamma]{} & D \end{array}$$

commutes, since ε is a natural transformation. Therefore a functor $Q : \mathscr{C} \longrightarrow \mathscr{A}^T$ is defined by $QC = (GC, G\varepsilon_C)$, $Q\gamma = G\gamma$ for every object C and morphism γ of $\mathscr{C}$. We see that $G^T Q = G$; also $QF = F^T$, since $QFA = (GFA, G\varepsilon_{FA}) = (FA, \mu_A) = F^T A$ and $QF\alpha = GF\alpha = T\alpha = F^T\alpha$ for every object A and morphism α of $\mathscr{A}$. $\qquad\square$

Exercises

1. Describe the triple on *Sets* induced by the forgetful functor from *R-Mods* to *Sets* and its left adjoint.

2. Describe the triple on *Abs* induced by the forgetful functor from *Rings* to *Abs* and its left adjoint.

3. Let I be a preordered set, viewed as a category. What is a triple on I? What is a T-algebra?

4. Verify that groups yield T-algebras.

5. Let G be a group. Show that there is a triple (T,η,μ) on *Sets* in which $TX = G \times X$, $\eta_X : X \longrightarrow G \times X$ sends $x \in X$ onto $(1,x)$, and $\mu_X : G \times (G \times X) \longrightarrow G \times X$ sends $(g,(h,x))$ onto (gh,x). Show that T-algebras coincide with group actions of G (on sets).

6. Let F and F' be two left adjoints of G. What can you say about the triples induced by (F,G) and (F',G)?

17.9. TRIPLEABILITY

In this section we prove Beck's Theorem which characterizes tripleable functors. With this theorem we can show that *Grps*, *Rings*, *R-Mods*, and so on, are tripleable over *Sets*, so that groups, rings, R-modules, and so on, can be defined as T-algebras. A general result of this nature is proved in Chapter 18.

17.9.a. Definition. A functor $G : \mathscr{C} \longrightarrow \mathscr{A}$ is **tripleable** (or **monadic**) in case it has a left adjoint and the functor Q in Proposition 17.8.3 is an isomorphism; equivalently, in case G is, up to isomorphism, the forgetful functor $\mathscr{A}^T \longrightarrow \mathscr{A}$ of a category of T-algebras over $\mathscr{A}$. A category $\mathscr{C}$ is **tripleable** over a category $\mathscr{A}$ in case there is a canonical functor $\mathscr{C} \longrightarrow \mathscr{A}$ which is tripleable; then $\mathscr{C}$ is isomorphic to a category of T-algebras over $\mathscr{A}$.

Tripleability does not always hold. Let *POS* be the category whose objects are all partially ordered sets and whose morphisms are all order-preserving mappings. The forgetful functor $POS \longrightarrow Sets$ has a left adjoint F, which assigns to each set X the **discrete** partially ordered set FX on X, in which $x \leqslant y$ if and only if $x = y$: indeed every mapping of X into a partially ordered set S "extends" uniquely to an order-preserving mapping of FX into S. The induced triple has $TX = X$ and $\eta_X = \mu_X = X$ for every set X. Hence T-algebras "are" sets; more precisely, F^T and G^T are isomorphisms. In Proposition 17.8.3, $G^T Q$ is the forgetful functor to *Sets*; therefore Q is not an isomorphism. Thus *POS* is not tripleable over *Sets*.

17.9.b. Split Coequalizers. A **split coequalizer** of $\alpha,\beta : A \longrightarrow B$ is a morphism $\sigma : B \longrightarrow C$ such that $\sigma\alpha = \sigma\beta$ and there exist morphisms $\mu : B \longrightarrow A$ and $\nu : C \longrightarrow B$ such that $\alpha\mu = 1_B$, $\sigma\nu = 1_C$, and $\beta\mu = \nu\sigma$.

$$A \underset{\mu}{\overset{\alpha}{\rightrightarrows}} B$$

$$\beta \downarrow \qquad \downarrow \sigma$$

$$B \underset{\nu}{\overset{\sigma}{\rightrightarrows}} C$$

Lemma 17.9.1. *Every split coequalizer is a coequalizer.*

Proof. Let $\alpha, \beta, \sigma, \mu, \nu$ be as above. Then $\sigma\alpha = \sigma\beta$. If $\varphi\alpha = \varphi\beta$, then $\varphi = \varphi\alpha\mu = \varphi\beta\mu = \varphi\nu\sigma$ factors through σ; moreover $\varphi = \psi\sigma$ implies $\psi = \psi\sigma\nu = \varphi\nu$. $\qquad\square$

Lemma 17.9.2. *Every functor preserves split coequalizers.*

Proof. If indeed $\sigma\alpha = \sigma\beta$, $\alpha\mu = 1_B$, $\sigma\nu = 1_C$, and $\beta\mu = \nu\sigma$, then $(F\sigma)(F\alpha) = (F\sigma)(F\beta)$, $(F\alpha)(F\mu) = 1_{FB}$, $(F\sigma)(F\nu) = 1_{FC}$, and $(F\beta)(F\mu) = (F\nu)(F\sigma)$. $\quad\square$

17.9.c. Beck's Theorem

Theorem 17.9.3 (Beck). *Let $\mathcal{A}$ and $\mathcal{C}$ have small Hom sets. A functor $G : \mathcal{C} \longrightarrow \mathcal{A}$ with a left adjoint is tripleable if and only if it creates coequalizers of pairs α, β such that $G\alpha, G\beta$ have a split coequalizer in $\mathcal{A}$.*

Proof. Let G be tripleable, F be a left adjoint of G, and (T, η, μ) be the triple on $\mathcal{A}$ induced by (F, G). We show that $G^T : \mathcal{A}^T \longrightarrow \mathcal{A}$ creates coequalizers of pairs α, β such that $G^T\alpha, G^T\beta$ have a split coequalizer in $\mathcal{A}$. This property extends to $G = G^T Q$ since $Q : \mathcal{C} \longrightarrow \mathcal{A}^T$ is an isomorphism.

Let $\alpha, \beta : (A, \varphi) \longrightarrow (B, \psi)$ be morphisms of T-algebras such that $G^T\alpha = \alpha$ and $G^T\beta = \beta$ have a split coequalizer $\sigma : B \longrightarrow C$. By Lemma 17.9.2, $T\sigma$ is a split coequalizer of $T\alpha$ and $T\beta$.

$$TA \underset{T\beta}{\overset{T\alpha}{\rightrightarrows}} TB \xrightarrow{\;T\sigma\;} TC$$
$$\varphi\downarrow \qquad \psi\downarrow \qquad \vdots\,\chi$$
$$A \underset{\beta}{\overset{\alpha}{\rightrightarrows}} B \xrightarrow{\;\sigma\;} C$$

Since $\sigma\psi(T\alpha) = \sigma\alpha\varphi = \sigma\beta\varphi = \sigma\psi(T\beta)$, there exists by Lemma 17.9.1 a unique morphism $\chi : TC \longrightarrow C$ such that $\sigma\psi = \chi(T\sigma)$.

We show that (C, χ) is a T-algebra. We have a commutative diagram

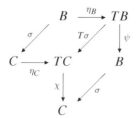

Hence $\chi \eta_C \sigma = \sigma \psi \eta_B = 1_C \sigma$; therefore $\chi \eta_C = 1_C$. Similarly there is a commutative diagram (to be completed by the reader)

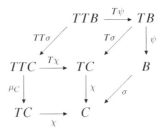

Hence $\chi(T\chi)(TT\sigma) = \chi \mu_C (TT\sigma)$; by Lemma 17.9.2, $TT\sigma$ is an epimorphism, so that $\chi(T\chi) = \chi \mu_C$.

We now have a unique T-algebra (C, χ) such that $\sigma : (B, \psi) \longrightarrow (C, \chi)$ is a morphism of T-algebras. It remains to show that σ is a coequalizer of α and β in $\mathscr{A}^T$. Let $\gamma : (B, \psi) \longrightarrow (X, \xi)$ be a morphism such that $\gamma\alpha = \gamma\beta$. By Lemma 17.9.1, $\gamma = \delta\sigma$ for some unique $\delta : C \longrightarrow X$. There is a commutative diagram

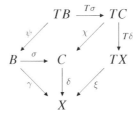

Hence $\xi(T\delta)(T\sigma) = \delta\chi(T\sigma)$ and $\xi(T\delta) = \delta\chi$ by Lemma 17.9.2. Thus δ is a morphism from (C, χ) to (X, ξ), and we have shown that σ is a coequalizer in $\mathscr{A}^T$.

The converse follows from Lemma 17.9.4 below. We have just shown that this Lemma can be applied to (F, G) and (F^T, G^T), in which case it yields the unique functor Q in Proposition 17.8.3 such that $F^T = QF$ and $G^T Q = G$. If G creates coequalizers as in the statement, then Lemma 17.9.4 applies to (F^T, G^T) and (F, G) and yields a unique functor R such that $F = RF^T$ and $GR = G^T$. Then $F^T = QRF^T$ and $G^T QR = G^T$; by the uniqueness in Lemma 17.9.4, QR is the identity on $\mathscr{A}^T$. Similarly RQ is the identity on $\mathscr{C}$. Therefore Q is an isomorphism. $\square$

Lemma 17.9.4. *Let* $(F, G, \eta, \varepsilon, \theta) : \mathscr{A} \longrightarrow \mathscr{C}$ *and* $(F', G', \eta', \varepsilon', \theta') : \mathscr{A} \longrightarrow \mathscr{C}'$ *be adjunctions that induce the same triple* $(T, \eta, \mu) = (GF, \eta, G\varepsilon F) = (G'F', \eta', G'\varepsilon'F')$ *on* $\mathscr{A}$. *Assume that* G' *creates coequalizers of pairs* α, β *such that* $G'\alpha, G'\beta$ *have a split coequalizer in* $\mathscr{A}$. *Then there is a unique functor* $R : \mathscr{C} \longrightarrow \mathscr{C}'$ *such that* $F = RF'$ *and* $GR = G'$.

Proof. In any T-algebra (A,φ), the morphism φ is a split coequalizer of μ_A and $T\varphi$: indeed $\varphi(T\varphi) = \varphi\mu_A$, and $\eta_A : A \longrightarrow TA$, $\eta_{TA} : TA \longrightarrow TTA$,

$$
\begin{array}{ccc}
TTA \underset{\eta_{TA}}{\overset{\mu_A}{\rightrightarrows}} TA \\[4pt]
T\varphi \downarrow \qquad \downarrow \varphi \\[4pt]
TA \underset{\eta_A}{\overset{\varphi}{\rightrightarrows}} A
\end{array}
$$

satisfy $\varphi\eta_A = 1_A$ since A is a T-algebra, $\mu_A\eta_{TA} = 1_{TA}$ since (T,η,μ) is a triple, and $(T\varphi)\eta_{TA} = \eta_A\varphi$ since η is a natural transformation.

Now assume that R exists. Since $\eta = \eta'$, $(1_{\mathscr{A}},R)$ is a functorial transformation of adjunctions; hence $R\varepsilon = \varepsilon'R$.

For each object C of $\mathscr{C}$ the diagram

$$
\begin{array}{ccc}
FGFGC & \xrightarrow{\varepsilon FGC} & FGC \\
FG\varepsilon_C \downarrow & & \downarrow \varepsilon_C \\
FGC & \xrightarrow{\varepsilon_C} & C
\end{array}
$$

commutes, since ε is a natural transformation. Appying R yields a commutative diagram

$$
\begin{array}{ccc}
F'G'F'GC & \xrightarrow{\varepsilon'_{F'GC}} & F'GC \\
F'G\varepsilon_C \downarrow & & \downarrow R\varepsilon_C \\
F'GC & \xrightarrow{R\varepsilon_C} & RC
\end{array}
$$

(note that $RFGFGC = F'GFGC = F'G'F'GC$). Applying G' yields the commutative diagram

$$
\begin{array}{ccc}
TTGC & \xrightarrow{\mu_{GC}} & TGC \\
TG\varepsilon_C \downarrow & & \downarrow G\varepsilon_C \\
TGC & \xrightarrow{G\varepsilon_C} & GC
\end{array}
$$

in which $G\varepsilon_C$ is a split coequalizer of μ_{GC} and $TG\varepsilon_C$, since $(GC, G\varepsilon_C)$ is a T-algebra by Proposition 17.8.3. By the hypothesis, there is exactly one morphism $\rho'_C : F'GC \longrightarrow C'$ such that $G'\rho'_C = G\varepsilon_C$, and it is a coequalizer of $\varepsilon'_{F'GC}$ and $F'G\varepsilon_C$. Therefore $RC = C'$ and $R\varepsilon_C = \rho'_C$. In particular, RC and $R\varepsilon_C$ are unique and $R\varepsilon_C$ is an epimorphism. If $\gamma : C \longrightarrow D$ is a morphism, then the diagram

$$
\begin{array}{ccc}
F'GC & \xrightarrow{R\varepsilon_C} & RC \\
F'G\gamma \downarrow & & \downarrow R\gamma \\
F'GD & \xrightarrow{R\varepsilon_D} & RD
\end{array}
$$

commutes, since $F' = RF$ and ε is natural; therefore $R\gamma$ is unique. Thus R is unique.

We now show that R exists. Let C be an object of $\mathscr{C}$. By the above, applying G' to $\varepsilon'_{F'GC}$ and $F'G\varepsilon_C$ yields morphisms μ_{GC} and $TG\varepsilon_C$ which have a split coequalizer $G\varepsilon_C$ in $\mathscr{A}$. By the hypothesis, $\varepsilon'_{F'GC}$ and $F'G\varepsilon_C$ have a coequalizer $\rho'_C : F'GC \longrightarrow C'$ which is the unique morphism such that $G'\rho'_C = G\varepsilon_C$ (and $G'C' = GC$). Define $RC = C'$. For each morphism $\gamma : C \longrightarrow D$ there is a diagram

$$
\begin{array}{ccccc}
F'G'F'GC & \overset{\varepsilon'_{F'GC}}{\underset{F'G\varepsilon_C}{\rightrightarrows}} & F'GC & \overset{\rho'_C}{\longrightarrow} & RC \\
{\scriptstyle F'G'F'G\gamma}\big\downarrow & & \big\downarrow{\scriptstyle F'G\gamma} & & \big\downarrow{\scriptstyle R\gamma} \\
F'G'F'GD & \underset{F'G\varepsilon_D}{\overset{\varepsilon'_{F'GD}}{\rightrightarrows}} & F'GD & \underset{\rho'_D}{\longrightarrow} & RD
\end{array}
$$

in which $\rho'_D(F'G\gamma)\varepsilon'_{F'GC} = \rho'_D(F'G\gamma)F'G\varepsilon_C$ (since ε is a natural transformation); therefore there is a unique morphism $R\gamma : RC \longrightarrow RD$ such that $(R\gamma)\rho'_C = \rho'_D(F'G\gamma)$. It is immediate that R is a functor. The definition of R shows that $G'R = G$.

Let $C = FA$. The image under G' of $\varepsilon'_{F'A} : F'GFA = F'G'F'A \longrightarrow F'A$ is $G'\varepsilon'_{F'A} = \mu_A$, which is a split coequalizer of $G'\varepsilon'_{F'GFA} = \mu_{TA}$ and $G'F'G\varepsilon_{FA} = T\mu_A$ since (TA, μ_A) is a T-algebra. Therefore $\varepsilon'_{F'A} : F'GFA \longrightarrow F'A$ coincides with $\rho'_{FA} : F'GFA \longrightarrow RFA$. It follows that $RF = F'$. $\qquad\square$

Some variants of Beck's Theorem are given in the exercises.

17.9.d. Examples. Using Beck's Theorem, we can show:

Proposition 17.9.5. *The forgetful functor from Grps to Sets is tripleable.*

Proof. We show that this forgetful functor creates coequalizers of pairs $\alpha, \beta : G \longrightarrow H$ of group homomorphisms which have a split coequalizer $\sigma : H \longrightarrow K$ in *Sets*.

Let $\mu_G : G \times G \longrightarrow G$ and $\mu_H : H \times H \longrightarrow H$ be the group operations on G and H. Since α and β are homomorphisms, we have $\alpha\mu_G = \mu_H(\alpha \times \alpha)$ and $\beta\mu_G = \mu_H(\beta \times \beta)$. By Lemma 17.9.2, $\sigma \times \sigma$ is a split coequalizer of $\alpha \times \alpha$ and $\beta \times \beta$. The diagram

$$
\begin{array}{ccccc}
G \times G & \overset{\alpha \times \alpha}{\underset{\beta \times \beta}{\rightrightarrows}} & H \times H & \overset{\sigma \times \sigma}{\longrightarrow} & K \times K \\
{\scriptstyle \mu_G}\big\downarrow & & \big\downarrow{\scriptstyle \mu_H} & & \big\downarrow{\scriptstyle \mu_K} \\
G & \underset{\beta}{\overset{\alpha}{\rightrightarrows}} & H & \underset{\sigma}{\longrightarrow} & K
\end{array}
$$

shows that $\sigma\mu_H(\alpha \times \alpha) = \sigma\mu_H(\beta \times \beta)$. Therefore there is a unique mapping $\mu_K : K \times K \longrightarrow K$ such that $\sigma\mu_H = \mu_K(\sigma \times \sigma)$. Since σ splits, it is surjective; hence K inherits associativity, identity element, and inverses from H, and is a group. Thus there is a unique group operation on the set K such that σ is a homomorphism.

It remains to show that σ is a coequalizer in *Grps*. Let $\varphi : H \longrightarrow L$ be a group homomorphism such that $\varphi\alpha = \varphi\beta$. Since σ is a coequalizer in *Sets*, there is a unique mapping $\psi : K \longrightarrow L$ such that $\varphi = \psi\sigma$. In the diagram

$$
\begin{array}{ccccc}
H \times H & \xrightarrow{\sigma \times \sigma} & K \times K & \xrightarrow{\psi \times \psi} & L \times L \\
\downarrow{\scriptstyle\mu_H} & & \downarrow{\scriptstyle\mu_K} & & \downarrow{\scriptstyle\mu_L} \\
H & \xrightarrow{\sigma} & K & \xrightarrow{\psi} & L
\end{array}
$$

the left square and the outer rectangle commute, since σ and φ are homomorphisms; hence the right square commutes, and ψ is a homomorphism.   □

Similar proofs show that *Rings*, *R-Mods*, *R-S-Bimods*, and so on, are tripleable over *Sets*. Other examples are given in the exercises.

Exercises

1. Prove that the forgetful functor from *Rings* to *Sets* is tripleable.
2. Prove that the forgetful functor from *R-Mods* to *Sets* is tripleable.
3. Prove that the forgetful functor from *Rings* to *Abs* is tripleable.
4. Prove that the forgetful functor from *R-Mods* to *Abs* is tripleable.
5. Topological spaces and continuous mappings are the objects and morphisms of a category *Tops*. Show that the forgetful functor from *Tops* to *Sets* has a left adjoint but is not tripleable. (On the other hand, the forgetful functor from compact Hausdorff spaces to sets is known to be tripleable.)

18

UNIVERSAL ALGEBRA

Universal algebra unifies the study of algebraic systems. It is more direct and less abstract than category theory and applies only to algebra, but we think it is almost as essential to its conceptual understanding. Its importance as a branch of algebra began with Birkhoff's work in the 1930s.

The main results of this chapter are that groups, rings, left R-modules, etc., constitute varieties, and that many properties of groups, rings, left R-modules, etc., are in fact true in all varieties.

Required results: this chapter draws examples from many of the previous chapters, particularly Chapters 1, 4, and 10, and category concepts from Chapter 17.

18.1. UNIVERSAL ALGEBRAS

In this chapter a **universal algebra** is a set endowed with any number of operations. This section give basic properties of universal algebras in general, such as the Homomorphism Theorem.

18.1.a. Definition. Recall that an *n*-**ary operation** on a set X is a mapping $\omega : X^n \longrightarrow X$, where n is a nonnegative integer and X^n is the cartesian product $X \times X \times \cdots \times X$ of n copies of X; n is the **arity** of ω. For instance, an operation of arity 2 is a **binary** operation. An operation of arity 1 on X is just a mapping of X into X and is a **unary** operation. By convention X^0 (the empty product) is your favorite one-element set (a terminal object of the category *Sets*); an operation of arity 0 on X, also called a **constant** operation, merely selects one element of X.

There are operations of infinite arity, which we do not use in this chapter because many later properties require finite arities. For the same reason we do not use partial operations (e.g., composition in a small category).

Universal algebras are first classified by their type, which specifies the operations on an algebra and their arities. Formally a **type** of algebra consists of a set T and a mapping $\omega \longmapsto n_\omega$ which assigns to each $\omega \in T$ a nonnegative integer n_ω, the (abstract) **arity** of $\omega \in T$.

A universal algebra **of type** T consists of a set A and a mapping, the type T algebra **structure** on A, which assigns to each $\omega \in T$ an operation ω_A on A (often denoted by ω) of arity n_ω.

18.1.b. Examples. Groups, rings, modules, etc., all fit into this definition.

A group can be viewed as a universal algebra with one binary operation (multiplication) which satisfies certain axioms; then T has just one element of arity 2. A group can also be viewed as a universal algebra with one constant operation (which selects the identity element), one unary operation ($x \longmapsto x^{-1}$), and one binary operation (the multiplication), which satisfy another set of axioms; then groups are algebras of type T, where T has one element of arity 0, one element of arity 1, and one element of arity 2.

A ring R [with an identity element] is a universal algebra with two binary operations (addition and multiplication) and a constant operation (which selects the identity element), which satisfy certain axioms. Thus, rings are of type, say, $\{a,m,u\}$, where a and m have arity 2 and u has arity 0. (As above, one may also view the additive group of a ring as an algebra with three operations.)

A left R-module is a universal algebra with one binary operation (addition) and one unary operation for each element r of R (the action of r on X), which satisfy certain axioms. Thus left R-modules are of type, say, $T = \{a\} \cup R$, where a has arity 2 and every $r \in R$ has arity 1.

These descriptions of groups, rings, and modules will be refined in Section 18.3 when we formally define identities.

On the other hand, partially ordered sets, small categories, or topological spaces are not readily described as universal algebras (assuming that one would want to); this will be explained to some extent by later results.

18.1.c. Homomorphisms. Homomorphisms are mappings which preserve all operations. In detail, let A and B be universal algebras of the same type T. A **homomorphism** of A into B is a mapping $\varphi : A \longrightarrow B$ such that the diagram

$$
\begin{array}{ccc}
A^n & \xrightarrow{\ \varphi^n\ } & B^n \\
{\scriptstyle \omega_A}\downarrow & & \downarrow{\scriptstyle \omega_B} \\
A & \xrightarrow[\varphi]{} & B
\end{array}
$$

commutes for every $\omega \in T$, where $n = n_\omega$ is the arity of ω; equivalently, such that

$$\varphi(\omega_A(x_1,\ldots,x_n)) = \omega_B(\varphi(x_1),\ldots,\varphi(x_n))$$

whenever $\omega \in T$ has arity n and $x_1,\ldots,x_n \in A$.

For instance, the identity mapping 1_A on an algebra A of type T is a homomorphism. If $\varphi : A \longrightarrow B$ and $\psi : B \longrightarrow C$ are homomorphisms of algebras of type T, then so is $\psi \circ \varphi : A \longrightarrow C$. Hence:

Proposition 18.1.1. *Universal algebras of type T and their homomorphisms are the objects and morphisms of a category Alg_T.*

An **isomorphism** of algebras of type T is a homomorphism $\theta : A \longrightarrow B$ for which there exists a homomorphism $\zeta : B \longrightarrow A$ such that $\zeta \circ \theta = 1_A$ and $\theta \circ \zeta = 1_B$; equivalently, a homomorphism which is bijective (see the exercises). As before we regard isomorphic algebras as essentially identical instances of the same "abstract" universal algebra.

18.1.d. Subalgebras. A **subalgebra** of an algebra A of type T is an algebra S of type T such that $S \subseteq A$ and the inclusion mapping $S \longrightarrow A$ is a homomorphism.

The definition of a subalgebra S of A implies that $\omega_S = (\omega_A)_{|S^n}$ for each $\omega \in T$ of arity n; therefore the algebra structure on S is completely determined by the subset S. In detail:

Proposition 18.1.2. *Let A be an algebra of type T. A subset S of A is [the underlying set of] a subalgebra of A if and only if it is closed under all operations on A: $\omega_A(x_1,\ldots,x_n) \in S$ for all $x_1,\ldots,x_n \in S$ whenever $\omega \in T$ has arity n. Then there is a unique type T algebra structure on S such that the inclusion mapping $S \longrightarrow A$ is a homomorphism, in which $\omega_S = (\omega_A)_{|S^n}$ when $\omega \in T$ has arity n.*

Thus a subalgebra of an algebra A may be identified with its underlying set; then a subalgebra of A is a subset of A which is closed under the operations on A. The word subalgebra then refers either to the subset S or to the algebra S.

The following properties are straightforward:

Proposition 18.1.3. *Let A be a universal algebra. Every intersection of subalgebras of A is a subalgebra of A.*

Recall that a **directed union** is the union of a family $(S_i)_{i \in I}$ such that for every S_i and S_j there is some $S_k \supseteq S_i, S_j$.

Proposition 18.1.4. *Let A be a universal algebra. Every directed union of subalgebras of A is a subalgebra of A. More generally, every directed union of algebras of type T is an algebra of type T.*

In particular, the union of a chain of subalgebras is a subalgebra. Proposition 18.1.4 no longer holds if infinitary operations are allowed.

18.1.e. Quotient Algebras. Universal algebras resemble sets in that their quotients must be constructed from equivalence relations rather than subalgebras.

Proposition 18.1.5. *For an equivalence relation $\mathscr{E}$ on an algebra A of type T the following conditions are equivalent:*

(1) *there is a type T algebra structure on $A/\mathscr{E}$ such that the projection $A \longrightarrow A/\mathscr{E}$ is a homomorphism;*

(2) *$x_1 \mathscr{E} y_1, \ldots, x_n \mathscr{E} y_n$ implies $\omega_A(x_1,\ldots,x_n) \mathscr{E} \omega_A(y_1,\ldots,y_n)$ for every $\omega \in T$, where n is the arity of ω.*

Then the algebra structure in (1) is unique.

Proof. Let E_x denote the $\mathscr{E}$-class of $x \in A$; the projection $\pi : A \longrightarrow A/\mathscr{E}$ assigns E_x to x. If there is an algebra structure on $Q = A/\mathscr{E}$ such that π is a homomorphism, and $\omega \in T$ has arity n, then $x_1 \mathscr{E} y_1, \ldots, x_n \mathscr{E} y_n$ implies

$$\pi(\omega_A(x_1,\ldots,x_n)) = \omega_Q(\pi(x_1),\ldots,\pi(x_n))$$

$$= \omega_Q(\pi(y_1),\ldots,\pi(y_n)) = \pi(\omega_A(y_1,\ldots,y_n));$$

thus (1) implies (2). Conversely, assume (2). Let $\omega \in T$ have arity n. If $E_1,\ldots,E_n \in Q$ are $\mathscr{E}$-classes, then the set

$$\omega_A(E_1,\ldots,E_n) = \{\omega_A(x_1,\ldots,x_n) ; x_1 \in E_1,\ldots,x_n \in E_n\}$$

is contained in a single $\mathscr{E}$-class, by (2). Therefore there is a mapping $\omega_Q : Q^n \longrightarrow Q$ which sends $(E_1,\ldots,E_n) \in Q^n$ to the $\mathscr{E}$-class which contains $\omega_A(E_1, \ldots,E_n)$. We see that the following square commutes:

$$\begin{array}{ccc} A^n & \xrightarrow{\pi^n} & Q^n \\ \omega_A \downarrow & & \downarrow \omega_Q \\ A & \xrightarrow{\pi} & Q \end{array}$$

Since π^n is surjective, ω_Q is the only mapping such that this square commutes. Thus we constructed an algebra structure on Q; it has type T and is the only algebra structure on Q such that the projection π is a homomorphism. $\square$

A **congruence** on an algebra A of type T is an equivalence relation $\mathscr{E}$ on A which satisfies the equivalent conditions in Proposition 18.1.5; then the unique algebra $A/\mathscr{E}$ in Proposition 18.1.5 is the **quotient** of A by $\mathscr{E}$. Proposition 1.1.5 is a particular case of Proposition 18.1.5.

We saw in Chapter 1 that all congruences on a group can be constructed from normal subgroups. This property extends to rings, modules, and so on, but does not hold in all universal algebras; the exercises give a counterexample.

Congruences have the following general properties.

Proposition 18.1.6. *Let A be a universal algebra. Every intersection of congruences of A is a congruence of A.*

Proposition 18.1.7. *Let A be a universal algebra. Every directed union of congruences of A is a congruence of A.*

The proofs are exercises.

18.1.f. Factorization. There are two basic results which allow one homomorphism of universal algebras to factor through another.

Proposition 18.1.8. *Let $\varphi : A \longrightarrow C$ and $\psi : B \longrightarrow C$ be homomorphisms of algebras of type T. If φ is injective, then ψ factors through φ ($\psi = \varphi \circ \chi$ for some homomorphism $\chi : B \longrightarrow A$) if and only if $\operatorname{Im} \psi \subseteq \operatorname{Im} \varphi$, and then ψ factors uniquely through φ (χ is unique).*

$$A \xrightarrow{\varphi} C$$
$$\chi \nwarrow \quad \nearrow \psi$$
$$B$$

Proposition 18.1.9. *Let $\varphi : C \longrightarrow A$ and $\psi : C \longrightarrow B$ be homomorphisms of algebras of type T. If φ is surjective, then ψ factors through φ ($\psi = \chi \circ \varphi$ for some homomorphism $\chi : A \longrightarrow B$) if and only if the congruence induced by φ is contained in the congruence induced by ψ ($\varphi(x) = \varphi(y)$ implies $\psi(x) = \psi(y)$); and then ψ factors uniquely through φ (χ is unique):*

$$C \xrightarrow{\varphi} A$$
$$\psi \searrow \quad \downarrow \chi$$
$$B$$

The proofs make good exercises for our diligent reader.

18.1.g. The Homomorphism Theorem. We can now prove:

Theorem 18.1.10. *When $\varphi : A \longrightarrow B$ is a homomorphism of algebras of type T:*

(1) *$\operatorname{Im} \varphi$ is a subalgebra of B;*
(2) *the equivalence relation $x \mathcal{E} y \iff \varphi(x) = \varphi(y)$ induced by φ is a congruence on A;*
(3) *$\operatorname{Im} \varphi \cong A/\mathcal{E}$; in fact there is a unique isomorphism $\theta : A/\mathcal{E} \longrightarrow \operatorname{Im} \varphi$ such that the square*

commutes, where $\pi : A \longrightarrow A/\mathscr{E}$ *is the projection and* $\iota : \mathrm{Im}\,\varphi \longrightarrow B$ *is the inclusion homomorphism.*

Proof. For every $\omega \in T$ of arity n and $b_1 = \varphi(a_1), \ldots, b_n = \varphi(a_n) \in \mathrm{Im}\,\varphi$, we have

$$\omega_B(b_1, \ldots, b_n) = \omega_B(\varphi(a_1), \ldots, \varphi(a_n)) = \varphi(\omega_A(a_1, \ldots, a_n)) \in \mathrm{Im}\,\varphi;$$

hence $\mathrm{Im}\,\varphi$ is a subalgebra of B. Similarly $\varphi(x_1) = \varphi(y_1), \ldots, \varphi(x_n) = \varphi(y_n)$ implies that

$$\varphi(\omega_A(x_1, \ldots, x_n)) = \omega_B(\varphi(x_1), \ldots, \varphi(x_n))$$

$$= \omega_B(\varphi(y_1), \ldots, \varphi(y_n)) = \varphi(\omega_A(y_1, \ldots, y_n))$$

for every $\omega \in T$ of arity n; hence the equivalence relation $\mathscr{E}$ induced by φ is a congruence on A.

From set theory we know that there is a unique bijection θ such that the square in the statement commutes; θ sends the $\mathscr{E}$-class E_x of $x \in A$ onto $\varphi(x)$. Let $Q = A/\mathscr{E}$, $x_1, \ldots, x_n \in A$, $\omega \in T$ have arity n, and $x = \omega_A(x_1, \ldots, x_n)$. Then $\omega_Q(E_{x_1}, \ldots, E_{x_n}) = E_x$ and

$$\theta(\omega_Q(E_{x_1}, \ldots, E_{x_n})) = \theta(E_x) = \varphi(x) = \varphi(\omega_A(x_1, \ldots, x_n))$$

$$= \omega_B(\varphi(x_1), \ldots, \varphi(x_n)) = \omega_{\mathrm{Im}\,\varphi}(\theta(E_{x_1}), \ldots, \theta(E_{x_n})).$$

Hence θ is an isomorphism. (We can also use Propositions 18.1.8 and 18.1.9.) $\square$

A **homomorphic image** of an algebra A of type T is an algebra B of type T for which there exists a surjective homomorphism $A \longrightarrow B$; equivalently, such that B is isomorphic to a quotient algebra of A.

Exercises

1. Prove that a homomorphism $\theta : A \longrightarrow B$ of algebras of type T is an isomorphism (there exists a homomorphism $\zeta : B \longrightarrow A$ such that $\zeta \circ \theta = 1_A$ and $\theta \circ \zeta = 1_B$) if and only if it is bijective.
2. Prove Proposition 18.1.2.
3. Let A be a universal algebra. Prove that every intersection of subalgebras of A is a subalgebra of A.
4. Let A be a universal algebra. Prove that every directed union of subalgebras of A is a subalgebra of A.
5. Prove that there is a one-to-one correspondence between congruences on a group G and normal subgroups of G.

6. Let A be an algebra with one operation (denoted multiplicatively) defined by $xy = y$ for all $x, y \in A$. Show that every equivalence relation on A is a congruence on A. If A has enough elements, show that there are congruences on A which are not determined by any one of their equivalence classes.

7. Let A be a universal algebra. Prove that every intersection of congruences on A is a congruence on A.

8. Let A be a universal algebra. Prove that every directed union on congruences on A is a congruence on A.

9. Show that an equivalence relation on a universal algebra A is a congruence on A if and only if it is a subalgebra of $A \times A$ (with the coordinatewise operations on $A \times A$).

10. Let $A = \mathbb{R} \cup \{\infty\}$ be the algebra with one infinitary operation which assigns to each infinite sequence its least upper bound. Show that a directed union of subalgebras of A need not be a subalgebra of A.

11. Let $\varphi : A \longrightarrow C$ and $\psi : B \longrightarrow C$ be homomorphisms of algebras of type T. If φ is injective, prove that ψ factors through φ if and only if $\operatorname{Im} \psi \subseteq \operatorname{Im} \varphi$; and then ψ factors uniquely through φ.

12. Let $\varphi : C \longrightarrow A$ and $\psi : C \longrightarrow B$ be homomorphisms of algebras of type T. If φ is surjective, prove that ψ factors through φ if and only if the congruence induced by φ is contained in the congruence induced by ψ, and then ψ factors uniquely through φ.

18.2. WORD ALGEBRAS

This section constructs word algebras and formally defines identities.

18.2.a. Generators. Let A be a universal algebra of type T. By Proposition 18.1.3, every intersection of subalgebras of A is a subalgebra of A. Hence there is for each subset X of A a subalgebra $\langle X \rangle$ of A **generated by** X which is the smallest subalgebra of A containing S, namely the intersection of all the subalgebras of A that contain X. Our eager reader will prove:

Proposition 18.2.1. *Let A be an algebra of type T and X be a subset of A. Define $S_k \subseteq A$ by induction: $S_0 = X$, and S_{k+1} is the union of S_k and $\{ \omega_A(x_1, \ldots, x_n) ; n \geqslant 0, x_1, \ldots, x_n \in S_k,$ and $\omega \in T$ has arity $n \}$. The subalgebra $S = \langle X \rangle$ of A generated by X is $S = \bigcup_{k \geqslant 0} S_k$.*

When A is generated by X, every element a of A can thus be calculated in finitely many steps from the operations on A and the elements of X. In general this calculation can be carried out in several different ways. The simplest way to construct an algebra which is generated by X is to ensure that different calculations yield different elements. The result is the word algebra on X. This algebra is conceptually similar to, say, the free group on X, except that, without associativity or inverses, there is no need to equate some words to others.

18.2.b. Construction. Let X be a set and T be a type of universal algebras. The set W_k of all T-**words** of **height** k in the alphabet X is constructed by induction on $k \geqslant 0$ as follows. A T-word of height 0 is an element of X or an element of T of arity 0. A T-word of height $k + 1 > 0$ is a sequence $(\omega, w_1, \ldots, w_n)$ in which $\omega \in T$ has arity $n > 0$, $w_1, \ldots, w_n$ are words of height at most k $(w_1, \ldots, w_n \in \bigcup_{\ell \leqslant k} W_\ell)$ and at least one of $w_1, \ldots, w_n$ has height k $(w_i \in W_k$ for some $i)$. The sets W_k are pairwise disjoint.

For example, let T have a single element m of arity 2. A T-word of height 0 is an element of X. The T-words of height 1 are all sequences (m, x, y) with $x, y \in X$. The T-words of height 2 are all $(m, x, (m, y, z))$, $(m, (m, x, y), z)$, and $(m, (m, x, y), (m, z, t))$, where $x, y, z, t \in X$.

In general, the **word algebra** of type T on the set X is the set $F = F_X^T = \bigcup_{k \geqslant 0} W_k$ of all T-words in the alphabet X, with the following type T algebra structure. Let $\omega \in T$ have arity n. If $n = 0$, then $\omega_F = \omega \in W_0$. If $n > 0$, then $\omega_F(w_1, \ldots, w_n) = (\omega, w_1, \ldots, w_n) \in W_{k+1}$, where k is the largest of the heights of $w_1, \ldots, w_n$. Hence $(\omega, w_1, \ldots, w_n) = \omega_F(w_1, \ldots, w_n)$ for all $(\omega, w_1, \ldots, w_n) \in W_{k+1}$.

We note the following property:

Proposition 18.2.2. F_X^T *is a directed union* $F_X^T = \bigcup_{Y \subseteq X, Y \text{ finite}} F_Y^T$.

Proof. It is immediate by induction on k that every $w \in W_k$ belongs to F_Y^T for some finite $Y \subseteq X$. $\square$

18.2.c. Universal Property. The universal property of F_X^T shows that it is indeed the "largest" algebra of type T generated by X.

Theorem 18.2.3. *The word algebra* F_X^T *of type* T *on the set* X *is generated by* X. *Moreover every mapping of* X *into an algebra* A *of type* T *can be extended uniquely to a homomorphism of* F_X^T *into* A.

Proof. $F = F_X^T$ is generated by X by Proposition 18.2.1.

Let A be an algebra of type T and $\varphi : X \longrightarrow A$ be a mapping. If $\psi : F \longrightarrow A$ is a homomorphism and extends φ, then $\psi(x) = \varphi(x)$ for all $x \in X$ and $\psi(\omega) = \psi(\omega_F) = \omega_A$ for every $\omega \in T$ of arity 0. Since $(\omega, w_1, \ldots, w_n) = \omega_F(w_1, \ldots, w_n)$ for all $(\omega, w_1, \ldots, w_n) \in W_{k+1}$, we must also have

$$\psi(\omega, w_1, \ldots, w_n) = \psi(\omega_F(w_1, \ldots, w_n)) = \omega_A(\psi(w_1), \ldots, \psi(w_n))$$

for all $(\omega, w_1, \ldots, w_n) \in W_{k+1}$. By induction, this determines the values of ψ on W_k. Hence there is at most one homomorphism $\psi : F \longrightarrow A$ which extends φ.

Conversely, a mapping $\psi : F \longrightarrow A$ is defined recursively by $\psi(x) = \varphi(x)$ for all $x \in X$ and

$$\psi(\omega, w_1, \ldots, w_n) = \omega_A(\psi(w_1), \ldots, \psi(w_n))$$

for all $(\omega, w_1, \ldots, w_n) \in W_{k+1}$. It is immediate that ψ is a homomorphism. $\square$

Theorem 18.2.3 shows that the forgetful functor from Alg_T to *Sets* has a left adjoint; η_X is the inclusion mapping $X \longrightarrow F_X^T$. In this sense word algebras are similar to free groups, free modules, and the like. However, free groups and free modules are not word algebras: indeed the conditions that characterize groups or modules do not hold in the corresponding word algebras.

By Theorem 18.2.3, a bijection $X \longrightarrow Y$ induces an isomorphism $F_X^T \cong F_Y^T$; hence, up to isomorphism, F_X^T depends only on T and $|X|$.

18.2.d. Identities. As in Section 2.6, a T-**relation** between the elements of a set X is a pair (u,v) of elements of F_X^T (often written as an equality $u = v$); a relation (u,v) **holds** in an algebra A of type T, **via** a mapping $\varphi : X \longrightarrow A$, in case the homomorphism $\psi : F_X^T \longrightarrow A$ which extends φ satisfies $\psi(u) = \psi(v)$.

An identity is a relation which holds via every mapping. In detail, let X be your favorite infinite set (e.g., $\mathbb{N}$). A T-**identity** is a pair (u,v) of elements of the word algebra F_X^T (often written as an equality $u = v$). A T-identity (u,v) **holds** in an algebra A of type T in case the relation (u,v) holds in A via *every* mapping $X \longrightarrow A$; equivalently, in case $\psi(u) = \psi(v)$ holds for every homomorphism $\psi : F_X^T \longrightarrow A$.

By Proposition 18.2.2, every identity (u,v), with $u,v \in F_X^T$, can be written using only finitely many elements of X. Hence our definition reaches all possible identities, regardless of the size of X, as long as X is at least countable.

For example, **associativity** is the formal T-identity

$$((m,x,(m,y,z)),\ (m,(m,x,y),z)),$$

where T has a single element m of arity 2 and x,y,z are distinct elements of X. Indeed an algebra A of type T is a set A with one binary operation m_A. For every $a,b,c \in A$ there is a homomorphism $\psi : F_X^T \longrightarrow A$ which sends x,y,z to a,b,c; then $\psi(m,x,(m,y,z)) = m_A(a,m_A(b,c))$ and $\psi(m,(m,x,y),z) = m_A(m_A(x,y),z)$. Therefore the identity $(m,x,(m,y,z)),(m,(m,x,y),z)$ holds in A if and only if $m_A(a,m_A(b,c)) = m_A(m_A(a,b),c)$ for all $a,b,c \in A$, if and only if m_A is associative.

Exercises

1. Prove Proposition 18.2.1.

2. Let $T = \{a\} \cup R$, where a has arity 2 and every $r \in R$ has arity 1. Describe all T-words of height $0,1,2$ in F_X^T.

3. Let $T = \{e,i,m\}$, where e has arity 0, i has arity 1, and m has arity 2. Describe all T-words of height $0,1,2$ in F_X^T.

4. Show that every algebra of type T is a homomorphic image of a word algebra of type T.

5. Write commutativity as a formal identity.

6. Write distributivity in a ring as a formal identity.

7. Let X be any infinite countable set and A be an algebra of type T. Prove that the set of all identities which hold in A is a congruence on F_X^T.

8. Show that every universal algebra is a directed union of finitely generated subalgebras.

18.3. VARIETIES

A variety consists of all algebras that satisfy a given set of identities. Most of the algebraic systems in this book (groups, rings, modules, etc.) constitute varieties; many of their properties extend to all varieties.

18.3.a. Definition. Let T be a type of algebra and X be your favorite infinite set (e.g., $X = \mathbb{N}$). For each set $R \subseteq F_X^T \times F_X^T$ of T-identities, let $\mathscr{V}(R)$ be the class of all algebras A of type T such that every identity in R holds in A. A **variety** of algebras of type T is a class $\mathscr{V}$ of the form $\mathscr{V}(R)$ for some set $R \subseteq F_X^T \times F_X^T$ of T-identities. A $\mathscr{V}$-algebra is an algebra $A \in \mathscr{V}$.

18.3.b. Examples. For any type T there are always two varieties. When $R = \varnothing$, then $\mathscr{V}(R)$ contains all algebras of type T; thus Alg_T is a variety. If $R = F_X^T \times F_X^T$ consists of all possible identities (including the identity $x = y$), then $\mathscr{V}(R)$ is the **trivial** variety, which consists of all the algebras of type T with one element (and, if there is no element of T of arity 0, the empty algebra).

We show that groups constitute a variety. The definition of groups as universal algebras with one binary operation is not suitable for this, since the existence of an identity element, or the existence of inverses, is not an identity. But we can regard groups as algebras with one constant operation 1, one unary operation $x \longmapsto x^{-1}$, and one binary operation $(x,y) \longmapsto xy$. (Then the subalgebras of a group are its subgroups, and the homomorphisms are the usual homomorphisms of groups.) An algebra G of this type is a group if and only if

$$1x = x \qquad \text{for all} \quad x \in G,$$

$$x1 = x \qquad \text{for all} \quad x \in G,$$

$$xx^{-1} = 1 \qquad \text{for all} \quad x \in G,$$

$$x^{-1}x = 1 \qquad \text{for all} \quad x \in G,$$

$$x(yz) = (xy)z \qquad \text{for all} \quad x,y,z \in G,$$

which means that certain identities hold in G. (Dedicated readers will write the corresponding formal identities.)

Abelian groups form a variety, which is characterized by the above and the additional **commutativity** identity $x + y = y + x$.

Rings [with identity element] are universal algebras with a constant operation 0, a unary operation $x \longmapsto -x$, a binary operation $(x, y) \longmapsto x + y$, another binary operation $(x, y) \longmapsto xy$, and another constant operation 1. Then rings are characterized by a set of identities, and constitute a variety.

For each ring R, left R-modules constitute a variety (of algebras with three operations for the additive group and one unary operation for each element of R).

18.3.c. Properties.

The next results state that many familiar constructions are possible in every variety.

Proposition 18.3.1. *Every variety is closed under direct products, subalgebras, homomorphic images, and directed unions.*

Proof. The **direct product** of a family $(A_i)_{i \in I}$ of algebras of type T is their cartesian product $A = \prod_{i \in I} A_i$ with the componentwise operations:

$$\omega_A((x_{1i})_{i \in I}, \ldots, (x_{ni})_{i \in I}) = (\omega_{A_i}(x_{1i}, \ldots, x_{ni}))_{i \in I}$$

whenever $\omega \in T$ has arity n and $(x_{1i})_{i \in I}, \ldots, (x_{ni})_{i \in I} \in \prod_{i \in I} A_i$. This is the only algebra structure on $\prod_{i \in I} A_i$ such that every projection $\pi_i : A \longrightarrow A_i$ is a homomorphism.

Let $R \subseteq F_X^T \times F_X^T$ be a set of identities and $\mathscr{V} = \mathscr{V}(R)$. Assume that $A_i \in \mathscr{V}$ for all i. If $(u, v) \in R$ and $\psi : F_X^T \longrightarrow A$ is a homomorphism, then $\pi_i \circ \psi : F_X^T \longrightarrow A_i$ is a homomorphism and $\pi_i(\psi(u)) = \pi_i(\psi(v))$, since $A_i \in \mathscr{V}$; since this holds for all i, it follows that $\psi(u) = \psi(v)$, which proves that $A = \prod_{i \in I} A_i \in \mathscr{V}$. Thus $\mathscr{V}$ is closed under direct products.

The reader will show that $\mathscr{V}$ is closed under subalgebras (every subalgebra of a $\mathscr{V}$-algebra is a $\mathscr{V}$-algebra), closed under homomorphic images (every homomorphic image of a $\mathscr{V}$-algebra is a $\mathscr{V}$-algebra), and closed under directed unions (every directed union of $\mathscr{V}$-algebras is a $\mathscr{V}$-algebra). $\square$

It follows from Proposition 18.3.1 that fields do not constitute a variety: there is no ingenious type of algebra T such that fields can be characterized as algebras of type T satisfying a set of identities. Indeed a direct product of two or more fields has zero divisors and is not a field.

In the next results, $\mathscr{C}$ is a class of algebras of type T (e.g., a variety). A $\mathscr{C}$-**algebra** is simply a member of $\mathscr{C}$. A homomorphism of $\mathscr{C}$-algebras is a homomorphism of algebras of type T (of one $\mathscr{C}$-algebra into another). By Proposition 18.1.1, $\mathscr{C}$-algebras and their homomorphisms are the objects and morphisms of a category, which we also denote by $\mathscr{C}$.

Proposition 18.3.2. *Let $\mathscr{C}$ be a class of algebras of type T which is closed under products and subalgebras (for instance, a variety). Then $\mathscr{C}$ is complete; in fact the forgetful functor from $\mathscr{C}$ to Sets creates limits.*

Proof. Let D be a diagram in $\mathscr{C}$ over a graph $\mathscr{G}$. Let $L : \lambda \longrightarrow D$ be a limit cone in *Sets* (which exists by Proposition 17.4.1). We have to show that there is exactly one algebra structure on L such that $\lambda : L \longrightarrow D$ is a cone in $\mathscr{C}$, and then λ is a limit cone in $\mathscr{C}$. It suffices to prove this for the standard limit cone, since every other limit cone is isomorphic to it.

In the standard limit cone $\lambda : L \longrightarrow D$,

$$ L = \{\, (x_i)_{i \in \mathscr{G}} \in \textstyle\prod_{i \in \mathscr{G}} D_i \; ; \; D_a(x_i) = x_j \text{ whenever } a : i \longrightarrow j \,\}, $$

and $\lambda_i : L \longrightarrow D_i$ sends $(x_i)_{i \in \mathscr{G}} \in L$ to x_i. There is a unique algebra structure on L such that every mapping λ_i is a homomorphism, namely the componentwise structure, whose operations are given by

$$ \omega_L((x_{1i})_{i \in \mathscr{G}}, \dots, (x_{ni})_{i \in \mathscr{G}}) = (\omega_{D_i}(x_{1i}, \dots, x_{ni}))_{i \in \mathscr{G}} $$

whenever $\omega \in T$ has arity n and $(x_{1i})_{i \in \mathscr{G}}, \dots, (x_{ni})_{i \in \mathscr{G}} \in L$; since every D_a is a homomorphism, $\omega_L((x_{1i})_{i \in \mathscr{G}}, \dots, (x_{ni})_{i \in \mathscr{G}})$ remains in L. (Equivalently, $(\omega_{D_i})_{i \in \mathscr{G}}$ is a homomorphism of diagrams; hence there is a unique mapping ω_L such that

$$
\begin{array}{ccc}
L^n & \xrightarrow{\;\lambda_i^n\;} & D_i^n \\[4pt]
\omega_L \big\downarrow & & \big\downarrow \omega_{D_i} \\[4pt]
L & \xrightarrow[\;\lambda_i\;]{} & D
\end{array}
$$

commutes for all i.) We see that L is a subalgebra of the direct product $\prod_{i \in \mathscr{G}} D_i$; hence $L \in \mathscr{C}$. Thus $\lambda : L \longrightarrow D$ is a cone in $\mathscr{C}$. It is immediate that λ is a limit cone in $\mathscr{C}$: if indeed $\varphi : A \longrightarrow D$ is any cone in $\mathscr{C}$, then the unique mapping $\overline{\varphi} : A \longrightarrow L$ such that $\varphi_i = \lambda_i \circ \overline{\varphi}$ for all i, given by $\overline{\varphi}(a) = (\varphi_i(a))_{i \in \mathscr{G}}$, is a homomorphism, and is the only homomorphism such that $\varphi_i = \lambda_i \circ \overline{\varphi}$ for all i. $\qquad\square$

The reader will show:

Proposition 18.3.3. *Let $\mathscr{C}$ be a class of algebras of type T which is closed under homomorphic images and directed unions (e.g., a variety). Then $\mathscr{C}$ has direct limits; in fact the forgetful functor from $\mathscr{C}$ to Sets creates direct limits.*

A $\mathscr{C}$-algebra F is **free** in $\mathscr{C}$ on a set X in case there exists a mapping $\eta : X \longrightarrow F$ such that, for every mapping φ of X into a $\mathscr{C}$-algebra A, there is a unique homomorphism $\psi : F \longrightarrow A$ such that $\varphi = \psi \circ \eta$.

For example, F_X^T is free on X in the class Alg_T of all algebras of type T. (Some definitions of free algebras also require the mapping η to be injective. This property holds in every nontrivial variety; see the exercises.)

Theorem 18.3.4. *Let $\mathscr{C}$ be a class of algebras of type T which is closed under direct products and subalgebras (e.g., a variety). In $\mathscr{C}$ there exists for every set X a free $\mathscr{C}$-algebra $\eta : X \longrightarrow F$ on X, and it is unique up to isomorphism and generated by $\eta(X)$.*

We denote the free $\mathscr{C}$-algebra on X by $F_X^{\mathscr{C}}$.

Proof. First, assume that there is a free $\mathscr{C}$-algebra $\eta : X \longrightarrow F$ on X. The universal property of F and η shows that they are unique up to isomorphism. Since $\mathscr{C}$ is closed under subalgebras, the subalgebra $G = \langle \eta(X) \rangle$ of F generated by $\eta(X)$ is a $\mathscr{C}$-algebra. If φ is a mapping of X into a $\mathscr{C}$-algebra A, there is a homomorphism $\psi : G \longrightarrow A$ induced by $F \longrightarrow A$ such that $\varphi = \psi \circ \eta$; ψ is unique, since G is generated by $\eta(X)$. Thus G is a free $\mathscr{C}$-algebra on X. Now G is generated by $\eta(X)$; this property is inherited by all free $\mathscr{C}$-algebras on X, which are isomorphic to G.

To prove the existence of free $\mathscr{C}$-algebras, we use the Adjoint Functor Theorem (Theorem 17.7.3) to show that the forgetful functor $G : \mathscr{C} \longrightarrow Sets$ has a left adjoint F; then $\eta_X : X \longrightarrow F(X)$ is a free $\mathscr{C}$-algebra on X. Both $\mathscr{C}$ and $Sets$ have small Hom sets. By Propositions 17.4.1 and 18.3.2, a limit can be assigned to every diagram in $\mathscr{C}$, and G preserves limits. It remains to verify the solution set condition: that to every set X can be assigned a set $\mathscr{S}$ of mappings $\sigma : X \longrightarrow C_\sigma$ such that C_σ is a $\mathscr{C}$-algebra and that every mapping α of X into a $\mathscr{C}$-algebra C factors as $\alpha = \gamma \circ \sigma$ through some homomorphism $\gamma : C_\sigma \longrightarrow C$.

We construct $\mathscr{S}$ as follows. Let $\mathscr{R}$ be the set of all congruences $\mathscr{E}$ on the word algebra F_X^T such that $F_X^T / \mathscr{E} \in \mathscr{C}$. For each $\mathscr{E} \in \mathscr{R}$, composing the inclusion $X \subseteq F_X^T$ and projection $F_X^T \longrightarrow F_X^T / \mathscr{E}$ yields a mapping $\sigma : X \longrightarrow C_\sigma = F_X^T / \mathscr{E}$ into a $\mathscr{C}$-algebra C_σ; $\mathscr{S}$ is the set of all such mappings. Let α be any mapping of X into a $\mathscr{C}$-algebra C. The subalgebra $\langle \mathrm{Im}\, \alpha \rangle$ of C generated by $\mathrm{Im}\, \alpha$ is a homomorphic image of F_X^T and is isomorphic to some C_σ:

$$
\begin{array}{ccc}
X & \xrightarrow{\ \sigma\ } & C_\sigma \\
{\scriptstyle \alpha}\big\downarrow & & \big\downarrow{\scriptstyle \cong} \\
C & \xleftarrow[\supseteq]{} & \langle \mathrm{Im}\, \alpha \rangle
\end{array}
$$

This provides a homomorphism $\gamma : C_\sigma \longrightarrow C$ such that $\alpha = \gamma \circ \sigma$. $\qquad\square$

Proposition 18.3.5. *Every variety $\mathscr{V}$ is cocomplete; in fact a colimit can be assigned to every diagram in $\mathscr{V}$.*

Proof. Let D be a diagram in $\mathscr{V}$ over some graph $\mathscr{G}$. Let X be the disjoint union of all the underlying sets D_i. By Theorem 18.3.4 there exists a free $\mathscr{V}$-algebra F on X; composing $D_i \subseteq X$ and $\eta : X \longrightarrow F$ yields a mapping $\eta_i : D_i \longrightarrow F$.

A cone $\varphi : D \longrightarrow C$ in $\mathscr{V}$ induces a mapping $X \longrightarrow C$ and a homomorphism $\psi : F \longrightarrow C$ unique such that $\varphi_i = \psi \circ \eta_i$ for all i. Since φ_i and ψ are homomorphisms we have

$$\psi(\eta_i(\omega_{D_i}(x_1,\ldots,x_n))) = \psi(\omega_F(\eta_i(x_1),\ldots,\eta_i(x_n))) \tag{1}$$

whenever $\omega \in T$ has arity n and $x_1,\ldots,x_n \in D_i$. Since $\varphi_i = \varphi_j \circ D_a$ whenever $a : i \longrightarrow j$ in $\mathscr{G}$, we also have

$$\psi(\eta_i(x)) = \psi(\eta_j(D_a(x))) \tag{2}$$

for all $a : i \longrightarrow j$ and $x \in D_i$. If, conversely, $C \in \mathscr{V}$ and $\psi : F \longrightarrow C$ is a homomorphism which satisfies (1) and (2), then each $\varphi_i = \psi \circ \eta_i$ is a homomorphism by (1), and $\varphi = (\varphi_i)_{i \in \mathscr{G}}$ is a cone from D to C.

We can therefore construct a colimit of D as follows. Let $\mathscr{L}$ be the intersection of all the congruences $\mathscr{E}$ on F such that

$$\eta_i(\omega_{D_i}(x_1,\ldots,x_n)) \, \mathscr{E} \, \omega_F(\eta_i(x_1),\ldots,\eta_i(x_n)) \tag{1'}$$

whenever $\omega \in T$ has arity n and $x_1,\ldots,x_n \in D_i$, and

$$\eta_i(x) \, \mathscr{E} \, \eta_j(D_a(x)) \tag{2'}$$

for all $a : i \longrightarrow j$ and $x \in D_i$. By Proposition 18.1.6, $\mathscr{L}$ is the smallest congruence on F with these properties. Then $L = F/\mathscr{L} \in \mathscr{V}$ by Proposition 18.3.1. Composing $\eta_i : D_i \longrightarrow F$ with the projection $F \longrightarrow L$ yields a mapping $\lambda_i : D_i \longrightarrow L$ which is a homomorphism by (1'). Any cone $\varphi : D \longrightarrow C$ in $\mathscr{V}$ induces a homomorphism $\psi : F \longrightarrow C$ with properties (1) and (2); ψ induces a congruence $\mathscr{E}$ on F with properties (1') and (2'), $\mathscr{L} \subseteq \mathscr{E}$, and ψ factors through the projection $F \longrightarrow L$.

This provides a homomorphism $\overline{\varphi} : L \longrightarrow C$ such that $\varphi_i = \overline{\varphi} \circ \lambda_i$ for all i; $\overline{\varphi}$ is unique since L is generated by $\bigcup_{i \in \mathscr{G}} \operatorname{Im} \lambda_i$. □

18.3.d. Birkhoff's Theorem. Birkhoff's Theorem states that the properties in Proposition 18.3.1 characterize varieties.

Theorem 18.3.6 (Birkhoff). *A class of universal algebras of the same type is a variety if and only if it is closed under direct products, subalgebras, and homomorphic images.*

Proof. By Proposition 18.3.1, every variety has these properties. Conversely, let $\mathscr{C}$ be a class of algebras of the same type T which is closed under direct products, subalgebras, and homomorphic images. Let R be the set of all T-identities that hold in every $\mathscr{C}$-algebra. Then $\mathscr{C}$ is contained in the variety $\mathscr{V} = \mathscr{V}(R)$.

Conversely, let $A \in \mathscr{V}$. There is a surjective homomorphism $\psi : F_X^T \longrightarrow A$ for some set X. By Theorem 18.3.4, there is a free $\mathscr{C}$-algebra $\eta : X \longrightarrow F$ on X, and it is generated by $\eta(X)$. Now the mapping $\eta : X \longrightarrow F$ extends to a homomorphism $\varphi : F_X^T \longrightarrow F$; φ is surjective, since F is generated by $\eta(X)$.

If $\varphi(u) = \varphi(v)$, then by Lemma 18.3.7 below the identity (u, v) holds in every $\mathscr{C}$-algebra; hence (u, v) holds in A and $\psi(u) = \psi(v)$. Therefore ψ factors through $\varphi : \psi = \chi \circ \varphi$ for some $\chi : F \longrightarrow A$. Now χ is surjective, since ψ is surjective; hence A is a homomorphic image of F, and the hypothesis yields $A \in \mathscr{C}$. $\square$

Lemma 18.3.7. *Let F be the free $\mathscr{C}$-algebra on a set X and $\varphi : F_X^T \longrightarrow F$ be the homomorphism which extends $\eta : X \longrightarrow F$. If $\varphi(u) = \varphi(v)$, then the identity (u, v) holds in every $\mathscr{C}$-algebra.*

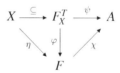

Proof. Let ψ be a homomorphism of F_X^T into a $\mathscr{C}$-algebra A. Since F is free on X, the mapping $\psi_{|X} : X \longrightarrow A$ factors through η: there is a homomorphism $\chi : F \longrightarrow A$ such that $\chi(\eta(x)) = \psi(x)$ for all $x \in X$. Then $\chi(\varphi(x)) = \psi(x)$ for all $x \in X$, $\chi \circ \varphi = \psi$, and $\varphi(u) = \varphi(v)$ implies $\psi(u) = \psi(v)$. $\square$

It follows from Birkhoff's Theorem 18.3.6 that every intersection of varieties of algebras of type T is a variety of algebras of type T. In particular, every class of algebras of type T generates a variety. Lemma 18.3.7 implies that any variety $\mathscr{V}$ is generated by the free $\mathscr{V}$-algebra on any infinite set.

A congruence $\mathscr{E}$ on a universal algebra A is **fully invariant** in case $x\,\mathscr{E}\,y$ implies $\xi(x)\,\mathscr{E}\,\xi(y)$ for every endomorphism ξ of A. The methods used in the proof of Theorem 18.3.6 yield the following result:

Proposition 18.3.8. *Let X be a countable infinite set. There is a one-to-one, order-reversing correspondence between varieties of algebras of type T and fully invariant congruences on F_X^T.*

Proof. To each fully invariant congruence $\mathscr{E}$ on F_X^T, we can assign the variety $\mathscr{V}(\mathscr{E})$ of all algebras of type T in which every identity $(u,v) \in \mathscr{E}$ holds. Conversely, to each variety $\mathscr{V}$ of algebras of type T we can assign the set $\mathscr{R}(\mathscr{V}) \subseteq F_X^T \times F_X^T$ of all T-identities which hold in every $\mathscr{V}$-algebra. We show that these are mutually inverse constructions.

Let $\mathscr{E}$ be a fully invariant congruence on F_X^T. We have $\mathscr{E} \subseteq \mathscr{R}(\mathscr{V}(\mathscr{E}))$. To prove the converse inclusion, we show that $F = F_X^T/\mathscr{E} \in \mathscr{V}(\mathscr{E})$. Let $\pi : F_X^T \longrightarrow F$ be the projection and $\varphi : F_X^T \longrightarrow F$ be any homomorphism. For each $x \in X$ there exists $w_x \in F_X^T$ such that $\pi(w_x) = \varphi(x)$. There is a homomorphism $\xi : F_X^T \longrightarrow F_X^T$ such that $\xi(x) = w_x$ for all x:

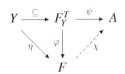

Then $\pi \circ \xi = \varphi$, since $\pi \circ \xi$ and φ agree on X. For all $(u,v) \in \mathscr{E}$ we have $(\xi(u), \xi(v)) \in \mathscr{E}$ and $\varphi(u) = \pi(\xi(u)) = \pi(\xi(v)) = \varphi(v)$. This proves that $F \in \mathscr{V}(\mathscr{E})$. (In fact F is the free $\mathscr{V}(\mathscr{E})$-algebra on X.) If now (u,v) is an identity which holds in every $\mathscr{V}(\mathscr{E})$-algebra, then it holds in F, and this implies $\pi(u) = \pi(v)$ and $(u,v) \in \mathscr{E}$. Thus $\mathscr{R}(\mathscr{V}(\mathscr{E})) \subseteq \mathscr{E}$.

Now let $\mathscr{V}$ be a variety of algebras of type T. First we show that $\mathscr{R}(\mathscr{V})$ is a fully invariant congruence on F_X^T. Let $\eta : X \longrightarrow F$ be the free $\mathscr{V}$-algebra on X. Let $\pi : F_X^T \longrightarrow F$ be the homomorphism which extends η. If an identity $(u,v) \in F_X^T \times F_X^T$ holds in every $\mathscr{V}$-algebra, then it holds in F, and $\pi(u) = \pi(v)$. The converse follows from Lemma 18.3.7. Thus $\mathscr{R}(\mathscr{V})$ is the equivalence relation induced by π and is a congruence on F_X^T. If now $(u,v) \in \mathscr{R}(\mathscr{V})$ and ξ is an endomorphism of F_X^T, then (u,v) holds in F, $\pi(\xi(u)) = \pi(\xi(v))$, and $(\xi(u), \xi(v)) \in \mathscr{R}(\mathscr{V})$. Thus $\mathscr{R}(\mathscr{V})$ is a fully invariant congruence.

We have $\mathscr{V} \subseteq \mathscr{V}(\mathscr{R}(\mathscr{V}))$. Conversely, let A be a $\mathscr{V}(\mathscr{R}(\mathscr{V}))$-algebra. As in the proof of Theorem 18.3.6, there is a surjective homomorphism $\psi : F_Y^T \longrightarrow A$ for some set Y. Let F be the free $\mathscr{V}$-algebra $\eta : Y \longrightarrow F$ on Y. The mapping $\eta : Y \longrightarrow F$ extends to a homomorphism $\varphi : F_Y^T \longrightarrow F$, which is surjective since F is generated by $\eta(Y)$ (Proposition 18.3.4):

$$Y \overset{\subseteq}{\longrightarrow} F_Y^T \overset{\psi}{\longrightarrow} A$$

If $\varphi(u) = \varphi(v)$, then by Lemma 18.3.7 the identity (u,v) holds in every $\mathscr{V}$-algebra; hence (u,v) holds in A and $\psi(u) = \psi(v)$. Therefore ψ factors through φ : $\psi = \chi \circ \varphi$ for some homomorphism $\chi : F \longrightarrow A$. Now χ is surjective, since ψ is surjective; hence A is a homomorphic image of F, and $A \in \mathscr{V}$. Thus $\mathscr{V}(\mathscr{R}(\mathscr{V})) \subseteq \mathscr{V}$. $\qquad\square$

18.3.e. Tripleability. Finally we show that every variety is tripleable over *Sets*. We saw several cases of this result in Chapter 17. On the other hand, ordered or topological structures yield categories that have many of the previous properties but are generally not tripleable over *Sets*.

Theorem 18.3.9. *For every variety $\mathscr{V}$, the forgetful functor from $\mathscr{V}$ to Sets is tripleable.*

Proof. We use Beck's Theorem 17.9.3. Let $\mathscr{V}$ be a variety of algebras of type T. By Theorem 18.3.4, the forgetful functor $G : \mathscr{V} \longrightarrow Sets$ has a left adjoint; we must show that G creates coequalizers of pairs $\alpha, \beta : A \longrightarrow B$ of homomorphisms of $\mathscr{V}$-algebras which have a split coequalizer $\sigma : B \longrightarrow C$ in *Sets*.

Let $\omega \in T$ have arity n. Since α and β are homomorphisms we have $\alpha \circ \omega_A = \omega_B \circ \alpha^n$ and $\beta \circ \omega_A = \omega_B \circ \beta^n$. Now σ^n is a split coequalizer of α^n and β^n. The diagram below shows that $\sigma \circ \omega_B \circ \alpha^n = \sigma \circ \omega_B \circ \beta^n$:

$$
\begin{array}{ccccc}
A^n & \underset{\beta^n}{\overset{\alpha^n}{\rightrightarrows}} & B^n & \overset{\sigma^n}{\longrightarrow} & C^n \\
\omega_A \downarrow & & \omega_B \downarrow & & \downarrow \omega_C \\
A & \underset{\beta}{\overset{\alpha}{\rightrightarrows}} & B & \underset{\sigma}{\longrightarrow} & C
\end{array}
$$

Therefore there is a unique mapping $\omega_C : C^n \longrightarrow C$ such that $\sigma \circ \omega_B = \omega_C \circ \sigma^n$. This provides an algebra structure on C such that σ is a homomorphism. Since σ is surjective, C is a homomorphic image of $B \in \mathscr{V}$ and $C \in \mathscr{V}$. Thus there is a unique $\mathscr{V}$-algebra on the set C such that σ is a homomorphism.

It remains to show that σ is a coequalizer in $\mathscr{V}$. Let $\varphi : B \longrightarrow D$ be a homomorphism of $\mathscr{V}$-algebras such that $\varphi \circ \alpha = \varphi \circ \beta$. Since σ is a coequalizer in *Sets*, there is a mapping $\psi : C \longrightarrow D$ unique such that $\varphi = \psi \circ \sigma$. For each $\omega \in T$ of arity n, we have a diagram

$$
\begin{array}{ccccc}
B^n & \overset{\sigma^n}{\longrightarrow} & C^n & \overset{\psi^n}{\longrightarrow} & D^n \\
\omega_B \downarrow & & \omega_C \downarrow & & \downarrow \omega_D \\
B & \underset{\sigma}{\longrightarrow} & C & \underset{\psi}{\longrightarrow} & D
\end{array}
$$

in which the left square and the outer rectangle commute, since σ and φ are homomorphisms; since σ^n is surjective, the right square commutes. Thus ψ is a homomorphism. $\qquad\square$

Exercises

1. Write a set of formal identities that characterize groups.

2. Write a set of formal identities that characterize rings [with identity elements].

3. Prove that every variety $\mathscr{V}$ is closed under subalgebras (every subalgebra of a $\mathscr{V}$-algebra is a $\mathscr{V}$-algebra).

4. Prove that every variety $\mathscr{V}$ is closed under homomorphic images (every homomorphic image of a $\mathscr{V}$-algebra is a $\mathscr{V}$-algebra).

5. Prove that every variety $\mathscr{V}$ is closed under directed unions (every directed union of $\mathscr{V}$-algebras is a $\mathscr{V}$-algebra).

6. Prove the following: if groups are defined as universal algebras with just one binary operation, then groups do not constitute a variety.

7. Let $\mathscr{C}$ be a class of algebras of type T which is closed under homomorphic images and directed unions (e.g., a variety). Show that $\mathscr{C}$ has direct limits; in fact the forgetful functor from $\mathscr{C}$ to *Sets* creates direct limits.

8. Let $\mathscr{V}$ be a variety and F be the free $\mathscr{V}$-algebra on some infinite set. Show that the variety generated by $\{F\}$ is $\mathscr{V}$.

9. Let $\mathscr{V}$ be a variety and $F_X^{\mathscr{V}}$ be the free $\mathscr{V}$-algebra on a set X. Assume that $\mathscr{V}$ is not the trivial variety (i.e., $\mathscr{V}$ contains an algebra with more than one element). Show that $\eta : X \longrightarrow F_X^{\mathscr{V}}$ is injective for every set X.

19

LATTICES

Lattice theory developed in the 1930s from the work of Birkhoff and Stone, culminating in 1940 with the publication of Birkhoff's *Lattice Theory*. Lattices are of interest as algebraic systems, and because there are so many examples.

Required results: this chapter draws examples from Chapters 1, 4, and 10, and occasional concepts from Chapters 17 and 18.

19.1. DEFINITION

This section contains the definition and some examples of lattices, and some basic properties.

19.1.a. Semilattices. Let S be a partially ordered set. A **lower bound** of a subset T of S is an element ℓ of S such that $\ell \leqslant t$ for all $t \in T$. A **greatest lower bound** or g.l.b. of T, also called a **meet** or **infimum**, is a lower bound g of T ($g \leqslant t$ for all $t \in T$) such that $\ell \leqslant g$ for every lower bound ℓ of T ($\ell \leqslant t$ for all $t \in T$ implies $\ell \leqslant g$).

It is immediate that greatest lower bounds are unique (a subset which has a g.l.b. has only one g.l.b.). Hence we speak of *the* greatest lower bound of a subset T. It is denoted by $\bigwedge_{t \in T} t$: if $T = (t_i)_{i \in I}$, then $\bigwedge_{t \in T} t$ is denoted by $\bigwedge_{i \in I} t_i$: if $T = \{t_1, \ldots, t_n\}$ is finite, then $\bigwedge_{t \in T} t$ is denoted by $t_1 \wedge t_2 \wedge \cdots \wedge t_n$ (see Proposition 19.1.1 below).

A **lower semilattice** (also called a **meet** or **inf** semilattice) is a partially ordered set $(S, \leqslant)$ in which any two elements a and b have a greatest lower bound $a \wedge b$. By definition, $a \wedge b \leqslant a$, $a \wedge b \leqslant b$, and $x \leqslant a$, $x \leqslant b$ implies $x \leqslant a \wedge b$.

For example, the set 2^X of all subsets of a set X, ordered by inclusion, is a lower semilattice, in which the g.l.b. of two subsets is their intersection. Therefore any set of subsets of X which is closed under intersections is a lower semilattice (when ordered by inclusion). Thus subgroups of a group, subrings of a ring, ideals of a ring, submodules of a module, all constitute semilattices.

Proposition 19.1.1. *The binary operation $\wedge$ on a lower semilattice $(S, \leqslant)$ is idempotent ($a \wedge a = a$ for all $a \in S$), commutative ($a \wedge b = b \wedge a$ for all $a, b \in S$), and associative ($a \wedge (b \wedge c) = (a \wedge b) \wedge c$ for all $a, b, c \in S$). Moreover every finite nonempty subset $\{a_1, \ldots, a_n\}$ of S has a greatest lower bound, namely $a_1 \wedge \cdots \wedge a_n = (\ldots((a_1 \wedge a_2) \wedge a_3) \wedge \cdots) \wedge a_n$.*

Proof. The first two properties are clear. We show by induction on $n > 0$ that $(\ldots((a_1 \wedge a_2) \wedge a_3) \wedge \cdots) \wedge a_n$ is the g.l.b. of $\{a_1, \ldots, a_n\}$. Certainly this holds if $n = 1$ or $n = 2$. If $n > 2$, the induction hypothesis provides a g.l.b. $b = (\ldots((a_1 \wedge a_2) \wedge a_3) \wedge \cdots) \wedge a_{n-1}$ of $a_1, \ldots, a_{n-1}$. Then $b \wedge a_n \leqslant a_i$ for all i; if, conversely, $x \leqslant a_i$ for all i, then $x \leqslant b$ (since $x \leqslant a_1, \ldots, a_{n-1}$), $x \leqslant a_n$, and $x \leqslant b \wedge a_n$. Thus $b \wedge a_n$ is a g.l.b. of $a_1, \ldots, a_n$. In particular, $(a \wedge b) \wedge c$ is the g.l.b. of $\{a, b, c\}$, and so is $a \wedge (b \wedge c) = (b \wedge c) \wedge a$. $\square$

Let S be a partially ordered set. An **upper bound** of a subset T of S is an element u of S such that $t \leqslant u$ for all $t \in T$. A **least upper bound** or l.u.b. of T, also called a **join** or **supremum** of T, is an upper bound ℓ of T ($t \leqslant \ell$ for all $t \in T$) such that $\ell \leqslant u$ for every upper bound u of T ($t \leqslant u$ for all $t \in T$ implies $\ell \leqslant u$). It is immediate that the least upper bound of $T \subseteq S$ is unique (when it exists). It is denoted by $\bigvee_{t \in T} t$; if $T = (t_i)_{i \in I}$, then $\bigvee_{t \in T} t$ is denoted by $\bigvee_{i \in I} t_i$; if $T = \{t_1, \ldots, t_n\}$ is finite, then $\bigvee_{t \in T} t$ is denoted by $t_1 \vee t_2 \vee \cdots \vee t_n$.

An **upper semilattice** is a partially ordered set $(S, \leqslant)$ in which any two elements a and b have a least upper bound $a \vee b$ (also called their **join** or **supremum**). By definition, $a \vee b \geqslant a$, $a \vee b \geqslant b$, and $x \geqslant a$, $x \geqslant b$ implies $x \geqslant a \vee b$.

Least upper bounds and greatest lower bounds are dual concepts in the following sense. The **dual** or **opposite** of a partially ordered set $(S, \leqslant)$ is the partially ordered set $(S, \leqslant)^{\mathrm{op}} = (S, \leqslant^{\mathrm{op}})$ on the same set with the opposite order relation, $x \leqslant^{\mathrm{op}} y$ if and only if $y \leqslant x$. We see that a least upper bound in $(S, \leqslant)$ is a greatest lower bound in $(S, \leqslant^{\mathrm{op}})$, and vice versa. Hence S is an upper semilattice if and only if S^{op} is a lower semilattice. By Proposition 19.1.1, every finite nonempty subset of an upper semilattice has a least upper bound.

19.1.b. Lattices. A **lattice** is a partially ordered set $(L, \leqslant)$ which is both a lower semilattice and an upper semilattice; equivalently, in which any two elements a and b of L have a greatest lower bound $a \wedge b$ (also called their **meet** or **infimum**) and a least upper bound $a \vee b$ (also called their **join** or **supremum**).

By Proposition 19.1.1, every finite nonempty subset of a lattice has a greatest lower bound and a least upper bound.

Reversing the order relation on a partially ordered set exchanges infimums and supremums (replaces $a \wedge b$ by $a \vee b$ and replaces $a \vee b$ by $a \wedge b$). Hence

L^{op} is a lattice if and only if L is a lattice. Therefore:

Metatheorem 19.1.2 (Duality Principle). *A theorem which holds in all lattices remains true if the order relation is reversed.*

As with the similar result in Chapter 17 (Metatheorem 17.1.4), one should be careful not to apply the Duality Principle to *specific* lattices: if Theorem T is true in L, then T^{op} is true in L^{op}, but it does not follow that T^{op} is true in L (unless T is also true in L^{op}).

19.1.c. Examples. Every **chain** (= totally ordered set) is a lattice: if, say, $a \leqslant b$, then $a \wedge b = a$ and $a \vee b = b$. For instance, $\mathbb{N}$, $\mathbb{Q}$, and $\mathbb{R}$, with their usual order relations, are lattices.

A finite lattice can be (like every finite partially ordered set) specified by a directed graph in which the elements are vertices and $x \leqslant y$ if and only if there is a path from x to y. It is understood that all arrows point upward and that arrow tips are omitted. For instance, in the graph

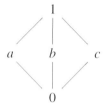

$0 < 1$, but $a \not\leqslant b$; the graph shows that $a \wedge b = b \wedge c = c \wedge a = 0$ and $a \vee b = b \vee c = c \vee a = 1$, and represents a lattice.

The set 2^X of all subsets of a set X, partially ordered by inclusion, is a lattice, in which the infimum of two subsets is their intersection, and the supremum of two subsets is their union.

Many examples of lattices arise from the following result.

Lemma 19.1.3. *Let $L \subseteq 2^X$. If $X \in L$ and L is closed under all intersections ($S \subseteq L$ implies $\bigcap S \in L$), then L, partially ordered by inclusion, is a lattice.*

Proof. When $A, B \in L$, then $A \cap B \in L$ is the infimum of A and B in L. The supremum $A \vee B$ in L is the intersection of all $C \in L$ containing $A \cup B$, which belongs to L by the hypothesis. $\square$

By Lemma 19.1.3, the subgroups of a group G constitute a lattice (when partially ordered by inclusion); the normal subgroups of G constitute a lattice; the subrings of a ring R constitute a lattice; so do the ideals of R; the submodules of an R-module constitute a lattice. In fact the subalgebras of any universal algebra A constitute a lattice; so do the congruences on A. In these lattices, infimums are intersections; supremums depend on the lattice.

19.1.d. Equivalent Definition. Unlike partially ordered sets in general, lattices and semilattices can be viewed as algebraic systems (as in Chapter 18).

Proposition 19.1.4. *A binary operation $\wedge$ is the infimum operation of a lower semilattice if and only if it is idempotent, commutative, and associative; then $\wedge$ is the infimum operation of a unique lower semilattice, in which $x \leqslant y$ if and only if $x \wedge y = x$.*

Proof. In a semilattice the operation $\wedge$ is idempotent, commutative, and associative by Proposition 19.1.1. Moreover $x \leqslant y$ implies $x \wedge y = x$; conversely, $x \wedge y = x$ implies $x = x \wedge y \leqslant y$.

Conversely, assume that $\wedge$ is an idempotent, commutative, and associative binary operation on a set S. Define a relation $\leqslant$ on S by $x \leqslant y$ if and only if $x \wedge y = x$. Then $x \leqslant x$ by idempotency, $x \leqslant y \leqslant x$ implies $x = y$ by commutativity, and $x \leqslant y \leqslant z$ implies

$$x \wedge z = (x \wedge y) \wedge z = x \wedge (y \wedge z) = x \wedge y = x$$

by associativity. Hence $\leqslant$ is an order relation. It remains to show that $a \wedge b$ is the greatest lower bound of a and b. First $a \wedge b \leqslant a$, since $(a \wedge b) \wedge a = a \wedge (a \wedge b) = (a \wedge a) \wedge b = a \wedge b$. Then $a \wedge b \leqslant b$ by (2). If, conversely, $x \leqslant a$ and $x \leqslant b$, then $x \wedge a = x \wedge b = x$, $x \wedge (a \wedge b) = (x \wedge a) \wedge b = x \wedge b = x$ by (3), and $x \leqslant a \wedge b$. ◻

Dually a binary operation $\vee$ is the supremum operation of an upper semilattice if and only if it is idempotent, commutative, and associative; then $x \leqslant y$ if and only if $x \vee y = y$.

Proposition 19.1.5. *Two binary operations $\wedge$ and $\vee$ on a set S are the infimum and supremum operations of a lattice L if and only if the identities*

(1) *(idempotency)* $a \wedge a = a$, $a \vee a = a$,

(2) *(commutativity)* $a \wedge b = b \wedge a$, $a \vee b = b \vee a$,

(3) *(associativity)* $a \wedge (b \wedge c) = (a \wedge b) \wedge c$, $a \vee (b \vee c) = (a \vee b) \vee c$,

(4) *(absorption laws)* $a \wedge (a \vee b) = a$, $a \vee (a \wedge b) = a$

hold for all $a, b, c \in S$; then $\wedge$ and $\vee$ are the infimum and supremum operations of a unique lattice L, in which $x \leqslant y$ if and only if $x \wedge y = x$, if and only if $x \vee y = y$.

Proof. We saw that (1), (2), and (3) hold in every lattice; (4) is immediate. Conversely, assume (1), (2), (3), and (4). By Proposition 19.1.4, S, partially ordered by $x \leqslant y$ if and only if $x \wedge y = x$, is a lower semilattice whose infimum operation is $\wedge$. Dually S, partially ordered by $x \leqslant y$ if and only if $x \vee y = y$, is an upper semilattice whose supremum operation is $\vee$. By (4), $x \wedge y = x$ implies

$y = y \vee (y \wedge x) = y \vee x$, and conversely, $x \vee y = y$ implies $x = x \wedge (x \vee y) = x \wedge y$. Thus the two order relations on S coincide, and S is a lattice. $\square$

19.1.e. Homomorphisms. Proposition 19.1.5 shows that lattices constitute a variety. From Chapter 18 lattices inherit a number of general definitions and properties, which we review quickly.

A **homomorphism** of a lattice A into a lattice B is a mapping $\varphi : A \longrightarrow B$ which preserves infimums and supremums: $\varphi(a \wedge b) = \varphi(a) \wedge \varphi(b)$ and $\varphi(a \vee b) = \varphi(a) \vee \varphi(b)$ for all $a, b \in A$. A homomorphism of lattices is order preserving: if $x \leqslant y$, then $x = x \wedge y$, $\varphi(x) = \varphi(x) \wedge \varphi(y)$, and $\varphi(x) \leqslant \varphi(y)$. The converse is false: an order-preserving mapping of one lattice into another need not be a homomorphism (see the exercises).

An **isomorphism** of lattices is a homomorphism which is bijective; then the inverse bijection is also a homomorphism. The reader will verify that a bijection θ between lattices is an isomorphism if and only if θ and θ^{-1} are order preserving.

A subset S of a lattice L is [the underlying set of] a **sublattice** of L if and only if it is closed under infimums and supremums (if $x, y \in S$, then $x \wedge y \in S$ and $x \vee y \in S$).

A **congruence** on a lattice L is an equivalence relation $\mathcal{E}$ on L such that $x \mathrel{\mathcal{E}} y$ and $z \mathrel{\mathcal{E}} t$ implies $x \wedge z \mathrel{\mathcal{E}} y \wedge t$ and $x \vee z \mathrel{\mathcal{E}} y \vee t$. Then there is a **quotient lattice** $L/\mathcal{E}$ such that the projection $L \longrightarrow L/\mathcal{E}$ is a lattice homomorphism.

The **Homomorphism Theorem** for lattices states:

Proposition 19.1.6. *When $\varphi : A \longrightarrow B$ is a homomorphism of lattices:*

(1) *Im φ is a sublattice of B;*

(2) *the equivalence relation $\mathcal{E}$ induced by φ is a congruence on A;*

(3) *$A/\mathcal{E} \cong$ Im φ.*

Exercises

1. Prove that greatest lower bounds have the following associativity property: if $I = \bigcup_{j \in J} I_j$ is a partition of I, then $\bigwedge_{i \in I} t_i = \bigwedge_{j \in J} (\bigwedge_{i \in I_j} t_i)$ provided that $\bigwedge_{i \in I} t_i$ exists and $\bigwedge_{i \in I_j} t_i$ exists for every $j \in J$.

2. Give an example of two lattices A and B and an order-preserving mapping of A into B which is not a lattice homomorphism.

3. Show that a bijection θ between lattices is an isomorphism if and only if θ and θ^{-1} are order preserving.

4. Give a direct proof that every intersection of sublattices is a sublattice.

5. Give a direct proof that every directed union of sublattices is a sublattice.

6. Let $\mathcal{E}$ be a congruence on a lattice L. Show that every $\mathcal{E}$-class C is a sublattice of L and is **convex** (if $x \leqslant y \leqslant z$ and $x, z \in C$, then $y \in C$).

19.2. COMPLETE LATTICES

A complete lattice is a lattice in which all infimums and supremums exist. Most of the examples in Section 19.1 have this property. This section contains basic examples and properties, and MacNeille's completion theorem.

19.2.a. Definition. A lower semilattice S is **complete** in case every nonempty subset of S has a greatest lower bound. In particular, S has a least element 0 (such that $0 \leqslant x$ for all $x \in S$), which is the greatest lower bound of S.

Dually an upper semilattice S is **complete** in case every nonempty subset of S has a least upper bound. In particular, S has a greatest element 1 (such that $x \leqslant 1$ for all $x \in S$), which is the least upper bound of S.

A **complete** lattice is a partially ordered set L (necessarily a lattice) in which every nonempty subset has a greatest lower bound and a least upper bound. In particular, a complete lattice L has a **least** element 0 ($0 \leqslant x$ for all $x \in L$) which is the greatest lower bound of L (and the least upper bound of the empty subset of L), and a **greatest** element 1 ($x \leqslant 1$ for all $x \in L$) which is the least upper bound of L (and the greatest lower bound of the empty subset of L).

Proposition 19.2.1. *A partially ordered set is a complete lattice if and only if it is a complete lower semilattice and has a greatest element.*

Proof. Every complete lattice has these properties. Conversely, let S be a complete lower semilattice with a greatest element 1. For each subset T of S, let

$$U = \{ x \in S ; t \leqslant x \text{ for all } t \in T \}$$

be the set of all upper bounds of T. Then $1 \in U$ and $U \neq \emptyset$. Hence U has a greatest lower bound a. We have $a \in U$: indeed $t \leqslant a$ for every $t \in T$, since t is a lower bound of U. But a is a lower bound of U; hence a is the least element of U, and a is a least upper bound of T. $\qquad\square$

19.2.b. Examples. Every finite lattice is complete.

The set 2^X of all subsets of a set X, partially ordered by inclusion, is a complete lattice.

By Proposition 19.2.1, a subset L of 2^X which contains X and is closed under intersections is a complete lattice. Thus Lemma 19.1.3 is a particular case of Proposition 19.2.1, and the lattices produced by Lemma 19.1.3 are complete lattices. Hence the lattice of subgroups of a group, the lattice of normal subgroups of a group, the lattice of subrings of a ring, the lattice of ideals of a ring, the lattice of submodules of a module, are all complete lattices. More generally, the subalgebras of a universal algebra A constitute a complete lattice; so do the congruences on A.

Not every lattice is complete. We saw that $\mathbb{N}$, $\mathbb{Q}$, and $\mathbb{R}$, with their usual order relations, are lattices; neither is complete.

19.2.c. Closure Mappings. A **closure mapping** on a partially ordered set X is a mapping $\gamma : X \longrightarrow X$ such that:

(a) γ is order preserving ($x \leqslant y$ implies $\gamma x \leqslant \gamma y$);
(b) γ is idempotent ($\gamma\gamma x = \gamma x$ for all x);
(c) $\gamma x \geqslant x$ for all x.

These properties are abstracted from well-known properties of the closure mapping $A \longmapsto \overline{A}$ on a topological space.

An element c of X is **closed** relative to a closure mapping $\gamma : X \longrightarrow X$ in case $\gamma c = c$. By (b), c is closed if and only if $c \in \mathrm{Im}\ \gamma$. For each $x \in X$, γx (the **closure** of x) is the least closed element $c \geqslant x$. Indeed γx is closed by (b), $x \leqslant \gamma x$ by (c), and $x \leqslant c = \gamma c$ implies $\gamma x \leqslant \gamma c = c$ by (a).

Proposition 19.2.2. *Let X be a complete lattice. For any closure mapping on X, the set of all closed elements of X is closed under infimums and is a complete lattice.*

Proof. Let $\gamma : X \longrightarrow X$ be a closure mapping on X. The greatest element 1 of X is closed by (c). When every element of $T \subseteq X$ is closed, then $c = \bigwedge_{t \in T} t$ is closed: indeed $\gamma c \leqslant \gamma t = t$ for all $t \in T$; therefore $\gamma c \leqslant c$, and $\gamma c = c$ by (c). Then the result follows from Proposition 19.2.1. ☐

Many of the previous examples of complete lattices arise from closure mappings. For example, let G be a group. The subgroup $\langle X \rangle$ of G generated by a subset $X \subseteq G$ consists of all finite products of elements of X and inverses of elements of X. This defines a closure mapping $\langle - \rangle$ on the complete lattice 2^G of all subsets of G; the closed subsets are the subgroups of G.

19.2.d. Galois Connections. A **Galois connection** between two partially ordered sets X and Y is an ordered pair (α, β) of mappings $\alpha : X \longrightarrow Y$ and $\beta : Y \longrightarrow X$ such that

(i) α and β are order reversing ($x \leqslant y$ implies $\alpha x \geqslant \alpha y$, and $y \leqslant z$ implies $\beta y \geqslant \beta z$);
(ii) $\beta\alpha x \geqslant x$ and $\alpha\beta y \geqslant y$ for all $x \in X$ and $y \in Y$.

Properties (i) and (ii) are abstracted from properties of the Galois group and fixed field constructions in field theory: given a Galois extension E of K with Galois group $G = \mathrm{Gal}(E/K)$, let X be the set of all intermediate fields $K \subseteq F \subseteq E$ and Y be the set of all subgroups of G, partially ordered by inclusion; then $\alpha(F) = \mathrm{Gal}(E/F)$ and $\beta(H) = \mathrm{Fix}_E(H)$ are order-reversing mappings with property (ii). The exercises give other examples.

Proposition 19.2.3. *Let (α, β) be a Galois connection between two partially ordered sets X and Y. Then α and β induce mutually inverse bijections between*

Im α and Im β; $\alpha\beta$ and $\beta\alpha$ are closure mappings; if X and Y are complete lattices, then Im $\alpha = $ Im $\alpha\beta$ and Im $\beta = $ Im $\beta\alpha$ are complete lattices.

An **anti-isomorphism** of a partially ordered set A onto a partially ordered set B is an isomorphism of A onto B^{op} (equivalently, an isomorphism of A^{op} onto B). In Proposition 19.2.3, Im α and Im β are anti-isomorphic.

Proof. In any Galois connection, (ii) implies $\alpha\beta\alpha x \geqslant \alpha x$, whereas $\beta\alpha x \geqslant x$ implies $\alpha\beta\alpha x \leqslant \alpha x$ by (i). Hence $\alpha\beta\alpha x = \alpha x$ for all $x \in X$. Similarly $\beta\alpha\beta y = \beta y$ for all $y \in Y$. Therefore the restrictions of α and β to Im β and Im α are mutually inverse bijections. Hence Im α and Im β are anti-isomorphic.

Next $\beta\alpha$ is order preserving, is idempotent by the above, satisfies $\beta\alpha x \geqslant x$ by (ii), and is a closure mapping on X. We have Im $\beta\alpha = $ Im β, since $\beta\alpha\beta = \beta$. If X is a complete lattice, then by Proposition 19.1.5 so is Im β. Similarly $\alpha\beta$ is a closure mapping on Y; if Y is a complete lattice, then so is Im $\alpha = $ Im $\alpha\beta$. $\square$

In a Galois extension, Proposition 19.2.3 provides the usual bijections between intermediate fields and closed subgroups of the Galois group.

19.2.e. MacNeille's Theorem. As another application of Galois connections we prove:

Theorem 19.2.4 (MacNeille). *Every partially ordered set X can be embedded into a complete lattice so that all existing greatest lower bounds and least upper bounds are preserved.*

Proof. For each subset Y of X, let

$$L(Y) = \{ x \in X \,;\, x \leqslant y \text{ for all } y \in Y \},$$

$$U(Y) = \{ x \in X \,;\, x \geqslant y \text{ for all } y \in Y \},$$

be the sets of all lower bounds and all upper bounds of Y. We see that $Y \subseteq Z$ implies $L(Y) \supseteq L(Z)$ and $U(Y) \supseteq U(Z)$; also $Y \subseteq U(L(Y))$ and $Y \subseteq L(U(Y))$ for all $Y \subseteq X$. Hence (L,U) is a Galois connection between 2^X and 2^X. Call a subset Y of X *closed* in case $L(U(Y)) = Y$. By Proposition 19.2.3, the closed subsets of X constitute a complete lattice C. By Proposition 19.2.2, every intersection of closed subsets is closed; thus infimums in C are intersections.

For each $t \in X$ let

$$\lambda(t) = \{ x \in X \,;\, x \leqslant t \} \qquad \text{and} \qquad \upsilon(t) = \{ x \in X \,;\, x \geqslant t \}.$$

We see that $U(\lambda(t)) = \upsilon(t)$ and $L(\upsilon(t)) = \lambda(t)$. Therefore $\lambda(t)$ is closed. Thus $\lambda : X \longrightarrow C$ is injective; and $x \leqslant y$ if and only if $\lambda(x) \subseteq \lambda(y)$.

Let Y be a subset of X. Then $L(Y) = \bigcap_{y \in Y} \lambda(y)$. If Y has a greatest lower bound t in X, then $L(Y) = \lambda(t)$, $\lambda(t) = \bigcap_{y \in Y} \lambda(y)$, and $\lambda(t)$ is the greatest lower bound of $\lambda(Y)$ in C.

Assume that $Y \subseteq X$ has a least upper bound t in X. Then $U(Y) = \upsilon(t)$ and $L(U(Y)) = \lambda(t)$. In particular, $\lambda(t)$ is a closed subset which contains $\lambda(y)$ for every $y \in Y$. If, conversely, Z is a closed subset which contains $\lambda(y)$ for every $y \in Y$, then $Y \subseteq Z$ and $\lambda(t) = L(U(Y)) \subseteq L(U(Z)) = Z$. Thus $\lambda(t)$ is the least upper bound of $\lambda(Y)$ in C. $\qquad\square$

The complete lattice C in this proof is the **MacNeille completion** of X. Some examples are given in the exercises, including a purely algebraic construction of $\mathbb{R}$ (which actually goes back to Dedekind).

19.2.f. Homomorphisms. Complete lattices do not constitute a variety (see the exercises). Nevertheless, a number of basic definitions and properties extend to complete lattices.

A **complete homomorphism** of a complete lattice A into a complete lattice B is a mapping $\varphi : A \longrightarrow B$ which preserves *all* infimums and supremums: $\varphi(\bigwedge_{t \in T} t) = \bigwedge_{t \in T} \varphi(t)$ and $\varphi(\bigvee_{t \in T} t) = \bigvee_{t \in T} \varphi(t)$ for every $T \subseteq A$. In particular, $\varphi(0) = 0$ and $\varphi(1) = 1$.

A **complete sublattice** of a complete lattice L is a subset S of L (and the corresponding complete lattice) which is closed under *all* infimums and supremums ($\bigwedge_{t \in T} t \in S$ and $\bigvee_{t \in T} t \in S$ for every $T \subseteq S$); in particular, $0, 1 \in S$.

A **complete congruence** on a lattice L is an equivalence relation $\mathscr{E}$ on L (necessarily a congruence) such that $x_i \mathrel{\mathscr{E}} y_i$ for all $i \in I$ implies $\bigwedge_{i \in I} x_i \mathrel{\mathscr{E}} \bigwedge_{i \in I} y_i$ and $\bigvee_{i \in I} x_i \mathrel{\mathscr{E}} \bigvee_{i \in I} y_i$. Then the quotient lattice $L/\mathscr{E}$ is complete, and the projection $L \longrightarrow L/\mathscr{E}$ is a complete homomorphism (see the exercises).

There is a Homomorphism Theorem for complete lattices, whose statement and proof are left to the reader.

Exercises

1. Prove that the sublattices of any lattice (partially ordered by inclusion) constitute a complete lattice.

2. Prove that the congruences on any lattice (partially ordered by inclusion) constitute a complete lattice. Describe suprema in this lattice.

3. Let L be a complete lattice and $C \subseteq L$ be a subset of L which is closed under infimums and contains the greatest element of L. Show that there is a closure mapping on L of which C is the set of closed elements.

4. Let X be a set and γ be a closure mapping on 2^X. When is γ is the closure mapping of a topology on X?

5. Let R be a ring. The **annihilator** of a left ideal L of R is the right ideal $\mathrm{Ann}(L) = \{x \in R; Lx = 0\}$ of R; the annihilator of a right ideal T of R is the left ideal $\mathrm{Ann}(T) = \{x \in R; xT = 0\}$ of R. Show that this defines a Galois connection between left ideals and right ideals of R.

6. Let R be a ring. The **centralizer** of a subring S of R is the subring $C(S) = \{x \in R; xs = sx \text{ for all } s \in S\}$. Show that this defines a Galois connection between subrings of R and subrings of R.

7. Find the MacNeille completion of $\mathbb{N}$.

8. Show that the MacNeille completion of $\mathbb{Q}$ is $\mathbb{R} \cup \{\infty\} \cup \{-\infty\}$.

9. Produce a complete lattice L and a chain of complete sublattices of L whose union is not a complete sublattice of L.

10. Let L be a complete lattice and $\mathscr{E}$ be a complete congruence on L. Show that $\mathscr{E}$ is a congruence, that $L/\mathscr{E}$ is complete, and that the projection $L \longrightarrow L/\mathscr{E}$ is a complete homomorphism.

11. State and prove a Homomorphism Theorem for complete lattices.

19.3. MODULAR LATTICES

Modular lattices have interesting chain properties. In this section we show that the length properties of abelian groups and modules are in fact properties of modular lattices.

19.3.a. Definition. In every lattice, $x \leqslant z$ implies $x \vee (y \wedge z) \leqslant (x \vee y) \wedge z$, since x and $y \wedge z$ are both $\leqslant x \vee y$ and $\leqslant z$. A lattice is **modular** in case $x \leqslant z$ implies $x \vee (y \wedge z) = (x \vee y) \wedge z$; equivalently, in case $x \leqslant z$ implies $x \vee (y \wedge z) \geqslant (x \vee y) \wedge z$.

Modular lattices are characterized by the identity

$$x \vee (y \wedge (x \vee t)) = (x \vee y) \wedge (x \vee t)$$

and therefore constitute a variety. Hence a sublattice of a modular lattice is modular; a homomorphic image of a modular lattice is modular.

Modularity is self-dual: a lattice L is modular if and only if L^{op} is modular.

19.3.b. Examples. Every chain is a modular lattice.
The following lattice, which we call N_5, is not modular:

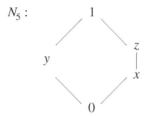

N_5 :

Indeed $x \leqslant z$, $x \vee (y \wedge z) = x \vee 0 = x$, but $(x \vee y) \wedge z = 1 \wedge z = z$.

Proposition 19.3.1. *The lattice of normal subgroups of a group is modular.*

Proof. In the lattice of normal subgroups of a group G, $A \wedge B = A \cap B$ and $A \vee B = AB$. If A, B, C are normal subgroups of G and $A \subseteq C$, then $x \in AB \cap C$ implies that $x \in C$, $x = ab$ for some $a \in A$ and $b \in B$, $b = a^{-1}x \in B \cap C$, and $x = ab \in A(B \cap C)$; thus $A(B \cap C) \supseteq AB \cap C$. Thus the lattice of normal subgroups of G is modular. $\qquad\square$

Modular lattices are named after the following class of examples:

Proposition 19.3.2. *The lattice of submodules of a module is modular.*

Proof. By Proposition 19.3.1, the lattice of subgroups of an abelian group is modular. A left (or right) R-module is in particular an abelian group; its lattice of submodules is a sublattice of its lattice of subgroups and is also modular. $\qquad\square$

19.3.c. Equivalent Definition. We give an equivalent definition of modular lattices which uses the lattice N_5 above.

Theorem 19.3.3. *A lattice is modular if and only if it contains no sublattice which is isomorphic to N_5.*

Proof. A sublattice of a modular lattice is modular and therefore cannot be isomorphic to N_5.

Conversely, assume that the lattice L is not modular. Then L contains elements a, b, c such that $a \leqslant c$ and $u = a \vee (b \wedge c) < (a \vee b) \wedge c = v$. We see that $a \leqslant u < v \leqslant c$. Hence $b \vee a \leqslant b \vee u \leqslant b \vee v \leqslant b \vee a$ and $b \vee u = b \vee v = b \vee a$. Similarly $b \wedge c \leqslant b \wedge u \leqslant b \wedge v \leqslant b \wedge c$ and $b \wedge u = b \wedge v = b \wedge c$.

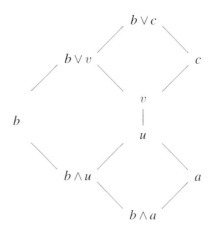

Thus b, u, v, $b \wedge u = b \wedge v$, and $b \vee u = b \vee v$ constitute a sublattice of L. We show that these five elements are distinct, so that our sublattice is isomorphic to N_5. We already have $u < v$. Also $b \nleqslant c$, otherwise $b = b \wedge c \leqslant u < v$ and $u = b \vee u = b \vee v = v$; and $a \nleqslant b$, otherwise $b = b \vee a \geqslant v > u$ and $u = b \wedge u = b \wedge v = v$. Hence

$v < b \vee v$; otherwise, $b \leqslant v \leqslant c$ and $u = a \vee b = v$;

$b \wedge u < u$; otherwise, $a \leqslant u \leqslant b$ and $u = b \wedge c = v$;

$b \wedge u < b$; otherwise, $b \wedge c = b \wedge u = b$ and $b \leqslant c$;

$b < b \vee v$; otherwise, $b \vee a = b \vee v = b$ and $a \leqslant b$. $\qquad\square$

19.3.d. Properties. In a partially ordered set X we say that an element b **covers** an element a in case $a < b$ and there is no $a < c < b$. This relationship is denoted by $b \succ a$ or by $a \prec b$.

Lemma 19.3.4. *In a modular lattice, $x \wedge y \prec x$ if and only if $y \prec x \vee y$.*

Proof. Assume that $x \wedge y \prec x$ and $y \not\prec x \vee y$. Then $x \wedge y < x$, $x \nleqslant y$, $y < x \vee y$, and $y < z < x \vee y$ for some z. Then $x \vee z = x \vee y$. Also $x \wedge y \leqslant x \wedge z \leqslant x$, and $x \wedge z = x \wedge y$, since $x \wedge z = x$ would imply $x \leqslant z$ and $x \vee z = z < x \vee y$. Hence $y \vee (x \wedge z) = y < z = (y \vee x) \wedge z$, contradicting modularity. Therefore $x \wedge y \prec x$ implies $y \prec x \vee y$; the converse implication is dual. $\qquad\square$

Theorem 19.3.5. *In a modular lattice, any two finite maximal chains have the same length.*

Proof. In a lattice L, a minimal element x is a least element: for any $y \in L$, $x \wedge y \leqslant x$ implies $x \wedge y = x$ and $x \leqslant y$. Dually a maximal element is a greatest element. If $x_0 < x_1 < \cdots < x_m$ is a maximal chain of L, then x_0 is a minimal element of L, $x_i \prec x_{i+1}$ for all $i < m$, and x_m is a maximal element of L; hence L has a least element 0 and a greatest element 1, and $x_0 = 0$, $x_m = 1$.

Now let L be modular with a finite maximal chain $0 = x_0 \prec \cdots \prec x_\ell = 1$ of length ℓ. We prove by induction on ℓ that in every such modular lattice every finite maximal chain has length ℓ. This is obvious if $\ell = 0$ or $\ell = 1$. Assume that $\ell > 1$, and let $0 = y_0 \prec \cdots \prec y_n = 1$ be a finite maximal chain.

If $x_{\ell-1} = y_{n-1}$, then $\ell = n$ by the induction hypothesis applied to the modular lattice $L(x_{\ell-1}) = \{x \in L ; x \leqslant x_{\ell-1} = y_{n-1}\}$.

Now assume that $x_{\ell-1} \neq y_{n-1}$. Then $x_{\ell-1} \nleqslant y_{n-1}$, $y_{n-1} < x_{\ell-1} \vee y_{n-1} \leqslant y_n$, and $x_{\ell-1} \vee y_{n-1} = y_n$. Let $z_{n-2} = x_{\ell-1} \wedge y_{n-1}$. By Lemma 19.3.4, $z_{n-2} \prec x_{\ell-1}$ and $z_{n-2} \prec y_{n-1}$. If $z_{n-2} = y_{n-2}$, we let $z_{n-3} = y_{n-3}$. If $z_{n-2} \neq y_{n-2}$, then as above $z_{n-2} \vee y_{n-2} = y_{n-1}$, and we continue with $z_{n-3} = z_{n-2} \wedge y_{n-2}$. By Lemma 19.3.4, $z_{n-3} \prec z_{n-2}$ and $z_{n-3} \prec y_{n-2}$. This constructs a maximal chain $0 = z_0 \prec \cdots \prec z_{n-2} \prec x_{\ell-1}$ of length $n - 1$ of $L(x_{\ell-1})$. By the induction hypothesis, this chain must have the same length as $0 = x_0 \prec \cdots \prec x_{\ell-1}$; hence again $\ell = n$. $\qquad\square$

Exercises

1. Show that a lattice is modular if and only if $x \leqslant t$ and $z \leqslant y$ implies $x \vee (y \wedge (z \vee t)) = ((x \vee y) \wedge z) \vee t$.

2. Show that a lattice is modular if and only if $a \wedge b = a \wedge c$, $a \vee b = a \vee c$, $b \leqslant c$ implies $b = c$.

3. Show that the lattice of all subgroups of a group need not be modular.

4. Show that there is a free modular lattice on three generators and that it has 28 elements.

In the following exercises, an **interval** of a lattice L is a sublattice

$$[a,b] = \{x \in L;\, a \leqslant x \leqslant b\},$$

where $a \leqslant b$ in L.

5. In a modular lattice L, $[a \wedge b, a] \cong [b, a \vee b]$ for all $a, b \in L$.

6. Let L be a lattice in which all maximal chains are finite and have the same length (the **length** of L). Further assume that $[a \wedge b, a]$ and $[b, a \vee b]$ have the same length for all $a, b \in L$. Prove that L is modular.

7. Let L be a lattice. Assume that $a \prec a \vee b$ and $b \prec a \vee b$ implies $a \wedge b \prec a$, $a \wedge b \prec b$. Prove that any two finite maximal chains of L have the same length.

19.4. DISTRIBUTIVE LATTICES

Distributive lattices are less general than modular lattices but still include some important examples. This section contains some simple structure results.

19.4.a. Definition

Proposition 19.4.1. *In any lattice L the following conditions are equivalent:*

(1) $x \wedge (y \vee z) = (x \wedge y) \vee (x \wedge z)$ *for all* $x, y, z \in L$.
(2) $x \vee (y \wedge z) = (x \vee y) \wedge (x \vee z)$ *for all* $x, y, z \in L$.

Proof. If (1) holds, then

$$(x \vee y) \wedge (x \vee z) = ((x \vee y) \wedge x) \vee ((x \vee y) \wedge z)$$

$$= x \vee ((x \wedge z) \vee (y \wedge z)) = x \vee (y \wedge z)$$

for all x, y, z and (2) holds. Dually (by Metatheorem 19.1.2) (2) implies (1). $\qquad\square$

A lattice is **distributive** when (1) and (2) hold. The exercises give equivalent definitions. Since (1) and (2) are identities, distributive lattices constitute a variety.

Distributive lattices are modular: if $x \leqslant z$, then (1) implies $x \wedge (y \vee z) = (x \wedge y) \vee (x \wedge z) = (x \wedge y) \vee z$. But the following lattice, which we call M_5,

$M_5:$

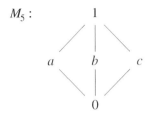

is modular (by Theorem 19.3.3) but not distributive, since $a \wedge (b \vee c) = a$, whereas $(a \wedge b) \vee (a \wedge c) = 0$.

The lattice 2^X of all subsets of a set X is distributive; so is every sublattice of 2^X. It can be shown that the lattice of congruences of any lattice is distributive. Other examples are given below and in the next section.

19.4.b. Equivalent Definition. The following result is similar to Theorem 19.3.3:

Theorem 19.4.2. *A lattice is distributive if and only if it contains no sublattice which is isomorphic to either M_5 or N_5.*

Proof. A sublattice of a distributive lattice is distributive, and cannot be isomorphic to either M_5 or N_5.

Conversely, assume that the lattice L is not distributive. If L is not modular, then L contains a sublattice which is isomorphic to N_5, and the theorem is proved. Hence we may assume that L is modular.

For any $a, b, c \in L$ we construct a sublattice S of L as follows: let

$$u = (a \wedge b) \vee (b \wedge c) \vee (c \wedge a),$$

$$v = (a \vee b) \wedge (b \vee c) \wedge (c \vee a).$$

We have $u \leqslant v$, since $a \wedge b$, $b \wedge c$, $c \wedge a$ are $\leqslant a \vee b$, $b \vee c$, $c \vee a$. Let

$$x = u \vee (a \wedge v) = (u \vee a) \wedge v,$$

$$y = u \vee (b \wedge v) = (u \vee b) \wedge v,$$

$$z = u \vee (c \wedge v) = (u \vee c) \wedge v.$$

Using modularity and $a \wedge v = a \wedge (b \vee c)$, $b \vee u = b \vee (c \wedge a)$, we obtain

$$x \wedge y = (u \vee (a \wedge v)) \wedge (u \vee (b \wedge v))$$

$$= u \vee ((a \wedge v) \wedge (u \vee (b \wedge v)))$$

$$= u \vee ((a \wedge v) \wedge ((u \vee b) \wedge v))$$

$$= u \vee ((a \wedge v) \wedge (u \vee b))$$

$$= u \vee ((a \wedge (b \vee c)) \wedge (b \vee (c \wedge a)))$$

$$= u \vee (((a \wedge (b \vee c)) \wedge b) \vee (c \wedge a)))$$

$$= u \vee (a \wedge b) \vee (c \wedge a) = u.$$

Permuting a, b, and c yields $y \wedge z = z \wedge x = u$. Dually $x \vee y = y \vee z = z \vee x = v$.
Thus $S = \{u, v, x, y, z\}$ is a sublattice of L.

Since L is not distributive, L contains elements a, b, c such that $a \wedge (b \vee c) \neq (a \wedge b) \vee (a \wedge c)$. Since $a \wedge b \leqslant a \wedge (b \vee c)$ and $a \wedge c \leqslant a \wedge (b \vee c)$, we have

$$p = (a \wedge b) \vee (a \wedge c) < a \wedge (b \vee c) = q.$$

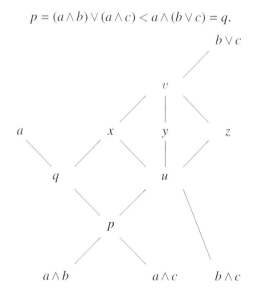

We show that $S \cong M_5$ when a, b, and c are chosen in this fashion. We have
$a \wedge v = a \wedge (b \vee c) = q$, and modularity yields

$$u \wedge a = (((a \wedge b) \vee (a \wedge c)) \vee (b \wedge c)) \wedge a$$

$$= ((a \wedge b) \vee (a \wedge c)) \vee ((b \wedge c) \wedge a) = (a \wedge b) \vee (a \wedge c) = p.$$

Since $p < q$, we have $u < v$. Hence x, y, and z are distinct (e.g., if $x = y$, then $u = x \wedge y = x \vee y = v$) and distinct from u and v (e.g., if $x = u$, then $y = x \vee y = v$ and $z = x \vee z = v = y$). Therefore $S \cong M_5$. □

19.4.c. Irreducible Elements. Our first structure result uses the following construction. Let X be any partially ordered set. An **order ideal** of X is a subset I of X such that $x \leqslant y \in I$ implies $x \in I$. Every union of order ideals of X is an order ideal of X, and every intersection of order ideals of X is an order ideal of X, so that the order ideals of X constitute a complete sublattice of 2^X. Therefore:

Proposition 19.4.3. *For any partially ordered set X, the order ideals of X, ordered by inclusion, constitute a distributive lattice* $\mathrm{Id}(X)$.

If, for instance, X is a **discrete** partially ordered set ($x \leqslant y$ if and only if $x = y$), then every subset of X is an order ideal and $\mathrm{Id}(X) = 2^X$. We now show that Proposition 19.4.3 yields all finite distributive lattices.

An element i of a lattice L is **irreducible** (short for sup irreducible) in case i is not the least element of L (if such exists) and $i = a \vee b$ implies $i = a$ or $i = b$. For example, the irreducible elements of 2^X are the one-element subsets of X.

Lemma 19.4.4. *In a finite lattice every element is the supremum of a set of irreducible elements.*

Proof. Let L be a finite lattice in which the statement is false (not every element is the supremum of a set of irreducible elements). Since L satisfies the descending chain condition, L contains an element m which is not the supremum of a set of irreducible elements and is minimal with this property. Then m is not the least element of L (which, if it exists, is the supremum of the empty set of irreducible elements), and m is not irreducible. Therefore $m = a \vee b$ for some $a, b \neq m$. Then $a, b < m$; by the choice of m, a and b are the supremums of sets of irreducible elements. But then $m = a \vee b$ is the supremum of a set of irreducible elements. This is the required contradiction. □

We denote by $\mathrm{Irr}(L)$ the set of all irreducible elements of L; $\mathrm{Irr}(L)$ is a partially ordered set, with $i \leqslant j$ in $\mathrm{Irr}(L)$ if and only if $i \leqslant j$ in L.

Theorem 19.4.5. *If L is a finite distributive lattice, then $L \cong \mathrm{Id}(\mathrm{Irr}(L))$. Moreover $\mathrm{Irr}(\mathrm{Id}(X)) \cong X$ for every finite partially ordered set X, so that $L \cong \mathrm{Id}(X)$ if and only if $X \cong \mathrm{Irr}(L)$.*

Proof. For each $x \in L$ let

$$\iota(x) = \{ i \in \mathrm{Irr}(L) ; i \leqslant x \}.$$

We see that $\iota(x)$ is an order ideal of $\operatorname{Irr}(L)$, that is, $\iota(x) \in \operatorname{Id}(\operatorname{Irr}(L))$. By Lemma 19.4.4, every $x \in L$ is the supremum of a set S of irreducible elements. Necessarily $S \subseteq \iota(x)$; hence $x = \bigvee_{i \in \iota(x)} i$. Hence ι is injective. Conversely, let S be an order ideal of $\operatorname{Irr}(L)$ and $x = \bigvee_{i \in S} i$. Then $S \subseteq \iota(x)$. If, conversely, $j \in \iota(x)$, then

$$j = j \wedge x = j \wedge (\bigvee_{i \in S} i) = \bigvee_{i \in S} (j \wedge i).$$

Since j is irreducible, we have $j = j \wedge i \leqslant i$ for some $i \in S$; hence $j \in S$, and $S = \iota(x)$. Thus ι is a bijection of L onto $\operatorname{Id}(\operatorname{Irr}(L))$. Clearly $x \leqslant y$ implies $\iota(x) \subseteq \iota(y)$; conversely, $\iota(x) \subseteq \iota(y)$ implies $x = \bigvee_{i \in \iota(x)} i \leqslant \bigvee_{i \in \iota(y)} i = y$. Thus ι and ι^{-1} are order preserving, and ι is a lattice isomorphism of L onto $\operatorname{Id}(\operatorname{Irr}(L))$. The isomorphism $\operatorname{Irr}(\operatorname{Id}(X)) \cong X$ for any partially ordered set X is an exercise. $\square$

19.4.d. Birkhoff's Theorem. Theorem 19.4.5 implies that a finite distributive lattice is isomorphic to a sublattice of 2^X for some finite (partially ordered) set X. We now extend this result to all distributive lattices.

Theorem 19.4.6 (Birkhoff). *A lattice is distributive if and only if it is isomorphic to a sublattice of the lattice of all subsets of a set.*

The proof of Theorem 19.4.6 uses the following definitions. An **ideal** of a lattice L is a subset I of L such that:

(1) $x \leqslant y \in I$ implies $x \in I$;
(2) $x, y \in I$ implies $x \vee y \in I$.

Thus a lattice ideal is an order ideal which is closed under finite supremums. For instance, the set

$$L(a) = \{x \in L; \ x \leqslant a\}$$

is a lattice ideal of L for every $a \in L$; ideals of this form are called **principal**. In general, L is an ideal of itself, and every intersection of ideals of L is an ideal of L. Hence the ideals of L constitute a complete lattice.

Lemma 19.4.7. *Let A and B be two ideals of a distributive lattice L. In the lattice of ideals of L,*

$$A \vee B = \{a \vee b; \ a \in A, \ b \in B\}.$$

Proof. Every ideal which contains both A and B must contain $C = \{a \vee b; a \in A, b \in B\}$. Hence it suffices to show that C is an ideal. Indeed (1) holds for C: if $a \in A$, $b \in B$, and $x \leqslant a \vee b$, then $x \wedge a \in A$ and $x \wedge b \in B$ by (1), and $x = x \wedge (a \vee b) = (x \wedge a) \vee (x \wedge b) \in C$. Also (2) holds for C: if $a, a' \in A$

and $b,b' \in B$, then $a \vee a' \in A$ and $b \vee b' \in B$ by (2), and $(a \vee b) \vee (a' \vee b') = (a \vee a') \vee (b \vee b') \in C$. □

An ideal P of L is **prime** in case $P \neq \varnothing, L$, and $x \wedge y \in P$ implies $x \in P$ or $y \in P$.

Lemma 19.4.8. *Let I be a nonempty ideal of a distributive lattice L and $c \in L$, $c \notin I$. There exists a prime ideal $P \supseteq I$ such that $c \notin P$.*

Proof. It is immediate that the union of a chain of ideals of L is an ideal of L. Hence the union of a chain of ideals which contain I but not c is an ideal which contains I but not c. By Zorn's Lemma, there exists an ideal P of L maximal such that $P \supseteq I$ and $c \notin P$.

We show that P is prime. Let $a,b \notin P$. Then $A = P \vee L(a)$ and $B = P \vee L(b)$ properly contain P. By the choice of P, $c \in A$ and $c \in B$. By Lemma 19.4.7, $c = p \vee x = q \vee y$ for some $p,q \in P$, $x \in L(a)$, and $y \in L(b)$, and

$$c = (p \vee x) \wedge (q \vee y) = (p \wedge q) \vee (p \wedge y) \vee (x \wedge q) \vee (x \wedge y).$$

Since P contains $p \wedge q$, $p \wedge y$, and $x \wedge q$, but not c, it follows that P does not contain $x \wedge y$. Therefore P does not contain $a \wedge b \geqslant x \wedge y$.

We can now prove Theorem 19.4.6. Let L be a distributive lattice and X be the set of all prime ideals of L. Define a mapping $\varphi : L \longrightarrow 2^X$ by

$$\varphi(a) = \{P \in X \, ; \, a \notin P\}.$$

We show that φ is a lattice homomorphism. Let $a,b \in L$. If P is a prime ideal, then $a \wedge b \in P$ if and only if $a \in P$ or $b \in P$; therefore $\varphi(a \wedge b) = \varphi(a) \cap \varphi(b)$. For any ideal P, $a \vee b \in P$ if and only if $a \in P$ and $b \in P$; therefore $\varphi(a \vee b) = \varphi(a) \cup \varphi(b)$.

We show that φ is injective. If $a \neq b$ in L, then $a \leqslant b$ and $b \leqslant a$ cannot both hold. If $a \nleqslant b$, then Lemma 19.4.8 provides a prime ideal P which contains $L(b)$ and b but not a. If $b \nleqslant a$, then Lemma 19.4.8 provides a prime ideal P which contains $L(a)$ and a but not b. In either case $\varphi(a) \neq \varphi(b)$. □

Exercises

1. Prove that a lattice is distributive if and only if $(x \wedge y) \vee (y \wedge z) \vee (z \wedge x) = (x \vee y) \wedge (y \vee z) \wedge (z \vee x)$ for all x,y,z.
2. Prove that a lattice is distributive if and only if $a \wedge b = a \wedge c$ and $a \vee b = a \vee c$ implies $b = c$.
3. Show that there is a free distributive lattice on three generators and that it has 18 elements.
4. Prove that $\mathrm{Irr}(\mathrm{Id}(X)) \cong X$ for every finite partially ordered set X.

5. Say that $x = i_1 \vee i_2 \vee \cdots \vee i_n$ is an irredundant supremum if $i_1 \vee \cdots \vee i_{j-1} \vee i_{j+1} \vee \cdots \vee i_n < x$ for all j. Show that every element of a finite distributive lattice can be written uniquely as an irredundant supremum of irreducible elements.

6. Prove that all maximal chains of a finite distributice lattice L have length $|\text{Irr}(L)|$.

19.5. BOOLEAN LATTICES

Boolean lattices are a generalization of the lattice of subsets of a set. They are used in mathematical logic, where the algebraic properties of conjunctions, disjunctions, and negations match those of infimums, supremums, and complements.

19.5.a. Definition. Let L be a lattice with a least element 0 and a greatest element 1. A **complement** of an element a of L is an element b of L such that $a \wedge b = 0$ and $a \vee b = 1$. For example, the usual complement $X \backslash Y$ of a subset Y of X is a complement in the lattice 2^X.

Proposition 19.5.1. *Let L be a distributive lattice with a least element 0 and a greatest element 1.*

(1) *An element of L has at most one complement.*

(2) *If a' is a complement of a and b' is a complement of b, then $a' \vee b'$ is a complement of $a \wedge b$ and $a' \wedge b'$ is a complement of $a \vee b$.*

Proof. (1) If b and c are complements of a, then

$$b = b \wedge (a \vee c) = (b \wedge a) \vee (b \wedge c) = b \wedge c \leqslant c;$$

exchanging b and c yields $c \leqslant b$.

(2) We have

$$(a \wedge b) \wedge (a' \vee b') = (a \wedge b \wedge a') \vee (a \wedge b \wedge b') = 0 \vee 0 = 0;$$

dually $(a' \vee b') \vee (a \wedge b) = 1 \vee 1 = 1$. Thus $a' \vee b'$ is a complement of $a \wedge b$. Dually $a' \wedge b'$ is a complement of $a \vee b$. $\qquad\square$

A **boolean lattice** is a distributive lattice with a least element and a greatest element, in which every element has a complement. For example, the lattice 2^X of subsets of any set X is boolean.

In a boolean lattice B, the complement of an element a is unique by Proposition 19.5.1, and is denoted by a'. Proposition 19.5.1 also shows that $a \longmapsto a'$ is an antiautomorphism of B.

Boolean lattices constitute a variety of universal algebras (with two binary operations $\wedge$ and $\vee$, two constant operations 0 and 1, and a unary operation $a \longmapsto a'$). Accordingly a **homomorphism** of boolean lattices is a lattice homomorphism φ which preserves 0 and 1 and preserves complements ($\varphi(0) = 0$, $\varphi(1) = 1$, and $\varphi(a') = \varphi(a)'$ for all a). A **boolean sublattice** of a boolean lattice B is a sublattice S which is contains 0 and 1 and is closed under complements ($x \in S$ implies $x' \in S$).

19.5.b. Boolean Rings. A **boolean ring** is a ring R [with an identity element] in which every element is idempotent ($x^2 = x$ for all $x \in R$).

Lemma 19.5.2. *A boolean ring is commutative and has characteristic* 2.

Proof. In a boolean ring R,

$$x + x = (x + x)(x + x) = x^2 + x^2 + x^2 + x^2 = x + x + x + x;$$

hence $x + x = 0$ for all $x \in R$. Then

$$x + y = (x + y)(x + y) = x^2 + xy + yx + y^2 = x + xy + yx + y$$

implies $xy + yx = 0$ and $yx = -xy = xy$ for all $x, y \in R$. □

For every boolean ring, a boolean lattice can be constructed as follows:

Proposition 19.5.3. *Let R be a boolean ring. Let $x \leqslant y$ in R if and only if $xy = x$. Then $L(R) = (R, \leqslant)$ is a boolean lattice, in which $x \wedge y = xy$, $x \vee y = x + y + xy$, and $x' = 1 - x$.*

Proof. By Proposition 19.1.4, $(R, \leqslant)$ is a lower semilattice, in which $x \wedge y = xy$. Let $x, y \in R$. Then $x(x + y + xy) = x + xy + xy = x$, so that $x \leqslant x + y + xy$. Similarly $y \leqslant x + y + xy$. Moreover $x, y \leqslant z$ implies $(x + y + xy)z = xz + yz + xyz = x + y + xy$ and $x + y + xy \leqslant z$. Thus $x + y + xy = x \vee y$ and $(R, \leqslant)$ is a lattice. $(R, \leqslant)$ is distributive, since

$$x \wedge (y \vee z) = x(y + z + yz) = xy + xz + xyxz = (x \wedge y) \vee (x \wedge z)$$

for all x, y, z. We see that $0 \leqslant x \leqslant 1$ for all x. Finally $x(1 - x) = x - x^2 = 0$ and $x + (1 - x) + x(1 - x) = 1$, so that $1 - x = x'$ is a complement of x for each $x \in R$. □

It is remarkable that Proposition 19.5.3 constructs all boolean lattices.

Theorem 19.5.4. *Let B be a boolean lattice. Define an addition and multiplication on B by*

$$x + y = (x' \wedge y) \vee (x \wedge y') \qquad and \qquad xy = x \wedge y.$$

Then $R(B) = (B, +, \cdot)$ *is a boolean ring and* $L(R(B)) = B$. *Moreover* $R(L(R)) = R$ *for every boolean ring* R.

In Theorem 19.5.4, L and R are isomorphisms of categories (see the exercises).

Proof. Readers who love computation will delight in showing that the addition $x + y = (x' \wedge y) \vee (x \wedge y')$ is associative. It is certainly commutative; furthermore $x + 0 = (x' \wedge 0) \vee (x \wedge 1) = x$ and $x + x = (x' \wedge x) \vee (x \wedge x') = 0$ for all x. Thus $(B, +)$ is an abelian group.

The multiplication $xy = x \wedge y$ is associative, commutative, and idempotent by Proposition 19.1.4. Also $1x = x$ and

$$xz + yz = ((x' \vee z') \wedge y \wedge z) \vee (x \wedge z \wedge (y' \vee z'))$$
$$= (x' \wedge y \wedge z) \vee (z' \wedge y \wedge z) \vee (x \wedge z \wedge y') \vee (x \wedge z \wedge z')$$
$$= (x' \wedge y \wedge z) \vee (x \wedge z \wedge y') = (x + y) \wedge z$$

for all x, y, z. Thus $R(B)$ is a boolean ring.

For all $x, y \in B$,

$$(x \vee y) + (x \wedge y) = (x' \wedge y' \wedge x \wedge y) \vee ((x \vee y) \wedge (x' \vee y'))$$
$$= (x \vee y) \wedge (x' \vee y')$$
$$= (x \wedge x') \vee (x \wedge y') \vee (y \wedge x') \vee (y \wedge y') = x + y;$$

therefore $x \vee y = x + y + (x \wedge y)$. Since $x \wedge y = xy$, it follows that $L(R(B)) = B$. If, conversely, R is a boolean ring, then, in $L(R)$, $x \wedge y = xy$ and

$$(x' \wedge y) \vee (x \wedge y') = (1 - x)y \vee x(1 - y)$$
$$= (1 - x)y + x(1 - y) + (1 - x)yx(1 - y)$$
$$= y - xy + x - xy = x + y;$$

thus $R(L(R)) = R$. □

19.5.c. Finite Boolean Lattices. An **atom** of a boolean lattice is a minimal nonzero element ($a > 0$, and there is no $a > x > 0$). In 2^X, the atoms are the subsets with one element.

Lemma 19.5.5. *In a finite boolean lattice, every element is the supremum of a set of atoms.*

Proof. Let B be a finite boolean lattice in which the statement is false. The descending chain condition holds in B. Since the statement is false, B contains an element m which is not the supremum of a set of atoms and is minimal with this property. Then $m \neq 0$ (since $0 = \bigvee \emptyset$) and m is not an atom. Hence there is an atom $a < m$. Then $m \wedge a' < m$, since $m \leqslant a'$ would imply $a = a \wedge m = 0$. By the choice of m, $m \wedge a'$ is the supremum of a set of atoms. But

$$m = m \wedge (a \vee a') = (m \wedge a) \vee (m \wedge a') = a \vee (m \wedge a').$$

Therefore m is the supremum of a set of atoms; this is the required contradiction.

$\square$

Lemma 19.5.5 also follows from Lemma 19.4.4, since the atoms of a finite boolean lattice are its irreducible elements (see the exercises). The following result can then be deduced from Theorem 19.4.5:

Theorem 19.5.6. *Every finite boolean lattice is isomorphic to the lattice of subsets of a finite set.*

Proof. Let B be a finite boolean lattice and A be the set of all atoms of B. For each $x \in B$ let

$$\alpha(x) = \{ a \in A; \, a \leqslant x \}.$$

By Lemma 19.5.5, every element $x \in B$ is the supremum of a set S of atoms. Necessarily $S \subseteq \alpha(x)$; therefore $x = \bigvee_{a \in \alpha(x)} a$. Hence α is injective. Conversely, let S be any subset of A and $x = \bigvee_{a \in S} a$. Then $S \subseteq \alpha(x)$. If, conversely, $b \in \alpha(x)$, then

$$b = b \wedge x = b \wedge (\bigvee_{a \in S} a) = \bigvee_{a \in S} (b \wedge a);$$

since b is an atom, we have $b = b \wedge a \leqslant a$ for some $a \in S$ and $b = a \in S$. Thus α is a bijection. Clearly $x \leqslant y$ implies $\alpha(x) \subseteq \alpha(y)$; conversely, $\alpha(x) \subseteq \alpha(y)$ implies $x = \bigvee_{a \in \alpha(x)} a \leqslant \bigvee_{a \in \alpha(y)} a = y$. Thus α and α^{-1} are order preserving, and α is a lattice isomorphism of B onto 2^A. $\square$

19.5.d. Birkhoff's Theorem. Finally, Birkhoff's Theorem (Theorem 19.4.6) can be specialized to boolean lattices.

Theorem 19.5.7. *Every boolean lattice is isomorphic to a boolean sublattice of the lattice of all subsets of a set.*

Proof. Let L be a boolean lattice and X be the set of all prime ideals of L. For each $a \in L$ let

$$\varphi(a) = \{ P \in X; \, a \notin P \}.$$

The proof of Theorem 19.4.6 shows that $\varphi : L \longrightarrow 2^X$ is a lattice homomorphism and is injective.

When P is a prime ideal of L, then $0 \in P$ (since $P \neq \emptyset$) but $1 \notin P$ (otherwise, $P = L$). Hence $\varphi(0) = \emptyset$ and $\varphi(1) = X$. Moreover $a \in P$ implies $a' \notin P$ (otherwise, $1 = a \vee a' \in P$) and $a \notin P$ implies $a' \in P$ (since $a \wedge a' = 0 \in P$). Thus $\varphi(a')$ is the complement of $\varphi(a)$, and φ is a homomorphism of boolean lattices. $\square$

Exercises

1. Recall that an interval of a lattice L is a subset

$$[a,b] = \{x \in L; \, a \leqslant x \leqslant b\}$$

where $a,b \in L$. Prove that every interval of a boolean lattice is a boolean lattice (though not a boolean sublattice).

2. Prove that the addition on $R(B)$ is associative.

3. In Theorem 19.5.4, show that L and R can be defined on homomorphisms, and yield isomorphisms of categories.

4. Show that the irreducible elements of a finite boolean lattice are its atoms.

5. A **generalized boolean lattice** is a lattice L with a least element such that the interval $[0,x]$ is a boolean lattice for every $x \in L$. Show that Theorem 19.5.4 extends to generalized boolean lattices, if rings are not required to contain an identity element.

6. Let B be a boolean lattice. Show that the $\mathscr{E}$-class of 0 is an ideal of B for every congruence $\mathscr{E}$ on B. Show that this defines a one-to-one correspondence between nonempty ideals of B and congruences on B.

7. Show that $(\bigvee_{i \in I} x_i) \wedge y = \bigvee_{i \in I} (x_i \wedge y)$ and $(\bigwedge_{i \in I} x_i) \vee y = \bigwedge_{i \in I} (x_i \vee y)$ hold for all $(x_i)_{i \in I}$ and y in a boolean lattice which is complete.

8. Show that $(\bigvee_{i \in I} x_i)' = \bigwedge_{i \in I} x_i'$ and $(\bigwedge_{i \in I} x_i)' = \bigvee_{i \in I} x_i'$ hold for all $(x_i)_{i \in I}$ in a boolean lattice which is complete.

9. Show that a complete boolean lattice is isomorphic to the lattice of all subsets of a set if and only if every element is the supremum of a set of atoms.

APPENDIX A

NUMBERS

This appendix contains standard constructions of all the familiar number systems (from natural numbers to complex numbers and quaternions). This is intended as review material for readers who have not seen these constructions before, and includes reviews of proof by induction (Section A.1) and basic properties of the integers (Section A.3).

A.1. NATURAL NUMBERS

Our construction of natural numbers is due to Peano. This construction provides a rigorous foundation for the natural numbers and their basic properties, and for subsequent number systems. For the reader this is also an opportunity to review or practice proofs by induction.

Recursion may be a bit much for the reader at this point, but this is how the operations on $\mathbb{N}$ are defined in Peano's construction. We have approached this difficulty frankly and proved (at the end of the section) that such recursive definitions are legitimate.

Another approach to natural numbers will be found in Section B.4.

A.1.a. Peano's Axioms. We want natural numbers so that we can count sequentially $(1, 2, 3, \ldots)$ and do proofs by induction. The definition of natural numbers must reflect these requirements: every natural number n must have a successor n^*; moreover a set S of natural numbers which contains 1 and is closed under successors (if $n \in S$, then $n^* \in S$) must contain all natural numbers.

Peano's Axioms define the set $\mathbb{N}$ of all natural numbers as a set with an element $1 \in \mathbb{N}$ and a mapping $n \longmapsto n^* = n + 1$ of $\mathbb{N}$ into $\mathbb{N}$, such that:

Axiom (1) (successor) $1 \in \mathbb{N}$ and $n \longmapsto n^*$ is a bijection of $\mathbb{N}$ onto $\mathbb{N} \setminus \{1\}$;

Axiom (2) (induction) If $S \subseteq \mathbb{N}$, $1 \in S$, and $n \in S$ implies $n^* \in S$, then $S = \mathbb{N}$.

The existence of a set $\mathbb{N}$ satisfying Axioms (1) and (2) follows from an axiom of set theory which implies the existence of an infinite set. Our first result is that Axioms (1) and (2) characterize $\mathbb{N}$ up to a $*$-preserving bijection; thereafter, $\mathbb{N}$ denotes any one of the suitable sets.

Proposition A.1.1. *If $\mathbb{N}$ and $\mathbb{N}'$ satisfy Axioms (1) and (2), then there is a unique bijection $\theta : \mathbb{N} \longrightarrow \mathbb{N}'$ such that $\theta(1) = 1'$ and $\theta(n^*) = \theta(n)^*$ for all $n \in \mathbb{N}$.*

This is proved at the end of the section (using recursion). The **natural numbers** are the elements of $\mathbb{N}$: thus $1, 2 = 1^*, 3 = 2^*$, and so on, are natural numbers.

Axiom (2) justifies proofs by induction in the usual format.

Proposition A.1.2 (Induction). *Let $P(n)$ be a property of the natural number n (a statement which is either true or false for each $n \in \mathbb{N}$). If $P(1)$ is true, and $P(n)$ implies $P(n + 1)$, then $P(n)$ is true for all $n \in \mathbb{N}$.*

(Proper definitions of "property" or "statement" are beyond the scope of this book.)

Proof. $S = \{ n \in \mathbb{N}; P(n) \text{ is true} \}$ has the properties in Axiom (2), hence is all of $\mathbb{N}$. $\qquad\square$

The **induction hypothesis** is the assumption that $P(n)$ is true in the proof that $P(n)$ implies $P(n + 1)$. This seems to assume what we are trying to prove. Actually we are only proving $P(n + 1)$, and we assume $P(n)$ only for the number n which precedes $n + 1$, not for all n.

A.1.b. Operations. The operations on $\mathbb{N}$ (addition and multiplication) are defined by induction. (Appendix B gives more elegant definitions.) The **sum** $a + b$ of two natural numbers a and b is defined by induction on b:

$$a + 1 = a^* \qquad \text{and} \qquad a + b^* = (a + b)^*$$

for all $a, b \in \mathbb{N}$. (This construction is discussed more fully at the end of the section.)

Proposition A.1.3. *Addition on $\mathbb{N}$ is associative ($a + (b + c) = (a + b) + c$), commutative ($b + a = a + b$), and cancellative ($a + b = a + c$ implies $b = c$).*

Proof. Mindful of the reader's need to practice proofs by induction, we only show associativity. The equality $a + (b + c) = (a + b) + c$ is proved by induction on c. First

$$a + (b + 1) = a + b^* = (a + b)^* = (a + b) + 1.$$

If $a + (b + c) = (a + b) + c$, then

$$a + (b + c^*) = a + (b + c)^* = (a + (b + c))^* = ((a + b) + c)^* = (a + b) + c^*.$$

$\square$

The **product** ab is defined by induction on b:

$$a1 = a \qquad \text{and} \qquad a(b^*) = ab + a$$

for all $a, b \in \mathbb{N}$.

Proposition A.1.4. *Multiplication on* $\mathbb{N}$ *is associative, commutative, cancellative* ($ab = ac$ *implies* $b = c$), *and distributive* ($a(b + c) = ab + bc$).

The proof is left to the reader. More elegant proofs of Propositions A.1.3 and A.1.4, using sets and counting rather than induction, will be found in the exercises.

A.1.c. Order

Proposition A.1.5. *A* total order relation *on* $\mathbb{N}$ *is defined by*

$$a \leqslant b \qquad \text{if and only if} \quad a = b \quad \text{or} \quad b = a + c \quad \text{for some} \quad c \in \mathbb{N}.$$

That $\leqslant$ is an order relation follows from properties of the addition (Proposition A.1.3). To show that $\leqslant$ is a total order relation, we prove a stronger property.

A **well-ordered** set is a partially ordered set in which every nonempty subset S has a least element (= an element ℓ of S such that $\ell \leqslant s$ for all $s \in S$). A well-ordered set is totally ordered, since every subset $\{x, y\}$ must have a least element.

Proposition A.1.6. $\mathbb{N}$ *is well-ordered.*

Proof. Let S be a subset of $\mathbb{N}$. We assume that S does not have a least element and show that $S = \emptyset$. Let

$$T = \{ t \in \mathbb{N}; \, t < s \text{ for all } s \in S \}.$$

We show that $1 \in T$. First $1 \leqslant n$ for all $n \in \mathbb{N}$, since by Axiom (1) $n \neq 1$ implies $n = p^* = p + 1 = 1 + p$ for some $p \in \mathbb{N}$. Therefore $1 \notin S$, otherwise S has a least element; hence $1 < s$ for all $s \in S$.

Let $t \in T$. For each $s \in S$, $t < s$ implies $s = t + c$ for some $c \in \mathbb{N}$. If $c = 1$, then $s = t^*$. If $c \neq 1$, then $c = d^*$ by Axiom (1), $s = t + d^* = t + d + 1 = t + 1 + d = t^* + d$ by Proposition A.1.3, and $s > t^*$. Thus $t^* \leqslant s$ for all $s \in S$. Again $t^* \notin S$, otherwise S has a least element; hence $t^* < s$ for all $s \in S$. Thus $t \in T$ implies $t^* \in T$.

By Axiom (2), $T = \mathbb{N}$. Hence $S = \emptyset$, since $s \in S$ implies $s \in T$ and $s < s$. □

Proposition A.1.6 allows a stronger form of proof by induction.

Corollary A.1.7 (Induction). *Let $P(n)$ denote a property of the natural number n (a statement which is true or false for each $n \in \mathbb{N}$). Assume that $P(n)$ is true whenever $P(k)$ is true for all $k < n$. Then $P(n)$ is true for all $n \in \mathbb{N}$.*

The hypothesis implies that $P(1)$ is true, since $P(k)$ is true, vacuously, for all $k < 1$.

Proof. Let $S = \{ s \in \mathbb{N};\ P(s)\text{ is false} \}$. Assume $S \neq \emptyset$. Then S has a least element ℓ ($\ell \in S$, and $\ell \leqslant s$ for all $s \in S$). We see that $P(\ell)$ is false (since $\ell \in S$), but $P(k)$ is true for all $k < \ell$ (since $k < \ell$ implies $k \notin S$). This contradicts the hypothesis on P. Therefore $S = \emptyset$ and $P(n)$ is true for all $n \in \mathbb{N}$. □

In a standard proof by induction, $P(n)$ is deduced from $P(n-1)$. Corollary A.1.7 allows us to prove $P(n)$ under the stronger **induction hypothesis** that $P(k)$ holds for all $k < n$.

A.1.d. Counting. Anticipating on Appendix B, we show that the natural numbers can be used to count the elements of finite sets.

A set S is **finite** in case there exists a bijection of S onto some natural number. A set which is not finite is **infinite**.

Lemma A.1.8. *When $m < n$ are natural numbers there exists no injection of $\{ 1,2,\ldots,n \}$ into $\{ 1,2,\ldots,m \}$.*

Proof. We prove by induction on n that no mapping $f : \{ 1,2,\ldots,n \} \longrightarrow \{ 1,2,\ldots,m \}$ is injective when $m < n$. This is vacuously true if $n = 1$. Assume that no mapping $f : \{ 1,2,\ldots,n \} \longrightarrow \{ 1,2,\ldots,m \}$ is injective when $m < n$, and let $m < n + 1$ and $f : \{ 1,2,\ldots,n + 1 \} \longrightarrow \{ 1,2,\ldots,m \}$ be a mapping.

First assume that $f(n+1) = m$. If $f(i) = m$ for some $i \leqslant n$, then f is not injective. If $f(i) < m$ for all $i \leqslant n$, then the restriction g of f to $\{ 1,2,\ldots,n \}$ is a mapping of $\{ 1,2,\ldots,n \}$ into $\{ 1,2,\ldots,m \}$; since $m < n$, g is not injective, and neither is f.

Now assume that $f(n+1) < m$. Then there is a permutation s of $\{ 1,2,\ldots,m \}$ such that $s(f(n+1)) = m$. By the above, the composition $s \circ f$ is not injective; hence neither is f. □

It follows from Lemma A.1.8 that there is for each finite set S only one natural number n with a bijection $S \longrightarrow \{ 1,2,\ldots,n \}$ (so that the number of elements of S does not depend on the order in which the elements are counted, despite some experimental evidence to the contrary). If indeed there are bijections $S \longrightarrow \{1,2,\ldots,n\}$ and $S \longrightarrow \{1,2,\ldots,m\}$, then there is a bijection $\{1,2,\ldots,n\} \longrightarrow \{1,2,\ldots,m\}$, Lemma A.1.8 rules out $n < m$ and $m < n$, and therefore $n = m$. Naturally n is the **number of elements** $n = |S|$ of S. (This definition applies only to finite sets. The number of elements of any set is defined in Section B.5.)

A.1.e. Recursion. A **finite** (nonempty) **sequence** of elements of a set S is a mapping of $\{1,2,\ldots,n\}$ into S (for some natural number n), usually written as a family $x = (x_i)_{i \leqslant n}$ or sequentially $x_1, x_2, \ldots, x_n$. An **infinite sequence** of elements of a set S is a mapping of $\mathbb{N}$ into S, usually written as a family $x = (x_n)_{n \in \mathbb{N}}$ or sequentially $x_1, x_2, \ldots, x_n, \ldots$.

Infinite sequences can be constructed by a process called **recursion**, which is similar to induction in that x_{n+1} is defined from x_n.

Proposition A.1.9 (Recursion). *Let S be a set, $a \in S$, and $f : S \longrightarrow S$ be a mapping. There exists a unique mapping $g : \mathbb{N} \longrightarrow S$ such that $g(1) = a$ and $g(n + 1) = f(g(n))$ for all $n \in \mathbb{N}$.*

Equivalently, there is a unique sequence $x_1, \ldots, x_n, \ldots$ of elements of S such that $x_1 = a$ and $x_{n+1} = f(x_n)$ for all n. It is natural to say that "x_n is defined by induction," but the fact is that *properties* of x_n can be proved by induction, whereas the existence or uniqueness of the infinite *sequence* $x_1, x_2, \ldots, x_n, \ldots$ is not the kind of property to which induction (Proposition A.1.2, or Corollary A.1.7) can be applied directly. On the other hand, one can prove, by induction on n, the existence and uniqueness of a *finite* sequence $x_1, x_2, \ldots, x_n$ such that $x_1 = a$ and $x_{i+1} = f(x_i)$ for all $i < n$; this provides the proof of Proposition A.1.9.

Proof. First we show, by induction on n, that there is at most one mapping $g_n : \{1,2,\ldots,n\} \longrightarrow S$ such that

$(C_n) \qquad g_n(1) = a \qquad$ and $\qquad g_n(i + 1) = f(g_n(i)) \quad$ for all $\quad i < n$.

This holds if $n = 1$, since (C_1) implies $g_1(1) = a$. If true for n and $g', g'' : \{1,2,\ldots,n, n + 1\} \longrightarrow S$ satisfy (C_{n+1}), then g' and g'' agree on $\{1,2,\ldots,n\}$ by the induction hypothesis, and agree on $n + 1$ by (C_{n+1}).

Next we show, by induction on n, that there exists a mapping $g_n : \{1,2, \ldots,n\} \longrightarrow S$ which satisfies (C_n). This holds if $n = 1$ (let $g_1(1) = a$). If $g_n : \{1,2,\ldots,n\} \longrightarrow S$ satisfies (C_n), we can define $g_{n+1} : \{1,2,\ldots,n + 1\} \longrightarrow S$ as follows: let $g_{n+1}(i) = g_n(i)$ for all $i \leqslant n$, and $g_{n+1}(n + 1) = f(g_{n+1}(n))$. Then g_{n+1} satisfies (C_{n+1}).

We now have, for each $n \in \mathbb{N}$, one mapping $g_n : \{1,2,\ldots,n\} \longrightarrow S$ which satisfies (C_n). By uniqueness, g_n and g_m agree on all $i \leqslant m, n$. Therefore there is a mapping $g : \mathbb{N} \longrightarrow S$ which extends every g_n; g may be defined by $g(i) = g_i(i)$ for all i, and then $g(i) = g_n(i)$ whenever $i \leqslant n$. Then $g(i + 1) = f(g(i))$ for all i; g is unique, since the restriction of g to $\{1,2,\ldots,n\}$ must satisfy (C_n). $\qquad \square$

The operations on $\mathbb{N}$ provide examples of recursive definitions. Proposition A.1.9, applied to $S = \mathbb{N}$, $a \in \mathbb{N}$, and $f(i) = i + 1 = i^*$, yields for each $a \in \mathbb{N}$ a unique mapping $n \longmapsto a + n$ of $\mathbb{N}$ into $\mathbb{N}$ such that $a + 1 = a^*$ and $a + (n + 1) = (a + n) + 1$ for all n. Similarly there is for each $a \in \mathbb{N}$ a unique mapping $n \longmapsto an$ of $\mathbb{N}$ into $\mathbb{N}$ such that $a1 = a$ and $a(n + 1) = an + a$ for all n.

As another application of recursion we prove Proposition A.1.1: if $\mathbb{N}$ and $\mathbb{N}'$ satisfy Axioms (1) and (2), then there is a unique bijection $\theta : \mathbb{N} \longrightarrow \mathbb{N}'$ such that $\theta(1) = 1'$ and $\theta(n^*) = \theta(n)^*$ for all $n \in \mathbb{N}$. By Proposition A.1.9 there is a unique mapping $\theta : \mathbb{N} \longrightarrow \mathbb{N}'$ such that $\theta(1) = 1'$ and $\theta(n^*) = \theta(n)^*$ for all $n \in \mathbb{N}$. We show that θ is bijective.

By the definition of θ, $1' \in \theta(\mathbb{N})$, and $n' \in \theta(\mathbb{N})$ implies $n'^* \in \theta(\mathbb{N})$. By Axiom (2) for $\mathbb{N}'$, $\theta(\mathbb{N}) = \mathbb{N}'$, and θ is surjective.

We prove by induction on $n \in \mathbb{N}$ that $\theta(m) \neq \theta(n)$ for all $m \neq n$. If $m \neq 1$, then $m = p^*$ for some p by Axiom (1), and $\theta(m) = \theta(p)^* \neq 1' = \theta(1)$. Now assume that $\theta(m) \neq \theta(n)$ for all $m \neq n$. Let $p \neq n^*$. If $p = 1$, then $\theta(p) \neq \theta(n^*)$ by the above, since $n^* \neq 1$. Otherwise, $p = m^*$ for some m, $m \neq n$ by Axiom (1) (since $m^* \neq n^*$), $\theta(m) \neq \theta(n)$ by the induction hypothesis, and $\theta(m^*) = \theta(m)^* \neq \theta(n)^* = \theta(n^*)$ by Axiom (1). Therefore $\theta(m) \neq \theta(n)$ for all $m \neq n$ holds for every $n \in \mathbb{N}$. $\qquad\square$

There are more complex forms of recursive definitions than are covered by Proposition A.1.9; for instance, the definition $x_{n+2} = x_{n+1} + x_n$, with $x_0 = x_1 = 1$, of the **Fibonacci sequence**. Products in Section 1.1 are another example. These examples are explored in the exercises.

Exercises

1. Prove that $2 + 2 = 4$.

2. Prove that $2 \times 3 = 6$.

3. Use induction to prove that the addition on $\mathbb{N}$ is commutative.

4. Prove that the addition on $\mathbb{N}$ is cancellative.

5. Use induction to prove that the multiplication on $\mathbb{N}$ is distributive.

6. Use induction to prove that the multiplication on $\mathbb{N}$ is associative.

7. Use induction to prove that the multiplication on $\mathbb{N}$ is commutative.

8. Prove that $ab = ac$ implies $b = c$ when $a,b,c \in \mathbb{N}$.

9. Prove the following: if X and Y are disjoint finite sets, then $|X \cup Y| = |X| + |Y|$.

10. Use the previous exercise to show that the addition on $\mathbb{N}$ is commutative and associative.

11. Prove the following: if X and Y are finite sets, then $|X \times Y| = |X||Y|$.

12. Use the previous exercise to show that the multiplication on $\mathbb{N}$ is associative, commutative, and distributive.

13. Let S be a set with a bijection θ of S onto a subset $T \neq S$ of S (the next two exercises show that this happens if and only if S is infinite). Let $1 \in S \backslash T$. Show that there is a smallest subset N of S such that $1 \in N$ and that $x \in N$ implies $\theta(x) \in N$. Show that N satisfies Peano's axioms (1) and (2) (with $n^* = \theta(n)$).

14. Let S be a set with a bijection θ of S onto a subset $T \neq S$ of S. Use the previous exercise to show that $S \neq \varnothing$ and that there is no natural number n for which there is a bijection of S onto $\{1,2,\ldots,n\}$.

15. Let S be a set. Assume that $S \neq \varnothing$ and that there is no natural number n for which there is a bijection of S onto $\{1,2,\ldots,n\}$. Show that there is an injection of $\mathbb{N}$ into S (this uses the Axiom of Choice in Appendix B). Then show that there is a bijection of S onto a subset $T \neq S$ of S.

16. (Fibonacci sequence) Prove that there is a unique sequence $x_1, x_2, \ldots, x_n, \ldots$ such that $x_1 = x_2 = 1$ and $x_{n+1} = x_n + x_{n-1}$ for all $n > 1$.

17. Let S be a set and f be a mapping which assigns an element of S to every finite subset of S. Show that there is a unique sequence $x_1, \ldots, x_n, \ldots$ of elements of S such that $x_{n+1} = f(x_1, \ldots, x_n)$ for all $n \in \mathbb{N}$.

18. (Products) Let X be a set with a binary operation (denoted multiplicatively). Prove that there is a unique mapping P which assigns to each finite nonempty sequence $x_1, \ldots, x_n$ of elements of X a subset $P(x_1, \ldots, x_n)$ of X, so that $P(x_1, \ldots, x_n) = \{x_1\}$ whenever $n = 1$, and

$$P(x_1, \ldots, x_n) = \bigcup_{1 < k < n} P(x_1, \ldots, x_k)\, P(x_{k+1}, \ldots, x_n)$$

for all $x_1, \ldots, x_n \in X$. (You may follow the proof of Proposition A.1.9; the set of all finite sequences of length at most n can serve as domain of g_n.)

A.2. INTEGERS

Natural numbers can be added and multiplied, but subtractions such as $1 - 1$ or $1 - 3$, and exact divisions such as $1 \div 2$ or $4 \div 3$, are not possible in $\mathbb{N}$. Additional numbers (integers and rational numbers) were invented to make these operations possible (to improve semigroups into groups). We begin by constructing the integers, with which subtraction is always possible.

A.2.a. Construction. We use an equivalence relation in the construction of integers. Intuitively an integer z is the difference $z = a - b$ between two natural numbers but can be written in this form in many ways. Accordingly we define an integer as an equivalence class of pairs of natural numbers, the equivalence of (a,b) and (c,d) being defined to eventually mean $a - b = c - d$. This construction depends on properties of the addition on $\mathbb{N}$ (Proposition A.1.3); a more general construction is indicated in the exercises.

Proposition A.2.1. *The binary relation $\equiv$ on $\mathbb{N} \times \mathbb{N}$ defined by*

$$(a,b) \equiv (c,d) \iff a + d = b + c$$

is an equivalence relation.

Proof. We have $(a,b) \equiv (a,b)$, since $a + b = b + a$. Also $(a,b) \equiv (c,d)$ implies $(c,d) \equiv (a,b)$, since $a + d = b + c$ implies $c + b = d + a$. Finally assume that $(a,b) \equiv (c,d) \equiv (e,f)$. Then $a + d = b + c$, $c + f = d + e$, $a + c + f = a + d + e = b + c + e$, $a + f = b + e$, and $(a,b) \equiv (e,f)$. $\qquad\square$

We denote by $\mathrm{cls}(a,b)$ the equivalence class of $(a,b) \in \mathbb{N} \times \mathbb{N}$. By definition, $\mathbb{Z} = (\mathbb{N} \times \mathbb{N})/\equiv$ is the set of all such equivalence classes; its elements are **integers**.

Lemma A.2.2. *An injective mapping* $\iota : \mathbb{N} \longrightarrow \mathbb{Z}$ *is defined by*

$$\iota(n) = \mathrm{cls}(a + n, a).$$

Proof. We have $(a + n, a) \equiv (b + n, b)$, so that all $(a + n, a)$ are contained in a single equivalence class. Hence there is a mapping $\iota : \mathbb{N} \longrightarrow \mathbb{Z}$ which assigns to each $n \in \mathbb{N}$ the equivalence class $\bar{n}$ that contains every $(a + n, a)$. Moreover $\mathrm{cls}(a + m, a) = \mathrm{cls}(b + n, b)$ implies $m = n$, since the addition on $\mathbb{N}$ is cancellative. $\qquad\square$

Lemma A.2.2 allows the common practice of regarding $n \in \mathbb{N}$ and $\iota(n) \in \mathbb{Z}$ as identical (so that natural numbers are integers).

A.2.b. Operations. The construction of the addition, multiplication, and order relation on $\mathbb{Z}$ is suggested by the following: $(a - b) + (c - d) = (a + c) - (b + d)$, $(a - b)(c - d) = (ac + bd) - (ad + bc)$, and $a - b \leqslant c - d \iff a + d \leqslant b + c$; these must eventually hold in $\mathbb{Z}$.

Proposition A.2.3. *An addition on* $\mathbb{Z}$ *is well-defined by*

$$\mathrm{cls}(a,b) + \mathrm{cls}(c,d) = \mathrm{cls}(a + c, b + d).$$

Proof. We follow the general procedure for inducing operations on quotient sets (Proposition 1.1.5). The statement defines at least one sum $\mathrm{cls}(a,b) + \mathrm{cls}(c,d)$. We must prove that there is only one sum $\mathrm{cls}(a,b) + \mathrm{cls}(c,d)$. We note that

$$(a,b) \equiv (a',b') \quad \text{and} \quad (c,d) \equiv (c',d') \quad \text{implies}$$
$$(a + c, b + d) \equiv (a' + c', b' + d'),$$

since $a + b' = b + a'$ and $c + d' = d + c'$ implies $a + c + b' + d' = b + d + a' + c'$. Hence the set

$$S = \{(a' + c', b' + d'); (a',b') \in \mathrm{cls}(a,b), (c',d') \in \mathrm{cls}(c,d)\}$$

is contained in a single equivalence class (i.e., $\mathrm{cls}(a + c, b + d)$). Therefore there is a mapping of $\mathbb{Z} \times \mathbb{Z}$ into $\mathbb{Z}$ which assigns to the pair $(\mathrm{cls}(a,b), \mathrm{cls}(c,d))$

the equivalence class that contains S; this equivalence class is $\operatorname{cls}(a+c,b+d)$.

□

Proposition A.2.4. *The addition on $\mathbb{Z}$ is associative, commutative, and cancellative $(x+y=x+z \Longrightarrow y=z)$. Furthermore:*

(1) $\iota(m)+\iota(n)=\iota(m+n)$ *for all* $m,n \in \mathbb{N}$;

(2) $x+0=x$ *for all* $x \in \mathbb{Z}$, *where* $0=\operatorname{cls}(1,1)=\operatorname{cls}(a,a)$;

(3) $x+(-x)=0$ *for all* $x \in \mathbb{Z}$, *where* $-x=\operatorname{cls}(b,a)$ *if* $x=\operatorname{cls}(a,b)$.

(Hence $\mathbb{Z}$ is an abelian group under addition.)

The straightforward proof is left to the reader. Proposition A.2.4 implies that a substraction is defined on $\mathbb{Z}$ by $x-y=x+(-y)$.

Proposition A.2.5. *A multiplication on $\mathbb{Z}$ is well-defined by*

$$\operatorname{cls}(a,b)\operatorname{cls}(c,d)=\operatorname{cls}(ac+bd, ad+bc).$$

Proof. We have

$$(a,b)\equiv(a',b') \quad \text{and} \quad (c,d)\equiv(c',d') \quad \text{implies}$$
$$(ac+bd, ad+bc)\equiv(a'c'+b'd', a'd'+b'c').$$

Indeed $a+b'=b+a'$ and $c+d'=d+c'$ implies $ac+bd+ad'+bc'=a(c+d')$ $+b(d+c')=a(d+c')+b(c+d')=ad+bc+ac'+bd'$ and similarly $ac'+bd'$ $+a'd'+b'c'=a'c'+b'd'+ad'+bc'$, so that $(ac+bd, ad+bc)\equiv(ac'+bd', ad'$ $+bc')\equiv(a'c'+b'd', a'd'+b'c')$. Hence the set

$$S=\{(a'c'+b'd', a'd'+b'c'); (a',b')\in\operatorname{cls}(a,b), (c',d')\in\operatorname{cls}(c,d)\}$$

is contained in a single equivalence class $\operatorname{cls}(ac+bd, ad+bc)$. Hence there is a mapping of $\mathbb{Z}\times\mathbb{Z}$ into $\mathbb{Z}$ which assigns to the pair $(\operatorname{cls}(a,b),\operatorname{cls}(c,d))$ the equivalence class that contains S, namely $\operatorname{cls}(ac+bd, ad+bc)$.   □

Proposition A.2.6. *The multiplication on $\mathbb{Z}$ is associative and commutative. Furthermore, for all $x,y,z \in \mathbb{Z}$:*

(1) $\iota(m)\iota(n)=\iota(mn)$ *for all* $m,n \in \mathbb{N}$;

(2) *if* $xz=yz$ *and* $z\neq 0$, *then* $x=y$;

(3) $1x=x$, *where* $1=\operatorname{cls}(2,1)=\operatorname{cls}(a+1,a)$;

(4) $x(y+z)=xy+xz$.

(Hence $\mathbb{Z}$ is a commutative ring with identity element.)

The straightforward proof is left to the reader.

A.2.c. Order. The next proofs are also left as exercises.

Proposition A.2.7. *A total order relation on $\mathbb{Z}$ is well-defined by $\mathrm{cls}(a,b) \leqslant \mathrm{cls}(c,d)$ if and only if $a + d \leqslant b + c$.*

Proposition A.2.8. *The order relation on $\mathbb{Z}$ has the following properties:*

(1) *$\iota(m) \leqslant \iota(n)$ if and only if $m \leqslant n$;*
(2) *$x \leqslant y$ implies $x + z \leqslant y + z$;*
(3) *$x \leqslant y$ implies $xz \leqslant yz$ if $z \geqslant 0$, $xz \geqslant yz$ if $z \leqslant 0$.*

We note that $\mathrm{cls}(a,b) > 0$ if and only if $\mathrm{cls}(a,b) = \iota(n)$ for some $n \in \mathbb{N}$. Hence natural numbers are identified with positive integers. (Equivalently, $\{z \in \mathbb{Z};\ z > 0\}$ satisfies Peano's axioms and may be used as $\mathbb{N}$.)

If $z = \mathrm{cls}(a,b) < 0$, then $-z = \mathrm{cls}(b,a) > 0$ and $z = -\iota(n)$ for some unique $n \in \mathbb{N}$. Hence an integer is either a natural number, or zero, or the opposite of a natural number.

Exercises

1. Prove Proposition A.2.4.
2. Prove Proposition A.2.6.
3. Prove Proposition A.2.7.
4. Prove Proposition A.2.8.
5. Show that $\mathrm{cls}(a,b) = \iota(a) - \iota(b)$ for all $a,b \in \mathbb{N}$.
6. Let S be a nonempty set with a binary operation which is associative, commutative, and cancellative. Show that the construction in this section embeds S into an abelian group.
7. Let $\iota : S \longrightarrow G$ be the embedding in the previous exercise. Let φ be a homomorphism of S into any abelian group A ($=$ a mapping φ such that $\varphi(ab) = \varphi(a)\,\varphi(b)$ for all $a,b \in S$). Show that there is a unique homomorphism $\psi : G \longrightarrow A$ (of groups) such that $\psi \circ \iota = \varphi$.

A.3. ARITHMETIC PROPERTIES OF THE INTEGERS

This section contains basic properties of integers: greatest common divisor, least common multiple, and prime factorization.

A.3.a. Integer Division

Proposition A.3.1. *Let $a,b \in \mathbb{Z}$ and $b > 0$. There exist $q,r \in \mathbb{Z}$ such that*

$$a = bq + r \qquad and \qquad 0 \leqslant r < b.$$

Furthermore q and r are unique.

Proof. (Existence) If $a \geqslant 0$, we proceed by induction on a. If $a < b$, then $q = 0$, $r = a$ serve. In general, assume that $a = bq + r$. Then $a + 1 = bq + (r + 1)$. If $r + 1 < b$, then we are done. If $r + 1 = b$, then $a + 1 = b(q + 1) + 0$.

If $a < 0$ then by the above $-a = bq + r$ with $0 \leqslant r < b$. If $r = 0$, then $a = b(-q) + 0$. If $r > 0$, then $a = b(-q - 1) + (b - r)$, with $0 < b - r < b$.

(Uniqueness) Assume that $a = bq + r = bq' + r'$ with $0 \leqslant r, r' < b$. Then $-b < r - r' < b$. Also $r - r' = b(q' - q)$ is a multiple of b. Now a nonzero multiple of b is either $\geqslant b$ or $\leqslant -b$. Therefore $r - r' = 0$. Then $bq = bq'$ yields $q = q'$. $\square$

The operation in Proposition A.3.1 is **integer division** of a by b; in the result, q is the **quotient** and r is the **remainder**.

A.3.b. Ideals. The remaining properties in this section follow from the properties of certain subsets of $\mathbb{Z}$. An **ideal** of $\mathbb{Z}$ is a subset I of $\mathbb{Z}$ such that:

(1) $0 \in I$;
(2) $x, y \in I$ implies $x - y \in I$.

Such a subset would normally be called a subgroup, but it fits the definition of an ideal due to the following properties.

Proposition A.3.2. *If I is an ideal of $\mathbb{Z}$ then:*

(3) $x, y \in I$ *implies* $x + y \in I$;
(4) $x \in I$ *implies* $xy \in I$ *for all* $y \in \mathbb{Z}$.

Proof. If $x, y \in I$, then $-y = 0 - y \in I$ and $x + y = x - (-y) \in I$, by (1) and (2). Now let $x \in I$. If $y > 0$, then $xy \in I$ by induction on y : $x1 = x \in I$, and $xy \in I$ implies $x(y + 1) = xy + x \in I$ by (3). Then $x0 = 0 \in I$ by (1), and $x(-y) = 0 - xy \in I$ by (1) and (2). $\square$

For each $n \in \mathbb{Z}$, the set $n\mathbb{Z} = \{nx; x \in \mathbb{Z}\}$ of all multiples of n is an ideal of $\mathbb{Z}$, the **principal ideal** of $\mathbb{Z}$ generated by n; n is a **generator** of $n\mathbb{Z}$. For example, $0\mathbb{Z} = \{0\}$ and $1\mathbb{Z} = \mathbb{Z}$. If $n > 0$, then n is the smallest positive element of $n\mathbb{Z}$.

Proposition A.3.3. *Every ideal of $\mathbb{Z}$ is principal and has a unique nonnegative generator.*

Proof. Let I be an ideal of $\mathbb{Z}$. If $I = \{0\}$, then $I = 0\mathbb{Z}$. Assume that $I \neq \{0\}$. Since $x \in I$ implies $-x \in I$, I contains a positive integer. Let n be the smallest positive element of I. Then $n\mathbb{Z} \subseteq I$ by (4). Conversely, let $x \in I$. By Proposition A.3.1, $x = nq + r$ with $0 \leqslant r < n$. Then $r = x - nq \in I$ by (2) and (4). Now $r > 0$ would contradict the choice of n. Hence $r = 0$ and $x = nq \in n\mathbb{Z}$. Thus $I = n\mathbb{Z}$.

Assume that $I = m\mathbb{Z} = n\mathbb{Z}$ with $m, n \geqslant 0$. Then $m = 0$ if and only if $n = 0$. If $m, n > 0$, then m is the smallest positive element of $m\mathbb{Z}$ and n is the smallest positive element of $n\mathbb{Z}$; hence $m = n$. This proves uniqueness. $\square$

A.3.c. Integers Modulo n. Let $n\mathbb{Z}$ be a nonzero ideal of $\mathbb{Z}$, with $n > 0$. **Congruence modulo n** is the relation

$$a \equiv b \ (\mathrm{mod}\ n) \iff a - b \in n\mathbb{Z} \iff a - b \quad \text{is a multiple of} \quad n.$$

The definition of an ideal shows that congruence modulo n is an equivalence relation. We denote by $\bar{a}$ the equivalence class of $a \in \mathbb{Z}$; thus $x \in \bar{a}$ if and only if $x \equiv a \ (\mathrm{mod}\ n)$. The quotient set is denoted by $\mathbb{Z}_n$; its elements are **integers modulo n**. By Proposition A.3.1, $a \in \bar{r}$ when r is the remainder in the integer division of a by n; hence $a \equiv b \ (\mathrm{mod}\ n)$ if and only if a and b have the same remainder when divided by n. Thus $\mathbb{Z}_n$ has n elements, namely $\bar{0}, \bar{1}, \ldots, \overline{n-1}$.

Lemma A.3.4. *If $a \equiv b \ (\mathrm{mod}\ n)$ and $c \equiv d \ (\mathrm{mod}\ n)$, then $a + c \equiv b + d \ (\mathrm{mod}\ n)$ and $ac \equiv bd \ (\mathrm{mod}\ n)$.*

The proof is an exercise. Lemma A.3.4 states that congruence modulo n is a congruence (for either operation) in the more general sense of Proposition 1.1.5. Hence the operations on $\mathbb{Z}$ induce operations on the quotient set $\mathbb{Z}_n$:

Proposition A.3.5. *An addition and a multiplication on $\mathbb{Z}_n$ are well-defined by*

$$\bar{a} + \bar{b} = \overline{a + b}, \quad \bar{a}\bar{b} = \overline{ab}.$$

Proof. By Lemma A.3.4, the set $\{x + y \in \mathbb{Z}; x \in \bar{a}, y \in \bar{b}\}$ is contained in a single equivalence class, namely $\overline{a + b}$. Therefore there exists a mapping $\mathbb{Z}_n \times \mathbb{Z}_n \longrightarrow \mathbb{Z}_n$ (the addition on $\mathbb{Z}_n$) which assigns to $(\bar{a}, \bar{b})$ the equivalence class $\bar{a} + \bar{b}$ that contains the set $\{x + y \in \mathbb{Z}; x \in \bar{a}, y \in \bar{b}\}$; and $\bar{a} + \bar{b} = \overline{a + b}$. The multiplication is defined similarly. $\square$

Proposition A.3.6. *The operations on $\mathbb{Z}_n$ have the following properties:*

(1) *The addition is commutative and associative, $\bar{0} + \bar{a} = \bar{a}$, and $\bar{a} + \overline{-a} = \bar{0}$ for all $\bar{a}$.*

(2) *The multiplication is commutative and associative, $\bar{1}\bar{a} = \bar{a}$, and $\bar{a}(\bar{b} + \bar{c}) = \bar{a}\bar{b} + \bar{a}\bar{c}$ for all $\bar{a}, \bar{b}, \bar{c}$.*

(Hence $\mathbb{Z}_n$ is a commutative ring with identity element.)

We will see (Proposition A.3.11) that $\mathbb{Z}_p$ is a field when p is prime.

A.3.d. L.c.m. and G.c.d. Recall that $a \in \mathbb{Z}$ **divides** $b \in \mathbb{Z}$ in case $b = aq$ for some $q \in \mathbb{Z}$ (in case b is a multiple of a). This relationship is denoted by $a \mid b$. It is equivalent to $b \in a\mathbb{Z}$ and to $b\mathbb{Z} \subseteq a\mathbb{Z}$. For example, 1 divides every integer, but only 1 and -1 divide 1.

Proposition A.3.7. *Let a and b be nonzero integers. There exists a positive integer m such that:*

(i) *$a \mid m$ and $b \mid m$;*

(ii) *if $a \mid c$ and $b \mid c$, then $m \mid c$.*

Furthermore m is unique.

Proof. (i) and (ii) state that $m\mathbb{Z} \subseteq a\mathbb{Z} \cap b\mathbb{Z}$, and that $c\mathbb{Z} \subseteq a\mathbb{Z} \cap b\mathbb{Z}$ implies $c\mathbb{Z} \subseteq m\mathbb{Z}$. Now $a\mathbb{Z} \cap b\mathbb{Z}$ has properties (1) and (2) and is an ideal of $\mathbb{Z}$. Also $a\mathbb{Z} \cap b\mathbb{Z} \neq \{0\}$, since $ab \in a\mathbb{Z} \cap b\mathbb{Z}$. Therefore $a\mathbb{Z} \cap b\mathbb{Z} = n\mathbb{Z}$ for some $n > 0$. Then (i) holds if and only if $m\mathbb{Z} \subseteq a\mathbb{Z} \cap b\mathbb{Z}$, and (ii) holds if and only if $c\mathbb{Z} \subseteq a\mathbb{Z} \cap b\mathbb{Z}$ implies $c\mathbb{Z} \subseteq m\mathbb{Z}$. Therefore (i) and (ii) hold if and only if $m\mathbb{Z} = a\mathbb{Z} \cap b\mathbb{Z} = n\mathbb{Z}$, which with $m, n > 0$ implies $m = n$. □

The positive integer m in Proposition A.3.7 is the **least common multiple** or **l.c.m.** of a and b, and is usually denoted by $m = [a, b]$; we prefer $m = \operatorname{lcm}(a, b)$.

A similar argument shows that every finite sequence $a_1, \ldots, a_n$ of nonzero integers has a l.c.m. $m = \operatorname{lcm}(a_1, \ldots, a_n)$.

Proposition A.3.8. *Let a and b be nonzero integers. There exists a positive integer d such that:*

(i) *$d \mid a$ and $d \mid b$;*

(ii) *if $c \mid a$ and $c \mid b$, then $c \mid d$.*

Furthermore d is unique, and there exists integers u, v such that $d = ua + vb$.

Proof. Parts (i) and (ii) state that $a\mathbb{Z} \subseteq d\mathbb{Z}$, $b\mathbb{Z} \subseteq d\mathbb{Z}$, and that $a\mathbb{Z} \subseteq c\mathbb{Z}$, $b\mathbb{Z} \subseteq c\mathbb{Z}$ implies $d\mathbb{Z} \subseteq c\mathbb{Z}$. Now

$$I = \{au + bv;\ u, v \in \mathbb{Z}\}$$

is an ideal of $\mathbb{Z}$. Also, $I \neq \{0\}$ since $a\mathbb{Z}, b\mathbb{Z} \subseteq I$. Hence $I = n\mathbb{Z}$ for some $n > 0$.

Since $d\mathbb{Z}$ is an ideal of $\mathbb{Z}$, (i) holds if and only if $n\mathbb{Z} = I \subseteq d\mathbb{Z}$. Similarly (ii) holds if and only if $n\mathbb{Z} = I \subseteq c\mathbb{Z}$ implies $d\mathbb{Z} \subseteq c\mathbb{Z}$. Hence (i) and (ii) are equivalent to $d\mathbb{Z} = n\mathbb{Z}$ and (since $n, d > 0$) to $d = n$. □

The positive integer d in Proposition A.3.8 is the **greatest common divisor** or **g.c.d.** of a and b, and is usually denoted by $d = (a, b)$; we prefer $d = \gcd(a, b)$.

A similar argument shows that every nonempty sequence $a_1, \ldots, a_n$ of nonzero integers has a g.c.d. $d = \gcd(a_1, \ldots, a_n)$, which can be written in the form $d = u_1 a_1 + u_2 a_2 + \cdots + u_n a_n$ for some $u_1, \ldots, u_n \in \mathbb{Z}$.

Two nonzero integers a and b are **relatively prime** in case $\gcd(a, b) = 1$.

Lemma A.3.9. *If $\gcd(a, b) = 1$ and $\gcd(a, c) = 1$, then $\gcd(a, bc) = 1$.*

Proof. If $sa + tb = ua + vc = 1$ for some $s, t, u, v \in \mathbb{Z}$, then

$$1 = (sa + tb)(ua + vc) = sua^2 + svac + tuab + tvbc = xa + ybc$$

for some $x, y \in \mathbb{Z}$, and $d = \gcd(a, bc) = 1$, since d must divide xa, ybc, and $xa + ybc = 1$. $\qquad\square$

A.3.e. Prime Factorization. A **prime** (short for **prime number**) is an integer $p > 1$ which has no divisor $1 < d < p$. For example, 2, 3, 5, 7 are primes, but 1, 4, 6, 8, 9 are not primes.

Lemma A.3.10. *Let p be prime. If $p \mid ab$, then $p \mid a$ or $p \mid b$. If $p \mid a_1 a_2 \ldots a_n$, then $p \mid a_i$ for some i.*

Proof. Assume that p divides neither a nor b. Then $\gcd(p, a) < p$; since p is prime, $\gcd(p, a) = 1$. Similarly $\gcd(p, b) = 1$. By Lemma A.3.9, $\gcd(p, ab) = 1$; hence p does not divide ab. This proves the first part of the statement. The second part follows by induction on n. $\qquad\square$

Proposition A.3.11. *When p is prime, there is for every $\overline{a} \in \mathbb{Z}_p$, $\overline{a} \neq \overline{0}$ some $\overline{b} \in \mathbb{Z}_p$ such that $\overline{a}\overline{b} = \overline{1}$. (Hence $\mathbb{Z}_p$ is a field.)*

Proof. If $\overline{a} \neq \overline{0}$, then p does not divide a and $\gcd(a, p) = 1$; hence $ab + pc = 1$ for some $b, c \in \mathbb{Z}$, and then $\overline{a}\overline{b} = \overline{1}$. $\qquad\square$

The main result in this section is:

Theorem A.3.12. *Every positive integer can be written uniquely (up to the order of the terms) as a product of positive powers of distinct primes.*

As in Chapter 1, we allow a product to have only one term (then it is equal to this term) or to be empty (then the product is 1; e.g., $0! = 1$).

Proof. First we prove by induction that every positive integer n is a product of primes. Assume that every integer $1 \leqslant k < n$ is a product of primes. If $n = 1$, or if n is prime, then n is an empty or one-term product product of primes. If $n > 1$ is not prime, then n has a divisor $1 < d < n$, and $n = cd$ with $c < n$. By the induction hypothesis, c and d are products of primes; hence so is $n = cd$.

In a product of primes, the order of the terms can be rearranged so that equal terms are next to each other. Then products of equal terms can be rewritten as positive powers. Hence every positive integer is a product of positive powers of distinct primes.

Uniqueness is proved by induction. A nonempty product of positive powers of primes must be greater than 1; hence the number 1 can be written only as an empty product. Now let

$$n = p_1^{k_1} p_2^{k_2} \dots p_r^{k_r} = q_1^{\ell_1} q_2^{\ell_2} \dots q_s^{\ell_s} > 1,$$

where $p_1, \dots, p_r$ are distinct primes, $q_1, \dots, q_s$ are distinct primes, and $k_1, \dots, k_r$, $\ell_1, \dots, \ell_s > 0$. Uniqueness for n means that $r = s$, $q_1, \dots, q_r = q_s$ can be renumbered so that $p_i = q_i$ for all i, and then $k_i = \ell_i$ for all i. Assume uniqueness for all $1 \leqslant m < n$.

Since $n > 1$, neither product is empty, and $r, s \geqslant 1$. By Lemma A.3.10, $p_1 \mid q_j$ for some j; since p_1 and q_j are prime, this implies $p_1 = q_j$. By changing the order of the terms in the second product, we can arrange that $p_1 = q_1$. Then

$$m = p_1^{k_1 - 1} p_2^{k_2} \dots p_r^{k_r} = q_1^{\ell_1 - 1} q_2^{\ell_2} \dots q_s^{\ell_s} < n,$$

with $p_1^{k_1 - 1}$ omitted if $k_1 = 1$, and similarly for $q_1^{\ell_1 - 1}$. By the induction hypothesis, applied to m, these two products contain the same primes with the same exponents. Thus either $p_1 = q_1$ appears in both products, and $k_1 - 1 = \ell_1 - 1$, or p_1 appears in neither, and $k_1 - 1 = \ell_1 - 1 = 0$; hence $k_1 = \ell_1$. Then

$$p_2^{k_2} \dots p_r^{k_r} = q_2^{\ell_2} \dots q_s^{\ell_s} < n;$$

by the induction hypothesis, $r = s$, we can permute $q_2, \dots, q_r$ so that $q_i = p_i$ for all $i \leqslant r = s$, and then $k_i = \ell_i$ for all $2 \leqslant i \leqslant r$.   □

We note some consequences of Theorem A.3.12. When $a = p_1^{a_1} p_2^{a_2} \dots p_r^{a_r}$ and $b = q_1^{b_1} q_2^{b_2} \dots q_s^{b_s}$ are products of positive powers of distinct primes, the sequences $p_1, \dots, p_r$ and $q_1, \dots, q_s$ can be merged into one sequence so that a and b are products of nonnegative powers of distinct primes $a = p_1^{a_1} p_2^{a_2} \dots p_t^{a_t}$ and $b = p_1^{b_1} p_2^{b_2} \dots p_t^{b_t}$. The following result is readily proved:

Proposition A.3.13. *Let $a = p_1^{a_1} p_2^{a_2} \dots p_r^{a_r}$ and $b = p_1^{b_1} p_2^{b_2} \dots p_r^{b_r}$ be products of nonnegative powers of distinct primes. Then:*

  (i) $\operatorname{lcm}(a, b) = p_1^{c_1} p_2^{c_2} \dots p_r^{c_r}$, *where $c_i = \max(a_i, b_i)$;*
  (ii) $\gcd(a, b) = p_1^{d_1} p_2^{d_2} \dots p_r^{d_r}$, *where $d_i = \min(a_i, b_i)$;*
  (iii) $\operatorname{lcm}(a, b) \gcd(a, b) = ab$.

Corollary A.3.14. *If $a \mid c$, $b \mid c$, and $\gcd(a, b) = 1$, then $ab \mid c$.*

Theorem A.3.15 (Euclid's Theorem). *There are infinitely many primes.*

Proof. The proof (also due to Euclid) shows that no finite sequence $p_1, p_2,$ $\dots, p_k$ can contain all primes. To see this, let $n = 1 + p_1 p_2 \dots p_k$. If $k = 0$, then the sequence $p_1,\ldots,p_k$, which is empty, does not contain the prime number 2. If $k > 0$, then $n > 1$; by Theorem A.3.12, some prime p divides n. Then $p \neq p_1, p_2, \ldots, p_k$; otherwise, p divides $p_1 p_2 \ldots p_k$ and divides $n - p_1 p_2 \ldots p_k = 1$, contradicting $p > 1$. □

Exercises

1. Prove Lemma A.3.4.

2. Prove that every finite sequence $a_1, \ldots, a_n$ of nonzero integers has a l.c.m.

3. Show that $\mathrm{lcm}(a, \mathrm{lcm}(b,c)) = \mathrm{lcm}(\mathrm{lcm}(a,b), c) = \mathrm{lcm}(a,b,c)$.

4. Prove that every nonempty sequence $a_1, \ldots, a_n$ of nonzero integers has a g.c.d. d, which can be written in the form $d = u_1 a_1 + u_2 a_2 + \cdots + u_n a_n$ for some $u_1, \ldots, u_n \in \mathbb{Z}$.

5. Show that $\gcd(a, \gcd(b,c)) = \gcd(\gcd(a,b), c) = \gcd(a,b,c)$.

6. Prove that $a \mid bc$ and $\gcd(a,b) = 1$ implies $a \mid c$.

7. Write 1650 as a product of powers of distinct primes.

8. Write 2310 as a product of powers of distinct primes.

9. Find $\mathrm{lcm}(1650, 2310)$ and $\gcd(1650, 2310)$.

10. Prove Proposition A.3.13.

A.4. RATIONAL NUMBERS

Exact division (with zero remainder) is generally not possible in $\mathbb{Z}$. Rational numbers were invented to make exact division possible without losing the properties of integers in Section A.2.

A.4.a. Construction. We use an equivalence relation in the construction of rational numbers. Intuitively, a rational number q is the quotient $q = a/b$ of one integer by another, but can be written in this form in many ways. Accordingly we define a rational number as an equivalence class of pairs of integers, the equivalence of (a,b) and (c,d) being defined to eventually mean $a/b = c/d$. It is convenient (but not necessary) to assume that the denominators b and d are positive. This construction depends on properties of the multiplication on $\mathbb{Z}$ (Proposition A.2.6). More general constructions will be found in Chapters 4 and 11.

Proposition A.4.1. *The binary relation $\equiv$ on $\mathbb{Z} \times \mathbb{N}$ defined by*

$$(a,b) \equiv (c,d) \iff ad = bc$$

is an equivalence relation.

Proof. We have $(a,b) \equiv (a,b)$, since $ab = ba$. Also $(a,b) \equiv (c,d)$ implies $(c,d) \equiv (a,b)$, since $ad = bc$ implies $cb = da$. Finally assume that $(a,b) \equiv (c,d) \equiv (e,f)$. Then $ad = bc$, $cf = de$, $acf = ade = bce$. If $c \neq 0$, then $af = be$. If $c = 0$, then $a = e = 0$ (since $ad = de = 0$ and $d \neq 0$), and again $af = be$. In either case $(a,b) \equiv (e,f)$. □

The equivalence class of $(a,b) \in \mathbb{Z} \times \mathbb{N}$ is a **fraction** and is denoted by a/b (or by a/b). By definition, $\mathbb{Q} = (\mathbb{Z} \times \mathbb{N})/\equiv$ is the set of all such equivalence classes. The elements of $\mathbb{Q}$ are **rational numbers**.

Lemma A.4.2. *An injective mapping $\iota : \mathbb{Z} \longrightarrow \mathbb{Q}$ is defined by*

$$\iota(x) = x/1 = ax/a .$$

Lemma A.4.2 allows the common practice of regarding $x \in \mathbb{Z}$ and $\iota(x) \in \mathbb{Q}$ as identical, so that integers are rational numbers.

A.4.b. Operations. The construction of the addition, multiplication, and order relation on $\mathbb{Q}$ is suggested by the following rules: $(a/b) + (c/d) = (ad + bc)/bd$, $(a/b)(c/d) = ac/bd$, and $a/b \leqslant c/d \iff ad \leqslant bc$ (since $b,d > 0$), which must eventually hold in $\mathbb{Q}$.

Proposition A.4.3. *An addition on $\mathbb{Q}$ is well-defined by*

$$a/b + c/d = (ad + bc)/bd .$$

Proof. We have

$$(a,b) \equiv (a',b') \qquad \text{and} \qquad (c,d) \equiv (c',d') \quad \text{implies}$$
$$(ad + bc, bd) \equiv (a'd' + b'c', b'd'),$$

since $ab' = ba'$ and $cd' = dc'$ implies $adb'd' + bcb'd' = a'd'bd + b'c'bd$. Hence

$$S = \{ (a'd' + b'c', b'd'); \ (a',b') \in a/b, \ (c',d') \in c/d \}$$

is contained in a single equivalence class, that of $(ad + bc, bd)$. Therefore there is a mapping of $\mathbb{Q} \times \mathbb{Q}$ into $\mathbb{Q}$ which assigns to the pair $(a/b, c/d)$ the equivalence class that contains the set S, namely $(ad + bc)/bd$. □

Proposition A.4.4. *The addition on $\mathbb{Q}$ is associative, commutative, and cancellative. Furthermore:*

(1) $\iota(x) + \iota(y) = \iota(x + y)$ *for all $x, y \in \mathbb{Z}$;*

(2) $x + 0 = x$ *for all $x \in \mathbb{Q}$, where $0 = 0/1 = 0/n$;*

(3) $x + (-x) = 0$ for all $x \in \mathbb{Q}$, where $-x = (-a)/b$ if $x = a/b$.

(Hence $\mathbb{Q}$ is an abelian group under addition.)

The straightforward proof is left as an exercise.

Proposition A.4.5. *A multiplication on $\mathbb{Q}$ is well-defined by*

$$(a/b)(c/d) = ac/bd.$$

Proof. We have

$$(a,b) \equiv (a',b') \qquad \text{and} \qquad (c,d) \equiv (c',d') \quad \text{implies} \quad (ac,bd) \equiv (a'c',b'd'),$$

since $ab' = ba'$ and $cd' = dc'$ implies $acb'd' = bda'c'$. Hence the set

$$S = \{ (a'c', b'd') \, ; \, (a',b') \in a/b, \, (c',d') \in c/d \}$$

is contained in a single equivalence class, namely, ac/bd. Hence there is a mapping of $\mathbb{Q} \times \mathbb{Q}$ into $\mathbb{Q}$ which assigns to the pair $(a/b, c/d)$ the equivalence class ac/bd that contains S. □

Proposition A.4.6. *The multiplication on $\mathbb{Q}$ is associative and commutative. Furthermore, for all $x, y, z \in \mathbb{Q}$:*

(1) $\iota(a)\iota(b) = \iota(ab)$ *for all* $a, b \in \mathbb{Z}$;
(2) $1x = x$, *where* $1 = 1/1$;
(3) *if* $x = a/b \neq 0$, *then* $xx^{-1} = 1$, *where* $x^{-1} = b/a$ *if* $a > 0$, $x^{-1} = -b/-a$ *if* $a < 0$;
(4) *if* $xz = yz$ *and* $z \neq 0$, *then* $x = y$;
(5) $x(y + z) = xy + xz$.

(Hence $\mathbb{Q}$ is a field.)

The straightforward proof is left to the reader. The inverse x^{-1} in the statement is often denoted by $1/x$ (a notation which would be confusing in this section).

We note that $(a/b)\iota(b) = ab/b = a/1 = \iota(a)$, so that $a/b = \iota(a)\,\iota(b)^{-1}$ and every rational number is indeed the quotient of two integers. Inverses also make exact division possible in $\mathbb{Q}$, with $x \div y = xy^{-1}$ as long as $y \neq 0$.

A.4.c. Order. The following results are straightforward:

Proposition A.4.7. *A total order relation on $\mathbb{Q}$ is well-defined by $a/b \leqslant c/d$ if and only if $ad \leqslant bc$.*

Proposition A.4.8. *The order relation on $\mathbb{Q}$ has the following properties:*

(1) $\iota(m) \leqslant \iota(n)$ *if and only if* $m \leqslant n$.
(2) $x \leqslant y$ *implies* $x + z \leqslant y + z$.
(3) $x \leqslant y$ *implies* $xz \leqslant yz$ *if* $z \geqslant 0$, $xz \geqslant yz$ *if* $z \leqslant 0$.

The order on $\mathbb{Q}$ is dense in the sense that between any two $x < y$ there is some $x < z < y$. But $\mathbb{Q}$ is not complete in the sense that a nonempty bounded subset of $\mathbb{Q}$ does not necessarily have a least upper bound or a greatest lower bound. This will be remedied in the next section.

Proposition A.4.9. *$\mathbb{Q}$ is archimedean: for every $\varepsilon > 0$ in $\mathbb{Q}$, there exists $N \in \mathbb{N}$ such that $1/n < \varepsilon$ for all $n \geqslant N$ in $\mathbb{N}$.*

Proof. Let $\varepsilon = a/b$, where $a, b \in \mathbb{N}$. Then $b\varepsilon = a \geqslant 1$ and $\varepsilon \geqslant 1/b$. Therefore $1/n < \varepsilon$ for all $n \geqslant b + 1$. $\qquad\square$

Exercises

1. Prove Lemma A.4.2.
2. Prove Proposition A.4.4.
3. Prove Proposition A.4.6.
4. Prove Proposition A.4.7.
5. Prove Proposition A.4.8.
6. Show that $x \leqslant y$ in $\mathbb{Q}$ implies $x + z \leqslant y + z$ and, if $z > 0$, $xz \leqslant yz$.
7. Repeat the construction in this section, allowing from the start all pairs (a, b) with $a, b \in \mathbb{Z}$ and $b \neq 0$. Show that, with suitable multiplications, this yields the same set $\mathbb{Q}$, with the same operations and order relation.

A.5. REAL NUMBERS

Rational numbers are not sufficient for mathematics. Geometry shows that algebraic equations such as $x^2 = 2$ should have solutions, none of which can be in $\mathbb{Q}$. Analysis requires that certain sequences and other constructions have limits, which $\mathbb{Q}$ mostly does not provide. Real numbers satisfy these needs.

Our construction of real numbers is the standard metric space completion of $\mathbb{Q}$. Section 9.2 contains a similar construction, which applies to all fields with absolute values. A totally different construction (based on the order relation, and essentially due to Dedekind) is given in Section 19.2.

A.5.a. Construction. The ideas behind the construction of real numbers come from analysis. Intuitively a real number such as $\pi = 3.1415926535\ldots$ is the limit of an infinite sequence of rational numbers $3, 3.1, 3.14, 3.141, \ldots$. But not every sequence of rational numbers ought to have a real number limit. Moreover a real number can be written as a limit of rational numbers in many different ways. This suggests that real numbers be defined as equivalence classes of suitable sequences of rational numbers.

In finding suitable sequences, we use the absolute value on $\mathbb{Q}$:

$$|x| = \begin{cases} x & \text{if } x \geqslant 0, \\ -x & \text{if } x \leqslant 0, \end{cases}$$

which measures how close a rational number x is to 0. A sequence $a = (a_1, \ldots, a_n, \ldots)$ of rational numbers is a **Cauchy sequence** in case for every $\varepsilon > 0$ in $\mathbb{Q}$ there exists $N \in \mathbb{N}$ such that $|a_n - a_m| < \varepsilon$ whenever $m, n \geqslant N$). Cauchy was first to observe that a sequence of real numbers has this property if and only if it has a limit. Thus Cauchy sequences of rational numbers are precisely the sequences that must have real limits.

In what follows, C denotes the set of all Cauchy sequences of rational numbers. It is easy to decide when two sequences should have the same limit:

Proposition A.5.1. *The following binary relation $\equiv$ is an equivalence relation on C: $a \equiv b$ if and only if for every $\varepsilon > 0$ in $\mathbb{Q}$ there exists $N \in \mathbb{N}$ such that $|a_n - b_n| < \varepsilon$ for all $n \geqslant N$.*

We denote by $\mathrm{cls}\, a$ the equivalence class of $a \in C$ modulo $\equiv$. By definition, $\mathbb{R} = C/\equiv$. The elements of $\mathbb{R}$ are **real numbers**. All we have to show now is that these are the usual real numbers. The proofs are left to those of our readers who need to practice their skill at analysis, or they can be looked up in standard texts.

Lemma A.5.2. *An injective mapping $\iota : \mathbb{Q} \longrightarrow \mathbb{R}$ is defined by*

$$i(x) = \mathrm{cls}\, \overline{x},$$

where $\overline{x}$ is the constant sequence $\overline{x} = (x, x, \ldots, x, \ldots)$.

Lemma A.5.2 allows the standard practice of regarding $x \in \mathbb{Q}$ and $\iota(x) \in \mathbb{R}$ as identical so that rational numbers are real numbers.

A.5.b. Operations. The operations on $\mathbb{R}$ can be defined from pointwise operations on Cauchy sequences.

Lemma A.5.3. *Let $a = (a_1, \ldots, a_n, \ldots)$ and $b = (b_1, \ldots, b_n, \ldots)$ be Cauchy sequence of rational numbers.*

(1) $a + b = (a_1 + b_1, \ldots, a_n + b_n, \ldots)$ *is a Cauchy sequence.*

(2) *Every Cauchy sequence is bounded (for each $a \in C$ there exists $A > 0$ in $\mathbb{Q}$ such that $|a_n| \leqslant A$ for all n).*

(3) $ab = (a_1 b_1, \ldots, a_n b_n, \ldots)$ *is a Cauchy sequence.*

Proposition A.5.4. *An addition on* $\mathbb{R}$ *is well-defined by:* $\mathrm{cls}\,a + \mathrm{cls}\,b = \mathrm{cls}\,(a + b)$. *It is associative, commutative, and cancellative. Furthermore:*

(1) $\iota(x) + \iota(y) = \iota(x + y)$ *for all* $x, y \in \mathbb{Q}$;
(2) $r + 0 = r$ *for all* $r \in \mathbb{R}$, *where* $0 = \iota(0)$;
(3) $r + (-r) = 0$ *for all* $r \in \mathbb{R}$, *where* $-r = \mathrm{cls}\,(-a_1, \ldots, -a_n, \ldots)$ *if* $r = \mathrm{cls}\,(a_1, \ldots, a_n, \ldots)$.

(Hence $\mathbb{R}$ *is an abelian group under addition.)*

Proposition A.5.5. *A multiplication on* $\mathbb{R}$ *is well-defined by* $\mathrm{cls}\,a\,\mathrm{cls}\,b = \mathrm{cls}\,(ab)$. *It is associative and commutative. Furthermore, for all* $r, s, t \in \mathbb{R}$:

(1) $\iota(x)\iota(y) = \iota(xy)$ *for all* $x, y \in \mathbb{Q}$;
(2) $1r = r$, *where* $1 = \iota(1)$;
(3) *if* $r \neq 0$, *then* $rr^{-1} = 1$ *for some* $r^{-1} \in \mathbb{R}$;
(4) *if* $rt = st$ *and* $t \neq 0$, *then* $r = s$;
(5) $r(s + t) = rs + rt$.

(Hence $\mathbb{R}$ *is a field.)*

The proof is straightforward except for (3). For this one can use:

Lemma A.5.6. *Let* $a \in C$. *If* $a \not\equiv \overline{0}$, *then there exist* $A > 0$ *in* $\mathbb{Q}$ *and* $N \in \mathbb{N}$ *such that either* $a_n \geqslant A$ *for all* $n \geqslant N$, *or* $a_n \leqslant -A$ *for all* $n \geqslant N$. *Hence any sequence* $b = (b_1, \ldots, b_n, \ldots)$ *with* $b_n = 1/a_n$ *for all* $n \geqslant N$ *is a Cauchy sequence.*

A.5.c. Order. With Lemma A.5.6 we can show:

Proposition A.5.7. *A total order relation on* $\mathbb{R}$ *is well-defined by* $\mathrm{cls}\,a < \mathrm{cls}\,b$ *if and only if there exist* $N \in \mathbb{N}$ *and* $A \in \mathbb{Q}$, $A > 0$ *such that* $b_n \geqslant a_n + A$ *for all* $n \geqslant N$. *For all* $r, s, t \in \mathbb{R}$:

(1) *when* $x, y \in \mathbb{Q}$, $\iota(x) \leqslant \iota(y)$ *if and only if* $x \leqslant y$;
(2) $r > 0$ *if and only if* $r > \iota(q)$ *for some* $q \in \mathbb{Q}$, $q > 0$;
(3) $r \leqslant s$ *implies* $r + t \leqslant s + t$;
(4) $r \leqslant s$ *implies* $rt \leqslant st$ *if* $t \geqslant 0$, $rt \geqslant st$ *if* $t \leqslant 0$.

$\mathbb{Q}$ is dense in $\mathbb{R}$: between any two real numbers $a < b$ there is a rational number $a < r < b$ (see the exercises).

A.5.d. Completeness. With the order on $\mathbb{R}$ the absolute value on $\mathbb{R}$ can be defined as usual: $|r| = r$ if $r \geqslant 0$, $|r| = -r$ if $r \leqslant 0$.

With this absolute value, limits in $\mathbb{R}$ can be defined. The reader will verify that every real number is the limit of a sequence of rational numbers.

The completeness properties of $\mathbb{R}$ are most often stated as follows:

Theorem A.5.8. *Every Cauchy sequence of real numbers converges.*

Theorem A.5.9. *Every nonempty set of real numbers with an upper bound has a least upper bound.*

Theorem A.5.9 implies that every nonempty set of real numbers with a lower bound has a greatest lower bound. Proofs of Theorems A.5.8 and A.5.9 can be found in analysis textbooks (see the bibliography). Section 9.2 also has a proof of Theorem A.5.8.

Exercises

1. Prove Proposition A.5.1.
2. Prove Lemma A.5.3.
3. Prove Lemma A.5.6.
4. Prove the following: if $r \neq 0$ in $\mathbb{R}$, then $rr^{-1} = 1$ for some $r^{-1} \in \mathbb{R}$. (Use Lemma A.5.6.)
5. Show that a total order relation on $\mathbb{R}$ is defined by $\operatorname{cls} a < \operatorname{cls} b$ if and only if there exist $N \in \mathbb{N}$ and $A \in \mathbb{Q}$, $A > 0$ such that $b_n \geqslant a_n + A$ for all $n \geqslant N$.
6. Prove that $|r + s| \leqslant |r| + |s|$ for all $r, s \in \mathbb{R}$.
7. Show that $\operatorname{cls} a = \lim_i (a_n)$ holds in $\mathbb{R}$ for every Cauchy sequence $a = (a_1, \ldots, a_n, \ldots)$ of rational numbers.
8. Show that $\mathbb{Q}$ is dense in $\mathbb{R}$ in the sense that $r < s$ in $\mathbb{R}$ implies $r < q < s$ for some $q \in \mathbb{Q}$.

A.6. COMPLEX NUMBERS

The field $\mathbb{C}$ of complex numbers is constructed so that every quadratic equation with real coefficients has solutions in $\mathbb{C}$. It turns out that in fact every polynomial equation with real or complex coefficients has solutions in $\mathbb{C}$. The algebraic properties of $\mathbb{C}$ come at a price: the order relation on $\mathbb{R}$ has no satisfactory expansion to $\mathbb{C}$, a fact which is further explained in Chapter 8.

A.6.a. Construction. Many quadratic equations lack solutions in $\mathbb{R}$ because negative real numbers have no square root. To remedy this, we construct $\mathbb{C}$ by adjoining to $\mathbb{R}$ an element i such that $i^2 = -1$. Then every negative real number r has a square root $i\sqrt{|r|}$ in $\mathbb{C}$. We see that $\mathbb{C}$ must contain all sums $a + bi$ with $a, b \in \mathbb{R}$, and since these sums can be added (by $(a + bi) + (c + di) = (a + c) + (b + d)i$) and multiplied (by $(a + bi)(c + di) = ac + (ad + bc)i + bdi^2 = (ac - bd) + (ad + bc)i$), they constitute all of $\mathbb{C}$.

A rigorous way to adjoin i to $\mathbb{R}$ is to define $\mathbb{C}$ as the vector space $\mathbb{R} \times \mathbb{R}$ over $\mathbb{R}$ which consists of all ordered pairs (a,b) of real numbers. (A more enlightened construction is given in Chapter 6). The elements of $\mathbb{C}$ are **complex numbers**. A complex number (a,b) can thus be visualized as the point with coordinates (a,b) in the euclidean plane.

The vectors $(1,0)$ and $(0,1)$ form a basis of $\mathbb{C}$ over $\mathbb{R}$ and are traditionally denoted by $1 = (1,0)$ and $i = (0,1)$. Then every complex number $z = (a,b)$ can be written in the form $a + bi = a1 + bi$ for some unique $a,b \in \mathbb{R}$. There is a mapping $\iota : \mathbb{R} \longrightarrow \mathbb{C}$ which sends $a \in \mathbb{R}$ to $a = a + 0i \in \mathbb{C}$. It is standard practice, suggested by the notation, to identify $a \in \mathbb{R}$ and $\iota(a) \in \mathbb{C}$, and to call a complex number $a + bi$ **real** when $b = 0$.

A.6.b. Operations. The vector space structure of $\mathbb{C}$ provides a coordinate-wise addition on $\mathbb{C}$: $(a,b) + (c,d) = (a + c, b + d)$ or

$$(a + bi) + (c + di) = (a + c) + (b + d)i.$$

The following properties of this addition are general properties of vector spaces, but can also be deduced from Proposition A.5.4 and the above.

Proposition A.6.1. *The addition on $\mathbb{C}$ is associative, commutative, and cancellative. Furthermore:*

(1) $\iota(a) + \iota(b) = \iota(a + b)$ *for all* $a,b \in \mathbb{R}$;
(2) $z + 0 = z$ *for all* $z \in \mathbb{C}$, *where* $0 = 0 + 0i = \iota(0)$;
(3) $z + (-z) = 0$, *where* $-(x + yi) = -x + (-y)i$;

(Hence $\mathbb{C}$ is an abelian group under addition.)

Since the addition on $\mathbb{C}$ is vector addition, it has the following geometric interpretation. Let $z = a + bi$, $t = c + di$, and A, B, C be the points (a,b), (c,d), $(a + c, b + d)$ corresponding to z, t, $z + t$. Then $\overrightarrow{OC} = \overrightarrow{OA} + \overrightarrow{OB}$ so that O, A, B, C are the vertices of a possibly squashed parallelogram.

The multiplication on $\mathbb{C}$ is defined by $(a,b)(c,d) = (ac - bd, ad + bc)$, or

$$(a + bi)(c + di) = (ac - bd) + (ad + bc)i.$$

The following properties are straightforward:

Proposition A.6.2. *The multiplication on $\mathbb{C}$ is associative and commutative. Furthermore, for all* $x,y,z \in \mathbb{C}$:

(1) $\iota(r)\iota(s) = \iota(rs)$ *for all* $r,s \in \mathbb{R}$;
(2) $1x = x$;
(3) *if* $z = a + bi \neq 0$, *then* $zz^{-1} = 1$, *where* $z^{-1} = a/(a^2 + b^2) - b/(a^2 + b^2)i$;

(4) *if* $xz = yz$ *and* $z \neq 0$, *then* $x = y$;

(5) $x(y + z) = xy + xz$.

(Hence $\mathbb{C}$ *is a field.)*

A.6.c. Conjugate and Modulus. The **conjugate** of a complex number $z = a + bi$ is $\bar{z} = a - bi$. We see that $\bar{z} = z$ if and only if z is real.

Proposition A.6.3. *For all* $x, y, z \in \mathbb{C}$, $\bar{\bar{z}} = z$, $\overline{x + y} = \bar{x} + \bar{y}$, *and* $\overline{xy} = \bar{x}\,\bar{y}$. *(Hence* $z \longmapsto \bar{z}$ *is an automorphism of the field* $\mathbb{C}$.*)*

In the geometric interpretation of complex numbers, z and $\bar{z}$ correspond to points that are symmetric to the x-axis $y = 0$.

The geometric interpretation of complex numbers also suggests the following definitions. The **modulus** of $z = a + bi$ is $|z| = \sqrt{a^2 + b^2} \in \mathbb{R}$; then $z\bar{z} = a^2 + b^2 = |z|^2$. The **argument** of $z = a + bi$, defined up to the addition of integer multiples of 2π, is any $\arg z = \theta \in \mathbb{R}$ such that $a = |z|\cos\theta$ and $b = |z|\sin\theta$. If A is the point (a, b) which corresponds to $z = a + bi$, then $|z|$ is the length of $\overrightarrow{OA}$, and $\arg z$ is the radian measure of the directed angle from the positive x-axis to $\overrightarrow{OA}$.

Proposition A.6.4. *For all* $x, y, z \in \mathbb{C}$, $|x + y| \leqslant |x| + |y|$, $|xy| = |x||y|$, *and* $\arg(xy) = \arg x + \arg y$.

Proof. Let $x = a + bi$, $y = c + di$, $r = |x|$, $s = |y|$, $\alpha = \arg x$, and $\beta = \arg y$ so that $r^2 = a^2 + b^2$, $s^2 = c^2 + d^2$, $a = r\cos\alpha$, $b = r\sin\alpha$, $c = s\cos\beta$, and $d = s\sin\beta$.

Since $a^2 d^2 + b^2 c^2 - 2abcd = (ad - bc)^2 \geqslant 0$, we have

$$(ac + bd)^2 = a^2 c^2 + b^2 d^2 + 2abcd \leqslant a^2 c^2 + b^2 d^2 + a^2 d^2 + b^2 c^2 = r^2 s^2$$

and $ac + bd \leqslant rs$. Hence

$$|x + y|^2 = (a + c)^2 + (b + d)^2 = a^2 + 2ac + c^2 + b^2 + 2bd + d^2 \leqslant r^2 + 2rs + s^2$$

and $|x + y| \leqslant |x| + |y|$. (This **triangle inequality** may be best proved by vector methods.)

Finally the addition formulas for sines and cosines yield

$$xy = r(\cos\alpha + i\sin\alpha)s(\cos\beta + i\sin\beta) = rs(\cos(\alpha + \beta) + i\sin(\alpha + \beta)).$$

Hence $|xy| = rs$ and $\arg(xy) = \alpha + \beta$. $\square$

The last two equalities in Proposition A.6.4 provide a geometric interpretation of the product of two complex numbers. Let A, B, C be the points that represent x, y, and xy, and U be the point which represents $1 = (1, 0)$. Since

$|xy| = |x||y|$ and $\arg(xy) = \arg x + \arg y$, $\overrightarrow{OC}$ is obtained from $\overrightarrow{OB}$ by multiplying $\overrightarrow{OB}$ by $|x|$ and rotating the resulting vector by $\arg x$. Applying the same construction to $\overrightarrow{OU}$ yields $\overrightarrow{OA}$. Therefore the triangles OUA and OBC are directly similar.

A.6.d. The Fundamental Theorem of Algebra

Theorem A.6.5 (Fundamental Theorem of Algebra). *Every nonconstant polynomial equation*

$$a_n x^n + a_{n-1} x^{n-1} + \cdots + a_1 x + a_0 = 0$$

with coefficients in $\mathbb{C}$ has a solution in $\mathbb{C}$.

Theorem A.6.5 received its traditional name at a time when algebra was mainly concerned with the solution of polynomial equations. An algebraic proof is given in Chapter 8. The author's favorite proof uses complex analysis. We know that a holomorphic function whose modulus has a maximum must be constant. Assume that $f(X) \in \mathbb{C}[X]$ has no zero in $\mathbb{C}$. Then the function $g(z) = 1/f(z)$ is holomorphic on all of $\mathbb{C}$, and its modulus $|g(z)|$ is continuous on all of $\mathbb{C}$. If f has degree 1 or more, then $|g(z)| \longrightarrow 0$ when $z \longrightarrow \infty$, so that $|g(z)|$ has a maximum value in $\mathbb{C}$; this also holds if f is constant. It follows that g is constant, and f is constant.

The following particular case of Theorem A.6.5 also follows from Proposition A.6.4:

Proposition A.6.6. *For each positive integer n, the equation $x^n = 1$ has n solutions in $\mathbb{C}$, namely $x = \cos(2k\pi/n) + i\sin(2k\pi/n)$ $(k = 0, 1, \ldots, n-1)$.*

The numbers $\cos(2k\pi/n) + i\sin(2k\pi/n)$ $(k = 0, 1, \ldots, n-1)$ are the complex n-th **roots of unity**. They are studied in more detail in Section 7.7.

Exercises

1. Prove Proposition A.6.2.

2. Prove Proposition A.6.3.

3. Let $z \neq 0$ be a complex number and n be a positive integer. Show that the equation $x^n = z$ has n solution in $\mathbb{C}$ (the complex n-th **roots** of z).

4. Prove that there exists no total order relation $\leqslant$ on $\mathbb{C}$ with the properties in Proposition A.5.7.

5. Let $j = \cos(2\pi/3) + i\sin(2\pi/3) = -1/2 + i\sqrt{3}/2$. Show that three complex numbers x, y, z form a counterclockwise equilateral triangle in the complex plane if and only if $x + jy + j^2 z = 0$.

A.7. QUATERNIONS

The quaternion algebra was devised not as an improvement on $\mathbb{C}$ but simply to see if $\mathbb{C}$ can be expanded to a still larger number system. The answer comes at a price: the quaternion algebra is not commutative.

It is shown in Chapter 17 that the quaternion algebra cannot be expanded to a larger number system (that would still be an associative division ring); hence this appendix must end here.

A.7.a. Construction. By definition, the **quaternion algebra Q** (also denoted by $\mathbb{H}$ after its discoverer Hamilton) is the vector space $\mathbb{R}^4$ over $\mathbb{R}$ which consists of all ordered quadruples (a,b,c,d) of real numbers. An element of **Q** is a **quaternion**. A quaternion (a,b,c,d) can thus be visualized (if that is the proper word) as the point with coordinates a,b,c,d in the euclidean 4 space. (Another geometric interpretation is given in the exercises.)

The vectors $(1,0,0,0)$, $(0,1,0,0)$, $(0,0,1,0)$, and $(0,0,0,1)$ form a basis of **Q** over $\mathbb{R}$ and are traditionally denoted by $1 = (1,0,0,0)$, $i = (0,1,0,0)$, $j = (0,0,1,0)$, and $k = (0,0,0,1)$. Then every quaternion $q = (a,b,c,d)$ can be written in the form $a + bi + cj + dk = a1 + bi + cj + dk$ for some unique $a,b,c,d \in \mathbb{R}$. There is a mapping $\iota : \mathbb{C} \longrightarrow \mathbf{Q}$ which sends $a + bi \in \mathbb{C}$ to $a + bi = a + bi + 0j + 0k \in \mathbf{Q}$. It is standard practice, suggested by the notation, to identify $a + bi \in \mathbb{C}$ and $\iota(a) \in \mathbf{Q}$, and to call a quaternion $a + bi + cj + dk$ **complex** when $c = d = 0$ and **real** when $b = c = d = 0$.

A.7.b. Operations. The vector space structure of **Q** provides a coordinate-wise addition on **Q**:

$$(a + bi + cj + dk) + (x + yi + zj + tk) = (a + x) + (b + y)i + (c + z)j + (d + t)k.$$

The following properties of this addition are general properties of vector spaces but can also be deduced from Proposition A.5.4 and the above:

Proposition A.7.1. *The addition on **Q** is associative, commutative, and cancellative. Furthermore:*

(1) $\iota(a) + \iota(b) = \iota(a + b)$ *for all* $a,b \in \mathbb{C}$;
(2) $q + 0 = q$ *for all* $q \in \mathbf{Q}$, *where* $0 = 0 + 0i + 0j + 0k = \iota(0)$;
(3) $q + (-q) = 0$, *where* $-(x + yi + zj + tk) = -x + (-y)i + (-z)j + (-t)k$.

*(Hence **Q** is an abelian group under addition.)*

Since the addition on **Q** is vector addition, it has a geometric interpretation by four dimensional parallelograms (possibly squashed).

To define the multiplication on **Q**, we observe the following. Let V be a vector space over a field K with a finite basis $e_1, e_2, \ldots, e_n$ (so that every element of V can be written in the form $x = \sum_{1 \leqslant i \leqslant n} x_i e_i$ for some unique $x_1, x_2, \ldots, x_n \in K$). Any bilinear multiplication on V satisfies

$$(\sum_{1 \leqslant i \leqslant n} x_i e_i)(\sum_{1 \leqslant j \leqslant n} y_j e_j) = \sum_{1 \leqslant i,j \leqslant n} x_i y_j e_i e_j \tag{A.1}$$

and is determined by the products $e_i e_j$ of the basis elements. Conversely, we can choose $e_i e_j \in V$ arbitrarily for all $1 \leqslant i, j \leqslant n$, and then (A.1) yields a bilinear multiplication on V for which all products $e_i e_j$ are the given products. For example, the multiplication on $\mathbb{C}$ is the bilinear multiplication such that $1^2 = 1$, $1i = i1 = i$, and $i^2 = -1$.

The multiplication on $\mathbf{Q}$ is the bilinear multiplication such that

$$11 = 1, \; 1i = i1 = i, \; 1j = j1 = j, \; 1k = k1 = k,$$

$$ii = jj = kk = -1, \; ij = k, \; jk = i, \; ki = j, \; ji = -k, \; kj = -i, \; ik = -j.$$

In detail,

$$(a + bi + cj + dk)(x + yi + zj + tk)$$
$$= (ax - by - cz - dt) + (ay + bx + ct - dz)i$$
$$+ (az + cx + dy - bt)j + (at + dx + bz - cy)k.$$

This multiplication is not commutative.

The following properties are straightforward:

Proposition A.7.2. *The multiplication on* $\mathbf{Q}$ *is associative. Furthermore, for all* $p, q, r \in \mathbf{Q}$:

(1) $\iota(a)\iota(b) = \iota(ab)$ *for all* $a, b \in \mathbb{C}$;

(2) $1p = p1 = p$;

(3) *if* $q = a + bi + cj + dk \neq 0$, *then* $qq^{-1} = q^{-1}q = 1$, *where*

$$q^{-1} = (a - bi - cj - dk)/(a^2 + b^2 + c^2 + d^2);$$

(4) $p(q + r) = pq + pr$, $(p + q)r = pr + qr$.

(Hence $\mathbf{Q}$ *is a division ring.)*

A.7.c. Conjugate and Modulus. Analogy with $\mathbb{C}$ suggests the following definitions. The **conjugate** of a quaternion $q = a + bi + cj + dk$ is $\overline{q} = a - bi - cj - dk$. We see that $\overline{q} = q$ if and only if q is real.

Proposition A.7.3. *For all* $q, r \in \mathbf{Q}$, $\overline{\overline{q}} = q$, $\overline{q + r} = \overline{q} + \overline{r}$, *and* $\overline{qr} = \overline{q}\,\overline{r}$. *(Hence* $q \longmapsto \overline{q}$ *is an automorphism of* $\mathbf{Q}$.)

The **modulus** of $q = a + bi + cj + dk$ is $|q| = \sqrt{a^2 + b^2 + c^2 + d^2} \in \mathbb{R}$. We see that $q\bar{q} = \bar{q}q = a^2 + b^2 + c^2 + d^2 = |q|^2$.

Proposition A.7.4. *For all* $q, r \in \mathbf{Q}$, $|q + r| \leqslant |q| + |r|$ *and* $|qr| = |q||r|$.

Exercises

1. Prove Proposition A.7.2.
2. Show that a quaternion $q = a + bi + cj + dk$ can be viewed as an ordered pair $(a, \vec{v})$ of a real number a and a three-dimensional real vector $\vec{v} = bi + cj + dk$, with operations

$$(a, \vec{v}) + (b, \vec{w}) = (a + b, \vec{v} + \vec{w}),$$

$$(a, \vec{v})(b, \vec{w}) = (ab - \vec{v} \cdot \vec{w}, \ a\vec{w} + b\vec{v} - \vec{v} \times \vec{w}).$$

   Use this interpretation to give a different proof of Proposition A.7.2.
3. Prove Proposition A.7.3.
4. Prove Proposition A.7.4.
5. Find the center of $\mathbf{Q}$ (the set $\{q \in \mathbf{Q}; \ qr = rq \text{ for all } r \in \mathbf{Q}\}$).

APPENDIX B

SETS AND ORDER

This appendix contains properties of sets and of partially ordered sets that are used in several chapters, including various methods of proof by induction and basic properties of cardinal and ordinal numbers.

B.1. SETS, MAPPINGS, AND EQUIVALENCE RELATIONS

This section reviews elementary set theory.

B.1.a. Sets. A **set** is a mathematical object S such that every mathematical object x is either an **element** of S (written $x \in S$) or not an element of S (written $x \notin S$).

Axiomatic set theory is the branch of mathematics which provides rigorous definitions of sets (not like the above). It is beyond the scope of this book. A rigorous treatment of sets is necessary because we wish to think of sets as collections, but contradictions arise if collections with too many elements (e.g., the collection of all sets) are treated as sets. These large collections are called **classes**.

One of the generally accepted axioms of set theory states:

Two sets are equal if and only if they have the same elements.

This axiom allows us to think of a set as the collection of its elements. It also provides a method for proving that two sets X and Y are equal: show that every element of X is an element of Y, and that every element of Y is an element of X.

A set S is a **subset** of a set T in case every element of S is an element of T. This relationship is called **inclusion** and is denoted by $S \subseteq T$ (often, by $S \subset T$), or by $T \supseteq S$. Thus $S = T$ if and only if $S \subseteq T$ and $T \subseteq S$. We write $S \subsetneqq T$ when $S \subseteq T$ and $S \neq T$.

The **empty** set $\emptyset$ is the set with no elements ($x \notin \emptyset$ for all x). It is a subset of every set.

B.1.b. Set Constructions. The **power set** of a set S is the set 2^S or $\mathscr{P}(S)$ whose elements are all subsets of S : $2^S = \{ X ; X \subseteq S \}$.

The **union** $S \cup T$, **intersection** $S \cap T$, **set difference** $S \backslash T$ (also denoted by $S - T$), and **cartesian product** $S \times T$ (also called **direct product**) of two sets S and T are the sets defined by

$$S \cup T = \{ x ; x \in S \text{ or } x \in T \text{ [or both]} \},$$

$$S \cap T = \{ x ; x \in S \text{ and } x \in T \},$$

$$S \backslash T = \{ x ; x \in S \text{ and } x \notin T \},$$

$$S \times T = \{ (x, y) ; x \in S \text{ and } y \in T \}.$$

The elements of $S \times T$ are **ordered pairs** or just **pairs** of an element of S and an element of T.

A **binary relation** between two sets S and T is a subset $\mathscr{R}$ of $S \times T$. Intuitively $x \in S$ is related to $y \in T$ (by $\mathscr{R}$) if and only if $(x, y) \in \mathscr{R}$; this is also written $x \mathscr{R} y$. If $S = T$, then $\mathscr{R}$ is a binary relation **on** S.

B.1.c. Mappings. A **mapping** $\varphi : S \longrightarrow T$ of a set S into a set T is a binary relation $\varphi \subseteq S \times T$ such that for each $x \in S$ there is exactly one $y \in T$ such that $(x, y) \in \varphi$. This element y is the **value** of φ at x and is commonly denoted by $\varphi(x)$. We say that the mapping φ **assigns** $\varphi(x)$ to x, and **sends** x to $\varphi(x)$; φ may be denoted by $\varphi : x \longmapsto \varphi(x)$. By definition, a mapping of S into T sends each element of S to exactly one element of T.

Other notations for $\varphi(x)$ are the **left operator** notation φx, the **right operator** notation $x \varphi$, the **exponential** notation x^φ, and the **family** notation φ_x.

When φ is a mapping of S into T, S is the **domain** of φ and T is a **codomain** of φ. The **range** or **image** of φ is the subset $\varphi(S) = \text{Im } \varphi = \{ \varphi(x) ; x \in S \}$ of T.) The mappings of S into T are the elements of a set T^S. (The reader will readily construct a one-to-one correspondence between subsets of S and mappings of S into $\{0, 1\}$; this explains the notation 2^S for the power set of S.)

The **identity** mapping 1_S on a set S is the mapping $1_S : S \longrightarrow S$ defined by $1_S(x) = x$ for all $x \in S$.

When $\varphi : S \longrightarrow T$ and $\psi : T \longrightarrow U$ are mappings, the **composition** of φ and ψ is the mapping $\psi \circ \varphi : S \longrightarrow U$ defined by $(\psi \circ \varphi)(x) = \psi(\varphi(x))$ for all $x \in S$. Composition is associative: if $\chi : U \longrightarrow V$, then $\chi \circ (\psi \circ \varphi) = (\chi \circ \psi) \circ \varphi$. When $\varphi : S \longrightarrow T$, then $1_T \circ \varphi = \varphi = \varphi \circ 1_S$.

A mapping $\varphi : S \longrightarrow T$ is **injective** or **one-to-one** or an **injection** in case $\varphi(x) = \varphi(y)$ implies $x = y$. A mapping $\varphi : S \longrightarrow T$ is **surjective** or a **surjection**, or is a mapping of S **onto** T, in case for every $t \in T$ there exists $x \in S$ such that $\varphi(x) = t$ (equivalently, in case $\varphi(S) = T$).

A mapping $\varphi : S \longrightarrow T$ is **bijective** or **one-to-one onto** or a **bijection** in case it is injective and surjective. Then for every $t \in T$ there exists a unique $s \in S$ such that $\varphi(s) = t$; the mapping $\varphi^{-1} : t \longmapsto s$ is the **inverse** bijection;

φ and φ^{-1} are **mutually inverse** bijections. Two mappings $\varphi : S \longrightarrow T$ and $\psi : T \longrightarrow S$ are mutually inverse bijections if and only if $\psi \circ \varphi = 1_S$ and $\varphi \circ \psi = 1_T$; such a pair (φ, ψ) is also called a **one-to-one correspondence** between S and T.

Let $\varphi : S \longrightarrow T$ be a mapping. When A is a subset of S, the **image** of A under φ is the subset

$$\varphi(A) = \{ \varphi(a) ; a \in A \}$$

of T. When B is a subset of T, the **inverse image** of B under φ is the subset

$$\varphi^{-1}(B) = \{ x \in S ; \varphi(x) \in B \}$$

of S. The notation $\varphi^{-1}(B)$ should not be taken to imply that φ is a bijection: inverse images are defined for all mappings.

Let $\varphi : S \longrightarrow T$ be a mapping. When A is a subset of S, the **restriction** of φ to A is

$$\varphi_{|A} = \{ (x,y) \in \varphi ; x \in A \};$$

it is a mapping of A into T. By definition, $\varphi_{|A}(x) = \varphi(x)$ for all $x \in A$. The range of $\varphi_{|A}$ is $\varphi(A)$.

B.1.d. Binary Operations. A **binary operation** on a set S is a mapping μ of $S \times S$ into S; $\mu(x,y)$ is most often denoted by $x + y$ or by xy.

B.1.e. Families. A **family** indexed by a set I is a set $S = (S_i)_{i \in I}$ together with a mapping $i \longmapsto S_i$ of I onto S. The set I is the **index set** of the family.

A (finite) **sequence** $s = s_1, s_2, \ldots, s_n$ is a family indexed by the set $I = \{ 1, 2, \ldots, n \}$ of the first n natural numbers (constructed in Appendix A). Letting $n = 0$ yields the **empty** sequence.

The **union** $\bigcup_{i \in I} S_i$, **intersection** $\bigcap_{i \in I} S_i$, and **cartesian product** $\prod_{i \in I} S_i$ of a family $S = (S_i)_{i \in I}$ of sets are defined by

$$\bigcup_{i \in I} S_i = \{ x ; x \in S_i \text{ for some } i \in I \},$$
$$\bigcap_{i \in I} S_i = \{ x ; x \in S_i \text{ for every } i \in I \},$$
$$\prod_{i \in I} S_i = \{ (x_i)_{i \in I} ; x_i \in S_i \text{ for every } i \in I \}.$$

These are sets if $I \neq \emptyset$. If $I = \emptyset$, then $\bigcap_{i \in I} S_i$ contains every mathematical object and is not a set; but $\bigcup_{i \in I} S_i = \emptyset$, and $\prod_{i \in I} S_i$ contains only the empty family and is your favorite one-element set (e.g., $\{\emptyset\}$, or $\{1\}$). The intersection of an empty family $(S_i)_{i \in I}$ of subsets of S is sometimes defined to be S itself. One form of the Axiom of Choice states that $\prod_{i \in I} S_i \neq \emptyset$ when $I \neq \emptyset$ and $S_i \neq \emptyset$ for all $i \in I$ (see Section B.3).

The union, intersection, and cartesian product of a sequence $S_1, S_2, \ldots, S_n$ of sets are also denoted by $S_1 \cup S_2 \cup \cdots \cup S_n$, $S_1 \cap S_2 \cap \cdots \cap S_n$, and $S_1 \times S_2 \times \cdots \times S_n$.

These constructions do not apply to classes in general: for instance, there is no such thing as the union of all sets.

B.1.f. Order Relations. An **order relation** on a set S (also called a **partial order relation** on S) is a binary relation (generally denoted by $\leqslant$) on S which is **reflexive** ($x \leqslant x$ for all $x \in S$), **antisymmetric** (if $x \leqslant y$ and $y \leqslant x$, then $x = y$), and **transitive** (if $x \leqslant y$ and $y \leqslant z$, then $x \leqslant z$). Then $x \geqslant y$ means that $y \leqslant x$; $x < y$ means that $x \leqslant y$ and $x \neq y$; and $x > y$ means that $x \geqslant y$ and $x \neq y$.

A **partially ordered** set is an ordered pair $(S, \leqslant)$ (usually called just S) of a set S and an order relation $\leqslant$ on S.

An order relation $\leqslant$ on a set S is **total** in case at least one of $x \leqslant y$ or $x \geqslant y$ holds for every $x, y \in S$. A set with a total order relation is a **totally ordered** set, or a **chain**.

B.1.g. Equivalence Relations. An **equivalence relation** on a set S is a binary relation $\mathscr{E}$ on S which is reflexive ($x \mathrel{\mathscr{E}} x$ for all $x \in S$), **symmetric** (if $x \mathrel{\mathscr{E}} y$, then $y \mathrel{\mathscr{E}} x$), and **transitive** (if $x \mathrel{\mathscr{E}} y$ and $y \mathrel{\mathscr{E}} z$, then $x \mathrel{\mathscr{E}} z$). Every mapping $f : S \longrightarrow T$ **induces** an equivalence relation on S, defined by $x \mathrel{\mathscr{E}} y$ if and only if $f(x) = f(y)$.

When $\mathscr{E}$ is an equivalence relation on S, every element of S has an **equivalence class** or $\mathscr{E}$-**class**, which we denote by E_x, namely

$$E_x = \{ y \in S ; x \mathrel{\mathscr{E}} y \}.$$

The equivalence classes constitute a **partition** of S: that is, S is the union of all the equivalence classes, and the equivalence classes are nonempty and **pairwise disjoint** (if $E_x \neq E_y$, then $E_x \cap E_y = \varnothing$). Thus each element x of S belongs to exactly one equivalence class (i.e., E_x); furthermore $E_x = E_y$ if and only if $x \mathrel{\mathscr{E}} y$. Conversely, every partition of S consists of the equivalence classes of some equivalence relation (i.e., $x \mathrel{\mathscr{E}} y$ if and only if x and y belong to the same member of the partition).

When $\mathscr{E}$ is an equivalence relation on S, the set of all equivalence classes of $\mathscr{E}$ is the **quotient set** $S/\mathscr{E}$ of S by $\mathscr{E}$. The quotient set comes with a **projection** $p : S \longrightarrow S/\mathscr{E}$, which is the mapping defined by $p(x) = E_x$; p is surjective and induces the given equivalence relation $\mathscr{E}$ ($p(x) = p(y)$ if and only if $x \mathrel{\mathscr{E}} y$).

B.2. CHAIN CONDITIONS

This section studies two useful finiteness conditions for partially ordered sets: the ascending chain condition and the descending chain condition.

B.2.a. The Ascending Chain Condition

Proposition B.2.1. *For a partially ordered set X the following conditions are equivalent:*

(1) *Every ascending sequence*

$$x_1 \leqslant x_2 \leqslant \cdots \leqslant x_i \leqslant x_{i+1} \leqslant \cdots$$

of elements of X terminates (there exists $n > 0$ such that $x_i = x_n$ for all $i \geqslant n$).

(2) *X contains no strictly ascending infinite sequence*

$$x_1 < x_2 < \cdots < x_i < x_{i+1} < \cdots.$$

(3) *Every nonempty subset S of X has a maximal element (an element m of S such that there is no $m < s \in S$).*

Proof. (1) implies (2), since a strictly ascending sequence cannot terminate.

(2) implies (3). Assume that a nonempty subset S of X has no maximal element. A strictly ascending sequence $x_1 < \cdots < x_i < \cdots$ of elements of S is constructed by induction, as follows. First x_1 is any element of $S \neq \varnothing$. If $x_i \in S$ has been constructed, then x_i is not a maximal element of S, and $x_i < x_{i+1}$ for some $x_{i+1} \in S$. (This proof makes implicit use of the Axiom of Choice.)

(3) implies (1). By (3), an ascending sequence $x_1 \leqslant x_2 \leqslant \cdots \leqslant x_i \leqslant \cdots$ must have a maximal element x_n. Then $x_i = x_n$ for all $i \geqslant n$, since $x_n \leqslant x_i$ and $x_n < x_i$ is impossible. $\qquad\square$

The **ascending chain condition** or **a.c.c.** is condition (2) in Proposition B.2.1; a partially ordered set with this property is sometimes called **noetherian**. The a.c.c. is a **finiteness** condition, which means that it holds whenever the partially ordered set X is finite. However, there are partially ordered sets which are noetherian and infinite (see the exercises).

A **chain** is a subset C of a partially ordered set X such that every $x, y \in C$ satisfy $x \leqslant y$ or $y \leqslant x$; that is, a totally ordered subset of X. Every ascending sequence is a chain, and the a.c.c. derives its name from this type of chain.

B.2.b. The Descending Chain Condition.
The **opposite** of an order relation $\leqslant$ is the relation $\leqslant^{\mathrm{op}}$ defined by

$$x \leqslant^{\mathrm{op}} y \iff y \leqslant x.$$

$\leqslant^{\mathrm{op}}$ is normally denoted by $\geqslant$ and is also an order relation. The **opposite** of a partially ordered set X (ordered by $\leqslant$) is the partially ordered set X^{op} which consists of the same set X ordered by $\leqslant^{\mathrm{op}}$.

A theorem which holds in *every* partially ordered set must also hold in its opposite, and will remain true if all inequalities are reversed. In particular,

Proposition B.2.1 implies:

Proposition B.2.2. *For a partially ordered set X the following conditions are equivalent:*

(1) *Every descending sequence*

$$x_1 \geqslant x_2 \geqslant \cdots \geqslant x_i \geqslant x_{i+1} \geqslant \cdots$$

of elements of X terminates (there exists $n > 0$ such that $x_i = x_n$ for all $i \geqslant n$).

(2) *X contains no strictly descending infinite sequence*

$$x_1 > x_2 > \cdots > x_i > x_{i+1} > \cdots.$$

(3) *Every nonempty subset S of X has a minimal element (an element m of S such that there is no $m > s \in S$).*

The **descending chain condition** or **d.c.c.** is condition (2) in Proposition B.2.2; partially ordered sets with this property are sometimes called **artinian**. Every finite partially ordered set satisfies the d.c.c.

Proposition B.2.3. *A partially ordered set X satisfies both the ascending chain condition and the descending chain condition if and only if every chain of X is finite.*

Proof. We assume that X satisfies the d.c.c. and contains an infinite chain C, and show that X does not satisfy the a.c.c. Construct a strictly ascending sequence $x_1 < x_2 < \cdots < x_i < \cdots$ by induction, as follows. By the d.c.c., C has a minimal element x_1. Since C is a chain, $x_1 < y$ for all $y \in C \backslash \{x_1\}$. Now assume that $x_1 < x_2 < \cdots < x_i$ have been constructed so that $x_1, \ldots, x_i \in C$ and $x_i < y$ for all $y \in C \backslash \{x_1, \ldots, x_i\}$. Since C is infinite, $C \backslash \{x_1, \ldots, x_i\}$ is not empty and has a minimal element x_{i+1}. Since C is a chain, $x_{i+1} < y$ for all $y \in C \backslash \{x_1, \ldots, x_i, x_{i+1}\}$. $\square$

B.2.c. The Ascending Chain Condition in Groups. In groups, rings, and other algebraic objects, the a.c.c and d.c.c. are applied to sets of subobjects partially ordered by inclusion. Our next result gives a typical characterization of the a.c.c.

A group G satisfies the **ascending chain condition on subgroups** in case the set of all subgroups of G, partially ordered by inclusion, satisfies the ascending chain condition.

Proposition B.2.4. *A group G satisfies the ascending chain condition on subgroups if and only if every subgroup of G is finitely generated.*

Proof. Assume that every subgroup of G is finitely generated. Let $H_1 \subseteq H_2 \subseteq \cdots \subseteq H_i \subseteq \cdots$ be an ascending sequence of subgroups of G. By Proposition 1.4.6, $H = \bigcup_{i \geq 1} H_i$ is a subgroup of G. Hence H is generated by finitely many elements $x_1, \ldots, x_k$ of H. Each x_j belongs to some H_{i_j}. If $n \geq \max(i_1, \ldots, i_k)$, then H_n contains $x_1, \ldots, x_k$, and $H_n = H$. Thus G satisfies the a.c.c. on subgroups.

Conversely, assume that G satisfies the a.c.c. on subgroups. Let H be a subgroup of G. Now H contains finitely generated subgroups (e.g., cyclic subgroups). By the a.c.c., H contains a maximal finitely generated subgroup M. If $M \neq H$, then for any $x \in H \backslash M$ the subgroup $M \cup \langle x \rangle$ of H is finitely generated (by x and the finitely many generators of M); this contradicts the maximality of M. Therefore $M = H$, and H is finitely generated. $\square$

Use of the a.c.c. to produce maximal elements (as in the previous proof) is sometimes called **noetherian induction**; similar use of the d.c.c. is **artinian induction**. Proposition A.1.6 shows that $\mathbb{N}$ satisfies the d.c.c. and that integer based induction, as in Corollary A.1.7, is a particular case of artinian induction.

Exercises

1. Show that the a.c.c. does not imply the d.c.c., and that the d.c.c. does not imply the a.c.c.

2. Construct a partially ordered set X which satisfies the a.c.c. and the d.c.c. and contains a finite chain with n elements for every positive integer n.

3. Show that the additive group $\mathbb{Z}$ does not satisfy the d.c.c. on subgroups.

4. Show that a partially ordered set X satisfies the a.c.c. if and only if every nonempty chain C of X has a greatest element (an element $m \in C$ such that $x \leq m$ for all $x \in C$).

B.3. THE AXIOM OF CHOICE

This section contains the Axiom of Choice and some of its most useful consequences, including Zorn's Lemma.

B.3.a. Choice Functions. A **choice function** on a set S is a mapping c which assigns to each nonempty subset T of S an element $c(T)$ of T ($=$ which chooses one element of every nonempty subset of S).

The **Axiom of Choice** states that there exists a choice function on every set. Though less intuitively "obvious" than other axioms, it has a number of useful consequences, and became one of the generally accepted axioms of set theory after Gödel proved that it is consistent with the other axioms, so that it may be assumed without generating contradictions.

This section contains several statements which are equivalent to the Axiom of Choice (assuming the other axioms). The first one is the easiest to accept as a necessary property of sets.

Proposition B.3.1. *The Axiom of Choice is equivalent to the following statement: when I is a nonempty set, and S_i is a nonempty set for each $i \in I$, then $\prod_{i \in I} S_i$ is nonempty.*

Proof. The Axiom of Choice implies that there is a choice function c on $\bigcup_{i \in I} S_i$. Then $(c(S_i))_{i \in I} \in \prod_{i \in I} S_i$.

Conversely, assume that $\prod_{i \in I} S_i$ is nonempty when I is a nonempty set and that S_i is a nonempty set for each $i \in I$. The empty mapping is a choice function on $\varnothing$. Let S be a nonempty set. Then $\prod_{T \subseteq S, T \neq \varnothing} T$ is nonempty. An element of $\prod_{i \in I} S_i$ is a mapping (usually written as a family) which assigns to each $i \in I$ an element of S_i. An element of $\prod_{T \subseteq S, T \neq \varnothing} T$ is a mapping which assigns to each $T \subseteq S$, $T \neq \varnothing$ an element of T; that's a choice function on S. $\square$

B.3.b. Zorn's Lemma. The next form of the Axiom of Choice is the most useful. A partially ordered set X is **inductive** in case every nonempty chain C of X has an upper bound in X (an element $b \in X$ such that $x \leqslant b$ for all $x \in C$).

Theorem B.3.2. *The Axiom of Choice is equivalent to the following statement, known as* **Zorn's Lemma**: *every nonempty inductive partially ordered set has a maximal element.*

We split the proof between two sections. Later in this section we show that Zorn's Lemma implies the Axiom of Choice. We prove the converse in the next section.

Zorn's Lemma provides a method of proof sometimes called **transfinite induction**, which is analogous to integer-based induction but more powerful. Consider the problem of proving that a partially ordered set $X \neq \varnothing$ has a maximal element. Using induction, we can proceed as follows. Start with any $x_1 \in X$. If x_1 is not maximal, then there exists $x_1 < x_2$. By induction, if x_n is not maximal, then there exists $x_n < x_{n+1}$. This constructs a strictly ascending sequence, which is sure to reach a maximal element only if X satisfies the ascending chain condition (equivalently, every nonempty chain of X has a greatest element). Zorn's Lemma yields a maximal element under the much weaker hypothesis that every nonempty chain of X has an upper bound. The proof in the next section reaches a maximal element by constructing a strictly ascending sequence which is indexed by ordinal numbers, so that it can be as long as necessary.

Our next result is a typical application of Zorn's Lemma; other applications are given in the exercises and throughout the text.

Proposition B.3.3. *Every equivalence relation $\mathscr{E}$ on a set X has a cross section (= a subset S of X such that every $\mathscr{E}$-class contains exactly one element of S).*

This can be derived directly from the Axiom of Choice (see the exercises).

Proof. Let $\mathscr{A}$ be the set of all subsets A of X such that every $\mathscr{E}$-class contains at most one element of A ($a\,\mathscr{E}\,b$ implies $a = b$ when $a, b \in A$). We see that $\emptyset \in \mathscr{A}$, so that $\mathscr{A} \neq \emptyset$. Now $\mathscr{A}$, partially ordered by inclusion, is inductive: if $(A_i)_{i \in I}$ is a nonempty chain in $\mathscr{A}$, then $A = \bigcup_{i \in I} A_i \in \mathscr{A}$ (and is an upper bound in $\mathscr{A}$). If $a, b \in A$, then $a, b \in A_i$ for some i and $a\,\mathscr{E}\,b$ implies $a = b$.

By Zorn's Lemma, $\mathscr{A}$ has a maximal element M, and M is a cross section of $\mathscr{E}$: indeed, if there is an $\mathscr{E}$-class C which contains no element of M, then for any $c \in C$ we have $M \cup \{c\} \in \mathscr{A}$, contradicting the maximality of M. $\square$

B.3.c. Well-Ordered Sets. A partially ordered set X is **well-ordered** in case every nonempty subset S of X has a least element (an element $s \in S$ such that $s \leqslant x$ for all $x \in S$). A well-ordered set X is totally ordered: for each $x, y \in X$, $x \leqslant y$ or $y \leqslant x$, since $\{x, y\}$ has a least element. A well-ordered set also satisfies the descending chain condition. For instance, $\mathbb{N}$ is well-ordered.

Theorem B.3.4. *The Axiom of Choice is equivalent to the following statement: every set can be well-ordered.*

Proof. A well-ordered set X has a choice function, which assigns to each nonempty subset S of X the least element of S. Hence the statement "every set can be well-ordered" implies the Axiom of Choice.

For the converse we show that Zorn's Lemma implies the statement "every set can be well-ordered." This also shows that Zorn's Lemma implies the Axiom of Choice, which is half of Theorem B.3.2. The proof of Theorem B.3.4 will be complete when we show in the next section that the Axiom of Choice implies Zorn's Lemma.

Let S be any set. Let $\mathscr{W}$ be the set of all ordered pairs $(X, \leqslant_X)$ such that $X \subseteq S$ and X is well-ordered by $\leqslant_X$. Note that $\mathscr{W} \neq \emptyset$, since $\emptyset \subseteq S$ is (trivially) well ordered. Let $(X, \leqslant_X) \leqslant (Y, \leqslant_Y)$ in $\mathscr{W}$ if and only if:

(1) $X \subseteq Y$;
(2) if $y \in Y$ and $y \leqslant_Y x \in X$, then $y \in X$; and
(3) when $x_1, x_2 \in X$, then $x_1 \leqslant_X x_2$ if and only if $x_1 \leqslant_Y x_2$.

(Thus $(X, \leqslant_X) \leqslant (Y, \leqslant_Y)$ when X is a whole lower chunk of Y with the induced order relation). It is an easy exercise that this defines an order relation on $\mathscr{W}$.

Let $(X_i, \leqslant_i)_{i \in I}$ be a nonempty chain in $\mathscr{W}$. Let $X = \bigcup_{i \in I} X_i$; let $x \leqslant_X y$ if and only if $x, y \in X_i$ and $x \leqslant_i y$ for some $i \in I$. It is clear that $\leqslant_X$ is reflexive.

If $x \leqslant_X y$ and $y \leqslant_X x$, then $x, y \in X_i$, $x \leqslant_i y$, $x, y \in X_j$, $y \leqslant_j x$ for some $i, j \in I$; since $(X_i, \leqslant_i) \leqslant (X_j, \leqslant_j)$ or $(X_j, \leqslant_j) \leqslant (X_i, \leqslant_i)$, we have $x, y \in X_k$ for some k, $x \leqslant_k y$, $y \leqslant_k x$, and $x = y$. Transitivity is proved similarly; likewise X is totally ordered by $\leqslant_X$.

Let T be a nonempty subset of X. Then $T \cap X_i \neq \emptyset$ for some i, and $T \cap X_i$ has a least element t. We show that t is the least element of T. Otherwise, $t > u$ for some $u \in T$; $u \in X_j$ for some j. If $(X_i, \leqslant_i) \leqslant (X_j, \leqslant_j)$, then $u < t \in X_i$ implies $u \in X_i$; $u \in X_i$ also holds if $(X_i, \leqslant_i) \geqslant (X_j, \leqslant_j)$. Hence $t > u$ is not the least element of $T \cap X_i$. Thus X is well ordered. Hence $(X, \leqslant_X)$ is an upper bound of $(X_i, \leqslant_i)_{i \in I}$ in $\mathcal{W}$ and we have shown that $\mathcal{W}$ is inductive.

By Zorn's Lemma there is a maximal element $(M, \leqslant_M)$ of $\mathcal{W}$. If $S \neq M$, then let $a \in S \backslash M$, $X = M \cup \{a\}$, and $x \leqslant_X y$ if and only if either $x, y \in M$ and $x \leqslant_M y$, or $y = a$. Then X is well ordered by $\leqslant_X$, $(X, \leqslant_X) \in \mathcal{W}$, and $(M, \leqslant_M) < (X, \leqslant_X)$, contradicting the maximality of $(M, \leqslant_M)$. Therefore $S = M$. $\qquad\square$

Exercises

1. Prove directly that the Axiom of Choice is equivalent to the following statement: every equivalence relation $\mathcal{E}$ has a cross section ($=$ a set S such that every $\mathcal{E}$-class contains exactly one element of S).

2. The **domain** of a binary relation R is the set $\{x; (x, y) \in R \text{ for some } y\}$. Show that the Axiom of Choice is equivalent to the following statement: every binary relation contains a mapping which has the same domain.

3. Let G be a group and $a \in G$, $a \neq 1$. Use Zorn's Lemma to prove that there is a subgroup M of G maximal such that $a \notin M$ (that is, $a \notin M$, and $M \underset{\neq}{\leqslant} H \leqslant G$ implies $a \in H$).

4. Let G be a group and A be a subgroup of G. Use Zorn's Lemma to prove that there is a subgroup M of G maximal such that $A \cap M = 1$ (that is, $A \cap M = 1$, and $M \underset{\neq}{\leqslant} H \leqslant G$ implies $A \cap H \neq 1$).

5. Use Zorn's Lemma to prove that every vector space contains a maximal linearly independent subset; show that the latter is a basis.

6. Use Zorn's Lemma to prove that every order relation is the intersection of total order relations.

B.4. ORDINAL NUMBERS

Ordinal numbers can be used instead of integers in inductive proofs and constructions; this method of proof, known as **ordinal induction**, is as powerful as Zorn's Lemma, and sometimes more convenient.

B.4.a. Definition. Recall that a partially ordered set X is well-ordered in case every nonempty subset of X has a least element; in particular, X is totally ordered. The Axiom of Choice implies that every set can be well-ordered; this will be proved at the end of this section.

Ordinal numbers are most naturally defined as isomorphy classes of well-ordered sets. Unfortunately, most of these classes are not sets and should not be collected into a class. To avoid this difficulty, ordinal numbers are defined as special well-ordered sets, so special that there is only one in each isomorphy class.

A set X is **transitive** in case $y \in Y$ and $Y \in X$ implies $y \in X$ (in particular, $\in$ is a transitive binary relation on X). For example, $\emptyset$ is transitive; so are $\{\emptyset\}$, $\{\emptyset, \{\emptyset\}\}$, $\{\emptyset, \{\emptyset\}, \{\emptyset, \{\emptyset\}\}\}$, etc.

An **ordinal number** (or just an **ordinal**) is a well-ordered transitive set in which $x < y$ if and only if $x \in y$. We denote by *Ord* the class of all ordinals.

The first ordinals are readily found. The empty set is an ordinal number. A nonempty ordinal α has a least element ζ, which must be empty since $x \in \zeta$ would imply $x < \zeta$ in α. If α has no other element, then $\alpha = \{\emptyset\}$. Otherwise, there is a least $\beta > \emptyset$ in α; necessarily $\beta = \{\emptyset\}$, since $\emptyset \in \beta$, and $x \in \beta$ implies $x < \beta$ in α and $x = \emptyset$. If α has no other elements, then $\alpha = \{\emptyset, \{\emptyset\}\}$. Similarly the next ordinals are

$$\{\emptyset, \{\emptyset\}, \{\emptyset, \{\emptyset\}\}\}, \{\emptyset, \{\emptyset\}, \{\emptyset, \{\emptyset\}\}, \{\emptyset, \{\emptyset\}, \{\emptyset, \{\emptyset\}\}\}\}, \quad \text{etc.}$$

The first ordinal numbers are generally identified with the first nonnegative integers:

$$0 = \emptyset,$$
$$1 = \{\emptyset\},$$
$$2 = \{\emptyset, \{\emptyset\}\},$$
$$3 = \{\emptyset, \{\emptyset\}, \{\emptyset, \{\emptyset\}\}\}, \quad \text{etc.}$$

We see that $1 = \{0\}$, $2 = \{0, 1\}$, $3 = \{0, 1, 2\}$, etc. This provides an alternate definition of the natural numbers (which does not require Peano's axioms).

In general, every element of an ordinal is an ordinal (see the exercises).

B.4.b. Ordering. *Ord* is partially ordered by inclusion.

Lemma B.4.1. *When α and β are ordinal numbers, then $\alpha \in \beta$ if and only if $\alpha \subsetneq \beta$. Hence Ord is totally ordered, with $\alpha < \beta$ if and only if $\alpha \in \beta$.*

Proof. Let α, β be ordinal numbers. If $\alpha \in \beta$, then $\alpha \subseteq \beta$ (since β is transitive), with $\alpha \subsetneq \beta$ since $\alpha \notin \alpha$ (otherwise, $\alpha < \alpha$ in β). Conversely, assume that $\alpha \subsetneq \beta$. Let γ be the least element of $\beta \backslash \alpha$. Then $x \in \gamma$ implies $x \in \alpha$ (otherwise, γ would not be least). If, conversely, $x \in \alpha$, then $x \neq \gamma$ (since $\gamma \notin \alpha$) and $\gamma \notin x$ (otherwise $\gamma \in \alpha$); hence $x \in \gamma$ (since β is totally ordered). Thus $\alpha = \gamma \in \beta$. Thus $\alpha \in \beta$ if and only if $\alpha \subsetneq \beta$ (if and only if $\alpha < \beta$ in *Ord*).

Now let α, β be any ordinal numbers. Then $\gamma = \alpha \cap \beta$ is transitive (since α and β are transitive) and is well ordered (as a subset of either α or β) with

$x < y$ in γ if and only if $x \in y$. Thus γ is an ordinal number. Also $\gamma \subseteq \alpha, \beta$. If $\gamma \neq \alpha, \beta$, then $\gamma \in \alpha$ and $\gamma \in \beta$ by the above, $\gamma \in \alpha \cap \beta = \gamma$, and $\gamma < \gamma$ in α, a contradiction. Therefore $\gamma = \alpha$ or $\gamma = \beta$; hence $\alpha \subseteq \beta$ or $\beta \subseteq \alpha$. $\qquad\square$

It follows from Lemma B.4.1 that $\alpha = \{\beta \in Ord; \beta < \alpha\}$ holds for every ordinal number α. Another consequence is that Ord is a well-ordered class:

Proposition B.4.2. *Every nonempty class of ordinal numbers has a least element.*

Proof. Let $\mathscr{S} \neq \varnothing$ be a class of ordinal numbers. Take any $\alpha \in \mathscr{S}$. If α is the least element of $\mathscr{S}$, then we are done. Otherwise, $\mathscr{S} \cap \alpha \neq \varnothing$, and $\mathscr{S} \cap \alpha$ has a least element γ. We show that γ is the least element of $\mathscr{S}$. Let $\sigma \in \mathscr{S}$. If $\sigma \geqslant \alpha$, then $\gamma < \alpha \leqslant \sigma$. Otherwise, $\sigma < \alpha$ by Lemma B.4.1, $\sigma \in \mathscr{S} \cap \alpha$, and $\gamma \leqslant \sigma$. Thus γ is the least element of $\mathscr{S}$. $\qquad\square$

B.4.c. Successor and Limit Ordinals. We now show that all ordinal numbers are generated by two constructions. First:

Proposition B.4.3. *If α is an ordinal number, then so is $\alpha \cup \{\alpha\}$; $\alpha \cup \{\alpha\}$ is the smallest ordinal number $\beta > \alpha$.*

Proof. $\beta = \alpha \cup \{\alpha\}$ is transitive. Order β as follows: $x < y$ in β if and only if either $x < y$ in α, or $x \in \alpha$ and $y = \alpha$. Then α is the greatest element of β. Also $x < y$ in β if and only if $x \in y$, and it is clear that β is well ordered. Finally $\alpha < \beta$, and β is the least such ordinal, since Ord is totally ordered and there is no set $\alpha \subsetneqq S \subsetneqq \beta$. $\qquad\square$

The ordinal number $\alpha \cup \{\alpha\}$ is the **successor** of α and is normally denoted by $\alpha + 1$. (The sum of any two ordinal numbers is defined in the exercises.)

The second construction is:

Proposition B.4.4. *The union of a set of ordinal numbers is an ordinal number.*

Proof. Let S be a set of ordinal numbers. Then $\tau = \bigcup_{\sigma \in S} \sigma$ is a set. By Lemma B.4.1, S is a chain, and any two elements of τ are elements of some $\sigma \in S$. Hence τ is transitive, since every $\sigma \in S$ is transitive; and τ is totally ordered with $x < y$ in τ if and only if $x \in y$ in some $\sigma \in S$, if and only if $x \in y$.

We show that τ is well ordered. Let T be a nonempty subset of τ. Then $T \cap \sigma \neq \varnothing$ for some $\sigma \in S$, and $T \cap \sigma$ has a least element γ. As in the proof of Proposition B.4.2, γ is the least element of T. Indeed let $\tau \in T$. If $\tau \geqslant \sigma$, then $\gamma < \sigma \leqslant \tau$. Otherwise, $\tau < \sigma$ by Lemma B.4.1, $\tau \in T \cap \sigma$, and $\gamma \leqslant \tau$. $\qquad\square$

Corollary B.4.5. *Ord is not a set.*

Proof. If *Ord* was a set, then $\bigcup_{\alpha \in} \alpha$ would be an ordinal number, would be the greatest ordinal number, and would be greater than his successor, which is possible for Louis XIV but not for an ordinal number. $\square$

Proposition B.4.6. *An ordinal number α is a successor if and only if it has a greatest element; then α is the successor of $\bigcup_{\gamma < \alpha} \gamma$. Otherwise, $\bigcup_{\gamma < \alpha} \gamma = \alpha$.*

Proof. A successor $\alpha = \beta + 1 = \beta \cup \{\beta\}$ has a greatest element β. Conversely, let the ordinal α have a greatest element β. Then $\beta = \bigcup_{\gamma < \alpha} \gamma = \bigcup_{\gamma \in \alpha} \gamma$. Also $\alpha > \beta$, and $\gamma > \beta$ implies $\gamma \geqslant \alpha$, since $\gamma < \alpha$ would imply $\gamma \in \alpha$ and and $\gamma \leqslant \beta$. Thus α is the least ordinal $\gamma > \beta$ and is the successor of β.

$\bigcup_{\gamma < \alpha} \gamma = \bigcup_{\gamma \subsetneqq \alpha} \gamma \leqslant \alpha$ holds for every ordinal α. If $\bigcup_{\gamma < \alpha} \gamma < \alpha$, then $\beta = \bigcup_{\gamma \in \alpha} \gamma \in \alpha$ is the greatest element of α. Therefore, if α has no greatest element, then $\bigcup_{\gamma < \alpha} \gamma = \alpha$. $\square$

An ordinal number α which is not zero or a successor is a **limit** ordinal; by Proposition B.4.6, α is the least upper bound or "limit" of all ordinals $\gamma < \alpha$.
Each ordinal begins an infinite sequence

$$\alpha = \alpha + 0 < \alpha + 1 < \alpha + 2 < \cdots < \alpha + n < \cdots$$

in which every term is the successor of the previous term ($\alpha + (n + 1) = (\alpha + n) + 1$ for all $n \geqslant 0$); $\alpha + n$ is the n-th **successor** of α. Since *Ord* is totally ordered, every ordinal is either less than or equal to α, or equal to an n-th successor of α, or greater than all successors of α. For example, 0 and its successors $1, 2, \ldots$ are the **finite** ordinals, which we saw may be identified with the nonnegative integers; the remaining ordinals are **infinite**. Then the sequence $\alpha, \alpha + 1, \ldots, \alpha + n, \ldots$ has a union β which is a limit ordinal since $\beta = \bigcup_{\gamma < \beta} \gamma$. In turn β is followed by its successors. This implies that *Ord* is a disjoint union of ascending sequences:

Proposition B.4.7. *Every ordinal number α can be written in the form $\alpha = \lambda + n$, where n is a nonnegative integer and λ is either 0 or a limit ordinal; furthermore n and λ are unique.*

Proof. Since $\alpha = \alpha + 0$, $\{\gamma \in Ord ; \alpha = \gamma + i$ for some $i \geqslant 0\}$ is not empty, and has a least element λ. Then $\alpha = \lambda + n$ for some $n \geqslant 0$, and λ is not a successor, since $\lambda = \gamma + 1$ would imply $\alpha = \gamma + (n + 1)$ with $\gamma < \lambda$, contradicting the choice of λ.

Assume that $\lambda + m = \mu + n$, where $m, n \geqslant 0$ and λ, μ are 0 or limit ordinals. We may assume that $\lambda \leqslant \mu$. Then $\lambda \leqslant \mu \leqslant \mu + n = \lambda + m$. Let k be the smallest nonnegative integer such that $\mu \leqslant \lambda + k$. If $k > 0$, then $\lambda + (k - 1) < \mu \leqslant \lambda + k$; since $\lambda + k$ is the successor of $\lambda + (k - 1)$, this implies that $\mu = \lambda + k$ is a successor, a contradiction. Therefore $k = 0$, so that $\lambda \leqslant \mu \leqslant \lambda$ and $\mu = \lambda$. Then $m = n$, since $m < n$ implies $\lambda + m < \lambda + n$. $\square$

Proposition B.4.7 shows that *Ord* is made of ascending sequences

$$0 < 1 < 2 < \cdots < \omega < \omega + 1 < \omega + 2 < \cdots < \omega + \omega < \omega + \omega + 1 < \cdots$$

which begin with zero and with limit ordinals $\omega, 2\omega, \ldots$. Next, zero and the limit ordinals constitute a well-ordered class and are (by Proposition B.4.8 below) similarly arranged into ascending sequences

$$0 < \omega < \omega + \omega < \cdots < \omega^2 < \omega^2 + \omega < \cdots < \omega^2 + \omega^2 < \omega^2 + \omega^2 + \omega < \cdots$$

whose starting points are similarly arranged, and so forth. (Sums and products of ordinal numbers are defined in the exercises.)

B.4.d. Well-Ordered Sets. We now prove that there is one ordinal number in every isomorphy class of well-ordered sets. (An **isomorphism** of a partially ordered set X onto a partially ordered set Y is a bijection $\theta : X \longrightarrow Y$ such that $x \leqslant y$ holds in X if and only if $\theta(x) \leqslant \theta(y)$ holds in Y.)

Proposition B.4.8. *Every well-ordered set is isomorphic to a unique ordinal number.*

For example, $\mathbb{N}$ is isomorphic to the first (smallest) limit ordinal ω.

To prove Proposition B.4.8, we define a **lower section** of a partially ordered set X as a subset S of X such that $x \leqslant s \in S$ implies $x \in S$. Lower sections are also called **order ideals**, and various other names.

Lemma B.4.9. *A subset S of a well ordered set X is a lower section of X if and only if $S = X$ or*

$$S = X(a) = \{ x \in X \,;\, x < a \}$$

for some $a \in X$.

Proof. It is clear that X and each $X(a)$ are lower sections. Conversely, let $S \neq X$ be a lower section. The nonempty subset $X \backslash S$ of X has a least element a. If $x < a$, then $x \in S$ (otherwise, a would not be the least element of $X \backslash S$). If $x \in S$, then $x < a$, since $a \leqslant x \in S$ would imply $a \in S$. Thus $S = X(a)$. $\square$

Lemma B.4.10. *Let S and T be lower sections of a well-ordered set X. If $S \cong T$, then $S = T$.*

Proof. Assume that $S \neq T$ and that there is an isomorphism $\theta : S \longrightarrow T$. Since $S \neq T$, we cannot have $\theta(x) = x$ for all $x \in S$. Hence the set $\{ x \in S \,;\, \theta(x) \neq x \}$ is nonempty and has a least element a. By definition of a, $\theta(x) = x$ for all $x \in X(a)$; hence $\theta(X(a)) = X(a)$ and $\theta(S \backslash X(a)) = T \backslash X(a)$. Now a is the least element of $S\backslash X(a)$. Since θ is an isomorphism, $\theta(a)$ is the least element of $\theta(S\backslash X(a)) = T\backslash X(a)$; that is, $\theta(a) = a$. This is the required contradiction. $\qquad\square$

We now prove Proposition B.4.8. Uniqueness follows from Lemma B.4.10: if, say, $\alpha < \beta$ in Ord, then $\alpha = \{\gamma \in \beta; \gamma < \alpha\}$ is a lower section of β, and $\alpha \ncong \beta$.

Now let X be a well-ordered set. Let φ consist of all ordered pairs (a, α) such that $a \in X$, α is an ordinal number, and $X(a) \cong \alpha$.

If (a, α), $(b, \beta) \in \varphi$, and $a = b$, then $\alpha \cong \beta$ and $\alpha = \beta$: hence φ is a mapping. Similarly, if (a, α), $(b, \beta) \in \varphi$, and $\alpha = \beta$, then $X(a) \cong X(b)$ and $a = b$ by Lemma B.4.10; hence φ is injective.

Assume that (a, α), $(b, \beta) \in \varphi$ and that $a < b$. Let $\theta : X(b) \longrightarrow \beta$ be an isomorphism. Then $\theta(a) < \beta$ and $\theta_{|X(a)}$ is an isomorphism of $X(a)$ onto $\{\gamma \in \beta; \gamma < \theta(a)\} = \theta(a)$. Therefore $(a, \theta(a)) \in \varphi$ and $\alpha = \theta(a) < \beta$. Thus φ is order preserving. This argument also shows that the domain $\text{dom}\,\varphi$ of φ is a lower section of X.

Similarly assume that (a, α), $(b, \beta) \in \varphi$ and that $\alpha < \beta$. Let $\zeta : \beta \longrightarrow X(b)$ be an isomorphism. Then $\zeta(\alpha) < b$ and $\zeta_{|\alpha}$ is an isomorphism of α onto $X(\zeta(\alpha))$. Therefore $(\zeta(\alpha), \alpha) \in \varphi$ and $a = \zeta(\alpha) < b$, since φ is injective. Thus φ is an isomorphism (of its domain onto its range). This argument also shows that the range $\text{ran}\,\varphi$ of φ is a lower section of Ord (if β is in $\text{ran}\,\varphi$, so is every $\alpha \leqslant \beta$).

Since Ord is not a set, $\text{ran}\,\varphi$ cannot be all of Ord, and there is a least ordinal $\gamma \notin \text{ran}\,\varphi$. Then $\text{ran}\,\varphi = \{\alpha \in Ord; \alpha < \gamma\} = \gamma$. Then $\text{dom}\,\varphi$ is all of X: otherwise $\text{dom}\,\varphi = X(c)$ for some $c \in X$, φ is an isomorphism of $X(c)$ onto γ, $(c, \gamma) \in \varphi$, and $c \in \text{dom}\,\varphi$, a contradiction. Thus $X \cong \gamma$. $\qquad\square$

A **transfinite sequence** is a family $(x_\alpha)_{\alpha \in Ord}$ or $(x_\alpha)_{\alpha < \sigma}$ indexed by Ord or indexed by an ordinal number σ. Proposition B.4.8 can be restated as follows:

Corollary B.4.11. *Every well ordered set X can be written as a transfinite sequence $X = (x_\alpha)_{\alpha < \sigma}$ indexed by an ordinal number σ so that $x_\alpha \leqslant x_\beta$ if and only if $\alpha \leqslant \beta$.*

The restriction $\alpha < \sigma$ is necessary in Corollary B.4.11:

Lemma B.4.12. *It is not possible to write a set X as a transfinite sequence $X = (x_\alpha)_{\alpha \in Ord}$ indexed by all ordinals so that $x_\alpha \neq x_\beta$ whenever $\alpha \neq \beta$.*

Proof. If this is possible then X is a well-ordered set, with $x_\alpha \leqslant x_\beta$ if and only if $\alpha \leqslant \beta$. (This makes X isomorphic to Ord!) By Corollary B.4.11, X can be written as a family $X = (y_\alpha)_{\alpha < \sigma}$ indexed by all ordinal numbers that are less than some ordinal number σ, so that $y_\alpha \leqslant y_\beta$ if and only if $\alpha \leqslant \beta$. Then Ord is isomorphic to σ, which will contradict Lemma B.4.10. We have $x_\sigma = y_\tau$ for some $\tau < \sigma$. For each $\alpha < \sigma$ we have $x_\alpha < y_\tau$, and there is a unique $\gamma < \tau$ such that $x_\alpha = y_\gamma$. If moreover $x_\beta = y_\delta$, then $\alpha \leqslant \beta$ if and only if $\gamma \leqslant \delta$. Thus σ and τ are isomorphic; since $\tau < \sigma$, this contradicts Lemma B.4.10.   □

In the next section we give a more natural proof of Lemma B.4.12, based on cardinality arguments. Lemma B.4.12 also follows from Corollary B.4.5.

B.4.e. Ordinal Induction. Integer based induction is based on the following property of natural numbers: if S is a subset of $\mathbb{N}$ such that $1 \in S$ and that $n \in S$ implies $n + 1 \in S$, then $S = \mathbb{N}$.

Ordinal numbers have a similar property.

Proposition B.4.13. *Let S be a class of ordinal numbers such that:*

(1) $0 \in S$;
(2) $\alpha \in S$ *implies* $\alpha + 1 \in S$;
(3) *if α is a limit ordinal and $\beta \in S$ for all $\beta < \alpha$, then $\alpha \in S$.*

Then $S = Ord$.

Proof. If S does not contain all ordinal numbers, then by Proposition B.4.2 there is a least ordinal number $\alpha \notin S$. Then $\beta \in S$ for all $\beta < \alpha$. If $\alpha = 0$, then $\alpha \in S$ by (1). If $\alpha = \beta + 1$ is a successor, then $\beta \in S$ and $\alpha \in S$ by (2). Otherwise, α is a limit ordinal, and $\alpha \in S$ by (3). This is the required contradiction.   □

Proposition B.4.13 can be stated more simply. Let S be class of ordinal numbers with the following property: if α is an ordinal number and $\beta \in S$ for all $\beta < \alpha$, then $\alpha \in S$. Then $S = Ord$.

The following result is proved similarly and left as an exercise:

Proposition B.4.14. *Let σ be an ordinal number and S be a class of ordinal numbers such that:*

(1) $0 \in S$;
(2) *if $\alpha \in S$ and $\alpha + 1 < \sigma$, then $\alpha + 1 \in S$;*
(3) *if $\alpha < \sigma$ is a limit ordinal and $\beta \in S$ for every $\beta < \alpha$, then $\alpha \in S$.*

Then S contains every ordinal number $\alpha < \sigma$.

Like Proposition B.4.13, Proposition B.4.14 can be stated more simply. Let σ be an ordinal number and S be class of ordinal numbers with the following property: if $\alpha < \sigma$ is an ordinal number and $\beta \in S$ for all $\beta < \alpha$, then $\alpha \in S$. Then S contains every ordinal $\alpha < \sigma$.

Ordinal induction is a method of proof based on Proposition B.4.13 or B.4.14. It resembles integer-based induction except for case (3). As a first example we show that transfinite sequences can be constructed by a process which is similar to integer-based recursion in that each x_α is constructed from $\{x_\gamma ; \gamma < \alpha\}$.

Proposition B.4.15 (Recursion). *Let S be a set and $f : 2^S \longrightarrow S$ be a mapping. There exists a unique transfinite sequence $(x_\alpha)_{\alpha \in Ord}$, indexed by all ordinals such that $x_\alpha = f(\{x_\gamma ; \gamma < \alpha\})$ for every ordinal number α.*

Informally we say that x_α is "defined by induction" by the formula $x_\alpha = f(\{x_\gamma ; \gamma < \alpha\})$.

Proof. First we show by induction on σ that there is at most one sequence $(x_\alpha)_{\alpha < \sigma}$, indexed by σ, such that

$$x_\alpha = f(\{x_\gamma ; \gamma < \alpha\}) \tag{$*$}$$

holds for every ordinal $\alpha < \sigma$. Assume that uniqueness has been proved for every $\tau < \sigma$. Let $(x'_\alpha)_{\alpha < \sigma}$ and $(x''_\alpha)_{\alpha < \sigma}$ satisfy $(*)$ for every $\alpha < \sigma$. If $\sigma = 0$, then $x'_\alpha = x''_\alpha$ holds vacuously for every $\alpha < \sigma$. If $\sigma = \tau + 1$ is a successor ordinal, then $x'_\alpha = x''_\alpha$ for all $\alpha < \tau$, and $(*)$ implies $x'_\tau = f(\{x'_\gamma ; \gamma < \tau\}) = f(\{x''_\gamma ; \gamma < \tau\}) = x''_\tau$. If finally σ is a limit ordinal, then $x'_\alpha = x''_\alpha$ for all $\alpha < \tau < \sigma$; since $\sigma = \bigcup_{\tau < \sigma} \tau$, this implies $x'_\alpha = x''_\alpha$ for all $\alpha < \sigma$.

Next we show by induction on σ that there exists a sequence $(x_\alpha)_{\alpha < \sigma}$, indexed by σ, such that $(*)$ holds for every ordinal $\alpha < \sigma$. This is vacuously true if $\sigma = 0$. Assume that $\sigma > 0$ and that existence has been proved for every $\tau < \sigma$. If $\sigma = \tau + 1$ is a successor ordinal, then there is a sequence $(x_\alpha)_{\alpha < \tau}$, indexed by τ, which satisfies $(*)$ for every $\alpha < \tau$; define $x_\tau = f(\{x_\gamma ; \gamma < \tau\})$; then $(x_\alpha)_{\alpha \leqslant \tau} = (x_\alpha)_{\alpha < \sigma}$ satisfies $(*)$ for every $\alpha < \sigma$. Finally let σ be a limit ordinal. For each $\tau < \sigma$ there is a sequence $(x^\tau_\alpha)_{\alpha < \tau}$, indexed by τ, which satisfies $(*)$ for every $\alpha < \tau$; by the first part of the proof, $x^\tau_\alpha = x^\upsilon_\alpha$ whenever $\alpha < \tau, \upsilon$; since $\sigma = \bigcup_{\tau < \sigma} \tau$, we can define x_α for all $\alpha < \sigma$ by $x_\alpha = x^\tau_\alpha$ whenever $\alpha < \tau < \sigma$; then $(x_\alpha)_{\alpha < \sigma}$ satisfies $(*)$ for every $\alpha < \sigma$.

We now have for each ordinal σ a sequence $(x^\sigma_\alpha)_{\alpha < \sigma}$, indexed by σ, which satisfies $(*)$ for every $\alpha < \sigma$. By the first part of the proof, $x^\sigma_\alpha = x^\tau_\alpha$ whenever $\alpha < \sigma, \tau$. Hence we can define x_α for all $\alpha \in Ord$ by $x_\alpha = x^\sigma_\alpha$ whenever $\alpha < \sigma$; then $(x_\alpha)_{\alpha \in Ord}$ satisfies $(*)$ for every α. This sequence is unique by the first part of the proof. $\square$

A more general version of Proposition B.4.15 is given in the exercises.

Ordinal induction is often used in proofs by contradiction. One method of proof uses recursion to construct a transfinite sequence (x_α) indexed by all ordinals, with $x_\alpha \neq x_\beta$ whenever $\alpha \neq \beta$ to contradict Lemma B.4.12.

B.4.f. Zorn's Lemma. As another example of ordinal induction we prove that the Axiom of Choice implies Zorn's Lemma. This completes the proof of Theorem B.3.2 (and the proof of Theorem B.3.4). Other examples will be found in the exercises.

Let X be a nonempty inductive partially ordered set. Assume that there is a choice function c on X, but that X has no maximal element. We use Proposition B.4.15 to construct a transfinite sequence $(x_\alpha)_{\alpha \in Ord}$ of elements of X, indexed by all ordinals, such that $\alpha < \beta$ implies $x_\alpha < x_\beta$; this contradicts Lemma B.4.12.

Since $X \neq \emptyset$, there is some $a \in X$. For each $S \subseteq X$ let

$$g(S) = \{ u \in X ; u > s \text{ for all } s \in S \}$$

and let $f(S) = c(g(S))$ if $g(S) \neq \emptyset$, otherwise let $f(S) = a$. By Proposition B.4.15, there is a transfinite sequence $(x_\alpha)_{\alpha \in Ord}$ such that

$$x_\alpha = f(\{x_\gamma ; \gamma < \alpha\})$$

for every ordinal α. We prove by induction on α that $x_\gamma < x_\alpha$ for all $\gamma < \alpha$. This is vacuously true if $\alpha = 0$. Assume that $\alpha > 0$ and that $x_\gamma < x_\beta$ whenever $\gamma < \beta < \alpha$. Let

$$S_\alpha = \{ x_\gamma ; \gamma < \alpha \},$$

so that $x_\alpha = f(S_\alpha)$. If $\alpha = \beta + 1$ is a successor, then x_β is not maximal, and there exists $u > x_\beta$; then $u > x_\gamma$ for all $\gamma < \alpha$ and $g(S_\alpha) \neq \emptyset$. If α is a limit ordinal, then the nonempty chain $(x_\gamma)_{\gamma < \alpha}$ has an upper bound u in X; since $\gamma < \alpha$ now implies $\gamma + 1 < \alpha$, we have $x_\gamma < x_{\gamma+1} \leqslant u$ for all $\gamma < \alpha$ and again $g(S_\alpha) \neq \emptyset$. In either case $x_\alpha = c(g(S_\alpha)) \in g(S_\alpha)$ and $x_\alpha > x_\gamma$ for all $\gamma < \alpha$. We now have $x_\alpha < x_\beta$ whenever $\alpha < \beta$, contradicting Lemma B.4.12.

For greater clarity this proof may be written more informally, as follows. Assume that X has no maximal element. Use ordinal induction to construct x_α for every ordinal α so that $x_\alpha > x_\beta$ whenever $\alpha > \beta$. Assume that x_γ has been constructed for every $\gamma < \alpha$, so that $x_\gamma < x_\beta$ whenever $\gamma < \beta < \alpha$. If $\alpha = 0$, this is vacuous, but since $X \neq \emptyset$, we can choose any $x_0 \in X$. If $\alpha = \beta + 1$ is a successor ordinal, then x_β is not maximal, and we can choose any $x_\alpha > x_\beta$. If α is a limit ordinal, then the nonempty chain $(x_\gamma)_{\gamma < \alpha}$ has an upper bound in X, and we can choose one such as x_α; then $x_\gamma < x_{\gamma+1} \leqslant x_\alpha$ for all $\gamma < \alpha$. As before we now have $x_\alpha < x_\beta$ whenever $\alpha < \beta$, contradicting Lemma B.4.12. It is understood that x_α is actually constructed by ordinal recursion, with a choice function providing each required choice.

Exercises

1. Prove the following: when the well-ordered sets X and Y are isomorphic, there is only one isomorphism of X onto Y.

2. Show that every element of an ordinal number α is an ordinal number.

3. Ordinal addition is defined as follows. Let X and Y be disjoint well-ordered sets. Let $Z = X \cup Y$, with $x \leqslant y$ in Z if and only if either $x \leqslant y$ in X, or $x \leqslant y$ in Y, or

$x \in X$ and $y \in Y$. Show that Z is well ordered. Show that $X \cong X'$, $Y \cong Y'$ implies $Z \cong Z'$. (By definition, the sum $\alpha + \beta$ of two ordinal numbers is isomorphic to Z when $\alpha \cong X$ and $\beta \cong Y$.)

4. Show that $\omega + 1 \neq 1 + \omega$.

5. Ordinal multiplication is defined as follows. Let X and Y be well-ordered sets. Let $Z = X \times Y$, with $(x_1, y_1) \leqslant (x_2, y_2)$ if and only if either $y_1 < y_2$, or $y_1 = y_2$ and $x_1 \leqslant x_2$. Show that Z is well ordered. Show that $X \cong X'$, $Y \cong Y'$ implies $Z \cong Z'$. (By definition, the product $\alpha\beta$ of two ordinal numbers is isomorphic to Z when $\alpha \cong X$ and $\beta \cong Y$.)

6. Show that $2\omega \neq \omega 2$.

7. Let σ be an ordinal number and S be a class of ordinal numbers such that:

   (1) $0 \in S$;
   (2) if $\alpha \in S$ and $\alpha + 1 < \sigma$, then $\alpha + 1 \in S$;
   (3) if $\alpha < \sigma$ is a limit ordinal and $\beta \in S$ for every $\beta < \alpha$, then $\alpha \in S$.

   Prove that S contains every ordinal number $\alpha < \sigma$.

8. Let S be a set and f be a mapping of $D \subseteq 2^S$ into S. Prove that there exists a unique transfinite sequence (x_α), indexed by all ordinals or by some ordinal, such that $x_\alpha = f(\{x_\gamma ; \gamma < \alpha\})$ for every ordinal number α for which $f(\{x_\gamma ; \gamma < \alpha\})$ is defined (=for which x_γ is defined for every $\gamma < \alpha$ and $\{x_\gamma ; \gamma < \alpha\} \in D$).

9. Let G be a group and $a \in G$, $a \neq 1$. Use ordinal induction to prove that there is a subgroup M of G maximal such that $a \notin M$.

10. Let G be a group and A be a subgroup of G. Use ordinal induction to prove that there is a subgroup M of G maximal such that $A \cap M = 1$.

11. Use ordinal induction to prove that every vector space contains a maximal linearly independent subset.

B.5. CARDINAL NUMBERS

Cardinal numbers are used to assign a number of elements to any set. This section also contains some notable properties.

B.5.a. Number of Elements. The number of elements of an arbitrary set is defined below. First we define what it means for one set to have as many elements as another, or fewer elements than another; this does not require actual counting.

By definition, a set X **has as many elements as** a set Y in case there exists a bijection of X onto Y; one also says that X and Y are **equipotent**.

For the next definition we note:

Lemma B.5.1. *For sets $X \neq \emptyset$ and Y, the following are equivalent:*

   (1) *There exists an injection $X \longrightarrow Y$.*
   (2) *There exists a surjection $Y \longrightarrow X$.*

Proof. Let $\iota : X \longrightarrow Y$ be an injection. Let $a \in X$. A mapping $\sigma : Y \longrightarrow X$ is defined by

$$\sigma(y) = \begin{cases} x & \text{if} \quad y = \iota(x), \\ a & \text{if} \quad y \notin \iota[X]. \end{cases}$$

Clearly $\sigma : Y \longrightarrow X$ is surjective.

The converse uses the Axiom of Choice. Let c be a choice function on Y. If $\sigma : Y \longrightarrow X$ is surjective, then $\sigma^{-1}(x) = \{ y \in Y ; \sigma(y) = x \} \neq \emptyset$ for all $x \in X$. Let $\iota(x) = c(\sigma^{-1}(x))$. Then $\sigma(\iota(x)) = x$. Hence $\iota : X \longrightarrow Y$ is injective. $\square$

A set X **has at most as many elements as** a set Y in case there exists an injection $X \longrightarrow Y$; equivalently, in case $X = \emptyset$ or there exists a surjection $Y \longrightarrow X$. For instance, a subset of X has at most as many elements as X.

Theorem B.5.2 (Cantor-Bernstein). *If there exist injections $X \longrightarrow Y$ and $Y \longrightarrow X$, then there exists a bijection $X \longrightarrow Y$.*

In other words, if X has at most as many elements as Y, and Y has at most as many elements as X, then X has as many elements as Y.

Proof. This proof is due to Halmos. We may assume that X and Y are disjoint. Let $\varphi : X \longrightarrow Y$ and $\psi : Y \longrightarrow X$ be injections.

The idea of the proof is to arrange $X \cup Y$ into families in which every element of either set begat (all by itself) one child in the other set. The *child* of $x \in X$ is $\varphi(x) \in Y$, and x is the *parent* of $\varphi(x)$; similarly the child of $y \in Y$ is $\psi(y) \in X$, and y is the parent of $\psi(y)$. Since φ and ψ are injective, an element of $X \cup Y$ has at most one parent; the elements of $\varphi(X)$ have a parent, and the elements of $Y \backslash \varphi(X)$ are *orphans*; similarly the elements of $\psi(Y)$ have a parent, and the elements of $X \backslash \psi(Y)$ are orphans. The *descendents* of $x \in X$ are x, $\varphi(x)$, $\psi(\varphi(x))$, $\varphi(\psi(\varphi(x)))$, etc.; the descendents of $y \in Y$ are y, $\psi(y)$, $\varphi(\psi(y))$, $\psi(\varphi(\psi(y)))$, etc. A descendent of a descendent of any $z \in X \cup Y$ is a descendent of z.

We classify the elements of $X \cup Y$ by tracing their ancestry as far back as possible. The ancestry of an element either extends indefinitely or ends (or rather, begins) with an orphan. Hence X can be partitioned into: the subset X_X of all descendents of elements of $X \backslash \psi(Y)$; the subset X_Y of all descendents of elements of $Y \backslash \varphi(X)$; and the subset X_∞ of all elements of X whose ancestry extends indefinitely. (Some of these sets may be empty.) Likewise Y is partitioned into: the subset Y_X of all descendents of elements of $X \backslash \psi(Y)$; the subset Y_Y of all descendents of elements of $Y \backslash \varphi(X)$; and the subset Y_∞ of all elements of Y whose ancestry extends indefinitely (is, like the author's, lost in the mists of time).

We see that $\varphi(X_X) = Y_X$, $\psi(Y_Y) = X_Y$, and $\varphi(X_\infty) = Y_\infty$ (also $\psi(Y_\infty) = X_\infty$). Thus there are bijections $X_X \longrightarrow Y_X$, $X_Y \longrightarrow Y_Y$, $X_\infty \longrightarrow Y_\infty$, which can be combined into a bijection $X \longrightarrow Y$. □

A set X **has fewer elements than** a set Y, and Y **has more elements than** X, in case X has at most as many, but not as many, elements as Y. For instance, the empty set has fewer elements than a nonempty set. For another example let $I_n = \{1, 2, \ldots, n\}$ be the set of all n first natural numbers. If $n > m$ then I_n has (by our definition) more elements than I_m:

Lemma B.5.3 (= Lemma A.1.8). *When $n > m \geqslant 0$ there is no injection $I_n \longrightarrow I_m$.*

Proof. We proceed by induction on n. If $n = 1$, then $m = 0$, and there is no injection $I_1 \longrightarrow I_0 = \emptyset$. Assume that $n > 1$, and let $\varphi : I_n \longrightarrow I_m$ be any mapping. If $\varphi(n) = \varphi(i)$ for some $i < n$, then φ is not injective. Now assume that $\varphi(n) \neq \varphi(1), \varphi(2), \ldots, \varphi(n-1)$. Let σ be a permutation of I_m such that $\sigma(\varphi(n)) = m$, and $\psi = \sigma \circ \varphi$. Then $\psi(1), \ldots, \psi(n-1) \neq \psi(n) = m$, and $\chi = \psi_{|I_{n-1}}$ is a mapping from I_{n-1} to I_{m-1}. By the induction hypothesis, χ is not injective; hence ψ is not injective, and neither is φ. □

B.5.b. Cardinal Numbers. Cardinal numbers are defined most naturally as equipotency classes of sets. Unfortunately, as with ordinal numbers, most of these classes are not sets and should not be collected into a class. To avoid this difficulty, cardinal numbers are defined as sets, chosen so that there is only one cardinal number in each equipotency class; in fact, as special ordinal numbers.

By definition, a **cardinal number** or **cardinal** is an ordinal number κ such that every ordinal number $\alpha < \kappa$ has fewer elements than κ. We denote by *Card* the class of all cardinal numbers. *Card* is (like *Ord*) ordered by inclusion and is (like *Ord*) a well-ordered class.

For example, every finite ordinal number is a cardinal number. Recall that the finite ordinals are $0 = \emptyset$ and its successors $1, 2, \ldots$, and that a finite ordinal number n is the set $\{0, 1, \ldots, n-1\}$ and may be identified with the nonnegative integer n; then Lemma B.5.3 shows that every $m < n$ has fewer elements than n.

The finite ordinal numbers are the **finite** cardinal numbers; every remaining cardinal number is **infinite**. For instance the first limit ordinal ω ($\cong \mathbb{N}$) is a cardinal number: indeed every ordinal $n < \omega$ is finite and has fewer elements than ω, for there is an injection $n \longrightarrow \omega$, but there is no injection $\omega \longrightarrow n$ since this would induce an injection $n + 1 \longrightarrow n$.

The reader will show that all infinite cardinals are limit ordinals, and can be arranged into a transfinite sequence, traditionally denoted by

$$\aleph_0 < \aleph_1 < \cdots < \aleph_\alpha < \aleph_{\alpha+1} < \cdots,$$

indexed by all ordinals. For instance $\aleph_0 = \omega$.

B.5.c. Cardinality. We now show that every equipotency class contains exactly one cardinal number. Hence cardinal numbers can be used to count the elements of any set.

Proposition B.5.4. *For every set X there is a unique cardinal number $|X|$ such that there exists a bijection $X \longrightarrow |X|$.*

Proof. This follows easily from the Axiom of Choice and the definitions. By Theorem B.3.4, every set X can be well ordered; by Proposition B.4.8, there is a bijection of X to some ordinal number α. The least such ordinal κ is a cardinal number. It is the only cardinal number κ with a bijection $X \longrightarrow \kappa$: if, say, $\kappa < \lambda$, then there is no bijection $\kappa \longrightarrow \lambda$, since κ has fewer elements than λ. □

The cardinal number $|X|$ in Proposition B.1.3 is the [cardinal] **number of elements** of X, or **cardinality** of X. We see that a set X has as many elements as a set Y (as defined above) if and only if $|X| = |Y|$; X has at most as many elements as Y if and only if $|X| \leqslant |Y|$; X has fewer elements than Y if and only if $|X| < |Y|$. In particular, an ordinal number κ is a cardinal number if and only if $|\alpha| < |\kappa|$ for all $\alpha < \kappa$.

Proposition B.5.5 (Cantor). $|2^X| > |X|$ *for every set X.*

Proof. First $|2^X| \geqslant |X|$, since there is an injection $x \longmapsto \{x\}$ of X into 2^X. To prove $|2^X| \neq |X|$, we show that there is no surjection of X onto 2^X. Let $f : X \longrightarrow 2^X$ be any mapping. Then $S = \{x \in X \,;\, x \notin f(x)\} \in 2^X$. If $x \in X$, then either $x \in S$ and $x \notin f(x)$, or $x \notin S$ and $x \in f(x)$; in either case $f(x) \neq S$. Thus $S \notin f(X)$ and f is not surjective. □

B.5.d. Countable Sets. A set X is **finite** in case $|X|$ is a finite cardinal; equivalently, in case there is a bijection $X \longrightarrow \{1,2,\ldots,n\}$ for some non-negative integer n (then $|X| = n$). A set X is **infinite** in case $|X|$ is an infinite cardinal; equivalently, in case X is not finite.

A set X is **countable** in case $|X| \leqslant \aleph_0$. (Countable sets are often defined by the equality $|X| = \aleph_0$; then a set X such that $|X| \leqslant \aleph_0$ is called finite or countable.) Since $\aleph_0 = |\mathbb{N}|$, a set X is countable if and only if there is a surjection $\mathbb{N} \longrightarrow X$, if and only if the elements of X can be arranged into a finite or infinite sequence $x_1, x_2, \ldots, x_n, \ldots$ (indexed by the natural numbers).

For instance, $\mathbb{N}$ is countable. The following properties yield more examples:

Proposition B.5.6. *A direct product of finitely many countable sets is countable.*

Proof. The elements of $\mathbb{N} \times \mathbb{N}$ can be arranged by increasing sums: $(1,1)$; $(1,2),(2,1)$; $(1,3),(2,2),(3,1)$;.... This arranges the elements of $\mathbb{N} \times \mathbb{N}$ into an infinite sequence and shows that $\mathbb{N} \times \mathbb{N}$ is countable. Hence a direct product $X \times Y$ of two countable sets is countable, since there is an injection $X \times Y \longrightarrow \mathbb{N} \times \mathbb{N}$. By induction on n, a direct product of n countable sets is count-able, for all $n \in \mathbb{N}$.   □

Proposition B.5.7. *A union of countably many countable sets is countable.*

Proof. A countable family of sets can be arranged into a finite or infinite sequence $A_1, A_2, \ldots, A_n, \ldots$. Let $B = A_1 \cup \cdots \cup A_n \cup \cdots$ be a union of countably many countable sets. Then B is a disjoint union of countably many countable sets $A_1' = A_1, A_2' = A_2 \setminus A_1, \ldots, A_n' = A_n \setminus (A_1 \cup \cdots \cup A_{n-1}), \ldots$. An injection of B into $\mathbb{N} \times \mathbb{N}$ can be constructed by combining injections $A_n' \longrightarrow \mathbb{N} \longrightarrow \mathbb{N} \times \{n\}$. Therefore B is countable, by Proposition B.5.6.   □

Proposition B.5.7 implies that $\mathbb{Z} = \{0\} \cup \mathbb{N} \cup -\mathbb{N}$ is countable, and that $\mathbb{Q}$ is countable, since $\mathbb{Q} = \bigcup_{n \in \mathbb{N}} \{a/n \in \mathbb{Q} ; a \in \mathbb{Z}\}$ is a union of countably many countable sets.

On the other hand:

Proposition B.5.8 (Cantor). $\mathbb{R}$ *is not countable.*

Proof. Let X be the set of all real numbers with a decimal expansion $0.d_1 d_2 \ldots d_n \ldots$ in which every digit $d_0, d_1, \ldots, d_n, \ldots$ is either 0 or 1. There is one such sequence for each subset S of $\mathbb{N}$, in which $d_n = 1$ if and only if $n \in S$. Hence $|\mathbb{R}| \geqslant |X| \geqslant |2^{\mathbb{N}}| > |\mathbb{N}| = \aleph_0$ by Proposition B.5.5.   □

The cardinal number $|\mathbb{R}|$ is often denoted by $\mathfrak{c}$ (for "continuum"). By Propo-sition B.5.5, $\mathfrak{c} > \aleph_0$, equivalently $\mathfrak{c} \geqslant \aleph_1$. The **Continuum Hypothesis** states that there is no cardinal number $\aleph_0 < \kappa < \mathfrak{c}$; equivalently, that $\mathfrak{c} = \aleph_1$. Gödel showed that this statement is unprovable: that is, one can add either the Contin-uum Hypothesis or its negation to the axioms of set theory without generating a contradiction.

A real number x is **algebraic** [over $\mathbb{Q}$] in case it satisfies a polynomial equa-tion $a_n x^n + \cdots + a_1 x + a_0 = 0$ with integer coefficients, and **transcendental** in case it is not algebraic. It can be shown, with some difficulty, that e and π are transcendental. It is much easier to prove, using the results in this section, the existence of transcendental numbers, even though the proof (which goes back to Cantor) does not actually construct any specific example. Indeed the set of all polynomials of degree n with integer coefficients is countable by Proposition B.5.6, since such a polynomial $a_n x^n + \cdots + a_1 x + a_0$ is determined by its sequence of coefficients $(a_n, \ldots, a_1, a_0) \in \mathbb{Z}^{n+1}$. Since a polynomial of degree n has finitely many real roots, the set A_n of all real roots of polyno-mials of degree n is the union of countably many finite sets and is countable by Proposition B.5.7. Then the set A of all algebraic real numbers, which is the union of $A_1, A_2, \ldots, A_n, \ldots$, is countable by Proposition B.5.7. Since $\mathbb{R}$ is not countable, there must be real numbers which are not algebraic. (In fact Corollary B.5.10 below implies that $|\mathbb{R} \setminus A| = |\mathbb{R}|$.)

B.5.e. Operations. Cardinal **addition** is defined by

$$\kappa + \lambda = |X \cup Y| \qquad \text{whenever} \quad |X| = \kappa, \quad |Y| = \lambda, \quad X \cap Y = \varnothing.$$

Addition is well-defined: if $|X'| = \kappa$, $|Y'| = \lambda$, and $X' \cap Y' = \varnothing$, then there are bijections $X \longrightarrow X'$ and $Y \longrightarrow Y'$ which can be combined into a bijection $X \cup Y \longrightarrow X' \cup Y'$, and $|X \cup Y| = |X' \cup Y'|$. Similarly cardinal **multiplication** and **exponentiation** are well-defined by

$$\kappa\lambda = |X \times Y| \qquad \text{and} \qquad \kappa^\lambda = |X^Y| \quad \text{whenever} \quad |X| = \kappa, \quad |Y| = \lambda,$$

where X^Y is the set of all mappings of Y into X. (When $Y = 2 = \{0,1\}$, there is a one-to-one correspondence between subsets of X and mappings of X into Y.)

The operations on cardinal numbers have a number of expected properties, which we leave as exercises. The jollities of cardinal arithmetic are best shown by unexpected results.

Proposition B.5.9. $\kappa + \kappa = \kappa$ *for every infinite cardinal number κ.*

Proof. For any set X, $|\{1,2\} \times X| = |X| + |X|$, since $\{1,2\} \times X$ is the disjoint union of $\{1\} \times X$ and $\{2\} \times X$. Also $\aleph_0 + \aleph_0 = \aleph_0$ by Proposition B.5.8.

Let A be a set with $|A| = \kappa$. Let $\mathscr{S}$ be the set of all ordered pairs (X, φ) such that $X \subseteq A$ and φ is a bijection of X onto $\{1,2\} \times X$. Since A is infinite, there is an injection $\mathbb{N} \longrightarrow A$, A contains an infinite countable subset X, there is a bijection of X onto $\{1,2\} \times X$ by Proposition B.5.7, and $\mathscr{S} \neq \varnothing$. Partially order $\mathscr{S}$ by $(X, \varphi) \leqslant (Y, \psi)$ in case $X \subseteq Y$ and $\varphi = \psi|_X$. It is immediate that every chain in $\mathscr{S}$ has an upper bound. By Zorn's Lemma, $\mathscr{S}$ has a maximal element (M, μ). We already have $|M| + |M| = |M|$ and show that $|M| = |A|$.

If $A \backslash M$ is infinite, then $A \backslash M$ contains an infinite countable subset Y, and bijections $\mu : M \longrightarrow \{1,2\} \times M$, $\xi : Y \longrightarrow \{1,2\} \times Y$ can be combined into a bijection $M \cup Y \longrightarrow \{1,2\} \times (M \cup Y)$ which extends μ, contradicting the maximality of (M, μ). Therefore $A \backslash M$ is finite. Now M is infinite, since $|M| + |M| = |M|$. Let X be any infinite countable subset of M. Then $X \cup (A \backslash M)$ is infinite countable by Proposition B.5.7, and

$$|A| = |X \cup (A \backslash M)| + |M \backslash X| = |X| + |M \backslash X| = |M|. \qquad \square$$

Corollary B.5.10. *If κ or λ is infinite, then $\kappa + \lambda = \max(\kappa, \lambda)$.*

Proof. Assume $\kappa \leqslant \lambda$. Then $\lambda \leqslant \kappa + \lambda \leqslant \lambda + \lambda = \lambda$; hence $\kappa + \lambda = \lambda$. $\qquad \square$

Proposition B.5.11. $\kappa\kappa = \kappa$ *for every infinite cardinal number κ.*

Proof. Proposition B.5.6 shows that $\aleph_0\aleph_0 = \aleph_0$. As in the proof of Proposition B.5.9, let A be a set with $|A| = \kappa$ and $\mathscr{S}$ be the set of all ordered pairs (X, φ) such that $X \subseteq A$ and φ is a bijection of X onto $X \times X$. Since A is infinite, A contains an infinite countable subset and $\mathscr{S} \neq \varnothing$. Partially order $\mathscr{S}$ by

$(X, \varphi) \leqslant (Y, \psi)$ in case $X \subseteq Y$ and $\varphi = \psi|_X$. It is immediate that every chain in $\mathscr{S}$ has an upper bound. By Zorn's Lemma, $\mathscr{S}$ has a maximal element (M, μ). Now $|M||M| = |M|$; we show that $|M| = |A|$.

Assume that $|M| < |A|$. Then $|A \backslash M| = |A|$ by Corollary B.5.10, since $|A| = |M| + |A \backslash M|$. Hence $|M| < |A \backslash M|$, there is an injection $M \longrightarrow A \backslash M$, and $A \backslash M$ has a subset X with $|X| = |M|$. Then

$$|X \times X| = |X \times M| = |M \times X| = |M \times M| = |M| = |X|,$$

$|(X \times X) \cup (X \times M) \cup (M \times X)| = |X|$ by Corollary B.5.10, and there is a bijection $(X \times X) \cup (X \times M) \cup (M \times X) \longrightarrow X$. Combining this with the bijection $\mu : M \times M \longrightarrow M$ yields a bijection $(M \cup X) \times (M \cup X) \longrightarrow M \cup X$. This contradicts the maximality of (M, μ), as required. □

Corollary B.5.12. *If κ or λ is infinite and $\kappa, \lambda \neq 0$, then $\kappa\lambda = \max(\kappa, \lambda)$.*

Proof. Assume $\kappa \leqslant \lambda$. Then $\lambda \leqslant \kappa\lambda \leqslant \lambda\lambda = \lambda$; hence $\kappa\lambda = \lambda$. □

Corollary B.5.13. *An infinite set X has $|X|$ finite subsets.*

Proof. The set X has $|X|$ subsets with one element and has at least $|X|$ finite subsets. On the other hand, X has $1 \leqslant |X|$ empty subset, $|X|$ subsets with one element, at most $|X||X| = |X|$ subsets with two elements (by Proposition B.5.11), and generally at most $|X||X| \ldots |X| = |X|^n = |X|$ subsets with n elements. By Corollary B.5.12, X has at most $|X| + |X| + \cdots = \aleph_0|X| = |X|$ finite subsets. □

Exercises

1. Prove that the union of a set of cardinal numbers is a cardinal number.
2. Prove that every infinite cardinal is a limit ordinal.
3. Prove that all infinite cardinals can be arranged into a transfinite sequence

$$\aleph_0 < \aleph_1 < \cdots < \aleph_\alpha < \aleph_{\alpha+1} < \cdots,$$

   indexed by all ordinals.
4. Prove that $2^{\aleph_0} = \mathfrak{c}$. (Hint: $\mathfrak{c} = |[0,1]|$.)
5. Prove directly that a subset of a countable set is countable.
6. Prove directly that the image of a countable set under a mapping is countable.
7. Prove directly that a countable set has countably many finite subsets.
8. Prove directly that $\mathbb{N}^{\mathbb{N}}$ is not countable.
9. Show that the addition of cardinal numbers is commutative and associative.
10. Show that the multiplication of cardinal numbers is commutative.
11. Show that the multiplication of cardinal numbers is associative.
12. Show that exponentiation of cardinal numbers is well defined.

13. Prove that $\kappa^{\lambda+\mu} = \kappa^{\lambda}\kappa^{\mu}$ for all cardinal numbers κ, λ, μ.

14. Prove that $\kappa^{\lambda\mu} = (\kappa^{\lambda})^{\mu}$ for all cardinal numbers κ, λ, μ.

15. Verify that infinite sums and products of cardinal numbers are well-defined: that is, if $|X_i| = |Y_i|$ for all $i \in I$, then $|\prod_{i \in I} X_i| = |\prod_{i \in I} Y_i|$, and if the sets $(X_i)_{i \in I}$ and the sets $(Y_i)_{i \in I}$ are pairwise disjoint, $|\bigcup_{i \in I} X_i| = |\bigcup_{i \in I} Y_i|$.

16. Show that $(\prod_{i \in I} \kappa_i)^{\lambda} = \prod_{i \in I} (\kappa_i^{\lambda})$.

17. Show that $\prod_{i \in I} (\kappa_i^{\lambda}) = \kappa^{(\sum_{i \in I} \lambda_i)}$.

18. Show that $\sum_{i \in I} \kappa_i = \sum_{j \in J} (\sum_{i \in I_j} \kappa_i)$ whenever $I = \bigcup_{j \in J} I_j$ is a partition of I.

19. Show that $\prod_{i \in I} \kappa_i = \prod_{j \in J} (\prod_{i \in I_j} \kappa_i)$ whenever $I = \bigcup_{j \in J} I_j$ is a partition of I.

BIBLIOGRAPHY

The following are some historically important books, and some books that I like; they do not constitute a complete bibliography.

GENERAL

VAN DER WAERDEN, B. L., *Moderne Algebra*, 2 vols., Springer, 1930, 1931.

LANG, Serge, *Algebra*, Addison-Wesley, 1965.

HUNGERFORD, Thomas W., *Algebra*, Holt, Rhinehart and Winston, 1974.

ISAACS, I. Martin, *Algebra, a Graduate Course*, Brooks/Cole, 1994.

KEMPF, George R., *Algebraic Structures*, Vieweg, 1995.

GROUPS

BURNSIDE, W., *Theory of Groups of Finite Order*, Cambridge, 1897.

ZASSENHAUS, Hans, *The Theory of Groups*, Chelsea, 1949.

SUSUKI, Michio, *Group Theory*, Springer-Verlag, 1986.

FIELDS

McCARTHY, Paul J., *Algebraic Extensions of Fields*, Chelsea, 1966.

RINGS AND MODULES

JACOBSON, Nathan, *Structure of Rings*, American Mathematical Society, 1956.

MACLANE, Saunders, *Homology*, Springer-Verlag, 1963.

FARB, Benson and DENNIS, R. Keith, *Noncommutative Algebra*, Springer-Verlag, 1993.

DAUNS, John, *Modules and Rings*, Cambridge, 1994.

COMMUTATIVE ALGEBRA

ZARISKI, Oscar and SAMUEL, Pierre, *Commutative Algebra*, 2 vols., van Nostrand, 1958, 1960.
EISENBUD, David, *Commutative Algebra with a View toward Algebraic Geometry*, Springer-Verlag, 1994.

OTHERS STRUCTURES

MACLANE, Saunders, *Categories for the Working Mathematician*, Springer-Verlag, 1971.
COHN, P. M., *Universal Algebra*, Harper and Row, 1965.
GRÄTZER, George, *Universal Algebra*, van Nostrand, 1968.
BIRKHOFF, Garrett, *Lattice Theory*, Amer. Math. Soc., 1940.
GRÄTZER, George, *General Lattice Theory*, Birkhäuser, 1978.

SET THEORY

HALMOS, Paul, *Naive Set Theory*, van Nostrand, 1964.
JECH, Thomas, *Set Theory*, Academic Press, 1978.

REAL NUMBERS

HEWITT, Edwin and STROMBERG, Karl, *Real and Abstract Analysis*, Springer-Verlag, 1969.

NOTATION

$A, B, \ldots$ sets, subsets, matrices, objects

$a, b, \ldots$ elements

$\mathscr{A}, \mathscr{B}, \ldots$ unusual sets, categories, classes

A_n, alternating group, 100

Abs, category of abelian groups, 506

R-Algs, category of *R*-algebras, 506

Alg_T, category of univ. alg. of type T, 554

Ann$(\ldots)$, annihilator, 305, 313

$\mathbb{C}$, complex number field

$C(A)$, coordinate ring of A, 377

$C_G(x)$, centralizer in G, 95

C_n, a cyclic group of order n

c_T, characteristic polynomial of T, 341

C_x, conjugacy class, 95

Diag $(\mathscr{G}, \mathscr{A})$, diagram category, 516

D_n, dihedral group, 10

$D(f)$, discriminant of f, 167

det, determinant

dim, dimension

$e(E \colon K)$, ramification index, 298

End$_R(A)$, ring of *R*-endomorph's, 133, 384

End$_R^{\text{Op}}(A)$, same, written on right, 306, 384

$f(E \colon K)$, residue class degree, 299

F_X, free something on X

Fix$_K(G)$, fixed field of G, 226

Func $(\mathscr{A}, \mathscr{B})$, functor category, 513

G_v, value group of valuation, 291

Gal(E/K), Galois group

gcd$(\ldots)$, greatest common divisor

$GF(q)$, finite field, 218

GL, general linear group, 10

Grps, category of groups, 506

Hom$_R(A, B)$, group of homomorph's, 433

Hom$_{\mathscr{C}}(A, B)$, class of morphisms, 511

Hom$_R(A, -)$, Hom$_R(-, B)$, functors

I, identity matrix or linear transformation

Im, image or range of a mapping, 18

Irr$_K(\alpha)$, irred. polynomial of α, 201

$\mathbb{J}_p$, *p*-adic integers

$J(R)$, Jacobson radical, 391

Ker, kernel of a homomorphism, 23

lcm$(\ldots)$, least common multiple

$\varinjlim$, direct limit, 442

$\varprojlim$, inverse limit, 449

M_5, a modular lattice, 583

$M_n(R)$, ring of $n \times n$ matrices

m_T, minimal polynomial of T, 338

R-Mods, category of *R*-modules, 506

$\mathbb{N}$, the natural numbers

$N_K^E(\alpha)$, norm of element, 244

$N_G(H)$, normalizer in G, 97

N_5, a non modular lattice, 579

$\ldots^{\text{op}}$, opposite something

Ord, class of all ordinals

$\mathbb{P}$, positive reals under multiplication, 276

$\mathscr{P}(X)$, power set of X, 622

PSL, projective special linear group, 120

Q, quaternion group, 73

$\mathbb{Q}$, rational number field

$\mathbb{Q}_n$, cyclotomic field

$\mathbb{Q}_p$, *p*-adic numbers

$\mathbf{Q}$, quaternion algebra

$Q(R)$, field of fractions of R, 152

$\mathbb{R}$, real number field

649

A/B, quotient thing
a/b, fraction
$\mathfrak{a}/c$, fractional ideal
$a|b$, divides
$a \nmid b$, does not divide
$|X|$, order or cardinality of X
$\|x\|$, norm of x, 285
$\widehat{F}$, completion of a field, 281
$\widehat{\mathscr{G}}$, category of paths, 507
F^*, $F \backslash \{0\}$ (F a field)
M^*, dual module, 467
φ^*, $\mathrm{Hom}_R(\varphi, A)$, 434
φ_*, $\mathrm{Hom}_R(A, \varphi)$, 434
x^{-1}, φ^{-1}, inverse
$\varphi^{-1}(A)$, inverse image

$\subseteq$, contained in
$\subsetneqq$, strictly contained in
$\leqslant$, order or preorder relation
$\leqslant$, subgroup or submodule of, 16
$\underset{\neq}{\leqslant}$, proper subgroup or submodule of, 16
$\vartriangleleft$, normal subgroup of, 23
$\underset{\neq}{\vartriangleleft}$, proper normal subgroup of, 23

$\cong$, isomorphic
$\langle X \rangle$, subgroup or submodule gen. by X
$\langle X; R \rangle$, gen'd by X subject to R, 71, 327
$\overline{x}$, $x + n\mathbb{Z}$ in $\mathbb{Z}_n$
$\circ$, composition of mappings
$\times$, cartesian or direct product
$\times_\varphi$, semidirect product, 51
$\otimes$, tensor product
$\bigotimes^n M$, tensor power, 481
$\prod$, product or direct product, 38, 316
$\prod$, composite field extension, 196
$\sum$, sum
$\oplus$, $\bigoplus$, direct sum, 42, 44, 316, 319
II, $\coprod$, coproduct or free product, 76
$\cap$, $\bigcap$, intersections
$\cup$, $\bigcup$, unions
$\backslash$, set difference
$\wedge$, $\bigwedge$, infimums, 570
$\wedge$, wedge product, 486
$\bigwedge(M)$, exterior algebra, 486
$\bigwedge^n(M)$, 486
$\vee$, $\bigvee$, supremums, 571

INDEX

PURE AND APPLIED MATHEMATICS
A Wiley-Interscience Series of Texts, Monographs, and Tracts

Founded by RICHARD COURANT
Editor Emeritus: PETER HILTON and HARRY HOCHSTADT
Editors: MYRON B. ALLEN III, DAVID A. COX, PETER LAX,
  JOHN TOLAND

*Now available in a lower priced paperback edition in the Wiley Classics Library.
†Now available in paperback.

DATE DUE

SEP 2 1 2008			
SEP 1 8 REC'D			
NOV 2 1 2005			
NOV 1 0 REC'D			